T0186758

Lecture Notes in Networks and Systems

Volume 482

Series Editor

Janusz Kacprzyk, Systems Research Institute, Polish Academy of Sciences, Warsaw, Poland

Advisory Editors

Fernando Gomide, Department of Computer Engineering and Automation—DCA, School of Electrical and Computer Engineering—FEEC, University of Campinas—UNICAMP, São Paulo, Brazil

Okyay Kaynak, Department of Electrical and Electronic Engineering, Bogazici University, Istanbul, Turkey

Derong Liu, Department of Electrical and Computer Engineering, University of Illinois at Chicago, Chicago, USA
Institute of Automation, Chinese Academy of Sciences, Beijing, China

Witold Pedrycz, Department of Electrical and Computer Engineering, University of Alberta, Alberta, Canada
Systems Research Institute, Polish Academy of Sciences, Warsaw, Poland

Marios M. Polycarpou, Department of Electrical and Computer Engineering, KIOS Research Center for Intelligent Systems and Networks, University of Cyprus, Nicosia, Cyprus

Imre J. Rudas, Óbuda University, Budapest, Hungary

Jun Wang, Department of Computer Science, City University of Hong Kong, Kowloon, Hong Kong

The series "Lecture Notes in Networks and Systems" publishes the latest developments in Networks and Systems—quickly, informally and with high quality. Original research reported in proceedings and post-proceedings represents the core of LNNS.

Volumes published in LNNS embrace all aspects and subfields of, as well as new challenges in, Networks and Systems.

The series contains proceedings and edited volumes in systems and networks, spanning the areas of Cyber-Physical Systems, Autonomous Systems, Sensor Networks, Control Systems, Energy Systems, Automotive Systems, Biological Systems, Vehicular Networking and Connected Vehicles, Aerospace Systems, Automation, Manufacturing, Smart Grids, Nonlinear Systems, Power Systems, Robotics, Social Systems, Economic Systems and other. Of particular value to both the contributors and the readership are the short publication timeframe and the world-wide distribution and exposure which enable both a wide and rapid dissemination of research output.

The series covers the theory, applications, and perspectives on the state of the art and future developments relevant to systems and networks, decision making, control, complex processes and related areas, as embedded in the fields of interdisciplinary and applied sciences, engineering, computer science, physics, economics, social, and life sciences, as well as the paradigms and methodologies behind them.

Indexed by SCOPUS, INSPEC, WTI Frankfurt eG, zbMATH, SCImago.

All books published in the series are submitted for consideration in Web of Science.

For proposals from Asia please contact Aninda Bose (aninda.bose@springer.com).

More information about this series at https://link.springer.com/bookseries/15179

Francesco Calabrò · Lucia Della Spina ·
María José Piñeira Mantiñán
Editors

New Metropolitan Perspectives

Post COVID Dynamics: Green and Digital
Transition, between Metropolitan and Return
to Villages Perspectives

Set 3

 Springer

Editors
Francesco Calabrò
Dipartimento PAU
Mediterranea University of Reggio Calabria
Reggio Calabria, Reggio Calabria, Italy

Lucia Della Spina
Mediterranea University of Reggio Calabria
Reggio Calabria, Italy

María José Piñeira Mantiñán
University of Santiago de Compostela
Santiago de Compostela, Spain

ISSN 2367-3370   ISSN 2367-3389 (electronic)
Lecture Notes in Networks and Systems
ISBN 978-3-031-06824-9   ISBN 978-3-031-06825-6 (eBook)
https://doi.org/10.1007/978-3-031-06825-6

© The Editor(s) (if applicable) and The Author(s), under exclusive license
to Springer Nature Switzerland AG 2022
This work is subject to copyright. All rights are solely and exclusively licensed by the Publisher, whether
the whole or part of the material is concerned, specifically the rights of translation, reprinting, reuse of
illustrations, recitation, broadcasting, reproduction on microfilms or in any other physical way, and
transmission or information storage and retrieval, electronic adaptation, computer software, or by similar
or dissimilar methodology now known or hereafter developed.
The use of general descriptive names, registered names, trademarks, service marks, etc. in this
publication does not imply, even in the absence of a specific statement, that such names are exempt from
the relevant protective laws and regulations and therefore free for general use.
The publisher, the authors, and the editors are safe to assume that the advice and information in this
book are believed to be true and accurate at the date of publication. Neither the publisher nor the
authors or the editors give a warranty, expressed or implied, with respect to the material contained
herein or for any errors or omissions that may have been made. The publisher remains neutral with regard
to jurisdictional claims in published maps and institutional affiliations.

This Springer imprint is published by the registered company Springer Nature Switzerland AG
The registered company address is: Gewerbestrasse 11, 6330 Cham, Switzerland

Preface

This volume contains the proceedings for the fifth International "NEW METROPOLITAN PERSPECTIVES. Post COVID Dynamics: Green and Digital Transition, between Metropolitan and Return to Villages' Perspectives", scheduled from May 25–27, 2022, in Reggio Calabria, Italy.

The symposium was promoted by LaborEst (Evaluation and Economic Appraisal Lab) of the PAU Department, Mediterranea University of Reggio Calabria, Italy, in partnership with a qualified international network of academic institution and scientific societies.

The fifth edition of "NEW METROPOLITAN PERSPECTIVES", like the previous ones, aimed to deepen those factors which contribute to increase citics and territories attractiveness, both with theoretical studies and tangible applications.

This fifth edition coincides with what is most likely the end of the COVID pandemic that began in 2020. The global health emergency, despite having been a phenomenon limited in time, has acted as an accelerator of some changes in behavior and in the organization of activities associated with the ever-increasing spread of ICT.

The phenomena are too recent and still ongoing to fully understand the implications they will have on settlement systems, but the conclusion reached at the previous edition of New Metropolitan Perspectives seems to be confirmed: from many of the works presented at the Symposium, a reduction in the relevance of the localization factor emerges with ever greater clarity, at least in the ways known so far from the times of the Industrial Revolution, bringing to light more and more a paradigm shift in the center-periphery dualism.

In fact, the phenomenon that in the past led to the birth of the modern city, the need to concentrate people and activities in small areas, seems to be decreasing: the progressive spread of smart working and the digital modality for the provision of services (just think, e.g., of the digital services of the Public Administration or online commerce) significantly reduces the gaps in terms of accessibility to goods and services between metropolitan cities and marginalized areas, such as inland areas.

But this edition of the symposium also coincides with the start of a new phase for European policies, guided toward the green and digital transition, for the period 2021-27, by the European Green Deal, especially through the tool of the Next Generation EU.

The links between new technologies and sustainability tend to focus on the role played and that can play the city at EU level in fighting climate change.

Many of the contributions collected in this volume address the issue of the green transition through multidisciplinary points of view, dealing with very different issues such as, for example: infrastructures and mobility systems, green buildings and energy communities, ecosystem services and the consumption of soil, providing interesting information on the main trends in progress.

The changes in individual behavior and social organization, associated with the digital transition, are illustrated by the contributions that have addressed the issue of rules and of social innovation practices that are prefiguring new forms of governance for the regeneration of settlement systems. In this context, the issues of the new declinations of the concept of citizenship were also addressed, also with reference to the need to create favorable contexts for individual initiative and entrepreneurship, especially for young people, as a possible response to the challenge of employability for the new generations.

In this context, territorial information systems take on a leading role, together with apps capable of making territories increasingly smart.

The substantial investments planned by the EU to support the green and digital transition in the coming years require multidimensional evaluation systems, capable of supporting decision makers in selecting the interventions most capable of pursuing the objectives. The financial resources used for the implementation of the policies are borrowed from future generations, to whom we will have the obligation to be accountable for our work.

Unfortunately, at the time of writing we must also register serious concerns for the future of humanity, stemming from the risks of the spread of the conflict between Russia and Ukraine. In addition to the obvious concerns about the suffering that wars always cause to civilian populations, this situation makes future scenarios even more uncertain: It is clear that the circulation of goods, people and ideas will be increasingly conditioned by future geopolitical balances.

The ethics of research, in the disciplinary sectors that the Symposium crosses, invites us to feed, with scientific rigor, policies and practices that make the territory more resilient and able to react effectively to catastrophic events such as the pandemic or the war: We hope to know the outcomes of these courses in the next editions of the New Metropolitan Perspectives symposium.

For this edition, meanwhile, the more than 300 articles received allowed us to develop 6 macro-topics, about "Post COVID Dynamics: Green and Digital Transition, between Metropolitan and Return to Villages' Perspectives" as follows:

1. Inner and marginalized areas local development to re-balance territorial inequalities

2. Knowledge and innovation ecosystem for urban regeneration and resilience
3. Metropolitan cities and territorial dynamics. Rules, governance, economy, society
4. Green buildings, post-carbon city and ecosystem services
5. Infrastructures and spatial information systems
6. Cultural heritage: conservation, enhancement and management.

And a Special Section, Rhegion United Nations 2020-2030, chaired by our colleague Stefano Aragona.

We are pleased that the International Symposium NMP, thanks to its interdisciplinary character, stimulated growing interests and approvals from the scientific community, at the national and international levels.

We would like to take this opportunity to thank all who have contributed to the success of the fifth International Symposium "NEW METROPOLITAN PERSPECTIVES. Post COVID Dynamics: Green and Digital Transition, between Metropolitan and Return to Villages' Perspectives": authors, keynote speakers, session chairs, referees, the scientific committee and the scientific partners, participants, student volunteers and those ones that with different roles have contributed to the dissemination and the success of the Symposium; a special thank goes to the "Associazione ASTRI", particularly to Giuseppina Cassalia and Angela Viglianisi, together with Immacolata Lorè, for technical and organizational support activities: without them the Symposium couldn't have place; and, obviously, we would like to thank the academic representatives of the University of Reggio Calabria too: the Rector Prof. Marcello Zimbone, the responsible of internationalization Prof. Francesco Morabito, the chief of PAU Department Prof. Tommaso Manfredi.

Thank you very much for your support.

Last but not least, we would like to thank Springer for the support in the conference proceedings publication.

Francesco Calabrò
Lucia Della Spina
Maria José Pineira Mantinan

Organization

Programme Chairs

Francesco Calabrò — Mediterranea University of Reggio Calabria, Italy
Lucia Della Spina — Mediterranea University of Reggio Calabria, Italy
María José Piñeira Mantiñán — University of Santiago de Compostela, Spain

Scientific Committee

Ibtisam Al Khafaji — Al-Esraa University College of Baghdad, Iraq
Shaymaa Fadhìl Jasim Al Kubasi — Koya University, Iraq

Pierre-Alexandre Balland — Universiteit Utrecht, Netherlands
Massimiliano Bencardino — Università di Salerno
Jozsef Benedek — RSABabes-Bolyai University, Romania
Christer Bengs — SLU/Uppsala Sweden and Aalto/Helsinki, Finland
Adriano Bisello — EURAC Research
Mario Bolognari — Università degli Studi di Messina
Nico Calavita — San Diego State University, USA
Roberto Camagni — Politecnico di Milano, Presidente Gremi
Sebastiano Carbonara — Università degli Studi "Gabriele d'Annunzio" Chieti-Pescara
Farida Cherbi — Institut d'Architecture de TiziOuzou, Algeria
Antonio Del Pozzo — Università degli Studi di MessinaUnime
Alan W. Dyer — Northeastern University of Boston, USA
Yakup Egercioglu — Izmir Katip Celebi University, Turkey
Khalid El Harrouni — Ecole Nationale d'Architecture, Rabat, Morocco
Gabriella Esposito De Vita — CNR/IRISS Ist. di Ric. su Innov. e Serv. per lo Sviluppo

Fabiana Forte	Università degli Studi della Campania "Luigi Vanvitelli"
Chro Ali HamaRadha	Department of Architectural Engineering/Faculty of Engineering/Koya University, Iraq
Christina Kakderi	Aristotelio Panepistimio Thessalonikis, Greece
Karima Kourtit	Open University, Heerlen, Netherlands
Olivia Kyriakidou	Athens University of Economics and Business, Greece
Ibrahim Maarouf	Alexandria University, Faculty of Engineering, Egypt
Lívia M. C. Madureira	Centro de Estudos Transdisciplinares para o DesenvolvimentoCETRAD, Portugal
Tomasz Malec	Istanbul Kemerburgaz University, Turkey
Benedetto Manganelli	Università degli Studi della Basilicata
Giuliano Marella	Università di Padova
Nabil Mohäreb	Beirut Arab University, Tripoli, Lebanon
Mariangela Monaca	Università di Messina
Bruno Monardo	Università degli Studi di Roma "La Sapienza"
Giulio Mondini	Politecnico di Torino
Pierluigi Morano	Politecnico di Bari
Grazia Napoli	Università degli Studi di Palermo
Fabio Naselli	Epoka University
Antonio Nesticò	Università degli Studi di Salerno
Peter Nijkamp	Vrije Universiteit Amsterdam
Davy Norris	Louisiana Tech University, USA
Alessandra Oppio	Politecnico di Milano
Leila Oubouzar	Institut d'Architecture de TiziOuzou, Algeria
Sokol Pacukaj	Aleksander Moisiu University, Albania
Aurelio Pérez Jiménez	University of Malaga, Spain
Keith Pezzoli	University of California, San Diego, USA
María José Piñera Mantiñán	University of Santiago de Compostela, Spain
Fabio Pollice	Università del Salento
Vincenzo Provenzano	Università di Palermo
Ahmed Y. Rashed	Farouk ElBaz Centre for Sustainability and Future Studies
Paolo Rosato	SIEV Società Italiana di Estimo e Valutazione
Michelangelo Russo	SIU Società Italiana degli Urbanisti
Helen Salavou	Athens University of Economics and Business, Greece
Stefano Stanghellini	INUIstituto Nazionale di Urbanistica
Luisa Sturiale	Università di Catania
Ferdinando Trapani	Università degli Studi di Palermo

Robert Triest	Northeastern University of Boston, USA
Claudia Trillo	University of Salford, UK
Gregory Wassall	Northeastern University of Boston, USA

Internal Scientific Board

Giuseppe Barbaro	Mediterranea University of Reggio Calabria
Concetta Fallanca	Mediterranea University of Reggio Calabria
Giuseppe Fera	Mediterranea University of Reggio Calabria
Massimiliano Ferrara	Mediterranea University of Reggio Calabria
Tommaso Isernia	Mediterranea University of Reggio Calabria
Giovanni Leonardi	Mediterranea University of Reggio Calabria
Tommaso Manfredi	Mediterranea University of Reggio Calabria
Domenico E. Massimo	Mediterranea University of Reggio Calabria
Marina Mistretta	Mediterranea University of Reggio Calabria
Carlo Morabito	Mediterranea University of Reggio Calabria
Domenico Nicolò	Mediterranea University of Reggio Calabria
Adolfo Santini	Mediterranea University of Reggio Calabria
Simonetta Valtieri	Mediterranea University of Reggio Calabria
Giuseppe Zimbalatti	Mediterranea University of Reggio Calabria
Santo Marcello Zimbone	Mediterranea University of Reggio Calabria

Scientific Partnership

SIEV - Società Italiana di Estimo e Valutazione, Rome, Italy
SIIV - Società Italiana Infrastrutture Viarie, Ancona, Italy
SIRD - Società Italiana di Ricerca Didattica, Salerno, Italy
SIU - Società Italiana degli Urbanisti, Milan, Italy
SGI, Società Geografica Italiana, Roma, Italy

Organizing Committee

ASTRI Associazione Scientifica Territorio e Ricerca Interdisciplinare
URBAN LAB S.r.l.
ICOMOS Italia, Rome, Italy

Contents

Green Buildings, Post Carbon City and Ecosystem Services

Infrastructures and Spatial Information Systems

Outdoor Green Walls: Multi-perspective Methodology for Assessing Urban Sites Based on Socio-environmental Aspects

Nicole Agnolio[1], Matilde Molari[2(✉)], Laura Dominici[2], and Elena Comino[2]

[1] Department of Architecture and Design (DAD), Politecnico di Torino, 10129 Turin, Italy
[2] Department of Environment, Land and Infrastructure Engineering (DIATI), Politecnico di Torino, 10129 Turin, Italy
`matilde.molari@polito.it`

Abstract. Green walls are adopted as technological nature-based solution to improve citizens' well-being through the naturalization of the built environment. The site selection is critical to the success of outdoor green walls due to their cross-functional application and to the urban grid's complexity. This process must consider environmental features, social needs, and citizens' habits, besides the urban context's morphology. In this framework, the adoption of a multi-perspective approach may deal with the design process of outdoor green walls as complex systems. This contribute presents the definition of a site selection's methodology for outdoor green walls based on the integration of multi-perspectives, from citizens to academic experts, applied on the case study of Biella municipality (Piedmont Region, Italy). The methodology, here presented as a *"work in progress"* tool, was interdisciplinary designed combining environmental and socio-cultural drivers for decision-making processes for outdoor green walls' applications. Preliminary results contribute to amplify and ease the debate between academic and non-academic stakeholders concerning the production of tangible and intangible benefits provided by outdoor green walls.

Keywords: Decision-making process · Ecosystem approach · Socio-cultural benefits

1 Introduction

In the framework of cities' densification, urban green infrastructures (UGIs) provide measures to maintain the quality of life and to improve human well-being in cities. This concept refers to the network of multifunctional traditional and technological green systems designed to spatially complement grey infrastructures, such as bioswales and constructed wetlands [1]. Outdoor vertical greening systems (VGSs) are included in the huge set of UGIs adopted to re-introduce vegetation and nature in the built environment and to increase the connectivity between green areas in densely built-up cities. VGSs integrate varied structures that allow vegetation to grow vertically, including green walls

© The Author(s), under exclusive license to Springer Nature Switzerland AG 2022
F. Calabrò et al. (Eds.): NMP 2022, LNNS 482, pp. 1905–1915, 2022.
https://doi.org/10.1007/978-3-031-06825-6_183

as well as green façades and self-standing systems [2]. VGSs are classified as nature-based solutions (NBSs), in particular as technological vertical units, that facilitate the transition towards greener, more resilient and socially inclusive cities through the creation of artificially-made ecosystems [3]. Going beyond mere ornamental and aesthetic purposes, NBSs employ vegetation and natural elements to face socio-ecological challenges. They provide benefits at local or city-scale with equal reliance upon social, environmental and economic domains to achieve city resilience [4]. Green walls are increasingly adopted as site- or location-specific systems that can promote important ecosystem services to enhance the quality of life and to improve citizens' well-being [5]. Indeed, the interdependence between high quality of life and the access to green areas in cities is well recognised and supported by experts and academics [6]. This aspect was particularly visible during pandemic with mobility restrictions that limited the access of citizens to natural ecosystems outside urban areas [7]. These technological units contribute to generate supporting, regulating, provisioning and cultural ecosystem services, such as carbon sequestration, habitat provision for insects, contrasting *urban heat island effect* (HUI) [8], and provide several social opportunities to engage citizens with nature.

In order to enhance the multi-functional potential of vertical greening, the site selection is a critical stage during the design process of successful outdoor green walls. Indeed, the process of site selection must simultaneously consider spatial, environmental and socio-cultural features of urban contexts as well as citizens. This contribute explores the interdisciplinary design process of a multi-perspective methodology for the assessment of the most appropriate site for outdoor green walls application. The methodology integrates data concerning spatial and environmental characteristics with subjective aspects acquired from the involvement of stakeholders. The article presents the application of the methodology to the case study of Biella municipality in the Piedmont Region (north-west of Italy). Despite the site-specific focus of this contribute, the methodology was defined to be applied to other urban areas as a strategic tool to identify environmental and socio-cultural drivers during preliminary stages of decision-making processes. Moreover, the engagement of citizens during the design process helped to spread the know-how about multiple benefits associated with outdoor green walls.

2 Methodology

A holistic context analysis was performed to characterise the selected sites in order to evaluate them through a multi-criteria model. The methodology follows a multi-perspective approach integrating information from literature, non-experts and experts from multi-disciplinary research fields. Criteria have been obtained combining data and information consequently derived from (1) spatial analysis, (2) environmental analysis and (3) social analysis [9]. The multiple-criteria decision analysis (MCDA) was adopted to tackle the multi-purpose nature of VGSs. It considers their benefits and criticalities in relation to a specific context and it aims to foster their functions in urban environment.

2.1 Step 1: Spatial Analysis

The spatial analysis consists of city's districts identification and characterisation carried out with onsite inspections. Close districts with similar features (e.g., residential areas) were grouped to facilitate the context analysis. Hubs of innovation, buildings with recreational or social-educational role, and historic or industrial value were identified as landmarks for each district. A simplified SWOT analysis was performed to highlight strengths, weaknesses, opportunities and threats of each district and building. During this stage particular attention focused on regulations that preserve historic buildings, as potential criticality for project realization. Then, green areas with recreational and ecological functions within each district (such as municipal gardens and woods) were mapped to discover the accessibility level of citizens to urban green areas.

2.2 Step 2: Environmental Analysis

Data concerning environmental features were collected consulting more recent information reported by national, regional, or municipal databases and sources, e.g., ARPA reports, ISPRA reports, Geoportale ARPA. Environmental data were mapped at city-scaled considering: (1) noise pollution level, identifying site variations (in decibel) of noise levels according to classes of sound emission; (2) temperature variation, considering temperature raises caused by microclimatic alteration of *urban heat island effect*; (3) and air quality, taking into account particularly concentrations of PM10.

2.3 Step 3: Social Analysis

The social analysis was mainly conducted using a survey distributed to citizens. Considering the scarce presence of information about socio-cultural ecosystem services related to VGSs [10], the survey investigated citizens' perception and satisfaction in response to an outdoor green wall. Moreover, it explored basic knowledge of citizens about green walls' benefits when implemented in urban areas. The structure of the survey consisted of: 30% personal questions on the participant's sample; 10% of questions on knowledge about outdoor green walls benefits; 35% questions on personal opinions about potential functions and scopes of outdoor green walls; and 25% questions about preferences for sites' applications, referring to districts type or condition (touristic, residential, historic, industrial, degraded) and buildings function (scholastic, sporting, industrial, cultural, commercial).

2.4 Step 4: Definition of Preliminary Multi-perspective Matrix

Data collected from spatial, environmental and social analysis were grouped in functional categories defining criteria and sub-criteria of a preliminary multi-perspective matrix. The matrix was used to conduct the multi-criteria analysis on selected buildings in order to rank the most suitable sites for the outdoor green wall insertion. The matrix was composed of five criteria and fifteen sub-criteria, as shown in Table 1. This tool deals with spatial, environmental and social aspects and evaluates site position, building conditions, environmental features of each district and possible citizen psycho-physical benefits. Moreover, expert and non-expert preferences and judgement about proposed buildings are reported in the matrix. Those data and observations are extracted from surveys answers and from a final interdisciplinary focus group with academics.

Table 1. List of 5 main criteria and short description with range of 15 sub-criteria.

Criterion 1: Building's condition	
Status of building	0. Recently constructed (last 5 years); 1. Recent building (from 10 to 30 years); 2. Old renovated building; 3. Old building (more 30 years)
Surface	0. Glazed surface more than 80%; 1. Glazed surface 50%–80%; 2. Glazed surface less than 80%; 3. No glazed surface
Maintenance	0. Difficulty in operational activities in front of the building's wall; 1. Ease in operational activities in front of the building's wall
Energy consumption	0. Excellent energy rating (class A-B); 1. High Energy rating (class C-D); 2. Medium energy rating (class E-F); 3. Low energy rating (class G)
Criterion 2: Position	
Green areas proximity	0. All typologies of green areas; 1. Parks, garden, sport green areas; 2. Urban green areas; 3. No green areas - *Considering an area of 500 m radius around the building*
Wall visibility	0. No visual accessibility from the street; 1. Visual accessibility from car; 2. Visual accessibility from pedestrian street; 3. Visual accessibility from meeting place
Area depth	0. Less than 2 m; 1. Between 2 to 5 m; 2. Between 5 to 10 m; 3. More than 10 m
Criterion 3: Environmental features	
Noise pollution	0. Noise emission in 1–2 classes; 1. Noise emission in 3–4 classes of standard table zoning; 2. Noise emission in 5 class; 3. Noise emission in 6 class - *(refers to standard table zoning)*
Temperature raises (caused by UHI)	0. No variation; 1. Temperature raise until 1 °C; 2. Temperature rise until 2 °C; 3. Temperature rise until 3 °C
Air quality	0. High concentration of PM10; 1. Low concentration of PM10

(*continued*)

Table 1. (*continued*)

Criterion 4: People wellbeing	
Social aspects	0. No educational and socio-cultural activities; 1. Less than 5 educational and socio-cultural activities; 2. 5–10 educational and socio-cultural activities; 3. More than 10 educational and socio-cultural activities - *Considering an area of 200 m radius around the building*
Psychological aspects	0. No presence of work activities; 1. Less than 5 work activities; 2. 5–10 work activities; 3. More than 10 work activities - *Considering an area of 200 m radius around the building*
Tourism	0. No presence of historic sites in the neighbourhood as elements of interest; 1. One historic site; 2. Two historic sites; 3. More than two historic sites
Criterion 5: Survey's preferences	
Non-experts	0. Number of site's preferences less than 80; 1. Between 80 and 100; 2. Between 100 and 110; 3. More than 110
Experts	0. Number of site's preferences less than 4; 1. Between 4 and 8; 2. Between 8 and 12; 3. More than 12

All criteria and sub-criteria were weighted in order to perform the multi-criteria analysis considering both quantitative data collected and citizens' priorities and perceptions. The weighting method combines the SMART and the swing technique [11]. Weights obtained by a preliminary evaluation of criteria and sub-criteria, were then discussed and validated during a focus group with stakeholders. The ranges of the "Survey's preferences" and "People wellbeing" sub-criteria could change respectively according to the number of surveys obtained and the city analysed.

2.5   Step 5: Focus Group with Expert Stakeholders

At the end, results reported by the preliminary multi-perspective matrix were discussed during an interdisciplinary focus group with stakeholders to integrate and validate the model. Experts in urban, environmental, and social research fields were involved into the focus group to obtain a robust multi-perspective model. During the meeting, the survey was distributed to participants and answers were analysed and integrated in the multi-perspective matrix adding data concerning the sub-criteria "Experts' preference". At the end of the meeting, stakeholders discussed weights attributed to criteria and sub-criteria in order to define the final ones.

3   Results

The methodology previously described was applied to the case study of the Biella Municipality to select the most appropriate site for the implementation of an outdoor green wall.

3.1 Step 1: Spatial Analysis of Biella's Urban Area

Biella is a medium-size city close to the Biellese Alps and several natural areas. Biella is also characterized by a significant industrial heritage derived from wool processing and textile manufacturing. The spatial analysis started with districts' analysis and definition, grouping some of them in order to obtain eight main areas. These areas present varied features that describe main districts' vocation, as shown in Table 2. Historical and central areas (Centro and Piazzo) were not considered in this study because changes of historical buildings' surfaces are not permitted according to local regulations. Eight buildings were also identified as significant potential sites for outdoor green wall application (see Fig. 1) and their main features are reported in Table 2. The Fig. 1 also shows the distribution of green areas around the city highlighting the presence of many public gardens and parks mainly in Vernato and San Paolo districts, and wooded area in Oremo and Piazzo districts.

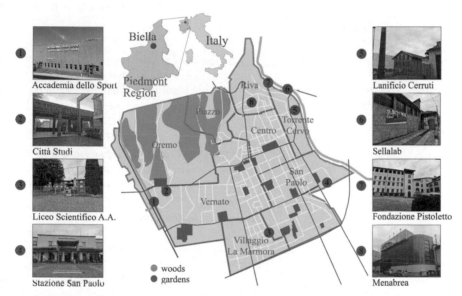

Fig. 1. Distribution of selected buildings on Biella's map.

3.2 Step 2: Environmental Analysis

Data analysis about the air quality showed that the concentration of PM10 in the urban area of Biella is lower than levels recorded in other cities of Piedmont Region [12]. Considering this aspect, "air quality" was not included as significant sub-criterion for this specific case study. Concerning the noise pollution, the analysis revealed that commercial and industrial areas, respectively Vernato and Cervo, presented the highest levels at the city scale [13].

Table 2. Districts and buildings features.

District	Features	Building	Features
Oremo	Peripheral district No services High risk of river's floods	1. Accademia dello Sport	Recent building Easily visible Popular by young
Vernato	Commercial district Several services University area	2. Città Studi	Recent building University campus and convention hall
Villaggio La Marmora	Peripheral district Active community of citizens	3. Liceo Scientifico A.Avogadro	Recent building Popular by young
San Paolo	Commercial district No services	4. Stazione S. Paolo	Good status of building Popular by citizens
Centro	Central district Several services	–	–
Torrente Cervo	Industrial district No services High risk of river's floods	5. Lanificio Cerrutti	Old and industrial building Not visible from the street Attention to sustainability
		6. Sellalab	Old and industrial building Innovative company Coworking space
		7. Fondazione Pistoletto	Creative hub Not visible from the street Attention to sustainability
Riva	Central district Full of services	8. Menabrea	Recently restructured building Easily visible from the street
Piazzo	Historical district No services Touristic area	–	–

According to the study conducted to investigate the relation between citizens' well-being and local climate, Biella reached the 79th position out of 107 municipalities of Italy, despite it is characterised by warm and temperate climate based on the evaluation of ten indicators [14]. Since data about climate present no significant variations between different areas of the city, they were not included into the matrix. To fill the lack of environmental quantitative data, according to the work of Musco et al. (2014), (1) the urban traffic, (2) the presence of industrial structures and (3) the residential density, were selected as qualitative indicators of the *urban heat island effect* [15].

3.3   Step 3: Social Analysis, Results of Survey

Results concerning the social analysis were outlined from about 300 surveys collected from citizens of different ages and cultural backgrounds. The survey has been prepared through a Google form and diffused via social media. Results provided information about urban greening's perception, green walls' appreciation, and level of knowledge about their benefits. They highlight that the 65% of participants have no previous knowledge of green walls and the percentage grows concerning their benefits. The most of them prefers to install green walls in busy and commercial areas (as Riva district) where there is a lack of green spaces, whereas the 55% of people would like to be involved in socio-cultural activities in which they can actively interact with the green wall. Furthermore, participants highlighted that green wall could encourage people to visit the buildings that host it. This result underlines that people prefer to install green walls in public and well visible sites than in private and more hidden places.

Following this preliminary screening, the top positions were occupied by "Città Studi", "Fondazione Pistoletto" and "Accademia dello Sport", followed by "Menabrea", "San Paolo" train station, "Sellalab", "Lanificio Cerruti" and "Liceo Scientifico Avogadro". Moreover, the analysis of survey's answers underlines that building's specific functions do not influence the choice of installing green walls, instead the visual impact of the building and the wall's visibility from the street are two significant characteristics that may influence the site's choice.

3.4   Step 4: Preliminary Multi-perspective Matrix

Data collected during previous stages were used to define criteria matrix and to assign specific weights (see Fig. 2). According to results elaborated from the survey, the highest weight was allocated to "People wellbeing" (25%), followed by "Building's condition", "Position", and "Environmental features" (20%), and at the end "Survey's preferences" (15%).

20 %		20 %		20 %		25 %		15 %	
BUILDING'S CONDITION		**POSITION**		**ENVIRONMENTAL FEATURES**		**PEOPLE WELLBEING**		**SURVEY'S PREFERENCES**	
STATUS OF BUILDING	6 %	GREEN AREAS	5 %	NOISE POLLUTION	10 %	SOCIAL ASPECTS	13 %	NON-EXPERTS	5 %
SURFACE	10 %	WALL VISIBILITY	5 %	TEMPERATURE RAISES	10 %	PSYCHOLOGICAL ASPECT	6 %	EXPERTS	10 %
MAINTENANCE	2 %	AREA DEPTH	10 %	AIR QUALITY	- %	TOURISM	6 %		
ENERGY CONSUMPTION	2 %								

Fig. 2. List of the main criteria with their weights

3.5   Step 5: Results from the Focus Group with Expert Stakeholders

Ten expert stakeholders were involved during the focus group for discussing technical aspects concerning outdoor applications of VGSs and focusing on the importance of "Energy consumption" for building's operations as one of the most significant sub-criteria of the matrix. Other sub-criteria were also widely discussed to integrate the perspective and the technical point of view of experts in socio-technical field of research. Stakeholders were asked to indicate three top choices of sites based on their personal considerations and to motivate it. Stakeholders expressed their preference for "San Paolo" train station, "Liceo Scientifico A. Avogadro", and "Menabrea". Experts highlight the relevance of the ornamental function of outdoor green wall as a tool to vehiculate eco-cultural message based on fostering ecological purposes of this type of NBSs among citizens. A green wall applied to the train station was considered as business card for the city of Biella. Moreover, the development of educational and recreational activities for young citizens, the creation of new economic opportunities and environmental compensation of a big industry were also discussed. Considerations extracted from the focus group were used to update the multi-perspective matrix and applied to obtain the final ranking of sites, as shown in the Table 3.

Table 3. Final ranking of sites' selection and evaluation combining results obtained from survey and focus group.

Building	Rank	Building	Rank
1. San Paolo	1,89	5. Sellalab	1,36
2. Menabrea	1,84	6. Liceo Scientifico A.Avogadro	1,35
3. Accademia dello Sport	1,42	7. Città Studi	1,27
4. Fondazione Pistoletto	1,36	8. Lanificio Cerruti	1,06

Finally, the "San Paolo" train station was selected for the installation of the outdoor green wall in the city of Biella.

4   Discussion and Final Remarks

The present contribute shows complex implications that may occur during the selection of the most appropriate site for installing an outdoor green wall in urban contexts. As demonstrated, the cross-functional application of this type of NBSs and the complexity of urban morphology require a multi-perspective approach able to assess tangible and intangible ecosystem services of outdoor VGSs. The spatial analysis was performed to preliminary screen an characterize suitable districts and buildings. This analysis led to identify a limited number of potential sites to be considered for outdoor green wall's installation. Considering these features, the methodology can be defined as a project-oriented process.

The simplified SWOT analysis applied to the proposed buildings highlighted their main criticalities and opportunities concerning the installation of an outdoor green wall. This preliminary buildings' characterization led to identify a set of criteria and sub-criteria that are not *"case study-specific"*, therefore they can be used to evaluate the buildings' suitability for VGSs application in other urban contexts. The decision to not consider the "Air quality" criterion and to adjust the "Temperature raises" one to better apply the multi-perspective matrix to the Biella's case study shows the opportunity to adapt the multi-criteria tool to a specific urban context. This characteristic also refers to the mixed quantitative and qualitative evaluation approach adopted to enhance the multi-perspective mindset of this methodology.

Dealing with social analysis, the multi-perspective approach that investigates non-expert opinions and preferences allowed to explore potential positive social effects of an outdoor green wall in the context of Biella. Considering the complexity in social evaluation criteria for urban greenery, survey enabled the identification of qualitative evaluation sub-criteria for intangible socio-cultural ecosystem services.

Information collected from citizens' survey highlighted the lack of knowledge about outdoor green walls and the desire to be involved in socio-cultural activities linked to them. The integration of these results influenced weights of matrix's criteria in which the "people wellbeing" criterion has the major influence, leading to a social-oriented building's selection.

The combination of personal non-experts and experts' preferences in building selection led to focus on the ornamental value of outdoor green wall as feature through which to activate socio-cultural ecosystem services. Citizens expressed that visual impact and surface' accessibility are the most significant aspect during the decision-making process. Moreover, experts' motivations at the end of the focus group revealed clear interest in outdoor green wall as a "performative ornament" that integrate ecological functions of VGSs with aesthetic purpose. Beyond considerations related to the environmental benefits, the selection of the "San Paolo" train station focuses on the aesthetic opportunity offered by outdoor green walls to improve the image of the city. This result highlights the relevance given to the socio-environmental message brought by this NBS typology. This feature is strictly related to the perception of technological greenery that drives socio-cultural ecosystem services acting through its aesthetic value. The citizen appreciation of a renovated space can foster a new concept of urban environment that integrate ecological aspects. This study presents work-in-progress results achieved in the definition of a multi-perspective tool for outdoor green wall application in urban context. The work fills the gap about site selection tools for vertical greenery system in outdoor context. The presented multi-perspective approach focuses on the development of a model that integrate environmental and socio-cultural ecosystem services. Although its academic nature, the present work aims to ease and amplify the collaboration between researchers and policymakers in the re-integration of natural elements in the urban environment.

References

1. Cameron, R.W.F., Blanuša, T.: Green infrastructures and ecosystem services – is the devil in the detail? Ann. Bot. **118**, 377–391 (2016). https://doi.org/10.1093/aob/mcw129

2. Radić, M., Dodig, M.B., Auer, T.: Green façades and living walls – a review establishing the classification of construction types and mapping the benefits. Sustainability **11**(17), 4579 (2019). https://doi.org/10.3390/su11174579

3. Castellar, J.A.C., et al.: Nature-based solutions in the urban context: terminology, classification and scoring for urban challenges and ecosystem services. Sci. Total Environ. **779** (2021). https://doi.org/10.1016/j.scitotenv.2021.146237

4. European Environment Agency, Castellari, S., Zandersen, M., Davis, M., et al.: Nature-based solutions in Europe policy, knowledge and practice for climate change adaptation and disaster risk reduction, Publications Office (2021)

5. Neonato, F., Tomasinelli, F., Colaninno, B.: Oro Verde. Quanto vale la natura in città. Il Verde Editoriale (2019)

6. Vujcic, M., Tomicevic-Dubljevic, J., Grbic, M., Lecic-Tosevski, D., Vukovic, O., Toskovic, O.: Nature based solutions to improve mental health and well-being in urban areas. Environ. Res. **158**, 385–392 (2017). https://doi.org/10.1016/j.envres.2017.06.030

7. Lu, Y., Zhao, J., Wu, X., Lo, S.M.: Escaping to nature during a pandemic: a natural experiment in Asian cities during the COVID-19 pandemic with big social media data. Sci. Total Environ. **777** (2021). https://doi.org/10.1016/j.scitotenv.2021.146092

8. Manso, M., Teotonio, I., Matos Silva, C., Oliveira Cruz, C.: Green roof and green wall benefits and costs: a review of the quantitative evidence. Renewable Sustain Energy Rev. **135** (2021). https://doi.org/10.1016/j.rser.2020.110111

9. Kremer, P., Hamstead, Z.A., McPhearson, T.: The value of urban ecosystem services in New York City: a spatially explicit multicriteria analysis of landscape scale valuation scenarios. Environ. Sci. Policy **62**, 57–68 (2016). https://doi.org/10.1016/j.envsci.2016.04.012

10. Cheng, X., Van Damme, S., Uyttenhove, P.: A review of empirical studies of cultural ecosystem services in urban green infrastructure. J. Environ. Manag. **293** (2021). https://doi.org/10.1016/j.jenvman.2021.112895

11. Németh, B., et al.: Comparison of weighting methods used in multicriteria decision analysis frameworks in healthcare with focus on low- and middle-income countries. J. Comp. Effectiveness Res. **8**(4) (2019). https://doi.org/10.2217/cer-2018-0102

12. Arpa Piemonte. Qualità dell'aria Biella. http://www.arpa.piemonte.it/approfondimenti/temi-ambientali/aria/aria/dati-giornalieri-di-particolato-pm10. Accessed 07 June 2021

13. Comune di Biella. http://sit2.comune.biella.it/maps/sitbiella.php?m=bi_prg&w=ks&o=0.8. Accessed 09 Dec 2021

14. Sole 24 ore, Qualità della vita (2018). https://lab24.ilsole24ore.com/indice-del-clima/?Biella. Accessed 09 Dec 2021

15. Musco, F., Fregolent, L., Magni, F., Maragno, D., Ferro, D.: Calmierare gli impatti del fenomeno delle isole di calore urbano con la pianificazione urbanistica: esiti e applicazioni del progetto UHI (Central Europe) in Veneto. In: Gaudioso, D., Giordano, F., Taurino, E. (eds.) Focus su Le città e la sfida dei cambiamenti climatici - Qualità dell'Ambiente Urbano X Rapporto. ISPRA, Roma (2014)

Systemic Decision Support Tool for Online Application to Aid NBS Co-creation

Fábio Matos[1]([⊠]), Rita Mendonça[1], Peter Roebeling[1,2], Piersaverio Spinnato[3],
Giovanni Aiello[3], Rúben Mendes[1], Maria Isabel Bastos[1], Max López-Maciel[1],
and Antonino Sirchia[3]

[1] CESAM & Department of Environment & Planning, Aveiro University (DAO-UA), Aveiro,
Portugal
{fabiomatos,ritaml1,peter.roebeling,ruben.tiago,mariaisabel,
max}@ua.pt
[2] Wageningen Economic Research, Wageningen University and Research (WUR), Wageningen,
Netherlands
[3] Engineering, Piazzale dell'Agricoltura, 24, 00144 Rome, Italy
{piersaverio.spinnato,giovanni.aiello,antonino.sirchia}@eng.it

Abstract. The world is facing numerous environmental challenges due to global change. These come in the form of air and water pollution, heat islands, floods, sprawl and gentrification issues, which stand as threats to citizen health, predominantly in urbanized areas, where most of the population is concentrated. It is now more than ever necessary to provide relevant information and help decision makers adopt measures for a green transition, which can be achieved by implementing nature-based solutions (NBS). The systemic decision support tool (SDST) and associated NBS simulation visualization tool (NBS-SVT) is a two-part online application that stores and manages technical multi-disciplinary data models concerning urban heat, air quality, flooding, and sprawl, gentrification and real-estate valuation, displaying them in the form of interactive and easy to digest geographical information. The application aims to help co-creation processes and decision making by showing the results of urban simulations with no action compared to NBS implementation across different climate scenarios. The aim of this paper is to present the SDST/NBS-SVT, explain its functioning and structure, as well as its capabilities for use in informed co-creation processes. The SDST/NBS-SVT has already been used in a number of workshops and online sessions, proving its potential as a means to transmit technical information in an easily understandable manner that can reach audiences with any level of technical knowledge.

Keywords: Decision support · Online application · Nature-based solutions · Decision making · Co-creation · Simulation

1 Introduction

It is estimated that over 75% of the European population currently lives in cities [1], and it is expected that up to 68% of the world's population will be as well by the year 2050

© The Author(s), under exclusive license to Springer Nature Switzerland AG 2022
F. Calabrò et al. (Eds.): NMP 2022, LNNS 482, pp. 1916–1925, 2022.
https://doi.org/10.1007/978-3-031-06825-6_184

[2]. With this the subsequent growth in urbanization, problems such as soil impermeabilization, loss of natural spaces, greenhouse gas (GHG) emissions and chemical releases to the environment, are expected to increase as well. In combination with climate change, this can lead to urban-environmental issues manifested in the form of urban heating, air and water pollution, flooding, ecosystem degradation, biodiversity losses, and sprawl, gentrification and real-estate devaluation [3].

In recent years, policy makers such as the European Commission started increasing their focus on urban environmental issues and searching for ways to mitigate them. One possible solution that has been the target of significant study is the incorporation of nature-based solutions (NBS) in city planning. NBS are actions inspired by, supported by or copied from nature, incorporating complex natural systems that enhance existing solutions to address in a sustainable way a variety of environmental, social and economic challenges [4]. NBS are energy and resource-efficient and can reduce disaster risk while promoting human well-being and socially inclusive green growth [4]. Moreover, there is emerging evidence that these actions can bring more natural features and processes into cities, providing effective solutions to multiple challenges [5].

Despite the growing scientific information at our disposal, the transition to more resilient and sustainable societies is slow and faced with several challenges. In order to create awareness and potentiate the rate of change, it is important to inform citizens, private sectors, non-government and government organizations to these urban issues, as well as potential solutions, and work to adopt more sustainable policies.

Decision support tools to help stakeholders (such as citizens, businesses, non-government organization and local governments) decide on what specific NBS suits their needs are scarce, even more so when considering tools that allow for the study of multiple benefits and services [6]. Such tools are necessary to aid co-creation and decision making in processes of participatory planning and public discussion, allowing for improved stakeholder awareness of the potential of NBS, and therefore acting as key components to changing the mentalities and incorporating these solutions in our society. Some tools of this type have been developed or are currently under development, as is the case of Deltares's Adaptation Support Tool (https://crctool.org/en/), Fregonese Associates' Envision Tomorrow (http://envisiontomorrow.org/) or Nature4Cities's NBenefit$ (https://mimes.list.lu/). However, these tools often come with several limitations, such as non context specificness, poor accessibility and presentation, limited scenario selection options, unclear modelling frameworks or lack of data resolution and interdisciplinarity.

The systemic decision support tool (SDST) and associated NBS simulation visualization tool (NBS-SVT) constitute a tool used to display a variety of environmental scenarios by simulating the effects of NBS implementation in different areas of the Urban Nature Labs (UNaLab; https://unalab.eu/en) front-runner cities (Eindhoven, Netherlands; Tampere, Finland; Genoa, Italy). These NBS scenarios were co-created with the input of the municipalities and citizens during workshops using the SDST/NBS-SVT. It is the first of its kind to offer spatially explicit and context-specific simulation results for multiple impacts of NBS across a variety of scenarios in an integrated manner. The aim of this paper is to present the SDST/NBS-SVT, explain its functioning and structure, as well as its capabilities for use in informed co-creation processes.

2 Methodology

2.1 Systemic Decision Support Tool

SDST Rationale and Objectives

The systemic decision support tool (SDST) integrates and builds upon data and information from disciplinary component models into a spatially explicit framework at the landscape scale [7]. The tool assesses the direct and indirect impacts of NBS on urban heat and air quality, flooding and water quality, and sprawl, gentrification and real-estate valuation [8]. Developed as part of the UNaLab ICT Framework, the SDST is a powerful tool for participatory planning, enabling stakeholders to visualize and discuss potential environmental, social and economic impacts of NBS and strategies in the face of local challenges [9].

The main goal of the SDST is to assist in co-creation ideation and design processes by providing the users with scenario simulations of the impacts of NBS. Its underlying principle is that NBS are co-created in a transparent, transdisciplinary, multi-stakeholder and participatory context as well as systematically incorporated into urban landscape planning. It aims to facilitate the participatory planning process and public discussion by improving stakeholder awareness about the multiple direct and indirect impacts of NBS. Hence, the SDST enriches public discussion, adds transparency and increases societal benefits. [8].

Fig. 1. SDST technical user flow diagram

From a technical point of view, SDST development starts with the selection of the disciplinary models to be used in the assessment of the different impacts of NBS (see Fig. 1). Statistical data used to feed these disciplinary models is analysed, prepared and stored in a common database depending on the type of information. Disciplinary models are then parameterized, calibrated and validated, such that reliable results are obtained [8]. Afterwards, NBS scenarios to be simulated are defined and characterized in detail, enabling the necessary changes to be made in disciplinary model input data (e.g. changes in land use or infiltration capacity). In turn, simulations of NBS scenarios are performed and the results organized and stored in a geodatabase (GeoDB). Then, the NBS scenario simulation results are uploaded onto the NBS simulation visualization tool (NBS-SVT), such that Baseline and NBS scenario simulation results can be viewed side-by-side [8]. Finally, internal feedback is collected regarding the data presented (colour codes, value ranges, zoom levels, coherence between baseline and scenario) to be used in its validation or, if necessary, readjustment and reuploading.

SDST Structure

The SDST, and its disciplinary component models, share a common database where the information is taken from. The UNaLab Knowledge Base (UKB) is a platform serving as the database to provide city data relative to the various aforementioned disciplinary models, such as meteorology, land use, emissions, air and water quality, demographics and socioeconomics. This data is then used alongside the specific scenario characteristics in modelling software to create the simulations (see Fig. 2) [9]. The state-of-the-art disciplinary component models constitute:

- Urban heat and air quality, assessed using the coupled WRF-CHEM model [10].
- Flooding and water quality, assessed using the InfoWORKS ICM [11] model (for Eindhoven) or the MIKE-Urban/Flood [12, 13] model (for Tampere and Genoa).
- Sprawl, gentrification and real-estate valuation, assessed using the Sustainable Urbanising Landscape Development (SULD) model [14].

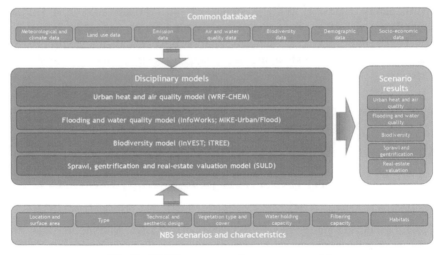

Fig. 2. SDST structure and component models

Note that the biodiversity model is currently under development and that other disciplinary component models can be used within the SDST – depending on the impacts to be assessed as well as the models available and generally used by local governments. It is, however, important to select disciplinary models that use similar temporal and spatial scales and to provide the output data in the GeoJSON (RFC7946) data format.

The simulations are performed for the Baseline scenario (2015), for the NBS scenarios under reference (Baseline) climate and population conditions (2030) and for the NBS scenarios under future climate and/or population conditions (2050). NBS scenarios and characteristics (e.g. NBS type, location and area, design and technical features), required by the disciplinary component models, are derived from local information (local knowledge; maps; design drawings; etc.) as well as from the NBS Technical Handbook [15].

Climate change projections for average temperature and cumulative precipitation by '2050' are estimated by applying the Weather Research and Forecasting (WRF) model forced by the Max Planck Institute Earth System Model (low-resolution MPI ESM-LR model results) and using the IPCC greenhouse gas concentration scenario RCP4.5 [16]. Population growth projections from '2015' to '2050' are based on municipal or regional projection data, in combination with extrapolations estimated using the Exponential Triple Smoothing (ETS) algorithm [16].

2.2 Nature-based Solution Simulation Visualization Tool

NBS-SVT Architecture and Implementation
The NBS-SVT was built using DigitalEnabler® (https://demo.digital-enabler.eng.it/ suite/), a powerful open source data-driven digital ecosystem platform provided by Engineering Ingegneria Informatica S.p.A.. DigitalEnabler provides mechanisms and features for data collection, standardisation and valorisation, including a dashboard modeler, which provides the capabilities to build self-contained dashboards presenting the data models through the use of interactive widgets, such as charts, diagrams and tables.

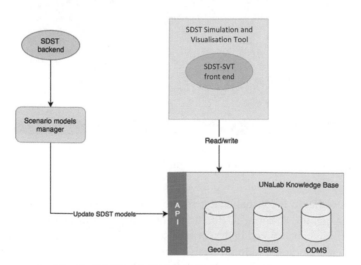

Fig. 3. SDST and NBS-SVT reference architecture

The reference architecture of the internet-based NBS-SVT is characterised by the decoupling of the data management from the components related to the front-end interaction (see Fig. 3). The 'Scenario models manager' will support the uploading of the SDST scenario simulation data models onto the GeoDB. The GeoDB, logically included in the UNaLab Knowledge Base (UKB), is the backend component in charge of the persistence and management of the NBS indicators and any other related information. The

NBS-SVT will interact with the UKB to retrieve the geodata models to be shown through the front-end user interfaces.

NBS-SVT Usage

The internet-based NBS simulation visualization tool (NBS-SVT; http://unalab.eng.it/ nbssvt_v4/) is used to present the NBS scenario simulation results managed by the SDST, acting as its front-end counterpart, in a graphic and intelligible way. The NBS-SVT supports users such as decision makers and citizens with the means to compare the potential environmental, social and economic impacts of non-action against the implementation of several possible NBS in one of the available cities (see Fig. 4). The goals of the NBS-SVT are to allow users to visualize and compare NBS scenarios relative to the Baseline scenario; evaluate alternatives, opportunities and feedbacks; identify future strategic objectives; and visualize geographical results.

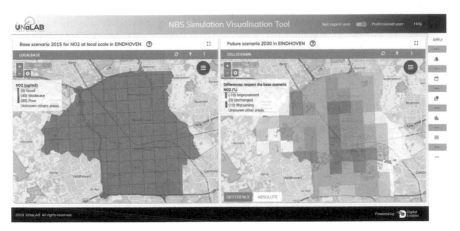

Fig. 4. Depiction of the NBS-SVT with the baseline scenario map (2015) on the left and one possible NBS scenario simulation map on the right for comparison.

The data available for visualization in the NBS-SVT encompasses four different indicators: air quality, urban heat, flooding, and sprawl, gentrification and real-estate valuation. These can be studied in the form of several parameters each, either at 'Local' (gridcell), 'Neighbourhood' (administrative unit), or City scale (municipal unit) for any of the three UNaLab front-runner cities: Eindhoven (Netherlands), Tampere (Finland) or Genoa (Italy). These parameters reflect the impact of the different NBS scenarios used in the models. Currently, between five and eight different scenarios can be visualized depending on the chosen city, including solutions such as river daylighting, grass block pavement, green roofs, green/blue spaces, street trees and bioswales, with some scenarios presenting variations in design and application or even mixed use of these solutions. By selecting the desired indicators and parameters, the user can then observe the data for the reference Base scenario (2015), the NBS scenario simulation under the reference climate and population conditions (2030) and for NBS scenarios under future climate and population conditions (2050) (Roebeling et al. 2021b), in the form of absolute values or percentage difference from the Baseline. The comparison between Baseline and Scenario simulations is done by visualization of the respective maps side by side in the same window, with each one being interactive and independent from the other.

The NBS-SVT comes with an intuitive and easy to use sliding menu containing the selection parameters the user wants to access, including NBS scenarios, climate condition scenarios (represented by the years 2030 and 2050), parameters for each indicator, return periods for specific indicators such as flooding, and scale options. An information button at the top of the page also provides information on every indicator parameter, while descriptions for singular parameters can be accessed by clicking a similar information button on the selection menu for the parameters themselves. The same goes for the NBS scenarios which all come with a detailed introduction and description of the city changes they represent and the issues they are expected to address. The data represented in the form of maps can also be visualized in the form of charts and numeric tables by clicking the appropriate button when selecting the desired indicator parameter. These as well as the maps themselves can also be individually exported as images and data sets. A user-experience tutorial detailing all of these options is displayed upon first entering the application to ensure the user is properly introduced to the NBS-SVT. A slider with "Professional" and "Non Expert" user option can also be toggled to open-up or limit, respectively, the amount of indicator parameters available for selection based on their relative simplicity and representativity.

3   Application: SDST/NBS-SVT for Co-creation

The SDST/NBS-SVT aims to provide stakeholders with different scenarios that are aligned with the cities' necessities, in order to tackle specific environmental, social and economic problems through the implementation of NBS. Hence, stakeholder workshops were held in the UNaLab front-runner cities (Eindhoven, Tampere and Genoa), to pinpoint their problems and visions for the future so that the proposed scenarios could represent them and contribute to the informed decision of NBS planning and strategies.

Co-creation workshops are acknowledged to improve the participatory character of the planning process, as the active participation of stakeholders from the beginning of the process is crucial to ensure the best possible outcomes, especially when a transdisciplinary problem is considered [8]. As proposed by Rios-White et al. [17], according to the Life Cycle Co-Creation Process (LCCCP) for NBS, the SDST/NBS-SVT contributes to the Co-Experiment stage. I.e., the SDST allows to experiment with different NBS measures (individual NBS) and strategies (suites of NBS), assessing their effectiveness and evaluating their multiple impacts, costs and benefits. This, in turn, will help to decide on the most desirable NBS to be implemented (Co-Implement stage).

The workshop model, following Löschner et al. [18] and Picketts et al. [19], envisioned to understand the main city problems and identify the best types and locations of NBS to address these problems in a co-creation setting, divided in four parts. The first entailed the SDST/NBS-SVT tool introduction and the establishment of the workshop objectives and activities. The following part included the problem identification and case study area definition, where groups discuss existing and expected environmental, social and economic problems in their city while indicating them on a map. To facilitate this process, a list of common urban problems (e.g. air quality, water quality, traffic, segregation, housing prices, etc.) was provided. The outcome of this part is a map with the city problems identified and spatially distributed and, consequently, the case study area.

Accordingly, the following workshop part, NBS mapping, used the city problem map as a base to brainstorm on NBS types and locations, considering their benefits and co-benefits. Again, to facilitate the discussion, a list of NBS definitions and benefits was provided. The objective was that stakeholders were able to make connections between the problems and the NBS that have the ability to tackle them, while considering the stakeholders' visions and possibilities. After the NBS identification, they were also spatially mapped, considering the problems' locations.

Following the creation of the city maps with the proposed NBS scenarios, the stakeholders explored the SDST/NBS-SVT tool to visualize different NBS scenarios already simulated to get a grasp on the NBS benefits and better understand the tool and how the results will be presented after the simulation of the workshop scenarios. The workshop is finalized with a plenary session to present and discuss city planning and design future strategies and, specifically, i) identified problems, ii) NBS chosen to tackle these problems, iii) SDST/NBS-SVT tool feedback.

On the one hand, as a general outcome of the workshops, stakeholders reflect on several city problems and possible NBS to tackle them, helping them to better integrate NBS for global change adaptation. On the other hand, and as an essential component for the development of the SDST/NBS-SVT, the necessary information was compiled to create and feed new NBS scenarios that are better adapted to the city's problems, capacities and expectations. Consequently, based on the identified city problems and the correspondingly identified NBS to address these problems, NBS scenarios were defined to be included for simulation using the SDST. In turn, results from these new NBS scenario simulations were uploaded onto the NBS-SVT and, thus, contribute to a more evidence-based and sustainable decision-making process.

4   Conclusions

This paper has succinctly explained the objectives as well as intricacies of the development of the application Systemic Decision Support Tool (SDST) and associated online NBS Simulation Visualization Tool (NBS-SVT). The two-part application is a powerful tool for co-creation and decision making focused on the geographical representation of the multiple direct and indirect impacts of nature-based solutions across several indicators and scenarios to better inform stakeholders of the value of these actions. The SDST acts as the data engine that manages the multidisciplinary simulation scenarios to be presented. The NBS-SVT is the interactive front-end component that displays all the scenarios and information in an intuitive way that intends to be easily operated and explored by users without requiring any sort of technical knowledge.

The simulations presented are a result of co-creation workshops with stakeholders discussing which solutions they envision to implement to address specific problems within their cities, and posterior data analysis by a team of experts using modelling software. The workshops as well as breakout sessions in which the SDST/NBS-SVT was displayed, have proved the potential of this application to reach a wider audience with varying degrees of technical expertise in the area. They also showed the tool is capable of transmitting information in a clear and concise manner that is easy to understand and compare across scenarios, both face-to-face and online, making it an invaluable tool to provide insight on the impacts, costs and benefits of NBS in the face of global change and in times of the global COVID pandemic.

The tool is, however, not without its limitations. It relies on specialized, data demanding and context-specific models that need to be applied by qualified staff. In addition, these models are generally large, requiring heavy duty computers to limit simulation time. As a consequence, scenario simulations are pre-run by the SDST and results uploaded onto the NBS-SVT for visualization – hence not allowing real-time scenario definition, simulation and visualization.

Future research will focus on expanding the range of NBS impact data, mainly by adding monetary model calculations and indicators (such as avoided flood damage or health costs). In addition, more NBS, including combined NBS strategy scenario simulations will be included as well as cost-benefit indicators. Refinement of the model presentation is also a constant until the final iteration of the tool is reached.

Acknowledgements. This work was supported by the UNaLab project (Grant Agreement No. 730052, Topic: SCC-2-2016-2017: Smart Cities and Communities Nature-based solutions). Thanks are also due to FCT/MCTES for the financial support to (UIDP/50017/2020 + UIDB/50017/2020), through national funds.

References

1. Oueslati, W., Alvanides, S., Garrod, G.: Determinants of urban sprawl in European cities. Urban Stud. **52**(9), 1594–1614 (2015). https://doi.org/10.1177/0042098015577773
2. United Nations Department of Economic and Social Affairs: World urbanization prospects: The 2018 revision. Working paper No. ESA/P/WP.252 (2018). https://population.un.org/wup/Publications/Files/WUP2018-Methodology.pdf
3. UN: World Population Prospects: Key Findings and Advance Tables (2017 Revision). Working Paper No. ESA/P/WP/248, Department of Economic and Social Affairs (DESA), United Nations (UN), New York, US, 46p. (2017)
4. EC: Towards an EU research and innovation policy agenda for nature-based solutions & re-naturing cities (2015). https://doi.org/10.2777/765301
5. European Commission (EC): Nature-Based Solutions page. https://ec.europa.eu/research/environment/index.cfm?pg=nbs. Accessed 10 Jan 2018
6. Croeser, T., Garrard, G., Sharma, R., Ossola, A., Bekessy, S.: Choosing the right nature-based solutions to meet diverse urban challenges. Urban For. Urban Green. **65**, 1–11 (2021). https://doi.org/10.1016/j.ufug.2021.127337
7. Bohnet, I.C., et al.: Landscapes toolkit – an integrated modeling framework to assist stakeholders in exploring options for sustainable landscape development. Landsc. Ecol. **26**, 1179–1198 (2011)
8. Roebeling, P.C., et al.: Systemic decision support tool user guide for municipalities. UNaLab project (https://www.unalab.eu), Deliverable D3.2 of 30-03-2021, CESAM – Department of Environment and Planning, University of Aveiro, Aveiro, Portugal (2021)
9. Roebeling, P.C., et al.: Internet-based systemic decision support tool application. UNaLab project (https://www.unalab.eu), Deliverable D4.6 of 28-05-2021, CESAM – Department of Environment and Planning, University of Aveiro, Aveiro, Portugal, 44p. (2021)

10. Grell, G.A., et al.: Fully coupled "online" chemistry within the WRF model. Atmos. Environ. **39**(37), 6957–6975 (2005)
11. Innovyze: InfoWorks ICM Help Documentation (2018). https://help.innovyze.com/display/infoworksicm
12. DHI: MIKE-Urban Modelling User Manual. Danish Hydraulic Institute (DHI), Horsholm, Denmark (2007). http://www.dhigroup.com/
13. DHI: MIKE-Flood 1D-2D Modeling User Manual. Danish Hydraulic Institute (DHI), Horsholm, Denmark (2007). http://www.dhigroup.com/
14. Roebeling, P.C., et al.: Assessing the socio-economic impacts of green/blue space, urban residential and road infrastructure projects in the Confluence (Lyon): a hedonic pricing simulation approach. J. Environ. Plann. Manag. **60**(3), 482–499 (2017)
15. Eisenberg, B., Polcher, V.: Nature-Based Solutions Technical Handbook. UNaLab project (https://www.unalab.eu/), Deliverable D5.1 of 31-05-2018, Institute of Landscape Planning and Ecology, University of Stuttgart, Stuttgart, Germany, 14p.+Appendices (2018)
16. Mendonça, R., et al.: Population growth and climate change scenarios for front-runner cities. Milestone Report M3.2, CESAM & Department of Environment and Planning, University of Aveiro, Aveiro, Portugal, 43p. (2020)
17. Ríos-White, M.I., Roebeling, P., Valente, S., Vaittinen, I.: Mapping the life cycle co-creation process of 408 nature-based solutions for urban climate change adaptation. Resources **9**(4), 39 (2020). https://doi.org/10.3390/resources9040039
18. Löschner, L., Nordbeck, R., Scherhaufer, P., Seher, W.: Scientist-stakeholder workshops: a collaborative approach for integrating science and decision-making in Austrian flood-prone municipalities. Environ. Sci. Policy **55**, 345–352 (2016)
19. Picketts, I.M., Werner, A.T., Murdock, T.Q., Curry, J., Déry, S.J., Dyer, D.: Planning for climate change adaptation: lessons learned from a community-based workshop. Environ. Sci. Policy **17**, 82–93 (2012)

An Evaluation Approach to Support Urban Agriculture Implementation in Post-covid19 Cities: The Case of Troisi Park in Naples

Marco Rossitti[1]([✉]) [iD], Chiara Amitrano[2] [iD], Chiara Cirillo[2] [iD],
and Francesca Torrieri[3] [iD]

[1] Politecnico di Milano, Via Bonardi 3, 20133 Milan, Italy
marco.rossitti@polimi.it
[2] Università degli Studi di Napoli Federico II, Via Università 100, 80055 Portici, Italy
[3] Università degli Studi di Napoli Federico II, Piazzale Vincenzo Tecchio 80,
80125 Naples, Italy

Abstract. The Covid-19 pandemic has had a dramatic impact worldwide by producing, especially in urban contexts, severe consequences not only in the healthcare field but also in socioeconomic terms with visible implications for food security. In this difficult context, Urban Agriculture (UA) stands as a valuable means to ensure social, environmental, and economic benefits for urban realities. Indeed, UA implementation's multi-dimensional opportunities can also be read in terms of ecosystem services. However, despite this wide acknowledgment of UA's multiple benefits, a gap exists between the number of policies already implemented to promote urban agriculture and the demand for these policies. The reasons for this gap can be found both in prejudice towards agriculture as a low-income activity, discouraging private investments, and in public administration's reduced financial capacity.

In this light, the paper proposes an evaluation approach based on a *Sustainability Assessment* to support agriculture-led implementation processes in urban spaces by dealing with financial constraints. This methodology is tested on the regeneration of Troisi Parks' greenhouses in Naples, which have recently been the subject of the Urbanfarm design challenge within the EU H2020 FoodE Project.

Thus, after describing the main features of the International Challenge and the project for Troisi Park, the paper delves into the application of the *Sustainability Assessment* to support Troisi Park's regeneration. Finally, the opportunities stemming from such an evaluation approach to UA and its possible room for improvement are discussed.

Keywords: Urban agriculture · Sustainability Assessment · Ecosystem services

1 Introduction

The Covid-19 pandemic has had a dramatic impact worldwide by producing, especially in urban contexts, severe consequences not only in the healthcare field but also

© The Author(s), under exclusive license to Springer Nature Switzerland AG 2022
F. Calabrò et al. (Eds.): NMP 2022, LNNS 482, pp. 1926–1936, 2022.
https://doi.org/10.1007/978-3-031-06825-6_185

in socio-economic terms [1]. More in detail, the lock-down adopted to tackle infections has resulted in a general drop in incomes and an increase in poverty levels with visible consequences for food security [2]. Furthermore, the pandemic's effects have been combined with other negative phenomena for food security as increases in the global population [3], climate change [4], and built areas expansion [5].

In this difficult context, Urban Agriculture (UA) stands as a valuable means to ensure social, environmental, and economic benefits for urban realities. Indeed, in social terms, especially in at-risk areas, UA can address urban poverty, improve local communities' well-being and promote social inclusion [6, 7]. From an environmental perspective, UA ensures several benefits: biodiversity protection, promoting organic waste cycles [8], tackling climate change and heat islands effects typical of urban areas by increasing the amount of urban green areas [9]. Finally, UA can produce several benefits even in economic terms, from meeting the urgent need for food security, exacerbated by the Covid pandemic [10], to promoting short food supply chains and creating new job opportunities at the local level [11].

The multi-dimensional opportunities from urban agriculture implementation can also be read in terms of provided Urban Ecosystem Services (UESs) [12]. Indeed, taking as reference the categorization of Ecosystem Services (ESs) provided by the Millennium Ecosystem Assessment (MEA) [13], the multi-dimensional benefits from UA implementation become even more evident (see Fig. 1). Indeed, UA implementation enhances several ecosystem services [14], primarily regulating and cultural services.

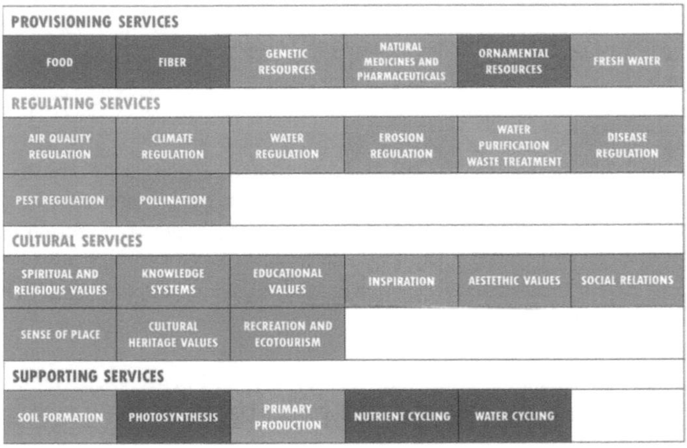

PROVISIONING SERVICES					
FOOD	FIBER	GENETIC RESOURCES	NATURAL MEDICINES AND PHARMACEUTICALS	ORNAMENTAL RESOURCES	FRESH WATER
REGULATING SERVICES					
AIR QUALITY REGULATION	CLIMATE REGULATION	WATER REGULATION	EROSION REGULATION	WATER PURIFICATION WASTE TREATMENT	DISEASE REGULATION
PEST REGULATION	POLLINATION				
CULTURAL SERVICES					
SPIRITUAL AND RELIGIOUS VALUES	KNOWLEDGE SYSTEMS	EDUCATIONAL VALUES	INSPIRATION	AESTETHIC VALUES	SOCIAL RELATIONS
SENSE OF PLACE	CULTURAL HERITAGE VALUES	RECREATION AND ECOTOURISM			
SUPPORTING SERVICES					
SOIL FORMATION	PHOTOSYNTHESIS	PRIMARY PRODUCTION	NUTRIENT CYCLING	WATER CYCLING	

Fig. 1. Ecosystem services provided by Urban Agriculture (colored cells) according to the MEA's categorization

However, despite this wide acknowledgment of UA's multiple benefits, a gap exists between the number of policies already implemented to promote urban agriculture and the demand for these policies from local communities and third sector entities. The reasons for this gap can be found both in prejudice towards agriculture as a low-income activity, discouraging private investments, and in public administrations' reduced financial capacity [15].

In this light, the paper proposes an evaluation approach to support agriculture-led implementation processes in urban spaces by dealing with financial constraints.

Such an approach is grounded on a *Sustainability Assessment* to define a priority order among the different designed interventions to be implemented within a UA project by considering the related multi-dimensional benefits and constraints.

This methodology is tested on the regeneration of Troisi Parks' greenhouses in Naples, which have recently been the subject of the Urbanfarm design challenge within the EU H2020 Project Food Systems in European Cities (FoodE).

Thus, after describing the Urbanfarm International Challenge requirements and the designed regeneration project for Troisi Park, the paper delves into applying the *Sustainability Assessment* to support Troisi Park's regeneration. Finally, the opportunities stemming from such an evaluation approach to UA and its possible room for improvement are discussed.

2 The Case Study: The Project for Troisi Park Regeneration Within the UrbanFarm2021 International Challenge

Urban Farm is an International Challenge funded within the EU H2020 Project Food Systems in European Cities (FoodE). Its objective is to promote the regeneration of neglected or vacant urban spaces through UA by embracing the FoodE mindset: "Think global, eat local" [16].

The Challenge asks participants to propose redesigning an abandoned agricultural area of 0.5 ha in Troisi Park in Naples. The site comprises four abandoned greenhouses (G1–G4), built in the 1980s and never used, and two open spaces (OS1–OS2). The Park belongs to Naples municipality, which allocated a budget of 70.000,00 €.

Since Troisi park is located in a tough neighborhood, victim of the 1960s' building speculation and affected by social instability, the Challenge requires creating a sustainable agricultural park able to flank production with social activities for the local communities.

In this light, the design of Troisi Park, based on a strongly participatory approach, leads to the creation of a multifunctional park, in which each space serves a different function (see Fig. 2) by adopting a multi-dimensional perspective towards sustainability. More in detail, Greenhouses G1, and G3 are devoted to crop production, G2 is dedicated to workshops and social events, and G4 is conceived as a community garden and a social garden for low-income people. The open spaces are designed to host a local market and other social events.

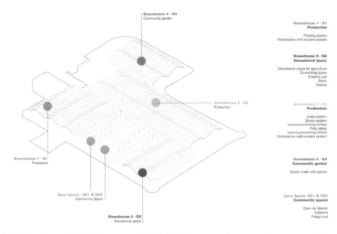

Fig. 2. Functional distribution of Troisi Park according to the proposed redesign

The architectural design adopts a circular economy perspective by promoting the reuse of recycled materials and the creation of multi-purpose furniture [17]. The agricultural practices in the production greenhouses aim to promote biodiversity, efficient application, and cycling water and nutrients within the growing systems. Finally, the project's environmental sustainability is addressed by fostering renewable energy sources, designing a rainwater collection, and setting zero-waste strategies.

3   *Sustainability Assessment* to Support the Agriculture-LED Regeneration of Troisi Park in Naples

Once the regeneration project for Troisi Park is designed, it is necessary to deal with its sustainability and feasibility, thus better investigating the opportunities and efforts for Naples municipalities related to its implementation.

More in detail, the need to immediately launch the project, thus contributing to the neighborhood's regeneration, combined with the budget constraints (the available budget accounts for 70,000.00 €, then increased by 35.000,00 € through a crowd-funding campaign) calls for an incremental approach to the project. This approach implies that the different designed interventions are developed according to specific timing, depending on the access to funding sources.

In this light, it is helpful to define an evaluation approach to support the project development by defining a priority order among the different designed interventions. Based on the key role of sustainability in the project, the proposed evaluation approach rests on a *Sustainability Assessment* (SA), performed through a Multi-Criteria Analysis.

SA is an approach to evaluation that «can help decision-makers decide what actions they should take […] in an attempt to make society more sustainable» [18]. In this context, considering the proposed project's incremental nature, this evaluation tool is geared towards understanding the most suitable intervention scenario for the project launch in light of the budget constraints (105.000,00 €).

The evaluation process is built referring to a solid conceptual framework (see Fig. 3), articulated in two parts: the SA principles and the SA procedure [19].

Thus, bearing in mind the SA principles [20], each step of the SA procedure is applied to the proposed project to accomplish the prefixed objective: to support the implementation of agriculture-led Troisi Park regeneration by selecting the project launch's favorable scenario.

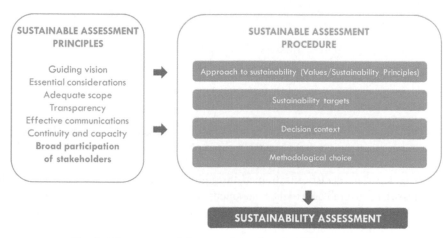

Fig. 3. The conceptual framework for Sustainability Assessment

3.1 Approach to Sustainability and Sustainability Targets

In performing the SA to support decisions for Troisi Park regeneration, the sustainability theme is approached through a broad perspective. Indeed, the concept of sustainability is not limited to the economic dimensions but opens to the other three sustainability pillars (social, environmental, and cultural).

3.2 Decision Context and Methodological Choice

The SA adopts a *scenario approach*: it identifies different scenarios and evaluates them by referring to criteria related to the four sustainability pillars.

More in detail, four different scenarios are defined by considering each of them as the recovery of a single greenhouse, together with the local market in the outdoor spaces.

Concerning the methodological choice, the decision issue to be faced calls for the Multi-Criteria Analysis (MCA) as the most suitable methodology to perform the *Sustainability Assessment* (see Fig. 4) [21, 22].

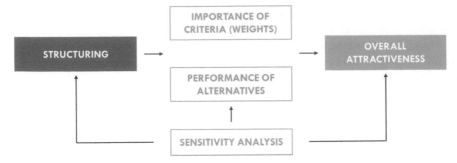

Fig. 4. Multi-Criteria Analysis (MCA) phases

3.3 Sustainability Assessment

Coming to the description of the assessment process to identify the favorable scenario for the project launch, it follows the MCA main phases:

1. *Structuring.* The decision is framed through a decision tree (see Fig. 5), which expresses: the overall goal of the assessment, the specific goals in the form of *criteria*, the *sub-criteria* to further detail the specific objectives, and the indicators to measure each sub-criteria.

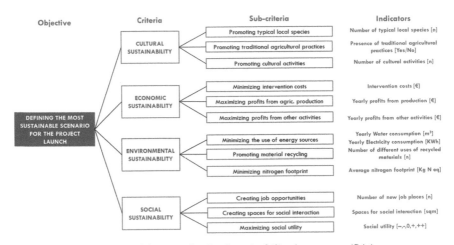

Fig. 5. Decision tree for the *Sustainability Assessment* (SA.)

In light of the approach to sustainability placed at the project's core, the selected criteria correspond to the four sustainability pillars as specific objectives to pursue. Each *criterion*, then, is better specified by defining through a literature review process [23] three relevant *sub-criteria* and the related *indicators* to measure the alternatives' performances.

2. *Performance of alternatives.* This phase is geared towards measuring the performance of each scenario in terms of each *sub-criteria.* Its output is the performance matrix displaying the four scenarios' performances (Table 1).

Table 1. Performance matrix for the *Sustainability Assessment*

Criteria	Sub-criteria	Indicator	C/B	U.m.	G1+ market	G2+ market	G3+ market	G4+ market
Cultural dimension	Promoting typical local species	Number of typical local species	B	n°	6	5	13	8
	Promoting traditional agricultural practices	Presence of typical agricultural practices	B	Yes/No	No	No	Yes	No
	Promoting cultural activities	Number of cultural activities	B	n°	1	5	1	2
Economic dimension	Minimizing intervention costs	Intervention costs	C	€	81.518,32	116.967,20	95.832,90	64.169,59
	Maximizing profits from agricultural production	Yearly profits from production	B	€	106.718,85	0,00	60.858,67	0,00
	Maximizing profits from other activities	Yearly profits from other activities	B	€	8.279,04	47.055,04	8.279,04	14.279,04
Environmental dimension	Minimizing the use of energy sources	Yearly water consumption	C	mc	20	18	1460	600
		Yearly electricity consumption	C	kWh	4949	2291	7713	2265
	Promoting material recycling	Number of different uses of recycled materials	B	n°	3	2	4	3
	Minimizing nitrogen footprint	Average nitrogen footprint	C	Kg N eq	75,1	4,7	92,1	124,3
Social dimension	Creating job opportunities	Number of new job places	B	n°	2	2	2	1

(*continued*)

Table 1. (*continued*)

Criteria	Sub-criteria	Indicator	C/B	U.m.	G1+ market	G2+ market	G3+ market	G4+ market
	Creating spaces for social interaction	Spaces for social interaction	B	sqm	435	900	410	450
	Maximizing social utility	Social utility	B	(−, 0, +)	Low (−)	Medium (0)	Low (−)	High (+)

3. *Importance of criteria (weights).* This phase aims to express the criteria' contribution in achieving the overall objective by defining *weights*. *Weights* are determined in a participatory way by the local community. During an online meeting held with local citizens and local associations, weights are generated by applying the *Swing Weights Method* in a card game format (Table 2) [24].

Table 2. Set of weights in the community's perspective

Cultural	Economic	Environmental	Social
0,250	0,201	0,273	0,277

4. *Overall attractiveness.* This phase defines a priority order among the different intervention scenarios. It relies on a weighted linear combination of the scenarios' performances related to each *sub-criteria* to obtain an overall score for each alternative. The alternatives' final ranking shows that the G2+market scenario represents the most favorable one for the project launch (see Fig. 6).

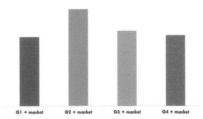

Fig. 6. Alternatives final ranking according to the weights defined by the local community

However, the G2+market scenario is incompatible with budget constraints (105.000,00 €). For this reason, it is worthy to start the project implementation from the runner-up scenario (G3 + market) and develop the other interventions according to the MCA results (see Fig. 7).

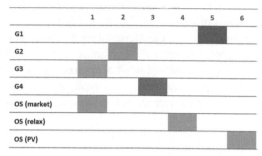

Fig. 7. Order of interventions suggested by the *Sustainability assessment*

5. *Sensitivity analysis*. The results' robustness is assessed through a sensitivity analysis based on repeating the assessment with two different sets of weights corresponding to other perspectives: an *expert* and a *political* perspective (Table 3). The sensitivity analysis shows that the obtained result is quite robust and, thus, the project development can follow the defined schedule.

Table 3. Comparison between the sets of weights defined by the two different stakeholders

Stakeholder	Cultural	Economic	Environmental	Social dimension
Political	0,143	0,357	0,286	0,214
Expert	0,231	0,215	0,246	0,308

4 Discussion and Conclusions

Based on recognizing the multi-dimensional benefits from UA implementation, the paper proposes an evaluation approach to support agriculture-led transformation processes in urban spaces by dealing with financial constraints and the sustainability requirements posed by the social and ecologic transition.

Indeed, such an approach allows grounding decisions for UA implementation on all the relevant dimensions for such a decisional issue without disregarding its economic feasibility. Furthermore, local communities' engagement enables decision-makers to build consensus around their choices by adopting a *place-based* approach.

Thus, the proposed evaluation approach can represent a valid support for private or public initiatives to capture the opportunities from UA implementation.

In this sense, the proposed approach can be further developed by integrating the *Sustainability Assessment* with a *Cost-Benefit Analysis*. Such an integration, indeed, would allow analyzing into detail the financial profitability of UA initiatives, thus encouraging private investments in this sector and overcoming public entities' skepticism. Furthermore, the evaluation approach could be enriched by measuring the impact produced by UA implementation in terms of Ecosystem Services (ESs).

Finally, it could be fruitful to test the replicability and adaptability of the proposed approach by applying it to other projects of urban areas' agriculture-led transformations.

Indeed, although the proposed methodological approach stands as a handy and robust tool to support UA implementation, the adaptation to other projects or territorial contexts requires a critical revision of the reference sub-criteria and indicators included in the *Assessment*.

References

1. Sharifi, A., Khavarian-Garmsir, A.R.: The COVID-19 pandemic: Impacts on cities and major lessons for urban planning, design, and management. Sci. Total Environ. **749**, 142391 (2020). https://doi.org/10.1016/j.scitotenv.2020.142391
2. Beland, L.P., Brodeur, A., Wright, T.: The short-term economic consequences of Covid-19: exposure to disease, remote work and government response (2020). https://ssrn.com/abstract=3584922. Accessed 27 Dec 2021
3. Pison, G.: The population of the world. Popul. Soc. **8**, 1–8 (2019). https://www.ined.fr/fichier/s_rubrique/29504/569.en.population.societies.world.2019.en.pdf. Accessed 27 Dec 2021
4. Pandya, A., Singh, S., Sharma, P.: Climate change and its implications for irrigation, drainage and flood management. In: Biswas, A.K., Tortajada, C. (eds.) Water Security Under Climate Change. WRDM, pp. 95–110. Springer, Singapore (2022). https://doi.org/10.1007/978-981-16-5493-0_6
5. Li, X., Zhou, Y., Eom, J., Yu, S., Asrar, G.R.: Projecting global urban area growth through 2100 based on historical time series data and future shared socioeconomic pathways. Earth's Future **7**(4), 351–362 (2019). https://doi.org/10.1029/2019EF001152
6. Rogge, N., Theesfeld, I., Strassner, C.: Social sustainability through social interaction—a national survey on community gardens in Germany. Sustainability **10**(4), 1085 (2018). https://doi.org/10.3390/su10041085
7. Poulsen, M.: Cultivating citizenship, equity, and social inclusion? Putting civic agriculture into practice through urban farming. Agric. Hum. Values **34**(1), 135–148 (2016). https://doi.org/10.1007/s10460-016-9699-y
8. Weidner, T., Yang, A.: The potential of urban agriculture in combination with organic waste valorization: assessment of resource flows and emissions for two European cities. J. Clean. Prod. **244**, 118490 (2020). https://doi.org/10.1016/j.jclepro.2019.118490
9. Bowler, D.E., Buyung-Ali, L., Knight, T.M., Pullin, A.S.: Urban greening to cool towns and cities: a systematic review of the empirical evidence. Landsc. Urban Plan. **97**(3), 147–155 (2010). https://doi.org/10.1016/j.landurbplan.2010.05.006
10. Viana, C.M., Freire, D., Abrantes, P., Rocha, J., Pereira, P.: Agricultural land systems importance for supporting food security and sustainable development goals: a systematic review. Sci. Total Environ. **806**, 150718 (2022). https://doi.org/10.1016/j.scitotenv.2021.150718

11. Medici, M., Canavari, M., Castellini, A.: Exploring the economic, social, and environmental dimensions of community-supported agriculture in Italy. J. Clean. Prod. **316**, 128233 (2021). https://doi.org/10.1016/j.jclepro.2021.128233

12. Dell'Ovo, M., Corsi, S.: Urban ecosystem services to support the design process in urban environment. a case study of the municipality of Milan. Aestimum (Special Issue), 219–239 (2020). https://doi.org/10.13128/aestim-9896

13. Millennium Ecosystem Assessment: Ecosystems and Human Well-Being. Synthesis. Island Press, Washington DC (2005). https://www.millenniumassessment.org/documents/document.356.aspx.pdf. Accessed 28 Dec 2021

14. Deksissa, T., Trobman, H., Zendehdel, K., Azam, H.: Integrating urban agriculture and stormwater management in a circular economy to enhance ecosystem services: connecting the dots. Sustainability **13**(15), 8293 (2021). https://doi.org/10.3390/su13158293

15. Vagneron, I.: Economic appraisal of profitability and sustainability of peri-urban agriculture in Bangkok. Ecol. Econ. **61**(2–3), 519–529 (2007). https://doi.org/10.1016/j.ecolecon.2006.04.006

16. Vittuari, M., et al.: Envisioning the future of European food systems: approaches and research priorities after COVID-19. Front. Sustain. Food Syst. **5**, 642787 (2021). https://doi.org/10.3389/fsufs.2021.642787

17. Geldermans, R.J.: Design for change and circularity - accommodating circular material & product flows in construction. Energy Procedia **96**, 301–311 (2016). https://doi.org/10.1016/j.egypro.2016.09.153

18. Devuyst, D.: How Green is the City? Sustainability Assessment and the Management of Urban Environment. Columbia University Press, New York (2001)

19. Sala, S., Farioli, F., Zamagni, A.: Progress in sustainability science: lessons learnt from current methodologies for sustainability assessment: Part 1. Int. J. Life Cycle Assess. **18**(9), 1653–1672 (2012). https://doi.org/10.1007/s11367-012-0508-6

20. Sala, S., Ciuffo, B., Nijkamp, P.: A systemic framework for sustainability assessment. Ecol. Econ. **119**, 314–325 (2015). https://doi.org/10.1016/j.ecolecon.2015.09.015

21. Tapia, C., Randall, L., Wang, S., Aguiar Borges, L.: Monitoring the contribution of urban agriculture to urban sustainability: an indicator-based framework. Sustain. Cities Soc. **74**, 103130 (2021). https://doi.org/10.1016/j.scs.2021.103130

22. Mareques-Perez, I., Segura, B.: Integrating social preferences analysis for multifunctional peri-urban farming in planning. An application by multi-criteria analysis techniques and stakeholders. Agroecol. Sustain. Food Syst. **42**(9), 1029–1057 (2018). https://doi.org/10.1080/21683565.2018.1468379

23. Caneva, G., Cicinelli, E., Scolastri, A., Bartoli, F.: Guidelines for urban community gardening: proposal of preliminary indicators for several ecosystem services (Rome, Italy). Urban For. Urban Green. **56**, 126866 (2020). https://doi.org/10.1016/j.ufug.2020.126866

24. Aubert, A., Esculier, F., Lienert, J.: Recommendations for online elicitation of swing weights from citizens in environmental decision-making. Oper. Res. Perspect. **7**, 100156 (2020). https://doi.org/10.1016/j.orp.2020.100156

Green Roof Benefits and Technology Assessment. A Literature Review

Astrid Carolina Aguilar Fajardo[1]([✉]), Gabriela Bacchi[2],
Jorge Alexis Cusicanqui Lopez[1], Giovanni Gilardi[1], Damodar Maggetti[2],
and Luca Tommasi[2]

[1] Politecnico di Torino, Corso Duca degli Abruzzi, 24, 10129 Turin, Italy
`astridcarolina.aguilar@asp-poli.it`
[2] Politecnico di Milano, Piazza Leonardo da Vinci, 32, 20133 Milan, Italy

Abstract. Amidst the growing awareness of environmental and sustainability concerns, and the emerging attention towards Ecosystem Services as the benefits supplied by the ecosystem, many researchers and designers are concentrating on the role of green roofs in the built environment. Although several contributions have been made to understand and quantify their benefits, researchers often tend to focus on their own field of study, be it energy, water management, air quality, or acoustic performance related. A further criticality was noticed in the existence of tools specifically designed to analyse green roofs, also embracing the different disciplinary aspects above. Given these premises, this contribution provides a systematic review and a citation network analysis to understand trends in the scientific literature about green roofs' benefits and existing methods to evaluate them, finding key performance indicators for quantifications. The literature review also supports an ongoing project to provide a user-friendly tool for supporting decision-makers in assessing green roofs performances.

Keywords: Green roof · Evaluation · Multi-dimensional performances

1 Introduction

Roofs compose a significant portion of the city environment. In Italy, roofs account for about 20–25% of the urban surfaces, and 60–70% of the building's envelope [1]. Furthermore, recent environmental concerns and an increase in energy cost in the last years have risen the interest in strategies regarding energy efficiency [2] and sustainability, and consequently, the attention given to green roofs, which involve the growth of plants on rooftops as a sustainable practice for typically unused roof spaces and covers the gap of benefit-cost [3]. Moreover, green roofs have been recognized by the European Commission in its EU Biodiversity Strategy for 2030 as one of the Green Infrastructure solutions for biodiversity by greening urban areas, which favour the pursue of the United Nations' Sustainable Development Goals (SDG) [4], precisely "Goal 11. Make cities and human settlements inclusive, safe, resilient and sustainable".

© The Author(s), under exclusive license to Springer Nature Switzerland AG 2022
F. Calabrò et al. (Eds.): NMP 2022, LNNS 482, pp. 1937–1946, 2022.
https://doi.org/10.1007/978-3-031-06825-6_186

The existing literature reveals that, although the benefits of green roofs are multidisciplinary, authors often focus their research on their own field [5]. Different scholars have focused on green roofs' thermal and energy performance and their capacity to reduce the Urban Island Heat effect [2, 6–16], their capacity of water retention acting on stormwater runoff [5, 17–21], the reduction of air pollution [3, 22–27], the increase of ecosystem services' multifunctionality [28], and even acoustical benefits [29]. However, it is worth mentioning that social benefits received significantly less attention than economic and environmental ones. In Perini and Rosasco's work [25] for example, social benefits were considered mainly connected to the quality of air improvement and CO_2 reduction, as they were able to be quantified.

Ecosystem Services (ESs), which gained more attention after initiatives such as the Millennium Ecosystem Assessment (MEA) [30–34] deal with ecosystem benefits deriving from several fields and dimensions, but multidisciplinary analyses of these services are still scarce [35]. The literature analysed also revealed a similar gap concerning ES provided by green roofs [5]. An exception among the analysed bibliography is for example the work of Clark et al. [36] which quantitively integrated stormwater, energy and air pollutant benefits of green roofs into an economic model.

A further shortage was noticed in the existence of tools specifically designed to analyse green roofs in a pre-construction phase for cost-benefit analysis. Analyses that embrace a multiplicity of aspects can help support the design phase [14, 35], allowing designers and clients to understand the impact and benefits of their choices while answering to demands of transparency and evidence [35].

The objective of the paper is to understand the trends in the scientific literature about green roofs' benefits and current methods to evaluate them, finding existing key performance indicators for such quantifications. The final purpose of the literature review is to aid in the elaboration of a user-friendly tool for supporting decision-makers in assessing green roofs considering their provision of ESs. Moreover, the research conducted through the Scopus database revealed no similar literature review done so far.

The paper is organized as follows: after the introduction, a section dedicated to the research methodology outlines the methods used to develop the literature review, in its different steps of analysis. Next, the results section reports the outcomes of the different steps of investigation, which are further analysed in the following discussion section. Finally, in the conclusions, the paper aims to explain the results obtained from the literature, pointing out also key points to support the ongoing development of the green roof assessment tool.

2 Research Methodology

The literature review aimed to achieve a comprehensive understanding of the research findings of the past years and to identify possible gaps or trends of interest in the context of green roofs, and their benefits and evaluation. The review was carried through the Scopus database in which multiple queries were inserted, obtaining hundreds of publications as a result of the research.

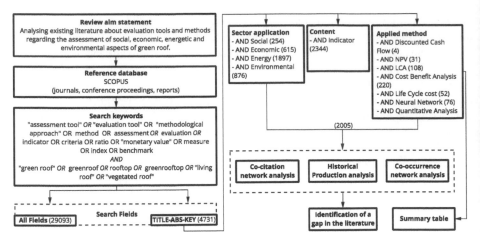

Fig. 1. Literature review framework.

Figure 1 shows the methodology adopted. Initially, a common set of keywords was identified, thus creating a basic query from which to start:

"assessment tool" OR "evaluation tool" OR "methodological approach" OR method OR assessment OR evaluation OR indicator OR criteria OR ratio OR "monetary value" OR measure OR index OR benchmark AND "green roof" OR greenroof OR rooftop OR greenrooftop OR "living roof" OR "vegetated roof".

Depending on whether the research was conducted on "All Fields" or "TITLE-ABS-KEY", the number of publications obtained varied considerably. The number of publications obtained from the base set of keywords using "All Fields" was 29,093, whereas using the "TITLE-ABS-KEY" research was 4,731. Leveraging on the latter type of Scopus query, a more specific investigation was conducted within the 4,731 results obtained, following a threefold perspective by attaching to the basic query some keywords representing the following three clusters:

- Sector application: investigating whether the publications focused on the social, economic, energy or environmental field.
- Content: identifying the papers that specifically provide quantitative indicators and formulas.
- Applied method: identifying the publications mentioning the methods used for the analysis (i.e., Cost-Benefit Analysis, Life Cycle [Cost] Assessment (LCA/LCC), etc.)

The number of articles obtained at each step is provided in brackets in Fig. 1. Summing up the number of articles obtained from the threefold analysis, a value higher than the initial 4,731 publications was obtained, since many articles appeared in more than one cluster. Therefore, after a data exploration activity characterized first by a data preparation work (dropping NAs and converting data types) and then by a full join process, the created database resulted in 2,005 articles sorted by descending number of citations, with no duplicates, and showing from which search query each publication was obtained by the adoption of as many Boolean columns as the queries of the threefold analysis.

Social	Economic	Energy	Environmental	Indicator	DCF	NPV	LCA	CBA	LFC	NN	QA	SOLVES	IT	Authors	Title	Year	Source title	Cited by	Doc-Typ
0	0	1	1	0	0	0	0	0	0	0	0	0	0	Santamouris	Cooling the cities - A review of ref ...	2014	Solar Energy	869	Article
0	0	1	0	1	0	0	0	0	0	0	0	0	0	Sadineni et al.	Passive building energy savings: A...	2011	Renew. Sustain. Energy Rev.	658	Review
0	0	0	1	0	0	0	0	0	0	0	0	0	0	Czemiel	Green roof performance towards ...	2010	Ecological Engineering	620	Review
0	0	1	0	0	0	0	0	0	0	1	0	0	0	Mellit and Pavan	A 24-h forecast of solar irradiance...	2010	Solar Energy	586	Article
0	0	1	0	0	0	0	0	0	0	0	0	0	0	Bacher et al.	Online short-term solar power for...	2009	Solar Energy	524	Article
0	1	1	1	0	0	0	0	0	0	0	0	0	0	Getter and Rowe	The role of extensive green roofs ...	2006	HortScience	422	Article
0	0	1	0	1	0	0	0	0	0	0	0	0	0	Sailor D.J.	A green roof model for building e...	2008	Energy and Buildings	422	Article
0	0	0	0	1	0	0	0	0	0	0	0	0	0	Susca et al.	Positive effects of vegetal on: Urb...	2011	Environmental Pollution	399	Article
0	0	0	1	0	0	0	0	0	0	0	0	0	0	VanWoert et al.	Green roof stormwater retention: ...	2005	J. Environ. Qual.	395	Article

Fig. 2. Sample of database created.

The resulting database was organized in a table as displayed above (see Fig. 2). The dummy variables created allowed to immediately understand the green roof aspect analysed in the paper (i.e. Social, Economic, Energy, Environmental), pointing out whether indicators are mentioned. Moreover, they also show in advance the applied method for evaluating impacts and economical effects of the green roof (e.g. Discount Cash Flow (DCF), Net Profit Value (NPV), LCA/LCC, etc.), and the presence of other specific topics of interest (Neural Network, Quantitative Analysis, Social Values for Ecosystem Services). The use of Boolean variables was essential to quickly identify the documents that needed to be explored in greater depth according to the themes being addressed.

3   Results

From the database obtained, further analyses were conducted. Co-citation and co-occurrence maps (see Fig. 3 and Fig. 4) were done through the VOSviewer software tool.

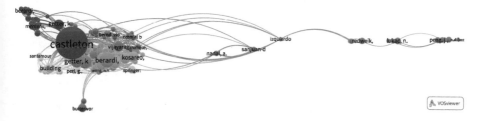

Fig. 3. Co-citation map.

The co-citation map (see Fig. 3), which tracks pairs of works that are cited together in the same paper, immediately shows how prominent Castleton's study [6] is. The nodes (circles) represent specific references cited and their size reflects the number of connections a reference has in the network. The edges (connection lines) signify that two references are cited together, and colours represent clusters of closely related publications.

More specifically, Castleton's study, titled "Green roofs; building energy savings and the potential for retrofit", is a prominent research in the energy field, providing a wide literature review of the current research activity. It also contains experimentally obtained results for U-Value (thermal transmittance rate) and thermal performances at varying moisture, thickness, and density of the soil.

The identification of papers, such as the aforementioned, by means of the co-citation map allows to investigate the publications that the papers in the database (see Fig. 2) adopt as references.

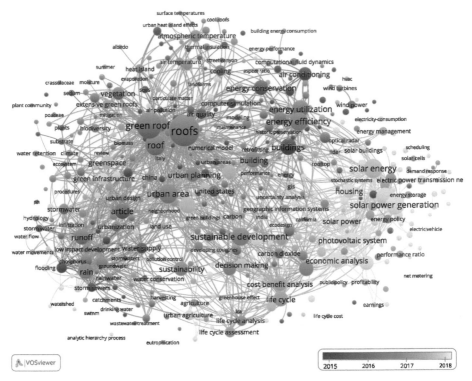

Fig. 4. Co-occurrence map.

A further step has been the plotting of the co-occurrence map (see Fig. 4), visualizing the keywords that occur repeatedly across several publications. The co-occurrence map includes only keywords that occurred at least 30 times to increase the robustness of the analysis. To further highlight the evolution in the studies fields, the timespan considered is 2015–2018. The most intense publication activity occurred in the last years (see Fig. 5) and a relative increase of publication year by year according to the field of interest has also been noted.

The map also shows that the initial studies were more focused on water quality and heat flux, aspects which in the following years have been further developed as confirmed by the larger diameter of the circle of "runoff" [5, 17–21], "rain" and "energy efficiency" [2, 6, 8–15]. In the same years, aspects such as "urban area" [14, 15, 37], "sustainable development" [7, 37, 38] and "urban planning" [35, 39] have been added to these themes, demonstrating the increasing interest in green roofs also from the point of view of city planning. In more recent years, however, aspects such as "climate change", "green infrastructure" and "solar power generation" have been of particular relevance, demonstrating the growing interest in sustainable development and reduction of environmental impacts.

These results are demonstrated in the following chart in Fig. 5, which shows the percentage increase in research activity year by year from 2005 to 2021. The chart is elaborated taking the number of publications in each year as a percentage of the total number in the 2005–2021 timeframe and adding this percentage to the previous year's percentage.

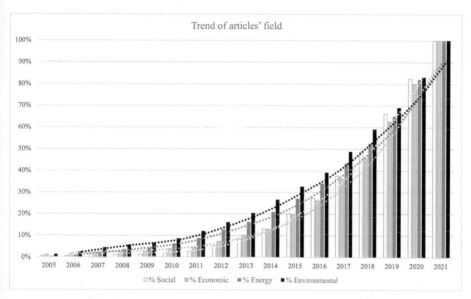

Fig. 5. The trend of articles according to the field of interest.

Visual inspection of the chart above (see Fig. 5), allowed to assess which historical period is linked to active research in each area. Indeed, from the analysis of the path of the trendlines, that were built using a moving average law with period equal to 2, for each topic of study and the keywords' occurrence by year, it was possible to confirm how topical is the research into the impacts of green roofs in social aspects.

4   Discussion

This research revealed the benefits of green roofs in energy, water management, air pollution, multifunctionality of ESs, and acoustic performance aspects. Moreover, since green roofs account for a large percentage (20–25%) of urban surfaces [1], they constitute an important strategy to replace urban green areas lost in construction.

A critical reading of the results underlines the already mentioned tendency of authors analysing green roofs focusing on their field of study [4], with particular attention often being given to energy, stormwater management, and air pollution reduction. Among the many benefits identified, it is possible to mention the thermal protection green roofs provide especially during summer [6]. Saiz et al. [12] in their research modelled three scenarios of roofs in Madrid (common flat roof, white roof, and intensive green roof) into an LCA, showing how during summer the green roof scenario outperformed the common flat one in solar radiation, reducing up to 25% the need for cooling energy. Vegetated roofs have mostly a positive impact on the total energy consumption of buildings implying a net reduction of the total annual energy demand compared to traditional roofs [6, 12–31].

An important gap noticed is in the assessment of green roofs' social benefits. Perini and Rosasco's [25] considered social benefits in connection to improvement of air quality and CO_2 reduction, as they were quantifiable. The literature review revealed many authors' association between air quality improvement and health [21, 23–26, 32], and this relation can help to bridge the gap of social assessment.

The gaps identified along the literature review can be summarized in:

1. The lack of a proper database of pertinent physical parameters.
2. A lack of LCA studies on green roofs and their components. Literature evidenced how some components of the green roof LCC analysis are often disconsidered. Specifically, the role of the disposal phase seems to be underestimated and/or lacking.
3. The lack of quantitative and qualitative data and methods of measurements of social benefits, losses, and impacts related to green roofs.

5   Conclusions and Future Implications

The present paper contributed to understanding the recent trends and discussions related to green roofs' benefits and evaluation. By critically reading the results obtained, it is possible to underline a scarcity of analysis that touch upon different fields of study simultaneously, as authors often focus either on energy, water management, air pollution, multifunctionality of ecosystem service, or acoustic properties at once. The work of various authors of different domains was gathered, creating a database of evaluation methods. Since a gap was also noticed in the social assessment of green roofs, the relation between air quality and health was underlined as a mean to bridge it.

Given the premises of the results discussed, as a future perspective, the final purpose of the analysis developed is the definition of an easy-to-use tool for assessing green roofs simultaneously regarding different domains and aiding decision-makers in their work. The existence of such tools can also contribute to a wider implementation of green roofs. Thuring et al. [20], when analysing green roofs from a water management perspective, mention limitations to their promotion due to the lack of design tools and models to anticipate its stormwater reduction effects.

Furthermore, the tool also aims to bridge the gap of a green roof analysis that involves different fields of study simultaneously. In fact, the research conducted so far resulted in a first selection of key performance indicators, concerning energy savings (annual cooling and heating savings), stormwater management (total runoff reduction), and air quality improvement (annual air pollutant uptake or deposition and amount of a particular air pollutant removed), the latter also relating to the social aspect of health. Moreover, indicators address more specifically ESs, e.g., Regulating Services of Air Quality, Climate Regulations, and Water Regulations [30]. In the end, the goal is to also provide a monetary outcome to the evaluation, helping designers and clients to understand the impacts not only in environmental terms but also economic. A first draft of guidelines for measuring the economic impact involves tax discounts related to the amount of pollutant removal and the reduction of energy consumption, within a certain analysis period. The future steps for the ongoing development of the tool must be accompanied, by additional datasets for review (Web of Science, Google Scholar, etc.) in order to further consolidate the indicators defined and better selecting the formulas for their quantification.

Acknowledgements. The research was developed in the context of the Green Roof Technology Assessment (GreenTA) project underway at the Alta Scuola Politecnica (ASP), Politecnico di Milano and Politecnico di Torino (Academic tutors: Cristina Becchio, Mauro Berta, Marta Bottero, Caterina Caprioli, Stefano Corgnati, Giulia Crespi, Federico Dell'Anna, Marta Dell'Ovo, Antonio Longo, Andrea Rebecchi, Leopoldo Sdino).

References

1. Mohajerani, A., Bakaric, J., Jeffrey-Bailey, T.: The urban heat island effect, its causes, and mitigation, with reference to the thermal properties of asphalt concrete. J. Environ. Manag. **197**, 522–538 (2017). https://doi.org/10.1016/j.jenvman.2017.03.095
2. Sadineni, S.B., Madala, S., Boehm, R.F.: Passive building energy savings: a review of building envelope components. Renew. Sustain. Energy Rev. **15**(8), 3617–3631 (2011). https://doi.org/10.1016/j.rser.2011.07.014
3. Rowe, D.B.: Green roofs as a means of pollution abatement. Environ. Pollut. **159**(8–9), 2100–2110 (2011). https://doi.org/10.1016/j.envpol.2010.10.029
4. European Commission, Directorate-General for Environment: Communication from the Commission to the European Parliament, the Council, the European Economic and Social Committee and the Committee of the Regions EU Biodiversity Strategy for 2030 Bringing nature back into our lives COM/2020/380 final. EUR-Lex - 52020DC0380 - EN - EUR-Lex, May 20 2020. https://eur-lex.europa.eu/legal-content/EN/ALL/?uri=CELEX%3A52020DC 0380. Accessed 27 Dec 2021
5. Czemiel Berndtsson, J.: Green roof performance towards management of runoff water quantity and quality: a review. Ecol. Eng. **36**(4), 351–360 (2010). https://doi.org/10.1016/j.ecoleng. 2009.12.014
6. Castleton, H.F., Stovin, V., Beck, S.B.M., Davison, J.B.: Green roofs; building energy savings and the potential for retrofit. Energy Build. **42**(10), 1582–1591 (2010). https://doi.org/10. 1016/j.enbuild.2010.05.004

7. Cirrincione, L., Peri, G.: Covering the gap for an effective energy and environmental design of green roofs: contributions from experimental and modelling researches. In: Andreucci, M.B., Marvuglia, A., Baltov, M., Hansen, P. (eds.) Rethinking Sustainability Towards a Regenerative Economy. FC, vol. 15, pp. 149–167. Springer, Cham (2021). https://doi.org/10.1007/978-3-030-71819-0_8

8. Gagliano, A., Detommaso, M., Nocera, F., Evola, G.: A multi-criteria methodology for comparing the energy and environmental behavior of cool, green and traditional roofs. Build. Environ. **90**, 71–81 (2015). https://doi.org/10.1016/j.buildenv.2015.02.043

9. Guattari, C., Evangelisti, L., Asdrubali, F., De Lieto Vollaro, R.: Experimental evaluation and numerical simulation of the thermal performance of a green roof. Appl. Sci. **10**(5), 1767 (2020). https://doi.org/10.3390/app10051767

10. Jaffal, I., Ouldboukhitine, S.E., Belarbi, R.: A comprehensive study of the impact of green roofs on building energy performance. Renew. Energy **43**, 157–164 (2012). https://doi.org/10.1016/j.renene.2011.12.004

11. Kumar, R., Kaushik, S.C.: Performance evaluation of green roof and shading for thermal protection of buildings. Build. Environ. **40**(11), 1505–1511 (2005). https://doi.org/10.1016/j.buildenv.2004.11.015

12. Luo, H., Wang, N., Chen, J., Ye, X., Sun, Y.-F.: Study on the thermal effects and air quality improvement of green roof. Sustainability **7**(3), 2804–2817 (2015). https://doi.org/10.3390/su7032804

13. Saiz, S., Kennedy, C., Bass, B., Pressnail, K.: Comparative life cycle assessment of standard and green roofs. Environ. Sci. Technol. **40**(13), 4312–4316 (2006). https://doi.org/10.1021/es0517522

14. Santamouris, M.: Cooling the cities – a review of reflective and green roof mitigation technologies to fight heat island and improve comfort in urban environments. Sol. Energy **103**, 682–703 (2014). https://doi.org/10.1016/j.solener.2012.07.003

15. Susca, T., Gaffin, S.R., Dell'Osso, G.R.: Positive effects of vegetation: urban heat island and green roofs. Environ. Pollut. **159**(8–9), 2119–2126 (2011). https://doi.org/10.1016/j.envpol.2011.03.007

16. Wong, N.H., Yok, T., Yu, C.: Study of thermal performance of extensive rooftop greenery systems in the tropical climate. Build. Environ. **42**, 25–54 (2007). https://doi.org/10.1016/j.buildenv.2005.07.030

17. Getter, K.L., Rowe, D.B., Andresen, J.A.: Quantifying the effect of slope on extensive green roof stormwater retention. Ecol. Eng. **31**(4), 225–231 (2007). https://doi.org/10.1016/j.ecoleng.2007.06.004

18. Metselaar, K.: Water retention and evapotranspiration of green roofs and possible natural vegetation types. Resour. Conserv. Recycl. **64**, 49–55 (2012). https://doi.org/10.1016/j.resconrec.2011.12.009

19. Stovin, V., Vesuviano, G., Kasmin, H.: The hydrological performance of a green roof test bed under UK climatic conditions. J. Hydrol. **414–415**, 148–161 (2012). https://doi.org/10.1016/j.jhydrol.2011.10.022

20. Thuring, C.E., Berghage, R.D., Beattie, D.J.: Green roof plant responses to different substrate types and depths under various drought conditions. HortTechnology **20**(2), 395–401 (2010). https://doi.org/10.21273/HORTTECH.20.2.395

21. VanWoert, N.D., Rowe, D.B., Andresen, J.A., Rugh, C.L., Fernandez, R.T., Xiao, L.: Green roof stormwater retention: effects of roof surface, slope, and media depth. J. Environ. Qual. **34**(3), 1036–1044 (2005). https://doi.org/10.2134/jeq2004.0364

22. Abhijith, K.V., et al.: Air pollution abatement performances of green infrastructure in open road and built-up street canyon environments – a review. Atmos. Environ. **162**, 71–86 (2017). https://doi.org/10.1016/j.atmosenv.2017.05.014

23. Berardi, U., GhaffarianHoseini, A., GhaffarianHoseini, A.: State-of-the-art analysis of the environmental benefits of green roofs. Appl. Energy **115**, 411–428 (2014). https://doi.org/10.1016/j.apenergy.2013.10.047

24. Currie, B.A., Bass, B.: Estimates of air pollution mitigation with green plants and green roofs using the UFORE model. Urban Ecosyst. **11**(4), 409–422 (2008). https://doi.org/10.1007/s11252-008-0054-y

25. Perini, K., Rosasco, P.: Cost–benefit analysis for green façades and living wall systems. Build. Environ. **70**, 110–121 (2013). https://doi.org/10.1016/j.buildenv.2013.08.012

26. Tomson, M., et al.: Green infrastructure for air quality improvement in street canyons. Environ. Int. **146**, 106288 (2021). https://doi.org/10.1016/j.envint.2020.106288

27. Yang, J., Yu, Q., Gong, P.: Quantifying air pollution removal by green roofs in Chicago. Atmos. Environ. **42**(31), 7266–7273 (2008). https://doi.org/10.1016/j.atmosenv.2008.07.003

28. Lundholm, J.T.: Green roof plant species diversity improves ecosystem multifunctionality. J. Appl. Ecol. **52**(3), 726–734 (2015). https://doi.org/10.1111/1365-2664.12425

29. Van Renterghem, T., Botteldooren, D.: Reducing the acoustical façade load from road traffic with green roofs. Build. Environ. **44**(5), 1081–1087 (2009). https://doi.org/10.1016/j.buildenv.2008.07.013

30. Millennium Ecosystem Assessment (Program) (ed.) Ecosystems and Human Well-Being: Synthesis. Island Press, Washington, DC (2005)

31. Caprioli, C., Bottero, M., Zanetta, E., Mondini, G.: Ecosystem services in land-use planning: an application for assessing transformation scenarios at the local scale. In: Bevilacqua, C., Calabrò, F., Della Spina, L. (eds.) NMP 2020. SIST, vol. 178, pp. 1332–1341. Springer, Cham (2021). https://doi.org/10.1007/978-3-030-48279-4_124

32. Caprioli, C., Bottero, M., Mondini, G.: Urban ecosystem services: a review of definitions and classifications for the identification of future research perspectives. In: Gervasi, O., et al. (eds.) ICCSA 2020. LNCS, vol. 12253, pp. 332–344. Springer, Cham (2020). https://doi.org/10.1007/978-3-030-58814-4_23

33. Caprioli, C., Oppio, A., Baldassarre, R., Grassi, R., Dell'Ovo, M.: A multidimensional assessment of ecosystem services: from grey to green infrastructure. In: Gervasi, O., et al. (eds.) ICCSA 2021. LNCS, vol. 12955, pp. 569–581. Springer, Cham (2021). https://doi.org/10.1007/978-3-030-87007-2_41

34. Sdino, L., Rosasco, P., Dell'Ovo, M.: Reclamation cost: an ecosystem perspective. In: Bevilacqua, C., Calabrò, F., Della Spina, L. (eds.) NMP 2020. SIST, vol. 178, pp. 1352–1358. Springer, Cham (2021). https://doi.org/10.1007/978-3-030-48279-4_126

35. Dell'Ovo, M., Oppio, A.: The role of the evaluation in designing ecosystem services. a literature review. In: Bevilacqua, C., Calabrò, F., Della Spina, L. (eds.) NMP 2020. SIST, vol. 178, pp. 1359–1368. Springer, Cham (2021). https://doi.org/10.1007/978-3-030-48279-4_127

36. Clark, C., Adriaens, P., Talbot, F.B.: Green roof valuation: a probabilistic economic analysis of environmental benefits. Environ. Sci. Technol. **42**(6), 2155–2161 (2008). https://doi.org/10.1021/es0706652

37. Buffoli, M., Rebecchi, A., Gola, M., Favotto, A., Procopio, G.P., Capolongo, S.: Green SOAP. A calculation model for improving outdoor air quality in urban contexts and evaluating the benefits to the population's health status. In: Mondini, G., Fattinnanzi, E., Oppio, A., Bottero, M., Stanghellini, S. (eds.) SIEV 2016. GET, pp. 453–467. Springer, Cham (2018). https://doi.org/10.1007/978-3-319-78271-3_36

38. Bianchini, F., Hewage, K.: Probabilistic social cost-benefit analysis for green roofs: a lifecycle approach. Build. Environ. **58**, 152–162 (2012). https://doi.org/10.1016/j.buildenv.2012.07.005

39. Mondini, G., Fattinnanzi, E., Oppio, A., Bottero, M., Stanghellini, S. (eds.): Integrated evaluation for the management of contemporary cities. In: SIEV 2016. GET, Springer, Cham (2018). https://doi.org/10.1007/978-3-319-78271-3

A Proposal to Assess the Benefits of Urban Ecosystem Services

Alessandra Oppio[1] (ID), Marta Dell'Ovo[1](✉) (ID), Caterina Caprioli[2] (ID),
and Marta Bottero[2] (ID)

[1] Department of Architecture and Urban Studies (DAStU), Politecnico di Milano, Via Bonardi, 3, 20133 Milan, MI, Italy
{alessandra.oppio,marta.dellovo}@polimi.it

[2] Interuniversity Department of Regional and Urban Studies and Planning (DIST), Politecnico di Torino, Castello del Valentino: Viale Pier Andrea Mattioli, 39, 10125 Turin, TO, Italy
{caterina.caprioli,marta.bottero}@polito.it

Abstract. The Millennium Ecosystem Assessment in 2005 defined and categorized the concept of Ecosystem Services and the strategic role of natural capital. The need to rethink our cities and public spaces is even more pressing in the COVID-19 era. In this context, green strategies could be the answer to the new demands raised by citizens about the built and natural environment. Green roofs, along with the other green spaces, form the city's green network, contribute to improving the quality of life and wellbeing of citizens. The present contribution aims to evaluate green roofs from an ecosystem perspective, by considering the evidence of their benefits on inhabitants' wellbeing, their ability to mitigate climate change and preserve biodiversity. A proposal for an integrated evaluation model is presented to take into account the different dimensions of value in the study of Ecosystem Services (ESs) and to support decision makers (DMs) in the definition of actions able to increase the quality of life in cities.

Keywords: Integrated evaluation model · Green strategies · Multicriteria Decision Analysis (MCDA)

1 Introduction

Globally, 55% of the population lives in urban areas, and this percentage is expected to increase to 68% by 2050 [1]. The current COVID-19 pandemic challenged cities that have shown their criticality since they were not equipped to meet the new needs that have emerged [2, 3]. This situation produced, at the same time, both negative and positive effects. On one hand, the socio-health crisis, while, on the other hand, the opportunity to rethink long-term strategies to design open urban spaces. In fact, considering the obvious vulnerability of economic, social, and environmental systems that cities are facing, there is the potential to account for all the issues that are affecting our society, i.e., demographic change, urbanization, and climate change [4]. Within this context, investing in green strategies brings multiple benefits, not only in terms of air quality, but also in terms of

© The Author(s), under exclusive license to Springer Nature Switzerland AG 2022
F. Calabrò et al. (Eds.): NMP 2022, LNNS 482, pp. 1947–1955, 2022.
https://doi.org/10.1007/978-3-031-06825-6_187

responsiveness to climate change, the creation of jobs, and local conditions for long-term economic growth. Moreover, considering the lockdown citizens have experienced, it has been demonstrated how people living in cities with a good availability of green spaces (accessible in that period), have incurred fewer negative impacts on their mental and physical health [5]. Today, the main reference that sets the principles of sustainable development is represented by the 2030 Agenda. With its 17 Sustainable Development Goals (SDGs) [6], it sets common goals that the member states of the United Nations have committed to achieving. The goals balance the three dimensions of sustainability: economic growth, social inclusion, and environmental protection.

Combining the concepts of sustainable planning and green strategies, brings to the notion of Ecosystem Services (ESs), defined as the benefits humans obtain from the action of natural capital [7] or as direct and indirect contributions of ecosystems to human well-being [8]. In fact, economic prosperity and well-being are closely dependent on the state of the natural resources around us, the so-called Natural Capital and Ecosystems that provide essential goods and services. Moreover, the demands on natural capital and ESs are increasing given their ability to support long-term conditions for life, health, security, good social relations, and other important aspects of human well-being. The importance of carrying out biophysical quantifications and monetary estimates to measure, on the one hand, the environmental costs associated with the loss of biodiversity and, on the other hand, the benefits obtained for human well-being has been recognized within the SDGs and by the Strategic Plan 2011–2020 of the Convention on Biological Diversity (CBD) with its Aichi Targets.

Given this context, the purpose of the contribution is to present an overview of ecosystem benefits provided by the natural environment and the green strategies (Sect. 2), with a focus on green roofs (Sect. 3). The objective is to propose an integrated approach (Sect. 5) which tries to overcome the criticalities and limits detected from the analysis of the existing evaluation methodologies (Sect. 4) and to combine both the evaluation of tangible and intangible values (Sect. 6).

2   Green Infrastructures in Urban Contexts

Urban green space is a component of "green infrastructure" and consists of a service that cities provide to citizens to promote their well-being [9]. The most common definition of urban green space has been given by the European Urban Atlas [10] which considers Urban Green Areas as public green areas mainly used for recreational purposes, such as gardens, zoo-gardens, parks, suburban natural areas, and forests, or green areas bordered by urban areas managed and used for recreational purposes.

In cities, there are few possibilities for citizens to get in touch with nature and experience biodiversity. In fact, green areas perform vital environmental and social functions that contribute to improving the quality of life and well-being [11]. Among the types of cities' green spaces, it is possible to recognize a variety of large and small, public and private, simple and complex natural spaces, that together form a green network (e.g., natural open spaces, river areas, forests, parks, gardens, squares, vegetable gardens, tree rows, urban greenery, ponds, green roofs, and green walls) [12].

We can list the following benefits produced by green areas [13]:

- Environmental: improvement of air pollution and the urban heat island effect.
- Social: natural features can play an important role in residents' feelings of belonging to the community and through the interaction.
- Health: individuals living in areas with a scarcity of green space may be more vulnerable to stress. As a positive consequence, there is a reduction in the number of hospitalizations caused by cardio-respiratory diseases.
- Physical: one of the main determinants of physical activity is access to green space.

Among the different types of green in urban contexts, green roofs represent a versatile and strategic solution, recognized by an increasing number of local governments to mitigate environmental impacts. Their installation can, in fact, improve the quality of air and life in cities, in the face of a reduction in green areas due to the densification of urban space [12]. This solution is innovative because it is adaptable to different technological systems, both for new and existing buildings. Additionally, the green roofs can improve the energy performance of the built environment, thus leading to economic benefits, as well as social impacts, i.e., perceptual, and visual quality [12].

3 A Special Green Infrastructure in Cities: The Green Roof

3.1 Best Practices

In the last decades, the number of green roofs that have been designed all over the world is intensively increased. This trend occurs both in highly dense places (such as cities) and sparse ones. Generally, in dense areas, they have the role of increasing the environmental quality of the site, particularly the recreational opportunities for the community. In the second case, green roofs are used to reduce the negative visual impact of buildings and infrastructures, creating a camouflage effect between the built environment and the surrounding natural areas.

Green roofs are now common in many countries, from Germany (where 35% of cities have integrated them into their regulations) to Denmark, from the United States (New York, Chicago, and Seattle) to Australia (Sydney, with also green wall policies) [14]. Since 2009, Toronto, the largest city in Canada, has adopted a law that obliges industrial, commercial, institutional, and residential buildings with a square footage gross exceeding 2,000 square meters to install green roofs [15, 16]. Since 2018, the European Commission has encouraged the spread of green roofs and walls, roof gardens, hedges, and trees in the city with EU Directive 2018/844 [17]. Then, in Italy, the 2018 Budget Law (extended in 2019 and 2021) provided for the so-called "Green Bonus", which makes possible a return on the capital invested due to the implementation and renovation of green spaces (e.g., hanging gardens and rooftops) [18].

To understand the main features of green roofs and their contribution to the ESs provision, a review of existing case studies was conducted. The analysis is also based on the grey literature. Specifically, the work was carried out by selecting the same number of cases for each localization category (i.e., Italy, Europe, and Outside Europe), to have a mix among building types, extensive/intensive roofs, functionalities of buildings and roofs, and extensions. For each case study, a detailed collection of data was carried out, starting from more general information (i.e., the designer/studio who carried out the project, the contractor, and year of construction) to more specific ones (i.e., the list of species, activities, and green roof management). Table 1 shows an example of the analysis developed for the green roofs. The main general characteristics are presented, while the specific features of each case study were useful for identifying the ES provided by each green roof project. The identification of ES provided by each project follows a qualitative procedure developed by the authors.

Table 1. Example of the analysis developed of green roofs' case studies based on the year, surface, type of green roof and the ES provided. (EXT: extensive; INT: intensive)

	Project - City	Year	Surface	Type of green roof	Ecosystem services*
IT	Cantina Le Mortelle - Castiglione della Pescaia (GR)	2010	360	EXT	CS; LC; PO; PH; AA
EU	The Städel Museum – Frankfurt (DE)	2012	3000	EXT	CS; LC; PO; PH; AA; TR; MH
Outside EU	Highline - New York (USA)	2009–2019	3.750.000	INT	CS; PO; PH; TR AA; MH

*ES abbreviations: FO (Food); RM (Raw Materials); MR (Medicine resources); FW (Fresh Water); CS (Carbon Sequestration); LC (Local Climate); WR (Water resources); SE (Soil Erosion); WM (Waste-water treatment); DP (regulation of diseases and pandemics); PO (Pollination); EX (Extreme Events); SF (Soil Formation); PH (Photosynthesis); NC (Nutrient Cycle); SE (Spiritual and religious values); TR (Tourism and Recreation); AA (Aesthetic values); MH (Mental and physic Health)

The green roofs analyzed show similar trends in the type of ESs produced. Carbon sequestration, Local climate amelioration, Pollination and Photosynthesis increase, and Aesthetic Appreciation are common to all projects under investigation. Additionally, the recreational activities provided by some intensive green roofs produce other ESs in the domains of cultural and social aspects, as well as in the well-being of people. Some projects then go on to collect and treat water resources. This analysis allows to understand the main trends in green roof design and characteristics, and eventually proposes challenging perspectives for future projects.

3.2 Economic and Financial Attractiveness

From the time when the environment was considered an externality in the urban economy, the presence, use, and modification of environmental systems have affected a range of economic aspects [19]. Many economic advantages are, in fact, linked to goods and environmental services, and particularly to urban green infrastructure in cities [20, 21].

A lot of research has highlighted the correlation between environmental benefits and the economic consequences, spanning the spatial value of environmental services, the value of green urban areas, the waterfront, and the perception of flood risks [22–28], among others.

Without considering the overall ESs provided by green roofs, this infrastructure can generate a set of monetary benefits. Firstly, the reduction of local taxes and the access to incentives. Reduced taxes for rainwater and waterproof coverings, as well as energy credits, subsidies, and tax incentives for green roofs, have been in place for decades in European countries such as Germany, the Netherlands, Switzerland, and Sweden. Also, cities in the United States and Canada have started to offer incentives [29]. Secondly, the reduction of energy costs: heat-insulating green roofs offer many energy savings. The benefits, of course, range from the geographic region, the type of insulating system installed or the thickness of the green roof, but authors generally agree on the reduction of peaks and the reduction of energy required for cooling and heating [30]. Thirdly, the increase in marketability of buildings with or surrounded by a green roof: skyscrapers, offices, and hotel rooms with natural views provided by green roofs can raise rents or room rates while maintaining occupancy levels. At the same time, the presence of green roofs on buildings can also produce an added value in real-estate prices [29].

Moreover, financial instruments, such as Green Bonds, can produce a return on the capital invested.

4   Existing Evaluation Methodologies: State-of-the Art

Three main values related to ESs assessment have informed the research: economic values, ecological/biophysical values, and socio-cultural values. It is not surprising that these three values exactly correspond to the three main pillars of the sustainable concept: economic, environmental, and social aspects.

Economic values stress how the loss of ecosystem services involves economic costs in terms of use (direct and indirect) or non-use (option and existence) [31–34]. Many reviews investigated and framed economic valuation approaches used to assess ESs (for an in-depth analysis, see [35]). Examples of economic valuation approaches in the context of ES are market valuation, revealed preference (Travel Cost Method, Hedonic Pricing Method), and Stated preferences (Conjoint Valuation Method, Choice Experiment, Group valuation).

The biophysical and ecological values of ES represent another huge group of values deeply analyzed in literature and research applications. A number of software packages have started to be implemented in order to simultaneously consider these two groups of values spatially distributed in a specific territory. Internationally, the most well-known software is INVEST [36]. It assesses 17 ecosystem services and, generally, works better at regional and national scales. I-Tree is another famous tool for the quantification of benefits and values produced by trees and can be used at smaller scales (streets, parcels, etc.). I-Tree provides different tools to assess specific benefits at different scales. Nationally, Simulsoil[1], developed by Politecnico di Torino during a LIFE project, quantifies

---

[1] http://www.sam4cp.eu/simulsoil/.

8 different ecosystem services both in biophysical/ecological and monetary terms and generally at territorial/national scales. Other tools are, for example, ARIES (Artificial Intelligence for Ecosystem Accounting)[2], ENVI-met[3], GI-VAL (Green infrastructure valuation toolkit)[4], ORVal (Outdoor Recreation Valuation Tool)[5].

Conversely, socio-cultural values still represent an underexplored and promising field of research for assessing a more comprehensive analysis of ESs and their values [37]. A few examples integrate these socio-cultural benefits provided by ESs into the evaluation [38]. Qualitative assessments, constructed scales, or narrations, deliberative processes and the use of locally defined metrics and guiding principles are some of the ways to capture and measure these values [39]. The use of Multicriteria Decision Analysis (MCDA) as an alternative approach for ESs value is a promising research direction [40], particularly when several features (e.g., multiple dimensions of well-being) should be taken into account and often not in monetary perspectives, as well as the creation of an open and transparent public debate of alternative courses of action among stakeholders. In the context of existing tools, an interesting example is SolVES (Social Values for Ecosystem Services)[6]. Its version 2.0 helps to assess, map, and quantify the perceived social values of ecosystem services, including food and fresh water, as well as cultural services like aesthetics and recreation.

5   Future Perspectives: A Proposal for an Integrated Approach

All these reflections emphasize the importance of incorporating multiple perspectives into the ESs assessment, as well as the foundation for framing our multidimensional assessment approach. In fact, as a result of the analysis carried out, with particular reference to the context of urban planning, it is as necessary as strategic the evaluation of green roofs according to an ecosystem perspective.

The objective of the proposal is to define an integrated method able to take into account the different dimensions of value in the study of ESs to support Decision Makers (DMs) in defining actions to increase the quality of life in cities. According to the analysis of the state of art previously provided, from the point of view of the evaluation, over the years, different approaches have been adopted to define the value of ESs. In particular, biophysical, ecological, and economic values have been extensively studied. On the other hand, there are critical issues in the assessment of intangible values, such as socio-cultural values, as well as the integrated consideration of all dimensions of value. In fact, a pluralistic vision of research on ecosystem services has become more important than ever, especially when it becomes a tool to support the planning and transformation of cities in a sustainable way. The proposed methodological approach considers not only the biophysical and ecological value provided by ESs, but also their translation into economic terms, integrated with values of a socio-cultural nature.

---

[2] https://aries.integratedmodelling.org/.

[3] https://www.envi-met.com/.

[4] https://ecosystemsknowledge.net/green-infrastructure-valuation-toolkit-gi-val#:~:text=Descri ption%3A,are%20given%20an%20economic%20value.

[5] https://www.leep.exeter.ac.uk/orval/.

[6] https://cities4forests.com/toolbox/tools/solves/.

To complete this framework, it is important to consider the costs of construction, maintenance, and management, aimed at having an exhaustive analytical picture to support the DMs. The model thus defined develops and applies cost-based methods, integrating them with value-based methods. The application of the proposed methodology provides an economic, biophysical, ecological, and socio-cultural evaluation of the performances provided by ESs. From the biophysical, ecological and economic point of view, existing tools will be tested (e.g., INVEST, Simulsoil, I-Tree); the assessment will then be integrated with a MCDA to take into account the more intangible aspects of ESs, such as recreational, social, cultural, perceptual values, etc. The qualitative or quantitative scales used (score-based) for the evaluation of intangible values can be defined with the support of case studies, literature [41] and panels of experts and will have as an output a performance score able to measure its intensity. The selection of the most suitable tool and the set of criteria will be guided by the scale of the intervention and the context of application in order to ensure the replicability of the method.

6 Conclusions

The current contribution is aimed at providing an overall overview of the current situation major cities are experiencing given the COVID-19 and how green actions could be a solution for long-term strategies given the multi-dimensional benefits provided [42]. In particular, due to urbanization and the cities growth, green roofs could support these perspectives, given the possibility to be installed both on new constructions and on the existing built environment. The benefits provided merge with the concept of ESs since they are able to meet multi-value services, both tangible and intangible. Their quantification and assessment may be required to support national and international policies [43], as well as to demonstrate how the initial, and sometimes individual, cost is offset by long-term collective benefits [44]. The definition of a comprehensive methodology to support and guide the decision-making phase becomes strategic as well as necessary in the urban planning context to provide common rules to follow, for common sense and to achieve the well-being of cities and citizens.

References

1. Affairs, U.N.D. of E and S: 68% of the world population projected to live in urban areas by 2050, says UN (2018)
2. Carozzi, F.: Urban Density and Covid-19. IZA Discussion Paper No. 13440 (2020)
3. Allain-Dupré, D., Chatry, I., Michalun, V., Moisio, A.: The territorial impact of COVID-19: managing the crisis across levels of government. OECD Tackling Coronavirus, pp. 2–44 (2020)
4. Capolongo, S., et al.: Healthy design and urban planning strategies, actions, and policy to achieve Salutogenic cities. Int. J. Environ. Res. Publ. Health **15** (2018). https://doi.org/10.3390/ijerph15122698
5. Stockholm Environment Institute: Lockdown highlights the value of green space in cities. https://www.york.ac.uk/sei/news/news-2020/lockdownhighlightsthevalueofgreenspaceincities/

6. United Nations: Transforming Our World: The 2030 Agenda for Sustainable Development United Nations United Nations Transforming Our World: The 2030 Agenda for Sustainable Development. A/RES/70/1. United Nations (2015)
7. United Nations: The Millennium Development Goals Report 2015. UN (2016). https://doi.org/10.18356/6cd11401-en
8. TEEB: The Economics of Ecosystems and Biodiversity: Ecological and Economic Foundations. Routledge (2012)
9. World Health Organization: Urban green spaces: A brief for action. Regional Office for Europe, p. 24 (2017)
10. European Commission: Building a Green for Europe Environment. European Union, p. 24 (2013). https://doi.org/10.2779/54125
11. Dell'Ovo, M., Corsi, S.: Urban Ecosystem Services to Support the Design Process in Urban Environment. A Case Study of the Municipality of Milan. Aestimum, vol. 2020, pp. 219–239 (2020). https://doi.org/10.13128/aestim-9896
12. Ajuntament de Barcelona: Barcelona green infrastructure and biodiversity plan 2020. Barcelona, Spain (in Spanish, English summary), p. 111 (2013)
13. D'Alessandro, D., Buffoli, M., Capasso, L., Fara, G.M., Rebecchi, A., Capolongo, S.: Green areas and public health: improving wellbeing and physical activity in the urban context. Epidemiol. Prev. **39**, 8–13 (2015)
14. City of Sydney: Green roofs and walls. https://www.cityofsydney.nsw.gov.au/environmental-support-funding/green-roofs-and-walls
15. City of Toronto: City of Toronto Green Roof Bylaw. https://www.toronto.ca/city-government/planning-development/official-plan-guidelines/green-roofs/green-roof-bylaw/
16. Feng, H., Hewage, K.N.: Economic benefits and costs of green roofs. In: Nature Based Strategies for Urban and Building Sustainability, pp. 307–318. Elsevier, Amsterdam (2018). https://doi.org/10.1016/B978-0-12-812150-4.00028-8
17. European Union: Directive (EU) 2018/844 of the European Parliament. Off. J. Eur. Union (2018)
18. Agenzia Entrate: Legge di bilancio 2018 (articolo 1, comma 12 della Legge n. 205 del 2017), Italy (2021)
19. Palmquist, R.B.: Property value models. In: Handbook of Environmental Economics (2005). https://doi.org/10.1016/S1574-0099(05)02016-4
20. Ambrey, C., Fleming, C.: Public greenspace and life satisfaction in urban Australia. Urban Stud. (2014). https://doi.org/10.1177/0042098013494417
21. Bertram, C., Rehdanz, K.: The role of urban green space for human well-being. Ecol. Econ. (2015). https://doi.org/10.1016/j.ecolecon.2015.10.013
22. Iqbal, A., Wilhelmsson, M.: Park proximity, crime and apartment prices. Int. J. Hous. Mark. Anal. (2018). https://doi.org/10.1108/IJHMA-04-2017-0035
23. Kim, H.S., Lee, G.E., Lee, J.S., Choi, Y.: Understanding the local impact of urban park plans and park typology on housing price: a case study of the Busan metropolitan region. Korea. Landsc. Urban Plann. (2019). https://doi.org/10.1016/j.landurbplan.2018.12.007
24. A. Samad, N.S., Abdul-Rahim, A.S., Mohd Yusof, M.J., Tanaka, K.: Assessing the economic value of urban green spaces in Kuala Lumpur. Environ. Sci. Pollut. Res. **27**(10), 10367–10390 (2020). https://doi.org/10.1007/s11356-019-07593-7
25. D'Acci, L.: Monetary, subjective and quantitative approaches to assess urban quality of life and pleasantness in cities (hedonic price, willingness-to-pay, positional value, life satisfaction, isobenefit lines). Soc. Indic. Res. **115**(2), 531–559 (2013). https://doi.org/10.1007/s11205-012-0221-7
26. Dahal, R.P., Grala, R.K., Gordon, J.S., Munn, I.A., Petrolia, D.R., Cummings, J.R.: A hedonic pricing method to estimate the value of waterfronts in the Gulf of Mexico. Urban Forestry Urban Green. (2019). https://doi.org/10.1016/j.ufug.2019.04.004

27. Sado-Inamura, Y., Fukushi, K.: Empirical analysis of flood risk perception using historical data in Tokyo. Land Use Policy (2019). https://doi.org/10.1016/j.landusepol.2018.11.031
28. Hiebert, J., Allen, K.: Valuing environmental amenities across space: a geographically-weighted regression of housing preferences in Greenville County. SC. Land. (2019). https://doi.org/10.3390/land8100147
29. Velazquez, L.S.: Organic green roof architecture: sustainable design for the new millennium. Environ. Qual. Manage. **14**, 73–85 (2005). https://doi.org/10.1002/tqem.20059
30. Urban Heat Island Initiative Pilot Project: Final Report (2000)
31. Union, E.: Mapping Ecosystem Services. Publications Office of the European Union, Luxembourg (2016)
32. Technical, E.E.A.: Mapping the impacts of natural hazards and technological accidents in Europe an overview of the last decade (2010)
33. Dobbs, C., Escobedo, F.J., Zipperer, W.C.: A framework for developing urban forest ecosystem services and goods indicators. Landsc. Urban Plann. **99**, 196–206 (2011). https://doi.org/10.1016/j.landurbplan.2010.11.004
34. Tyrväinen, L., Miettinen, A.: Property prices and urban forest amenities. J. Environ. Econ. Manag. (2000). https://doi.org/10.1006/jeem.1999.1097
35. Pascual, U., et al.: The economics of valuing ecosystem services and biodiversity. In: The Economics of Ecosystems and Biodiversity: Ecological and Economic Foundations (2012). https://doi.org/10.4324/9781849775489
36. Sharp, R., et al.: InVEST User's Guide (2018). https://doi.org/10.13140/RG.2.2.32693.78567
37. Caprioli, C., Bottero, M., Zanetta, E., Mondini, G.: Ecosystem services in land-use planning: an application for assessing transformation scenarios at the local scale. In: Bevilacqua, C., Calabrò, F., Della Spina, L. (eds.) NMP 2020. SIST, vol. 178, pp. 1332–1341. Springer, Cham (2021). https://doi.org/10.1007/978-3-030-48279-4_124
38. Caprioli, C., Bottero, M., Mondini, G.: Urban ecosystem services: a review of definitions and classifications for the identification of future research perspectives. In: Gervasi, O., et al. (eds.) ICCSA 2020. LNCS, vol. 12253, pp. 332–344. Springer, Cham (2020). https://doi.org/10.1007/978-3-030-58814-4_23
39. Gómez-Baggethun, E., de Groot, R., Lomas, P.L., Montes, C.: The history of ecosystem services in economic theory and practice: from early notions to markets and payment schemes. Ecol. Econ. (2010). https://doi.org/10.1016/j.ecolecon.2009.11.007
40. Saarikoski, H., et al.: Multi-criteria decision analysis and cost-benefit analysis: comparing alternative frameworks for integrated valuation of ecosystem services. Ecosyst. Serv. **22**, 238–249 (2016). https://doi.org/10.1016/j.ecoser.2016.10.014
41. Dell'Ovo, M., Oppio, A.: The role of the evaluation in designing ecosystem services. a literature review. In: Bevilacqua, C., Calabrò, F., Della Spina, L. (eds.) NMP 2020. SIST, vol. 178, pp. 1359–1368. Springer, Cham (2021). https://doi.org/10.1007/978-3-030-48279-4_127
42. Dell'Anna, F., Bravi, M., Bottero, M.: Urban green infrastructures: how much did they affect property prices in Singapore?. Urban Forestry Urban Green. 127475 (2022). https://doi.org/10.1016/j.ufug.2022.127475
43. Assumma, V., Bottero, M., Caprioli, C., Datola, G., Mondini, G.: Evaluation of ecosystem services in mining basins: an application in the piedmont region (Italy). Sustainability **14**(2), 872 (2022). https://doi.org/10.3390/su14020872
44. Sdino, L., Rosasco, P., Dell'Ovo, M.: Reclamation cost: an ecosystem perspective. In: Bevilacqua, C., Calabrò, F., Della Spina, L. (eds.) NMP 2020. SIST, vol. 178, pp. 1352–1358. Springer, Cham (2021). https://doi.org/10.1007/978-3-030-48279-4_126

Intra-scale Design and Benefit Assessment of Green Stormwater Infrastructures

Roberto Bosco[1]([⊠]), Savino Giacobbe[2], Salvatore Losco[1] [iD], and Renata Valente[1] [iD]

[1] Università degli Studi della Campania "Luigi Vanvitelli", Aversa, CE, Italy
roberto.bosco@unicampania.it
[2] Caserta, Italy

Abstract. By considering the effects of climate change in urban areas, the paper investigates the benefits of environmental design strategies as green stormwater infrastructures. The environmental dimension of spatial planning sits alongside and interfaces with other sustainability challenges in the urban environment. Moreover, the advent of the COVID-19 pandemic focused their crucial importance in the reorganization of places that are "safe" because they allow movement through cities with minimal risk of contagion, providing an adequate opportunity for safe socialization. The research studied the design of green streets equipped with GSI devices in a portion of the town of Aversa (CE), to mitigate flooding, improve the microclimatic conditions and the quality of public spaces. The work was carried out with the support of software simulation of the microclimatic and physical behavior of urban areas (ENVI-Met) and evaluation of the benefits of ecosystem services (iTree Eco), performing in a cyclic way the phases of Analysis, Design and Verification. The approach allowed to refine the results obtained by selecting the most appropriate solutions for the project. The computer simulation of microclimatic conditions is a useful tool for the environmental assessment of limited urban ecosystems, where it would be particularly costly to install a series of small stations to detect the local weather conditions. The holistic approach to urban regeneration planning and project leads to a constant assessment of the increase of the ecosystem services resulting from the solutions chosen in the cyclical process of research through design.

Keywords: Environmental design · Green stormwater infrastructure · Ecosystem services

1 Background and Goals

According to the latest report of the Intergovernmental Panel on Climate Change (IPCC), the changes in the Earth's climate are unprecedented; some of its effects will be irreversible for hundreds or thousands of years. Strong production of greenhouse gases caused global temperatures to rise by about 1.1 °C since the last century, but the effects of climate change are also seen in significant changes in humidity levels, winds and precipitation. In cities, some aspects of the climate will be amplified, including heat waves and flooding due to heavy rainfall. The COVID-19 pandemic focused the crucial

© The Author(s), under exclusive license to Springer Nature Switzerland AG 2022
F. Calabrò et al. (Eds.): NMP 2022, LNNS 482, pp. 1956–1965, 2022.
https://doi.org/10.1007/978-3-031-06825-6_188

importance in the reorganization of places that are "safe" because they allow movement through cities with minimal risk of contagion. While addressing the need for social distancing, open air exercise, and mobility without use of public transport, these measures resulted in other environmental and social benefits [1]. Green infrastructures provide an optimal solution for stormwater management in highly urbanized areas: several U.S. cities have successfully tested technical solutions involving the use of green stormwater infrastructures, gathering results in manuals, which provide useful resources for implementing these practices in Mediterranean latitudes [2, 5]. In this context, this research design proposes a sustainable *green streets* network to inform the future, permanent street redesign. A comparison of the climatic conditions incurred in the various case studies made it possible to select the best strategies suitable for the area in question. Based on topographic, urban pattern, morphologic, and climatic data, it evaluates a series of contiguous road sections, defining redesign capacities and critical conditions to implement sustainable interventions to manage urban runoff, mitigate extreme heat events, expand pedestrian paths and provide a bicycle network. Moreover, the holistic approach to sustainable urban planning and design, supported by reproducible data and parameters, serves as a replicable model for the roads redesign in similar urban settings. This kind of approach derives from the idea that a precondition for solving complex problems is the awareness that they are interconnected, therefore they cannot be solved in a disjointed way [3]. The extent, integration, and complexity of the study triggered an interdisciplinary framework, which made it easier to plan and design detailed and quantified assessments of environmental outcomes. This urban regeneration approach is intrinsically intra-scale, since the functional sum of the parts is always greater than the same parts taken individually in a systemic approach. In fact, if locally the redevelopment brings benefits in terms of fruition of specific areas, replicating the interventions on multiple urban and peri-urban contexts triggers wide-ranging benefits, including ecosystemic effects. A key objective of the work was the assessment of the benefits provided by the selected solutions, both with reference to software modeling and calculations for the considered metrics, also comparing the costs of traditional infrastructure with the planned implementations.

The paper is structured in four parts respectively analyzing the design and evaluation methodology adopted, urban planning implications of the intra-scale approach, the application of the method on a pilot area and the expected enhancement of the environmental and economic indicators considered.

2   Survey and Assessment Methodology

The interdisciplinary research through design work, driven with the idea of learning by doing and gaining further awareness, guided the study of green streets equipped with GSI devices in Aversa (Ce) precisely at the South of the historic center, where several buildings with public destinations are located. The process consisted in the cyclic execution of the three phases of Analysis, Design and Verification, allowing to refine the results obtained by selecting the most appropriate solutions for the project. All the information obtained in the Analysis phase made it possible to identify the areas where the microclimatic conditions are more adverse, such as those with an increased possibility of having heat islands or where wind speeds constitute a risk to the stability of trees, signs or scaffolding. Consequently, the appropriate indicators, grouped by hydraulic, urban, microclimatic, and socio-economic performance were defined to effectively assess the improvements provided by the hypothesis [4]. Data obtained from ENVI-Met software, able to simulate and reproduce the microclimatic and physical behavior of urban areas, allowed to verify the presumed climatic change of the area according to the design choices, for summer and winter conditions. The study of the current state of the area suggests performing a requalification based on the implementation of green streets, to mitigate flooding, improving microclimatic conditions and quality of public spaces [5]. To this purpose the redevelopment of two areas has been tested in detail: Piazza Bernini, which has a great potential attractiveness due to the various commercial activities that overlook it, but which is currently constantly occupied by parked cars; the area destined for parking between this square and Viale degli Artisti, where a highly adverse microclimatic condition was verified. Given the lack of public green spaces in close proximity, it was decided to convert the parking lot into a neighborhood *mini-park*, for citizens' benefit both in terms of climate and social life.

The results obtained from the hydraulic analysis showed an excessive water inflow due to stochastic events of short duration. The total stormwater volume (calculated through the method of runoff) for the five micro-watersheds of the project sites is about 605 m^3, on a public surface of about 20.278 m^2. This allowed to derive the minimum drainage infrastructure requirement of approximately 1.272 m^2. The ENVI-Met fluid dynamic models show a very poor microclimatic quality in the area, particularly in the PMV values, which are very high and denote a strong climatic discomfort for users. Wind, quite present in the streets of the area, represents a conditioning factor too. Particularly high UAR values, detected from the analysis of eleven street cross sections, indicate that the distance between buildings determines Urban Canyon conditions, also detectable in ENVI-Met through the parameter of the Average Radiant Temperature. Similarly unsatisfactory results are obtained by applying the indicators of RIE and BAF, whose values are much lower than the minimum suggested. Surveys of the area revealed a strong presence of vehicular traffic, cause of both air and noise pollution. It was also observed a clear state of degradation, due to poor maintenance of the building and the action of time. The existing vegetation has several criticalities, including the scarcity of space available for the growth of tree roots, which cause disruption to roadways.

3   From Traditional Spatial Planning Towards Eco-planning

Traditional spatial planning fragments and compromises, sometimes irreversibly, ecosystems. Eco-Planning [6] is an innovative approach to master-planning based on the ecological concept of physical planning as a bio-integration between the built environment and natural systems. It structures the natural and man-made environment into a single system consisting of four infrastructures: green, blue, gray and red. The green one is nature eco-infrastructure; the blue is water eco-infrastructure (natural drainage, water conservation systems and hydrological management in general); the grey is the engineering infrastructure, (roads, sewers, drainpipes, etc. as support systems sustainable for urban development); the red is the human infrastructure meaning the built context, including human activities and economic, legislative and social systems [7]. This contribution seeks to highlight that the macro-categories listed as separate are in fact very interconnected, as demonstrated by the study summarized in the paper. The management of water resources will be a major concern in the 21st century. The Water Sensitive Urban Design [8, 9] method provides valid techniques to plan and design a sustainable drainage scheme to manage storm water, ensuring that it remains within the area and is managed and conserved within the built environment. Projects will need to introduce effective strategies to conserve and make best use of this precious resource. The design and redevelopment of environmentally sustainable neighborhoods will need to integrate efficient water demand management. In this general interpretative framework, GSI can be regarded as a bio-integration technique between green and blue infrastructures with interconnections with red and grey infrastructures too. The sample area is part of a recently expanded neighbourhood of Aversa (Ce), the city has been characterized in the last fifty years by a strong increase in settlement (Fig. 1), with a consequent increase in impermeable surfaces, thus facilitating a possible condition of water stress in the sewer collector.

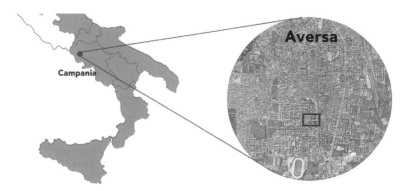

Fig. 1. Site location.

The study is representative of European city typical expansion after the Second World War and can be used to test the application of a GSI in this type of urbanization. From 2001 the Municipality of Aversa adopted the Land Use Plan, approved by the Provincial Administration in 2004 and still in force. The pilot area is zoned B1-Saturated built-up and is regulated by Article 38 of the Technical Implementation Rules that does not impose urban transformations with a low environmental impact. The reduction of impervious surfaces in the Aversa neighborhood highlights the fact that, although only the green and blue infrastructures of a complex system are affected, it is possible to achieve significant results. This leads to interesting thoughts about interconnections between various infrastructures through which the complexity of a territory can be delineated and the interrelations between different elements of the same infrastructure. It reminds of the interconnections between blue and green infrastructure in the design of urban spaces or the cause-effect link between gray and red infrastructure (land consumption) and blue infrastructure. The reduction of the first two has beneficial effects on the third one and brings about great improvements in the first one, so much that it is not necessary a re-planning and subsequent construction of a new hydraulic infrastructure. Urban (eco) planning turns to be the most suitable technical tool to manage and organize land transformation actions, both in the structural and operational form of the municipal plan and building regulations. GSI use could be useful also to transform into technical choices some addresses present in the Strategic Environmental Assessment (SEA) of plans and in the Environmental Impact Assessment (EIA) of the projects.

4 Pilot Area Research Through Design

The research's intra-scale dynamics allowed for the identification of areas where Green Stormwater Infrastructures should be placed through the study of spatial-level water flow lines. Given the morphological characteristics of the area, the objective is to prevent runoff from invading the lowest points; the planned infiltration areas (GSI) (*Stormwater Bump-out, Stormwater Planter, Stormwater Rain Garden*) therefore aim to increase local permeability, exploiting the mechanism of lamination to store the volume of rainwater with controlled and deferred release of flow. As active components of these infrastructures, plants purify runoff water, which brings along many pollutants from vehicular traffic in an urban environment [10]. To this end, species in the *genus Juncus, Salir*, and *Populus* have been selected for their phytodepuration properties. For the urban green, according to the analyses carried out with the ENVI-Met software, which reported a worsening of winter effects due to the presence of evergreen trees, it was decided to plant new species of *Tilia sp.pl.* and *Cercis siliquastrum L.* It was then planned to rearrange the existing trees in pots more suitable for their roots. The activities illustrated below have been elaborated by the research through design method; they are the product of the iterative, explorative and constructive approach that characterized the reflection on the possible project outcomes, representing an effective research strategy aimed at improving the values of the indicators considered in the design process. Pedestrian areas were rethought by placing vegetation strips alongside the sidewalks when feasible, to improve visual and sensory perception of pathways. In addition, changing traffic directions was assessed to encourage the smoothest and safest vehicular circulation possible, reducing traffic and noise. Finally, the route of a bike path has been planned in continuity with the one created by the Municipality of Aversa (Fig. 2).

The green area between Via Vanvitelli and Viale degli Artisti was rearranged by adding play equipment for children and providing seating in shaded areas to protect users from high summer temperatures (Fig. 3). Studies show that case rates of COVID-19 may be reduced among individuals who spend more time in parks compared to other recreational areas, especially in urbanized, high-density areas [11]. These findings suggest that is advisable to use the parks for recreational activities in these scenarios, demonstrating an additional utility of green spaces beyond the known health and wellness benefits under normal non-pandemic conditions. Within the area of Piazza Bernini, among the problems found in the analysis are the high temperatures and PMV values in summer and the high wind speed in winter. To reduce the value of microclimatic parameters, it has been planned an increase in the number of trees, whose *quinconce* arrangement creates a barrier against the strong wind. Asphalt paving will allegedly be replaced by a light-colored stone one, to reduce the amount of heat stored in the area during the day.

To increase soil permeability and mitigate stormwater runoff, four Stormwater Planters were included (Fig. 4). Data extrapolated from the software simulation for indicator results show that these interventions are successful in addressing the area's deficits.

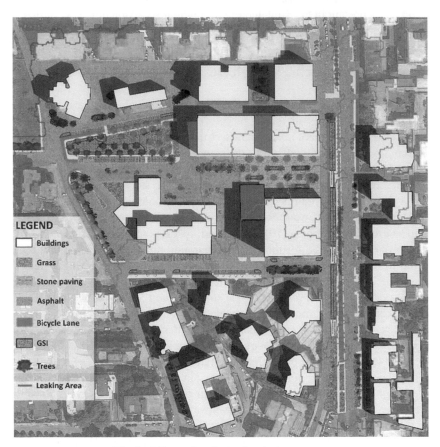

Fig. 2. Pilot project in the South Aversa area.

At the urban level, the area can be linked to the whole city through pedestrian and bicycle paths and with vegetation, so that "the human work becomes solidary with the work of nature" [12]. The study carried out on Viale degli Artisti shows severe summer microclimatic conditions due to high temperatures; also, the location, coincident with the intersection of several runoff lines, recommends redesigning the site as a large rainwater harvesting area to avoid downstream runoff. The research process therefore suggested the development of the area around the large Rain Garden, which collects most of the rainwater of the basin in which it is placed (with a requirement of about 730 m^2). New landscaping and the inclusion of drainage paving have been foreseen to further support rainwater harvesting. Finally, it has been tested the planting of 24 new trees positioned in *quinconce*, in addition to the 9 already present (Fig. 4).

Fig. 3. Mini-park demonstrator project (left). Detail of the Rain Garden (right).

Fig. 4. Demonstrator project in Bernini Square (left). Detail of the Stormwater Planter (right).

5   Findings About Benefits Assessment

Through the software ENVI-Met it was possible to verify the presumed climate mutation of the area with reference to the days 31/07/2018 and 26/02/2019, respectively the hottest and the coldest of the last two years[1]. In the summer period the project interventions result in a significant improvement of the critical parameters. This resulted from both the beneficial effects of increased green surfaces and the replacement of street paving with a lighter colored one. In addition, shadows generated from the large amount of new trees planted reduce the extent of area directly exposed to the sun. Substantially satisfactory values were also obtained for the simulation of the winter period: thanks to the special attention brought to the design and layout of urban green areas, wind speed is reduced in areas where it represented a danger for users and the stability of urban furniture. Nevertheless, it should be noted that the vegetation, while producing beneficial effects, at the same time creates a slightly colder and more humid habitat; in any case, the parameters of PMV and Relative Humidity are within tolerable standards. The benefits of ecosystem services (€) derived from trees were calculated through the software i-Tree ECO, adding the input values parametrized by the software itself. The pilot project of study area counts 179 trees (compared to 83 of the actual state) and the most present species are *Cercis Siliquastrum* (38%), *Tilia Platyphyllos* (16%) and *Pinus Pinea* (13%).

---

[1] Data detected by the meteorological stations of Aversa and Casoria.

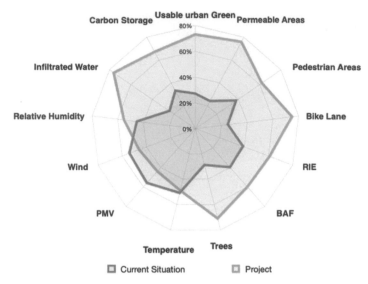

Fig. 5. Graph of the improvements in the indicators considered (percentage).

The transport systems theory allowed to derive the benefits obtained from the variation of the project vehicle flows, by adopting the concept of capacity, that is the maximum flow of vehicles users who can use an element of the transport offer in a given time interval[2] [13]. After calculating the cost-benefit ratio[3] for a 30-year time span, performance indicators were derived, leading to the following considerations: the demonstration project presented entails a positive NPV of 1.27 million euros; the B/C ratio is 4.10, attesting to the convenience of any investment; the *PayBack Period* is 5 years with a not-too-distant return time, therefore the risk inherent in the investment can be considered low [14]. A particularly relevant datum is the one obtained by calculating the difference in costs between the realization of GSI and the adaptation of the existing sewer network for the same water flow rate. Using the Veneto Region's price list parameters, it was possible to estimate the cost of the structural adjustment (about €350,000), which would be considerably higher than that envisaged for the implementation of the GSI, (about €115,000, three times lower). The operation would therefore result in a virtual saving of over 230.000 euros, without considering the other accessory works. With referring to the studies conducted on the relationships between air pollution [15], park use [11] and COVID-19 virus transmissibility, the results obtained from the software modeling show an improvement of the indicators, with the increase of usable green area (+45%) and carbon storage (+33%) and the reduction of PMV values (−11%). From the economic standpoint, savings related to the decrease in hospital admissions should be considered, knowing that each occupied bed has a daily cost ranging from 10 to 20.000€

---

[2] The mistake that must be avoided using this method is not considering the real vehicular flow of the area, but only the maximum flow.

[3] Cost-benefit analysis involves flaws, such as the non-summitability of effects and the limitation to economic benefits only.

[16]. Ultimately, we can classify the benefits obtainable from the project in *monetized benefits* and *other benefits*. In the first category are those obtained from the CBA, which demonstrate that the design developed by the research work is also cost-effective. In the second category we find all those benefits whose effects are not easily quantifiable, such as: public and private expenses resulting from flood damages, social and health benefits, presence of recreational spaces, creation of *Green Jobs*.

6   Concluding Remarks

Large-scale replicability of the described practices can bring the necessary improvements to mitigate the effects of climate change on urban areas. In order to manage the volume of rainwater in the pilot area, the implementation of GSIs for a total of 1615 m^2 has been studied, which would allow to absorb more than 100% of the volume of rainwater, allowing to meet the needs of even longer return times. Their placement at critical locations would allow about 90% of the water to infiltrate the surface aquifer, leaving only 10% to flow into the sewer. In addition to the mitigation of meteoric effects, the research work focused on the revaluation of the places subject to intervention, identifying places where it is possible to configure urban spaces capable of satisfying even the soul of people [17], in which the city of asphalt must disappear [18]. In fact, one of the purposes of the redevelopment is to recreate the true concept of a square in Piazza Bernini and at the same time to turn Viale degli Artisti into a neighborhood urban park. In both places the presence of GSI devices for water collection also represents a mean to make the population aware of the advantages of this environmental design approach.

Local microclimatic conditions can differ even drastically from the broader conditions of the large area in which they occur. Sunlight, ventilation, and relative humidity may vary locally; the latter, for example, can be affected by the presence of groups of trees that create small, localized urban forests. As vegetation brings great benefits to the city and its users, it is increasingly necessary to quantify what until now has been defined on a perceptive basis. This need also stems from the fact that wrong choices may lead to plant associations (in urban areas) that are particularly expensive to manage and not very environmentally useful [19]. It is therefore desirable to test with appropriate simulations the effectiveness of vegetation, not forgetting fundamental design and executive aspects such as the choice of suitable materials for paving and equipment.

The microclimate simulation software proves to be a reliable tool for the environmental assessment of limited urban ecosystems, where it would be particularly expensive to install a series of small stations to detect local weather conditions, also considering that it would be necessary to repeat the monitoring for adequately extended time intervals to obtain consistent results. The holistic approach to urban regeneration planning and project leads to a constant assessment of the increase of the ecosystem services resulting from the solutions chosen in the cyclical process of research through design. The economic benefits identified are deemed to be key factors for the effective applicability of the technical solutions hypothesised.

Acknowledgements. The contribution is based on the master thesis work by Savino Giacobbe and is the result of a shared evolution of research by the four authors. Valente R. wrote paragraph 1, Valente R. and Giacobbe S. paragraph 2, Losco S. paragraph 3, Bosco R. paragraph 4, Giacobbe S. paragraph 5.

References

1. Valente, R., et al.: Environmental regeneration integrating soft mobility and green street networks: a case study in the metropolitan periphery of Naples. Sustainability **13**, 8195 (2021)
2. Valente, R.: Water sensitive urban open spaces: comparing North American best management practices. UPLanD-Journal of Urban Planning, Landscape & environmental Design **2**, 285–297 (2017)
3. Bell, W.: The sociology of the future and the future of sociology. Sociological Perspectives **39**(1) (1996). https://doi.org/10.2307/1389342
4. D'Ambrosio, V., Leone, M.F.: Progettazione ambientale per l'adattamento al Climate Change. Strumenti e indirizzi per la riduzione, Clean Editori, Napoli (2017)
5. Valente, R., Cozzolino, S., Ferrara, P.: Enforceability and benefits of Mediterranean green street, Sustainable Mediterranean Construction (2019)
6. Yeang, K.: Ecomasterplanning, p. 167. Wiley, London (2009)
7. Yang, Z. (ed.): Eco-Cities. A Planning Guide. Routledge, London (2013)
8. Hoyer, J., et al.: Water Sensitive Urban Design. Jovis, Berlin (2011)
9. Larry, W., Mays, L.W.: Stormwater Collection Systems Design Handbook. McGraw-Hill, USA (2001)
10. Losasso, M., Lucarelli, M.T., Rigillo, M., Valente, R.: Adattarsi al clima che cambia. Innovare la conoscenza per il progetto ambientale, Maggioli Editore, Santarcangelo di Romagna (RN) (2020)
11. Johnson, T.F., Hordley, L.A., et al.: Associations between COVID-19 transmission rates, park use, and landscape structure, Science of The Total Environment, vol. 789 (2021)
12. Saragosa, C.: Città tra passato e futuro. Donzelli Editore, Roma (2011)
13. Cascetta, E.: Modelli per i sistemi di trasporto: Teoria e applicazioni. UTET, Napoli (2006)
14. Farris, P.W., et al.: Marketing Metrics: The Definitive Guide to Measuring Marketing Performance. Pearson Education, Inc., Upper Saddle River (2010). ISBN 0-13-705829-2
15. Travaglio, M., Yu, Y., et al.: Links between air pollution and COVID-19 in England, Environ. Pollut., 268 (2021)
16. https://www.liucbs.it/ricerca-applicata-e-advisory/centro-sulleconomia-e-il-management-nella-sanita-e-nel-sociale/osservatori-e-club/healthcare-datascience-lab-hd-lab/
17. Bosco, A., Rinaldi, S., Valente, R.: Strumenti di progetto per il microlandscape urbano, Alinea Editrice, Firenze (2012)
18. Giedion, S.: Breviario di architettura, Bollati Boringhieri, Milano 2008 (1956)
19. http://urbanbo.urbanit.it/wp-content/uploads/2017/05/5-capitolo4.pdf

Urban Green Space to Promote Urban Public Health: Green Areas' Design Features and Accessibility Assessment in Milano City, Italy

Maddalena Buffoli[1] , Francesco Villella[2], Nasko Stefanov Voynov[2], and Andrea Rebecchi[1](✉)

[1] Department of Architecture, Built Environment and Construction Engineering (ABC), Politecnico di Milano, via G. Ponzio 31, 20133 Milan, Italy
{maddalena.buffoli,andrea.rebecchi}@polimi.it
[2] School of Architecture Urban Planning Construction Engineering (AUIC), Politecnico di Milano, Piazza Leonardo da Vinci 32, 20133 Milan, Italy
{francesco.villella,naskostefanov.voynov}@mail.polimi.it

Abstract. During the first waves of the Covid-19 pandemic period, urban environments were stressed; the resilience of our cities were tested, highlighting the strengths and weaknesses of the urban contexts, not always capable to promote and protect the population health status. Urban Green Spaces (UGS) have proved essential role as "tools" to improve Urban Public and Mental Health. Unfortunately, the heterogeneous distribution of UGS inside the contemporary cities, together with the disparity in quality of such spaces, led to some exclusion phenomena. The paper would describe a research experience based on four consequential phases: theoretical background update; tool definition phase (Quantitative assessment: Proximity of the UGS in Milano, and Qualitative assessment: RECITAL 2.0 Milano); application phase in the urban context of Milano city and findings analysis. About the application phase, 24 parks were evaluated: by the comparison of the "RECITAL 2.0 Milano" results with the UGS surface data, no significant pattern emerges, that means that environmental quality is not linked to the extension of the UGS. By the comparison of the overall score with the average real estate values in the analyzed area, emerges an easily readable and expected correlation: the top-performing parks are in the most exclusive areas of the city, often in or near the city center, easily reachable by public transportation alternatives. Qualitative assessments can detect criticalities in-side the urban environment, while quantitative assessments can find areas of the cities deprived of the benefits of UGS. The overlap of both findings could be an indicator of the presence of some form of exclusion phenomena, thus requiring attention both of Urban Planners and Policy Makers to ensure healthier and more equal urban environments.

Keywords: Urban Green Spaces (UGS) · Green areas and infrastructures · Urban Health · Healthy cities · Environmental assessment

© The Author(s), under exclusive license to Springer Nature Switzerland AG 2022
F. Calabrò et al. (Eds.): NMP 2022, LNNS 482, pp. 1966–1976, 2022.
https://doi.org/10.1007/978-3-031-06825-6_189

1   Theoretical Background

1.1   Mental Health and Urban Green Spaces

The urbanization phenomenon has changed the living habits and lifestyles of urban dwellers at an unprecedented speed, even more, if compared with past years. For the first time, in 2007, more than half of the global population was living in cities, in 2050 cities population is estimated to grow to two-thirds of the total [1]. Cities are not only the place where most people live but also the urban environment where most of the key determinants for Public Health can be found. If the urban contexts are the most impactful to collective health, it appears obvious that the relationship between individuals and public space can have a significant effect, both positive and negative, on the physical and mental health status of the urban populations [2].

Urbanization phenomena, both in developing countries, emerging economies and the densification of already established cities with excellent level in term of the urban quality, suggest crucial challenges and new threats to Public Health emerge: air-noise-visual pollution, social inequalities, accessibility to services, increased incidence of mental pathologies. In most contemporary cities, Public Health is not exclusively caused by the efficiency of sanitary systems, but is also influenced by urban planning initiatives and healthy design actions. Although, Mental Health is dependent on some individual factors, like genetics, the characteristics of the individual or his social-economic status, there are others risk factors that depend on behavioral and environmental factors such as lifestyle, amount of Physical Activity, quality of the space in which population live or work, the persistence or view of green areas, and numerous other variables often connected to the design of the urban spaces [3]. These factors can be controlled to some extent to improve the quality of life and to give citizens the possibility to access the restorative effects of well-planned, socially inclusive public spaces.

During the first waves of the Covid-19 sanitary crisis, urban systems were stressed. During this period, the resilience of our cities and population were tested, highlighting the criticalities in both sanitary systems and urban planning. Urban Green Spaces (UGS) have proved essential role, not only as a safe recreational space but also as an irreplaceable tool to improve Urban Public and Mental Health due to its formidable restorative capabilities [4–6]. Unfortunately, the heterogeneous distribution of UGS inside of urban environments, and the disparity in quality of such spaces, led to some exclusion phenomena, further underlining their relevance in egalitarian and democratic cities.

Evidence/experience-based research strongly demonstrated the positive effects on Public Health of the UGS, and for this reason, they are now becoming the strategic and challenging issue of many urban regeneration programs. From 2008 to 2020 the number of publications researching the effects of green space on Public Health has increased five-folds. In an ever-growing multi-disciplinarity, environmental and urban sciences are frequently encompassed in the research, suggesting, to a certain degree that health factors were increasingly considered in urban planning and design [7]. The importance of UGS as a key infrastructure has generated the necessity of developing new health-centered design criteria able to conform to their new role in urban environments. The augmentation of UGS surface alone, does not necessarily make cities more livable. An increase in area and surfaces does not translate in ease of accessibility from all social groups or from all the cities' neighborhoods, or not does it give data on the qualities of such areas, like potential for social engagement or Physical Activity.

How to determine the effectiveness of UGS in providing Mental Health benefits for its users is currently the subject of much research [8]. It can however be claimed that accessibility and environmental quality (safety, biodiversity, paths, presence of recreational areas, etc.) are effective indicators that can be used to assess the current state of UGS in cities, as well as finding criticalities in urban contexts.

1.2 Urban Green Spaces of Milan

The city of Milan (Northern Italy, Lombardy Region) is one of the 14 Metropolitan Cities of Italy; it's a compact city, with few wide-open areas, where green spaces are not governed by a matrix or recognizable pattern [9]. Aside some historical examples, Milan's green spaces are fragmented into a multitude of medium to small-sized areas, with various level of accessibility, vastly heterogeneous in quality and characterization. Since the early 2000s the municipality of Milan has invested important resources to increase the amount of green surfaces.

According to the current Sustainable Development Goals (SDGs), Milan is supposed to double its 2010 surface of green areas, by 2030, reaching 67 million of m^2. Being the second most populous city in Italy, the second most densely populated, and given the relatively rapid pace of the greening effort, the issues of green spaces' accessibility and quality are particularly relevant for the objective of improve Physical and Mental Health of its inhabitants.

The following research focuses on the methodological tools capable to evaluate and map both accessibility and quality of the UGS of Milano. By investigating and mapping both of these issues, it is possible to highlight, in a complex urban environment, both the most virtuous and the most critical areas. The data would allow, through further analysis, to frame best practices and could help Policy Makers and Urban Planners to strategically prioritize their actions like urban planning and renewal initiative that will be capable to integrate urban greenery and healthy purposes.

2 Methodology

The research methodology is divided into 4 consequential phases: intelligence phase, tool definition phase, application phase and analysis (Fig. 1).

The proposed analysis starts with a selection of scientific research that highlights significant correlations between UGS and the citizens' psychological health status. After reviewing the papers, the need to have parameters capable of describing the state of urban green spaces emerges. UGS's potential can be expressed in function of accessibility and quality. For each of these two parameters, there are tools capable of detecting them. In an urban setting, buffer areas (isochronous) evaluate accessibility based on pedestrian walking time and therefore on the distance that citizens need to cross to reach green areas. It can be argued that, although it is not the only tool available for accessibilities/distribution studies, this constitutes a sort of known and shared standard among several studies.

Fig. 1. Methodological flowchart

The same cannot be upheld for qualitative assessments. Among the multitude of existing quality assessment tools, a shared conceptual framework can be recognized: the qualities of public space are divided into various domains, which are sub-divided into individual determinants of the aforementioned domains (or items). These can be subsequently evaluated and contribute to defining a score that can be adopted as an index of environmental quality. The number of domains identified, the number of items, the scoring methods, and the way of direct inspection varies from study to study. The tool chosen for this survey is the RECITAL developed by the Universitat Autonoma de Barcelona (UAB) in 2019 [10]. Subsequently, the presented tools, after being re-elaborated and adapted to the objectives of the research and to the specific context's features of Milan. The results of both assessments (with Green Areas Distance Method and with Recital 2.0 Milan Tool) are then analyzed, crossed with each other and with other parameters like real estate values of the neighborhoods.

3 Results

3.1 Quantitative Assessment - Proximity of the UGS in Milano

One of the first phases of the research, preliminary to the further assessment phases, was to draw up - referring to the GIS-based software QGIS - dynamic and descriptive maps of the relationships between density of population, density of the urban fabric, UGS (in all its forms) availability and accessibility, and city users' perception of those green areas and infrastructures existing in the context of Milano city.

Only the significant green areas for the research purposes were identified and selected, therefore poor-quality green areas and those insignificant have been eliminated, as they are not accessible, private and not suitable for recreational/sporting activities, or not capable of creating opportunities in terms of healthy life-styles' promotion.

Among the green areas previously identified, it was decided to carry out a further selection that allowed to analyze only the areas with a size greater than 15,000 m², as they were considered the most representative for their relevant size and sufficient for outdoor Physical Activity promotion.

Subsequently, three buffer zones were defined, respectively of 250, 500 and 750 m, which allow to evaluate the distance from the green areas (Fig. 2). These distances can be easily reached on foot from 3 to 10 min and are significant for assessing the quality of the available green areas into the neighborhoods.

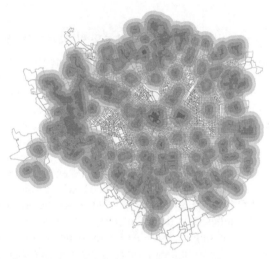

Fig. 2. Milan: buffer for analysis and definition of the distance from public green areas with an area greater than 15,000 m²

By combining the elaborations carried out with the information relating to the resident population in the several census sections (ISTAT 2011) it was possible to quantify the population included in the three buffer zones (250, 500 and 750 m) and therefore evaluate the availability of green areas in the cities of Milan, considering the served population VS excluded from the three defined buffer zones.

From a first analysis it is evident that about 90% of the population is served by a quality green area within a buffer area of 750 m; the buffer zone of 500 m corresponds to 78% of the resident population; further reducing the distance to a buffer of 250 m, a clear decrease of 49% is observed (Fig. 3).

Both the elaborated maps and the graphs obtained show how the resident population in urban areas is not equally served by quality green areas, in particular the central areas are those with a higher population density, with a high consumption of land and less availability of green areas.

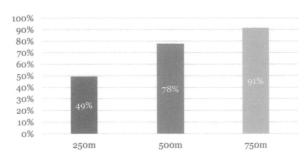

Fig. 3. Milan: quantification of the population included in the three buffer zones (250, 500 and 750 m)

3.2   The Tool RECITAL (uRban grEen spaCe quallTy Assessment tooL).

As previously discussed, for Public Health purposes, an accessibility (proximity) analysis is not sufficient to describe the state of Milan's UGS, it is therefore paired with a qualitative assessment. RECITAL (uRban grEen spaCe quallTy Assessment tooL) is a tool designed to evaluate the qualitative aspects of UGS relevant to human health.

The aforementioned tool is constituted of 90 evaluated items, scored by a 0 to 4 (Likert scale) and grouped into 11 "dimensions". The eleven "dimensions" in which the UGS are deconstructed are: "Surrounding", "Access", "Facilities", "Amenities", "Aesthetics and attraction", "Incivilities", "Safety", "Potential Usage", "Land Covers", "Animal Biodiversity", "Bird Biodiversity". RECITAL is based on previously developed tools, such as BRAT-DO [11], NEST [12], POST [13]. As the previously cited studies, the assessment tool also relies on a direct inspection from an operator on the field to determine the scores of the single items. Although detailed, the instrument does not give any prescription on how to use the collected data, which can be used as single-item scores, scores by dimensions, or as an overall global score, which would allow to compare directly different UGS. The major limitation of the tool derives from the partial subjectivity of the evaluation, the capabilities of the technician and the time frame of the inspection.

3.3   Qualitative Assessment - RECITAL 2.0 Milano

To better adapt the tool to the technician skills and to better tailor it to the qualities of the surveyed areas (the city of Milan, norther Italy, Lombardy Region), it undergoes some alterations. Without modifying the Likert scale system, or the single items, some dimensions were removed, and others were merged because they assess connected aspects.

Both the "Animal Biodiversity" and the "Bird Biodiversity" dimensions were omitted from the evaluated parameters as they require a too high level of monitoring to be evaluated (for example they ask the number of each type of bird, animals or insect). The "Surroundings" and "Access" dimensions were combined in the "Access and surrounding" macro-category; "Aesthetics and Attractions" and "Land covers" are merged into the "Environmental and Aesthetics Qualities" macro-category while "Safety" and "Incivilities" are merged into the "Security and Incivilities" macro-category. To sum up, the changes from the original eleven dimensions: two were removed and the remaining nine dimensions are merged into 6, reducing the number of items from 90 to 75 (Fig. 4).

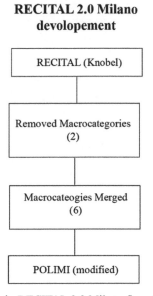

RECITAL 2.0 Milano devolopement

RECITAL (Knobel)

Removed Macrocategories (2)

Macrocateogies Merged (6)

POLIMI (modified)

Fig. 4. RECITAL 2.0 Milano flowchart

After tuning the RECITAL tool on Milan, it will be conventionally named "RECITAL 2.0 Milano" to distinguish it from the original version. The Recital 2.0 Milano was tested in some areas previously analyzed with the Green Areas Distance Method, described into the Sect. 3.1 paragraph. The sample of UGS encompasses green areas homogeneously distributed in the municipality, with different characteristics and inauguration years. By exploiting the administrative division of the city of Milan in radial sections (municipals) and sampling two to three parks from each one of the sections, to guarantee to have city center, intermediate and outskirts city example. The UGS were surveyed in the months of July and August of 2021, between 10 a.m. and 4 p.m., filling the evaluation table on site, following an observational approach. No areas were visited during rainy days or festivities.

At the end of the survey, 24 parks were evaluated (Fig. 5). Each UGS generated a global score which allowed direct comparison between UGS. Partial scores, one for each macro-category, allows comparison by domain between the various parks. By the comparison the "RECITAL 2.0 Milano" results with the UGS surface data, no significant pattern emerges, suggesting that environmental quality is not linked to the extension of the UGS (Fig. 6). By the comparison the overall score with the average real estate values in the area emerges an easily readable and expected correlation: the top-performing parks are in the most exclusive areas of the city, often in or near the city center, easily reachable by public transportation alternatives (Fig. 7).

Fig. 5. Map of the UGS tested using the RECITAL 2.0 Milano tool

The restrictions on the use of public space during the first wave Covid-19 have highlighted the importance of having quality green spaces easily accessible to the citizens. The RECITAL tool and other quality assessment systems can detect criticalities inside the urban environment, while Accessibilities assessment tools can find areas of the cities deprived of the benefits of UGS [14]. The overlap of both shortcomings could be an indicator of the presence of some form of exclusion phenomena, thus requiring attention from urban planners and Policy Makers to ensure a healthier and more equalitarian urban environment.

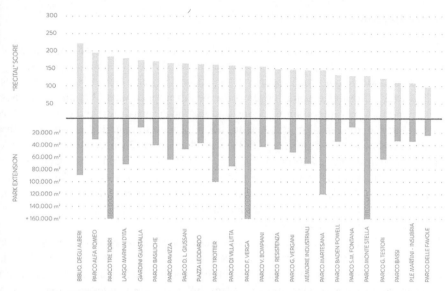

Fig. 6. Comparison between overall score and surface of the UGS

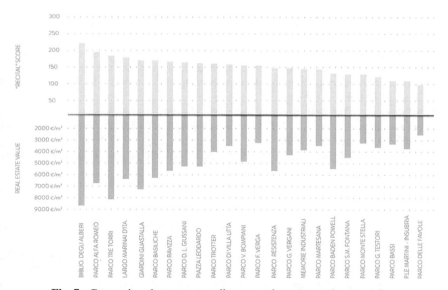

Fig. 7. Comparison between overall score and average real estate values

4 Conclusions

The UGS have multiple positive effects on Public Health as well as on the environmental quality and Climate Change reduction strategies, and for this reason, nowadays, they are a crucial strategies in urban regeneration actions [15] and Urban Health purposes [16]. However, in order to better develop the health implications [17] and outcomes of the green areas, it's important to define and guarantee proximity and quality of the green areas for all citizens. For this reason, it's important to map and geo-localize the green areas (process started in Italy with the introduction of recent regulations that ask to the Municipalities to develop and apply tools such as the green areas' Census, the Mapping, the Regulations and the Green Strategic Plan), but it's even more crucial to evaluate the quality of the aforementioned green areas in terms of Inclusive Design, Safety and Security features, equipment and provision of services (training, culture or entertainment), paths, biodiversity, etc.

The stand-alone provision of green areas into neighborhoods, does not guarantee the effectiveness of the green strategy for the purposes of Physical and Mental Health protection and promotion, if they are not carefully designed and maintained. Obviously, each urban context could have specific needs linked to the catchment area and the environmental and social context, but tools such as RECITAL can be a valid and effective support for identifying areas characterized by less Environmental Justice [18]. This would be a valid support in guiding strategic actions to protect and promote Public Health, in areas with greater needs related to environmental, social and health inequalities.

References

1. Talukder, S., Capon, A., Nath, D., Kolb, A., Jahan, S., Boufford, J.: Urban health in the post-2015 agenda. Lancet **385**, 769 (2015)
2. D'Alessandro, D., et al.: Instruments for promoting health and for assessing the hygienic and sanitary conditions in urban areas. Strategies for disease prevention and health promotion in urban areas: the erice 50 charter. Annali di Igiene **29**(6), 481–493 (2017). https://doi.org/10.7416/ai.2017.2179
3. Capolongo, S., et al.: Healthy design and urban planning strategies, actions, and policy to achieve salutogenic cities. Int. J. Environ. Res. Public Health **15**(12), 2698 (2018). https://doi.org/10.3390/ijerph15122698
4. Harting, T.: Tracking restoration in natural and urban field setting. J. Environ. Psychol., 109–123 (2003)
5. Ulrich, R.S.: Stress recovery during exposure to natural and urban environments. J. Environ. Psychol., 11 (1991)
6. Capolongo, S., et al.: COVID-19 and cities: From urban health strategies to the pandemic challenge. a decalogue of public health opportunities. Acta Biomedica **91**(2), 13–22 (2020). https://doi.org/10.23750/abm.v91i2.9515
7. Zhang, J., Yu, Z.: Links between space and public health: a bibliometric review of global research trends and future prospects from 1901 to 2019. Environ. Res. Lett. **15** (2020)
8. Gianfredi, V., et al.: Association between urban greenspace and health: a systematic review of literature. Int. J. Environ. Res. Public Health **18**, 5137 (2021). https://doi.org/10.3390/ijerph18105137
9. Comune di Milano: il verde del comune di Milano, Analisi del sistema paesistico-ambientale. Quadro di riferimento ambientale. Rapporto di riferimento Ambientale (2019)

10. Knobel P., Dadvand, P.: Development of the urban green space quality assessment tool (RECITAL). J. Urban Forestry Urban Greening, 1–8 (2020)
11. Bedimo-Rung, A., Gustat, J.: Developement of a direct observation instrument to measure environmental characteristics of parks for physical activity. Human Kintetc Journal
12. Gidlow, C., Van Kempen, E.: Development of the natural environment scoring tool (NEST). J. Urban Forestry Urban Greening, 322–333 (2017)
13. Giles-Corti, B., Broomhall, M.H.: Increasing walking: how important is distance to, attractiveness, and size of public open space? Am. J. Preventive Med., 169–176 (2005)
14. D'Alessandro, D., Buffoli, M., Capasso, L., Fara, G.M., Rebecchi, A., Capolongo, S.: Green areas and public health: Improving wellbeing and physical activity in the urban context. Epidemiol. Prev. **39**(4), 8–13 (2015)
15. Capolongo, S., Buffoli, M., Brambilla, A., Rebecchi, A.: Healthy urban planning and design strategies to improve urban quality and attractiveness of places. TECHNE **19**, 271–279 (2020). https://doi.org/10.13128/techne-7837
16. Capolongo, S., Buffoli, M., Mosca, E.I., Galeone, D., D'Elia, R., Rebecchi, A.: Public health aspects' assessment tool for urban projects, according to the urban health approach. In: Della Torre, S., Cattaneo, S., Lenzi, C., Zanelli, A. (eds.) Regeneration of the Built Environment from a Circular Economy Perspective. RD, pp. 325–335. Springer, Cham (2020). https://doi.org/10.1007/978-3-030-33256-3_30
17. Buffoli, M., Rebecchi, A., Gola, M., Favotto, A., Procopio, G.P., Capolongo, S.: Green SOAP. a calculation model for improving outdoor air quality in urban contexts and evaluating the benefits to the population's health status. In: Mondini, G., Fattinnanzi, E., Oppio, A., Bottero, M., Stanghellini, S. (eds.) SIEV 2016. GET, pp. 453–467. Springer, Cham (2018). https://doi.org/10.1007/978-3-319-78271-3_36
18. Lenzi, A., et al.: New competences to manage urban health: health city manager core curriculum. Acta Biomed. **91**, 21–28 (2020). https://doi.org/10.23750/abm.v91i3-S.9430

Social Environmental Profitability Index (SEPI) and BIM to Support Decision-Making Processes in Public Green Infrastructure Investments

Marcellina Bertolinelli, Lidia Pinti$^{(\boxtimes)}$, and Serena Bonelli

Department of Architecture, Built Environment and Construction Engineering, Polytechnic of Milan, Via Ponzio 31, 20133 Milan, Italy

{marcellina.bertolinelli,lidia.pinti,serena.bonelli}@polimi.it

Abstract. Building Information Modelling (BIM) is a method for managing data of an object for the overall life cycle: from the design to the Facility management phase till disposal. Thanks to parameters, it is possible to carry out scenario comparisons, to perform checking activities and simulations in order to make conscious decisions. Nowadays BIM is largely applied in the built environment field, while its application to outdoor areas is rare. Green space assets are subject to modifications and changes since they are by nature in a state of constant evolution: vegetal entities are characterized by most frequent transformations of considerable extent that should be properly evaluated and managed. From a public perspective, the evaluation of external factors in the selection of an appropriate landscape project, constitutes the basis for successful decisions and environmental improvement. For this reason, a suitable approach for driving aware investment actions is required. In this context, the research aims, through BIM, at defining a method for measuring the social-environmental profitability index (SEPI) to support decision making processes in public school green-outdoor infrastructure investments. A proper project appraisal should relate expenses (*starting costs, design, construction,* and *operating costs*) to the impact of the project's potential future benefits; thus, SEPI is set as the ratio of benefits/costs. In the end, research results show how BIM and SEPI should be used to classify projects, defining a ranking of alternatives, and to select among options for a sustainable public resources allocation aside from the investment scale.

Keywords: Social Environmental Profitability Index (SEPI) · Public investments · Building information modelling · Digital outdoor learning school space estimate · Evaluation tools · School green infrastructures

1 Introduction

The world population lives mainly in urban areas: in 2010, it was 50% of the global population and the forecast promised to rise to 68% by 2050 [1]. In this kind of context, the major environmental, social and economic challenges have arisen [2]. Moreover, Global Sustainable Development Goals (GSDG) outlined within the 2030 Agenda by

© The Author(s), under exclusive license to Springer Nature Switzerland AG 2022
F. Calabrò et al. (Eds.): NMP 2022, LNNS 482, pp. 1977–1989, 2022.
https://doi.org/10.1007/978-3-031-06825-6_190

the United Nations (2015), encourage inclusive and sustainable urbanization, promoting the social participation in planning and management activities [3].

Indeed, green public areas are considered a representative element of the quality of life and green infrastructures in an urban environment increasingly represent a heritage of natural and semi-natural areas [2]. Therefore, these represent a crucial issue to pursue these objectives. Thus, an effective methodology for approaching the theme of the outdoor green spaces investments and management can provide multiple environmental and social benefits [2]. At the macro level, landscape or the outdoor environment in general, constitutes an evolving context both for its physical extension and for its environmental and ornamental value; This value requires regular and appropriate methodologies for driving Decision Makers (DMs) actions.

1.1 Green Infrastructures

The landscape term represents a wide concept since it embraces environmental features with different characteristics. The landscape includes several components such as rivers, trees, and plants but also buildings etc. In this research work the interpretation is reduced to the micro-level, corresponding to an outdoor environment made of green areas and located in the medium-high density urban background.

Green spaces as assets that can be built, maintained/preserved are subject to modifications and changes since they are by nature in a state of constant evolution. Materials in the field of landscape design can be considered "animated" and their life cycle like that of human beings; vegetal materials are affected by degradation phenomena on the basis of their chemical components.

Furthermore, the positive externalities such as oxygen production, pollution reduction, climatic mitigation etc., generated by the presence of green infrastructures, can be extended over time through the optimization of project choices from the first stages of the design [4]. Therefore, the economic evaluation of an outdoor space project with reference to its Quantity Take-off, should take into account costs of the execution of works such as the cost of regeneration or renovation, but also should include management costs such as the operational ones in a ten-year period which can be considered an effective time span in a public perspective for measuring reliable data. This is fundamental to avoid squandering of public resources and benefits losses. In the light of the longer horizon (10 years), an accurate planning strategy is demanded [5, 6]; this needs also to consider the economic and financial liquidity of a Public Administration (PA) in a medium/long temporal view. The ex-ante analysis of benefits/costs of alternatives asks for transparent and impartial criteria that need to be communicated effectively.

1.2 BIM Application to the Outdoor Environment

Building Information Modelling (BIM) comes from the evolution of the construction industry [7–9] since design methods and the whole sector are progressively following technological innovation [9]. According to Babatunde et al. [9] and Lee et al. [7], BIM is a process of creating an intelligent virtual model which integrates the project data from design to construction and operation.

Looking at BIM evolution it is possible to confirm that it was introduced first to improve Architecture, Engineering, and Construction (AEC) industry productivity but nowadays its application has rapidly grown worldwide thanks to governmental support [10]. Indeed, BIM diffusion is also related to government actions in demanding BIM standards [10]. After the implementation of BIM in construction and in designing urban venues in the US, BIM was largely debated in Europe at the academic level [11] but without an in-depth analysis of the outdoor environment theme [12]. Therefore, there is a gap concerning BIM application in this field [11]. Regarding the literature and research contributions, very few studies approach the subject of BIM in green open spaces [12]; this is likely due to the fact that BIM was intended for an application in construction [11]. Therefore, nowadays BIM is largely applied in the built environment field while its implementation in dealing with the outdoor space or green infrastructure is rare [12]. As it is stated above, BIM is a method for managing data of a building or a component or an object in general for the overall life cycle: from the initial design stage to the Facility Management one (FM) and to the disposal phase. The BIM digital twin is based on the concept of 3D parametric elements, representing real life entities and related information. Thanks to parameters, it is possible to carry out analysis according to the field of application. Looking at the major differences between building elements and landscape components it should be underlined that building elements during *useful life* are influenced by degradation and obsolescence phenomena but in general they do not change substantially their geometrical attributes (apart from in cases of structural collapses etc.); on the other hand, vegetal entities, as reported above, are characterized by most frequent transformations of considerable extent that should be properly evaluated and managed. In line with these aspects, recently specific research works or analyses on outdoor infrastructures have been moving towards digital technologies and tools for innovation [13].

According to Eastman et al. [14] support of digital techniques, especially regarding information modelling, which is based both on graphics and attributes could provide possible benefits in outdoor spaces planning and design.

BIM also incorporates dynamic processes to design and enables visualization of several different possibilities before reaching the desired result. This dynamic aspect arises from parametric modelling since each component is built with entities, whose attributes may be fixed or variable [15]. The method of setting parameters for the project components, allows objects to automatically change according to the context transformations. This matter is of primary importance within the field of the outdoor environment since, as it is stated above, vegetal elements are by nature affected by strong modifications over time. Indeed, the parametric modelling of green infrastructure components (plant objects) enables simulations of real transformations affecting the outdoor environment and at the same time allows for visualize real-time evolutions at the graphical and informative level.

1.3 Paper Contribution

This research work aims, through BIM, at defining a methodology to support public decision-making processes in the field of the outdoor environment. In detail, through the Case study development, the paper concentrates on the evaluation of school green infrastructures renovation projects. During the Master's Degree Course of Economic Evaluations of Projects (2020 session) carried out at Polytechnic of Milan, students were required to develop a restoration project of outdoor school grounds located in different urban areas. At first, a qualitative analysis was done to evaluate the needs of the people directly involved: a questionnaire is sent to the community that in the future may be engaged in the schoolyard regeneration project. Then, BIM was applied to design the outdoor environment and to derive costs (both design and management costs in 10 years). In this context, BIM is a promising tool not only to represent the dynamism of the outdoor environment but also to extract data useful for the next in-depth evaluations; BIM enables real time simulations to assess the related results. Since a proper economic analysis should relate expenses (starting costs, design, construction, and operating costs) of the regeneration project to the impact of the potential future benefits, the research continues with the definition of the SEPI index.

In the end, research results show how BIM and SEPI should be used to classify projects, defining a ranking of alternatives, and to select among options for a sustainable public resources allocation aside from the investment scale.

Concerning the article structure, it is organized as follows: Sect. 2 describes the research methods and tools implemented; Sect. 3 presents the application of the research methods through a Case study approach; Sect. 4 on the basis of qualitative and quantitative data of the Case Study, defines the Social Environmental Profitability Index (SEPI) as an economic indicator for DMs in the choice of the best project alternative within the field of green infrastructures; Sect. 5 discusses the results and finally Sect. 6 presents the conclusions.

This work, through a step-by-step methodology, provides a methodology that enables the public DMs to consider social, environmental, and cost issues in doing investments. The last part of the research concerning the definition of the SEPI indicator needs further improvements since it is at the embryonic stage. Therefore, future developments and in-depth analysis are required to fine-tune this Index This method could be tested on different outdoor infrastructures, and it could be the baseline for future improvements in the area.

2 Research Methodology

This section starts by introducing the concepts applied in the study and selected as the most appropriate for evaluating public investments in green infrastructures and then for providing appropriate data for the next analysis: SEPI definitions for driving the most efficient choice among alternatives. Thus, the following paragraphs illustrate the tools then applied in the Case study section. The purpose of this part is to give a technical and global vision of the essential aspects addressed in the research. Therefore, the research's main concepts are reviewed in the next paragraphs and summarized below:

– Definition of an Information system (Work Breakdown Structure);
– Parametric modelling of green infrastructures;
– BIM application for evaluating design and management costs of the outdoor spaces.

2.1 Work Breakdown Structure (WBS)

One of the most relevant tasks, able to provide a remarkable benefit over the whole project life cycle, is the definition of a WBS. Since the fundamental point of a feasible integration among disciplines and professionals requires an adequate communication code based on a standardized language, the WBS is the key element to govern the overall process; it gives a proper framework and a clear codification which is strategic for the project development. A classification plan is a regulated standard which is the background of a WBS [16]. The definition of an adequate WBS based on a classification plan determines the goal achievement of a project.

The creation of a WBS is the first step for organising the project's information during the whole project life cycle. In particular, all the data characterizing each element of the outdoor environment must be structured in such a way that it can be uniquely identified and read by all the actors involved at any time. The WBS is a management document that, through a hierarchical breakdown into codes, allows to identify the specific information of an element in any project or management document [17]. Furthermore, it enables information to be aggregated or disaggregated according to needs or to the corresponding project phase.

The WBS is used as a project management technique for setting an efficient management and also for transmitting information to all the actors involved in the building process. Thus, the WBS represents the basis of BIM implementation also for the outdoor environment.

2.2 How to Model the Peculiarities of Green Areas

The research starts from the need to represent the peculiarities of urban green areas, both from a geometric and from an informative point of view, which inherently are characterized by two main elements (categories): all alike and static objects, which do not mutate over time their shape but which could deteriorate if not properly maintained (such as sidewalks, benches, street lamps, etc.); all different and dynamic objects, which change continuously over time (e.g. trees, lawns, bushes, etc.), both for their growth and as a result of maintenance work, or partial or non-partial removal, etc.

The representation of these two aforementioned categories of elements is still a challenging matter because of the tools currently available, since the software dedicated to construction can better simulate inanimate objects rather than vegetation [12].

This research work has faced this challenge of trying to fill the gaps of available applications on the market in the best possible way, managing the "dynamic" information of vegetal elements by filling parameters specifically created by the user.

As a matter of fact, if the parametric modelling and the graphical representation of the static elements do not seem to be a problem, the parametric modelling and the graphical representation of dynamic elements (vegetation) must take into consideration the additional parameters that allow the virtual representation of their dynamism over time (such as, for example the growth).

Focusing the attention on the most "critical" category corresponding to the dynamic element, the research work has provided two different paths in order to satisfy the graphical and informational needs of these elements. The first way, which satisfies the graphical representation and also the rendering needs of the elements (Fig. 1), goes through the insertion of Autodesk Revit RPC families for vegetation elements.

Fig. 1. Autodesk Revit RPC families for vegetation elements

Using this path, it is possible to obtain a virtual representation of the reality from a graphical point of view, but not to associate all the other information useful for the life cycle of the vegetal element. For this reason, the second option, involves the modelling of the vegetation elements through "masses" (Fig. 2). This second path, which is more schematic from a graphical point of view, provides the possibility to simulate the growth of the vegetation over the years and to enter all the parameters related to the design and maintenance cost data of the modelled element.

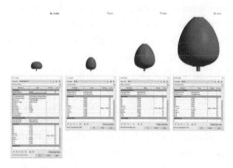

Fig. 2. Vegetation elements through "masses"

2.3 Digital Evaluation of Design and Management Costs of Outdoor Spaces

BIM is a methodology based on a flexible system that can be implemented continuously and uploaded; it can be applied in several contexts and fields from the design to the management stage and represent a suitable tool for the landscape environment purposes. In most cases, characteristics and extension of green areas, connected with the absence of a reliable mapping of underground utilities can lead to a complete review of the technical and economic feasibility study previously drafted. The process of data integration within the model in case of dealing with the outdoor environment, is made through the association of technical information and cost data of the individual component (the plant object). This integration allows an instant simulation of the possible incoming expense related to a precise design choice; Moreover, through BIM it is possible to mitigate or cancel interferences among the elements after their aggregation and to verify the impact of the green infrastructures in the surrounding context, taking in consideration also the built environment.

In the light of all of the above, the described solutions were then applied to the modelling designs of green areas in school building projects.

The green areas were then modeled taking into consideration the dynamic elements (trees, bushes, turf, etc.) and the static ones (benches, flooring, lighting, etc.) of the project, characterizing them with parameters not only containing cost data related to the project but also using cost data related to future maintenance of the modelled elements.

In this way it was possible to proceed with the preparation of the Bill of Quantity of the project and to create a simulation of the management costs of the green area for the next 10 years.

In fact, the possibility of inserting construction costs and management data within the parametric model makes it possible to know, from the earliest stages of the process, the costs that will be incurred for the project to realize and to keep under control also the cash flows related to future interventions.

3   Case Study: Outdoor Areas of Educational Buildings

The outdoor spaces in educational buildings such as schools, universities and in particular in primary or elementary schools perform a crucial role. Indeed, green areas are recognized as "open spaces classrooms": special places for doing learning and experimental activities outside. Moreover, the theme of outdoor areas, especially connected to public buildings, acquires a primary role in the light of the Covid-19 pandemic. Indeed, Covid-19 has emphasized the need of having adequate external areas available and areas suitable for outdoor education. Therefore, community participation and in general participatory processes become of strategic importance in renovation and redevelopment interventions of this kind of area. In the Master's Degree Course of the Economic Evaluations of Projects (2020 session) carried out at Polytechnic of Milan, students were required to develop a project of schoolyard restoration. In particular, the design point was to concentrate on green school infrastructures by applying BIM to the outdoor environment to derive design and maintenance costs.

Concerning the case study, the school ground requalification projects are distributed in the following regions: Lombardy (18), Veneto (3), Tuscany (1), Lazio (1), Apulia (1). Indeed, the schools are 24 in total: kindergarten (8), primary school (8), school complex gathering more levels of education (8).

3.1   Qualitative Analysis: The Participation Approach

This section illustrates the participatory process undertaken in the case of dealing with the regeneration of external educational spaces of public buildings. First of all, in order to understand the needs of children, living in the school spaces, a questionnaire was sent to families (having children at school age and directed interested in the transformation process) and then submitted online.

Regarding the survey, 518 questionnaires are delivered in total: 120 for kindergartens, 398 for primary schools.

The questionnaire answers drove the case study design through BIM. Therefore, it is possible to confirm that BIM allows the integration and test of the survey results in the model in order to obtain and evaluate the schoolyards projects considering also the users 'needs satisfaction.

3.2   Quantitative Analysis: Cost Estimate Through BIM Application

BIM as defined before, was born firstly for an application in the AEC environment and thus for being implemented in building projects; now, its use in the field of outdoor-green infrastructure needs future improvements and advancements in technologies and software tools. Considering the Course's assignment, students created a parametric model of the context using a BIM system; then the model was enriched with the individual instances of plants characterizing the project. The real growth of vegetal objects was reproduced through an algorithm introduced as input data within the model and the associated to the corresponding parameter. Thanks to this operation, the parametric 3D plant was able to show its condition in different phases of its useful life simulating different steps of development. Each vegetal component was characterized by specific data in the form of a Plant ID card, fundamental for defining the proper redevelopment strategy. Then, from this model it is possible to derive datasheets (through the Abacus tool) for evaluating in qualitative and quantitative terms maintenance interventions [5].

Indeed, each maintenance work was identified through its price and the extrapolation of the Quantity Take-off became dynamic, accelerating all the processes related to costs analysis. According to this method, intervention costs can be easily classified. Course results are summarized in the Table below (Table 1).

Each project is the outcome of different preliminary evaluations made between the initial investment and related management costs in ten years. The choice among project options was done by comparing costs of different projects, area extension and number of students.

Looking at the data, the incidence of management costs (in ten years) over the initial costs for regenerating the schoolyards is strongly different as highlighted in Table 2.

The lowest incidence is observed in project n. 5, which presents at the same time the highest project cost, while project n. 3 shows the most relevant incidence. Moreover, the case n. 5 has the most significant initial cost, followed by project n.1 which takes up the second position for the incidence of its operational costs.

The differences in cost values depend on the heterogeneity of the school ground areas and of the design choices in the study.

At this stage, through the BIM application, each schoolyard regeneration project was evaluated from the point of the users 'needs and of the starting costs, design, construction, and operating costs actualized over a 10-year span. Since the evaluation of an investment should require in addition of the analysis of the related expenses, that of the impact of potential future benefits, the research continues with the elaboration of the SEPI index.

Table 1. School ground design data implemented and extracted from a BIM model

Tipologia di scuola	Nr. Alunni	Area spazio esterno [m²]	Costo Progetto [€]	Costo Progetto / Area [€/m²]	Costo Progetto / Alunni [€/alunno]	Costo Gestione (10 anni)	Costo Gestione / Area [€/m²]	Costo Gestione / Alunni [€/alunno]	Progetto
Scuola dell'Infanzia, Primaria e Secondaria di primo Grado	233	5811	145.293 €	24,98	623,53	80.824 €	15 €	351 €	
Scuola Primaria e Secondaria di primo grado	430	2109	106.739 €	50,83	248,23	35.444 €	17 €	82 €	
Scuola dell'Infanzia, scuola primaria, scuola secondaria, Liceo	642	2658	121.848 €	45,57	189,79	91.658 €	39 €	151 €	
Scuola materna, scuola elementare	154	2581	78.051 €	30,24	506,84	36.096 €	14 €	240 €	
Scuola Elementare	210	2300	162.442 €	70,63	773,53	19.378 €	8 €	92 €	

Table 2. Initial design costs and operational/management costs

School ground Project	Design costs (€)	Operational/management costs (in 10 years) (€)	Incidence of the management costs over the total project expense
1	145.283	88.824	38%
2	106.739	35.444	25%
3	121.843	96.958	44%
4	78.053	36.996	32%
5	162.442	19.378	11%
Total	**614.360**	**277.600**	**31%**

4   Social Environmental Profitability Index (SEPI)

The main objective of a Public Administration lies in the fact to guaranteeing the positive effects generated by conscious choices in the regeneration of public areas. Moreover, a public body has the important assignment of providing a sustainable allocation of resources [18, 19]. The selection of the most effective investment should be carried out considering costs but also the impact of potential future benefits. Indeed, the analysis that evaluates only cost impacts could be considered incomplete since an appropriate approach needs also to measure the benefits. This is the point for driving aware investment actions. Considering the Investment Profitability Index (IPE), in the application to the outdoor infrastructure context, the research defines the Social and Environmental Profitability Index (SEPI).

$$\text{SEPI} = \frac{\alpha \Delta V_1 + \beta \Delta V_2 + \gamma \Delta V_3 + \varepsilon [\Delta V_4 + (\Delta V_5) * k] + \zeta (S_a - S_p) * V_{pav}}{\eta KC_p + \delta * \sum_{(t=0)}^{(t=n)} [(KM_p - KM_a) * (1 + r)^{-n}]} \tag{1}$$

Considering the numerator, this includes the so-called *stock values* such as that of the carbon stocked in the lawn and the *flow values* (for example that of carbon dioxide (CO_2) absorbed per year); Flow values need to take into account the discount rate for actualization.

$\Delta V_x = V_{xp} - V_{xa}$ (x = 1;5)
V_{xa} = cost value for managing the asset at the state of the art
V_{xp} = cost value related to the regeneration project
V_1 = square meters of green area (market value)/individual student
V_2 = square meters of equipped area (for sport, recreational or educational activities) in addition to V_1; (market value)
V_3 = trees ornamental value
V_4 = value of stocked CO_2[*1]
V_5 = value of absorbed CO_2 per year[*1]

---

[1] *The value is obtained multiplying Kg of stocked/absorbed CO_2 by its price/Kg on the European market (Emissions Trading Scheme - ETS).*

r = discount rate corresponding to the profit tax rate of 10-year government bonds (BTP)

$\alpha, \beta, \gamma, \varepsilon, \zeta, \eta, \delta$ = weighting factors.

S_a = square meters of the waterproofed area (considering the unequipped area) of the state of the art

S_p = square meters of the waterproofed area (considering the unequipped area) of the renovation project

V_{pav} = market value of the paved area.

k = financial coefficient of actualization for constant annual value: $(1 + r)n - 1/r (1 + r)n$

Considering the denominator, the factors composing the equation are as follows:

K_{Cp} = cost of the school ground renovation project at time t_0 (starting, design, construction costs)

K_{Mp} = cost for managing the school ground per year[*2].

K_{Ma} = actual cost for managing the school ground per year at time t_0 (state of the art)[*2]

n = time span (10 years).

r = discount rate corresponding to the profit tax rate of 10-year government bonds (BTP)

5 Results Discussion

The index (expressed as a ratio) takes into account not only the initial expenses and management costs of a project but also the overall social and environmental benefits. Regarding the SEPI structure, it should be highlighted that the denominator is obtained through the BIM application which is capable of providing the individual values of the related costs identified as K_C and K_M. Furthermore, concerning the first part of the formula and in particular the numerator composition, it is important to underline that this research work has identified the components (economically objective and identifiable) that contribute to the overall "investment benefit" since each of them generates an effect in terms of advantages. Therefore, all the variables are selected and calculated through the application of specific coefficients ($\alpha; \delta$) that operate as weighting factors.

Indeed, apart from the denominator's values which can be considered solid, further developments are underway to fine-tune numerator weighting coefficients in order to reach a higher significance level even though the data obtained till this point are enough for validating the general SEPI structure. Comparing SEPI to indicators such as the so-called Social Return On Investment (SROI), it should be said that the benefits in the SEPI structure are quantifiable and assessable through market values other than the participation approach which characterizes the SROI index. In detail, the SROI parameter is built on the stakeholder's identification and the social benefits first; on the other hand, the SEPI tool considers both environmental and social benefits and stakeholders have a marginal role. Moreover, as a predictive index, it can be applied both to the monitoring phase and to the final financial statement stage.

Thus, the SEPI approach can represent a suitable tool for driving decision making processes in public green infrastructure investments (apart from the scale).

---

[2] K_{Mp}; K_{Ma} = *this is intended as a precise cost value and not as a mean value per year.*

6 Conclusions

All the projects dealing with urban regeneration dedicate particular attention to sustainability and green infrastructure matters. Within the public context or in the field of private-public partnerships, the actors involved are demanded to evaluate several alternatives and projects. For example, they can be involved in selecting the best option in cases of redevelopment landscape projects or in deciding the area in which a precise project should be taken.

Regarding the theme of resources, it is important to highlight that even though relevant forms of public funding exist (as the case of PNRR), resources are exhaustible; thus, it is fundamental to define indicators which are easy to be deduced and communicated.

Considering the investment profitability index, obtained through the ratio of the project cash flows and related initial costs, a Social Environmental Profitability Index is defined. A proper project appraisal should relate expenses (*starting costs, design, construction,* and *operating costs*) to the impact of the project's potential future benefits; thus, SEPI is set as the ratio of benefits/costs. Then, operating costs distributed over time, are also discounted. The SEPI index is capable of highlighting benefits concerning expenses. Indeed, it should be used to classify projects, defining a ranking of alternatives, and to select among options for a sustainable public resources allocation aside from the investment scale.

Considering the limitations, it should be said that in order to obtain the most representative SEPI, further efforts and research should be done to achieve the highest completeness of the numerator composing the SEPI index. Therefore, future research directions could be addressed to benefits identification and measurement. Indeed, future studies should test BIM and SEPI application in different fields and cases.

References

1. United Nations, Transforming Our World: The 2030 Agenda for Sustainable Development Available online. https://sdgs.un.org/2030agenda. Accessed 1 Mar 2022
2. Dell'Anna, F., Bravi, M., Bottero, M.: Urban green infrastructures: how much did they affect property prices in Singapore? Urban For. Urban Greening **68**, 127475 (2022). https://doi.org/10.1016/j.ufug.2022.127475
3. Bottero, M., Assumma, V., Caprioli, C., Dell'Ovo, M.: Decision making in urban development: the application of a hybrid evaluation method for a critical area in the City of Turin (Italy). Sustain. Cities Soc. **72**, 103028 (2021). https://doi.org/10.1016/j.scs.2021.103028
4. Caprioli, C., Oppio, A., Baldassarre, R., Grassi, R., Dell'Ovo, M.: A multidimensional assessment of ecosystem services: from grey to green infrastructure. In: Gervasi, O., et al. (eds.) ICCSA 2021. LNCS, vol. 12955, pp. 569–581. Springer, Cham (2021). https://doi.org/10.1007/978-3-030-87007-2_41
5. Pinti, L., Bonelli, S., Brizzolari, A., Mirarchi, C., Dejaco, M.C.: Kiviniemi A, pp. 21–28. Building information modelling and database integration for the Italian public administration, Integr. Inf. Manage. Fm (2018)
6. Mirarchi, C., Pinti, L., Munir, M., Bonelli, S., Brizzolari, A., Kiviniemi, A.: Understand the value of knowledge management in a virtual asset management environment, pp. 13–20 (2018)

7. Lee, N., Dossick, C., Foley, S.: Guideline for building information modeling in construction engineering and management education. J. Prof. Issues Eng. Educ. Practice **139**, 266–274 (2013). https://doi.org/10.1061/(ASCE)EI.1943-5541.0000163

8. Liu, R., Hatipkarasulu, Y.: Introducing Building Information Modeling Course into a Newly Developed Construction Program with Various Student Backgrounds, 15 June 2014, p. 24.806.1–24.806.8

9. Babatunde, S.O., Ekundayo, D., Babalola, O., Jimoh, J.A.: Analysis of the drivers and benefits of BIM Incorporation into quantity surveying profession: academia and students' perspectives. J. Eng. Des. Technol. **16**, 750–766 (2018). https://doi.org/10.1108/JEDT-04-2018-0058

10. Pinti, L., Codinhoto, R., Bonelli, S.: A Review of Building Information Modelling (BIM) for Facility Management (FM): implementation in public organisations. Appl. Sci. **2022**, 12 (2022). https://doi.org/10.3390/app12031540

11. Kim, B.Y., Son, Y.: The current status of BIM in the field of landscape architecture and the issues on the adoption of LIM. J. Korean Inst. Landscape Archit. **42**, 50–63 (2014). https://doi.org/10.9715/KILA.2014.42.3.050

12. Zahradkova Zajickova, V.; Achten, H. Landscape Information Model: Plants as the Components for Information Modelling. In Proceedings of the Computation and Performance – Proceedings of the 31st eCAADe Conference; Delft, The Netherlands, 2013; Vol. 2

13. Fukuda, K.: Science, technology and innovation ecosystem transformation toward society 5.0. Int. J. Prod. Econ. **220**, 107460 (2020). https://doi.org/10.1016/j.ijpe.2019.07.033

14. Eastman, C., Teicholz, P., Sacks, R., Liston, K.: BIM Handbook: A Guide to Building Information Modeling for Owners, Managers, Designers, Engineers, and Contractors. Wiley (2008). ISBN 978-0-470-18528-5

15. Moura, N., et al.: Landscape information modelling to improve feedback in the geodesign international collaboration for carbon credit enhancement in metropolitan regions – the case study of Fortaleza, Brazil. In: Gervasi, O., et al. (eds.) ICCSA 2021. LNCS, vol. 12954, pp. 405–419. Springer, Cham (2021). https://doi.org/10.1007/978-3-030-86979-3_29

16. Utica, G., Pinti, L., Guzzoni, L., Bonelli, S., Brizzolari, A.: Integrating laser scanner and BIM for conservation and reuse: "The Lyric Theatre of Milan." vol. 4, pp. 77–82 (2017)

17. Utica, G.: Tecniche avanzate di analisi e gestione dei progetti. McGraw-Hill Education (2011). ISBN 88-386-6569-9

18. Bertolinelli, M., Guzzoni, L., Masseroni, S., Pinti, L., Utica, G.: Innovative participatory evaluation processes: the case of the ministry of defense real-estate assets in Italy. In: Mondini, G., Fattinnanzi, E., Oppio, A., Bottero, M., Stanghellini, S. (eds.) SIEV 2016. GET, pp. 547–557. Springer, Cham (2018). https://doi.org/10.1007/978-3-319-78271-3_43

19. Carbonara, S., Stefano, D.: An operational protocol for the valorisation of public real estate assets in Italy. Sustainability **12** (2020). https://doi.org/10.3390/su12020732

Assessing Tangible and Intangible Values of Cultural Ecosystem Services for Sustainable Regeneration Strategies

Maria Cerreta, Eugenio Muccio, and Giuliano Poli[(✉)]

Department of Architecture (DiARC), University of Naples Federico II, Via Toledo 402, 80134 Naples, Italy
giuliano.poli@unina.it

Abstract. The exceptional circumstances we are going through, due to the Covid-19 pandemic, request a reflection on the need for new methods and tools for sustainable urban development implementation. Culture is one of the critical components to promote new approaches for a more sustainable circular city, able to activate and support innovative processes for urban regeneration. The knowledge framework can be effectively considered in terms of Ecosystem Services, with specific attention to Cultural Ecosystem Services through their tangible and intangible dimensions, able to describe the cultural resources and their valorization potentials. The case study of the city of Naples (Italy) represents the opportunity to test the proposed classification of Cultural Ecosystem Services to effectively identify and and spatially representing enabling cultural contexts and provide an innovative methodological process to support decision-makers, able to identify enabling development policies and strategies for urban cultural regeneration.

Keywords: Complex values · Ecosystem services · Cultural regeneration

1 Introduction

Culture plays an essential role in regulating human relations, shaping institutions and organizational rules of our societies and urban ecosystem [1]. In a context of growing social inequalities accentuated by the pandemic scenario, it is essential to support policy-makers in assessing the potential of culture and creativity to promote social inclusion and cities' resilience. The synergy between cultural processes and urban regeneration has become the main topic in many European cities over the last decades, underlining the role of culture as one of the main drivers of urban and economic upgrowth [2–5]. The analysis of the different components of a territorial and urban landscape, intended as a socio-ecological system [6], can be usefully developed through the Ecosystem Services (E.S.) framework and Cultural Ecosystem Services specifically (C.E.S.) [7]. Indeed, C.E.S. has the great potential to contribute to biodiversity and local economies promotion [8] and the development of environmental policies and plans to enhance local ecosystem services [9]. Moreover, they express the benefits linked to the direct use of

© The Author(s), under exclusive license to Springer Nature Switzerland AG 2022
F. Calabrò et al. (Eds.): NMP 2022, LNNS 482, pp. 1990–1999, 2022.
https://doi.org/10.1007/978-3-031-06825-6_191

both ecological resources of a landscape [10–12] and recreational, aesthetic, educational, psychological values, and those related to cultural heritage [13–15].

C.E.S. are one of the four categories of E.S., representing the characteristics, ecological functions, or processes that directly or indirectly contribute to human well-being and the benefits that people derive from functioning ecosystems. C.E.S. refers to the non-material benefits that people get from ecosystems [7] and directly influences the quality of life. Over the last twenty years, multiple tools and frameworks for the E.S. classification have been developed [16, 17]. Among the most recent ones, the Common International Classification of Ecosystem Services (C.I.C.E.S.) [18], in its 4.3 version, has been considered a starting point for an urban and territorial reflection. This classification consists of a matrix articulated into "Divisions", "Groups", and "Classes" of services, relevant to identify the related components. The introduction of the Divisions represents a significant new factor, allowing the declination of C.E.S. in terms of tangible and intangible services. It is necessary to take this duality into account to detect an urban ecosystem's heterogeneous set of cultural values and describe it more refinedly. In this regard, this contribution aims to evaluate the cultural resources of the city of Naples, in Italy, to propose innovative models of sustainable development and urban cultural regeneration.

The paper introduces "Materials and methods" in Sect. 2, the methodological approach for the Cultural Ecosystem Services classification, elaborated for Naples to identify sustainable regeneration strategies in Sect. 3, and some concluding remarks in Sect. 4.

2 Materials and Methods

Once the general goal was set, a methodological process was structured (see Fig. 1) starting from the selection of the two criteria for the mapping of services: a classification of C.E.S. has been proposed starting from C.I.C.E.S. framework's Divisions, taking into account "tangible interactions" and "intangible interactions", analyzed through hard data and soft data respectively (Table 1).

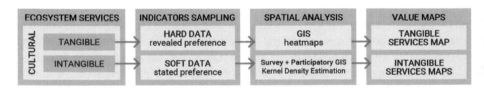

Fig. 1. The methodological approach

Tangible services have been reconsidered by integrating the three domains (and their relative dimensions) into the Groups, borrowed from the Cultural and Creative Cities Monitor proposed by the European Commission's J.R.C. [19] and identified as:

1. Cultural Vibrancy, expressed in terms of infrastructure and cultural participation;
2. Creative Economy, related to the ability to guarantee employment and innovation in the cultural and creative sectors;

3. Enabling Environment, identified with the resources that facilitate cultural processes to be activated.

Intangible services classification was identified by combining the more consolidated approach of C.E.S., deduced from the literature of C.I.C.E.S. one [18] and selected for the city of Naples, in Italy [20]. The Classes of services thus obtained were subsequently expressed through a selection of indicators that took into account the specificities of Naples and the availability of data.

The spatial distribution of information was expressed through concentration maps of C.E.S. Therefore, the most suitable methods for evaluating C.E.S. in the context of a spatial analysis appear to be non-monetary methods in most cases, which allow to express them in alternative terms to the economic ones [21].

In the assessment phase, the selected indicators were identified considering:

– hard data, obtained with the revealed preferences method through the use of national databases and local censuses;
– soft data, obtained with the stated preferences method expressed by an analysis of perceptions of a group of selected stakeholders.

Table 1. Cultural Ecosystem Services classification proposal

Services	Divisions	Groups	Classes
Cultural ecosystem services	Tangible Interactions	Cultural Vibrancy	Cultural Venues & Facilities
			Cultural Participation & Attractiveness
		Creative Economy	Creative Jobs & Activities
		Enabling Environment	Human Capital & Education
			Openness, Tolerance & Trust
			Local Connections
			Quality of Governance
	Intangible Interactions	Spiritual	Symbolism
			Spirituality
		Emblematic	Inspiration
			Sense of place
			Beauty
			Identity

The cartograms were elaborated through the support of Geographic Information Systems (G.I.S.), obtaining tangible and intangible C.E.S. heat maps, spatially represented indicator values. The maps of single services allowed us to analyze the phenomena' spatial distribution and get the final outputs through a weighted overlap. The results consist of two maps of values, which express the concentration of tangible and intangible services and thus allow the identification of enabling cultural areas or potentially sensitive to the activation of cultural regeneration processes.

3   The Cultural Ecosystem Services for the City of Naples Urban Cultural Regeneration Strategies

The C.E.S. classification proposal was tested on the case study of the city of Naples, working simultaneously on the tangible and intangible services assessment to achieve the expected results.

Table 2. Tangible C.E.S. indicators matrix

Groups	Code	Indicators	Unit	Year	Sources
Cultural Vibrancy		Museums, monuments and archaeological areas	n	2015	ISTAT
	H1	Museum, gallery or collection	n		
	H2	Monument or monumental complex	n		
	H3	Archaeological area or park	n		
		Innovative museums	n	2015	ISTAT
	H4	Digitization of assets and collections	n		
	H5	Shows and cultural entertainment	n		
	H6	Collaboration with other cultural institutions	n		
	H7	Evening openings	n		
	H8	Abandoned heritage	n	2012	Archeologia Attiva
	H9	Libraries	n	2018	Regione Campania
	H10	Theaters	n	2019	teatri.it
	H11	Parishes	n	2019	italia.indettaglio.it
Creative Economy		Cultural operators	n	2014	Comune di Napoli

(*continued*)

Table 2. (*continued*)

Groups	Code	Indicators	Unit	Year	Sources
	H12	Cinema	n		
	H13	Music	n		
	H14	Cultural promotion	n		
	H15	Social promotion and recreational activities	n		
	H16	Territorial promotion	n		
	H17	Theatre and entertainment	n		
	H18	Enhancement of artistic and archaeological heritage	n		
	H19	Social innovators	n	2018	Censimento NapoliAttiva
Enabling Environment	H20	Train stations	n	2016	OpenStreetMap
	H21	Bus stops	n	2016	OpenStreetMap

The tangible services evaluation was developed using the revealed preference method by observing the phenomena expressed by quantitative indicators, which reflect the values of C.E.S. Groups. The sources for collecting such data were databases of national and local authorities, digital databases and independent or developed censuses of research.

The indicators matrix expressed by hard data was built by considering the identification code, the name of the data, the unit of measurement, the year and the source for each indicator (Table 2).

The values of the selected indicators have been categorized into thematic maps using the Q.G.I.S. software, showing the intensity of concentration of the phenomena through a colour gradient.

The overlap of the selected indicators maps led to a preliminary map of tangible cultural values (see Fig. 2). The map shows that the most significant concentration of these services happens in the city's heart, specifically within the Historic Centre perimeter. Outside this area, the concentration decreases sharply, then slightly grows in the hilly areas and along the coast. The result indicates a strong potential in cultural terms of the Historical Centre and, secondarily, of the western coastal strip, due to the coexistence of cultural infrastructures, cultural and creative associations and companies, and transport systems.

Specific indicators have been built up referring to the number of places in the city connected to six intangible C.E.S. to assess the consistency of intangible Cultural Services. They specify the value of each Class of intangible services (Table 3).

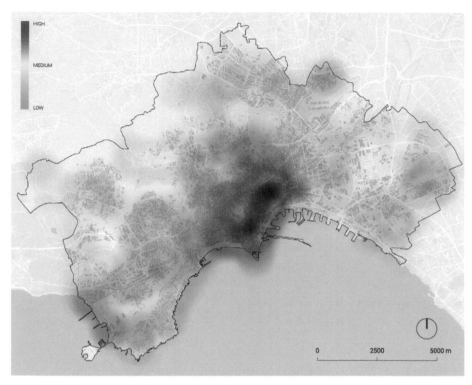

Fig. 2. Tangible cultural values map

The data collection was developed with the stated preference method through the elaboration and diffusion of a questionnaire. The survey was addressed to a significant sample of "social innovators" to evaluate the perception of intangible Cultural Services.

Table 3. Intangible C.E.S. indicators matrix

Groups	Classes	Indicators
Spiritual	Symbolism	Places connected to popular traditions, folklore, myth and other symbolic characters
	Spirituality	Places that express the spiritual and religious character of the city
Emblematic	Inspiration	Places that stimulate creative thoughts, ideas or expressions
	Sense of place	Places that express a sense of belonging and authentic attachment to the city
	Beauty	Places of particular beauty
	Identity	Places that represent the identity of the community

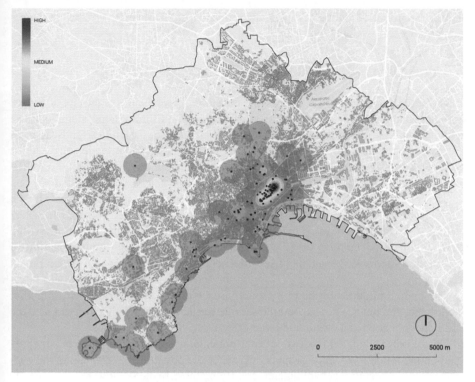

Fig. 3. Intangible cultural values map

This sample - identified through a stakeholder analysis - is made up of 44 associations and companies which in the last ten years have taken over a site of cultural interest, which guarantee the public function and which have an economic sustainability model. The ability of these organizations to respond to the social needs of the community and the effectiveness of the activated processes with a bottom-up approach make the sample significant for the purposes of the survey.

The Intangible C.E.S. Classes and related indicators therefore have become the structure of a six-question questionnaire. The sample associates three places of Naples with each of the six Classes [22] and also assigns a score that expresses for each site a judgment about the quality of the underlying service, on a Likert scale from 1 to 5, in which "1" shows a "very low" score and "5" a "very high" score.

The stated preferences were represented through concentration maps, allowing the identification of areas with the highest density of intangible Cultural Services according to their scores.

The overlap of the service maps led to a new density map of cultural values, which represents the concentration of intangible C.E.S. (see Fig. 3). Once more, the most significant concentration of intangible C.E.S. occurs within the perimeter of the Historic Centre but limited between Duomo cathedral and the "decumani" streets.

In this circumscribed area it is possible to use and benefit from the highest number of functioning Cultural Services connected to symbolism, spirituality, inspiration, a sense of place, beauty, and identity of Naples, according to the opinion expressed by the sample of interviewees.

4   Conclusions

The proposed methodology allows showing the cultural specificities of the city of Naples through a mapping of the concentration areas of the main Cultural Ecosystem Services. In its dual tangible and intangible meaning, the complexity of the cultural landscape was effectively represented in its spatial distribution through G.I.S. software. The outputs of this process consist of two maps of tangible and intangible values, which return the current picture of cultural strengths and weaknesses connected to the places of Naples.

The comparative analysis of the two maps also allows identifying the areas in which it is possible to benefit from the C.E.S. that can generate new values. Starting from the above results, it is possible to develop circular regeneration strategies that include interstitial areas identified as "cultural waste" [23, 24], thus supporting the implementation of autopoietic systems [25], capable of self-regenerating and activating new development processes.

The main limitations of the proposed approach are due to the relative availability of hard data and the need to expand the sample for the soft data collection to make the C.E.S. identification and evaluation more effective.

On the other hand, the most significant potential consists of eliciting and spatially representing intangible values, which concretizes the possibility of enhancing the consolidated areas of cultural enabling and identifying those in the shadow, recognizing the most suitable places-opportunities to the activation of new regeneration processes.

Future developments of this research will be oriented to test other methods and tools able to spatially evaluate the characteristics of cultural landscapes, considering the potential of a Spatial Decision Support System (S.D.S.S.) and Multi-Stakeholder Decision Analysis (M-SDA) for the construction of complex values maps [26–28]. Its efficacy could be tested in other urban landscapes, adapting the selection of the indicators to the peculiarities of specific urban and metropolitan contexts and verifying its replicability to establish a common basis for comparison and provide policy-makers with an effective monitoring tool for supporting territorial policies in the cultural field.

Author Contributions.   The authors jointly conceived and developed the approach and decided on the overall objective and structure of the paper: Conceptualization, M.C. and G.P.; Methodology, M.C., E.M. and G.P.; Software, E.M.; Validation, M.C. and G.P.; Formal Analysis, G.P. and E.M..; Investigation, E.M.; Resources, M.C. and G.P.; Data Curation, E.M..; Writing-Original Draft Preparation, E.M.; Writing-Review and Editing, M.C., E.M. and G.P.; Visualization, E.M.; Supervision, M.C. All authors have read and agreed to the published version of the manuscript.

References

1. Fusco Girard, L., Nijkamp, P.: Le valutazioni per lo sviluppo sostenibile della città e del territorio. Milano, Franco Angeli (1997)
2. Cerreta, M., Daldanise, G., Sposito, S.: Culture-led regeneration for urban spaces Monitoring complex values networks in action. Urbani Izziv **29**, 9–28 (2018)

3. Mondini, G., Oppio, A., Stanghellini, S., Bottero, M., Abastante, F. (eds.): Values and Functions for Future Cities. GET, Springer, Cham (2020). https://doi.org/10.1007/978-3-030-237 86-8
4. Caprioli, C., Oppio, A., Baldassarre, R., Grassi, R., Dell'Ovo, M.: A multidimensional assessment of ecosystem services: from grey to green infrastructure. In: Gervasi, O., et al. (eds.) ICCSA 2021. LNCS, vol. 12955, pp. 569–581. Springer, Cham (2021). https://doi.org/10. 1007/978-3-030-87007-2_41
5. Torres, A.V., Tiwari, C., Atkinson, S.F.: Progress in ecosystem services research: A guide for scholars and practitioners. In: Ecosystem Services, vol. 49 (2021)
6. Consiglio d'Europa, Convenzione europea del paesaggio. Firenze, 20 ottobre 2000 (2000). http://www.convenzioneeuropeapaesaggio.beniculturali.it/index.php?id=2&lang=it
7. M.E.A., Millennium Ecosystem Assessment, Ecosystems and Human Well-being: Synthesis. Island Press, Washington, DC (2005)
8. Milcu, A.I., Hanspach, J., Abson, D., Fischer, J.: Cultural ecosystem services: a literature review and prospects for future research, Ecology and society, vol. 18, n. 3 (2013)
9. Lee, J.H., Park, H.J., Kim, I., Kwon, H.S.: Analysis of cultural ecosystem services using text mining of residents' opinions. Ecol. Ind. **115**(106368), 1–7 (2020)
10. Katz-Gerro, T., Orenstein, D.E.: Environmental tastes, opinions and behaviors: social sciences in the service of cultural ecosystem service assessment. Ecology Soc. **20**(3) (2015)
11. Andersson, E., Tengö, M., McPhearson, T., Kremer, P.: Cultural ecosystem services as a gateway for improving urban sustainability. In: Ecosystem Services, vol. 12, pp. 165–168 (2015)
12. Rall, E., Bieling, C., Zytynska, S., Haase, D.: Exploring city-wide patterns of cultural ecosystem service perceptions and use. In: Ecological Indicators, vol. 77, pp. 80–95 (2017)
13. Hernández-Morcillo, M., Plieninger, T., Bieling, C.: An empirical review of cultural ecosystem service indicators. Ecol. Ind. **29**, 434–444 (2013)
14. Attardi, R., Cerreta, M., Franciosa, A., Gravagnuolo, A.: Valuing cultural landscape services: a multidimensional and multi-group SDSS for scenario simulations. In: Murgante, B., et al. (eds.) ICCSA 2014. LNCS, vol. 8581. Springer, Cham (2014). Doi: https://doi.org/10.1007/ 978-3-319-09150-1_29
15. Perchinunno, P., Rotondo, F., Torre, C.M.: The evidence of links between landscape and economy in a rural park. Int. J. Agric. Environ. Inf. Syst. (IJAEIS) **3**(2), 72–85 (2012)
16. Costanza, R., et al.: Twenty years of ecosystem services: How far have we come and how far do we still need to go? Ecosyst. Serv. **28**, 1–16 (2017)
17. Vallés-Planells, M., Galiana, F., Van Eetvelde, V.: A classification of landscape services to support local landscape planning. Ecology Soc. **19**(1), 1–11 (2014)
18. Haines-Young, R., Potschin, M.: Common international classification of ecosystem services (CICES) V5.1. Guidance on the Application of the Revised Structure. Fabis Consulting Ltd. The Paddocks, Chestnut Lane, Barton in Fabis, Nottingham, UK (2017)
19. Montalto, V., Tacao Moura, C., Panella, F., Alberti, V., Becker, W., Saisana, M.: The Cultural and Creative Cities Monitor: 2019 Edition, EUR 29797 EN, Publications Office of the European Union, Luxembourg (2019)
20. Cerreta, M., Muccio, E., Poli, G.: ValoreNAPOLI: la Valutazione dei Servizi Ecosistemici Culturali per un Modello di Città Circolare. In: BDC. Bollettino Del Centro Calza Bini, 20(2), pp. 277–296 (2020)
21. Cheng, X., Van Damme, S., Li, L., Uyttenhove, P.: Evaluation of cultural ecosystem services: a review of methods. Ecosyst. Serv. **37**(100925), 1–10 (2019)
22. Plieninger, T., Dijks, S., Oteros-Rozas, E., Bieling, C.: Assessing, mapping, and quantifying cultural ecosystem services at community level. Land Use Policy **33**, 118–129 (2013)
23. Hawkins, G., Muecke, S.: Culture and Waste: The Creation and Destruction of Value. Rowman & Littlefield, Lanham, Maryland, U.S.A. (2002)

24. Cerreta, M., Savino, V.: Circular enhancement of the cultural heritage: an adaptive reuse strategy for ercolano heritagescape. In: Gervasi, O., et al. (eds.) ICCSA 2020. LNCS, vol. 12251, pp. 1016–1033. Springer, Cham (2020). https://doi.org/10.1007/978-3-030-58808-3_72

25. Fusco Girard, L.: Toward a smart sustainable development of port cities/areas: the role of the "historic urban landscape" approach. Sustainability **5**, 4329–4348 (2013)

26. Cerreta, M., Mele, R.: A landscape complex values map: Integration among soft values and hard values in a spatial decision support system. In: Murgante, B. (eds.) ICCSA 2012. LNCS, Part 2, vol. 7334, pp. 653–669. Springer, Heidelberg (2012). Doi: https://doi.org/10.1007/978-3-642-31075-1_49

27. Cerreta, M., Inglese, P., Malangone, V., Panaro, S.: Complex values-based approach for multidimensional evaluation of landscape. In: Murgante, B., et al. (eds.) ICCSA 2014. LNCS, vol. 8581, pp. 382–397. Springer, Cham (2014). https://doi.org/10.1007/978-3-319-09150-1_28

28. Cerreta, M., Panaro, S.: From Perceived Values to Shared Values: A Multi-Stakeholder Spatial Decision Analysis (M-SSDA) for Resilient Landscapes. Sustainability 2017, 9, 1113 (2017)

The Use of the Adoption Prediction Outcome Tool to Help Communities Improve the Transition Towards the Implementation of Nature-Based Solutions

Max López-Maciel[1]([envelope]), Peter Roebeling[1,2], Rick Llewellyn[3], Elisabete Figueiredo[4], Rita Mendonça[1], Rúben Mendes[1], Fábio Matos[1], and Maria Isabel Bastos[1]

[1] CESAM and Department of Environment and Planning, Aveiro University (DAO-UA), Aveiro, Portugal
{max,peter.roebeling,ritam11,ruben.tiago,fabiomatos, mariaisabel}@ua.pt
[2] Wageningen Economic Research, Wageningen University and Research (WUR), Wageningen, Netherlands
[3] Commonwealth Scientific and Industrial and Research Organisation (CSIRO), Agriculture and Food, Waite Campus, Urrbrae, Australia
rick.llewellyn@csiro.au
[4] Department of Social, Political and Territorial Sciences and Research Unit on Governance, Competitiveness and Public Policies (GOVCOPP), Aveiro University, Aveiro, Portugal
elisa@ua.pt

Abstract. There is a general trend of increasing urban population in cities worldwide in the coming decades, leading to an intensification of urban environmental impacts as well as the probability of cities becoming vulnerable to climate change impacts. In this sense, it is essential to increase their resilience, to better cope with climate change effects. A strategy that allows taking common measures of adaptation and mitigation approaches is the implementation of nature-based solutions (NBS). However, its level of success in urban communities depends on a series of factors related to their level and time of adoption. This study adjusts the Adoption Prediction Outcome Tool (ADOPT) to the case of NBS, as to evaluate the potential level of adoption and rate of diffusion of NBS. To test the adaptation and application of this tool, the potential for implementation of green roofs in the city of Eindhoven (Netherlands) was used as a case study. Stakeholder workshops with participants from academia and local governments were held to apply ADOPT and discuss results. It is shown that the potential level of adoption of green roofs varies between 12% and 68%, while the rate of diffusion varies between 16 and 19 years. The level of adoption could be significantly improved by enhancing the relative advantage of green roofs, while the rate of diffusion is relatively robust and, thus, difficult to improve upon.

Keywords: Prediction tool · Nature-based solutions · Green infrastructure · Climate resilience · Stakeholder's engagement · Participatory approach

© The Author(s), under exclusive license to Springer Nature Switzerland AG 2022
F. Calabrò et al. (Eds.): NMP 2022, LNNS 482, pp. 2000–2011, 2022.
https://doi.org/10.1007/978-3-031-06825-6_192

1 Introduction

The effects of climate change (CC) have been widely studied at a global scale, and according to scenarios developed by the United Nations' Intergovernmental Panel on Climate Change [1], their impacts on the natural and urban ecosystems could highly increase over the next years. Moreover, cities will also grow, being expected that 68% percent of the global population will live in cities by 2050 [2]. Hence, the ecological risk is expected to increase [3], including heat islands, flooding, and water scarcity that, in turn, affect infrastructure of cities, local energy use, water management, air quality and human health [4].

In this sense, while urban areas become one of the most vulnerable types of human communities [6], they not only become receptors but also important generators of climate change impacts. This contrast allows cities to be the place where a significant response to reduce global warming effects can be achieved [7]. Cities must become more flexible in their dynamics within as well as with the external environment by improving their resilience. In this sense, infrastructure including the healthy design and urban planning strategies [8] plays a key role from an environmental as well as a social perspective [9]. Following the previous statement, the association between green spaces and health have been analysed by some authors founding those green spaces potentially benefits physical and metal heat and wellbeing in the population [10]. For instance, during the COVID pandemic the limited availability of and access to green space emphasized the inequality in some cities, highlighting the importance of cities green infrastructure implementation [11]. In that sense, some authors like Capolongo et al. [12], propose a decalogue of public health opportunities which includes the re-thinking of building typologies, fostering the presence of semi-private or collective spaces, which considers the conversion of roofs into green roofs.

One option to improve the urban resilience could be the use of nature-based solutions (NBS) [13]. NBS are generally defined as actions taken in a systemic view, which use specific solutions to diverse societal challenges, inspired, supported, or copied by or from nature, [14]. Some benefits of those include carbon extraction and storage, local climate and air quality, slopes and shoreline stabilizations, flood control, and erosion prevention, which can help with CC mitigation, adaptation, and disaster risk reduction [15]. Some examples of NBS are green roofs, street trees, parks, and artificial wetlands.

Examples of multiple impacts and benefits from NBS and green roofs can be found for instance in the cities of Seattle and Chicago including the promotion of environmental justice with the potential to redefine practices of green economic growth incorporating social equity and community coherence [16].

However, to achieve the successful implementation of joint mitigation and adaptation NBS strategies as part of urban infrastructure policies, synergies have to be addressed with a cost-efficiency perspective and a political view allowing for the strengthening of the frameworks used by decision-makers in the cities [17] – including the willingness of the community to accept them as a new implementation [18]. Thus, to improve the rate of success it is essential to assess the probability of the adoption of NBS in each city specific context, addressing the different factors that influence their level adoption and rate of diffusion.

Adoption prediction outcome tools create, based on perceived barriers and drivers, scenarios for measuring the potential level of adoption and rate of diffusion of innovations. The Adoption and Diffusion Outcome Prediction tool (ADOPT) is such a tool that creates scenarios for measuring the potential level of adoption and rate of diffusion of innovations in the agricultural sector [19]. The tool focuses on measuring the interrelationships between a targeted population and a specific innovation, based on concepts of the diffusion of innovations theory [20]. Based on variables associated with Relative advantage for the population, Learnability Characteristics of the innovations, Population-specific influences on the ability to learn about innovations, Relative advantage of innovations, ADOPT determines the peak adoption level (PAL) and the time to the peak adoption level (TPAL) of the innovation. The ADOPT has been used for instance in case studies in North America, South Asia, and Oceania [22–24].

This study is focused on the adaptation an application of the ADOPT tool to assess the factors influencing the level of adoption and rate of diffusion of NBS in urban areas. The ADOPT tool helps encouraging stakeholder participation in the decision making when adopting NBS. To this end, ADOPT was adapted to the case of urban NBS – in particular for the case of green roofs in Eindhoven (Netherlands). Section 2 describes the methodology used in this article regarding the use of ADOPT and the development of the workshop settings. The following Sect. 3 shows the results of the cited workshops; Sect. 4 discusses and analyses the results obtained. Finally, the conclusions are presented in Sect. 5.

2   Methodology

2.1   Adoption and Diffusion Outcome Prediction Tool (ADOPT)

This study carried out a qualitative and quantitative approach, orientated towards the adaptation and application of the ADOPT [24] to the case of NBS (green roofs) in Eindhoven (Netherlands).

ADOPT is centred around four interconnected quadrants. The first quadrant relates to the relative advantage of green roofs for the population, it includes the variables: Profit orientation, environmental orientation, risk orientation, enterprise scale, management horizon and short-term constraints. The second quadrant relates to the learnability characteristics of the innovation, it describes the observability of the innovation, their trialability and complexity. The third relates to the specific influences on the ability of the population to learn about the innovation, the main variables are advisory support, group involvement, relevant existing skills and knowledge and innovation awareness. Finally, the fourth quadrant relates to the relative advantage of the innovation, including the relative upfront costs of the innovation, reversibility of the innovation, profit benefit in the years it is used, profit benefit in the future, time for profit benefits to be realised, environmental benefits, time for environmental benefits to be realised, risk and ease and convenience [24].

ADOPT uses interacting variables, including networks, profit expectations, property size, the short-term costs of adoption, the innovation´s impacts on profits, impacts on risk, the complexity of the innovation, perceived environmental impact of the practice, triability on a small scale, observability, and readily apparent effects [24].

The adaptation to NBS, consisted in the modification of the original questions in ADOPT towards the adoption of green roofs. Table 1 describes the 22 variables, along with the questions associates. The answers in each question were set into ADOPT to be evaluated.

Table 1. 22 variables and questions associated in ADOPT-NBS

Variable	Question associated	Variable	Question associated
1. Profit/utility orientation	What proportion of the target population (governments, private sector and/or property owners) has maximizing profit/utility as a strong motivation for implementing green roofs?	12. Relevant existing skills and knowledge	What proportion of the target population (governments, private sector and/or property owners) will need to develop substantial new skills and knowledge to use the green roofs?
2. Environmental orientation	What proportion of the target population (governments, private sector and/or property owners) has protecting the natural environment as a strong motivation for implementing green roofs?	13. Innovation awareness	What proportion of the target population (governments, private sector and/or property owners) would be aware of the use or trailing of green roofs in their urban area?
3. Risk orientation	What proportion of the target population (governments, private sector and/or property owners) has risk minimisation as a strong motivation for implementing green roofs?	14. Relative upfront cost of innovation	What is the size of the of the initial investment relative to the implementation of green roofs?
4. Enterprise scale	What proportion of the target population (governments, private sector and/or property owners) could benefit from the green roofs?	15. Reversibility of innovation	To what extent is the adoption of green roofs able to be reversed?
5. Management horizon	What proportion of the target population (governments, private sector and/or property owners) has a long-term (more than 10 years) planning horizon?	16. Profit benefit in years that it is used	To what extent is the use of green roofs likely to affect the profitability of the target population (governments, private sector and/or property owners) in the years during its implementation and use?

(*continued*)

Table 1. (*continued*)

Variable	Question associated	Variable	Question associated
6. Short terms constraints	What proportion of the target population (governments, private sector and/or property owners) is under conditions of severe financial constraints?	17. Future profit benefit	To what extent is the use of green roofs likely to have additional effects on the future profitability of the target population?
7. Triable	How easily can the green roof (or significant components of it) be trialled on a small scale before a decision is made to adopt it on a larger scale?	18. Time until any future profit benefits are likely to be realized	How long after green roofs are first adopted would it take for effects on future profitability to be realised?
8. Innovation complexity	Does the complexity of green roofs and its components allow effects of their use to be easily evaluated when they are used?	19. Environmental costs and benefits	To what extent would the use of green roofs have net environmental benefits or costs?
9. Observability	To what extent would green roofs be observable to those in the target population (governments, private sector and/or property owners) who are yet to adopt them when they are used in their urban area?	20. Time to environmental benefit	How long after the green roofs are first adopted would it take for the expected environmental benefits or costs to be realised?
10. Advisory support	What proportion of the target population (governments, private sector and/or property owners) uses paid advisors capable of providing advice relevant to the implementation of green roofs?	21. Risk exposure	To what extent would the use of green roofs affect the net exposure of the owners' properties to risk?
11. Group Involvement	What proportion of the target population (governments, private sector and/or property owners) participates in groups that discuss this type of NBS (Green roofs)?	22. Ease and convenience	To what extent would the use of green roofs affect the ease and convenience of the management of the properties where they are applied, during the years that they are used?

The outcome of ADOPT is a report that includes a table containing the PAL and TPAL, the expected adoption level at five and ten years from the start, and adoption curves that display the number of years and the percentage of adoption of an innovation. It also includes a series of graphs representing the adoption level as well as a sensitivity analysis, which help to point out the opportunity areas for the improvement of the innovation adoption [19].

2.2 Elicitation Process and Workshop Setting

As part of the ADOPT implementation in Eindhoven, a structured elicitation process was developed. The structured elicitation process was proposed considering general concepts explained by Aspinall and Cook [25], the community readiness concept and a structured expert elicitation protocol developed by Hemming et al. [26]. Hence, a participative workshop setting for the different types of stakeholders was developed, consisting of two sessions orientated to the assessment of the possible adoption of green roofs in the city. The first one was performed at the Eindhoven University of Technology (TU/e), with 9 foreign postgraduate students holding a background orientated to urban planning and architecture; the second one was performed at the Municipality of Eindhoven, with 7 staff members of the Urban Planning department. The workshop and the questionnaire were both in English.

The performing steps in the case study were the following ones: (1) The participant institutions were contacted and invited beforehand by email, with a date proposed for each workshop. (2) During the workshops, an introduction of the general topic to the audience was performed along with the delivery of a consent letter for the participation in the workshop in which the participants agreed to contribute with the study. (3) The workshop was divided into two rounds. (3.1.) The first round was performed with the participants filling the survey of twenty-two questions adapted for ADOPT-NBSs and giving a brief explanation of the reasons for setting the score they chose. (3.2.) The second round, consisted of a similar activity, but this time the participants were conformed into heterogenic discussion groups, giving them the possibility to exchange different types of opinions, with the purpose of determining if their view changed or not respecting their first individual answers about the perception in the adoption of green roofs in the city of Eindhoven. (4) After that activity, the answers of all the participants were calculated as an average per question using Microsoft Excel and then set into the ADOPT for analysis, resulting in the PAL and TPAL graphs. (5) A group presentation and discussion about the results was conducted. (6) In the end, the participants filled a feedback form, with their comments for the continuous improvement of this type of workshops.

3   Results

The results are described regarding the two participatory workshop sessions in Eindhoven, derived from the results given by ADOPT. As stated in the previous section, the individual results of each session were evaluated along with the comparison of the questionnaire answers collected from both groups that participated. The analysis was done per question and in a global form, which allowed to have an average result that was used to have a generalization of the view of the participants.

3.1 Results from the TU/e and the Municipality of Eindhoven

Table 2 shows the results comparison between the first and second round PAL and TPAL of green roofs for the city of Eindhoven, according to participants from the Municipality of Eindhoven and the Eindhoven University of Technology (Tu/e). It is shown that the predicted PAL in the municipality was 49% for the first round and 68% for the second round, increasing the PAL in 18%, whereas in the Tu/e the PAL was 25% in the first round and 12% in the second round, reducing the perception of adoption in 13%. On the other hand, the rate of diffusion of green roofs in the municipality was 18 and 19 years in the first and second round respectively in each round and, for the Tu/e, from 18 to 16 years respectively in each round.

Table 2. Predicted adoption levels based on 1st and 2nd round responses from the Municipality of Eindhoven and the Eindhoven University of Technology (Tu/e)

	Eind. municipality		Tu/e	
	1st round	2nd round	1st round	2nd round
Predicted peak level of adoption	49.0%	68.0%	25%	12%
Predicted years to peak adoption	18	19	18	16
Predicted years to near peak adoption	14	15	14	13
Predicted adoption level in 5 years from start	16.3%	20.4%	8.30%	5%
Predicted adoption level in 10 years from start	42.6%	56.8%	21.70%	11.10%

Table 3 shows the questions in ADOPT divided by quadrants and the city of Eindhoven participants' approximate sensitivity, which indicates how the change in one of the variables could affect the PAL and TPAL of the green roof adoption. It is shown that the level of adoption could be significantly improved by enhancing the relative advantage of green roofs, specifically the key factors from variables: Profit benefit in years that it is used, which is focused only in the active profit during the green roofs use; Future profit benefit, related to the profit or loss after the green roofs are used; Environmental costs, which evaluates the environmental costs and benefits of the green roofs adoption; Risk exposure, that analyses if the adoption of green roofs and, Ease and convenience, which evaluates the management perspectives of the green roofs adoption. Finally, the TPAL resulted to be relatively robust – i.e. the variation in any of the answers had relatively little impact on the rate of diffusion of green roofs in Eindhoven.

Table 3. ADOPT final sensitivity analysis for PAL and TPAL (Tu/e and Eindhoven municipality). Adapted from CSIRO (2017)

Variable	Eind. municipality		Tu/e		Eind. municipality		Tu/e	
	Change in PAL (% step down)	Change in PAL (% step up)	Change in PAL (% step down)	Change in PAL (% step up)	Change in TPAL (years, step down)	Change in TPAL (years, step up)	Change in TPAL (years, step down)	Change in TPAL (years, step up)
Relative advantage for the population								
1. Profit/utility orientation	−7.5	7.5	−2.5	2.5	0.2	−0.2	0.1	−0.1
2. Environmental orientation	−14	11	−2.5	2.5	0.25	−0.25	0.1	−0.1
3. Risk orientation	N/a	N/a	N/a	N/a	N/a	N/a	N/a	N/a
4. Enterprise scale	−15.5	N/a	−3	4	0.25		0.17	−0.17
5. Management horizon	−4	4	−2	2	0.05	−0.05	0.05	−0.05
6. Short terms constraints	N/a	N/a	N/a	N/a	1	−1	1	−1
Learnability Characteristics of the Green Roofs								
7. Triable	N/a	N/a	N/a	N/a	1.5	−1.5	1.5	−1.5
8. Innovation complexity	N/a	N/a	N/a	N/a	1.5	−1.5	1.5	−1.5
9. Observability	N/a	N/a	N/a	N/a	0.65	−0.65	0.45	−0.45
Population-specific influences on the ability to learn about Green Roofs								
10. Advisory support	N/a	N/a	N/a	N/a	1.5	−1.5	1.1	−1.1
11. Group Involvement	N/a	N/a	N/a	N/a	1.05	−1.05	0.75	−0.75
12. Relevant existing skills and knowledge	N/a	N/a	N/a	N/a	N/a	−2	1.49	−1.49
13. Innovation awareness	N/a	N/a	N/a	N/a	0.5	−0.5	0.4	−0.4

(*continued*)

Table 3. (*continued*)

Variable	Eind. municipality		Tu/e		Eind. municipality		Tu/e	
	Change in PAL (% step down)	Change in PAL (% step up)	Change in PAL (% step down)	Change in PAL (% step up)	Change in TPAL (years, step down)	Change in TPAL (years, step up)	Change in TPAL (years, step down)	Change in TPAL (years, step up)
Relative advantage of Green Roofs								
14. Upfront costs	−7	6	−2.5	3	1.5	−1.5	1.05	−1.05
15. Reversibility of innovation	−2.5	2.5	−1	1	0.05	−0.05	0.025	−0.025
16. Profit benefit in years that it is used	−22.5	16	−6	11	0.4	−0.4	0.25	−0.25
17. Future profit benefit	−16	12.5	−5	8	0.3	−0.3	0.2	−0.2
18. Time until any future profit benefits are likely to be realized	−9	4.9	−1	0.8	0.2	−1.5	0.05	−0.05
19. Environmental costs and benefits	−20.5	15	−6	10.5	0.3	−0.3	0.25	−0.25
20. Time to environmental benefit	−4.9	4.9	−3	2	0.1	−0.1	0.1	−0.1
21. Risk exposure	−22	15	−5.5	0.95	0.4	−0.4	0.25	−0.25
22. Ease and convenience	−20.5	15	−5.3	10	0.35	−0.35	0.25	−0.25

4 Discussion

The experiment for the adaptation of ADOPT for the use of NBS revealed two different views in the adoption of green roofs in the city of Eindhoven. Even when the TPAL was slightly different (16 and 19 years, respectively), the potential level of adoption was quite different (12% and 68%, respectively). In this sense, the results show that the different workshops should be analysed in their respective context. It is important to express again that in the workshop realized in the Tu/e, the participants were all foreign students. In that sense, their opinion is the view for specialists that are outsiders to the city, and their perception of green roofs adoption is the result of their expectations with their knowledge of the city. On the other hand, the personnel interviewed in the municipality corresponds to participants that work with the public sector and are more related to the context and capabilities of the city from the local government's perspective.

The elicitation process resulted in a more inclusive exercise, allowing to reduce the degree of uncertainty when making decisions due to consensus generated. With the case study it was confirmed that a change in individual answers of the participants was observed after the exchange of knowledge and information regarding green roofs in Eindhoven. This resulted in a decrease in the general perception of adoption (Tu/e workshop) or the increase of it (Municipality of Eindhoven).

In order to improve the methodology of the workshop setting and following the comments in the feedback survey It was suggested to set the answers in the local language of the participants to make it easier to them to understand the content, in the same way making the question less technical was also a suggestion to reach a broader level of participants.

In this case with the information generated, it is possible to make an average of the results of both workshops, giving a general value of 40% of adoption in 17 years for the city of Eindhoven. This type of result does not represent a negative or positive value. The case study survey was broad, intending to show the potential of the whole city. But the adoption of NBS such as green roofs, will depend in the intentions and planning of the decision makers, a project in a neighborhood could have an objective of 100% of adoption, but NBS introduced to a bigger urban area such as the whole metropolitan area of the city could have a low level of adoption proportionally to the size of the urban area required,

Another aspect related to the use of ADOPT was the sensitive analysis of the answers. This is also an important part of the tool. Because once the stakeholders have a result of the PAL and TPAL, then they can identify in which of the 22 independent variables should put more effort in order to improve the probability of adoption and consequently the rate of success when implementing a new NBS. Which could be improving the design of the innovation, making more extension effort (in the promotion of the implementation) or checking the orientation of the target population (which is difficult to change).

5   Conclusions

The ADOPT tool has a good potential for the assessment of the level of adoption of NBS. However, considering that it is a predictive scenario generator tool, it is essential to be cautious when reviewing results. Since the scenarios do not represent an absolute truth but a prospection and are generated through an elicitation process, this means that the diversification of participants is vital to get a more generalized overview when analysing a determined urban area. Nevertheless, if it is considered as a pre-filter that gives a first approach to decision makers when thinking about adopting NBS, the answers given by ADOPT can offer a good idea regarding which type of NBS could be easier to implement in a particular urban community. Also, to determine which opportunity areas they should tackle for improving the rate of success when implementing NBS. Furthermore, the workshop setting methodology, gives a fast-track response; this makes ADOPT a practical tool for improving the rate of success when analysing potential adoption of NBS, saving time and resources for the NBS promoters and enchasing the citizen participation in the decision making. The case study in Eindhoven represented the first approach in the use of this tool for NBS along with the methodology proposed for assessing its potential.

It is suggested for future studies to test the tool in more communities, also with different scales of projects, and in parallel, to follow the implementation of NBS through the time, to generate a database that could strengthen the credibility and sharpness of the results given in the scenarios, and for possible tool calibration purposes.

The methodology proposed could serve as a first filter to the decision makers in urban areas to differentiate projects related to NBS, before taking an investment decision, and promotes the citizen participation which allows having a better idea of the expectations and acceptance level of the population to the potential strategy implementation.

For future studies, it is recommended to involve an even more diverse group in the same workshop and make a comparison of those scenarios with the separated scenarios created.

Lastly, the use of this digital tool has the potential to facilitate the transition to the adoption of green infrastructure, allowing the possibility to perform it in diverse circumstances.

References

1. Masson-Delmotte, V., et al.: Working Group I Contribution to the Sixth Assessment Report of the Intergovernmental Panel on Climate Change (2021)
2. OurWorldInData.org. https://ourworldindata.org/urbanization?source=content_type%3Ar eactlfirst_level_url%3Aarticlelsection%3Amain_contentlbutton%3Abody_link#citation. Accessed 16 Dec 2021
3. Luo, F., et al.: Assessing urban landscape ecological risk through an adaptive cycle framework. Landsc. Urban Plan. **180**, 125–134 (2018)
4. Rosenzweig, C., et al.: ARC3.2 Summary for City Leaders — Climate Change and Cities: Second Assessment Report of the Urban Climate Change Research Network. Cambridge University Press, New York (2015)
5. UNEP. https://www.unisdr.org/files/5453_092UNE.pdf. Accessed 16 Dec 2021
6. UN News. https://news.un.org/en/story/2019/09/1046662. Accessed 16 Dec 2021
7. Capolongo, S., et al.: Healthy design and urban planning strategies, actions, and policy to achieve salutogenic cities. Int. J. Environ. Res. Public Health **15**(12), 2698 (2018)
8. Friend, R., Moench, M.: What is the purpose of urban climate resilience? Implications for addressing poverty and vulnerability. Urban Clim. **6**, 98–113 (2013)
9. Gianfredi, V., et al.: Association between urban greenspace and health: a systematic review of literature. Int. J. Environ. Res. Public Health **18**(10), 5137 (2021)
10. Curran, W., Hamilton, T.: Nature-based solutions in hiding: goslings and greening in the still-industrial city. Socio-Ecol. Pract. Res. **2**(4), 321–327 (2020)
11. Capolongo, S., et al.: COVID-19 and cities: from urban health strategies to the pandemic challenge A Decalogue of Public Health opportunities. Acta bio-medica: Atenei Parmensis **91**(2), 13–22 (2020)
12. Raymond, C.M., et al.: An impact evaluation framework to support planning and evaluation of nature-based solutions projects. Report prepared by the EKLIPSE Expert Working Group on Nature-based Solutions to Promote Climate Resilience in Urban Areas, Food as Medicine, pp. 51–71 (2017)

13. Kabisch, N., et al.: Nature-based solutions to climate change mitigation and adaptation in urban areas: perspectives on indicators, knowledge gaps, barriers, and opportunities for action. Ecol. Soc. (2016)
14. Seddon, N., et al.: Understanding the value and limits of nature-based solutions to climate change and other global challenges. Philos. Trans. R. Soc. B Biol. Sci. **375**(1794), 20190120 (2020)
15. McKendry, C., Janos, N.: Greening the industrial city: equity, environment, and economic growth in Seattle and Chicago. Int. Environ. Agreem. Politics Law Econ. **15**(1), 45–60 (2014)
16. Landauer, M., et al.: The role of scale in integrating climate change adaptation and mitigation in cities. J. Environ. Planning Manag. **62**(5), 741–765 (2019)
17. Ruth, W., et al.: Community readiness: research to practice. J. Community Psychol. **28**, 291–307 (2000)
18. Commonwealth Scientific and Industrial Research Organisation, "CSIRO ADOPT". https://adopt.csiro.au/. Accessed 16 Dec 2021
19. Rogers, E.M.: Diffusion of Innovations, 5th edn. The Free Press, New York (2003)
20. Natcher, D., et al.: Assessing the constraints to the adoption of containerized agriculture in Northern Canada. Front. Sustain. Food Syst. **5**, 134 (2021)
21. Monjardino, M.: Quantifying the value of adopting a post-rice legume crop to intensify mixed smallholder farms in Southeast Asia. Agric. Syst. **177**, 102690 (2020)
22. James, A.R., Harrison, M.: T: Adoptability and effectiveness of livestock emission reduction techniques in Australia's temperate high-rainfall zone. Anim. Prod. Sci. **56**(3), 393–401 (2016)
23. CSIRO: Adoption Prediction Outcome Tool. CSIRO (2017)
24. Kuehne, G., et al.: Predicting farmer uptake of new agricultural practices: a tool for research, extension and policy. Agric. Syst. **156**, 115–125 (2017)
25. Aspinall, W.P., Cooke, R.M.: Quantifying scientific uncertainty from expert judgement elicitation. Risk and Uncertainty Assessment for Natural Hazards, vol. 9781107006195, pp. 64–99 (2011)
26. Hemming, V., et al.: A practical guide to structured expert elicitation using the IDEA protocol. Methods Ecol. Evol. **9**(1), 169–180 (2018)

Traffic Monitoring Using Intelligent Video Sensors to Support the Urban Mobility Planning

Domenico Gattuso and Domenica Savia Pellicanò[✉]

Mediterranea University, 89124 Reggio Calabria, Italy
domenica.pellicano@unirc.it

Abstract. The vehicular traffic is an extremely topical issue given its influence on the economic, social and environmental system. The mobility of people and goods is a significant factor of dynamism, but the congestive phenomena, caused by the greater vehicular flows, can be reflected in negative terms on the quality of citizens life. Smart management of transport networks is necessary and this, also and above all, means a traffic monitoring on the road. The systematic collection of traffic data is important for the safety/security related to the traffic phenomena, but also in support of urban mobility planning, for a more efficient management of infrastructures, to ensure appropriate traffic safety levels and the best possible levels of service. There are now monitoring tools, such as those based on video sensors, which can facilitate the task of transportation planners and managers and which can also provide opportunities as those relating to monitoring the security of citizens with respect to acts of vandalism or violence (security).

In the first part, the paper proposes a synthetic framework on technologies used for traffic monitoring, with a description of the sensors and variables likely to be captured. Attention is then focused on the monitoring technologies using video sensors, bringing out the greater wealth of information that can derive from it and the positive effects for the purposes of traffic planning. Finally, a proposal for a methodological approach is put forward to configure a continuous monitoring system based on a distribution of permanent video sensors on the urban road network, useful for the effective management of mobility phenomena and for planning activities.

Keywords: Traffic monitoring · Smart sensors · UAS · Urban mobility · Transportation planning

1 Introduction

Urban mobility planning is a key tool to have sustainable transport infrastructures with high quality and services, responding to an evolving demand and, at the same time, reducing the impacts produced by the vehicular traffic. Efficient mobility is the basis for improving citizens' quality of life. The traffic management has aroused interest, in recent decades. It gives a fundamental support in the collection and management of data in real time as it not only allows to quickly ascertain any situation that may involve people in motion, vehicles or the road, for safety and security purposes, but also provides useful indicators to improve mobility planning. An adequate monitoring system can guarantee

© The Author(s), under exclusive license to Springer Nature Switzerland AG 2022
F. Calabrò et al. (Eds.): NMP 2022, LNNS 482, pp. 2012–2023, 2022.
https://doi.org/10.1007/978-3-031-06825-6_193

a better knowledge of the mobility and its critical situations; thus, it can guide decision makers and technicians in defining the objectives to be pursued, the indicators to be considered, the strategies to be adopt.

Nowadays, the knowledge on the mobility system is often fragmented and incomplete. The limit of the detection methods is in the costs of investigation and processing of the collected data. In recent years, different approaches have been developed to try to obtain information characterizing the mobility in an indirect way. For example, the use of big data coming from people's mobile phones or from vehicle tracking devices connected to the telecommunication network has become established; with the aim of tracing the behavior of users in urban areas, reducing costs to a minimum. However, the goodness of this type of approach is questionable, and it is not always possible to obtain reliable data of interest. The work points to the opportunity to use video sensors for the purpose of effective traffic planning; they allow continuous monitoring over time and can guarantee control of the planning process through periodic updating phases. The paper presents a methodological approach to configure an organic monitoring system through an orderly distribution on the urban road network of intelligent video sensors, useful for the effective management of mobility phenomena and for planning activities.

2   A Synoptic Framework Related to the Most Diffused Technologies for the Vehicular Traffic Monitoring

Many traffic monitoring technologies are on the market today. They can be differentiated in relation to the sensors used to detect and recognize vehicles on a carriageway; there are "point" sensors, which allow to detect the traffic parameters at a punctual road section, and "area" sensors that focus on a longer spatial domain. The sensors can be intrusive, i.e. arranged on the pavement, or no intrusive, i.e. placed under the pavement or on the edge of the roadway. A brief description of the most common methods and tools for traffic monitoring is below; Table 1 shows a synoptic list.

Table 1. Traffic monitoring technologies and measurable traffic variables

Variables	Axes	Vehicles	Speed	Mass	Time occup.	Length	Veh. class
Pneumatic tubes	X		X				
Triboelectric cables	X						
WIM sensors				X			
Inductive loops		X					
Piezoelectric sensors	X			X^a			
Magn.-dynamic sensors		X	X		X	X	
Microwave radars		X	X				
Infrared sensors		X	X				X^b
Acoustic sensors		X	X		X		

[a]with class I devices; [b]with type of true-presence microwave radar; [b]with a pair of type active sensors [1]

Pneumatic tubes represent one of the oldest automated vehicle flow measurement systems; they are positioned on the carriageway, in a direction orthogonal to the flow direction and act as sensors as the passage of the vehicle's wheels generates a pressure wave propagating at the ends up to a counter device. The latter one, consisting of a membrane switch that vibrates and generates a pulse in the counter, allows the detector device to count the vehicle axes. Associating an equivalent car to the pass of two axes on the tubes, it is possible to derive, with sufficient approximation, the vehicular flow rate. The tubes can also be used to detect the vehicles speed.

Triboelectric cables have the same operating principle as the pneumatic tubes but they use the electrification by rubbing of a dielectric material. When a vehicle moves on the sensor, the steel wires of the external ring rub the surface of the dielectric material, electrifying it, causing an accumulation of electric charge; an electrical signal is then sent and therefore the recording of the passage of the vehicle axis. Compared to pneumatic tubes, triboelectric sensors have the advantage of being more resistant and robust.

Weigh In Motion (WIM) sensors are pressure detectors which, when vehicles transit, perform dynamic weighing, thus identifying the classes of vehicles. The technological evolution in recent years has led to the creation of a wide range of WIM sensors including piezoelectric cables, capacitive plates and load cell sensors. These tools can be used both for temporary surveys, fixing the cables above the road surface, and for permanent surveys, after their installation inside the bituminous surface. The advantages are the easy cables installation and a wide range of information; the disadvantages are related to the higher cost compared to other pressure sensors, the poor reliability and the risk of mechanical breakage of the cables.

Inductive loops are most widely used techniques, still today. The sensor is composed by one or more cables of electric wire, arranged to form a square or rectangular section with a side of 2–3 m. The loops can be installed above the road surface in the case of temporary surveys or embedded in the pavement with the function of permanent detectors. The measurement of the vehicular presence on each loop is linked to the variation of the magnetic field produced by the presence of the vehicle.

Magneto-dynamic sensors are punctual detectors, capable of detecting traffic parameters by analysing the variations in the earth's magnetic field as a vehicle passes. The most common magneto-dynamic sensors are VMI (Vehicle Magnetic Imaging), small rectangular plates consisting of a microprocessor powered by batteries, positioned at the middle of the lane. The measurement is based on the variation of the earth's magnetic field induced by the interference of the vehicle metal components, translated in an electrical signal, directly proportional to the magnetic mass; this signal is analysed by the micro-processor and encoded in its internal memory. The stored data can be used by specific software to obtain a wide range of information (counts, speed measurements, occupancy times, vehicle lengths, time distances, etc.).

Microwave radars (Fig. 1) are modern above-ground technologies that detect the passage and speed of vehicles on a road section. They can be of two types: "Doppler microwave radars" and "Truepresence microwave radars". In the first, the operating principle of the sensor is based on the Doppler-Fizeau effect, consisting in the frequency modification of an electromagnetic wave in the presence of relative motion between source and receptor. True-presence microwave radars differ in the frequency of the electromagnetic wave emitted; not more a constant frequency, but a continuous wave with a modulated frequency over time. This kind of sensor also allows to detect stationary vehicles, thus allowing any accidents to in real time. Both types of microwave radars have the advantage of not being affected by atmospheric conditions, avoiding problems of performance degradation due to bad weather or fog. However, they are more expensive than traditional detectors installed on the road surface, even if in the long term they can be cheaper in relation to the modest maintenance cost.

Infrared sensors can be divided into passive and active. Passive sensors have a receiver device capable of detecting the energy of infrared radiation emitted by the pavement when vehicles pass; therefore, they only detect the passage of vehicles. The sensor is installed above the carriageway and on the middle of each lane. Active devices detect the vehicles and their speed. The sensor consists of a source and a receiver of infrared rays. The two types of infrared sensors have the advantage of not causing disturbance to road traffic during the installation.

Fig. 1. Microwave sensors and Infrared sensors

Acoustic sensors can be ultrasonic (or active acoustic) or passive acoustic. The first are the most widely used and allow the detection of the traffic volume, occupancy rate, transit speed. An ultrasonic sensor is a small-sized instrument consisting essentially of a generator and a receiver of sound waves (ultrasounds) with a frequency between 25 and 60 kHz. This tool does not require works on the road surface and is normally installed on a portal or on an overpass above the carriageway. The operating principle is based on the reflection of sound waves phenomenon, according to which the time taken by a wave to leave the source, bounce off a reflecting surface and return to the source is directly proportional to the distance between the source and the reflection surface. Passive acoustic sensors are based on the measurement of the noise produced by each vehicle; when a vehicle crosses the detection zone, an increase in sound energy occurs, which corresponds to a signal of presence of the vehicle.

3   Monitoring Technologies by Using Video Sensors

Video sensors are no-intrusive tools capable to detect the vehicular traffic continuously and to operate on a wide spatial domain. A video monitoring system is based on the use of cameras that offer a spatial-temporal representation of the vehicular flow and can also be used by administrations for issues relating to road safety and security.

To determine the traffic parameters, it is necessary to process the images of the video through methods able both to interpret the content of each image (spatial analysis) and to sequentially correlate their contents (temporal analysis). Image processing can be done manually or automatically. The Automatic Image Processing (AIP) is more advantageous in terms of accuracy, reliability and costs; it is performed using specialized software. The monitoring system includes even a video recorder, an analogical-digital converter which transforms the electrical signal coming out of the camera (or video recorder) into digital form; a processing unit equipped with specialized software, which processes the digital images supplied by the converter and consequently extracts the traffic variables of interest. The cameras must be implanted on stable supports, in such a way as to prevent dynamic stresses caused by wind or other factors causing oscillations of the optical sensor, which could negatively affect the quality of the shooting and image processing. Video recordings are a very rich source of information; they allow:

- detection and tracking of vehicles and people;
- detection of unwanted events, such as accidents and congestive phenomena;
- quite precise position and speed measurements;
- a space-time reading of individual behaviours and interactions.

There are today several commercial products on the market that operate the AIP for traffic and the performances are still subject to refinement [2–4].

In general, there are two approaches in AIP (Fig. 2): *Tripwire*, or recognition of vehicles crossing a line, and *Tracking*, or trajectory drawing of moving vehicles.

The Tripwire approach can be used profitably for gate or intersection control, for example for safety control (crossing a gate, crossing an intersection with red traffic lights, etc.). Some references for specific applications are proposed [5–7]. Tracking appears more interesting for drawing the space-time trajectory of vehicles along a section of road, and for obtaining a greater number of information on vehicular flow.

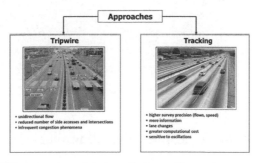

Fig. 2. Video sensors and Automatic Image Processing systems (AIP)

A particular type of video sensors are the drones (Unmanned Arial Systems - UAS), light aircraft, controlled remotely through radio waves or autonomously on a predetermined path [8]. Drones can cover several applications in hard-to-reach areas, such as landslides, degraded and polluted areas, flooded areas, collapsed infrastructures; detection of critical congestive conditions; monitoring of the decay of important structures such as tunnels, bridges, viaducts.

UAS (Fig. 3) offer interesting opportunities in specialized monitoring operations, scientific activity, experimentation and research. They allow to overcome some of the main limitations of traditional technologies; for instance:

- compared to fixed video sensors, installation costs are lower and operating costs are significantly reduced; moreover, they have the advantage of being able to be used in different time contexts, in relation to specific needs;
- if the drone is difficult to use in urban areas (due to regulatory constraints), it is very advantageous for surveys on motorways or extra-urban routes as it allows to follow the vehicles in space and time;
- the quality of the images is generally high and it is possible to save the videos for subsequent checks; it is possible to use an automatic detection system starting from the videos (deferred or in real time), with high reliability.

The use of UAS is bound by the rules on air traffic, which prevent the flight in urban contexts and in restricted areas as airport traffic zones. It is forbidden to fly over crowded areas, sporting events, public events. For the traffic survey on urban roads a buffer area around the operating site is required, in order to ensure appropriate levels of safety; the dimensions of the area are determined by evaluating the possible behaviour of the drone in the eventuality of malfunctions.

Fig. 3. UAS for traffic control (Ph.1 www.coverdrone.com; Ph.2. Authors)

4   Detection Tools for Transportation Planning

In the transportation planning, traffic detection tools are often used to acquire useful data for the knowledge of mobility phenomena in a land, the demand estimation and the calibration of modelling simulation models. There is a wide range of transport planning tools such as:

- Urban Traffic Plan (UTP), an operational management plan (few years);

- Sustainable Urban Mobility Plan (SUMP), a medium-long term strategic plan (ten years), for large urban areas;
- Public Transport Plan, generally on a sub-regional scale;
- Regional Transportation Plan, Extra-urban Road Plan, Road Safety Plan, etc.

The objectives of the plans are generally multiple like the improvement of accessibility, the reduction of congestion, the modal rebalancing, the harmonious development of transport/land use system, the reduction of road accidents, atmospheric and acoustic pollution, the socio-economic sustainability, the social inclusion and equity, in a nutshell the eco-sustainable mobility.

Considering the importance of an articulated cognitive framework and the opportunity to relate trend analyses and scenario simulations to real mobility data, the significance of appropriate traffic surveys in the field, in particular on the road network, is comprehensible. The main variables of the vehicular flow subject to monitoring are the flow rate, speed and density of traffic on the road, vehicle classification (mass, length, height, type); other typical variables are related to road safety (accidents and related characteristics, infringements), to the behaviour on the road (turning manoeuvers, lane changes, prohibited manoeuvers, etc.), to meteorological and environmental conditions (such as the presence of fog, ice, wind, rain, snow; pollutants concentration, noise levels). The information acquired can therefore be used for multiple purposes. Some significant ones are recalled here.

An application is linked to traffic regulation; monitoring activities are aimed at optimizing the operation of traffic lights in a network in real time (to minimize vehicle travel time, to reduce waiting time at intersections or the extent of congestion). The monitoring system is useful for the management of the streets and the control of parking in urban areas; data acquired in real time can suggest measures to better manage the capacity of roads and street parking spaces.

Traffic data on freeways and motorways are very useful to maintain a good level of service and a good response capacity in relation to unwanted events such as accidents or slowdowns due to traffic peaks, which can alter normal outflow conditions. Monitoring, integrated with an appropriate user information system (Variable Message Panels, radio information emissions or via App/navigators) can induce actions such as speed regulation or the diversion of users to other routes. The integrated traffic monitoring and user information system is essential for the management of urban and extra-urban mobility; in fact, users, informed about the traffic conditions in the roads they are travelling on or towards, can re-evaluate their route choices.

Other uses of the monitoring system are related to safety and security; specifically, it is possible to automatically detect infringements of the Highway Code in order to dissuade users from non-regulatory behaviours that can result in dangerous situations.

On the other hand, mobility data represent a very important element even in the context of territorial planning, as mobility phenomena are a primary component of the socio-economic activities system. And therefore, also, the integrated transport/land use management.

An important area of application, even if less visible, is related to the use of traffic data for studies and research. The data acquired in the field can be aimed, for example, at calibrating mathematical models and feeding traffic simulation packages on networks or analysing the driving behaviour of drivers.

For the mobility plans elaboration, data are required which must be appropriately retrieved whenever there is a need to elaborate or update the plan tools, with consequent surveying costs which can be considerable. To overcome this problem, nowadays, there is a tendency to resort to shortcuts or other knowledge tools such as data deriving from the connection of mobile devices (mobile phone, vehicle location systems), through software applications able to give traffic information. This approach seems interesting for its low costs and the ease-of-use thanks also to the Internet. Some examples are Google Maps Traffic, Waze, Michelin Guide. However, the derived data do not easily translate into traffic variables and real quantitative measures.

5   Operational Proposal for an Effective Monitoring System

The elaboration of urban mobility plans is essential for the development of cities but requires data expensive to find in terms of cost and time. For this reason, information on the mobility system is often fragmented and incomplete. To overcome this issue, in recent years, some analysts prefer to use "big data"; the trend of citizens to use the telephone and internet to interact and exchange information makes possible to acquire at low cost a large amount of data on users behaviour concerning the mobility. But the reliability of this type of approach is doubtful so, to ensure continuous monitoring over time and therefore facilitate the control of the planning process, it is necessary to have an effective traffic data detection system that gives elements of knowledge in realistic terms. The analysis and modelling of mobility phenomena require anchoring to primary real data such as traffic volumes and related characteristics (direction, vehicle classes, distribution over time, origin/destination, etc.).

There is no absolute detection system preferable to the others; rather, the best solution consists of an integrated mix of devices and survey techniques. The choice of the final solution will ultimately be determined by the objectives of the traffic monitoring and a careful evaluation of the cost/benefit ratio. A basic solution is proposed here, in relation to the needs of a mobility plan in a limited area such as an urban context. It is necessary to state at the outset some basic principles:

- it is better to have a monitoring system allowing continuous traffic measurements, to install sensors capable of providing significant data in relation to singular events, for example when some changes occur in the transportation network or when the plan is updated;
- a monitoring system should give easy surveillance functions, for the purposes of road safety and security;
- it is advisable to adopt systematic forms of data recording, in order to build significant, homogeneous and historicized databases;
- traffic detection devices could be accompanied by other devices able to measure impacts caused by traffic such as detectors of polluting gases concentration, of noise (phonometer), of accidents;

- it is advisable to have a Traffic Monitoring Centre that manages the collected information in a unified and coordinated approach, also equipped to activate alarms and emergency actions by police.

With reference to an enclosed area (Fig. 4), significant lines can be identified, such as the cordon line (or belt line) that delimits the area of interest (perimeter of the inhabited city, municipal border, etc.), one or several crossing lines identifying significant macro-areas of territory (they can correspond to a railway line, a river, a mountain relief), but also lines that make it possible to divide the land in relation to main traffic currents (e.g. North/South, East/West) or the urban fabric morphology (city centre, urban area); if extended in terms of corridor it may be useful to divide the urban land into bands, if divided into detached urban areas it may be appropriate to consider crossing sections on the connecting axes (Fig. 5).

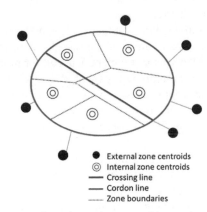

Fig. 4. Cordon line of a study area and internal crossing line

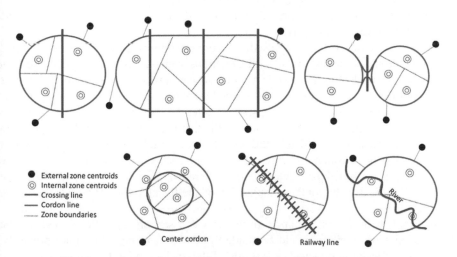

Fig. 5. Examples of crossing lines in relation to the land morphology

Significant network nodes can still be identified, considered strategic in terms of the role of meeting and sorting traffic, such as intersections, roundabouts, important junctions in the vehicle circulation dynamics, stations, toll gates.

The cordon and crossing lines allow to identify road sections; so, if the perimeter of the area closed by the belt is circular, the belt sections will correspond to the main radial traffic routes; if the crossing line divides the land into two parts, the corresponding crossing sections will be located on the streets cut by the crossing line.

The traffic measurements at the cordon are essential for a reading of the exchange flows between the area of interest and the outside, and therefore for the construction of the origin/destination matrices of external/internal exchange, internal/external exchange, crossing (flows with origin and destination outside the cordon, but in transit through the area). The traffic measurements along the crossing lines are even interesting for understanding the extent of internal exchanges between macro-sectors or neighbourhoods of the urban organism, and therefore contribute to the better determination of the matrix of internal movements within the area.

The measurements of traffic volumes at the strategic nodes then take on great importance for the verification and calibration of the displacement matrices. The wealth of information provided by video sensors (cameras), useful for monitoring, traffic control, safety and security, suggests equipping cities with this type of tools which, moreover, do not require road works, are not invasive, and are easy to maintain. It is also necessary to be aware that the video sensors allow the recording of videos and the possibility of subsequent verification. They also allow continuous measurements even for long periods of investigation.

The reading of traffic data (counting and classification of vehicles, turning manoeuvers) with reference to adequate time intervals (5, 15, 60 min) can be done "manually" by operators in the central office, or be "automated" through AIP specialised software. The reliability check of the data resulting from the AIP can be performed through manual reading on sample intervals. It will eventually be possible to make corrections to the measures where systematic percentage errors are found between real flows and flows returned by the software or to provide for a recalibration of the AIP system with the support of the software company.

In particular cases, UAS can be used; the Traffic Monitoring Centre can be equipped with small drones to be used in specific situations such as difficulties in accessing an area due to an accident or a landslide, relief of decay forms in infrastructural artworks such as bridges or viaducts that are not immediately accessible, states of emergency or collapse of the road system. In such cases they offer the possibility to operate otherwise very expensive kind of monitoring such as those usually operated by helicopters. In special cases, sensors embedded in the asphalt can be used (e.g. magnetic loops); this eventuality can be considered in cases of peripheral road sections, or in sites where the camera view is prevented, or at intersections where the sensor can be aimed at the dual function of traffic detector for the management of intelligent traffic lights and monitoring of vehicle flows.

To supplement the data coming from the video sensors, specific surveys on the characteristics of mobility should be made systematic and periodic; more precisely, surveys by means of interviews: on the side of the road, both on the cordon and on the crossing sections, through the provision of permanent and well-sized parking areas, to accommodate vehicles and personnel assigned to operations in safe conditions; at home, identifying an appropriate sample, representative and stable, of citizens.

These surveys can integrate the Data Base on mobility characteristics with three kinds of information: data relating to mobility in the day (origin, destination, motivation, mode of transport, revealed preferences, others), behavioural data (stated preferences) in relation to plan scenarios, data relating to customer satisfaction. It will be better to make use of specialized employees and tools for compiling and quickly recording the answers. The ideal would be that this staff is an integral component of the Traffic Monitoring Centre and is also responsible for the management and statistical analysis of traffic, pollution, road accident, safety data. In Italy, in the norm establishing the Urban Traffic Plans (UTP) stable expert figures are provided, within a "Technical Traffic Office".

6   Conclusions

Congestion due to vehicular traffic is one of the main factors negatively affecting people's quality of life. A continuous and real-time monitoring of road traffic is necessary, which is important not only for safety and security but as fundamental tool for urban planning activities. Traffic monitoring allows to obtain a lot of information on mobility and to identify the main criticalities, guiding the strategies of the decision makers. To date, knowledge on mobility is often fragmented and incomplete due to costs and times to acquire and process data, leading to innovative solutions.

It is considered appropriate to use video sensors for continuous monitoring over time that can guarantee control of the planning process through periodic updating phases. This is fundamental for the development of sustainable and quality transport systems. The paper presented a methodological approach to configure an organic monitoring system through an orderly distribution on the urban road network of intelligent video sensors, useful for the effective management of mobility phenomena and for planning activities.

References

1. MIT - Ministero delle Infrastrutture e dei Trasporti: Sistemi di monitoraggio del traffico -Linee guida per la progettazione (2002). https://trafficlab.eu/bfd_download/linee-guida-del-monito raggio-del-traffico/
2. Gulati, I., Rajendran, S.: Image processing in intelligent traffic management. Int. J. Recent Technol. Eng. (IJRTE) 8(2S4), 213–218 (2019)
3. Rhodes, A., Smaglik, E.J., Bullock, D.: Vendor comparison of video detection systems. Joint Transp. Res. Program 279 (2006). https://docs.lib.purdue.edu/cgi/viewcontent.cgi?art icle=1750&context=jtrp
4. Grenard, J., Bullock, D., Tarko, A.: Evaluation of selected video detection systems at signalized intersections. Joint Transportation Research Program, Indiana Department of Transportation and Purdue University, West Lafayette (2001)
5. Zender, R.L., Chang, K., Abdel -Rahim, A.: Evaluation of vehicle detection systems for traffic signal operations. Report Number RP 236, FHWA-ID-16-236 (2016). https://rosap.ntl.bts.gov/ view/dot/34949

6. Medina, J.C., Benekohal, R.F., Chitturi, M.: Evaluation of video detection systems. In: Effects of Configuration Changes in the Performance of Video Detection Systems, Illinois Center for Transportation Series N. 08-024, vol. 1 (2008)
7. VIMAR: DVR/NVR video analysis configuration. https://download.vimar.com/irj/go/km/docs/z_catalogo/DOCUMENT/49401250A0_R02.93059.pdf
8. Kardasz, P., Doskocz, J.: Drones and possibilities of their using. J. Civil Environ. Eng. **6**, 1–7 (2016)

Cultural Heritage Recovery Interventions Through Steel Endoskeletons: A Case Study

Antonino Fotia(✉) [iD], Francesco Caccamo, and Rocco Buda

PAU – Heritage Architecture and Urbanism Department, Mediterranea University,
Via dell'Università 23, 89124 Reggio Calabria, Italy
`antonino.fotia@unirc.it`

Abstract. The existing buildings in Italy represent most of the building heritage, with significantly different characteristics in relation to the type and structural scheme used, the activities that take place in them, the materials used and the current state of conservation. Recent and past seismic events have highlighted the high vulnerability of the existing building heritage, as well as the need to identify intervention strategies for the adaptation, improvement, or seismic strengthening of these buildings (or portions of them) characterized by high versatility, ease of application, reversibility and, where possible, economy.

In this context, the application of structural strengthening/consolidation systems in steel is presented as an optimized and performing solution able to merge the excellent mechanical performance of the material, in strength and ductility, with the ability to meet the current regulatory requirements for buildings in seismic areas through simple and easily applicable and installable systems.

The article in question studies an intervention of improvement and reorganization of a masonry building or historian, a Saracen tower in Cardeto (RC, Southern Italy), in order to expose a methodology of analysis and intervention aimed at improving and consolidating it.

Keywords: Steel endoskeletons · Monitoring · Cultural heritage

1 Introduction

Italy has a rich building heritage particularly varied. Since the beginning of the 80s we have witnessed an abrupt stop of new construction and urban planning has increasingly turned to the reuse of buildings already built that otherwise would have remained in a state of degradation, with a consequent waste of built space.

The particular geographical position of our country, located near the convergence between the African and European plates, make it very dangerous from a seismic point of view. This factor, combined with the considerable population density and the presence of infrastructure, determines a high seismic risk for large portions of the Italian territory. In this framework, it is evident that most of heritage building (masonry buildings) need to be recovered or safeguarded with prevention works [1, 2].

The difficulty of analyzing and reading masonry structures often stems from their empirical design. All the great masonry achievements, from antiquity to the Renaissance,

© The Author(s), under exclusive license to Springer Nature Switzerland AG 2022
F. Calabrò et al. (Eds.): NMP 2022, LNNS 482, pp. 2024–2034, 2022.
https://doi.org/10.1007/978-3-031-06825-6_194

were made based principally on a solid basis of structural sensitivity in the use of the material, and not on scientific support. The knowledge of masonry is still ongoing. In recent years the interest in masonry has been rekindled, because it is better fire resistant than other materials such as wood and steel.

From a seismic point of view, it has been observed that, if masonry buildings are well designed, they have a greater seismic-resistant capacity than modern reinforced concrete framed buildings; in fact, unlike framed buildings, that are subject to global collapses, they have partial collapses, keeping intact the bearing capacity to the vertical loads of the undamaged portions.

The aim of this note is to analyze a possible approach to improve the masonry building's visitor's safety thanks a seismic improvement through a endoskeleton system. The intervention is also aimed at restoring the functionality of the building, lost over time mainly due to abandonment, with effective consolidation methods that do not distort its original conception. The application in made on the ruins of an ancient fortification, a watchtower called by the Cardeto inhabitants "Saracen tower" (Fig. 1), located in a strategic position on the Messina strait, ideal for preventing continuous Saracen assaults. The site is located. Cardeto is a small village of the south Italy, its origins probably date back to between the 10th and 11th centuries. The town owes its name to the presence, throughout the area surrounding the village, of the thistle plant (lat. Carditum "place of thistles") or, according to another version to the Roman thistle.

Fig. 1. Saracen tower

2   Materials and Methods

2.1   UAV Survey and 3D Modelling

UAVs coupled to the use of digital photogrammetry have a fundamental role in the field of Cultural Heritage. In fact, they grant the access on dangerous or inaccessible areas, with very low costs and short times activities.

In the case study, indeed, given the particular tower location, the UAV survey is the only methodology applicable for the return of a 3D model [3, 4].

For the aerial photogrammetric survey, we used a UAV, model DJI Phantom 4 Pro equipped with the high resolution OcuSync transmission system, with a $1''$, 20-megapixel sensor capable of making videos in 4K at 60fps and shooting photographs at 14 fps. We designed a double-grid flight plan with a camera angle of 70 degrees, so that between one acquisition and the other an 80% overlap was guaranteed in both directions [5, 6]. A flight height of 20 m has been set to obtain a GSD on the ground equal to 0.68 cm/pixel. The acquisition time was 17 min, taking 181 images.

The flight was carried out ensuring the optimal light conditions for the subsequent processing phase.

For the image processing we used the Metashape software, a suitable software to create 3D models starting from a set of images.

The automatic photogrammetry has allowed us to automatically identify well-recognized key points in three or more images. Starting from them through a triangulation process, the camera was automatically calibrated, and the shooting position of the individual camera was recognized. Finally, a dense cloud, a mesh, a solid and a texture are restituted (Fig. 2) [7, 8].

Fig. 2. a) Sparse cloud, b) dense cloud, c) 3D elaboration.

2.2 HBIM

Usually, throw HBIM technology, the dense cloud allows a reverse engineering process on the existing building, because modelling the buildings, in the most likely way possible, need the knowledge of its characteristics, geometric, materials, historical, etc. The presence of defects and diversity in the forms makes a historic building unique. In this regard, the methodology we followed for the realization of this model is no longer based on point clouds but directly on the 3D model. The 3D model was cut in several elements and then imported as individual element in the BIM software as obj file, using the entire cloud as a guideline for the placement of the various objects. This process guarantees to obtain the geometric and material appearance closer to the real conformation of each component (also ensuring an information management). In relation to the insertion of the data flow (that from different sources must flow into the model and become an integral part of the model), we proceeded by adding to the geometry technical information, description of the materials, historical information, cadastral data, presence of constraints, interventions maintenance (Fig. 3).

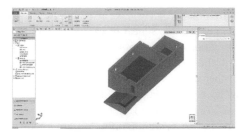

Fig. 3. HBIM reconstruction

3   Structural Analysis

3.1   Safety Assessment: Methodological Approach

Faced with a very large series in which it is necessary to intervene on the existing to reduce the seismic risk, it is important to immediately notice a great difference, from the design point of view, in the approach adopted for existing constructions compared to that for new interventions. In new buildings it is the designer who governs the process, deciding to create dissipative or non-dissipative structures, thus being able to control in some way the damage to the structure, while for existing buildings, however, the approach changes completely, in fact at moment that the damage is formed in the most fragile points the designer must be able to identify the fragility of the existing structure trying to understand the problems and events that, starting from the original project, have conditioned the structure during its life[9].

The first step to be taken is certainly the assessment of the structural safety of existing buildings, to be carried out in accordance with the provisions of Chapter 8 of the NTC 2018 and the Application Circular, which can be schematized in the following operational phases:

Historical-Critical Analysis (NTC 2018, § 8.5.1)
The historical-critical analysis allows to acquire most of the information on the basis of the existing documentation, including the elaborations and the projectual reports of the first realization of the construction and of any subsequent interventions, test reports.

It is aimed at understanding the following aspects:

period of construction; techniques, construction rules and technical standards of the time of construction; original form and subsequent modifications; trauma suffered and alterations in boundary conditions; deformations, instability and cracking pictures, with indications, where possible, of their evolution over time; any irregularities in the building; lack of construction details that guarantee the connection between structural elements, previous consolidation interventions and urban and historical aspects that have regulated building development.

Geometric-Structural Survey and Construction Details (NTC 2018, § 8.5.2)
Investigation tool very often used, with the aim of identifying the resistant organism of the construction, in particular:

- Global and local geometry and structural organization.
- Types of floors, warping, vertical section.
- Types and dimensions of non-structural elements.
- State of conservation of the structural elements.
- Possible cracking pictures and damage mechanisms

Mechanical Characterization of Materials (NTC 2018, § 8.5.3)

The mechanical characterization of materials has as its purpose the definition of the strength and deformability of materials as well as their degradation (Fig. 4).

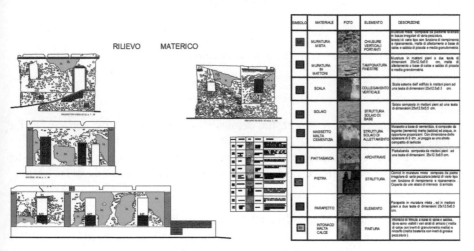

Fig. 4. Material analysis

Knowledge Levels and Confidence Factors (NTC 2018, § 8.5.4)

Based on the level of in-depth analysis achieved in the previously mentioned cognitive phases, appropriate levels of knowledge are identified (ordered by increasing information: LC1, LC2 and LC3) for the different parameters involved in the model and the related confidence factors are defined, to be used in safety checks.

Structural Modeling and Analysis

At this point the designer has the possibility to transform the physical model, reconstructed on the basis of the information available, into an analytical model for the subsequent structural analysis. Below are the types of analysis possible:

- Linear static analysis with elastic spectrum
- Linear static analysis/modal dynamic analysis with behavior factor q, a value that must be representative of the current ductility of the structure, difficult to determine a priori.
- Nonlinear static/nonlinear dynamic analysis, the most effective tool to make a diagnosis on the state of health of an existing building.

It is important to underline that, depending on the level of LC knowledge achieved in the cognitive phase, it is possible to use only some types of analysis [10, 11], as in the following figure (Fig. 5).

Fig. 5. Analysis model

Verification

Following the analysis carried out, the demand is compared with the capacity of the structure in terms of forces.

For the evaluation of the level of performance of the structures, it is possible to refer to the methods proposed in § C7.3.4.2 of the Application Circular of the NTC 2018.

Synthesis

Finally, the safety index ζ E is calculated, i.e. the ratio between demand and capacity.

Traditional strategies for the recovery of existing historic buildings, can increase capacity in terms of lift (strength and stiffness) and/or ductility towards lateral actions.

These strategies are implemented through two different classes of interventions: global and local.

The global interventions provide for the introduction inside or outside of the existing structural organism of resistant seismic elements made with the use of special bracing, which can be divided into two macro-categories: Endoskeletons and Exoskeletons, structures that modify the static dynamic behavior of the existing building.

Instead, the local interventions are instead applied to portions of existing buildings and aim to increase the plastic capacity of the system.

3.2 Interventions on the Existing

The N.T.C 2018 of Construction (point 8.4) currently in force provide for the use of three types of interventions to be carried out on existing structures:

1) **Repairs or premises** (NTC 2018, § 8.4.1):

Interventions that concern individual parts/elements of the structure that must not significantly change the overall behavior of the structure.

They are mainly intended to: Restore damage; Improve strength/ductility of parts/elements of the structure even if not damaged; Prevent local collapse mechanisms and modify the element/limited part of the structure.

The assessment of the safety and the increase in the level of safety refers only to the element/part of the structure considered (local).

2) **Seismic improvement interventions** (NTC 2018, § 8.4.2):

Interventions aimed at achieving an increase in the safety of the entire construction, without necessarily reaching safety levels of current regulations.

Also, significant changes in the local/global structural behavior in terms of stiffness/strength, introduction of new structural elements.

$$\zeta E < 1$$

3) **Seismic retrofitting interventions** (NTC 2018, § 8.4.3):

Interventions with the aim of achieving the safety levels required for new buildings specified in the NTC 2018.

Required for:

− Elevations: $\zeta E \geq 1$
− Extensions with structures connected to the existing such as to significantly alter the response: $\zeta E \geq 1$
− Change of intended use involving an increase in vertical global loads in the foundation > 10% in the characteristic combination of the actions [2.5.2] of the NTC 2018: $\zeta E \geq 0.8$

Structural system transformation: $\zeta E \geq 0.8$
Variations in use class leading to class III for school use or to CLASS IV: $\zeta E \geq 0.8$

3.3 Intervention Strategy

There are different strategies of intervention on existing structures, starting from the assumption in which the demand is always higher than the capacity, in fact it is possible to classify these strategies according to whether the interventions act on the reduction of demand and/or on the increase in capacity [12].

It is possible, on the one hand, to intervene by increasing the capacity of existing structures through a global reinforcement through the use of steel bracing that increases their strength or by acting locally by increasing the ductility and/or resistance even of individual components through, for example, the use of steel bandages.

On the other hand, it is possible to adopt intervention strategies that reduce demand, for example by reducing the seismic action on the building or through the insulation at the base that increases the period of the structure or the damping of the seismic action with the use of viscous dissipation devices, going to reduce the spectrum [13].

Steel is a material that, unlike others, provides different solutions based on the type of existing building on which it is necessary to intervene. It is the most widely used above all for the strengthening of existing buildings through local interventions, improvement, or adaptation, according to the strategies of both increasing capacity and reducing demand [14].

For global interventions, stiffening structures are usually made, made with the use of special bracing, which can be divided into two macro-categories: Endoskeletons and Exoskeletons [15, 16].

The Exoskeletons

Seismo-resistant system always of additive type that can also be adaptive, that is, not necessarily bound to the mesh of the existing frame.

The steel structure is external to the existing building [17].

Exoskeletons are solutions known for their low impact.

They can be classified in turn into:

– 2D exoskeletons, in which bracing is realized and work in the plane both parallel and orthogonally.
– 3D exoskeletons, which envelop the entire building, potentially even the co-pertura, giving the structure a boxy or shell behavior.

You can take advantage of the connection of the exoskeleton to the existing structure by inserting an additional dissipation element to the interface.

The Endoskeletons

They are typically used for the adaptation and seismic improvement of existing buildings; they are interventions in which metal bracings are inserted inside the frame of the existing structure or along its entire inter-no perimeter.

The endoskeleton itself is a seismo-resistant system of the "additive" type, which increases the strength of the frame, but not "adaptive" that is bound by the mesh of the existing frame. Among its advantages it can be highlighted that the resistance of the existing structure to horizontal loads can therefore be improved by adding cutting walls with a lattice scheme, this solution also allows to balance the distribution of stiffness with respect to the cutting center, in order to minimize torsional effects.

These systems must in any case be connected to the existing decks (Fig. 6).

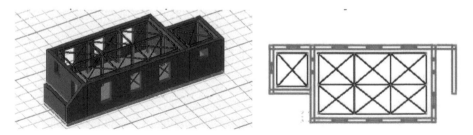

Fig. 6. Improvement intervention

4   Conclusion

The intervention suggested in this article belongs to the category of global interventions, as its application induces an overall increase in resistance and stiffness.

The proposal of intervention under consideration provides for the use of an endoskeleton, applied to the case study of Torre Saracena located in Serra within the municipality of Cardeto, in the province of Reggio Calabria. The building built around 1500 is composed of two structural blocks in stone aggregated according to a rectangular plan, the same is composed on a single level, and is oriented in an East-West direction.

The information obtained from the surveys mentioned above in relation to geometry, construction details and material properties has enabled an adequate level of knowledge to be defined (LC2) in accordance with current technical regulations for NTC buildings 2018.

The proposal of intervention under consideration provides for the use of an endoskeleton, applied to the case study of Torre Saracena located in Serra within the municipality of Cardeto, in the province of Reggio Calabria. The building built around 1500 is composed of two structural blocks in stone aggregated according to a rectangular layout, the same is composed on a single level, and is oriented in an East-West direction.

The information obtained from the surveys mentioned above in relation to geometry, construction details and material properties has enabled an adequate level of knowledge to be defined (LC2) in accordance with current technical regulations for NTC buildings 2018. The type of Endoskeleton hypothesized is realized with the use of metallic materials installed through a dry typology but research in place do not exclude future applications with the use of alloys of other nature. The same technological choice makes the solution reversible and adaptable to different contexts. This 'double skin' applied and connected along the entire internal perimeter of the existing structure is equipped with its own foundations, joined to the existing ones, because unlike the traditional systems of wind, it has a large spread of modules along the internal perimeter of the structure to ensure the correct transfer of actions and consequently its operation. The Endoskeleton itself consists of cutting walls arranged orthogonally or perpendicularly to the internal facade of the artifact to be protected.

It is a modular system, from which it is possible to obtain different configurations for example, extending the application to the roofing deck you get an integral endoskeleton which facilitates the realization of the roof that is currently absent.

In addition, in order to preserve the historical integrity of the building, Thanks to their frame configuration, the same steel profiles act as a support for the installation of glass sheets in order to allow the visitor to be able to observe the historical originality of the artifact and everything that is beyond this 'double skin'.

Acknowledgements. We thank MTC srl for the help and availability for field operations.

References

1. Kurz, J.H.: Monitoring of timber structures. J. Civil Struct. Health Monit. **5**, 97–98 (2015). https://doi.org/10.1007/s13349-014-0075-6
2. Cavalli, A., Togni, M.J.: Monitoring of historical timber structures: state of the art and prospective. J. Civil Struct. Health Monit. **5**, 107–113 (2015). https://doi.org/10.1007/s13349-014-

3. Fotia, A., Modafferi, A., Nunnari, A., D'amico, S.: From UAV survey to 3D printing, geomatics techniques for the enhancement of small village cultural heritage. WSEAS Trans. Environ. Dev. **17**, 479-489 (2021). Article number 46. Open Access. ISSN 17905079. https://doi.org/10.37394/232015.2021.17.46

4. Barrile, V., Fotia, A., Bernardo, E., Bilotta, G., Modafferi, A.: Road infrastructure monitoring: an experimental geomatic integrated system. In: Gervasi, O., et al. (eds.) ICCSA 2020. LNCS, vol. 12252, pp. 634–648. Springer, Cham (2020). https://doi.org/10.1007/978-3-030-58811-3_46

5. Barrile, V., Bilotta, G., Fotia, A., Bernardo, E.: Road extraction for emergencies from satellite imagery. In: Gervasi, O., et al. (eds.) ICCSA 2020. LNCS, vol. 12252, pp. 767–781. Springer, Cham (2020). https://doi.org/10.1007/978-3-030-58811-3_55

6. Barrile, V., Leonardi, G., Fotia, A., Bilotta, G., Ielo, G.: Real-time update of the road cadastre in GIS environment from a MMS rudimentary system. In: Calabrò, F., Della Spina, L., Bevilacqua, C. (eds.) ISHT 2018. SIST, vol. 101, pp. 240–247. Springer, Cham (2019). https://doi.org/10.1007/978-3-319-92102-0_26

7. Barrile, V., Bilotta, G., Fotia, A.: Analysis of hydraulic risk territories: comparison between LIDAR and other different techniques for 3D modeling. WSEAS Trans. Environ. Dev. **14**, 45–52 (2018). ISSN 17905079

8. Barrile, V., Fotia, A., Ponterio, R., Mollica Nardo, V., Giuffrida, D., Mastelloni, M.A.: A combined study of art works preserved in the archaeological museums: 3D survey, spectroscopic approach and augmented reality. ISPRS Ann. Photogramm. Remote Sens. Spatial Inf. Sci. **42**(2/W11), 201-207 (2019). Open Access. 2nd International Conference of Geomatics and Restoration, GEORES 2019, Milan, 8 May 2019 through 10 May 2019

9. Tondini, N., Zanon, G., Pucinotti, R., Di Filippo, R., Bursi, O.S.: Seismic performance and fragility functions of a 3D steel-concrete composite structure made of high-strength steel. Eng. Struct. **174**, 373–383 (2018). https://doi.org/10.1016/j.engstruct.2018.07.026

10. Barrile, V., Fotia, A., Leonardi, G., Pucinotti, R.: Geomatics and soft computing techniques for infrastructural monitoring. Sustainability **12**(4), 1606 (2020). https://doi.org/10.3390/su12041606

11. Lorenzoni, F., Casarin, F., Modena, C., Caldon, M., Islami, K., da Porto, F.: Structural health monitoring of the Roman Arena of Verona, Italy. J. Civil Struct. Health Monit. **3**, 227–246 (2013). https://doi.org/10.1007/s13349-013-0065-0

12. Cristofaro, M.T., Pucinotti, R., Tanganelli, M., De Stefano, M.: The dispersion of concrete compressive strength of existing buildings. In: Cimellaro, G.P., Nagarajaiah, S., Kunnath, S.K. (eds.) Computational Methods, Seismic Protection, Hybrid Testing and Resilience in Earthquake Engineering. GGEE, vol. 33, pp. 275–285. Springer, Cham (2015). https://doi.org/10.1007/978-3-319-06394-2_16

13. Pucinotti, R., Tripodo, M.: The Fiumarella bridge: concrete characterisation and deterioration assessment by nondestructive testing. Int. J. Microstruct. Mater. Prop. **4**(1), 128 (2009). https://doi.org/10.1504/IJMMP.2009.028438

14. Bergamasco, I., Marzo, A., Marghella, G., Carpani, B.: In-situ experimental campaign of the covering on the structures of Villa dei Misteri in Pompeii. J. Civil Struct. Health Monit. (2017). https://doi.org/10.1007/s13349-018-0274-7

15. Mesquita, E., et al.: Structural health monitoring of the retrofitting process, characterization and reliability analysis of a masonry heritage construction. J. Civ. Struct. Heal. Monit. **7**(3), 405–428 (2017). https://doi.org/10.1007/s13349-017-0232-9

16. Pucinotti, R., Tondini, N., Zanon, G., Bursi, O.S.: Tests and model calibration of high-strength steel tubular beam-to-column and column-base composite joints for moment-resisting structures. Earthq. Eng. Struct. Dyn. **44**(9), 1471–1493 (2015). ISSN 00988847. https://doi.org/10.1002/eqe.2547

17. Pucinotti, R.: Cyclic mechanical model of semirigid top and seat and double web angle connections. Steel Compos. Struct. **6**(2), 139–157 (2006). https://doi.org/10.12989/scs.2006.6.2.139

Sant'Aniceto Castle from the Survey to the Enhancement

Francesco Amodeo, Davide Rocco Castagnoli, Daniele Marino, Pasquale Repaci[✉],
and Antonino Siclari

Dipartimento DICEAM di Ingegneria dell'Università degli Studi Mediterranea (RC), Università
Mediterranea di Reggio Calabria, via Graziella, Feo di Vito, 89128 Reggio Calabria, Italy
pasquale.repaci@unirc.it

Abstract. The Scan to BIM method applied to the field of cultural heritage,
today represents an opportunity for public administrations in order to enhance
and disseminate the cultural heritage present in their territories, building models
able to connect different databases and, consequently, allow a multidisciplinary
approach in such a way as to guarantee the transition and the transformation from
an unsustainable production system from the point of view of the use of resources,
to a model that instead has its own strength in sustainability, environmental, social
and economic, especially in existing buildings.

The proposed method was applied to the case study of the Castle of
Sant'Aniceto (often referred to as S. Niceto) located in Motta San Giovanni (RC).
To obtain the 3D model of the structure under examination, it was necessary to
carry out, as a first step, an investigation with drone photogrammetry that allowed
the construction of a dense cloud of points and high-resolution orthophotos useful
for the study of the architectural and historical analysis of the elements present
within the rocky site. Subsequently, the individual objects identified using Edi-
ficius were modeled allowing to reconstruct the objects in a virtual environment
that can be fully visited through virtual reality applications.

Keywords: Drone · BIM · Cultural heritage · Virtual reality

1 Introduction

The purpose of this document is to highlight the potential of Scan toBIM, applied to
the field of cultural heritage, to build a 3D model able to connect different databases
and, consequently, allow a multidisciplinary approach in such a way as to guarantee the
transition and transformation from an unsustainable production system from the point
of view of the use of resources, to a model that instead has its own strength in sustainability, environmental, social and economic. in new buildings and in existing ones,
the combination and choice of materials, construction technologies, energy sources,
their control and regulation bring considerable benefits in terms of sustainability and
living comfort. The possibility of applying the BIM (Building Information Modelling)
approach to the historical building heritage (HBIM) is an interesting challenge in the
management/evaluation of the buildings and their 3D modeling. HBIM allows you to

© The Author(s), under exclusive license to Springer Nature Switzerland AG 2022
F. Calabrò et al. (Eds.): NMP 2022, LNNS 482, pp. 2035–2044, 2022.
https://doi.org/10.1007/978-3-031-06825-6_195

represent historic buildings in a digital environment, using all updatable documentation in addition to geometric information, together with the various restoration activities carried out. The survey of the existing cultural heritage requires a first phase to acquire geometric information useful for design (also through photogrammetry or terrestrial laser scanner). Cultural heritage buildings are also unique thanks to the friezes and the particular "presence of construction defects", so accurate modeling requires importing all these details into a digital model. As already mentioned, BIM is a creative methodology for the design of new buildings; therefore the processes for the reproduction of those elements are not immediate. The digitization of Cultural Heritage is an essential prerequisite for the proper conservation, management and enhancement of the structures and landscapes present in the territory. In order to achieve the objective of documenting and cataloguing the architectural elements present within the site under consideration and to obtain a high degree of detail, both from a geometric and semantic point of view, a new approach was proposed [1, 2].

Starting from a historical-bibliographical reconstruction followed by an accurate detailed inspection, it was possible to obtain a three-dimensional model of the site integrated with a series of heterogeneous information that will form the basis of an RDBMS (Relational Database Management System) prepared for continuous implementation. The metric and documentary analysis of the construction of a particular historical, architectural or archaeological interest can be divided into the following phases:

– Historical analysis;
– Detection and processing of metric data3D;
– Identification of the characteristic elements and their modeling;
– Digitization of semantic, documentary and graphic information.

For the three-dimensional survey of architectural heritage, as in the case of the investigation of the ancient village, the modern techniques that lead to an excellent result in terms of accuracy and quality of the data are essentially attributable to photogrammetric survey and TLS surveying. Therefore, in the process of importance of cultural and architectural heritage, generally characterized by complex geometries, the integration of the two different acquisition techniques and in particular the use of active sensors (high data density and acquisition speed) and passive sensors (texture quality and low cost) allow to obtain a model three-dimensional accurate and detailed. The proposed method was applied to the case study of the Castle of Sant'Aniceto located in Motta San Giovanni (RC). To obtain the 3D model of the structure under examination, it was necessary to carry out, as a first step, an investigation with a drone photogrammetry that allowed the construction of a dense cloud of points and high-resolution orthophotos useful for the study of the architectural and historical analysis of the elements present within the rocky site. Subsequently, the individual identified objects were modeled from the point cloud developing an original procedure based on the use of some tools developed in the Rhinoceros and MeshLAB software that allowed to reconstruct the objects geometrically with precision and therefore to import them into the BIM environment.

In fact, the difference between the point cloud and the BIM model was a few millimeters, demonstrating the high quality of the proposed method [3].

1.1   The Castle of Sant'Aniceto

Built in the Byzantine era to overcome the Saracen raids, the Castle of Motta S. Aniceto is a Norman fortification built in the first half of the eleventh century that protects the entire Calabrian territory.

Located on a rocky hill, oaks and olive trees protect the Castle, a visit stimulates the references of the past, when inside the walls the activity of one of the best equipped fortresses in the whole of Calabria trembled. But at the same time visiting the castle of S. Niceto offers the discovery of extraordinary natural beauty, with views that range over the last Aspromontane appendages up to Capo D'Armi, Mount Etna and the Strait of Messina. A rare example of early medieval architecture in Calabria, it is one of the few Byzantine fortifications subjected to restoration and recovery work to allow good future conservation, the Castle of San Niceto, has been for centuries a place of sighting and shelter for the people of Reggio, during the Saracen raids. You can then visit the walls, about 3.5 m high, the imposing entrance door and two square towers. There are also ruins of a water cistern and other watchtowers. At the foot of the climb, there is the church of SS. Annunziata with the dome frescoed in a traditionally Byzantine way: the subject is the Christ Pantocrator. The irregular plan of the structure has the shape of a ship: the stern to the sea and the bow to the mountain. During the thirteenth century the castle became the command center of the flourishing fief of Santo Niceto that in 1200 was tormented by the wars between the Angevins and the Aragonese who alternated on the territory of Reggio and, like many other areas of Calabria, passed into different hands; in 1321 it was handed over to the Angevins. In 1434 Santo Niceto became a barony and dominated the territories of Motta San Giovanni, Montebello and Paterriti. In the fifteenth century Santo Niceto, like the other pro-Angevin mottes, came into conflict with the city of Reggio supported by the Aragonese. In 1459, with the good of Duke Alfonso of Calabria, the People of Reggio conquered Santo Niceto through a stratagem: during a dark night, the teams of Armigers of Reggio lurk in a valley near the castle, and on the opposite side of it they let wander a flock of goats to which candles lit on the horns had been applied. The castellans, mistaking the flock for an enemy army, launched themselves on it leaving Sant'Aniceto unguarded: the Reggio soldiers, taking advantage of the situation, assaulted and invaded the castle, putting it to fire and sword. In a document of 1604 Santo Niceto is said to belong to the Barony of Motta San Giovanni. The dedication of the castle to Santo Niceto betrays the Sicilian origin of part of the founders: in those years, in fact, in Sicily the devotion to the Byzantine admiral San Niceta, who lived between the seventh and eighth centuries, was particularly widespread. Landed in Calabria with the support of the Byzantine government, the Sicilian refugees participated with the local populations in the construction calling him by the name of their patron saint (Fig. 1).

Fig. 1. Case study: Sant'Aniceto caste.

2 Materials and Methods

2.1 Drone Survey

For the planning of the flight it was instead necessary to know a priori the final representation scale, since it determines the tolerance limit within which the uncertainty of the measurements must reside. In this case, after establishing a final representation scale 1:100, it was possible to define the accuracy within which to maintain this voltage: scale 1:100 □ σ = 100 * 0.2 mm = 20 mm.

Based on this, some of the fundamental parameters for planning the flight scheme (GSD, relative height of hpro design, average frame scale mb, etc.) were established and it was therefore possible to determine the start and end points of the strip. The drone's flight was planned thanks to the Drone Harmony program, an open source software, which can be used both as a flight configuration tool and as a dynamic control tool of the APR. For the acquisition of photos to be used in 3D reconstruction, the DJI Mavic Pro drone equipped with a stabilized 4k was used a camera with a 12 Mpx sensor that, thanks to its technical characteristics and handling, lends itself well to use of this type. The flights for the different surveys were set in automatic mode thanks to the drafting of flight plans at 30 m, with an orthogonal and inclined 70° camera [4, 5]. The possibility of real-time control of the UAV through a ground station allowed to have detailed and precise shots of the investigated works, monitoring the position, altitude and status of the device. We made a number of shots consistent with the generation of the model, integrating, where necessary, the shots taken by the UAV with further shots from the ground. The tables show the data related to the model of the works detected (Fig. 2).

For the work in question (the 3D reconstruction in Fig. 3), the software used is Agisoft Metashape. The workflow is fully automatic both in terms of image orientation and model generation and reconstruction. This condition has led to an optimization of the processing times ensuring good performance of the machine/complex software.

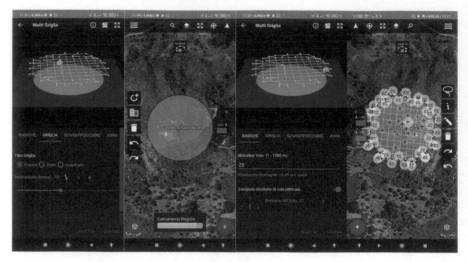

Fig. 2. Flight plan.

2.2 BIM

The acronym for BIM is Building Information Modeling: that is, model of a building with project information. BIM is therefore a methodology that if anything includes more software suitable for design and also for construction, used for its various stages of development [6].

Everything related to the construction and construction sector is therefore collected and combined digitally through the BIM.

BIM is not a software, but a methodology, complex and innovative, essential for the construction, architectural and infrastructure sector mainly public, because of strategic interest for governments but also for private individuals and for all those large projects. BIM is also an integrated design method whose uniqueness lies in the ability to collect, unify and combine all the data concerning the planning of the design of a building. The field of use of BIM concerns both the ex novo for what concerns times and methods of execution and the future life of the building itself through management and facility management.

The objective of this work is to provide a contribution to the metric-critical documentation of an asset of historical-architectural interest using new generation technologies, and therefore to be able to evaluate the potential of the same in a terrain of such high interest for the whole community of specialists who deal with the conservation of cultural heritage. The methodology used is called BIM (Building Information Modeling), which involves the creation of a three-dimensional model of the asset being analyzed to which to accompany information of different types. BIM technology was used to document and analyze an existing building. The property taken into consideration is the Castle of Sant'Aniceto in Motta San Giovanni (RC).

3   Application Case

3.1   From Capture to Point Cloud: Photogrammetry

Aerial photogrammetry is a system of observation and acquisition of spatial data through the use of a remotely piloted drone, which can generally be equipped with a digital camera and other sensors, such as the lidar camera or spectrometers [7, 8]. The direct results obtained by photogrammetric procedure are the spatial arrangement of the pixels derived from the analyzed images. Thus, the fundamental result consists of the well-known point cloud, represented in 3D space by a series of X, Y, Z coordinates. As explained in several documents available in the literature, the technique of the movement structure requires the realization of several shots from different points of view and directions, in order to locate each observed point. Importing point clouds within the project environment is a relatively simple operation, but some precautions must be used so that the georeferencing of the model is not lost. In fact, it is necessary to set the parameter "Positioning - From source to origin" during the import of the first point cloud and for the subsequent "From origin to last positioning" [9–11]. This allows you to maintain a local reference system, which allows you to have a georeferencing of the model in case of addition of other processes of the same type. Compared to standard photogrammetry, working with UAV-based photogrammetry [12, 13], additional parameters must be considered, such as the shutter speed of the drone's camera, the number of photos per second, and the flight speed. These parameters must be defined according to the desired percentage of overlap or the space between two shots [14]. The definition of the scale depends on the Ground Sample Distance (GSD), which represents the distance between the center of two consecutive pixels or, alternatively, the amount of the observed structure included in a pixel. After the definition of the GSD, the second phase of aerial photogrammetric survey is the definition of a polygonal lattice that is synchronized from a remote device to the drone. In order to obtain a correct georeferencing of the processed points, the position of each frame is generally improved by the placement of physical markers on the structure that are located via a Global Position System (GPS) station [15]. The GPS coordinates of each marker are used as reference points in the processing of all images [16, 17]. Agisoft Metashape software was used in this work. The elaboration process is composed of different phases:

1. Photo alignment: a phase that identifies the tie points with common characteristics in order to be adequately superimposed in the various photos.
2. Build dense cloud: this phase allows to build a dense cloud using dense image matching algorithms.
3. Build mesh: it consists in generating a polygonal model based on the newly created dense cloud.
4. Build texture: this phase allows to generate a twodimensional raster image in a way to apply a photorealistic colour to the mesh.

Fig. 3. 3d model.

3.2 From Cloud to HBIM

Currently starting from a point cloud, it is possible to generate an intelligent and parametric model that contains the highest possible degree of knowledge, however the process takes a long time to complete. Starting from the detection of the building, the process must determine as much information as possible and transfer it back to a model. This procedure may vary depending on the purpose for which the investigation is carried out [18, 19]. Current research is focusing on the development of standard procedures and automatic process to perform this operation faster than manual methods, avoiding the loss of information and ins and outs. Actually, two main methods are used: the first compares the point cloud, with a database of objects already present in the libraries in search of the most similar, the second Procedures use surface information (from point clouds) to perform another type of classification [20–23]. The intelligent model contains within its parametric information regarding geometry, physical properties, constituent materials. The process of moving from an existing object to an intelligent object is to follow a process in which starting from the capture data you can get a digital model with and how it is displayed in the model depends on the use. Therefore, it is possible to generate a congruent and consistent with a real estate model, in which most of the information collected up to that point is contained, as if it were a digital "catalog" that could be queried in case of need [24–26] (Fig. 4).

Fig. 4. Representation in BIM

4 Conclusions

The method described in this document has made it possible to obtain a parameterized 3D model with the enormous potential of integrating this model within the 3D software, where each object can be associated with more information. Point cloud modeling is a key role in building object surfaces. In this environment, in fact, it was possible to have a faithful reconstruction of the elements through the various tools developed within the software. To achieve this, a new procedure has been developed. The method proposed and described in the document allows to overcome the limits present in the management of information in BIM software, such as the management of the materials of the individual structural elements; currently materials with physical/thermal properties can be assigned in the BIM platform. In addition, the number of meshes integrated into the environment is less than the meshes generated in the SfM-MVS environment or generated by a TLS survey. In fact, this approach led to the construction of a parametric model much lighter in handling than that obtained from the mesh model built on the entire point cloud. The subdivision of the structures was then parameterized and defined individually. Each layer was deduced using the stratigraphic method, that is, the assignment of a unique code (USM, USR, etc.). This code represents, within 3D, the key field for the relational database; with each element, you can associate additional information.

Acknowledgements. The author thank MTC srl for the help and availability on the elaboration operations.

References

1. Barrile, V., Bilotta, G.: An application of remote sensing: object-oriented analysis of Satellite data. Int. Arch. Photogramm. Remote Sens. Spatial Inf. Sci. **XXXVII**, 107–113 (2008)
2. Barrile, V., Bilotta, G.: Object-oriented analysis applied to high resolution satellite data. WSEAS Trans. Signal Process. **4**(3), 68–75 (2008)

3. Barrile, V., Fotia, A., Ponterio, R., Mollica Nardo, V., Giuffrida, D., Mastelloni, M.A.: A combined study of art works preserved in the archaeological museums: 3D survey, spectroscopic approach and augmented reality. ISPRS Ann. Photogramm. Remote Sens. Spatial Inf. Sci. **42**(2/W11), 201–207 (2019). 2nd International Conference of Geomatics and Restoration, GEORES 2019, Milan, 8 May 2019 through 10 May 2019

4. Barrile, V., Bilotta, G., Fotia, A., Bernardo, E.: Road extraction for emergencies from satellite imagery. In: Gervasi, O., et al. (eds.) ICCSA 2020. LNCS, vol. 12252, pp. 767–781. Springer, Cham (2020). https://doi.org/10.1007/978-3-030-58811-3_55

5. Barrile, V., Leonardi, G., Fotia, A., Bilotta, G., Ielo, G.: Real-time update of the road cadastre in GIS environment from a MMS rudimentary system. In: Calabrò, F., Della Spina, L., Bevilacqua, C. (eds.) ISHT 2018. SIST, vol. 101, pp. 240–247. Springer, Cham (2019). https://doi.org/10.1007/978-3-319-92102-0_26

6. López, F., Lerones, P., Llamas, J., Gómez-García-Bermejo, J., Zalama, E.: A review of heritage building information modeling (H-BIM). Multimodal Technol. Interact. **2**(21), 1–29 (2018)

7. Fotia A, Modafferi A, Nunnari A, D'amico S: From UAV survey to 3D printing, geomatics techniques for the enhancement of small village cultural heritage. WSEAS Trans. Environ. Dev. **17**, 479–489 (2021). Article number 46

8. Barrile, V., Fotia, A., Bernardo, E., Bilotta, G.: Road cadastre an innovative system to update information, from big data elaboration. In: Gervasi, O., et al. (eds.) ICCSA 2020. LNCS, vol. 12252, pp. 709–720. Springer, Cham (2020). https://doi.org/10.1007/978-3-030-58811-3_51

9. Bernardo, E., Bonfa, S., Anderson, J.: Implementation of the digital inland water smart strategy using geomatics instruments and the big data SmartRiver platform. In: Borgogno-Mondino, E., Zamperlin, P. (eds.) ASITA 2021. CCIS, vol. 1507, pp. 83–94. Springer, Cham (2022). https://doi.org/10.1007/978-3-030-94426-1_7

10. Bernardo, E., Bonfa, S., Calcagno, S.: Techniques of geomatics and soft computing for the monitoring of infrastructures and the management of big data. WSEAS Trans. Environ. Dev. **17**, 371–385 (2021). https://doi.org/10.37394/232015.2021.17.37

11. Barrile, V., Fotia, A., Bernardo, E., Candela, G.: Geomatics techniques for submerged heritage: a mobile app for tourism. WSEAS Trans. Environ. Dev. **16**, 586–597 (2020). https://doi.org/10.37394/232015.2020.16.60

12. Barrile, V., Bilotta, G., Nunnari, A.: 3D Modeling with Photogrammetry by UAVs and model quality verification. ISPRS Ann. Photogramm. Remote Sens. Spatial Inf. Sci. **IV-4/W4**, 129–134 (2017). https://doi.org/10.5194/isprs-annals-IV-4-W4-129-2017

13. Barrile, V., Gelsomino, V., Bilotta, G.: UAV and computer vision in 3D modeling of cultural heritage in Southern Italy. IOP Conf. Ser. Mater. Sci. Eng. **225**, 012196 (2017)

14. Barrile, V., Fotia, A., Bernardo, E.: The submerged heritage: a virtual journey in our seabed. Int. Arch. Photogram. Remote Sens. Spatial Inf. Sci. **XLII-2/W10**, 17–24 (2019). https://doi.org/10.5194/isprs-archives-XLII-2-W10-17-2019

15. Barrile, V., Fotia, A., Bernardo, E., Bilotta, G., Modafferi, A.: Road infrastructure monitoring: an experimental geomatic integrated system. In: Gervasi, O., et al. (eds.) ICCSA 2020. LNCS, vol. 12252, pp. 634–648. Springer, Cham (2020). https://doi.org/10.1007/978-3-030-58811-3_46

16. Barrile, V., Cacciola, M., Morabito, F.C., Versaci, M.: TEC measurements through GPS and artificial intelligence. J. Electromagn. Waves Appl. **20**, 1211–1220 (2006). https://doi.org/10.1163/156939306777442962

17. Barrile, V., Bilotta, G., Fotia, A.: Analysis of hydraulic risk territories: comparison between LIDAR and other different techniques for 3D modeling. WSEAS Trans. Environ. Dev. **14**, 45–52 (2018)

18. Dore, C., Murphy, M.: Historic building information modeling (HBIM). In: Brusaporci, S. (ed.) Handbook of Research on Emerging Digital Tools for Architectural Surveying, Modeling, and Representation, pp. 239–279. IGI Global (2015)

19. Barrile, V., Bernardo, E., Candela, G., Bilotta, G., Modafferi, A., Fotia, A.: Road infrastructure heritage: from scan to Infrabim. WSEAS Trans. Environ. Dev. V **16**, 633–642 (2020)
20. Oreni, D., Brumana, R., Cuca, B.: Towards a methodology for 3D content models: the reconstruction of ancient vaults for maintenance and structural behaviour in the logic of BIM management. In: 2012 18th International Conference on Virtual Systems and Multimedia, Milan, pp. 475–482 (2012). https://doi.org/10.1109/VSMM.2012.6365961
21. Barrile, V., Meduri, G.M., Bilotta, G.: Comparison between two methods for monitoring deformation with laser scanner. WSEAS Trans. Signal Process. **10**(1), 497–503 (2014)
22. Barrile, V., Meduri, G.M., Bilotta, G.: Laser scanner technology for complex surveying structures. WSEAS Trans. Signal Process. **7**(3), 65–74 (2011)
23. Barrile, V., Meduri, G.M., Bilotta, G.: Laser scanner surveying techniques aiming to the study and the spreading of recent architectural structures. In: Proceedings of the 2nd WSEAS International Conference on Engineering Mechanics, Structures and Engineering Geology, EMESEG 2009, pp. 25–28 (2009)
24. Özdemir, E., Remondino, F.: Segmentation of 3D photogrammetric point cloud for 3D building modeling. Int. Arch. Photogramm. Remote Sens. Spatial Inf. Sci. **XLII-4/W10**, 135–142 (2018)
25. Paris, L., Wahbeh, W.: Survey and representation of the parametric geometries in HBIM. Disegnarecon **9**(16), 12.1–12.9 (2016)
26. Barrile, V., Candela, G., Fotia, A.: Point cloud segmentation using image processing techniques for structural analysis. Int. Arch. Photogramm. Remote Sens. Spatial Inf. Sci. **XLII-2/W11**, 187–193 (2019). https://doi.org/10.5194/isprs-archives-XLII-2-W11-187-2019

GIS Roads Cadastre, Infrastructure Management and Maintenance

Silvia Simonetti(✉), Agostino Currà, Salvatore Minniti, and Maurizio Modafferi

Mediterranea University of Reggio Calabria, Località Feo di Vito, 89124 Reggio Calabria, Italy
silvia.simonetti@unirc.it

Abstract. The updating of the Italian Road Cadastre is regulated by the D.M. 3484/2001 and currently takes place through the manual insertion on the WebGIS platform detected by geo-referenced territorial data detected thanks also to the Mobile Mapping System (MMS) technology.

This research illustrates an automated way of updating a rudimentary MMS system (which involves the installation of a camera on a vehicle equipped with a GPS system) and a UAV (Unmanned Aerial Vehicle). Images processing takes place through specific algorithms (YOLO, SVM, Edge Detection, etc.). The test carried out highlighted some advantages in the use of drones, in particular a reduction in data acquisition and processing times.

Keywords: Mobile Mapping System · UAV · SVM · YOLO · Road Cadastre

1 Introduction

The article 13, paragraph 6 of the Italian Legislative Decree no. 285 of 30/04/1992 (New Highway Code) provides the obligation for road owners to establish and keep up-to-date maps and the cadastre of roads and related appurtenances, according to the procedures established with a specific decree issued by the Ministry of Public Works (Ministerial Decree no. 3484 of 01/06/2001).

To comply with the legislation, the Road Cadastre was established, a GIS platform for the collection, updating and consultation of all data relating to the road network and structures of interest for traffic management. The information it contains is extremely important for traffic management, road maintenance, and the maintenance of adequate road safety standards.

The need to keep the system constantly updated does not comply with the difficulty of manually entering a large amount of data. The MMS (Mobile Mapping System) methodology, despite not solving this problem, allows to acquire most of the data required by law, with excellent levels of detail and accuracy [1].

However, the MMS survey is a long procedure; for this reason, an attempt was made to automate it through the use of UAVs, instruments equipped with geo-referenced and automated photographic sensors. The images, thanks to the georeferencing parameters with which they are accompanied, provide the archive from which to extract, even at

© The Author(s), under exclusive license to Springer Nature Switzerland AG 2022
F. Calabrò et al. (Eds.): NMP 2022, LNNS 482, pp. 2045–2053, 2022.
https://doi.org/10.1007/978-3-031-06825-6_196

different and subsequent times, the location, qualitative and metric data relating to the different classes of territorial elements of specific interest.

The proposal developed in this note aims to integrate the results of the MMS methodology, now well established, with innovative tools (UAVs) associated with soft computing methodologies for processing images and extracting the information contained therein. Once processed, these data can be integrated into the Road Cadastre, allowing the search on a geographical basis of all territorial information (cartographic, graphic and alphanumeric), thus allowing the analysis of the information distributed on the territory, exploiting the spatial relationships and the overlapping of information layers (cadastral cartography, census data, planimetric and real estate market observatory data, orthophotos and road graphs).

1.1 Identification of the Study Area

The study area is located in the historic center of the metropolitan city of Reggio Calabria (Fig. 1). The area is characterized by roads with different geometric sections and with evident deterioration; there are also horizontal and vertical road signs, manholes under study. This set of features lends itself well to system prototyping activities. However, the traffic becomes particularly intense during rush hours due to the proximity of schools and offices. In this regard, for safety reasons, the inspections were carried out exclusively during the early hours of the day.

Fig. 1. Aerial view of the study area.

2   Materials and Methods

In order to process the data extrapolated from the images acquired by MMS and drone, and intended for inclusion in the Road Cadastre database, it is necessary to perform the following steps.

Once the data has been acquired, a road graph conforming to the GDF standard on a relational database is obtained [2]. This activity requires the determination of the position of the acquisition points and the subsequent saving of the coordinates.

To identify the width of the carriageway, the right-hand half-carriageways were considered, obtained from the measurements resting on the outward and return trajectories, and the points thus obtained were transformed into flat cartographic coordinates. Once the two margin polylines were drawn, a third polyline was generated for each vertex, the distances of which were calculated from the other two polylines, adding them and saving the result in the value field.

The identification of horizontal and vertical road signs occurs through algorithms of recognition of the elements of the image, based on neural networks that compare the portion of the image analyzed with training datasets. Once the element of interest has been identified, it is signaled through special bounding boxes and correlated the acquisition position to the size of the bounding box itself to correct the relative positioning.

The D.M. 3484/2001 explicitly identifies the composition of the Road Cadastre in a relational database and in a GIS application in which to represent the cartography of the territory and the road graph and select the individual elements to display the associated attributes. The compliance check of the database generated with the described procedures was performed with the GIS ArcMap 8.2 desktop. A further check was performed by generating the sequential ASCII file in compliance with the GDF 3.0 standard starting from the relational database of the Road Cadastre (DBMS, Database Management System) compliant with the Ministerial Decree 3484/2001.

2.1   Image Acquisition Through Rudimentary MMS and UAVs

A MMS is a vehicle equipped with various types of sensors (high resolution photogrammetric cameras, laser scanners, cameras, georadars, DGPS, inertial systems, odometers) which allow the acquisition of georeferenced images of all the territorial elements existing along the route of the vehicle and underground (with the use of the georadar) [3].

For the application in question, a rudimentary MMS was created by installing a Canon EOS 1100D camera with a resolution of 12 Mp on the front roof of a car for the acquisition of digital images, and a GPS RTK system for georeferencing the points. of grip. The acquired data were automatically saved inside a PC to which the two systems were connected [4].

The same trajectories performed with MMS were re-executed using a MAVIC AIR 2 UAV (Fig. 2), equipped with a 48 MP camera stabilized on three axes suitable for shooting in 4K/60 fps [5–7].

Fig. 2. UAV MAVIC AIR 2 (Barrile et al. 2020).

2.2 Processing

The acquired images were used to create orthophotos and as a database for processing with soft computing techniques aimed at extracting data of interest [8–14].

The reconstruction of the orthophotos is carried out using Agisoft Metashape, a standalone software capable of performing photogrammetric processing of digital images and generating data that can be used by GIS applications. The automatic steps to be followed in the software for the creation of the orthophoto include the following phases: loading and aligning the images, identifying homologous points and creating a scattered cloud, alignment correction, thickening of the points homologues, creation of a dense cloud, creation of a 3D model through the generation of a mesh (and the consequent creation of the texture) [7] and, finally, creation of the orthophoto. The orthophoto was generated by setting the "Type" parameter to "Planar". By setting the "Projection plane" parameter on the reference plane of the elevation from which the orthophoto can be generated [15–18].

For object recognition, the You Only Look Once (YOLO) algorithm was used, which uses a neural network, this neural network provides delimitation frames and the probabilities of associating the image pixels to a class directly from the complete images (avoiding the partition of the image) in a single evaluation for the recognition of the road signs. YOLO's base model processes images in real-time at 45 frames per second and is particularly suitable for use on MMS and drones for real-time, motion, road signs and artifact detection [2]. The proposed system identifies the elements of interest, memorizing the frame, acquiring it at a given distance and georeferencing it [19–21].

3 Application Case

The survey in the study area was initially carried out by means of a rudimentary MMS; the subsequent use of the UAV has made it possible to acquire all the missing data upon completion of the survey. 75% of the longitudinal overlap with respect to the flight direction and 60% of the lateral overside between the strips were obtained. The photos were taken following a regular grid. The camera was kept at a constant height above the ground as much as possible to ensure the desired GSD [22–24].

The acquired images were pre-processed, segmented and finally classified (SVM, Support Vector Machine). From the analysis of the images it was possible to extrapolate the geometric characteristics of the infrastructures and detect the degradation of the road surface, the presence and maintenance status of the horizontal road signs, the presence of manholes (Fig. 3).

The training set was divided into 3 equal subsets. To ensure correct learning, a subset was tested using the trained classifier on the remaining two subsets. The goal was to identify good parameters so that the classifier could accurately predict the test data.

Fig. 3. Vertical road signs within the study area.

3.1 Data Processing and Road Cadastre Updating

The database for the application was built thanks to the images captured by the UAV using a special unit containing the ASCII files necessary to connect the information between the trajectory data and the images [25].

The Geodatabase was created on pgAdmin and the PostGIS spatial extension was inserted inside it. A connection to the database was made on QGIS in pgAdmin and through the "DB manager" plugin PostGIS was chosen among the available spatial extensions (thus guaranteeing the connection between the database and the layers).

A first map was created showing the updated road network using the data collected by the proposed experimental system (Fig. 4) [26, 27].

Fig. 4. Updated road network map.

The experience of a combined use of MMS and UAV allows a comparison between the two systems, with advantages and disadvantages summarized in Table 1.

UAVs saves time and resources; automated data acquisition is a significant point in favor of this technology. Furthermore, the acquisition time of an UAV does not depend on vehicular traffic, as for a MMS. It is also necessary to consider the meteorological limitations, for which the UAV cannot be used in rainy or windy days, and law limitations, for which flying in busy areas is not allowed; to overcome this latter problem, a parachute has been set up on the UAV in order to avoid impacts on vehicular traffic.

Table 1. Comparison between UAV and MMS.

	UAV	MMS
Operators	1	2
Acquisition	28 min	1 h 33 min
Trasmission	Real time	Real time
Processing	Real time	Post processing

4 Conclusions

The Road Cadastre was created with the aim of taking a census of the existing road patrimony and is configured as a fundamental tool for the management of road infrastructures. Improved data collection and classification systems can considerably facilitate the planning of maintenance activities. This work illustrates an innovative system for updating information on the Road Cadastre from Big Data Elaboration. From the results of the case examined it emerged that UAV is well suited to replace the surveys carried out with MMS, allowing a saving of time and economic resources.

More generally, UAV technology and mobile mapping systems (MMS) can be considered complementary solutions for modeling the road network using photogrammetric techniques [28, 29]. The UAV equipped with standard digital cameras can collect high quality aerial optical images with centimeter accuracy; however, it presents difficulties in acquiring information from objects on the ground. The MMS system is particularly suitable for the acquisition of geometric data correlated to the road axis, for which an accuracy of one decimeter is required. Elements are assigned a progression and an offset. In addition, the range of tools on board the MMS makes it possible to obtain thematic information relating to road signs and street furniture.

Future developments are aimed at the implementation of functions that can allow the planning of the maintenance of additional elements such as pavements, vertical road signs and works of art related to roads.

References

1. Barrile, V., Meduri, G.M., Critelli, M., Bilotta, G.: MMS and GIS for self-driving car and road management. In: Gervasi, O., et al. (eds.) ICCSA 2017. LNCS, vol. 10407, pp. 68–80. Springer, Cham (2017). https://doi.org/10.1007/978-3-319-62401-3_6
2. Barrile, V., Meduri, G.M., Bilotta, G.: Comparison between two methods for monitoring deformation with laser scanner. WSEAS Trans. Sign. Proces. 10, 497–503 (2014)
3. Barrile, V., Bernardo, E., Fotia, A., Candela, G., Bilotta, G.: Road safety: road degradation survey through images by UAV. WSEAS Trans. Environ. Dev. 16, 649–659 (2020). https://doi.org/10.37394/232015.2020.16.67. Article number 67 Open Access ISSN 17905079
4. Gruen, A., Li, H.: Road extractions from aerial and satellite images by dynamic programming. ISPRS J. Photogram. Remote Sens. 50(4), 111–120 (1995)
5. Barrile, V., Fotia, A., Bernardo, E.: Non-invasive geomatic techniques to study and preserve cultural heritage: the case study of the Cardeto tower. Non-invasive techniques to study and preserve cultural heritage. Springer, Cham (2020)

6. Barrile, V., Fotia, A.: Active faults: Geomatics and soft computing techniques for analysis, monitorig and risk prevention in central tyrrhenian Calabria (Italy). WSEAS Trans. Environ. Dev. **17**, 436–448 (2021). https://doi.org/10.37394/232015.2021.17.43. Article number 43 Open Access ISSN 17905079
7. Fotia, A., Modafferi, A., Nunnari, A., D'amico, S.: From UAV survey to 3D printing, geomatics techniques for the enhancement of small village cultural heritage. WSEAS Trans. Environ. Dev. **17**, 479–489 (2021). https://doi.org/10.37394/232015.2021.17.46. Article number 46 Open Access ISSN 17905079
8. Pucinotti, R., Tondini, N., Zanon, G., Bursi, O.S.: Tests and model calibration of high-strength steel tubular beam-to-column and column-base composite joints for moment-resisting structures. Earthq. Eng. Struct. Dyn. **44**(9), 1471–1493 (2015). https://doi.org/10.1002/eqe.2547. ISSN 00988847
9. Cristofaro, M.T., Pucinotti, R., Tanganelli, M., De Stefano, M.: The dispersion of concrete compressive strength of existing buildings. In: Cimellaro, G.P., Nagarajaiah, S., Kunnath, S.K. (eds.) Computational Methods, Seismic Protection, Hybrid Testing and Resilience in Earthquake Engineering. GGEE, vol. 33, pp. 275–285. Springer, Cham (2015). https://doi.org/10.1007/978-3-319-06394-2_16
10. Pucinotti, R.: Cyclic mechanical model of semirigid top and seat and double web angle connections. Steel Compos. Struct. **6**(2), 139–157 (2006). https://doi.org/10.12989/scs.2006.6.2.139. ISSN 12299367
11. Pucinotti, R., Tripodi, M.: The fiumarella bridge: concrete characterisation and deterioration assessment by nondestructive testing. Int. J. Microstruct. Mater. Prop. **4**(1), 128–139 (2009). https://doi.org/10.1504/IJMMP.2009.028438. ISSN 17418410
12. Tondini, N., Zanon, G., Pucinotti, R., Di Filippo, R., Bursi, O.S.: Seismic performance and fragility functions of a 3D steel-concrete composite structure made of high-strength steel. Eng. Struct. **174**, 373–383 (2018). https://doi.org/10.1016/j.engstruct.2018.07.026
13. Barrile, V., Meduri, G.M., Bilotta, G.: Laser scanner technology for complex surveying structures. WSEAS Trans. Signal Process. **7**(3), 65–74 (2011)
14. Barrile, V., Meduri, G.M., Bilotta, G.: Laser scanner surveying techniques aiming to the study and the spreading of recent architectural structures. In: Proceedings of the 2nd WSEAS International Conference on Engineering Mechanics, Structures and Engineering Geology, EMESEG 2009, pp. 25–28 (2009)
15. Hinz, S., Baumgartner, A., Mayer, H., Wiedemann, C., Ebner, H.: Road extraction focusing on urban areas. In: Baltsavias, E.P., Gruen, A., Gool, L.V. (eds.) Automatic Extraction of Manmade Objects from Aerial and Space Images (III), pp. 255–265. A.A. Balkema Publishers (2001)
16. Barrile, V., Fotia, A., Bernardo, E.: The submerged heritage: a virtual journey in our seabed. Int. Arch. Photogramm. Remote Sens. Spatial Inf. Sci. **XLII-2/W10**, 17–24 (2019). https://doi.org/10.5194/isprs-archives-XLII-2-W10-17-2019
17. Barrile, V., Candela, G., Fotia, A.: Point cloud segmentation using image processing techniques for structural analysis. Int. Arch. Photogram. Remote Sens. Spatial Inf. Sci. **XLII-2/W11**, 187–193 (2019). https://doi.org/10.5194/isprs-archives-XLII-2-W11-187-2019
18. Barrile, V., Bilotta, G.: An application of remote sensing: object-oriented analysis of Satellite data. Int. Arch. Photogramm. Remote Sens. Spatial Inf. Sci. **XXXVII**, 107–113 (2008)
19. Grüen, A., Li, H.: Linear feature extraction with 3-D LSB-snakes. In: Automatic Extraction of Man-Made Objects from Aerial and Space Images (II), pp. 287–298. BirkhäuserVerlag, Basel (1997)
20. Barrile, V., Bilotta, G., Fotia, A., Bernardo, E.: Road extraction for emergencies from satellite imagery. In: Gervasi, O., et al. (eds.) ICCSA 2020. LNCS, vol. 12252, pp. 767–781. Springer, Cham (2020). https://doi.org/10.1007/978-3-030-58811-3_55

21. Barrile, V., Fotia, A., Bernardo, E., Bilotta, G., Modafferi, A.: Road infrastructure monitoring: an experimental geomatic integrated system. In: Gervasi, O., et al. (eds.) ICCSA 2020. LNCS, vol. 12252, pp. 634–648. Springer, Cham (2020). https://doi.org/10.1007/978-3-030-58811-3_46

22. Barrile, V., Fotia, A., Candela, G., Bernardo, E.: Integration of 3D model from UAV survey in BIM environment. Int. Arch. Photogram. Remote Sens. Spatial Inf. Sci. **XLII-2/W11**, 195–199 (2019). https://doi.org/10.5194/isprs-archives-XLII-2-W11-195-2019

23. Barrile, V., Cacciola, M., Morabito, F.C., Versaci, M.: TEC measurements through GPS and artificial intelligence. J. Electromagn. Waves Appl. **20**, 1211–1220 (2006). https://doi.org/10.1163/156939306777442962

24. Barrile, V., Bilotta, G., Fotia, A.: Analysis of hydraulic risk territories: comparison between LIDAR and other different techniques for 3D modeling. WSEAS Trans. Environ. Dev. **14**, 45–52 (2018)

25. Barrile, V., Fotia, A.: Seismic risk: GPS/GIS monitoring and neural network application to control active fault in Castrovillari area (South Italy) [Rischio sismico: Monitoraggio GPS/GIS e applicazioni di reti neurali per il controllo di una faglia attiva nell'area di Castrovillari] ArcHistoRIssue **6**, 570–583 (2019). https://doi.org/10.14633/AHR182

26. Barrile, V., Leonardi, G., Fotia, A., Bilotta, G., Ielo, G.: Real-time update of the road cadastre in GIS environment from a MMS rudimentary system. In: Calabrò, F., Della Spina, L., Bevilacqua, C. (eds.) ISHT 2018. SIST, vol. 101, pp. 240–247. Springer, Cham (2019). https://doi.org/10.1007/978-3-319-92102-0_26

27. Barrile, V., Fotia, A., Bernardo, E., Bilotta, G.: Road cadastre an innovative system to update information, from big data elaboration. In: Gervasi, O., et al. (eds.) ICCSA 2020. LNCS, vol. 12252, pp. 709–720. Springer, Cham (2020). https://doi.org/10.1007/978-3-030-58811-3_51

28. Soilán, M., Riveiro, B., Martínez-Sánchez, J., Arias, P.: Automatic road sign inventory using mobile mapping systems. In: The International Archives of the Photogrammetry, Remote Sensing and Spatial Information Sciences, 2016 XXIII ISPRS Congress, vol. XLI-B3, Prague, Czech Republic, 12–19 July 2016 (2016)

29. Barrile, V., Fotia, A., Ponterio, R., Mollica Nardo, V., Giuffrida, D., Mastelloni, M.A.: A combined study of art works preserved in the archaeological museums: 3D survey, spectroscopic approach and augmented reality. ISPRS Ann. Photogramm. Remote Sens. Spatial Inf. Sci. **42**(2/W11), 201–207 (2019). 2nd International Conference of Geomatics and Restoration, GEORES 2019, Milan, 8 May 2019 through 10 May 2019

OBIA to Detect Asbestos-Containing Roofs

Giuliana Bilotta[✉] [ID]

Department of Planning, IUAV University of Venice, Santa Croce 191 Tolentini, 30135 Venice, Italy
giuliana.bilotta@iuav.it

Abstract. Asbestos was primarily a problem in certain industrial areas of Italy. A link between asbestos and pleural mesothelioma was discovered in the 1960s. Italy is not the only country with an asbestos problem. International interest in the subject has been heightened by studies conducted in the aftermath of the September 11, 2001, World Trade Center attacks in New York, in which asbestos, primarily used for fire protection of large metal structures, spread over a region of several square kilometers in the form of fine dust because of the Twin Towers' explosive collapse, inhaled by thousands of people.

Despite long-known dangers, the number of asbestos-containing building roofs remains high in many countries. Many buildings in Italy still have roofs made of a cement-asbestos mixture, which was once popular due to the material's mechanical strength and low cost.

In this work, we used object classification to solve the problem of the true location of asbestos cement roofs, using cadastral data in vector format as a thematic layer within the segmentation procedure to obtain objects complete with property attributes as a result of segmentation, useful in determining who owns the buildings.

Keywords: Asbestos roofs · Satellite imagery · Object based image analysis

1   Introduction

1.1   Preserving the Health of Our Lungs

In this pandemic period, we have learned how important it is to preserve the health of our lungs, which easily become prey to deadly diseases.

This applies to infection by viruses with a certain degree of lethality, such as SARS-CoV-2, which is responsible for the current COVID-19 pandemic, but also to other dangerous threats, some of which come from human activity itself, asbestos being one of them.

1.2   Asbestos and Health Risks

Asbestos is present in nature, for example, in Val di Susa, Italy (Fig. 1), where the population led the No-TAV protests because of the high health risks posed by drilling

© The Author(s), under exclusive license to Springer Nature Switzerland AG 2022
F. Calabrò et al. (Eds.): NMP 2022, LNNS 482, pp. 2054–2064, 2022.
https://doi.org/10.1007/978-3-031-06825-6_197

tunnels rich in this mineral, tunnels needed for the high-speed railway line; but it has long been used in the construction industry.

Long-term asbestos exposure has been linked to the emergence of dangerous diseases affecting the respiratory system. The most dangerous element is friable asbestos, because its fibres are easily dispersed and have a unique ability to penetrate the respiratory system. Despite the long-known risks, the number of roofs of buildings containing asbestos remains high in many countries. In Italy, there are still many buildings with roofs made of a mixture of cement and asbestos, which was once popular due to the mechanical strength of the material and its low cost.

The asbestos problem is not limited to Italy. At the international level, interest in the subject has been heightened by studies conducted in the aftermath of the World Trade Center attacks in New York on September 11, 2001, in which asbestos, which was primarily used for fire protection of large metal structures, spread over a region of several square kilometers in the form of fine dust as a result of the Twin Towers' explosive collapse. This dust, which was made up of the pulverized remains of the Towers' non-metallic components, contained significant amounts of asbestos and was undoubtedly inhaled by thousands of people.

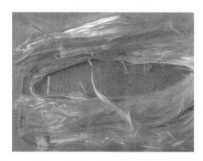

Fig. 1. Asbestos (tremolite) silky fibres from Val di Susa, Italy.

Asbestos was primarily a problem in certain industrial areas in Italy. Since the Swiss company Eternit obtained a patent in 1906, the Piedmontese town of Casale Monferrato has housed a factory that produced fiber cement until 1986, when it was closed down, despite the fact that evidence of a link between asbestos and mesothelioma of the pleura had already been discovered in the 1960s.

In addition to contamination from the factory, which was reduced in the 1970s after the installation of filtering systems, there was contamination transmitted to the workers' families and contamination caused by Eternit's distribution of asbestos processing scraps (the so-called 'powder') in the Casale area, which was given to citizens to make flooring and insulation. Two-thirds of the 20–30 people who die from mesothelioma each year are citizens who have never had any contact with the factory.

Asbestos cement materials are made up of a cement matrix that contains 6 to 12 percent asbestos fibres (Fig. 2). This material was primarily used to make sheets for use as external roofing for industrial and civil buildings. After a few years, the sheets tend to release a significant amount of asbestos fibres into the atmosphere, which can be harmful to human health.

Fig. 2. Anthophyllite asbestos, Scanning electron microscope picture.

Asbestos' fibrous consistency, which is the basis of its excellent technological properties, also confers material risk characteristics, as it is the cause of serious diseases, primarily affecting the respiratory system [1, 2]. Asbestos materials degrade over time, releasing potentially inhalable fibres: they tend to split longitudinally into increasingly fine fibrils with diameters small enough to be breathed in and penetrate deeply into the pulmonary alveoli. As a result of respiratory movements, finer, needle-like fibres pass through the lung tissue and easily reach the pleura [3, 4].

1.3 Searching Asbestos Roofs

Because the airborne payload is typically a hyperspectral sensor, environmental monitoring for detecting the presence of this material in urban areas is characteristically accomplished through the use of expensive dedicated air travel.

Since mapping is a regional competence, like other environmental competences, many regions have organised themselves to carry it out in order to monitor the development of this health risk. The first to do so was Val d'Aosta in 2001 (using hyperspectral imaging of the entire regional territory completed in October 1999, but many others have joined the ranks, especially since 2011 (Fig. 3). In general, mapping was carried out from airborne MIVIS (Multispectral Infrared and Visibile Imaging Spectrometer) sensors.

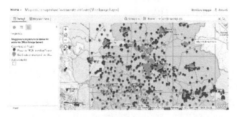

Fig. 3. Asbestos survey in Casale Monferrato, Piedmont [5].

Legislation requires that asbestos roofing be cleaned up, but in many countries construction companies continue to use this material despite being banned. It is required intervention to replace these roofs, and among the possible replacements is the smart substitution of asbestos roofs with PV roofs for energy production as in the Veneto Region since 1 January 2012 [6]. The survey is being completed throughout Italy (Fig. 4) [7]. Law n. 257 of 1992 banned all products containing asbestos and banned the production and use of asbestos cement. It also required regions to register buildings where materials or products containing asbestos were present. Decree no. 101 of 18 March 2003 of the Ministry of the Environment and Protection of the Territory also laid down the rules for the complete mapping of the areas of the national territory affected by the presence of asbestos and for carrying out reclamation work of particular urgency, identifying the criteria for attributing the character of urgency to reclamation work, the persons responsible for carrying out the mapping and the tools and methods with which it must be done.

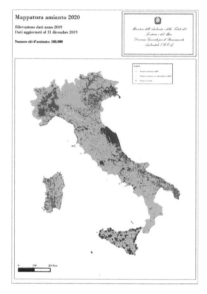

Fig. 4. Asbestos survey in Italy (Ministry of the Environment): ●Asbestos survey 2019 ●Asbestos survey preceding 2019 ●Natural asbestos.

The aim of this work is to provide a low-cost tool for the automatic identification of buildings that still use this very dangerous technology as a roof covering. This is possible thanks to the asbestos spectral profile (Fig. 5).

This research, similarly to others already carried out [8] also with the use of different algorithms and software tools compared with each other [9–11], proposes the use of remote sensing data from satellites, which are less expensive than data from airborne sensors.

Satellites, unlike air missions, continuously orbit the Earth. Some commercial optical satellites in sun-synchronous orbit that monitor at very high geometric resolution are

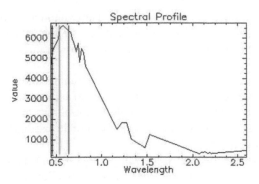

Fig. 5. The spectral profile of a light-grey Eternit roof in the wavelength range 0.5–2.5 mm.

especially well suited for this type of search. This paper proposes the use of optical satellite data, which have proven to be effective for detecting asbestos roofs through object-based analysis.

What is proposed could become a more common and less expensive practice than the current practice, which involves the use of airborne sensors. Object-based techniques for classifying satellite data enable the identification of buildings with asbestos-containing roofs. In this paper, satellite imagery is integrated from the start with cadastral data in vector format (shapefile) to maintain the cadastral information relating to property through the various procedures of segmentation and classification. This is important in order to identify the owners of the buildings.

The object-based method employs a Nearest Neighbor classification tool in conjunction with a multi-resolution segmentation of the entire scene.

2   Materials and Methods

In this study we proceed with a structural technique, through a multiresolution segmentation of the whole scene - which automatically creates vector polygons directly extracted from the raster, with a perfect coincidence in the overlap - and a classification using as test areas some buildings with known asbestos cement covering.

The aim of this contribution is to illustrate an application of Object Based Image Analysis on high resolution data, such as Ikonos multispectral data, with the objective of showing how automatic analysis - with minimal manual intervention - can facilitate the recognition of roofs made of materials containing asbestos.

As data we used European Space Imaging satellite images acquired by the Ikonos-2 satellite on 21 December 2002 at 11:00 local time, georeferenced in WGS84 (Datum WGS84, UTM 33 N). The cadastral data is in shapefile format.

2.1   Object-Based Image Analysis

Traditional image processing and interpretation techniques are based on statistical oper-ations on data extracted from the intrinsic characteristics of the single pixel. This is low-level semantic information - the amount of energy emitted by the pixel - in which context plays no role. In object-based image analysis, the semantic level is raised; topological, statistical and context-related information is stored among the attributes of the created objects.

It introduces further rules for the location of the context and relationships between objects, while significantly increasing the probability of automatic recognition of objects on the earth's surface. The choice of the scale factor allows the size of the resulting polygons to be calibrated, and its definition is linked to the cartographic reference scale that the end user must obtain.

With this structural analysis it is possible to obtain from the remote sensing data information that can be immediately integrated into GIS, allowing the direct creation of vector maps. Starting from the same image it is possible to generate various hierarchical levels of polygons with different scale factors, therefore the segmentation process is multiresolution.

Recognition is based on concepts of mathematical morphology applied to image analysis [12] and on fuzzy logic for classification, similar to that used by the human photo-interpreter. Unlike the work that can be obtained by the latter, however, the processes guarantee qualities of uniformity, standardisation and reproducibility of the results, characteristics that cannot be ensured by a human photo-interpreter.

The inclusion of the cadastral shapefile in the segmentation procedure also allows information to be organised in such a way as to preserve, among the attributes of the vectorial objects that are created, information relating to the property and other cadastral data.

2.2   Segmentation

With the software used, eCognition by Definiens Imaging, now released by Trimble, the image pixels are gradually aggregated in a series of steps until the resulting polygons have the characteristics desired by the user [13].

The procedure leads to the minimisation of the spectral heterogeneity of each polygon derived from the digital number values of the included pixels and on the basis of the geometric heterogeneity dependent on the shape of the polygons created. The spectral heterogeneity h_s of each polygon generated with the segmentation process is obtained as a weighted sum of the standard deviations of the digital number values of each spectral band obtained for each of the pixels included in the polygon:

$$h_s = \sum_{c=1}^{q} w_c \sigma_c \tag{1}$$

h_s = spectral heterogeneity of the polygon; q = number of spectral bands; σ_c = standard deviation of the digital number values of the c-th spectral band; w_c = weight assigned to the c-th band.

The segmentation algorithm merges adjacent polygons from each pixel in the image until the change in observable heterogeneity between the two primitive polygons and the new polygon obtained exceeds a threshold assigned by the user (scale factor). If the assigned threshold is not exceeded, the merger takes place, otherwise the polygons remain distinct. The difference in heterogeneity (*overall fusion value*) between the potential fused object and the two original polygons is equal to:

$$f = w_f \Delta h_s + (1 - w_f)\Delta h_g \tag{2}$$

where: f = overall fusion value; w_f = weight assigned according to the relative importance of spectral heterogeneity to geometric heterogeneity (h_g), ranging from 1 (only geometric heterogeneity is considered) to 0 (only spectral heterogeneity is considered).

The difference in spectral heterogeneity (Δh_s) between the potentially merged polygon and the two original polygons is calculated as:

$$\Delta h_s = \sum_{c=1}^{q} w_c \left[n_{merge}\sigma_{merge_c} - \left(n_{obj1}\sigma_{obj1_c} + n_{obj2}\sigma_{obj2_c} \right) \right] \tag{3}$$

where: n_{merge} = number of pixels included in the polygon generated by the merger; σ_{merge_c} = standard deviation of the digital number values of the c-th spectral band in the merged polygon; n_{obj1} = number of pixels included in the first of the two polygons before merging; σ_{obj1_c} = standard deviation of the digital number values of the c-th spectral band in the first of the two polygons before merging; n_{obj2} = number of pixels included in the second of the two polygons before merging; σ_{obj2c} = standard deviation of the digital number values of the c-th spectral band in the second of the two polygons before merging.

The theory of fuzzy logic, developed to deal with imprecise information, can provide a more appropriate solution [14–16] when each region has partial membership in a class or multiple memberships in several classes. Partial membership allows better representation and use of information related to more complex situations. Moreover, fuzzy logic emulates human thought by using its linguistic rules.

2.3 Procedure

The subject of this study is the recognition of materials containing asbestos in the roofs of buildings in Melito di Porto Salvo (in the Province of Reggio Calabria, Italy).

In Fig. 6 the flowchart of the procedure.

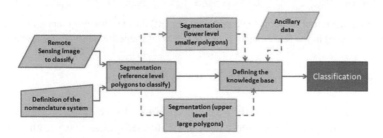

Fig. 6. Flowchart from segmentation to classification.

For the multiresolution segmentation we assigned greater weights to the Red and IR bands, while setting shape parameter to 0.5, compactness factor is set to 0.5, the scale factor is 15/Fig. 7). The scale parameters assigned here are reduced because the parcel division, relating to the oldest urban nucleus, is very fragmented (Fig. 8).

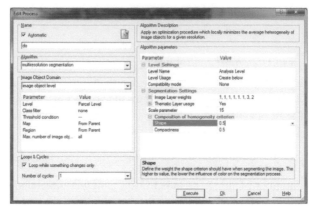

Fig. 7. Multiresolution segmentation with assignment of greater weights to the Red and IR bands, the shape parameter is 0.5, the compactness factor is set to 0.5, the scale factor is 15.

Fig. 8. The segmentation shows in the Ikonos image (left) the extreme fragmentation of the oldest urban core, on the right the representation of the cadastral shapefile with the segmented area.

3 Results

In the first phase, the area was firstly segmented in order to create objects corresponding to the cadastral polygons that preserve their attributes, assigning a very high scale factor. This procedure proved effective in a study on the discrimination of illegal dumps [17] and other applications [18–42]. In contrast to a rural area, in this case the parcel division, relating to the oldest urban nucleus known as "Paese Vecchio", is very fragmented; therefore the scale parameters assigned here are reduced.

A simple classification was then carried out based on test areas consisting of a few buildings whose roofs are already known to contain asbestos cement. The classes are: Asbestos and Other (Fig. 9).

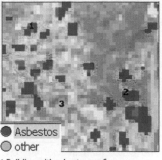

1 Building with asbestos roofing
2 Building with probable asbestos roofing
3 Building without asbestos roofing

Fig. 9. Detection of buildings with asbestos roofing, with probable asbestos roofing and without asbestos roofing.

4   Conclusions

In this work we have applied object classification to find a solution to the problem of the real position of asbestos cement roofs, using the cadastral data in vector format as a thematic layer as part of the segmentation procedure in order to obtain, as a product of segmentation, objects complete with property attributes. This is important in order to identify the owners of the buildings.

This procedure has proved to be applicable in other contexts and is still open to further improvement.

One of the possible improvements that can be implemented in the future is the monitoring of changes and transformations in the roofs already identified, through comparison with historical series of images, in order to verify over time the response to one of the environmental hazards that have proven to be very dangerous to human health.

References

1. Mollo, F., et al.: Lung adenocarcinoma and indicators of asbestos exposure. Int. J. Cancer **60**, 289–293 (1995)
2. Weill, H.: Occupational Lung Diseases: Research Approaches and Methods, pp. 220–228. Marcel Dekker, New York (1981)
3. Rinaudo, C., Gastaldi, D., Belluso, E.: Characterization of chrysotile, antigorite and lizardite by FT-Raman Spectroscopy. Can. Mineral. **4**, 883–890 (2003)
4. Cherry, K., Cherry, C.: Asbestos: Engineering, Management and Control, pp. 127–150. CRC Press, Boca Raton (1988)
5. ARPA Piemonte. https://webgis.arpa.piemonte.it/agportal/home/webmap/viewer.html?use Existing=1&layers=10c8528222f042dbb1b868471c7911e5. Accessed 22 Dec 2021
6. "Veneto 2050" - regional law no. 14 of 4 April 2019, approved by the Veneto Region. https://www.veneto2050.it/category/commento-lr-14-19/. Accessed 21 Dec 2021
7. INSIC. https://www.insic.it/tutela-ambientale/sostanze-pericolose/mappatura-amianto-la-guida-regione-per-regione/. Accessed 23 Dec 2021

8. Barrile, V., Cacciola, M., Cotroneo, F.: An advanced method to detect asbestos coverages in building roofs using ikonos images. Remote Sensing Applications for a Sustainable Future. ISPRS Archives 36, Part 8 (2006)

9. Barrile, V., Bilotta, G., Pannuti, F.: A comparison between methods – a specialized operator, object oriented and pixel-oriented image analysis – to detect asbestos coverages in building roofs using remotely sensed data. ISPRS Archives **37**, 427–433 (2008)

10. Altshuller, G.: Innovation Algorithm, pp. 31–40. Technical Innovation Center Inc., Worcester (1999)

11. Sangwine, S.: The Colour Image Processing Handbook, pp. 67–105. Springer, Berlin (1998). https://doi.org/10.1007/978-1-4615-5779-1

12. Serra, J.: Image Analysis and Mathematical Morphology. Theoretical Advances, vol. 2. Academic Press, London (1988)

13. Baatz, M., et al.: eCognition 4.0 professional user guide. Definiens Imaging GmbH, Monaco (2004)

14. Soille, P., Pesaresi, M.: Advances in mathematical morphology applied to geoscience and remote sensing. IEEE Trans. Geosci. Remote Sens. **40**(9), 2042–2055 (2002)

15. Small, C.: Multiresolution analysis of urban reflectance. In: Proceedings of IEEE/ISPRS Joint Workshop on Remote Sensing and Data Fusion over Urban Areas, pp. 15–19 (2001)

16. Tso, B., Mather, P.: Classification Methods for Remotely Sensed Data, pp. 309–326. Taylor & Francis, London (2001)

17. Barrile, V., Bilotta, G., Meduri, G.M.: Recognition and classification of illegal dumps with object based image analysis of satellite data. In: Proceeding of the Third Annual Hyperspectral Imaging Conference, vol. 2, pp. 12–17, INGV, Rome (2012)

18. Barrile, V., Bilotta, G.: An application of remote sensing: object-oriented analysis of Satellite data. Int. Arch. Photogram. Remote Sens. Spatial Inf. Sci. **XXXVII**, 107–113 (2008)

19. Bilotta, G., Nocera, R., Barone, P.M.: Cultural Heritage and OBIA. WSEAS Trans. Environ. Dev. **17**, 449–465 (2021)

20. Bilotta, G., Calcagno, S., Bonfa, S.: Wildfires: an application of remote sensing and OBIA. WSEAS Trans. Environ. Dev. **17**, 282–296 (2021). https://doi.org/10.37394/232015.2021. 17.29

21. Barrile, V., Meduri, G.M., Bilotta, G.: Comparison between two methods for monitoring deformation with laser scanner. WSEAS Trans. Signal Process. **10**(1), 497–503 (2014)

22. Barrile, V., Meduri, G.M., Bilotta, G.: Laser scanner technology for complex surveying structures. WSEAS Trans. Signal Process. **7**(3), 65–74 (2011)

23. Barrile, V., Meduri, G.M., Bilotta, G.: Laser scanner surveying techniques aiming to the study and the spreading of recent architectural structures. In: Proceedings of the 2nd WSEAS International Conference on Engineering Mechanics, Structures and Engineering Geology, EMESEG 2009, pp. 25–28 (2009)

24. Leonardi, G., Barrile, V., Palamara, R., Suraci, F., Candela, G.: 3D mapping of pavement distresses using an unmanned aerial vehicle (UAV) system. In: Calabrò, F., Della Spina, L., Bevilacqua, C. (eds.) ISHT 2018. SIST, vol. 101, pp. 164–171. Springer, Cham (2019). https://doi.org/10.1007/978-3-319-92102-0_18

25. Barrile, V., Candela, G., Fotia, A.: Point cloud segmentation using image processing techniques for structural analysis. Int. Arch. Photogramm. Remote Sens. Spatial Inf. Sci. **XLII-2/W11**, 187–193 (2019). https://doi.org/10.5194/isprs-archives-XLII-2-W11-187-2019

26. Barrile, V., Fotia, A., Bernardo, E.: The submerged heritage: a virtual journey in our seabed. Int. Arch. Photogramm. Remote Sens. Spatial Inf. Sci. **XLII-2/W10**, 17–24 (2019). https://doi.org/10.5194/isprs-archives-XLII-2-W10-17-2019

27. Barrile, V., Fotia, A., Candela, G., Bernardo, E.: Integration of 3D model from UAV survey in BIM environment. Int. Arch. Photogramm. Remote Sens. Spatial Inf. Sci. **XLII-2/W11**, 195–199 (2019). https://doi.org/10.5194/isprs-archives-XLII-2-W11-195-2019
28. Barrile, V., Cacciola, M., Morabito, F.C., Versaci, M.: TEC measurements through GPS and artificial intelligence. J. Electromagn. Waves Appl. **20**, 1211–1220 (2006). https://doi.org/10.1163/156939306777442962
29. Barrile, V., Bilotta, G., Fotia, A.: Analysis of hydraulic risk territories: comparison between LIDAR and other different techniques for 3D modeling. WSEAS Trans. Environ. Dev. **14**, 45–52 (2018)
30. Barrile, V., Bernardo, E., Bilotta, G., Fotia, A.: "Bronzi di Riace" geomatics techniques in augmented reality for cultural heritage dissemination. In: Borgogno-Mondino, E., Zamperlin, P. (eds.) ASITA 2021. CCIS, vol. 1507, pp. 195–215. Springer, Cham (2022). https://doi.org/10.1007/978-3-030-94426-1_15
31. Bernardo, E., Bonfa, S., Anderson, J.: Implementation of the digital inland water smart strategy using geomatics instruments and the big data SmartRiver platform. In: Borgogno-Mondino, E., Zamperlin, P. (eds.) ASITA 2021. CCIS, vol. 1507, pp. 83–94. Springer, Cham (2022). https://doi.org/10.1007/978-3-030-94426-1_7
32. Barrile, V., Bilotta, G., Fotia, A., Bernardo, E.: Road extraction for emergencies from satellite imagery. In: Gervasi, O., et al. (eds.) ICCSA 2020. LNCS, vol. 12252, pp. 767–781. Springer, Cham (2020). https://doi.org/10.1007/978-3-030-58811-3_55
33. Barrile, V., Fotia, A., Bernardo, E., Bilotta, G.: Geomatic techniques: a smart app for a smart city. In: Bevilacqua, C., Calabrò, F., Della Spina, L. (eds.) NMP 2020. SIST, vol. 178, pp. 2123–2130. Springer, Cham (2021). https://doi.org/10.1007/978-3-030-48279-4_200
34. Barrile, V., Bilotta, G., Fotia, A., Bernardo, E.: Integrated Gis system for post-fire hazard assessments with remote sensing. Int. Arch. Photogramm. Remote Sens. Spatial Inf. Sci. **XLIV-3/W1-2020**, 13–20 (2021). https://doi.org/10.5194/isprs-archives-XLIV-3-W1-2020-13-2020
35. Barrile, V., Fotia, A., Bernardo, E., Bilotta, G., Modafferi, A.: Road infrastructure monitoring: an experimental geomatic integrated system. In: Gervasi, O., et al. (eds.) ICCSA 2020. LNCS, vol. 12252, pp. 634–648. Springer, Cham (2020). https://doi.org/10.1007/978-3-030-58811-3_46
36. Barrile, V., Bernardo, E., Candela, G., Bilotta, G., Modafferi, A., Fotia, A.: Road infrastructure heritage: from scan to Infrabim. WSEAS Trans. Environ. Dev. V **16**, 633–642 (2020)
37. Barrile, V., Fotia, A., Bernardo, E., Candela, G.: Geomatics techniques for submerged heritage: a mobile app for tourism. WSEAS Trans. Environ. Dev. **16**, 586–597 (2020). https://doi.org/10.37394/232015.2020.16.60
38. Barrile, V., Fotia, A., Bernardo, E., Bilotta, G.: Road cadastre an innovative system to update information, from big data elaboration. In: Gervasi, O., et al. (eds.) ICCSA 2020. LNCS, vol. 12252, pp. 709–720. Springer, Cham (2020). https://doi.org/10.1007/978-3-030-58811-3_51
39. Bilotta, G., Bernardo, E.: UAV for precision agriculture in vineyards: a case study in Calabria. In: Borgogno-Mondino, E., Zamperlin, P. (eds.) ASITA 2021. CCIS, vol. 1507, pp. 28–42. Springer, Cham (2022). https://doi.org/10.1007/978-3-030-94426-1_3
40. Barrile, V., Bilotta, G., Nunnari, A.: 3D modeling with photogrammetry by UAVs and model quality verification. ISPRS Ann. Photogramm. Remote Sens. Spatial Inf. Sci. **IV-4/W4**, 129–134 (2017). https://doi.org/10.5194/isprs-annals-IV-4-W4-129-2017
41. Barrile, V., Gelsomino, V., Bilotta, G.: UAV and computer vision in 3D modeling of cultural heritage in Southern Italy. IOP Conf. Ser. Mater. Sci. Eng. **225**(1), 012196 (2017)
42. Barrile, V., Bilotta, G.: Object-oriented analysis applied to high resolution satellite data. WSEAS Trans. Signal Process. **4**(3), 68–75 (2008)

Geomatic Techniques: A Smart App for Cultural Heritage

Ernesto Bernardo[1](✉) ⓘ, Giuliana Bilotta[2] ⓘ, and Adila Sturniolo[1]

[1] DICEAM Department, Faculty of Engineering, University "Mediterranea" of Reggio Calabria, 89100 Reggio Calabria, Italy
ernesto.bernardo@unirc.it

[2] Planning Department, University IUAV of Venice, Santa Croce 191 Tolentini, 30135 Venice, Italy

Abstract. Accelerated by the COVID-19 pandemic, global digital transformation plays an important role in the green transition. Effective coordination between green and digital transitions is therefore currently considered to be crucial for the European recovery.

Specifically, with the advent of Smart Cities, the transposition and consultation of large amounts of data on digital media, the creation of new methods of interaction with places of interest and services, through the use of apps and digital platforms, considerably reduces emissions and climate impact, thus promoting the green transition.

In this context, the aim of this research is to help clarify (for cultural heritage and tourism) how these major changes in global systems can shape these transitions and positively influence people and places, in particular by helping to break down territorial inequalities and the phenomena of social exclusion, useful objectives in the strategy of the European Green Deal (EGD).

Keywords: Cultural heritage · Tourism App · Unity 3D

1 Introduction

Today tourism plays a fundamental role in the economy of our country. With the advent of Smart Cities there is a growing tendency to introduce urban planning strategies aimed at optimizing and innovating public services in order to relate the material infrastructures of cities "with the human, intellectual and social capital of those who live there" [1–3].

The introduction of intelligent systems that communicate with each other and interact with the outside world has therefore facilitated the spread of apps that make it possible to make cultural heritage accessible. In a broader project for the dissemination and enhancement of the cultural heritage of the city of Reggio Calabria, the Geomatics laboratory of the Mediterranean University is developing a tourist app for educational purposes that allows you to interact directly with the surrounding environment and explore models 3D reconstructed with photogrammetric techniques [4–7].

The app is developed in the Unity environment. This work illustrates the various operating models of the app, as well as the assembly process of the different scenes

© The Author(s), under exclusive license to Springer Nature Switzerland AG 2022
F. Calabrò et al. (Eds.): NMP 2022, LNNS 482, pp. 2065–2072, 2022.
https://doi.org/10.1007/978-3-031-06825-6_198

and the various techniques used to reconstruct the three-dimensional models inserted in them.

2 Materials and Methods

2.1 Context and Case Study

The tourism sector is very important for Italy and for individual communities, both from an economic and employment point of view (thanks also to the business and to the related industries it creates).

Shortly before the pandemic, in Italy the data recorded positive numbers, with tourist flows (in search of natural, cultural, historical, artistic and food and wine experiences) increasing throughout the peninsula. A trend that seemed destined to increase also in the following years.

Due to the recent pandemic, the tourism sector has been one of the hardest to be hit by the spread of the virus, however, it is a sector that is slowly recovering, and for this reason these technologies should accelerate his recovery time. In a market rich in offers, especially in the field of food and wine and archaeological tourism, we believe it is important to find experimental solutions to offer a different and new tourist experience in order to increase the appeal to less known archaeological sites. It is a known fact that Italy, especially southern Italy, does not provide enough visibility to the multitude of ancient historical sites that rest in the deep part of the Aspromonte.

Tourists, in fact, explore places unknown to them, but there are many more which they neglect the existence because they are not properly sponsored. That's why they appreciate having easy access to all the information about their destination: wifi hotspots, weather, must-see tourist attractions. There are numerous "classic" applications that provide this information, but augmented reality has the advantage of making their consultation playful and attractive. For many tourists, learning about history and discovering other cultures is one of the main purposes of the trip. Augmented reality (AR) allows us to learn information and concepts in an engaging way. Through interactivity, it sets in motion the mental and motor functions, as it stimulates the users' curiosity towards the information presented through the smartphone interface. With augmented reality, the tourist will be able to see 3D models of archaeological sites and historical monuments in different eras. It allows users to experience these experiences firsthand, rather than just reading and looking at the images. AR is a new form of information and learning [8–13].

Visitors are engaged and make the most of their visit. Even museums, which by many are considered boring or banal, finally have the opportunity to surprise their visitors.

Another important aspect to cover, is how the city is perceived by someone how lives there against someone who is just a visitor. There are many hidden gems in the city that are only acknowledged by the inhabitants, such as restaurants, astonishing landscapes and tiny, interesting history points which are full of history. For this reason, our app will develop an entire area dedicated to these places. This will be done thanks to the Peer-to-Peer technology. Anyone who has a registered and approved account, can propose a place which characterizes the city they live and after being approved by a moderator, it can be reviewed by others and seen by tourists. In this way, even a tourist can live part of the life of someone who already lives there [14]. We are also considering the possibility to offer a some kind of discount for those who show up in these restaurants or museum and are displaying this app, which will generate a unique qr code, linked to the user. It is considered a win-win situation both for us developers and the owner of the restaurant/the museum [15]. Our app will be gladly used by the users, while the restaurant/museum will have more clients/visitors, so a higher profit.

2.2 The Developed App

In a larger project, the Geomatics Laboratory of the Mediterranean University of Reggio Calabria has developed for mobile devices, in the Unity environment, a tourist app for academic purposes (this is a first version of an app for tourist purposes, developed through unconventional, which require special processing for the use of models acquired with geomatic techniques and therefore difficult to use because they are rather heavy).

The app allows the user to have more information (historical, multimedia…) through Augmented Reality (AR). It also allows, in real time, to take a virtual tour with the 3D model as the main scene and to interact with the surrounding environment.

The app is currently being defined and allows users to access different services based on the choices made previously, and in the future, we hope to add location-related tags, such as offering a range of services related to where you are. find or marketing-oriented events.

The app was tested on the case study of S. Pietro di Deca in Torrenova (ME), Sicily, Italy.

In general, the app is able to:

- show different information regarding the history of the artifacts that the smartphone camera is capturing at that moment.
- view multimedia content, such as video or audio, associated with the object framed by the user through their device.
- highlight the details of interest directly on the object we are framing and of which we want to carry out a more in-depth study, making learning easier and more immediate.
- view the 3D model of the product with the possibility of transforming or disassembling and assembling it,
- take a virtual tour of the museum or cultural property as a passive spectator, or take the same tour but as an active spectator interacting through the device screen.
- offer the user a virtual guide that accompanies him inside the museum or cultural property and connects him with the surrounding environment.
- report the position of the visitor on the floor plans of the museum or cultural property with the names of the rooms in which they are located.
- Build a rating system, based on the reviews of the place the visitors attend.

Main and innovative features of the app:

a) The developed app, exploiting the connection with the GPS coordinates of the device on which it is installed, allows you to:

- Identify important archaeological sites in the surroundings (even those that are not adequately valued).
- Send a notification of the presence of places of historical interest and indicate the shortest route.
- Send notifications about nearby events of interest.
- Promote and enhance the understanding of archaeological sites, even in less developed areas, by improving their ability to attract visitors and thus creating an economic link from which the resident population can also benefit.
- Explore the 3D model with VR and AR.
- View movies with virtual guides.
- The VR system is a virtual diving exhibit that allows users to explore the 3D reconstruction of important heritage site and receive historical and archaeological information about the exhibits and structures of the site. The VR system can also be used by tourists because of its capability to make a detailed planning of the operations to improve and preserve the site it is linked to. The system indeed represents a reliable instrument to plan and simulate the tourist itinerary that is performed at a later time in the real environment, such as mimicking a foggy or sunny weather, depending on the real condition of the site. For this reason, the software architecture plays a major role in the VR system. It consists of five main elements: a database, a web service, a scene editor module, a visualization module, and the controller module. In particular, the database manages all of the data of the virtual scene. The web service provides a bilateral communication between the database and the other modules. The scene editor, visualization, and controller modules have been implemented by means of the cross-platform game engine, Unity for making the experience as real as possible. The scene editor module allows to compose the virtual scene by integrating 3D objects and multimedia information stored in the database. Once the scene is created, the interaction module is adopted to implement the logics of the virtual scenario defining the physics and behaviors of the elements that belong to the virtual environment. As a next step, it could be considered to insert weather change system, for realizing a unique experience every time you take part to a VR exploration.
- We should also pay attention to the presence of the local fauna and flora, which are an integrated part of the whole experience of the exploration of these beautiful hidden sites here in Italy. Thus, we will record in high quality, the sounds of the many bird species, and we will consider the possibility to create an augmented encyclopedia of the local wildlife.

b) It allows to defend and enhance the cultural heritage, placing all the problems related to its conservation at the center, through the creation of 3D models, and therefore through the creation of digital archives that report and provide in addition to the various information also sufficient information for a possible faithful reconstruction (for example through 3D printing) in case of destructive events.

c) The main scenes and 3D models inserted within the app have been reconstructed from images treated with innovative techniques that enhance their contours and shapes, allowing us to have excellent results in terms of quality.

The real-time rendering process allows the user to navigate between objects qualitatively representative of reality, albeit considerably "lightened" in terms of points and edges.

In the research, particular attention was given to the development of new unique and customizable functions such as:

1) A section that provides notifications of places of historical and cultural interest in the vicinity.
2) A section that provides real-time notifications on authorized cultural events in the specific territory (in collaboration with the municipality or private events).
3) A section that provides the shortest routes to reach the place of interest.
4) The ability to customize (through selectable filters) tourist routes based on certain parameters (historical period, opening hours to the public, reaching by public transport, etc.).
5) The introduction of virtual guides of places of interest.
6) The possibility to leave comments and an evaluation of the places or museums or cultural works visited.
7) The possibility of visiting places also with Virtual Reality.

Thus, giving the tourist not only the complete perception of the cultural offer that each individual territory can offer but also the possibility of designing do-it-yourself itineraries according to their tastes.

These routes and this information can be conveniently viewed from the screen of our smartphone while we are walking.

Thanks to our app, the protagonist is not only the cultural heritage but also the user.

2.3 Unity

Unity is a cross-platform tool for creating interactive 3D content, such as architectural visualizations or real-time 3D animations, using two different programming languages: NET and C++.

His programming work is based on the use of so-called "Game Objects". These elements, which may or may not have a graphic representation, can be associated with scripts, all extensions of the "Mono-Behavior" base class. This Unity script, linked to a Game Object running both on the server and on the client, has the task of verifying in each frame the presence of new data inside the Buffer and therefore, based on these, of updating its representation on both server and on the customer's side.

This effect is achieved by using the Remote Procedure Calls (RPC) functions.

Therefore, based on the content of the buffer, the action to be taken on the object in question is chosen and performed both on the version of the object kept on the server and on those maintained on the clients.

3 Discussion

Vuforia is the most used platform in the world for the development of augmented reality thanks to its compatibility with the main mobile devices. Simplify your work when it comes to solving complex problems in virtual reality development.

From an operational point of view, this SDK (Software Development Kit) is based on real-time tracking of images and objects. On these targets or "Markers", it is possible to position virtual elements (3D models for example), which are then displayed on the screen following the observer's perspective. The freedom to choose these Markers is so important because the framework allows the recognition of words, objects and images, even more than one at the same time, previously scanned through special tools, made available to developers.

The app, created with Unity 3D, is structured through a series of scenes.

Precisely, there is an initial scene, a loading scene and an exploration scene, the sequence of which depends on the choices made by the user.

Thus, a virtual tour is conceived as a sequence of scenes presented in succession over time, creating what is the sensation of movement. Motion tracking allows us to activate animations or videos based on the position and orientation of the device, move around an object and interact with it as if it were part of the environment and also allows us to track the position of the device with respect to the environment. surrounding environment. This type of tracking has already been implemented in our app, as it allows us to let the app know when we move and accordingly react, to achieve this ARCore is used, a software development kit made by Google that is implemented within the Unity environment.

4 Conclusions

With the advent of Smart Cities, the transposition and consultation of large amounts of data on digital media, the creation of new methods of interaction with places of interest and services, through the use of apps and digital platforms, considerably reduces emissions and climate impact, thus favoring the green transition.

Nevertheless, we should not forget about the importance of the rural areas, which are one of the sharpest spear southern Italy has to offer, with its multitude of historical and interest points.

Making cities and communities safe, inclusive, resilient and sustainable is also one of the goals of the 2020–2030 Urban Agenda.

The environment around us can drastically affect our habits and lifestyles [16–19]. The use of technology can help users change their routines. A city to be defined as smart can only keep up with technology [20–22]. The development of digital telecommunications in a smart way sees the availability of platforms and applications for the optimization of city life. The app described, once started, allows the suppliers of offers and information on nearby shops and attraction points thanks to the GPS of the device, without detecting personal data.

Virtual Reality and Augmented Reality allow you to integrate the real world with information from different sources enriched by graphic, textual and media information.

The purposes can be both historical research and studies related to the history of the territory, whether they are for educational or cultural purposes. Virtual models with their realism become community heritage also allowing tourists and citizens to be made aware of the cultural value of the territory.

References

1. Barrile, V., Meduri, G.M., Bilotta, G.: Comparison between two methods for monitoring deformation with Laser Scanner. WSEAS Trans. Signal Process. **10**(1), 497–503 (2014)
2. Barrile, V., Meduri, G.M., Bilotta, G.: Laser scanner technology for complex surveying structures. WSEAS Trans. Signal Process. **7**(3), 65–74 (2011)
3. Barrile, V., Meduri, G.M., Bilotta, G.: Laser scanner surveying techniques aiming to the study and the spreading of recent architectural structures. In: Proceedings of the 2nd WSEAS International Conference on Engineering Mechanics, Structures and Engineering Geology, EMESEG 2009, pp. 25–28 (2009)
4. Leonardi, G., Barrile, V., Palamara, R., Suraci, F., Candela, G.: 3D mapping of pavement distresses using an unmanned aerial vehicle (UAV) system. In: Calabrò, F., Della Spina, L., Bevilacqua, C. (eds.) ISHT 2018. SIST, vol. 101, pp. 164–171. Springer, Cham (2019). https://doi.org/10.1007/978-3-319-92102-0_18
5. Barrile, V., Fotia, A., Bernardo, E.: The submerged heritage: a virtual journey in our seabed. Int. Arch. Photogramm. Remote Sens. Spatial Inf. Sci. **XLII-2/W10**, 17–24 (2019). https://doi.org/10.5194/isprs-archives-XLII-2-W10-17-2019
6. Barrile, V., Bilotta, G.: An application of remote sensing: object-oriented analysis of Satellite data. Int. Arch. Photogramm. Remote Sens. Spatial Inf. Sci. **XXXVII**, 107–113 (2008)
7. Barrile, V., Fotia, A., Candela, G., Bernardo, E.: Integration of 3D model from UAV survey in BIM environment. Int. Arch. Photogramm. Remote Sens. Spatial Inf. Sci. **XLII-2/W11**, 195–199 (2019). https://doi.org/10.5194/isprs-archives-XLII-2-W11-195-2019
8. Barrile, V., Cacciola, M., Morabito, F.C., Versaci, M.: TEC measurements through GPS and artificial intelligence. J. Electromagn. Waves Appl. **20**, 1211–1220 (2006). https://doi.org/10.1163/156939306777442962
9. Barrile, V., Bilotta, G., Fotia, A.: Analysis of hydraulic risk territories: comparison between LIDAR and other different techniques for 3D modeling. WSEAS Trans. Environ. Dev. **14**, 45–52 (2018)
10. Barrile, V., Bernardo, E., Bilotta, G., Fotia, A.: "Bronzi di Riace" geomatics techniques in augmented reality for cultural heritage dissemination. In: Borgogno-Mondino, E., Zamperlin, P. (eds.) ASITA 2021. CCIS, vol. 1507, pp. 195–215. Springer, Cham (2022). https://doi.org/10.1007/978-3-030-94426-1_15
11. Cannistraro, M., Bernardo, E.: Monitoring of the indoor microclimate in hospital environments a case study the Papardo Hospital in Messina. Int. J. Heat Technol. **35**(Special Issue 1), S456–S465 (2017). https://doi.org/10.18280/ijht.35Sp0162
12. Bernardo, E., Musolino, M., Maesano, M.: San Pietro di Deca: from knowledge to restoration. studies and geomatics investigations for conservation, redevelopment and promotion. In: Bevilacqua, C., Calabrò, F., Della Spina, L. (eds.) NMP 2020. SIST, vol. 178, pp. 1572–1580. Springer, Cham (2021). https://doi.org/10.1007/978-3-030-48279-4_147
13. Barrile, V., Bilotta, G., Fotia, A., Bernardo, E.: Road extraction for emergencies from satellite imagery. In: Gervasi, O., et al. (eds.) ICCSA 2020. LNCS, vol. 12252, pp. 767–781. Springer, Cham (2020). https://doi.org/10.1007/978-3-030-58811-3_55
14. Barrile, V., Fotia, A., Bernardo, E., Bilotta, G.: Geomatic techniques: a smart app for a smart city. In: Bevilacqua, C., Calabrò, F., Della Spina, L. (eds.) NMP 2020. SIST, vol. 178, pp. 2123–2130. Springer, Cham (2021). https://doi.org/10.1007/978-3-030-48279-4_200
15. Barrile, V., Fotia, A., Bernardo, E., Candela, G.: Geomatics techniques for submerged heritage: a mobile app for tourism. WSEAS Trans. Environ. Dev. **16**, 586–597 (2020). https://doi.org/10.37394/232015.2020.16.60

16. Barrile, V., Bilotta, G., Fotia, A., Bernardo, E.: Integrated GIS system for post-fire hazard assessments with remote sensing. Int. Arch. Photogramm. Remote Sens. Spatial Inf. Sci. **XLIV-3/W1-2020**, 13–20 (2021). https://doi.org/10.5194/isprs-archives-XLIV-3-W1-2020-13-2020

17. Barrile, V., Fotia, A., Bernardo, E., Bilotta, G., Modafferi, A.: Road infrastructure monitoring: an experimental geomatic integrated system. In: Gervasi, O., et al. (eds.) ICCSA 2020. LNCS, vol. 12252, pp. 634–648. Springer, Cham (2020). https://doi.org/10.1007/978-3-030-58811-3_46

18. Barrile, V., Bernardo, E., Candela, G., Bilotta, G., Modafferi, A., Fotia, A.: Road infrastructure heritage: from scan to InfraBIM. WSEAS Trans. Environ. Dev. V **16**, 633–642 (2020)

19. Barrile, V., Fotia, A., Bernardo, E., Bilotta, G.: Road cadastre an innovative system to update information, from big data elaboration. In: Gervasi, O., et al. (eds.) ICCSA 2020. LNCS, vol. 12252, pp. 709–720. Springer, Cham (2020). https://doi.org/10.1007/978-3-030-58811-3_51

20. Bilotta, G., Bernardo, E.: UAV for precision agriculture in vineyards: a case study in Calabria. In: Borgogno-Mondino, E., Zamperlin, P. (eds.) ASITA 2021. CCIS, vol. 1507, pp. 28–42. Springer, Cham (2022). https://doi.org/10.1007/978-3-030-94426-1_3

21. Bilotta, G., Nocera, R., Barone, P.M.: Cultural heritage and OBIA. WSEAS Trans. Environ. Dev. **17**, 449–465 (2021). https://doi.org/10.37394/232015.2021.17.44

22. Bilotta, G., Calcagno, S., Bonfa, S.: Wildfires: an application of remote sensing and OBIA. WSEAS Trans. Environ. Dev. **17**, 282–296 (2021). https://doi.org/10.37394/232015.2021.17.29

Use of Big Data and Geomatics Tools for Monitoring and Combating Pandemics

Vincenzo Barrile[1]([✉]), Ernesto Bernardo[1], and Stefano Bonfa[2]

[1] DICEAM, Mediterranea University of Reggio Calabria, Reggio Calabria, Italy
vincenzo.barrile@unirc.it
[2] UK Economic Interest Grouping (UKEIG) N. GEE000176, 87 Hungerdown, London E4 6QJ, UK

Abstract. The rapid spread of COVID-19 has highlighted the need for rapid and digital monitoring means for managing information relating to the spread of the virus. The researchers have produced some innovative experiments in this sense, using geomatics technologies and technologies for the treatment of Big Data. In this research, we used an integrated platform (still under experimentation and validation) that allows the treatment of Big Data to obtain forecast models that can be viewed in a GIS. The advantages of the proposed methodology are the ability to use and correlate a lot of data together through a single platform and view the results on a WebGIS platform.

Keywords: Big Data · Machine Learning · Artificial Intelligence · Remote sensing · GIS

1 Introduction

The study of Background showed how the rapid spread of COVID-19 has produced disastrous consequences in many nations, highlighting the need for fast and digital monitoring means for the communication and management of information relating to the spread of the virus, useful for planning interventions aimed at combating its diffusion and its negative impact on the territory. Hence the need to use suitable tools to manage and correlate a huge and heterogeneous amount of data (Big Data) obtained both through geomatics tools (images from earth observation satellites, UAVs, etc.), and from other sources such as the statistics of infected, healed or deceased people (age, residence, previous health status, etc.), from the climatic conditions (temperature, humidity, etc.), and from the environmental situation (pollution level, presence of fine particles in the air, etc.). All these data are combined within a properly implemented COVID RISK experimental platform that uses Machine Learning, Artificial Intelligence and Analytics techniques to try to map the pandemic risk to study its spread, evaluating adequate containment measures in relation to the different territories [1–4]. This map, being in constant evolution, requires continuous updates, and effective tools such as Geographic Information Systems (GIS) for the analysis and processing of geolocated data over time

© The Author(s), under exclusive license to Springer Nature Switzerland AG 2022
F. Calabrò et al. (Eds.): NMP 2022, LNNS 482, pp. 2073–2082, 2022.
https://doi.org/10.1007/978-3-031-06825-6_199

that can be integrated with other types of data. Up to now these systems have been effectively used for monitoring risk situations, such as volcanic, geological or environmental ones, but in the world of research, GIS solutions are also being tested in emergency situations such as the spread of an epidemic [5]. The management of emergencies and disasters has in fact, in the correlation and treatment of large amounts of different data (big data), and in their spatial-temporal and geolocalized contextualization, the essential tools, useful for responding adequately and quickly to problems arising (being them of a health, social, institutional, training, legal, economic, inclusion, productive nature, or in our specific case for the purpose of interpreting the pandemic situation [6, 7]). The real value associated with the availability of big data does not lie only in the quantity, but in the ability to use them to process, analyze and find objective evidence and correlations between various sectors, in the specific example in favor of the monitoring of pandemic risk through big data [8]. As known, Big Data means the processing of large amounts of data and useful methodologies for their correlation, to be used for research, management and forecasting applications [9]. To manage large amounts of data, basic technologies and specific technologies are required. Recently, we have experienced a strong increase in all three main dimensions of big data: volume, velocity, and variety, with tremendous amount of new developments in many fields, made possible through new technologies and new hardware and software, multi-temporal data analytics, data management, and information mining technologies [10]. Moreover, also recently, we have had a momentum in the field of Big Data from Earth observation, thanks to the recent proliferation of open access initiatives that offer new opportunities to researchers and companies [11]. In particular for Space Science and Earth Observation data with the release to the public of the complete archive of Landsat data by the United States Geological Survey and the European Copernicus program, whose Sentinel missions managed by the European Space Agency will provide free and open access to global data in the microwave and optical/infrared ranges [12]. The first one, Sentinel-1A, launched on April 3, 2014, and the second, Sentinel-1B, launched on April 25, 2016, are delivering high-resolution Synthetic Aperture Radar (SAR) global data every 12 days at a daily rate of 2.5 TB. The Sentinel-2A acquires optical data with high revisit frequency, coverage, timeliness and reliability with MSI Spectral Bands span from the Visible and the Near Infra-Red to the Short Wave Infra-Red: 4 bands at 10m, 6 bands at 20m, 3 bands at 60m. It operates in a reference sun-synchronous orbit with a repeat cycle of 10 days for the overall duration of the mission. Sentinel-2B in 2017 will be in the same orbit, allowing a ground-track revisit frequency of 5 days for the dual-spacecraft constellation [13]. Considering the existence of this pandemic risk, the availability of Big data, the existence of algorithms that allow their correlation and fusion, the current existence of very performing predictive systems in the field of machine learning, and finally GIS systems that allow the visualization of the results, it was therefore decided to utilize a platform opendatacube by implementing the appropriate functions useful for combining the various data sources available. We have thus created a "SmartCovidRisck" platform which represents a unique and complete solution, that, although still using rudimentary algorithms and incomplete data, returns the first results on the evolution of the pandemic (which can be viewed through a GIS platform), e consequently on the socio-economic effects of the spread of the pandemic and its correlation with some parameters. It underlines and points out that the results obtained are forcely partial and reported only for demonstrative purposes to highlight the potential of the developed methodology, implemented and proposed (and useable for provisional purposes in the fields of competitions) analyzed and used, are currently limited and the methodology is still in the testing and validation phase, precisely wanting to represent when completed a useful system for the prediction/monitoring of territorial and environmental events of different types. The experimental pilot project has been tested in the Italian territory, in the regions of Calabria and Sicily, where there are frequent movements between regions. The trials carried out were only experimental to validate the goodness of the proposed research procedure, as only partial data were used, we don't have a long historical series and, in the future, the algorithms currently used will certainly be implemented, improved and updated with other algorithms more performant.

2 Methodology

In relation to the resolution of the problem on the spread of the virus, several studies have been addressed individually.

The most recent bibliography investigates the use of traditional geomatics tools (GIS, Earth Observation satellites, etc.) for monitoring Covid 19, for monitoring atmospheric pollution, for monitoring population movements using mobile devices generated (big) data, and the critical study of the most emerging technologies to address the COVID-19 pandemic (including geospatial technology, artificial intelligence (AI), big data, telemedicine, blockchain, 5G technology, smart applications, Internet of Medical Things (IoMT), robotics, and additive manufacturing). However, researchers have almost always used only specific methods and specific data to obtain specific results, thus not attempting to correlate the large amount of available data and obtain more types of results. In our work, we have correlated all the data within a platform whose advantages are those of managing large amounts of data and obtaining more results. As free Big Data and numerous correlation and prediction algorithms are available, we decided to group and correlate this huge amount of data within a platform, in order to visualize first results about the possible evolution of a pandemic, the socioeconomic risks and the correlation between some phenomena and the spread of the pandemic. The trials carried out were only experimental to validate the goodness of the proposed research procedure, as only partial data were used, we don't have a long historical series data and we used algorithm that, in the future, will certainly be implemented, improved and updated with other algorithms more performant. We have thus realized a platform called SmartCovidRisk (The flowchart in Fig. 1 explains the SmartCovidRisk platform) wich is a platform able to process large amounts of data (Big Data) providing predictive models thanks to Artificial Intelligence and, thanks to the END-USER API, it allows through a WEB GIS graphical interface the interaction and participation of people from various communities and supervisory bodies, i.e."Big Data Living Labs".

This platform allows the interaction between the stakeholders related to the management of health crises, together with the transfer of knowledge between communities and bodies of expertise.

In particular, The SmartCovidRisk platform consists of three main pillars:

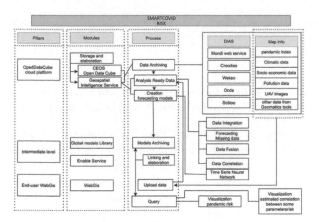

Fig. 1. Platform explanation

i.   The OpenDataCube cloud platform, which serves as a data-lake for ARD data (storage) and enables data processing and predictive modeling with algorithms and Artificial Intelligence (processing).
ii. An intermediate level of data services organization to interface the Open Data Cube with END-USER APIs (interaction between storage and visualization).
iii. The end-user WebGis visualization APIs to be implemented for the case study for the "Big Data Living Labs" communities (end-user visualization).

Specifically, the operation of the platform includes the following modules:
1) Storage and Elaboration (Opendatacube; Geospatial intelligent Service), 2) Global Models Library, 3) Enabled services, and 4) WebGIS application.

The various phases of the process will be better detailed directly through the explanation of the proposed case study.

3 Case Study

The experimentation was carried out in the territory between Calabria and Sicily (Fig. 2), two areas affected by a high flow of vehicular and pedestrian traffic and by a high flow of movements between regions because of their geographical position. Initially these two regions were free from a high number of cases of Covid 19, then, precisely because of the outbreak of the pandemic and the high number of movements between regions (returns of students and workers away from home) the cases have increased exponentially.

3.1 Storage and Elaboration: OpenDataCube

The CEOS OpenDataCube: The opendatacube works as a multidimensional value matrix to manage arrays, multi-terabyte/petabyte data warehouses and time series of image data, thus acting as a data-lake storage: the CEOS opendatacube platform automatically and strategically (depending on the case study) takes data from ESA's DIAS (services provided on accessing data from Copernicus satellites) that are of interest to our case study. We chose DIAS because it can provide ready data, easily integrated and usable in the CEOS Opendatacube platform.

3.2 Storage and Elaboration: Geospatial Intelligence Services (1. Analysis Ready Data, 2. Creation of Forecast Models (Correlation, Machine Learning))

Once the automatic storage phase in the CEOS Open Data Cube was completed, these data were implemented (merged and integrated with other data provided by geomatics tools), transformed and manually loaded into the platform, creating an Analysis ready data (ARD) with various other data (in the specific case Sicilia – Calabria were used: data from earth observation satellites (Copernicus Landsat, Sentinel 1, Sentinel 2, Sentinel-5P); climatic data (wind, temperature, atmospheric humidity, etc. provided by Eumetsat satellites or weather stations or historical data); socio-economic data, distribution and composition of the population (population per municipality, population density; old age index), statistical data Tourism; data on air pollution (2019 and 2020 comparison); electromagnetic pollution, total difference in deaths per municipality between 2020 and average 2016–2019 (ISTAT data); hospitals and service centers for the elderly; UAV images (GB multispectral and thermal); rainfall data (stations or historical data); data on population movements; total covid-19 cases; population by municipalities and provinces (source: civil protection); enrolled at the University of Messina (Unime) and Reggio Calabria (UNIRC) in 2019 and related Region of origin) providing, where necessary, to their transformation and to their loading also manually in the platform, paying particular attention to the Open Data Cube (ODC) format and procedures, and allowing the Analysis Ready data to make them ready for the next steps (correlation, prediction).

More specifically, as regards the Analysis Ready data, we distinguish the case in which the data is in the form of an image or not. If the data is in the form of a satellite image, the upload process within the OpenDataCube (ODC) involves the following steps (Indexing – Ingestion – Load Data and eventually Data Fusion); If the data to be entered in the OpenDatacube is not an image, it is transformed into an image (In particular, through the maps software it was possible to locate the numerical data available, creating thematic maps). In this way we have comparable data all in the same format. Once the parameter to be estimated has been identified (In the specific case the pandemic forecast index), its correlation with all the other data (parameters) available has been assessed by evaluating their correlation trying to maximize it (possibly aggregating the variables) if it is low, in order to be able to use them with a neural network or a predictive system. In this regard, the data are correlated with each other (we use Pearson Correlation), possibly integrated with missing data and then processed through Machine Learning techniques (Support Vector Regression, Random Forest, Neural Network, Time Series Neural Network) to create a predictive model and choose from time to time the best in relation to the one that best "explains" the results of regression between the various data first found. In order to predict a value from the observation of other values in fact it is necessary that the values under observation are correlated with the value to be predicted (spread of the pandemic and socio-economic effects). As an example, we report the correlations found related to the 5 parameters reported (Table 1).

Fig. 2. Case study: territory between Calabria and Sicily

Table 1. Correlations found related to the 5 parameters reported.

Variables	Correlation
Temperature	−0.080423
PM10	1.000000
Pressure	−0.036171
Wind	−0.236895
Rain	−0.041258

To implement the aforementioned network, the choices listed below were made in relation to the following parameters that characterize the construction of the above mentioned Neural Network:

1. *The temporal subdivision of the database.*
2. *The number of hidden layers and the number of neurons to be inserted in each layer.*
3. *The connective mechanisms between the different layers.*
4. *The activation function.*
5. *Learning rules.*
6. *Update of connection weights of neurons.*

Therefore, for the forecasting phase, machine learning techniques were used (Support, Vector Regression (SVR), Random Forest (RF), Neural Network Time Series (NNTS), Neural Network (NN)), in order to evaluate and make decisions based on the test results and observing as the most important value the mean square error deviation on the forecasts. Thanks to the libraries the implementation was reduced to understanding how to use the classes of data made available.

Table 2 shows the mean square deviation results obtained in relation to the different forecasting models used.

Table 2. Standard deviation.

Model	Mean square error deviation (standard deviation)
SVR	0.107838
RF	0.110606
NNTS	0.107838
NN	0.001012

Table 3. Standard deviation variability.

Number of neurons	Mean square error deviation (standard deviation)
1, 12, 24	0.001062, 0.001031, 0.001062

As it can be observed from the table, the most performing model has resulted the Neural Network with temporal series. In Table 3 we see how the mean square deviation varies, when the number of neurons varies for the chosen neural network. In our application the most performing model has resulted in the Neural Network with temporal series.

4 Global Models Library/Enabled Services

The generated forecast models are loaded into this library (this is where forecast model creation ends) to be available for Enabled services. Once the models are obtained, the operators (citizens or administrators) query the library thanks to tools that allow them to query the forecasting model or extrapolate the information that best meets their needs. In our specific case the information on the pandemic risk of the territory between Calabria and Sicily.

5 WebGis Application

We have therefore implemented a WebGis application that allows us to consult the Global Models Library thanks to tools that allow us to query the forecast model.

The WebGis application created, allows end users, or data scientists, citizens and stakeholders, respectively:

a) To upload (additional INPUT) land mapping data and to simulate scenarios, e.g. possible virus outbreaks using artificial intelligence predictive models, leveraging shared data made available both locally and globally.
b) To input data and opinions through targeted surveys, so that the various stakeholders can participate in the policy making process related to the solutions to be adopted for the pandemic risk.
c) To visualize on ad-hoc dashboards the set of scenarios simulated by data-scientists on the basis of data made available also by citizens, for a decision-making process based on data. This open data can be processed and displayed in different ways, for example as text or on an interactive map, on the website, in the mobile application or in the software being developed.

Our application allows end users to:

1) Visualize the pandemic risk on the Calabrian and Sicilian territory and the socio-economic impact that these two events bring to the economy and to the population residing in the territory.
2) Visualize the estimated correlation between some parameters (presence of NO2, PM10, little wind, little vegetation) and the possible spread of Covid 19.
3) Upload any additional INPUT data to interact with the data already in the Open Data Cube.
4) Input data and opinions through targeted surveys so that the various stakeholders can participate in the policy making process related to countermeasures to be taken.

In the developed WebGis, the user interface (dashboard) allows for customization (e.g., can include information for local authorities) and visualization of the set of simulated scenarios. Multi-layer visualization capabilities, earth mapping data loading, and algorithm simulation are provided to users. Such open data can be processed and displayed in different ways, such as text or on an interactive map, on the website, in the mobile application, or in the software being developed [14–16].

6 Results of the SmartCovidRisk

The experimentation of SmartCovidRisk thus realized and applied to the Sicilian-Calabrian territory has allowed to obtain, visualize and manage on the GIS predictive results (virus propagation and socio-economic effects) and correlations between the evolution of the pandemic and some parameters (NO2, PM10, little wind, little vegetation). In relation to the prediction on the possible propagation of Covid 19, Fig. 3 shows some results on the possible evolution of the virus. The study carried out shows that the cities and towns most affected seem to be those in the vicinity of ports and airports and that those in the mountains and more isolated have less effect. Having a tool that provides predictive models and a correlation between phenomena and pandemic risk at the local level, provides significant benefits to populations and local governments. In fact, a better awareness of the evolution of the possible epidemic and of the possible correlations between phenomena and pandemic risk will allow the elaboration of suitable political choices to contain harmful effects. In relation to the estimation of socioeconomic effects, Fig. 4 shows the predictive models of possible socioeconomic effects that the territories affected by the propagation of pandemic risk could have.

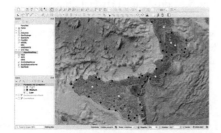 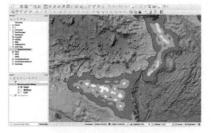

Fig. 3. Case study, Sicily and Calabria, the software implemented shows the propagation of the virus, 3 months after the data acquired without having adopted countermeasures. In black the cities and towns with high effects, in dark gray with medium effects, in light gray with low effects.

Fig. 4. Case study, Sicily and Calabria. The implemented software shows the areas affected by the effects (socioeconomic on the territory) of the pandemic emergency. High effects in black, medium effects in dark gray, low effects in light gray.

As we can see from the images, the greatest socio-economic effects occur in agricultural and industrial areas as well as in large cities. During this period, in fact, the sector that usually is the first to be affected is the agricultural and industrial one as well as stores, restaurants, bars, hotels, tourism, etc. and all the induced activities.

A short-term spread of the virus that persists from three to six months can have small and medium impacts (direct and indirect) on these sectors.

Through the GIS realized it is also possible to estimate the dependence between some parameters (NO2 Fig. 5, PM10 Fig. 6, little wind Fig. 7, little vegetation Fig. 8) and the possible spread of Covid. The results obtained allow to highlight a significant correlation with the presence of these parameters.

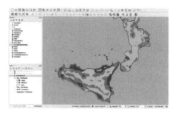

Fig. 5. Correlation between NO2 concentration and pandemic spread. In the presence of important values of smog and dust the spread of the virus is higher.

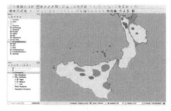

Fig. 6. Correlation between PM10 and pandemic spread. In the presence of high PM10 concentration, the spread of pandemic risk is higher.

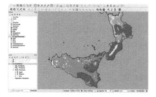

Fig. 7. Correlation between little wind and pandemic spread. In the presence of little wind, the spread of the virus is higher.

Fig. 8. Correlation between little vegetation and pandemic spread. In areas with less forest and vegetation, the spread of pandemic risk is higher.

7   Conclusion

The danger level of a pandemic risk is high among many countries. For governments/regions/municipalities and people, the value of the proposed platform is very significant especially because it can be replicated in areas all over the territories with countless benefits. For this experimentation we have used the CEOS OpenDatacube, a product already on the market, however, in the future, we will try to use an OpenDataCube made by us and use new algorithms more performing.

The experiments carried out were only experimental to validate the validity of the proposed research procedure. In the future, in order to have more reliable values it will be necessary to have homogeneously distributed data, time series and certain and complete parameter values. The algorithms currently used, in the future, can be implemented, improved, and updated with other algorithms more performing and possibly more relevant.

References

1. Bello-Orgaz, G., Jung, J.J., Camacho, D.: Social big data: recent achievements and new challenges. Inf. Fusion **28**, 45–59 (2016)
2. Barrile, V., Fotia, A., Bernardo, E., Bilotta, G.: Road cadastre an innovative system to update information, from big data elaboration. In: Gervasi, O., et al. (eds.) ICCSA 2020. LNCS, vol. 12252, pp. 709–720. Springer, Cham (2020). https://doi.org/10.1007/978-3-030-58811-3_51
3. Barrile, V., Bilotta, G.: An application of remote sensing: object oriented analysis of satellite data. Int. Arch. Photogramm. Remote Sens. Spatial Inf. Sci. **37**, 107–114 (2008)
4. Bernardo, E., Musolino, M., Maesano, M.: San Pietro di Deca: from knowledge to restoration. Studies and geomatics investigations for conservation, redevelopment and promotion. In: Bevilacqua, C., Calabrò, F., Della Spina, L. (eds.) NMP 2020. SIST, vol. 178, pp. 1572–1580. Springer, Cham (2021). https://doi.org/10.1007/978-3-030-48279-4_147
5. Barrile, V., Meduri, G.M., Bilotta, G.: Comparison between two methods for monitoring deformation with laser scanner. WSEAS Trans. Signal Process. **10**(1), 497–503 (2014)
6. Barrile, V., Meduri, G.M., Bilotta, G.: Laser scanner technology for complex surveying structures. WSEAS Trans. Signal Process. **7**(3), 65–74 (2011)
7. Barrile, V., Bernardo, E., Bilotta, G., Fotia, A.: "Bronzi di Riace" geomatics techniques in augmented reality for cultural heritage dissemination. In: Borgogno-Mondino, E., Zamperlin, P. (eds.) ASITA 2021. CCIS, vol. 1507, pp. 195–215. Springer, Cham (2022). https://doi.org/10.1007/978-3-030-94426-1_15
8. Barrile, V., Bilotta, G., Fotia, A., Bernardo, E.: Road extraction for emergencies from satellite imagery. In: Gervasi, O., et al. (eds.) ICCSA 2020. LNCS, vol. 12252, pp. 767–781. Springer, Cham (2020). https://doi.org/10.1007/978-3-030-58811-3_55
9. Barrile, V., Fotia, A., Bernardo, E., Bilotta, G.: Geomatic techniques: a smart app for a smart city. In: Bevilacqua, C., Calabrò, F., Della Spina, L. (eds.) NMP 2020. SIST, vol. 178, pp. 2123–2130. Springer, Cham (2021). https://doi.org/10.1007/978-3-030-48279-4_200
10. Barrile, V., Bilotta, G., Fotia, A., Bernardo, E.: Integrated GIS system for post-fire hazard assessments with remote sensing. Int. Arch. Photogramm. Remote Sens. Spatial Inf. Sci. **XLIV-3/W1-2020**, 13–20 (2021)
11. Barrile, V., Fotia, A., Bernardo, E., Bilotta, G., Modafferi, A.: Road infrastructure monitoring: an experimental geomatic integrated system. In: Gervasi, O., et al. (eds.) ICCSA 2020. LNCS, vol. 12252, pp. 634–648. Springer, Cham (2020). https://doi.org/10.1007/978-3-030-58811-3_46
12. Bilotta, G., Bernardo, E.: UAV for precision agriculture in vineyards: a case study in Calabria. In: Borgogno-Mondino, E., Zamperlin, P. (eds.) ASITA 2021. CCIS, vol. 1507, pp. 28–42. Springer, Cham (2022). https://doi.org/10.1007/978-3-030-94426-1_3
13. Barrile, V., Bernardo, E., Candela, G., Bilotta, G., Modafferi, A., Fotia, A.: Road infrastructure heritage: from scan to Infrabim. WSEAS Trans. Environ. Dev. V **16**, 633–642 (2020)
14. Bilotta, G., Nocera, R., Barone, P.M.: Cultural heritage and OBIA. WSEAS Trans. Environ. Dev. **17**, 449–465 (2021)
15. Barrile, V., Fotia, A., Bernardo, E., Candela, G.: Geomatics techniques for submerged heritage: a mobile app for tourism. WSEAS Trans. Environ. Dev. **16**, 586–597 (2020)
16. Bilotta, G., Calcagno, S., Bonfa, S.: Wildfires: an application of remote sensing and OBIA. WSEAS Trans. Environ. Dev. **17**, 282–296 (2021)

Modern Tools of Geomatics as an Indispensable Support to the Public Administration for the Protection, Restoration, Conservation and Dissemination of Cultural Heritage

Ernesto Bernardo[✉] [iD]

Department of Civil, Energetic, Environmental and Material Engineering -DICEAM- Geomatics Lab, Mediterranea University of Reggio Calabria, 89123 Reggio Calabria, Italy
ernesto.bernardo@unirc.it

Abstract. This research demonstrates the utility of some modern Geomatics tools (UAV, aerial photogrammetry, 3D models, Augmented and Virtual Reality, GIS) as an indispensable support to the public administration for the protection, restoration, conservation and dissemination of cultural heritage. The solution proposed by us is not as complete as [3], however it is faster, cheaper and more easily repeatable over time. Specifically, for the protection, restoration and conservation of the artifact detected, in order to obtain a 3D model (with geolocalized information and with precision of detail of a geometric-material survey), we initially created an aerial photogrammetric survey using of UAVs; subsequently we used the images acquired for the creation of a 3D model through the SFM, using the canonical method described in the manuals, continuing we made a comparison of the model previously obtained with a second 3D model (made using some tricks for noise reduction). The repetition of the photographic survey at fixed intervals, possibly to be included in the maintenance program, can allow a comparison over time of the evolution of deterioration and damage, and therefore establish a ranking of intervention priorities. Finally, for the dissemination, knowledge and enhancement of the artifact detected, we have created an AR/VR app and a GIS system.

Keywords: UAV · Aerophotogrammetry · 3D model · Augmented reality · Virtual reality · GIS

1 Introduction

"Take diligent care of your monuments, and you will have no need to restore them […] And all this, do it lovingly, with reverence and continuity, and more than one generation can still be born and die in the shadow of that building", John Ruskin, *The seven lamps of architecture* [1]. The Italian territory is one of the richest countries in cultural heritage, they produce 1.6% of the national GDP, about 27 billion euros, 117 thousand employees [2] and represent a great opportunity for the whole country but especially for individual local communities. Taking care of historic buildings, ancient and therefore very delicate

© The Author(s), under exclusive license to Springer Nature Switzerland AG 2022
F. Calabrò et al. (Eds.): NMP 2022, LNNS 482, pp. 2083–2092, 2022.
https://doi.org/10.1007/978-3-031-06825-6_200

artifacts represents a very difficult challenge for public administrations. The obstacles for scheduled maintenance are enormous: the seismic risk of the territory (we know that the entire Italian territory is at seismic risk) and the consequent damage it produces to the building such as injuries, weakening of the structure - or worse - the total collapse of the same with all that it holds; hydrogeological risk (flooding, landslides, subsidence, with all the consequent damages, in most cases irreparable); and even the environmental context (this strongly conditions the state of conservation of the artefacts, in particular of the wall faces: the growth of spontaneous vegetation on the wall structures, whose roots, entering into the interstices, cause the enlargement to the point of causing disconnection and expulsion of the facing stones, causing very serious damage).

Although ordinary maintenance can be adequately planned, and kept under control at an adequate frequency, however, the repair of the damage produced by the phenomena previously described can be much more difficult and costly. Therefore, in contrast to the practice carried out so far by the public administration which included sporadic interventions dictated by an emergency or linked to the occasional availability of funds, a new awareness linked to the need to carry out more organic maintenance interventions, or better conservation scheduled. This new approach favors the execution of small preventive and monitoring interventions, minimal and repetitive operations that in the long term have undoubted economic advantages compared to occasional interventions aimed at restoring an assumed initial condition of the architectural asset following serious damage. To deal comprehensively and solve all the problems concerning the protection, restoration, conservation and dissemination of cultural heritage, the ideal solution would be [3]. However, this solution is very expensive, both in terms of time and money. So we present in this paper a less complete but cheaper, faster and easily repeatable solution over time that can be useful to public administrations to solve many of the problems mentioned above. A solution that uses some of the tools of Geomatics (UAV, aerial photogrammetry, 3D models, Augmented and Virtual Reality, GIS). Useful tools for urgent maintenance operations, for the preparation of a periodic control and maintenance program to be repeated over time, and finally for the dissemination of the product. Specifically, for the protection, restoration and conservation of the artifact detected, in order to obtain a 3D rendering with precision of detail of a geometric-material relief, we initially created an aerial photogrammetric survey using UAVs; subsequently we used the acquired images for the creation of a 3D model through the SFM, using the canonical method described in the manuals; continuing we made a comparison of the model previously obtained with a second 3D model (made using some tricks for noise reduction). The repetition of the photographic survey at fixed intervals, to be included in the maintenance program if possible, can allow a comparison over time of the evolution of degradation and damage, for example the state of the lesions or the presence of new lesions, the speed of vegetation regrowth pest, possible humidity problems, etc., and therefore establish a ranking of intervention priorities. Finally, for the dissemination, knowledge and enhancement of the artifact detected, we have created an AR / VR app and a GIS system.

2 Materials and Methods

2.1 Study Area

We detected the ruins of the Grangia di San Leonte (or San Leonzio) in Stilo, province of Reggio Calabria, Calabria, (Italy), (Fig. 1). The grangia was a former Basilian monastery, belonging until the time of its suppression, to the administrative jurisdiction of the Certosa di S. Stefano del Bosco. The first historical references (1059; 1093; 1184) relating to the monastery of San Lorenzo are due to some claims of possessions and donations by some "powerful" of the period; subsequently, in 1191 the monastery was donated to the Certosa di S. Stefano del Bosco, in exchange for the Casale del Conte. There are no precise data regarding the suppression of the grange of S. Leonzio, however the hypothesis that this suppression took place with the institution of the Cassa Sacra, following the earthquake of 1783, is certainly valid. Another hypothesis may be that due to the need to give renewed spiritual content to the various religious congregations, which led the Pope of the time to suppress the small convent communities in 1652. In a manuscript by D. Martire we learn that the hamlet of S. Leonte had already disappeared in the 1600s. The 4 sides of the complex were surrounded by a wall more than 3 m high, inside there were the factories of the grange whose ground floors, consisting of large rooms served as granaries, cellars and shelters for livestock. At the center of the complex there was a building 13 m high consisting of numerous loopholes, with the possible function of a watchtower and armory. The only access to the tower was the elevated passage that connected it to the rest of the complex. Today only the ruins remain of the building complex, which clearly show the defensive function of the grange (Fig. 2).

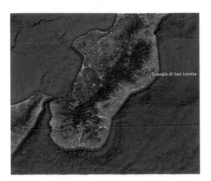

Fig. 1. Geographical location of the case study.

Fig. 2. Ruins of the Grangia of San Leonte.

2.2 Survey by UAV and 3D Reconstruction

As previously written, for the protection, restoration and conservation of cultural heritage, the use of aerial photogrammetry and SFM is essential in order to obtain a detailed 3D rendering of a geometric-material relief. The repetition of the photographic survey at fixed intervals, possibly to be included in the maintenance program, can allow a comparison over time of the evolution of degradation and damage (for example the state of the lesions or the presence of new lesions, the speed of vegetation regrowth pest, possible humidity problems, etc.) and therefore establish a ranking of intervention priorities. An agile and precise tool like the UAV is very useful in the field of cultural heritage maintenance, as it also allows you to monitor the current situation; coordinate interventions; obtain the state of affairs before and after the interventions carried out and promptly identify any necessary repairs such as the removal and restoration of damaged parts. The UAV has the flexibility of use of a geographic-territorial database (SIT, Geographic Information System, or GIS, Geographic Information System) together with the precision of detail of a geometric-material survey; and at the same time allows a sufficiently fast and repeatable realization over time. Thanks to the use of increasingly advanced software we are able to obtain three-dimensional models with increasingly precise "textures" (projections of the surfaces obtained from photographs), thus increasing the detail and accuracy of the model.

We have created a topographic support network using dual frequency GPS stations, in RTK (Real Time Kinetec) mode, connected to the regional support network, using the VRS-Nearest-DGPS differential correction. 65 topographic points were detected, with a mean square error (SQM) of less than 3 cm planimetric and 4 cm altimetric, fixed to the ground with topographic nails and chosen on the top and foot of the walls in order to cover with at least 5 points, allowing the appropriate overlaps, each of the different portions of the circuit that is expected to be detected with a single flight session. A DJI Mavic 2 Pro UAV was used for the acquisition phase of the photographic data set (Fig. 3), equipped with omnidirectional vision sensors and infrared sensors, of DJI brand technologies, such as the obstacle detection system, intelligent features such as Hyperlapse, Point of Interest, ActiveTrackTM 2.0, TapFly and Quick-Shots, as well as assisted piloting systems (APAS). The Mavic 2 Pro features a fully stabilized three-axis gimbal camera, with a 1-inch CMOS sensor (developed in collaboration with Hasselblad) for recording 4K video and taking 20-megapixel photos. Mavic 2 Pro uses the latest technologies to increase shooting stability and quality by reducing the range of angular vibrations to $\pm$ 0.01°.

Fig. 3. UAV used.

Flight Plan Set-up. To obtain images with high quality centimeter accuracy, the image acquisition plane has been divided into three phases: 1) definition of the image acquisition plane type, 2) definition of the Ground Sampling Distance (GSD), and 3) definition overlay of images. The type of image acquisition has been set on the waypoint automatic flight mission for the UAV to perform an automatic flight. The passage points have been placed around the artifact. The software automatically calculates the image acquisition plan and mission settings, the following parameters have been defined: flight height (consequently GSD), overlap (%) and area to map. To obtain centimeter accuracy (GSD < 1) the legal flight height, considering the specifications of the DJI Mavic 2 Pro camera the maximum flight height has been set at 10 m for vertical flight. Several missions were automatically performed along the route using the Pix4d flight planning software (Pix4d, Lausanne). To collect photographic information, flight plans were organized with the camera tilt set to 70°. During the flight, the drone always maintained a constant height with respect to the take-off point, therefore initial tests were necessary to verify that the difference in height could not in any way compromise the safety of flight operations. The absence of any obstacles (electric cables, etc.) along the flight plan was also checked. Apps such as Flight plan, on the other hand, allowed us to modify the various heights of the stages, thus optimizing the procedures necessary on the fly.

3D Model Construction. 3D reconstruction was performed using the Agisoft Photoscan photogrammetry software (Agisoft, St. Petersburg). The elaboration process consists of three different phases: 1. Photographic alignment: identification of the connection points through operators of interest. The points chosen in the various photos must have common characteristics in order to be adequately superimposed. For a good result, the image quality must be high; you need to have few shaded areas and adequate lighting; 2. Construction of the dense cloud. During this phase, a dense cloud is built using image matching algorithms. These are divided into algorithms that use a stereo pair to find the matches (stereo matching) and those that identify them in multiple images (stereo multi-view); 3. Build mesh, which consists in generating a polygonal model based on the newly created dense cloud. Mesh is a subdivision of a solid into smaller polyhedral-shaped solids; 4. Build texture, allows to obtain the 3D representation of the work in question. As an experiment, for the 3d reconstruction of the artefact, 369 images were purchased from different gripping distances (with circular flights) which allowed to build different chunks for the metrically more defined reconstruction of the details. It should be noted that the processing of the different chunks and consequently their union despite having tripled the processing times (18 h and 45 m) generated a model disturbed by noise (point cloud of 8707014 points), and of unsatisfactory quality, waste root mean square equal to 0.326 (calculated on the basis of comparison with real data), Fig. 4.

In this regard, it should be noted that, in general, the creation of a 3d model obtained starting from a circular flight, may also include the need to return a part of it in greater detail (element of interest). In these cases, it is necessary, consequently, to proceed with the creation of a more "dense" relief in correspondence with the element of interest (creation of more chunks). In the restitution phase of the entire model, however, it is observed that the union of different chunks generates noise if the images used for the restitution have aligned grip points, thus resulting redundant (overlapping and with different resolutions). Although for the construction of the single point clouds (of the various chunks) the optimal conditions suggested by the software manual are used (same grip distance and orthogonality of the camera with respect to the product), the above problem continues to exist in relation to accuracy global of the entire model [4–11].

Fig. 4. Dense cloud from the union of several chunks.

Consequently, in order to solve the highlighted problem, an own algorithm has been created that identifies the frames that have an overlap equal to or greater than 90% (in fact, variable threshold values are used), so that they can be discarded and eliminated from processing. In the specific case, the use of the proposed algorithm (using various threshold parameters) allows you to automatically reconstruct the global model using various combinations of grip points and number of images.

The optimal result (Fig. 5) was obtained with the configuration relating to the use of a single gripping distance and 119 images (point cloud of 10731638). The result of the processing (performed with the same processing parameters), in addition to reducing the processing times (6 h and 22 min), returned a model without noise, and a root mean square of 0.168 (calculated on the basis of the comparison with the real data).

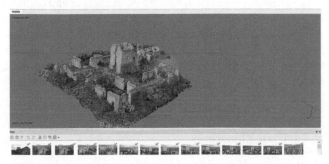

Fig. 5. Dense cloud from a single chunk (reduced number of images and single grip point).

3   The App Created

In order to spread and enhance the cultural heritage, the Geomatica laboratory of Reggio Calabria has developed a mobile tourist application in an Augmented and Virtual reality environment for research purposes. The app created is a tourist app for mobile device that allows the user to obtain in real time additional information on the object of the survey and to be able to view the 3D model in AR and VR (Fig. 6). The user can thus take a virtual tour inside the building (with a viewer or simply through the device screen) and be "accompanied" during the visit by a virtual guide who interacts with the surrounding environment or projects informative videos. The app allows you to project the reconstruction of the exterior and interior of the building, and to highlight the details.

Fig. 6. App made.

For the programming of the app we used Unity, based on the use of so-called "Game Objects". These elements (which may or may not have a graphical representation) can be associated with scripts, all extensions of the base class "Mono-Behavior". This Unity script, connected to a Game Object running both on the server and on the client, has the task of verifying in each frame the presence of new data inside the Buffer and therefore, based on these, of updating its representation. both on the server side and on the client side. This effect is achieved by using the RPC functions. Therefore, based on the content of the buffer, the action to be performed on the object in question is chosen and performed both on the version of the object kept on the server and on those maintained on the clients [12–19].

3.1   3D Model Analysis: Rhinoceros

To verify the cohesion of the 3D model and possibly intervene on the "open" or "missing" parts, we used the software Rhinoceros 3D ver.6 (Robert McNeel & Associates, USA). We initially used the edge analysis tool, once an open polysurface was detected, the open edges were joined with a joint operation. We joined the separated surface to the main body of the model, closing the object. In the case of missing surfaces, we proceeded by creating a new surface that was in perfect contact with the edges of the surrounding surface, or if the distance was less than the Rhinoceros tolerance, a forced union was used. Subsequently, using the "Select damaged objects" tool we have identified the damaged objects. Once the defective surfaces were isolated, we proceeded to restore the edges of the surfaces to their original state before they were joined with the "Rebuild Edges" command. This obviously led to differences between the scanned object and the model, but the differences found were very small.

4 GIS

For the purposes of dissemination, knowledge and enhancement of the cultural heritage of Calabria, the use of the GIS tool is of enormous importance.

For this reason we have created a GIS platform (Fig. 7) by inserting the location of the artifact detected, the 3D model that can be recalled through a link, the archival and literary material, and all the most significant information of the artifact, implementing it with other sites archaeological sites we have detected, and creating a tool - a sort of modern "map" - which allows both scholars and the most demanding citizens to know the location of most of the cultural heritage of the Calabrian territory.

This system can be integrated into information totems installed in the points of greatest interest (Fig. 8). Furthermore, in our opinion, it is essential for every public administration to have a geolocalized inventory of cultural assets containing all the information useful for the management of the cultural asset (maintenance plan, interventions carried out, materials used, restorations, historical and archaeological information, etc.): in this case the use of GIS is fundamental not only as a modern tool for tourists but also as a tool for managing cultural heritage for the entrants in charge [20–24].

Fig. 7. Visualization of the georeferenced archaeological map.

Fig. 8. Totem test in the Geomatics laboratory.

5 Conclusions

The work we have carried out has the potential to become an important resource for the Public Administration, offering the possibility of obtaining a large amount of detailed information on an architectural asset quickly and inexpensively.

The repetition of the photographic survey at fixed intervals, to be included in the maintenance program if possible, can allow a comparison over time of the evolution of the degradation, for example the state of the lesions or the presence of new ones, the regrowth speed of the weeds, any problems humidity, etc. and then establish a ranking of intervention priorities. The geometric model obtained can be integrated with other information (restoration or repairs carried out, the chronology of the interventions, documentation, materials used and their possible origin, etc.), as well as with other relevant technologies (laserscanner, georadar, close-up photogrammetry, etc.) in order to increase the knowledge and quality of details. Finally, tools such as apps and GIS appear to be indispensable to allow the dissemination of the wealth of knowledge acquired regarding a little-known and difficult to reach asset.

Future research will focus on the automation and simplification of the processing procedures of the acquired images in order to reduce the time.

References

1. Ruskin, J.: Le sette lampade dell'architettura. Trad. it., JACA BOOK, Milano (1849)
2. MIBACT. https://storico.beniculturali.it/mibac/export/MiBAC/sitoMiBAC/Contenuti/Mib acUnif/Comunicati/visualizza_asset.html_1125367599.html. Accessed 13 Sept 2020
3. Barrile, V., Fotia, A., Bernardo, E.: Non-invasive geomatic techniques to study and preserve cultural heritage: the case study of the Cardeto tower. Non-invasive techniques to study and preserve cultural heritage. Springer, Cham (2020)
4. Bernardo, E., Musolino, M., Maesano, M.: San Pietro di Deca: from knowledge to restoration. Studies and geomatics investigations for conservation, redevelopment and promotion. In: Bevilacqua, C., Calabrò, F., Della Spina, L. (eds.) NMP 2020. SIST, vol. 178, pp. 1572–1580. Springer, Cham (2021). https://doi.org/10.1007/978-3-030-48279-4_147
5. Menna, F., Nocerino, E., Remondino, F., Dellepiane, M., Callieri, M., Scopigno, R.: 3D digitization of an heritage masterpiece-a critical analysis on quality assessment. ISPRS Int. Arch. Photogramm. Remote Sens. Spat. Inf. Sci. **XLI-B5**, 675–683 (2016)
6. Tucci, G., Bonora, V., Conti, A., Fiorini, L.: High-quality 3D models and their use in a cultural heritage conservation project. ISPRS–Int. Arch. Photogramm. Remote Sens. Spat. Inf. Sci. **XLII-2/W5**, 687–693 (2017)
7. Apollonio, F.I., Ballabeni, M., Bertacchi, S., Fallavollita, F., Foschi, R., Gaiani, M.: From documentation images to restauration support tools: a path following the neptune fountain in bologna design process. ISPRS–Int. Arch. Photogramm. Remote Sens. Spat. Inf. Sci. **XLII-5/W1**, 329–336 (2017)
8. Green, S., Bevan, A., Shapland, M.: A comparative assessment of structure from motion methods for archaeological research. J. Archaeol. Sci. **46**, 173–181 (2014)
9. Chandler, J.H., Lane, S.N.: Structure from motion (SFM) photogrammetry. In: Geomorphological Techniques, pp. 1–12. Routledge, Abingdon (2015)
10. Bilotta, G., Bernardo, E.: UAV for precision agriculture in vineyards: a case study in Calabria. In: Borgogno-Mondino, E., Zamperlin, P. (eds.) ASITA 2021. CCIS, vol. 1507, pp. 28–42. Springer, Cham (2022). https://doi.org/10.1007/978-3-030-94426-1_3
11. Redweik, P., Cláudio, A.P., Carmo, M.B., Naranjo, J.M., Sanjosé, J.J.: Digital preservation of cultural and scientific heritage: involving university students to raise awareness of its importance. Virtual Archaeol. Rev. **8**, 22 (2017)
12. Bae, H., Golparvar-Fard, M., White, J.: High-precision vision-based mobile augmented reality system for context-aware architectural, engineering, construction and facility management (AEC/FM) applications. Vis. Eng. **1**, 1–13 (2013)
13. Bedford, L.: Storytelling: the real work of museums. Curator Mus. J. **44**, 27–34 (2001)
14. Barrile, V., Fotia, A., Costa, G., Bernardo, E.: Geomatics techniques for cultural heritage dissemination in augmented reality: Bronzi di Riace case study. Heritage **2**(3), 2243–2254 (2019)
15. Lowe, D.G.: Distinctive image features from scale-invariant keypoints. Int. J. Comput. Vis. **60**, 91–110 (2004). Heritage **2**, 2254 (2019)

16. Caspani, S., Brumana, R., Oreni, D., Previtali, M.: Virtual museums as digital storytellers for dissemination of built environment: possible narratives and outlooks for appealing and rich encounters with the past. ISPRS–Int. Arch. Photogramm. Remote Sens. Spat. Inf. Sci. **XLII-2/W5**, 113–119 (2017)

17. Barrile, V., Bernardo, E., Bilotta, G., Fotia, A.: "Bronzi di Riace" geomatics techniques in augmented reality for cultural heritage dissemination. In: Borgogno-Mondino, E., Zamperlin, P. (eds.) ASITA 2021. CCIS, vol. 1507, pp. 195–215. Springer, Cham (2022). https://doi.org/10.1007/978-3-030-94426-1_15

18. Barrile, V., Fotia, A., Bernardo, E., Bilotta, G.: Geomatic techniques: a smart app for a smart city. In: Bevilacqua, C., Calabrò, F., Della Spina, L. (eds.) NMP 2020. SIST, vol. 178, pp. 2123–2130. Springer, Cham (2021). https://doi.org/10.1007/978-3-030-48279-4_200

19. Barrile, V., Fotia, A., Bernardo, E., Candela, G.: Geomatics techniques for submerged heritage: a mobile app for tourism. WSEAS Trans. Environ. Dev. **16**, 586–597 (2020). https://doi.org/10.37394/232015.2020.16.60

20. Barrile, V., Bilotta, G., Fotia, A., Bernardo, E.: Road extraction for emergencies from satellite imagery. In: Gervasi, O., et al. (eds.) ICCSA 2020. LNCS, vol. 12252, pp. 767–781. Springer, Cham (2020). https://doi.org/10.1007/978-3-030-58811-3_55

21. Barrile, V., Bilotta, G., Fotia, A., Bernardo, E.: Integrated GIS system for post-fire hazard assessments with remote sensing. Int. Arch. Photogramm. Remote Sens. Spatial Inf. Sci. **XLIV-3/W1-2020**, 13–20 (2021). https://doi.org/10.5194/isprs-archives-XLIV-3-W1-2020-13-2020

22. Barrile, V., Fotia, A., Bernardo, E., Bilotta, G., Modafferi, A.: Road infrastructure monitoring: an experimental geomatic integrated system. In: Gervasi, O., et al. (eds.) ICCSA 2020. LNCS, vol. 12252, pp. 634–648. Springer, Cham (2020). https://doi.org/10.1007/978-3-030-58811-3_46

23. Barrile, V., Bernardo, E., Candela, G., Bilotta, G., Modafferi, A., Fotia, A.: Road infrastructure heritage: from scan to Infrabim. WSEAS Trans. Environ. Dev. V **16**, 633–642 (2020)

24. Barrile, V., Fotia, A., Bernardo, E., Bilotta, G.: Road cadastre an innovative system to update information, from big data elaboration. In: Gervasi, O., et al. (eds.) ICCSA 2020. LNCS, vol. 12252, pp. 709–720. Springer, Cham (2020). https://doi.org/10.1007/978-3-030-58811-3_51

Virtual Reality Approach for Geoparks Fruition During SARS-Cov2 Pandemic Situation: Methodological Notes and First Results

Salvatore Praticò[1]([email]) [iD], Marco Neves[2], Angelo Merlino[3], Raimondo Tripodi[1], Peppuccio Bonomo[3], Paulo Barreira[2], and Giuseppe Modica[1] [iD]

[1] Dipartimento di Agraria, Mediterranea University of Reggio Calabria, loc. Feo di Vito snc, Reggio Calabria, Italy
salvatore.pratico@unirc.it
[2] Interact Ideas, Rua do Areeiro n.°150, 2415-315 Leiria, Portugal
[3] Madonie UNESCO Global Geopark, Corso Paolo Agliata 16, Petralia Sottana, PA, Italy

Abstract. In recent decades there has been a growing interest in geological heritage conservation and management. In 2015, UNESCO (United Nations Educational, Scientific and Cultural Organization) created the new label UNESCO Global Geoparks to identify all areas of international geological significance where landscapes and single geosites are managed with a holistic, bottom-up approach to ensure protection, sustainable management, and education activities. With the advent of the SARS-CoV2 pandemic, many countries and organisations have put lockdown and other mobility restrictions in place. During this period, the need for new methodologies for the study was perceived and the fruition of the geosites and, more in general, of geological processes and their visible signs. This paper presents the first results of an ongoing activity carried out in several geosites of different Geoparks and aimed at providing virtual-reality tools to assist in studying the geological heritage. To this end, we propose an integrated approach between different sensors to the survey of geosites, 3D modelling, and VR reconstruction. Moreover, we proposed the development of a platform for the exploitation of the obtained data.

Keywords: 3D modelling · 360 photos · Augmented Reality (AR) · Geomatics

1 Introduction

In recent decades there has been a growing interest in cultural and geological heritage conservation and management [1–8]. In 2015, the UNESCO (United Nations Educational, Scientific and Cultural Organization) created the new label UNESCO Global Geoparks (UGGp) to identify all those geographical areas of international geological significance where landscapes and single sites (i.e., geosites) are managed with a holistic, bottom-up approach to ensure protection, sustainable management and education activities [9, 10]. One of the primary purposes of a Geopark is to help transfer knowledge in geosciences and environmental sciences [11]. To achieve these objectives of the

© The Author(s), under exclusive license to Springer Nature Switzerland AG 2022
F. Calabrò et al. (Eds.): NMP 2022, LNNS 482, pp. 2093–2103, 2022.
https://doi.org/10.1007/978-3-031-06825-6_201

UGGp, a deep knowledge of geological processes and their appearance on Earth surface is needed. This can be reached through studies and researches, mainly conducted with field activities to observe the geosites and collect data.

With the advent of the pandemic situation, due to the SARS-CoV2 (severe acute respiratory syndrome coronavirus-2), lockdown and other restrictions on mobility have been put in place by many countries and organisations. During this period, the need for new methodologies for the study and the fruition of natural areas [12], especially geosites and, more in general, geological processes and their visible signs, was perceived. In this framework, a key role is played by the use of virtual reality (VR) and augmented reality (AR) techniques that allow 3D reconstruction of details of interest and remote viewing of the areas [13–16]. These techniques have been used in teaching and learning activities and have proved helpful in several subjects [17, 18].

This paper presents the first results of an ongoing activity carried out in several geosites of different Geoparks located in Italy, Croatia, Poland, Hungary and Turkey, and aimed at providing VR tools to assist in studying the geological heritage. To this end, we propose an integrated approach between different sensors like 360 cameras, RGB cameras mounted onboard unmanned aerial vehicles (UAVs) and terrestrial RGB cameras to survey geosites, 3D modelling and VR reconstruction. Moreover, we propose the development of a platform for the exploitation of the data. Within the scope of technological development for exploration in an educational environment, an application was developed for an Android mobile device that will include the possibility of exploring the various Geoparks in a virtual reality environment through exploration with 360 photos.

2   Materials and Methods

To achieve our goal, we present a method based on two main phases: i) data survey and processing and ii) VR platform implementation (see Fig. 1).

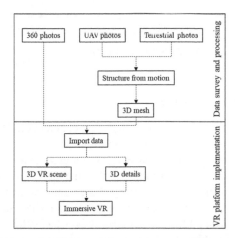

Fig. 1. Synthetic workflow of the proposed method

The first one consists of surveying geosites with various instruments, while the second of creating and implementing a VR platform to explore the data created in the previous step.

2.1   Study Sites

The proposed methodology was tested in several geosites of two different Geoparks (see Fig. 2): Madonie UGGp in Italy and Papuk UGGp in Croatia. The first one was founded in 2004, the first in Italy, and falls within a natural protected area of regional interest. It is located in Sicily, in the middle of the Mediterranean basin, covering almost 400 km^2. The second one, established in 2007, is located in the eastern part of Croatia, in Slavonia Region, encompassing the whole area of a Natural Park of national interest. It covers a surface of about 524 km^2, in the middle of the area occupied by the Pannonian Sea.

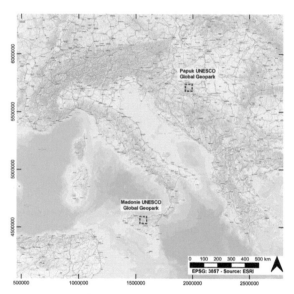

Fig. 2. Study sites geolocalisation.

2.2   Data Surveys and Processing

In order to better detect the different geomorphological characteristics of the two Geoparks and the geosites inside them, we adopted a multi-source survey approach (see Fig. 3 a1-a2-a3). We used a 360 camera to have a 3D immersive view of each geosite. Several 360 photos have been shot accordingly to the characteristics and dimensions of the site, having one photo with the entire surveyed area and one or more other photos for details. For this purpose, we used the 360 camera Ricoh Theta model SC2 (https://theta360.com/en/about/theta/sc2.html - last access 2021/12/27), the main characteristics of which are summarised in Table 1. The camera is composed of two opposite objectives that simultaneously shoot two 180 photos that are automatically merged to obtain a complete 360 view that can be shown as spherical or developed in the plane (see Fig. 3 b1). The acquired 360 photos have been used to reconstruct geosites in an immersive way inside the VR platform using VR headsets.

To have a deeper view of the geosites, details of particular interest for teaching and learning activities have been reconstructed as 3D mesh after a structure from motion (SfM) process (see Fig. 3 b2–b3).

Table 1. Main characteristics of Ricoh Theta SC2 360 camera.

Dimensions	4.52 × 13.06 cm
Weight	104 g
Image resolution	5376 × 2688 pxl
Image compression method	JPEG (Exif Ver2.3)
Video resolution	4K
Video compression method	MP4
Internal storage	14 GB
Lens focus value	2.0
Image sensors size	1/2.3

Depending on the dimension of the target to be acquired, the images were taken with an RGB camera mounted onboard a quadcopter DJI Phantom 4 Pro Plus, for bigger targets, or with a terrestrial RGB full-frame mirrorless Sony α7 II, for smaller targets or specific details. The main characteristics are synthesised in Table 2 and Table 3.

Table 2. Main characteristics of DJI Phantom 4 Pro plus.

Dimensions	35 cm (diagonal)
Weight	1388 g
Image resolution	5472 × 3648 pxl
Sensor	CMOS 1″
Focal length	8.6 mm
Ground sample distance (GSD)	2.74 cm at 100 m of flight height
Field of view (FOV)	84°

Table 3. Main characteristics of the Sony α7 II.

Dimensions	12.69 × 9.57 × 5.97 cm
Weight	556 g
Image resolution	6000 × 4000 pxl
Sensor	CMOS full-frame 35 mm
Focal length	35 mm–55 mm
Lens focus value	2.8

At the basis of the SfM process is the concept that an object can be reconstructed in 3D using a series of overlapping images taken from a different angle and covering the whole object surface [19]. Regarding UAV-based surveys, considering the typical irregular surface of geomorphological details, UAV flight was conducted entirely in manual mode, ensuring different flight heights above ground level (AGL) and camera angles. Once acquired the images, all photogrammetric processes to obtain the SfM-based 3D mesh were conducted in Pix4Dmapper (Pix4D SA, Switzerland) environment for the images taken with the UAV camera (see Fig. 3 b2) and in Agisoft Metashape [20] environment for the terrestrial ones (see Fig. 3 b3). The processing time depends on the size of the dataset and the characteristics of the hardware. For this work we used a HP-Z820 workstation with the following characteristics: CPU Intel Xeon E5–2697 v2, 64 GB RAM DDR3 1866 MHz, GPU NVIDIA K5000.The obtained 3D meshes have been put in a public repository to be recalled as hyperlinks in the platform environment.

2.3 Virtual-Reality (VR) Platform Implementation

The development of the application is structured in the first process of analysis and investigation of User Experience/User Interface (UX/UI), where the user interaction process with the application was designed in order to provide the best possible experience in these environments. App was developed for Android devices. Tools to develop Android applications are freely available and can be downloaded from the Web. In this project is used the Android Studio IDE that provides the fastest tools for building apps on every type of Android device. Android Studio is the official integrated development environment (IDE) for Google's Android operating system. It is available for download on Windows, macOS and Linux based operating systems. Concerning the data it is collected trough a 360 camera, Google VR SDK for Android that supports the creation of the 360 immersive experience has been used. The Cardboard SDK allows us to build immersive cross-platform VR experiences for Android, with essential VR features such as motion tracking, stereoscopic rendering, and user interaction.

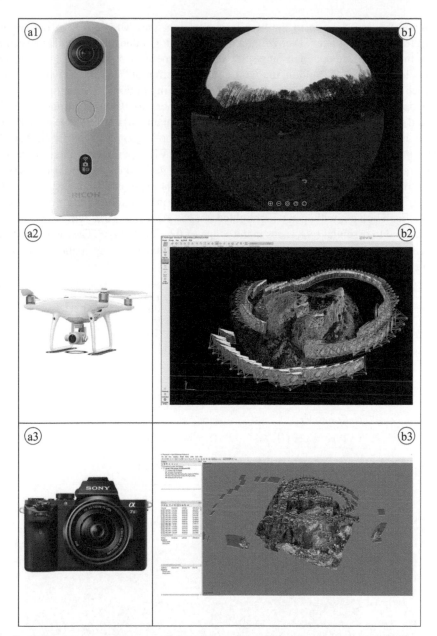

Fig. 3. Different sources of surveyed data (Ricoh Theta SC2 360 camera – a1; DJI Phantom 4 Pro plus – a2; Sony α7 II – a3), 360 photo spherical view (b1) and structure from motion (SfM) processing phases (Pix4D mapper environment - b2; Agisoft Metashape environment - b3)

When the user logs in to the mobile application, they will have two options for use: i) for navigation in a VR environment; ii) to explore educational contents, such as geocaching games, lesson plans and scientific experiments according to each Geopark. Once selecting the "VR" option, the choice between different Geoparks will be possible. For each of them, it is possible to pick the five following options: i) geomorphology of the Geopark (rocks, soil, fair chimneys, cliffs, volcanoes, fossils) and a virtual trip to the selected geosite; ii) flora of the Geopark and endemic plants iii) fauna of the Geopark and extinct animals; iv) minerals of the Geopark; v) geofoods/geo products/geo-heritage of the Geopark. For each option, will be available a 360° VR experience in three degrees of freedom (3DOF) environment. During the experience itself, the user always can search, return to the main menu and go back to the previous experience. In each of the 360 images, when explored, the user will have the option to interact with hotspots (see Fig. 4). Each hotspot will open a highlighted image (2D format) that will allow the user to be alerted to essential references in the scope of that 360° image. They will also have in each 360° experience a short questionnaire that validates the exploration carried out in line with the learning developed (Table 4).

Table 4. Main characteristics of Google Cardboard V2 headset.

Refresh rate	60 Hz or above
Field of view (FOV)	80°
Dimension of VR scene	150 × 90 × 55 mm
Weight	96 g
Controller	No
Type screen	Depends on smartphone

Fig. 4. Example of a 360 image, in planar view, with highlighted hotspots.

3   Main Surveys Results

To achieve the project's goal, 360 photos and the SfM process have been used to recreate VR geosites scenes and 3D mesh of interesting details. The number of 360 photos changes concerning the dimension of the surveyed geosite and the number of details that should be highlighted. Due to its dimension, its position without near obstacles and its distance from public roads, about 350 UAV-based images have been used to obtain the 3D mesh reconstruction of the geosite "Rocca di Sant'Otiero" (see. Figure 5a), covering a surface around 5 ha in Madonie UGGp (Italy). Concerning the details of other geosites, due to their position and dimension, limitations in using drones in foreign countries, and the project's specific objectives, 3D meshes have been reconstructed using terrestrial RGB photos. Figures 5b and 5c show two examples of terrestrial RGB-based 3D meshes. Details of columnar stones of geosite Rupnica (see Fig. 5b) and details of a magma and volcanic rocks fracture of geosite Trešnjevica (see Fig. 5c) have been reconstructed starting from images acquired in Papuk UGGp (Croatia).

As mentioned in the previous section, the developed application (still in beta version) consists of different menus that allow the users to make certain choices to personalise their experience. With the first menu (see Fig. 6a), users can decide to navigate the immersive VR environment or access other content (e.g., caching games, quizzes, learning data, etc.). If VR experience has been selected, users can choose which Geoparks want to visit (see Fig. 6b) and which aspects want to be highlighted (see Fig. 6c). Once made these choices, users can navigate in an immersive VR environment and always

Fig. 5. Examples of SfM-based 3D meshes of "Rocca di Sant'Otiero" (a), details of columnar rocks of "Rupnica" (b) and magmatic and volcanic rocks fracture of Trešnjevica (c) geosites.

have the option to search, return to the main menu and go back to the previous experience (see Fig. 6d).

Fig. 6. Main menus of the developed application: a) choice between virtual reality (VR) experiences or learning experiences; b) choice between different Geoparks; c) choice of the aspect of the Geopark to be investigated; d) immersive VR navigation menu.

4   Conclusions

In this work, we illustrated an integrated approach for making Geoparks fruition available virtually. We have shown our proposed method step by step, beginning with the surveys until the implementation of the VR application platform. The developed application represents a reasonable effort to fruition of places during restricted periods (e.g., during a pandemic situation) and for people with limited mobility who cannot easily reach places like Geoparks. Moreover, concerning teaching and learning activities, this application does not intend to replace the entire educational process, but rather a complement that validates the pedagogical interactions through specific moments of VR experimentation where students, with the proper educational framework, will reinforce their learning through moments of almost direct contact with different park structures.

This work is part of an ongoing project, so all data presented should be assumed as a first attempt to achieve our goal of providing virtual-reality tools to assist in the study of the geological heritage.

Acknowledgements. This work was funded by Erasmus + Programme 2014-2020 project "VR@Geoparks" under the registration 2020-1-IT02-KA227-SCH-095493.

References

1. Wimbledon, W., et al.: The development of a methodology for the selection of British geological sites for conservation: part 1. Mod Geol. **20**, 159–202 (1995)

2. Gray, M.: Geodiversity: Valuing and Conserving Abiotic Nature. Wiley and sons, Hoboken (2004)

3. Feuillet, T., Sourp, E.: Geomorphological heritage of the pyrenees national park (france): assessment, clustering, and promotion of geomorphosites. Geoheritage **3**(3), 151–162 (2010). https://doi.org/10.1007/s12371-010-0020-y

4. Brilha, J.: Inventory and quantitative assessment of geosites and geodiversity sites: a review. Geoheritage **8**(2), 119–134 (2015). https://doi.org/10.1007/s12371-014-0139-3

5. Reynard, E., Perret, A., Bussard, J., Grangier, L., Martin, S.: Integrated approach for the inventory and management of geomorphological heritage at the regional scale. Geoheritage **8**(1), 43–60 (2015). https://doi.org/10.1007/s12371-015-0153-0

6. Mariotto F.P., Bonali, F.L., Venturini, C.: Iceland, an open-air museum for geoheritage and earth science communication purposes. Resources **9**(2), 14 (2020). https://www.mdpi.com/2079-9276/9/2/14

7. Lasaponara, R., et al.: Spatial open data for monitoring risks and preserving archaeological areas and landscape: case studies at Kom el Shoqafa, Egypt and Shush. Iran. Sustain. **9**(4), 572 (2017)

8. Calabrò, F., Iannone, L., Pellicanò, R.: The historical and environmental heritage for the attractiveness of cities. the case of the umbertine forts of pentimele in Reggio Calabria, Italy. In: Bevilacqua, Carmelina, Calabrò, Francesco, Spina, Lucia Della (eds.) NMP 2020. SIST, vol. 178, pp. 1990–2004. Springer, Cham (2021). https://doi.org/10.1007/978-3-030-48279-4_188

9. UNESCO. https://en.unesco.org/global-geoparks

10. Pásková, M., Zelenka, J.: Sustainability management of Unesco global geoparks. Sustain. Geosci. Geotour. **2**, 44–64 (2018). https://www.scipress.com/SGG.2.44

11. de Grosbois, A.M., Eder, W.: International viewpoint and news. Environ. Geol. **55**(2), 465–466 (2008). https://doi.org/10.1007/s00254-008-1340-y

12. Di Fazio, S., Vivona, S., Veltri, A., Luzzi, G., Modica, G.: Tranquillity areas mapping : a project in sila national park: first results and importance in the covid-19 era. LaborEst **22**, 75–85 (2021)

13. Bailey, J.E., Chen, A.: The role of Virtual Globes in geoscience. Comput Geosci. **37**(1):1–2 (2011). https://linkinghub.elsevier.com/retrieve/pii/S0098300410001597

14. Mel, K., et al.: Workflows for virtual reality visualisation and navigation scenarios in earth sciences. In: Proceedings of the 5th International Conference on Geographical Information Systems Theory, Applications and Management. SCITEPRESS - Science and Technology Publications, pp. 297–304 (2019). http://www.scitepress.org/DigitalLibrary/Link.aspx?doi=10.5220/0007765302970304

15. Tibaldi, A., et al.: Real world–based immersive Virtual Reality for research, teaching and communication in volcanology. Bull. Volcanol. **82**(5), 1–12 (2020). https://doi.org/10.1007/s00445-020-01376-6

16. Fragomeni, P., Lorè, I.: VR as (in)tangible representation of cultural heritage. scientific visualization and virtual reality of the doric temple of punta stilo: interference ancient-modern. In: Bevilacqua, Carmelina, Calabrò, Francesco, Spina, Lucia Della (eds.) NMP 2020. SIST, vol. 178, pp. 1851–1861. Springer, Cham (2021). https://doi.org/10.1007/978-3-030-48279-4_175

17. Kesim, M., Ozarslan, Y.: Augmented reality in education: current technologies and the potential for education. Procedia Soc Behav Sci. **47**, 297–302 (2012). https://linkinghub.elsevier.com/retrieve/pii/S1877042812023907

18. Majeed, Z.H., Ali, H.A.: A review of augmented reality in educational applications. Int J Adv Technol Eng Explor. **7**(62), 20–27 (2020). https://www.accentsjournals.org/paperInfo.php?journalPaperId=1186

19. Westoby, M.J., Brasington, J., Glasser, N.F., Hambrey, M.J., Reynolds, J.M.: 'Structure-from-Motion' photogrammetry: a low-cost, effective tool for geoscience applications. Geomorphology **179**, 300–314 (2012). https://doi.org/10.1016/j.geomorph.2012.08.021
20. Agisoft. Agisoft Metashape User Manual, p. 187 (2021)

Geographic Information and Socio-Economic Indicators: A Reading of Recent Territorial Processes in the Test Area of Basilicata Region

Valentina Santarsiero[1,2]([⊠]) [iD], Gabriele Nolè[1], Francesco Scorza[2], and Beniamino Murgante[2]

[1] IMAA-CNR C.da Santa Loja, Zona Industriale, Tito Scalo, Italy
valentina.santarsiero@unibas.it, {Valentina.santarsiero, gabriele.nole}@imaa.cnr.it
[2] School of Engineering, University of Basilicata, Viale dell'Ateneo Lucano 10, 85100 Potenza, Italy
{francesco.scorza,beniamino.murgante}@unibas.it

Abstract. This work proposes the development of a methodology directed to analyze the potential of the territory in terms of local development and the evolution that this has undergone in the historical period between 2017 and 2021. Taking into account the social and environmental realities present, the study also developed an in-depth analysis of the evolution of the urban and rural fabric and socio-economic characteristics, paying particular attention to the multifunctional role played within the territory.

Analyses of urban and rural environments typically focus on the application of methodologies that assess quality objectives at the environmental and urban levels. Research has shown that a system of indicators can be useful in developing qualitative and quantitative descriptors of the area, investigating urban and inland areas. The first step was to formulate a methodology to measure the quality of life in these areas based on lists of objective indicators, aggregated to develop a framework to assess the level of quality and spatial development of the areas investigated. The second step was to apply this methodology to evaluate the settlements of tested area of Basilicata Region, characterized by a multiplicity of different environments that make possible the coexistence of a great variety of environmental and territorial phenomena. The main results of this research concern the opportunity to measure numerically the objective aspects that influence the development of the territory in urban and rural areas. In this way, the most critical areas to be upgraded have been highlighted in order to prepare policies congruent with the local context.

Keywords: Geographic information · Urban and Rural environment · Territorial processes

© The Author(s), under exclusive license to Springer Nature Switzerland AG 2022
F. Calabrò et al. (Eds.): NMP 2022, LNNS 482, pp. 2104–2111, 2022.
https://doi.org/10.1007/978-3-031-06825-6_202

1 Introduction

Define a territorial model represents a constant challenge to the systematization of data and spatial information able to understand the mechanisms that, at the local scale, determine the organization of the demand and, consequently, of the offer of services and equipment's [1–4].

This consists in an interpretative approach to the dynamics of settlement, territorial, infra-structural endowments and organizational models that condition, for example, territorial accessibility and that lead citizens to self-determine the residence and systematic movements according to criteria of optimization of the modalities of use of space and territory. The research for rules and criteria that define the settlement pattern finds utility in the planning of sustainable forms of territorial development. The search for rules and criteria that contribute to the definition of the settlement model is useful in the planning of sustainable forms of territorial development. This is particularly critical in the management of territories with low settlement density in which the rules and standards defined for the organization of large metropolitan aggregates lose effectiveness. Basilicata represents this criticality in the management of territories, being one of the regions with the lowest settlement density, conditioned by a delay in development, which derives from a secular infrastructural deficit. The search for rules and criteria that contribute to the definition of the settlement model is useful in the planning of sustainable forms of territorial development. The region still bases its strategic line of territorial development on the use of autochthonous resources linked to the system of diffused naturalness, to the quality of the productions of the agricultural sector, to the uniqueness of the values - historical and cultural.

The purpose of this work was to analyze the trend of territorial development in terms of public and private services present between 2017 and 2021.

The article presents a methodology to investigate the relationships between socio-economic phenomena present in the territory and its spatial distribution, developed in a Geographic Information System (GIS) environment using a free and open-source data and software (FoSS). The integration of different integrated datasets in a GIS environment, combined with the application of new territorial analysis models, represent a fundamental tool for studying and monitoring the evolution and development of the territory. The interoperability of the various territorial data and analysis models represent an important tool for the planner and the territorial government bodies, and contribute to improving the definition of plans and strategies consistent with the real needs and territorial problems [5–10].

2 Materials and Methods

2.1 Study Area

The territorial area studied includes thirty-one municipalities of Basilicata Region (Fig. 1) that, according to the SNAI classification [11, 12], all fall within inland areas. The study area is characterized by a multiplicity of different natural environments, characterized by wooded and mountainous areas, hilly areas and plains with the presence of agricultural areas, badlands and barren areas [13, 14]. This variety of environment makes possible the coexistence of a great variety of plant and animal species. The urban and rural settlements are mainly located in the hills and mountains, some are also located in the flat coastal area of the Metapontum plain. The study area can be divided into three macro areas: a Tyrrhenian mountainous area bordering the Campania region, an internal hilly area and finally an Ionian area falling partly in the Metapontum plain and partly in the territory of the Matera mountains (Fig. 1). The climate is Mediterranean, with a marked two-season regime characterized by hot and dry summers and wet and cold winters. From a geological and geomorphological point of view, the whole of Basilicata is among the regions with a high seismic and geomorphological risk, it is in fact affected by the presence of innumerable landslides [14–17].

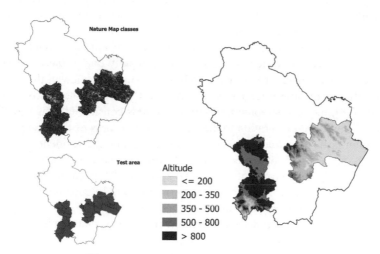

Fig. 1. Territorial framework of study area.

The main road system is represented by a few arteries that cross the whole region and from which develops a system of secondary roads that have the function of connecting the individual centers to roads of higher level, allowing access to suburban roads and agricultural land. From the demographic point of view, by the steady reduction of the regional resident population since the 50's; currently, there are about 545 thousand inhabitants. The structure of the economic system of the study area shows a prevalence of agricultural enterprises, followed by the commercial sector and the tertiary sector. The prevalence of the agricultural sector does not correspond to an occupational structure composed in terms of employees and collaborators of the companies. This denotes a system linked to individual and family businesses with a low level of industrialization of production processes.

2.2 Methodology

In this paper we consider two main informative data for the territorial framework of some regional centers chosen as test areas: the demographic structure of the resident population and the endowment of services and equipment. The framing of the demographic structure refers to ISTAT data [18], while the equipment of services and equipment derives from a work of reconnaissance and detailed mapping of the offer of public and private services, which together determine the different types of territorial equipment.

The criteria for evaluating the evolution of the development of the 31 municipalities in the period studied are analyzed through the development of indicators in a GIS environment. Numerical assessments and maps were extracted to allow an objective and quantifiable comparative analysis of their transformation between 2017 and 2021.

The first methodological approach was to reconstruct the stock of services (public and private) and equipment (economic activities, associations, etc.) in the years 2017 and 2021, with the use of open data [16, 18–22] processed in the QGIS software [23], it was possible to define a map of local services relating to the entire test area. A large part of Italian small urban centers, despite having potential for development linked to cultural tourism, suffer from serious economic and above all social hardship, above all due to the processes of depopulation. The processes of abandonment have significant effects on the landscape heritage which deteriorates more rapidly due to the absence of any maintenance practice. In order to frame a preliminary summary view of the main socio-economic variables of the area read with respect to the trends that emerge from the ISTAT census data [18], the demographic trend in the periods considered, was subsequently analyzed. A significant elaboration to describe the seriousness of the aging process of the population is the construction of indicators that frame the resident population in percentage terms, divided into three classes: Youth population (aged between 14 and 35 years), adult population (over 55 years) and aged population (over 80 years).

The purpose is to create a territorial framework that will be used for spatial processing, with the aim of integrating traditional methodologies and geostatistical approaches in the definition of the territorial socio-economic framework, that represent the basis of future development planning.

3 Results and Discussion

In this preliminary work, two main information layers were compared from the point of view of regional territorial development: the demographic structure of the settled population and the provision of services and equipment. The purpose of this preliminary work was to first create a dataset of data useful for the calculation of various socio-economic indicators, in order to frame the demographic evolution and the evolution of the stocks of public and private services.

Preliminary phase of the study was to identify the stock of services (public and private) and equipment, the data collection phase allowed the identification of approximately 3600 in 2017 and 5600 in 2021 activities and services in all the municipalities studied (Table 1). Globally there is a general increase in 2021 in all classes of services and equipment surveyed.

Table 1. Service and equipment classes in 2017 and 2021.

Services and equipment	2017	2021
Education	222	246
Commerce	768	1395
Culture, art, publishing	279	495
Health	370	508
Services	843	1342
Public services	113	135
Financial services	161	212
Safety	96	95
Sport and Free time	395	560
Tourism	388	577

Analyzing the supply of services and equipment in a territory is a parameter against which to evaluate the quality of life in a specific territory, also through comparison with reference realities. On the other hand, it can be understood as a deficit assessment, or the absence of minimum requirements for the offer of services and equipment in reference to the urban functions exercised by each territorial unit.

The demographic structure of the municipalities analyzed reflects the regional trend towards depopulation, in fact the resident population has gone from 131.425 (2017) to 125.202 (2021) inhabitants. The results of the calculation of the indicators divided by age groups for the year 2021 are very significant. The indicators measure, as a percentage, the population in the three age groups considered (youth (age 14–35), age over 55 and age over 80) out of the total resident population (Fig. 2). These values were compared to the values of the national average. The indicators have been constructed and classified into 5 value classes, with reference to the national average of southern Italy. As regards the indicator relating to the percentage of the resident population in the age group 14 - 35, the class with the value 5 represents a resident youth population rate greater than 27% until it decreases to the value 1, which represents the percentage of resident youth population less than 18%. What emerges from this first analysis is that the resident population for this age group in the areas considered is of the order of 21% - 18% class 3 and 2. With regard to the adult population over 55, the indicator measures, also in this case, the resident population in percentage terms compared to the average of southern Italy. Class 5 is attributed with a percentage of the population over 55 residents of less than 30% while class 1 represents a resident population over 55 years of greater than 40%. From the analyzes it emerges that the population residing in the test area, in the over 55 age group is greater than 40%. Similar argument was made for the resident population in the age group over 80, where in class 5 a resident population of less than 3% is represented while in class 1 a resident elderly population value greater than 10%. In this case it emerges that the percentage of the population residing in this age group in the area is between 7% and 10%.

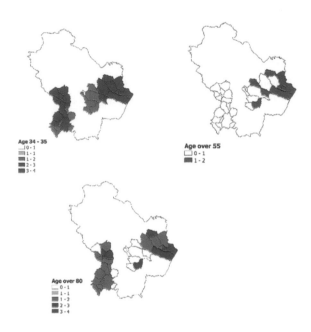

Fig. 2. Indicators of the distribution of the resident population by age group.

4   Conclusion

The situation of the studied area, and in general the entire Basilicata region, raises the need for a new multi-year regional strategic planning that allows the region to implement the public policies necessary to trigger balanced development. The results of demographic trends highlight the structural territorial weaknesses linked above all to depopulation and abandonment of smaller towns. This work represents the main elements of a first experimentation, at the municipal level, of the methodology described which achieves preliminary results. Further studies and analyzes are needed to provide a socio-economic geographical picture of the area considered. Future developments concern the implementation of new indicators regarding the accessibility of these areas to the nearest school complexes and also to medical health centers. A further object of analysis will be to study the degree of infrastructure of the road and railway network and assess its exposure to hydrogeological risk.

References

1. Danese, M., Nolè, G., Murgante, B.: Visual impact assessment in urban planning. Stud. Comput. Intell. **176**, 133–146 (2009). https://doi.org/10.1007/978-3-540-89930-3_8
2. Murgante, B., Borruso, G., Lapucci, A.: Geocomputation and urban planning. Stud. Comput. Intell. **176**, 1–17 (2009). https://doi.org/10.1007/978-3-540-89930-3_1

3. Fortunato, G., Scorza, F., Murgante, B.: Hybrid oriented sustainable urban development: a pattern of low-carbon access to schools in the city of Potenza. In: Gervasi, O., et al. (eds.) ICCSA 2020. LNCS, vol. 12255, pp. 193–205. Springer, Cham (2020). https://doi.org/10. 1007/978-3-030-58820-5_15

4. Scorza, F., Saganeiti, L., Pilogallo, A., Murgante, B.: Ghost planning: the inefficiency of energy sector policies in a low population density region. Archivio di Studi Urbani e regionali **127**, 34–55 (2020). https://doi.org/10.3280/ASUR2020-127-S1003

5. Scorza, F., Casas, G.L., Murgante, B.: Overcoming interoperability weaknesses in e-government processes: organizing and sharing knowledge in regional development programs using ontologies. In: Lytras, M.D., Ordonez de Pablos, P., Ziderman, A., Roulstone, A., Maurer, H., Imber, J.B. (eds.) WSKS 2010. CCIS, vol. 112, pp. 243–253. Springer, Heidelberg (2010). https://doi.org/10.1007/978-3-642-16324-1_26

6. Las Casas, G., Scorza, F., Murgante, B.: New urban agenda and open challenges for urban and regional planning. In: Calabrò, F., Della Spina, L., Bevilacqua, C. (eds.) ISHT 2018. SIST, vol. 100, pp. 282–288. Springer, Cham (2019). https://doi.org/10.1007/978-3-319-92099-3_33

7. Pontrandolfi, P., Dastoli, P.S.: Comparing impact evaluation evidence of eu and local development policies with new urban agenda themes: the agri valley case in Basilicata (Italy). Sustainability **13**(16), 9376 (2021). https://doi.org/10.3390/su13169376

8. De Toni, A., Vizzarri, M., Di Febbraro, M., Lasserre, B., Noguera, J., Di Martino, P.: aligning inner peripheries with rural development in italy: territorial evidence to support policy contextualization. Land Use Policy **100**, 104899 (2021). https://doi.org/10.1016/J.LANDUS EPOL.2020.104899

9. Saganeiti, L., Pilogallo, A., Scorza, F., Mussuto, G., Murgante, B.: Spatial indicators to evaluate urban fragmentation in basilicata region. In: Gervasi, O., et al. (eds.) ICCSA 2018. LNCS, vol. 10964, pp. 100–112. Springer, Cham (2018). https://doi.org/10.1007/978-3-319-95174-4_8

10. Murgante, B., Borruso, G., Lapucci, A.: Sustainable development: concepts and methods for its application in urban and environmental planning. Stud. Comput. Intell. **348**, 1–15 (2011). https://doi.org/10.1007/978-3-642-19733-8_1

11. Ministro per il Sud e la Coesione territoriale - Strategia Nazionale Aree Interne. https:// www.ministroperilsud.gov.it/it/approfondimenti/aree-interne/strategia-nazionale-aree-int erne/, Accessed 03 Feb 2022

12. Strategia nazionale per le Aree interne: definizione, obiettivi, strumenti e governance* *Documento tecnico collegato alla bozza di Accordo di Partenariato trasmessa alla CE il 9 dicembre 2013

13. Carta della Natura—Italiano. https://www.isprambiente.gov.it/it/servizi/sistema-carta-della-natura, Accessed 28 Dec 2021

14. RSDI. https://rsdi.regione.basilicata.it/, Accessed 24 Nov 2021

15. IFFI - Inventario dei fenomeni franosi in Italia—Italiano. https://www.isprambiente.gov.it/it/progetti/cartella-progetti-in-corso/suolo-e-territorio-1/iffi-inventario-dei-fenomeni-franosi-in-italia, Accessed 28 Dec 2021

16. Home. http://www.distrettoappenninomeridionale.it/, Accessed 03 Feb 2022

17. Bentivenga, M., Grimaldi, S., Palladino, G.: Caratteri geomorfologici della instabilità del versante sinistro del fiume Basento interessato dalla grande frana di Brindisi di Montagna Scalo (Potenza, Basilicata). G. di Geol. Appl. **4**, 123–130 (2006). https://doi.org/10.1474/GGA.2006-04.0-16.0144

18. Popolazione residente al 1° gennaio : Basilicata. http://dati.istat.it/Index.aspx?QueryId= 18564, Accessed 23 Nov 2021

19. Portale Unico dei Dati della Scuola I MIUR. https://dati.istruzione.it/opendata/, Accessed 03 Feb 2022

20. Open Data - Dati - Posti letto per struttura ospedaliera. https://www.dati.salute.gov.it/dati/dettaglioDataset.jsp?menu=dati&idPag=18, Accessed 03 Feb 2022
21. Catalogo Geodati I RSDI. https://rsdi.regione.basilicata.it/catalogo-geodati/, Accessed 03 Feb 2022
22. Welcome   page.   https://registroimprese.infocamere.it/nmov/welcome.jsp?CdC=FR, Accessed 03 Feb 2022
23. https://qgis.org/it/site/, Accessed 24 Nov 2021

Methods and Tools for a Participatory Local Development Strategy

Priscilla Sofia Dastoli[1](✉) ⓘ and Piergiuseppe Pontrandolfi[2] ⓘ

[1] School of Engineering, Laboratory of Urban and Regional Systems Engineering, University of Basilicata, Viale dell'Ateneo Lucano, 10, 85100 Potenza, Italy
priscillasofia.dastoli@unibas.it
[2] DiCEM, Architecture, University of Basilicata, Via Lanera, 20, 75100 Matera, Italy
piergiuseppe.pontrandolfi@unibas.it

Abstract. The paper is based on the activities promoted and developed within the project "Rehabilitating Countries. Operational Strategies for the Valorisation and Resilience of Inner Areas" (RI.P.R.O.VA.RE). It describes the first results of the activities to define a local development strategy based on participatory practices and processes, in which institutional and non-institutional actors of the Media Val d'Agri territory in Basilicata (Italy) were involved. The so-called inland areas, such as the one under investigation, are involved in a twofold challenge: on the one hand, to comply with the issues common to the marginal territorial realities of our country (accelerated depopulation, poor infrastructure network and public transport, poor accessibility to essential services), and on the other hand, to cope with the consequences of the recent pandemic and its impacts.

The participatory process approach refers to a combination of traditional and innovative tools, from the classic interview to online questionnaires, from SWOT analysis to the AHP (Analytic Hierarchy Process) method, from Living Lab to Geodesign, up to completing the phases of the Logical Framework Approach (LFA) hoping for future implementation by local actors. Therefore, a so-called "engaged research" has been launched, which - starting from the involvement of stakeholders - tries to promote a more resilient and responsible territorial development process, based on the Sustainable Development Goals (SDGs) and that can be financed by the excellent public investments that are involving Italy and Europe in this historical moment.

Keywords: Inland areas · Participatory urban planning · Local development strategy

1 Introduction

The project "Rehabilitating Countries. Operational Strategies for the Valorisation and Resilience of Inland Areas" (RI.P.R.O.VA.RE)[1] [1] is part of the broader context of

---

[1] The Departments involved are the Department of Architecture and Industrial Design, University of Campania Luigi Vanvitelli (Lead Partner), the Department of Civil Engineering, University of Salerno, and the Department of European and Mediterranean Cultures, University of Basilicata (Partners).

© The Author(s), under exclusive license to Springer Nature Switzerland AG 2022
F. Calabrò et al. (Eds.): NMP 2022, LNNS 482, pp. 2112–2121, 2022.
https://doi.org/10.1007/978-3-031-06825-6_203

inland areas which, thanks to the National Strategy for Inland Areas (SNAI) [2–5], have been re-evaluated in terms of potential and development opportunities [6–8]. The SNAI is an innovative national policy for territorial development and cohesion to counter Italian inland areas' marginalization and demographic decline. The total national resources made available amount to almost 600 million euro, in addition to the allocations coming from the Operational Programs of the SIE Funds and other public and private funds, to face the pursuit of social cohesion objectives aimed at slowing down and reversing the depopulation phenomena of the Inner Areas [9, 10].

The RI.P.R.O.VA.RE project identified two regions of Southern Italy, Campania, and Basilicata, as focus areas. It was structured around three research objectives, all aimed at providing support to policies and strategies for the "re-centralization" of inland areas:

1. Redefine the current geography of inland areas based on an integrated approach, capable of adding to the parameters already used by the SNAI parameters capable of describing more effectively the fragility and potential of these territories;
2. Develop operational tools to understand, on the one hand, the main pressure factors that hinder the development of inland areas and, in some cases, threaten their very survival (threats); on the other hand, the characteristics of these systems that determine their greater or lesser capacity to respond to these threats (resilience), in support of the definition of future development policies;
3. To outline, through co-design processes and concerning selected focus areas, Integrated Strategies and pilot projects able to act on the resilience characteristics of the systems under study, combining, for example, measures aimed at risk reduction and actions oriented to regenerate and enhance the potential, in terms of natural and cultural resources and productive capacities of these territories and integrating, in this way, the themes of the SNAI with the priorities of the National Strategy for Sustainable Development (SNSvS) [11].

The paper focuses on the methodology, the activities carried out, and the results that were obtained through the achievement of the last objective of the project, which cannot disregard the analyses and evaluations that were carried out in the two previous phases in the restricted territory composed of three pilot areas (Agri, Ufita, and Matese), for a total of about 60 municipalities. Further checks on the resilience values of the municipalities concerned, as well as the analysis of the single indicators that contributed to composing the Resilience Matrix, led to the selection of a restricted set of municipalities that, in the current case, corresponds to a group of six municipalities in the Medio Agri area, in Basilicata.

2   The Context of the Middle Agri Valley

The Medio Agri area covers the central sector of the basin of the Agri river, one of the five rivers that cross the Basilicata region and flows into the Ionian Sea. The area has a mainly mountain morphology in the western part, with the Monte Raparo Site of Community Interest (SCI); most of the remaining area has a hilly morphology, with sandy and conglomeratic hills, which characterize it for its high hydrogeological risk.

According to the SNAI classification, the six municipalities that are part of the survey group (Gallicchio, Missanello, Roccanova, San Chirico Raparo, San Martino d'Agri, and Sant'Arcangelo) fall into the class F - ultra-peripheral, i.e., those municipalities of the internal areas that are more than 75 min away from a pole that has at the same time a complete upper secondary school offer, at least one hospital with a level I d.e.a. (Emergency and Acceptance Department) and at least a silver railway station. An evident migratory phenomenon in the sample area leads to progressive depopulation. In the last decade, almost one thousand units have left the area; about 8% of the population currently stands at 10.634 inhabitants [12].

In describing the knowledge framework, a section was reserved for studying the spatial distribution of current policies and project impacts, both in terms of total quantification of public investment and impact evaluation [13, 14]. It was found that the municipalities of the Medio Agri area, except Sant'Arcangelo, were less able to apply for and carry out projects for Community and national funding. These municipalities do not have an adapted structure to play a role within the development programs. For this reason, it is believed that the development program itself should adapt the structures of the weaker territories to be competitive as the others.

The previous activities of the project focused on the elaboration of 84 indicators that, combined adequately in a resilience matrix, could express a resilience value for each municipality; in a range from 120 to 185, in the municipalities of interest, there are low (133, 137), medium (146, 148, 156) and high values only for Sant'Arcangelo (166). The latter municipality (6,267 inhabitants) plays a role in polarizing services and population for the whole area.

Mainly to overcome the lack of services, which compounds the marginal character of the area, the Union of Municipalities [15] of Medio Agri (Missanello, Roccanova, San Chirico Raparo, and Sant'Arcangelo) was established in 2017, recently expanded with the inclusion of the municipalities of Gallicchio and Armento. This Union aims to face up jointly to the difficulties affecting the area, starting with accessibility to essential services (education, health, transport). The setting up of the Union of Municipalities affected the choice of the area to be researched because it is believed that the smaller centers should join institutional forms of association between municipalities, both to ensure sustainable management of services and functions and to guarantee more opportunities for citizens.

The Medio Agri area has significant potential, especially in cultural and natural heritage, with a high ecological value of the ecosystems [16, 17]. In particular, the area is affected by the perimeter of the Lucano Val d'Agri-Lagonegrese Apennines National Park, by a ZSC 'Murge di S. Oronzo', by a ZPS 'Lucano Apennines, Agri Valley, Monte Sirino, Monte Raparo' and by two SIC 'Lago Pertusillo' and 'Monte Raparo'.

3 Methodology

The logical ordering procedure of the Logical Framework Approach (LFA), which is part of Project Cycle Management (PCM), refers to the cyclical nature of planning and is organized into an Analysis and a Synthesis phase [18]. At first, it involves the organization and planning of all activities; each context is different from the others and requires a specific strategic plan. A second activity focuses on evaluating the context through an internal and external diagnosis, which is usually carried out through the technique of SWOT analysis. In the third part, the most important decisions are taken; in fact, the diagnosis developed makes it possible to define the strategic objectives as strong ideas to form the basis of the intervention plan and from which the corresponding strategic lines to be adopted can be deduced. In this phase, the problem tree and the objective tree techniques are prepared to organize problems from causes to effects and objectives from means to ends. The last activity, which is part of the Synthesis phase, concerns the detailed organization of the objectives for each strategic theme and the identification of the actions that will need to be implemented to achieve them [19]. In this last phase of the co-design activities, the Geodesign platform [20] was used as a tool to support the identification of actions.

Geodesign can be defined as the integration process of methods, techniques, and tools of Geo spatial information sciences to support the design and planning of physical development Design. It can be described as a multidisciplinary collaboration with direct interaction between design professionals, spatially oriented scientists, and local people, using available information technologies [21].

3.1 Preliminary Actions

After defining the area of the middle Val d'Agri where the methodology was to be applied, the first step was to involve the municipal administrations to explain the purpose of the project and ask for their collaboration in the interlocution with the different actors the community. In addition, a questionnaire was submitted to some administration members concerning the problems, characteristics, projects, and traditional features of their municipality to verify and extend the preliminary knowledge acquired from other sources. The University of Basilicata research group made up of planners, architects, and anthropologists was able to refer to these latter figures for ethnographic research in the field, capable of involving a wider audience of citizens willing to participate in the subsequent phases of the project.

To involve also young stakeholders, it was decided to start a dialogue with the Sant'Arcangelo-based Higher Education Institute, IIS "Carlo Levi." In this respect, an agreement was signed. The research group was allowed to involve six fifth classes, for about 120 students, in the fields of Science, Applied Sciences, Linguistics, and Administration-Finance-Marketing.

To avoid crowds, in compliance with anti-Covid regulations, the presentation of the project to the students and their involvement took place purely at a distance; in the first instance, a questionnaire was submitted as a "google form," from which needs, perspectives, and judgments on the various sectors of the quality of life could emerge.

Further preliminary actions concerning the organization and involvement of stakeholders led to the drafting of a calendar of activities for the students of the IIS "Carlo Levi" and a calendar of activities for the citizens of the six municipalities, whose involvement was structured in the form of a Living Lab. The Living Lab activities were organized in six meetings, one for each municipality, where citizens, enterprises, research centers, and institutions could meet and exchange views.

3.2 Context Evaluation

The study area has been the subject of advanced, in-depth studies, mainly carried out by the lead research group of the University of Campania 'Luigi Vanvitelli':

– In the two focus areas, the Geographical Atlas was implemented, representing the first helpful tool for understanding the internal areas of Campania and Basilicata. In the digital version [1], it is possible to view information on the six themes addressed in the research for all 575 municipalities (446 out of 550 in the Campania region and 129 out of 131 in the Basilicata region). The Geographical Atlas is structured in six sections (Geographies of Contraction, Geographies of Fragility, Geographies of Marginalisation, Geographies of Quality, Geographies of Innovation, and Geographies of Relationships) from which the relevant maps and data can be accessed. The Atlas is designed to be implemented over time with the contribution of researchers, administrators, and citizens. The database structure that constitutes the Atlas of Geographies is built from a selection of the primary data available on the currently existing public and private national platforms and aims to provide a cognitive framework of the municipalities of Campania and Basilicata [22].
– In the three pilot areas (Ufita, Agri, and Matese), for a total of about 60 municipalities, an attempt was made to calculate the value of resilience; this operation involved the implementation of a two-level resilience matrix, with the search for indicators covering the physical-functional, economic-productive, natural and socio-institutional sub-sectors.
– In the three restricted study areas, which in the case of the Agri sample area corresponds to the middle Val d'Agri, a detailed analysis was carried out in addition to the previous ones. The information for this cognitive level comes from interviews and questionnaires to institutional subjects and local stakeholders, municipal urban planning instruments, and the field observation carried out by the Lucanian research group.

The three analysis levels provided a clear and detailed knowledge of the area in question. The research team has systematically organized the information with a context assessment through an internal and external diagnosis using the SWOT analysis technique.

3.3 Problems and Objectives of the *Medio Agri* Area

The stakeholders' active participation was encouraged to bring out the territory's problems and critical issues first. The problem tree technique was used in the schools and the Living Lab experience. In particular, the problems indicated or deduced are acquired with the problem tree technique. The main problem is identified, and all issues are organized, dividing them into ones that cause the main problem and the effects of the main problem. In the study area, due to the careful assessment of the context, the issues were placed into five themes:

1. *Quality of life*: under the quality of life theme, issues relating to mobility, social and welfare services, school services, sports facilities, and cultural facilities of various kinds were discussed.
2. *Economic activities and employment*: employment is an essential issue on which various phenomena depend, including migration and the consequent depopulation of the area.
3. *Protection and enhancement of the natural heritage risks*: inland areas are characterized by various elements of marginality, but there is a widespread natural heritage that must be protected and enhanced.
4. *Cultural heritage and identity*: the topic investigated the main aspects of cultural heritage in terms of protection, valorization, fruition, and, above all, the ability and the will to hand down the inestimable value of this heritage to future generations.
5. *Hospitality*: this topic dealt with a critical issue for inland areas: the depopulation phenomenon and the need to support initiatives for a widespread reception of new residents. To remedy the abandonment of the territory, virtuous mechanisms must be set in motion so that those who live in these places remain and can welcome the new population so that services, shops, and other commercial activities remain active. There is a permanent presence in the territory.

Stakeholders provided input on numerous problems according to their interests and experience; the research team reorganized them to compose thematic problem trees. The logic of problem-cause - main problem - problem-effect was clear.

The same procedure was carried out for the Tree of objectives, which is to put the tree of problems in a positive way; it is necessary to express the problems in the form of objectives where the main problem becomes the main objective, the cause-problems become the means to reach the main objective. The effect-problems become the objectives to be achieved.

Within urban planning, the identification of objectives plays a decisive role. Of the objectives, which have a different impact on the achievement of the overall objective, it is possible to identify which are the priority ones, and thus which have the most influence if achieved. Among the evaluation techniques that can be useful for this purpose, Multi-Criteria Analysis (MCA) tries to rationalize the policy maker's choice process by optimizing a vector of several criteria, weighted according to the priorities declared by the policymaker.

The AMC represents a broad family of techniques whose structure can depend on whether the evaluation criteria used are compensatory (rebalancing) or not. Among the techniques with a compensatory character is the Analytic Hierarchy Process (AHP).

The AHP method, developed by Thomas L. Saaty in the late 1970s, has become internationally popular thanks to its (relative) simplicity of application; it is based on an analytical and synthetic approach that facilitates communication between the actors involved in the decision-making process [23]. The Hierarchical Analysis procedure is based on binary comparisons (pairwise comparison), which allows the decision-maker to make judgments based on comparing two elements at a time. In general, this procedure is developed through three logical operations:

- Hierarchical structuring: The problem to be assessed is structured hierarchically, with the objectives at the highest level and the criteria and alternatives at successive levels;
- Comparative judgment: all elements of each level are compared in pairs according to each element of the next level;
- Comparative judgment: all elements of each level are compared in pairs according to each element of the next level.

The Living Lab participants tried to find priority objectives through the AHP technique, starting from 11 summary objectives that emerged from the objective trees, which are listed below according to the final order resulting from the negotiation and sharing process:

1. Stopping the territory's abandonment by current residents and encouraging the migrants' return;
2. Promoting more progressive forms of governance, more widespread and permanent participation of citizens in decision-making processes;
3. Building an efficient network of essential services according to a grid organization of the primary services;
4. Promoting initiatives to raise awareness and improve approaches to local development issues;
5. Enhancing the natural and cultural heritage by promoting the development of tourism;
6. Reinforcing the economic system with interventions in the leading agricultural and craft sectors, as well as in services;
7. Making the local public transport system efficient to promote exchanges and relations between the municipalities;
8. Creating the necessary conditions for the area to be attractive to non-residents and groups;
9. Improving the existing infrastructure in safety and practicability;
10. Promote the physical renovation of existing settlements;
11. Promote participatory and shared planning to end oil extraction. The priority objectives were the guidelines to identify the actions to be developed.

3.4 The Setting of Actions to Increase the Resilience of the Territory

The Logical Framework Approach (LFA) procedure requires the elaboration of the Logframe Matrix in this phase. The conclusions of the analysis phase are summarised, and the strategy's general objectives and specific objectives, actions to be undertaken, products, and economic resources are indicated. However, within the Living Lab of Medio Agri, whose participants belonged to different professional fields, it was decided to use the tool of Geodesign [20, 24] (through the online platform of Geodesign Hub) to focus the attention on mapping the interventions over the whole area of Medio Agri and to make concrete the ideas expressed in the objectives description phase. A simplified Geodesign approach was used in this case, as the activities were concentrated in two days. In particular, all participants were asked to locate on a map the interventions or policies that could realize the 11 objectives of the previous phase. Each intervention or policy had to have a title, a location, a description, an estimated total amount and had to belong to one of the seven systems into which the context analysis was summarised (agriculture, natural heritage, industry-trade, cultural heritage, institutional services, infrastructure and mobility, reception) (Fig. 1).

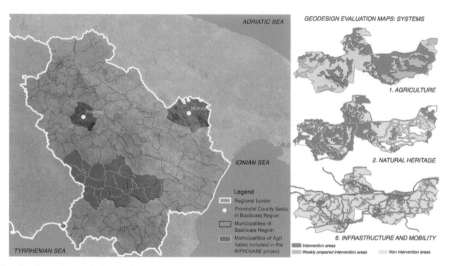

Fig. 1. Overview of the study area and assessment maps by systems to define interventions on Geodesign.

In this phase, 86 actions in the seven systems were located and described, divided into interventions and policies. On the second day, the participants were divided into two groups, each of which selected useful actions to build a strategy shared by the participants. Afterward, the two groups were asked to negotiate the two scenarios until a single shared strategic development scenario was reached.

Geodesign proved to be a useful tool to accompany a negotiation process between the stakeholders, in a very short period, with the participants' achievement of a shared development scenario.

4   Results and Conclusions

The Living Lab experience in the middle of Val d'Agri was able to take place in the presence during a lull in the pandemic between July and December 2021. The research group's commitment was constant and of great support in all work phases to guarantee the success of the activities in the field; the synthesis, elaboration, and systematization activity is still in progress. It is possible to discuss the first results, especially from a methodological point of view, of approaches and tools to set up a participatory local development strategy.

Currently being studied and clarified, the development strategy is based on proposals for action for the seven themes identified. The following is a summary by theme:

1. Agriculture system (Objectives 4-5-6-8-9-11): to promote the revitalization of the agricultural economy through tax exemption for labor; to promote local production and cultivation (olive groves, vineyards, dairy products, hazelnut groves, poultry, cattle, and pig farms) and to guarantee their quality, including through associations.
2. Natural heritage system (Objectives 4-5-7-8): recovery and enhancement of the secondary road system connecting the centers to guarantee an attractive and enhanced tourist itinerary; recovery and maintenance of nature trails and camper van areas; promotion of the use of sustainable means of transport (bike-sharing, eco-car-sharing).
3. Industry-Trade system (Objectives 4-6-9-11): strengthening the production of typical products and the protection of handicraft products;
4. Cultural heritage system (Goals 1-2-5-8-10): recovery and systematization of the territory's main cultural assets (Sant'Angelo Abbey, Baronial Palace, historical centers); inter-library project between the six municipalities.
5. Institutional services system (Objectives 2-3-5-6-8): promoting and guaranteeing health care, especially for the elderly; improving the management of sports facilities and ensuring their use by the entire district; innovating the education system and providing adequate training courses.
6. Infrastructure and mobility system (Goals 2-3-4-6-7): implement an innovative public road transport system and adapt it to the population's needs (shuttle bus, on-call service, social taxi, eco-car-sharing, bike sharing); improve and secure specific critical points of the road infrastructure.
7. Reception system (Goals 1-3-4-8-10): encourage and systematize reception activities for the elderly, migrants, and tourists.

The broad participation in the Living Lab created an open innovation environment, thanks to the concerted efforts of public bodies, businesses, universities, research centers, and citizens' groups. The active involvement of the end-users allowed the realization of co-creation paths of new services, products, and social infrastructures (based on the needs of the end-users) that could be tested in a circumscribed geographical context and in a defined time frame. The Union of Municipalities of the Medio Agri represents the ideal institutional entity that can benefit from the outcomes of the participatory local development strategy to implement it in the short term, concerning the exceptional public investments that concern the country.

References

1. RIPROVARE. https://www.riprovare.it/. Accessed 20 Dec 2021
2. Monaco, F.: Il ruolo dei Comuni ed il requisito associativo nella strategia nazionale "aree interne" (Snai). Agriregionieuropa. 12, n° 45 (2016)
3. Punziano, G.: Health, mobility, education: Strategies for inner areas. Sci. Reg. **18**, 65–92 (2019). https://doi.org/10.14650/92353
4. Di Giusy, P., Laura, S.: Toward an Italian National Strategy for Inner Areas 2.0, as an opportunity of institutional learning. Lessons from an action-research process (2020). https://doi.org/10.3280/ASUR2020-129003

5. Cardillo, G., Fusco, C., Mucci, M.N., Occhino, T., Picucci, A., Xilo, G.: Associazionismo e attuazione. I comuni alla prova della realizzazione della Strategia per le Aree Interne (2021)
6. Mallamace, S., Calabrò, F., Meduri, T., Tramontana, C.: Unused real estate and enhancement of historic centers: legislative instruments and procedural ideas. In: Calabrò, F., Della Spina, L., Bevilacqua, C. (eds.) ISHT 2018. SIST, vol. 101, pp. 464–474. Springer, Cham (2019). https://doi.org/10.1007/978-3-319-92102-0_49
7. Calabrò, F.: Integrated programming for the enhancement of minor historical centres. The SOSTEC model for the verification of the economic feasibility for the enhancement of unused public buildings. ArcHistoR. **13**(7), 1509–1523 (2020)
8. Oteri, A.M., Scamardì, G. (eds.): Un paese ci vuole. Studi e prospettive per i centri abbandonati e in via di spopolamento (2020)
9. Agenzia per la Coesione Territoriale. https://www.agenziacoesione.gov.it/strategia-nazion ale-aree-interne/. Accessed 23 Dec 2021
10. Di Fazio, S., Modica, G.: Trasformazione del paesaggio, sistemi insediativi e borghi rurali. ArcHistoR **7** (2020). https://doi.org/10.14633/AHR232
11. Galderisi, A., Fiore, P., Pontrandolfi, P.: Strategie Operative per la valorizzazione e la resilienza delle Aree Interne: il progetto RI.P.R.O.VA.RE. BDC. Boll. Del Cent. Calza Bini. **20**, 297–316 (2020). https://doi.org/10.6092/2284-4732/7557
12. ISTAT. http://dati.istat.it/Index.aspx?QueryId=18564. Accessed 22 Dec 2021
13. Pontrandolfi, P., Dastoli, P.S.: Comparing impact evaluation evidence of EU and local development policies with New Urban Agenda themes: the Agri Valley case in Basilicata, Italy (2021)
14. Nolè, G., Murgante, B., Calamita, G., Lanorte, A., Lasaponara, R.: Evaluation of urban sprawl from space using open source technologies. Ecol. Inform. **26**, 151–161 (2015). https://doi.org/10.1016/j.ecoinf.2014.05.005
15. Camera dei deputati. https://www.camera.it/temiap/documentazione/temi/pdf/1105809.pdf?_1555520990223. Accessed 26 Dec 2021
16. Saganeiti, L., Pilogallo, A., Faruolo, G., Scorza, F., Murgante, B.: Territorial fragmentation and renewable energy source plants: which relationship? Sustainability **12**, 1828 (2020). https://doi.org/10.3390/SU12051828
17. Las Casas, G., Murgante, B., Scorza, F.: Regional local development strategies benefiting from open data and open tools and an outlook on the renewable energy sources contribution. In: Papa, R., Fistola, R. (eds.) Smart Energy in the Smart City. GET, pp. 275–290. Springer, Cham (2016). https://doi.org/10.1007/978-3-319-31157-9_14
18. Las Casas, G., Sansone, A.: Un approccio rinnovato alla razionalità nel piano. In: Politiche e strumenti per il recupero urbano. Edicomedizioni, Monfalcone (GO) (2004)
19. Dastoli, P.S., Pontrandolfi, P.: strategic guidelines to increase the resilience of inland areas: the case of the Alta Val d'Agri (Basilicata-Italy). In: Gervasi, O., et al. (eds.) ICCSA 2021. LNCS, vol. 12958, pp. 119–130. Springer, Cham (2021). https://doi.org/10.1007/978-3-030-87016-4_9
20. Steinitz, C., Campagna, M.: Un Framework per il Geodesign: Trasformare la Geografia con il Progetto (2017)
21. Nyerges, T., et al.: Geodesign dynamics for sustainable urban watershed development. Sustain. Cities Soc. **25**, 13–24 (2016). https://doi.org/10.1016/j.scs.2016.04.016
22. Riprovare. https://www.riprovare.it/geografie-mappe.html. Accessed 24 Dec 2021
23. Stanghellini, S.: Valutazione multicriteriale. Corso di valutazione economica del progetto. Clamarch. A.A. 2013/2014
24. Scorza, F.: Training decision-makers: GEODESIGN workshop paving the way for new urban agenda. In: Gervasi, O., et al. (eds.) ICCSA 2020. LNCS, vol. 12252, pp. 310–316. Springer, Cham (2020). https://doi.org/10.1007/978-3-030-58811-3_22

On the Use of Big Earth Data in the Copernicus Era for the Investigation and the Preservation of the Human Past

Rosa Lasaponara[1](✉), Carmen Fattore[1](✉), Nicodemo Abate[2], and Nicola Masini[2]

[1] National Research Council (CNR) - Institute of Methodologies for Environmental Analysis, Contrada S. Loja - Zona Industriale C.P. 27, 85050 Tito Scalo, PZ, Italy
{rosa.lasaponara,carmen.fattore}@imaa.cnr.it
[2] National Research Council (CNR) - Institute of Heritage Science, Contrada S. Loja-Zona Industriale, C.P. 27, 85050 Tito Scalo, Italy

Abstract. Earth Observation (EO) Big data have emerged in the past few years providing opportunities to improve and/or enable research and decision support applications with unprecedented value for digital CH and archaeology.

The currently available digital data, tools and services with particular reference to Copernicus initiatives make possible to characterize and understand the state of conservation of CH and opened up a frontier of possibilities for the discovery of archaeological sites from above also using data available free of charge. In this paper an overview of the state of the art in the field of EO Big data for CH is brief summarized.

Keywords: Cultural heritage · Archaeology from space · Copernicus big data · Sentinel data · Optical images · Radar data · Environmental risk · Discovering

1 Introduction

In the recent decades, the availability of Earth Observation technologies (from satellite, aerial and ground) for archaeology is stepping into a golden age characterized by an increasing growth of both classical and emerging multidisciplinary methodologies, addressed to the study and documentation of the human past [1–4]. The main critical, challenging aspect of the use of EO in archaeology is a lack of correspondence between the great amount of data and information from diverse technologies (satellite, aerial, ground RS) and effective methods to extract information linked to the study of the human past [5]. So that, today, archaeologists have the possibility and opportunity to use an ensemble of diverse technologies, which however require, as preliminary steps, the setting up of ad hoc data processing and integration methodologies for archaeological investigations, analysis and interpretation. All of these aspects must to be tackled by the scientific community in conjunction with the "end uses needs" to ensure an effective and reliable applicability.

The use of space technologies, big data, artificial intelligence (AI) for the study of the human past highlights the multi-trans and inter-disciplinary scientific and technical

© The Author(s), under exclusive license to Springer Nature Switzerland AG 2022
F. Calabrò et al. (Eds.): NMP 2022, LNNS 482, pp. 2122–2131, 2022.
https://doi.org/10.1007/978-3-031-06825-6_204

aspects not only for the novel concepts and approaches proposed, but also for the development between and across diverse disciplines. EO data are today also available free of charge, as in the case of Copernicus initiative [6], which can suitably support new operational applications but pose several critical issues, as those linked to data processing and interpretation, for transforming data into useful information. Satellite Copernicus data can be free download from the ESA web site Copernicus Open Data Hub) since 2014 where open software tools are also available for the data processing SNAP [7, 13]. Nevertheless, it must be considered that all the satellite Copernicus are big data and this means that for example for a square of 100 km per 100 km Sentinel-2, is around 600 to 800 megabytes. Therefore, analyses on extended areas or over time of the same areas can be prohibitive. For these reasons, in 2017, EC funded the set-up of cloud and edge computing to make available an open European cloud to deploy applications and data processing centralized with no need to transfer and duplicate petabytes of data. So that EC and ESA supported the developments of the so-called Data and Information Access Service (DIAS), in service since 2018 (See Fig. 1).

Fig. 1. Data and information access services. They are five cloud-based platforms funded by EU to provide centralised access to Copernicus data and information, as well as to processing tools, thus facilitating and standardizing access to data [8].

DIAS platforms provide the "foundations" on which build applications exploiting Copernicus data, along with other tools developed ad hoc to enable a data manipulation easier also for user without technical expertise of the earth observation domain Sentinel Hub initiative [9], Data Cube Facility Service which provides fast access to a considerable amount of EO information in order to establish a "bridge from Space to Applications" [10]. Moreover, additional open tools are also available in Google Earth Engine [11].

2 Semantic Technologies for Big Data: Volume, Velocity, Variety, Veracity and V's

Big data are generally characterized using the so-called V's to capture their complex nature, that is generally not an absolute "categorization" but strongly linked and depending on a given application:

- Volume – amount of data
- Velocity – generation, or analyzed
- Variety – differences in data sources
- Veracity – uncertainty of data
- Validity – the suitability
- Volatility – temporal validity
- Value – usefulness of the information
- Visualization – displaying and showcasing
- Vulnerability – security and privacy
- Variability – the changing meaning of data

On the other side, the main technological challenges related to Big data are:

- Heterogeneity – differences in structure
- Uncertainty – data reliability
- Scalability – sizing the workflow and infrastructure Timeliness - real-time requirements
- Fault tolerance – sensitivity to errors
- Data security – privacy issues, data leaks
- Visualization – displaying of information

The main "Big" challenges related to Big data are related to analytics i.e. the analyses and knowledge extraction from enormous amount of data (Fig. 3). In other words, Big data processing and storing are challenging but, it is really important to highlight that they exist only to serve the knowledge extraction. To this aim, several tools in terms of algorithms/methodological approaches and their implementations are available today in open and commercial software. For example, focusing supervised learning methodologies, also known as supervised learning.

3 State of the Art: BIG EO Data for Archaeology

The availability of big and open satellite data, such as those from Copernicus Sentinel-1 (S-1) and Sentinel-2 (S-2) [9, 12] offers big opportunities and big challenges, relevant for multi- and inter-disciplinary studies as "space archaeology" (Fig. 2) including the site discovery, monitoring and preservation, today considered a priority at European and international level with important cultural, social, and economic repercussions. This is clearly highlighted by the increasing number of papers [14–17] and dedicated conferences and workshops on archaeological object detection in remote sensing data and from the fact that in 2017, the European Commission (EC) organized a specific workshop "to assess the potential of Copernicus in support of cultural heritage preservation and management" [18]. Subsequently, a "Copernicus Cultural Heritage Task Force" was also established in 2019, but, up to now only a few studies have very recently assessed the potentiality of Copernicus S-1 and S-2 data for archaeological research [19–24].

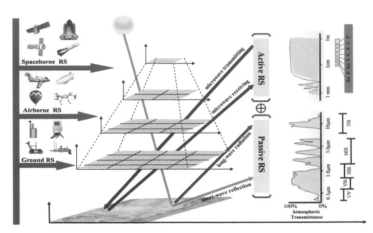

Fig. 2. Multiscale and multi sensor remote sensing based approach to archaeology and cultural heritage management (courtesy of Lei Luo et al. [3]).

For most of the 20th century aerial photography has been the unique remote sensing tool adopted for landscape archaeology and for detecting buried archaeological structures through the visual interpretation based on archaeological proxy indicators [5] (Fig. 3).

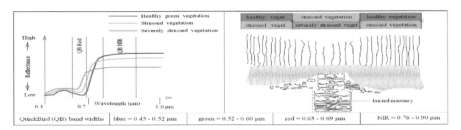

Fig. 3. Example of archaeological proxy indicators: crop marks.

The most common archaeological proxy indicators are generally known as crop, soil, shadow, and damp marks and are caused by the presence of buried remains and traces of ancient environs still fossilized in the modern landscape. These features induce spatial anomalies (in vegetation growth and/or status, surface moisture content, micro-reliefs) that are generally not visible in situ but only evident from above [19] even if are very subtle not permanent signals. In the last decades, significant improvements have been obtained in the identification of archaeological proxy indicators from VHR satellite optical and SAR data [24–26].

Starting from this heritage, new applications and developments are expected particularly from the use of open data as satellite S-1 and S-2 (see Fig. 4) that are part of the Europe's ambitious Copernicus Programme.

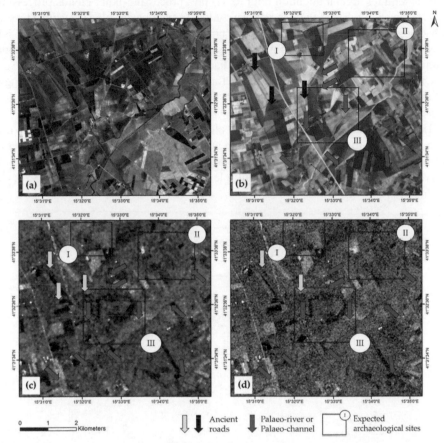

Fig. 4. Satellite based study of palaeoenvironment: identification of palaeorivers in Northern Apulia (South Italy) using Sentinel-1 and 2 [19].

All the Sentinel missions release under an Open Data policy to foster knowledge, innovative applications, and advanced developments. Pioneering studies highlighted that the multi-temporal S-2 data can be useful to extract information about the contemporary and past landscape, detect buried archaeological remains, infer changes in the current and former environment. Moreover, also the use of open satellite hyperspectral data, as PRISMA mission (which offers 239 spectral channels at 30 m, plus a panchromatic at 5 m) can provide (integrating and fusing pan and spectral bands) a suitable tool for both the site discovery and monitoring.

3.1 State of the Art: LiDAR Based Archaeology

The difficulties in the documentation, survey, and material collection increase in areas characterized by the presence of dense vegetation [27]. In these cases, the LiDAR-based analysis requires special attention both to: (i) pre- processing-point clouds processing and classification to avoid that the removal of low vegetation can also determine loss of archaeological information, and (ii) enhancement of LiDAR-Derived Models (LDMs), based on topographical modeling parameters (slope, convexity) and relief visualization techniques (Local Relief model, Sky View Factor, Openness) [28–30].

Over the past two decades, LiDAR has found increasing popularity in archaeology, and many archaeological landscape projects across Europe, America and Asia [31–33] focused on/or incorporated LiDAR.

The available data set types (from point cloud to DTM and orthophoto) allow to use more approaches to extract features, also in automated way, linked referable to diverse archaeological proxy indicators, including those linked to changes in microtopography and vegetation growth [34, 35] and multiple archaeological feature classes [36] (see Fig. 5).

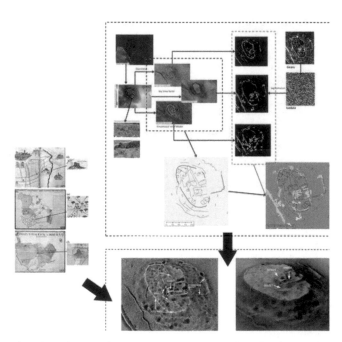

Fig. 5. LiDAR based semiautomatic extraction and interpretation of archaeological features. Case study related to a medieval village including a castle in Basilicata (South of Italy) [34].

3.2 State of the Art: UAV Archaeology

Photographs from UAV has been and is today widely used in archaeology also because available at low cost, in therefore drones often generate large amounts of data – sometimes more than we can handle, the use of AI can be the answer to this challenge.

The most recent UAV systems offer advanced equipment ranging from (i) multi to hyperspectral cameras (including the thermal bands), to (ii) LIDAR and (iii) geophysical prospection based on Ground Penetrating Radar (GPR) that can be faster and cheaper acquired compared to the traditional ground RS. Nevertheless, all of these technologies are rarely used today in archaeological applications, considered still quite expensive and complex.

3.3   State of the Art: Ground Remote Sensing for Archaeology

Ground Penetrating Radar (GPR), Magnetic surveys, along with other geophysical prospection have been (since long time) and are today widely used in archaeology; but except a few examples [37] rarely fully integrated with satellite and aerial data sets. This may be due to the fact that geophysics as well as aerial and satellite remote sensing belong to diverse scientific communities (experts and specialists) and there is an urgent need of sharing experience and background to facilitate the development between and across the diverse disciplines. However, when geophysics is functionally integrated with archaeological research and excavation, it can become a valid predictive tool, even in a diachronic key (Fig. 6).

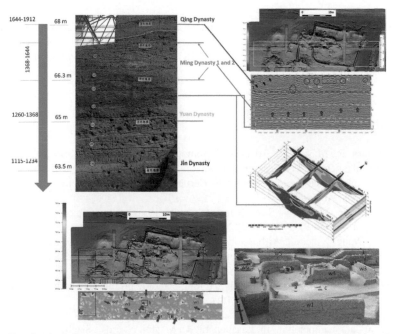

Fig. 6. Geophysical data integration for the diachronic detection of anthropogenic layers in Kaifeng (Henan, China) [37].

An important contribution can come from AI in order to maximize the information content of archaeological geophysics. The large spatial extent of areas covered by landscape surveys using available multi-sensor arrays and subsequently the large volume of collected data offer challenging opportunities for the development and use of Machine and Deep learning based automated analysis methods of geophysycal imagery [38].

4 Conclusion

Big data has emerged in the past few years providing opportunities to improve and/or enable research and decision support applications with unprecedented value for digital CH and archaeology. The possibility to fast analyses relatively large, varied and rapidly changing huge quantities of data has sped up the work during the diverse phases of application ranging from survey, mapping, documentation, exploitation and monitoring at diverse scales of interest, moving from small artefacts to architectural structures and landscape scale. There is, therefore, no doubt that EO big data will significantly change the scientific approach, data analysis and methodologies as well as discoveries of unknown archaeological sites. Moreover, these technologies are non- invasive survey methods very reliable not only for the discovery of lost archaeological remains/sites, but also useful to investigate cultural landscapes, assess the condition of archaeological features that is a mandatory step for the preservation and management of Cultural properties and historic sites.

Acknowledgements. The authors wish to thank the anonymous reviewers for their comments and suggestions that were constructive and which significantly improved the quality of this manuscript. The authors would also like to thank the SMART Heritage project for cooperation in this research study.

References

1. Lasaponara, R., Masini, N.: Advances in remote sensing for archaeology and cultural heritage management. In: Proceedings of International EARSeL Workshop "Advances in Remote Sensing for Archaeology and Culturale Heritage Management", Rome, 30 September–4 October 2008. Aracne (2008)
2. Lasaponara, R., Masini, N.: Satellite remote sensing in archaeology: past, present and future. J. Archaeol. Sci. **38**(9), 1995–2002 (2011)
3. Luo, L., Wang, X., et al.: Airborne and spaceborne remote sensing for archaeological and cultural heritage applications: a review of the century (1907–2017). Remote Sens. Environ. **232**, 111280 (2019)
4. Opitz, R., Herrmann, J.: Recent trends and long-standing problems in archaeological remote sensing. J. Comput. Appl. Archaeol. **1**(1), 19–41 (2018)
5. Lasaponara, R., Masini, N.: Remote sensing in archaeology: from visual data interpretation to digital data manipulation. In: Lasaponara, R., Masini, N. (eds.) Satellite Remote Sensing, vol. 16, pp. 3–16. Springer, Dordrecht (2012). https://doi.org/10.1007/978-90-481-8801-7_1
6. Tapete, D., Cigna, F.: Appraisal of opportunities and perspectives for the systematic condition assessment of heritage sites with copernicus Sentinel-2 high-resolution multispectral imagery. Remote Sens. **10**, 561 (2018)

7. https://step.esa.int/main/download/snap-download/
8. https://www.copernicus.eu/en/access-data/dias. Accessed 02 Feb 2022
9. https://www.sentinel-hub.com/
10. https://eo4society.esa.int/2019/05/21/european-data-cube-facility-service-an-eo-resource-factory/
11. Amani, M., et al.: Google Earth Engine cloud computing platform for remote sensing big data applications: a comprehensive review. IEEE J. Sel. Top. Appl. Earth Obs. Remote Sens. **13**, 5326– 5350 (2020)
12. Copernicus Open Access Hub. https://scihub.copernicus.eu/dhus/#/home. Accessed 11 Mar 2020
13. https://step.esa.int/main/download/snap-download/. Accessed 02 Feb 2022
14. Kvamme, K.L.: An examination of automated archaeological feature recognition in remotely sensed imagery. In: Bevan, A., Lake, M. (eds.) Computational Approaches to Archaeological Spaces, pp. 53–68. Left Coast Press, Walnut Creek (2013)
15. Schneider, A., Takla, M., Nicolay, A., et al.: A template-matching approach combining morphometric variables for automated mapping of charcoal kiln sites. Archaeol. Prospect. **22**(1), 45–62 (2018)
16. Orengo, H.A., Conesa, F.C., et al.: Automated detection of archaeological mounds using machine-learning classification of multisensor and multitemporal satellite data. PNAS **117**(31), 18240–18250 (2020)
17. Trier, Ø.D., Cowley, D.C., Waldeland, A.U.: Using deep neural networks on airborne laser scanning data: results from a case study of semi-automatic mapping of archaeological topography on Arran, Scotland. Archaeol. Prospect. **26**, 165–175 (2019)
18. Copernicus services in support to Cultural Heritage (2019). https://www.copernicus.eu/sites/default/files/201906/Copernicus_services_in_support_to_Cultural_heritage.pdf.   Accessed 20 Aug 2020
19. Abate, N., Elfadaly, A., Masini, N., Lasaponara, R.: Multitemporal 2016–2018 Sentinel-2 data enhancement for landscape archaeology: the case study of the Foggia Province, Southern Italy. Remote Sens. **12**, 1309 (2020)
20. Elfadaly, A., Abate, N., Masini, N., Lasaponara, R.: SAR Sentinel 1 imaging and detection of palaeo-landscape features in the mediterranean area. Remote Sens. **12**, 2611 (2020)
21. Agapiou, A., Alexakis, D.D., Sarris, A., Hadjimitsis, D.G.: Evaluating the potentials of Sentinel-2 for archaeological perspective. Remote Sens. **2014**(4), 2176–2194 (2014)
22. Zanni, S., De Rosa, A.: Remote sensing analyses on Sentinel-2 images: looking for Roman roads in Srem region (Serbia). Geosciences **9**, 25 (2019)
23. Agapiou, A., Alexakis, D., Hadjimitsis, D.G.: Potential of virtual earth observation constellations in archaeological research. Sensors **19**, 4066 (2019)
24. Kalayci, T., Lasaponara, R., Wainwright, J., Masini, N.: Multispectral contrast of archaeological features: a quantitative evaluation. Remote Sens. **11**, 913 (2019)
25. Masini, N., Lasaponara, R.: Sensing the past from space: approaches to site detection. In: Masini, N., Soldovieri, F. (eds.) Sensing the Past. GE, vol. 16, pp. 23–60. Springer, Cham (2017). https://doi.org/10.1007/978-3-319-50518-3_2
26. Jiang, A., Chen, F., Masini, N., et al.: Archeological crop marks identified from Cosmo-SkyMed time series: the case of Han-Wei capital city, Luoyang, China. Int. J. Digit. Earth **10**(8), 846–860 (2016)
27. Lasaponara, R., Masini, N.: Special issue on "remote sensing for cultural heritage management and documentation". J. Cult. Heritage **10S** (2009)
28. Hesse, R.: LiDAR-derived local relief models a new tool for archaeological prospection. Archaeol. Prospect. **17**(2), 67–72 (2010)
29. Zakšek, K., Oštir, K., Kokalj, Ž: Sky-view factor as a relief visualization technique. Remote Sens. **3**, 398–415 (2011)

30. Doneus, M.: Openness as visualization technique for interpretative mapping of airborne LiDAR derived digital terrain models. Remote Sens. **5**(12), 6427–6442 (2013)
31. Chase, A.F., et al.: Airborne LiDAR, archaeology, and the ancient Maya landscape at Caracol, Belize. J. Archaeol. Sci. **38**, 387–398 (2011)
32. Evans, D.H., et al.: Uncovering archaeological landscapes at Angkor using LiDAR. Proc. Natl. Acad. Sci. USA **110**, 12595–12600 (2013)
33. Masini, N., Lasaponara, R.: On the reuse of multiscale LiDAR data to investigate the resilience in the late medieval time: the case study of Basilicata in South of Italy. J. Archaeol. Method Theory **28**(4), 1172–1199 (2020). https://doi.org/10.1007/s10816-020-09495-2
34. Masini, N., et al.: Medieval archaeology under the canopy with LiDAR. The (re)discovery of a medieval fortified settlement in Southern Italy. Remote Sens. **10**, 1598 (2018)
35. Trier, O.D., Reksten, J.H., Løseth, K.: Automated mapping of cultural heritage in Norway from airborne lidar data using faster R-CNN. Int. J. Appl. Earth Obs. Geoinform. (2021). https://doi.org/10.1016/j.jag.2020.102241
36. Verschoof-van der Vaart, W.B., Lambers, K.: Learning to look at LiDAR: the use of R-CNN in the automated detection of archaeological objects in LiDAR data from The Netherlands. J. Comput. Appl. Archaeol. **2**(1), 31–40 (2019)
37. Masini, N., Capozzoli, L., et al.: Towards an operational use of geophysics for archaeology in Henan (China): archaeogeophysical investigations, approach and results in Kaifeng. Remote Sens. **9**(8), 809 (2017)
38. Küçükdemirci, M., Sarris, A.: Deep learning based automated analysis of archaeo-geophysical images. Archaeol. Prospect. **27**(2), 107–118 (2020)

FIRE-SAT System for the Near Real Time Monitoring of Burned Areas and Fire Severity Using Sentinel-2: The Case Study of the Basilicata Region

Rosa Lasaponara[1](✉), Carmen Fattore[1](✉), Nicodemo Abate[2], Angelo Aromando[1], Gianfranco Cardettini[1], Guido Loperte[3], and Marco Di Fonzo[4]

[1] National Research Council (CNR) - Institute of Methodologies for Environmental Analysis, Contrada S. Loja - Zona Industriale C.P. 27, 85050 Tito Scalo, PZ, Italy
{rosa.lasaponara,carmen.fattore}@imaa.cnr.it
[2] National Research Council (CNR) - Institute of Heritage Science, Contrada S. Loja-Zona Industriale, C.P. 27, 85050 Tito Scalo, Italy
[3] Ufficio per la Protezione Civile della Regione Basilicata, Potenza, Italy
[4] Comando Carabinieri per la Tutela Forestale, Rome, Italy

Abstract. Since the 2007, the FIRE-SAT system has been developed for the operational monitoring of fires in the Basilicata Region, as the results of a systematic collaboration between the Civil Protection of the Basilicata Region and the Argon Laboratory of the Institute of Methodologies for Environmental Analysis (IMAA) of the CNR-National Council of Research.

FIRE-SAT has a modular structure defined ad hoc for the different steps of the risk management: from the dynamic forecast of the wildfire hazard, to the mapping of the areas crossed by the wildfire, from the assessment of the fire impact (on vegetation, soil and atmosphere), to the estimation of the post-fire risk, such as erosion, increased hydrogeological crisis, biodiversity loss.

This paper provides an overview of the FIRE-SAT capability to mapping wildfires in the immediate aftermath of the event, to support post-event mitigation decisions. The system is based on the acquisition of satellite Sentinel-2 leveraged for the immediate mapping of burned areas, the estimation of fire severity and the long term post fire monitoring to assess the vegetation recovery capability.

Keywords: Wildfire · ΔNBR · Natural risk · Rural landscape · Earth Observation (EO)

1 Introduction

Wildfires are considered one of the most significant causes of environmental disturbance and damage, capable of affecting the functionality of the ecosystems and generating complex socio-economic impacts at local and global levels. The phenomenon of wildfires is extremely complex as it depends on several aspects among the other those related to the characteristics of vegetation, morphology, meteorological conditions, and, above all,

© The Author(s), under exclusive license to Springer Nature Switzerland AG 2022
F. Calabrò et al. (Eds.): NMP 2022, LNNS 482, pp. 2132–2145, 2022.
https://doi.org/10.1007/978-3-031-06825-6_205

anthropic factors. The use of satellite data makes it possible to support the monitoring and management of forest fire before, during and after the emergency and at different spatial/temporal scales (from the local to the global scale).

Satellite data are the core of the FIRE-SAT system developed the for monitoring of forest fires as a result of systematic long term collaboration (since 2007) between the civil protection of the Basilicata Region and the Argon Laboratory (directed by Dr. Rosa Lasaponara) of the Institute of Methodologies for Environmental Analysis (IMAA) of the CNR (National Research Council).

FIRE-SAT is systematically updated and based on a modular structure designed ad hoc for the different phases of risk management: from the dynamic prediction of forest fire danger, to the mapping of areas affected by fire, from the assessment of fire impact (on vegetation, soil and atmosphere), to the estimation of post-fire risk, such as erosion, increasing hydrogeological risk, and biodiversity loss, some of the details in [1–5]. FIRE-SAT is a system founded on the acquisition of satellite data made available at no cost by NASA and ESA, combined with ancillary data, weather parameters obtained by the regional network and weather forecast provided from COSMO-2.

In particular, FIRE-SAT provides fire danger maps that are updated and published on the region's website on a day-to-day basis (during the fire season, generally from June to September). Moreover, in the event of a maximum alert, the system automatically transmits the fire danger map via certified email on a town-by-town scale to the mayors of the given municipalities. This allows that, in the case of fire event, the fire fighting can be organized (ground, with or without mechanical means, aerial/Canadair) according to the fire severity levels (low, medium, high, extreme) obtained from several parameters, among which: vegetation type and status, morphology, meteorological conditions and weather forecasts. which correspond to the expected speed of propagation.

FIRE-SAT is focused on a systematic (year by year) and continuous updating: over the years, starting from 2007 the operational application is systematically supported by new research activities aimed at improving performance for the specific fire management phases (pre-, during and post-event) through the assimilation of new and/or updated data (such as the new Sentinel satellite data or weather forecasts available over the years at increasingly refined spatial/temporal resolutions). New data and/or ancillary information are comprehensively tested and then progressively incorporated into the pre-operational phase and subsequently into the operative one.

One of the latest improvements in particular is related to near real-time (weekly) "post-event" damage updates based on the use of Sentinel-1 and 2. The system, which is currently undergoing experimentation, has been implemented in the Google Earth Engine (GEE) platform to facilitate the access to large computational resources (necessary for the purpose), in order to:

1. delineate the fire-affected areas immediately after the fire event, based on the availability of new weekly acquisition of Sentinel-1 and 2 data;
2. analyse the degree of fire intensity (Fire Severity) to support mitigation activities (required in the immediate post-event period);
3. monitor the recovery of vegetation through the processing of multi-temporal images.

Throughout the years, FIRE-SAT has been adopted as best-practice in several European projects, such as, for example, SERV_FOR_FIRE (2017–2021) [6], FirEUrisk "Developing A Holistic, Risk-Wise Strategy For European Wildfire Management" H2020 (2021–2025).

2   Materials and Methods

2.1   Google Earth Engine for Perimeter and Monitoring of Burned Areas

Google Earth Engine (GEE) is a platform which facilitates the access to remotely available computing resources and enables manipulation, analysis and visualisation of geospatial data reducing cost and computation time without the need to download massive quantities (petabytes) of data and the use of a supercomputer [7]. GEE provides high performance for the processing of Big Data, including satellite data, land use information and maps, topography and socio-economic datasets, at no cost for research activities (or low cost for different uses) [8, 9]. The applications of GEE are numerous and in recent years its employment has increased within the scientific community [8] and it has been ascertained that it is efficient as a monitoring tool for environmental hazards of both archaeological areas, rural landscapes and forested areas damaged by the fire passage [10, 29, 30]. In particular, GEE [11] allows access to the data archives provided by several satellite platforms including the European Copernicus platforms with particular reference to Sentinel-1 [12] and Sentinel-2 data [13]. GEE archives are constantly updated with new satellite acquisitions (in the case of Sentinel-2 data, every 5 days). Sentinel-2 data are acquired in different spectral bands and with a spatial resolution ranging from 10 m to 60 m, as reported in Table 1.

Table 1. Sentinel-2 spectral bands.

Bands	Central wavelength (nm)	Spatial resolution
Band 1 – Coastal aerosol	0.443	60 m
Band 2 – Blue	0.490	10 m
Band 3 – Green	0.560	10 m
Band 4 – Red	0.665	10 m
Band 5 – Vegetation Red Edge	0.705	20 m
Band 6 – Vegetation Red Edge	0.740	20 m
Band 7 – Vegetation Red Edge	0.783	20 m
Band 8 – NIR	0.842	10 m
Band 8A – Vegetation Red Edge	0.865	20 m
Band 9 – Water vapor	0.945	60 m
Band 10 – SWIR – Cirrus	1.375	60 m
Band 11 – SWIR	1.610	20 m
Band 12 – SWIR	2.190	20 m

The GEE platform can be accessed via a website [11] and allows the processing of satellite data and images through appropriate programming and the creation of user-friendly interfaces for users not familiar with remote sensing techniques. GEE, therefore, allows the use of the Sentinel-2 database and the potential that this satellite offer: (i) a systematic and regular acquisition over time (every 5 days); (ii) a spatial resolution of the data of 10–20 m/pixel; (iii) a good spectral resolution that allows to calculate indices or maps obtained from appropriate mathematical combinations of the different bands to emphasize the presence of vegetation and any signs left by the passage of the fire. Thanks to the computing power of GEE you can get in a few seconds:

1. pre- and post-event RGB images, for visual discrimination of the event itself;
2. indices and therefore maps that emphasise the damage caused by the fire.

The approach has been developed completely within the platform of GEE (see Fig. 1).

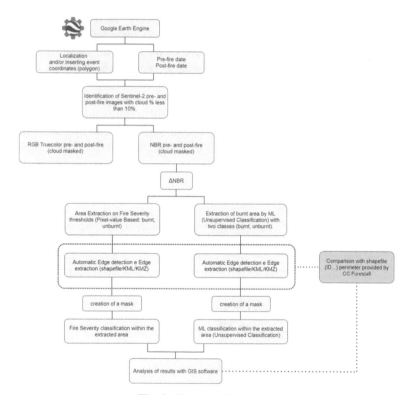

Fig. 1. Sentinel-2 flowchart.

As already extensively discussed in the literature and used for the monitoring of burnt areas, the bands involved in the process were mainly those of Blue, Green and Red (B2, B3, B4) for the creation of the truecolor RGB images, and the bands of Nir and Swir (B8, B12) for the creation of the index NBR (Normalized Burnt Ratio) [14–19] (1).

$$NBR = (B8 - B12)/(B8 + B12) \tag{1}$$

The NBR index is an index of the water content of plants that reveals the presence of burnt vegetation [20–22]. It is most commonly used as a difference detected before and after the ΔNBR event (2).

$$\Delta NBR = NBRpre - NBR\ post \tag{2}$$

The ΔNBR is extremely efficient in identifying the burnt area, thanks to the evident change in pixel reflectance between the pre and post-event NBR index, and it is also useful to understand the effect produced by the fire in terms of vegetation loss [2]. One of the advantages that ΔNBR offers is the possibility of classifying the different degrees of fire severity on the basis of several thresholds, moving from high severity to low severity. A widely used classification is given in Table 2.

Table 2. Classification by Fire Severity thresholds on ΔNBR, in accordance with the categorization suggested by USGS (from Lasaponara et al. 2018 [2], Table 1).

ΔNBR	Severity
<-0.25	Hight post-fire regrowth
-0.25 to -0.1	Low post-fire regrowth
-0.1 to 0.1	Unburned
0.1 to 0.27	Low-severity burn
0.27 to 0.44	Moderate-low severity burn
0.44 to 0.66	Moderate-high severity burn
>0.66	High-severity burn

Fire Severity is used to classify the intensity of fire, and hence its impact on vegetation and soil, and, with appropriate models, to estimate atmospheric emissions and the capacity of vegetation to recover. Furthermore, it is essential to consider that the effects of fire on the ecosystem generally appear at different spatial/temporal scales. For example, according to the time with which they occur, it is possible to divide them into three macro categories [23, 24]:

- immediate effects: these effects can be observed in the short time, from the first weeks of the event up to a few months, considered in relation to the amount of vegetation lost, the consumption of biomass within the affected area and erosive processes in the case of heavy rains immediately after the event;
- short-term effects: these occur a few months to years after the event, and involve a change in the structure and composition of the ecosystem, reducing the availability of nutrients in the soil;
- long-term effects: these persist many years after the fire and cause changes in the vegetation and structure of the affected area, thereby exposing it to other natural hazards such as soil erosion, landslides and flooding. These effects are caused by the fire which, upon damaging the surface, releases a coat of ash that compacts to create a water-repellent patina, increasing surface flow, and thus through runoff and erosion due to the mechanical action of water, there is a loss of organic substances.

Therefore, Fire Severity analysis becomes a fundamental tool to support the assessment of the impact of the given fire event, at different spatial/temporal scales. The methodology developed by the ARGON lab of the CNR-IMAA not only allows to identify and map the fire affected areas and its updated on a weekly frequency, but also to have an estimation of fire severity useful and need to support the identification and definition of mitigation strategies, where required.

3 Results and Discussion

Below are three examples, selected because significant and representative of different types of cover (forest and mixed) and characterised by areas affected by fire of variable extension from a few hectares (in areas with mixed cover) to more extensive events in forested areas. In the following examples, the perimeter obtained by satellite immediately after the event is compared with the perimeter obtained and provided by the Comando dei Carabinieri Forestali conducted in the field generally at the end of the fire season. The fire of 25 July 2018 in Grottole (MT) involved a total area of 4.04 ha and the event of 16 July 2018 in Sant'Arcangelo (PZ) affected 9.56 ha, shown in Fig. 2. However, the methodology used results to be performant even with small fires with different vegetation cover, as can be evident from Fig. 2.

In particular, the image shows how the ΔNBR allows a very detailed perimeter, despite the 10 m spatial resolution of the Sentinel-2 satellite images.

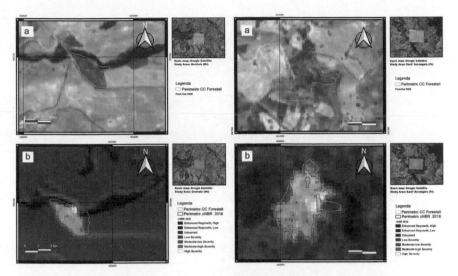

Fig. 2. Left: fire of 25 July 2018 near Grottole (MT) affecting an area of 4.04 ha with mainly agricultural use. (A) Sentinel-2 post-fire satellite image; (B) ΔNBR 2018 and comparison between the perimeters obtained from the ΔNBR (red) and the perimeter provided by CC Forestali (cyan). Right: fire of 16 July 2018 near Sant'Arcangelo (PZ), which affected a total area of 9.56 ha, mainly in agricultural use (C) Sentinel-2 pre-fire satellite image; (D) ΔNBR 2018 and comparison between the perimeters obtained from the ΔNBR (red) and the perimeter provided by CC Forestali (cyan).

3.1   The Fire in Metaponto (MT) on 13 July 2017

The fire occurred on the 13th July 2017 affected a large wooded area in the Metaponto pine forest (Fig. 3) - *Pinus halepensis* - present along the Ionian coast of the Basilicata region. The fire event affected both part the pine forest and an agricultural area close to. The flame front was estimated to be about 400 m wide and required several hours of work to extinguish and caused the evacuation of 600 tourists staying at three camp sites in the area. To extinguish the fire was required the intervention of the Fire Department of Bari and Metaponto, two teams of the Program Area and the Civil Protection of the Basilicata Region and the Carabinieri.

The analysis of the fire that occurred in Metaponto (MT) was oriented towards the identification of the burnt areas and fire severity automatically extracted from Sentinel-1 and Sentinel-2 data [25].

3.1.1   On the Mapping of Burned Area and Fire Severity

The classification of Fire Severity and the perimeter of the burnt area within the pine forest of Metaponto (MT) were conducted using multispectral satellite data from the Sentinel-2 mission and the Google Earth Engine computational tool [11]. The priority was to extract information and automatically map the effects of the event (fire) on the vegetation, with the hypothesis of creating a fast and repeatable method for the calculation of Fire Severity. Throughout the entire process, the main objective was to minimise operator intervention and input, focusing mainly on automating the process

Fig. 3. Territorial overview of the area affected by the fire.

and obtaining valid data. The operations performed manually include (i) the input of a geographical reference, related to the area of the event, and (ii) the input of a pre- and post-event temporal reference, in order to identify and search the Sentinel-2 images useful for the process (one for the pre-event and one for the post-event).

The Sentinel-2 L2A data have the advantage of being provided already corrected for atmospheric effects (Bot of the Atmosphere or BOA) and incorporate useful data for soil classification and the development of cloud, snow and water masks (SLC map). The SCL map, at 20 m/pixel resolution, distinguishes 11 classes [26] and is also relevant for land use classification within the image.

Upon choosing the dates of the event, the script searches the S2-L2A catalogue for the first available pre- and post-fire images, filtering them on the area of interest and the dates themselves, as well as on the percentage of cloudiness indicated in the S2 image metadata (less than 20%). Subsequently, using the data contained in the SLC map, the process provides for the subtraction of clouds, if present, from the pre- and post-event images.

The automatic edge extraction procedure was performed through the (Edge Detection and Edge Extraction, functions contained within GEE) and vectorization of the perimeter of the burned area. The perimeter thus obtained automatically from satellite data through GEE (Fig. 4) gives an area covered by fire of approximately 131 ha, while the area calculated through the perimeter provided by the CC Forestali is 138 ha.

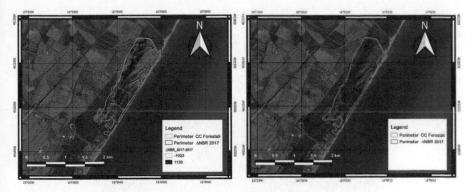

Fig. 4. Perimeter of the fire area calculated from the ΔNBR in blue and comparison with the perimeter provided by the CC Forestali (cyan).

A slight difference that demonstrates the accuracy of the sensor and the spectral index. In addition, the classification of the satellite-derived index (ΔNBR) permitted the discrimination of fire severity (Fig. 6) using thresholds (shown in Table 2) provided by the United States Geological Survey (USGS) [27] (Fig. 5).

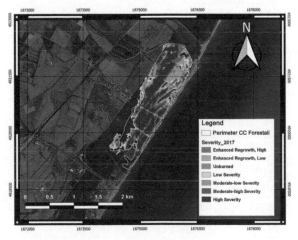

Fig. 5. Classification of Fire Severity within the area identified and circumscribed as in Fig. 2.

These thresholds are currently being investigated by researchers at the Argon Laboratory of the CNR-IMAA to optimise their use for both Lucanian and European ecosystems. In particular for this purpose, analyses and development of methodologies based on artificial intelligence approaches [1] validated with continuous in situ testing are ongoing (Fig. 6).

Fig. 6. From post-fire field surveys conducted in July and August 2017) in the Metaponto pine forest area (photo by: R. Lasaponara).

3.2 On the Monitoring of Post-event Vegetation Recovery

The similar methodology was successively applied to satellite images acquired in the years following the 2017 fire (Table 3).

Table 3. S2-A processed images for monitoring over the years.

S2-A	Pre-fire	Post-fire
2017	04/07/2017	24/07/2017
2018	04/07/2017	14/07/2018
2019	04/07/2017	14/07/2019
2020	04/07/2017	14/07/2020

In these years no further events took place and the analysis of the spectral indices of the NBR and ΔNBR showed some recovery of the less damaged vegetation, as highlighted in Fig. 7.

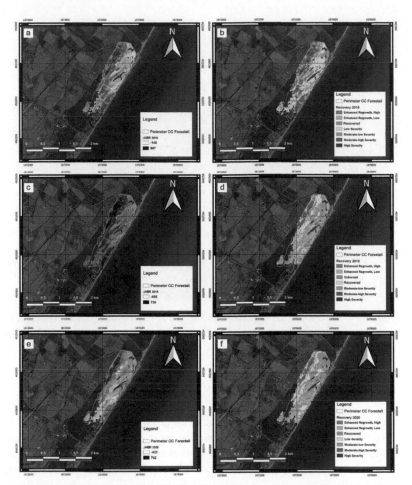

Fig. 7. Multi-temporal analysis of vegetation regrowth: (a) and (b) ΔNBR and Recovery of 2018; (c) and (d) ΔNBR and Recovery of 2019; (e) and (f) ΔNBR and Recovery of 2020. Vegetation recovery can be seen based on the change in spectral index, which detects positive values over the years, decreasing the area of medium to high severity.

A visual and qualitative comparison of the vegetation recovery trend in the years following the 2017 event was attempted with the aid of Google Earth, which allows high-resolution satellite imagery to be displayed. The area most affected by the event (circled in red) is also clearly visible in the 2018 images, as can be seen in Fig. 8. The analyses performed show that the vegetation is able to re-grow in a non-homogeneous pattern, in relation to the severity class.

Fig. 8. From the left a Google Earth image of 8 July 2017 before the event, on the right a zoom of the most affected area in 2017. Below left Google Earth image acquired in July 2018, exactly one year after the event. On the right a zoom of the damaged area.

4   Conclusions

In recent decades, both on a global and national scale, and therefore also on a local scale, the problem of forest fires has profoundly increased in intensity and frequency, due to anthropogenic and natural phenomena, including climate change, the abandonment of rural areas, and the constant loss of biodiversity that makes agro-forestry systems increasingly fragile. Hence, the fire season has broadened its spatial/temporal range of action, accentuating even in remote areas, such as the boreal forests that were once little affected by the problem, and committing the institutions and bodies involved in combating the problem on the various continents in a more intensive way. Preserving the "Natural Capital" and adopting good practices to monitor and support the sustainability of ecosystem goods and services is a priority especially for regions like Basilicata [28] and countries like Italy with the highest biodiversity in Europe. The use of advanced systems for operational observation can provide a valuable support to improve both contrast and risk mitigation activities as well as the management of the different phases of the emergency before, during and after the event. In particular, satellite data can provide synoptic information that is systematically updated and, with modern facilities, also available free of charge. For these reasons, the use of satellite systems for the monitoring of forest resources can in some aspects be considered a consolidated practice, although obviously, the continuous increasing availability of information and data and the latest technological developments have opened up new frontiers that make it necessary to develop Sentinel or weather forecasts available over the years at increasingly refined spatial/temporal resolutions). In addition, the FIRE-SAT system has been further enriched with the development of methodologies, presented in this article, based on the use of Sentinel-1 and 2 data and platforms, such as Google Earth Engine (GEE), that facilitate access to remotely usable computing resources, allowing the manipulation, analysis and visualisation of geospatial data without the need for supercomputers and, above all, without having to download huge amounts of data, reducing calculation times and optimising information extraction. The tool developed by the CNR-IMAA exploits the great capabilities of the GEE platform, managing to optimise and automate the processing phase relating to the perimeter of the areas affected by fire, the analysis of the impact (Fire Severity) and the ability to restore the vegetation after the passage of the fire. Three examples were presented, selected because they were significant and representative of different types of cover (forest and mixed) and characterised by areas affected by fire ranging from a few hectares (in areas with mixed cover) to more extensive events in forested areas. Comparison with the data provided by the Carabinieri Forestali showed the excellent performance obtained from satellite data even for fires of a few hectares in mixed-cover areas. In addition, the system also gave excellent results for the identification of areas affected by stubble burning from which, in all likelihood, the fire then spread to the forested area.

Acknowledgements. The authors are grateful to the anonymous reviewers for their constructive observations. In addition, the authors gratefully acknowledge the FirEUrisk project funded from the European Union's Horizon 2020 research and innovation programme for their encouragement and continued support in this research study.

References

1. Lasaponara, R., Tucci, B.: Identification of burned areas and severity using SAR Sentinel-1. IEEE Geosci. Remote Sens. Lett. **16**(6), 917–921 (2019)
2. Lasaponara, R., Tucci, B., Ghermandi, L.: On the use of satellite Sentinel 2 data for automatic mapping of burnt areas and burn severity. Sustainability **10**(11), 3889 (2018)
3. Pourghasemi, H.R., Gayen, A., Lasaponara, R., Tiefenbacher, J.P.: Application of learning vector quantization and different machine learning techniques to assessing forest fire influence factors and spatial modelling. Environ. Res. **184**, 109321 (2020)
4. Lasaponara, R., Lanorte, A.: Patent an integrated system for fire monitoring patent prot. 408719 del 24/08/2009 sistema di lotta attiva agli incendi boschivi, n. 2008 A0016 (2009)
5. Li, X., Lanorte, A., Lasaponara, R., Lovallo, M., Song, W., Telesca, L.: Fisher-Shannon and detrended fluctuation analysis of MODIS normalized difference vegetation index (NDVI) time series of fire-affected and fire-unaffected pixels. Geomatics Nat. Hazards Risk **8**(2), 1342–1357 (2017)
6. Project reports of the SERV-FORFIRE project (ERA4CS EU). https://servforfire-era4cs.eu/
7. Tamiminia, H., Salehi, B., Mahdianpari, M., Quackenbush, L., Adeli, S., Brisco, B.: Google Earth Engine for geo-big data applications: a meta-analysis and systematic review. J. Photogramm. Remote Sens. **164**, 152–170 (2020)
8. Amani, M., et al.: Google Earth Engine cloud computing platform for remote sensing big data applications: a comprehensive review. IEEE J. Sel. Top. Appl. Earth Obs. Remote Sens. **13**, 5326–5350 (2020)

9. Gorelick, N., Hancher, M., Dixon, M., Ilyushchenko, S., Thau, D., Moore, R.: Google Earth Engine: planetary-scale geospatial analysis for everyone. Rem. Sens. **202**, 18–22 (2017)
10. Fattore, C., Abate, N., Faridani, F., Masini, N., Lasaponara, R.: Google earth engine as multi-sensor open-source tool for supporting the preservation of archaeological areas: the case study of flood and fire mapping in Metaponto, Italy. Sensors **21**, 1791 (2021)
11. https://earthengine.google.com/. Accessed 08 Jan 2022
12. https://developers.google.com/earth-engine/datasets/catalog/COPERNICUS_S1_GRD. Accessed 08 Jan 2022
13. https://developers.google.com/earth-engine/datasets/catalog/sentinel-2. Accessed 08 Jan 2022
14. Miller, J.D., Thode, A.E.: Quantifying burn severity in a heterogeneous landscape with a relative version of the delta Normalized Burn Ratio (dNBR). Remote Sens. Environ. **109**, 66–80 (2007)
15. Huang, H., et al.: Separability analysis of Sentinel-2A multi-spectral instrument (MSI) data for burned area discrimination. Remote Sens. **8**, 873 (2016)
16. Filipponi, F.: Exploitation of Sentinel-2 time series to map burned areas at the national level: a case study on the 2017 Italy wildfires. Remote Sens. **11**, 622 (2019)
17. Verhegghen, A., et al.: The potential of Sentinel satellites for burnt area mapping and monitoring in the Congo Basin forests. Remote Sens. **8**, 986 (2016)
18. Murphy, K.A., Reynolds, J.H., Koltun, J.M.: Evaluating the ability of the di erenced normalized burn ratio (dNBR) to predict ecologically significant burn severity in Alaskan boreal forests. Int. J. Wildl. Fire **17**, 490–499 (2008)
19. Key, C.H., Benson, N.: The normalized burn ratio (NBR): a landsat TM radiometric measure of burn severity. United States Geological Survey, Northern Rocky Mountain Science Center (1999)
20. van Leeuwen, W.J.D.: Monitoring the effects of forest restoration treatments on post-fire vegetation recovery with MODIS multitemporal data. Sensors **8**, 2017–2042 (2008)
21. Keeley, J.E., Brennan, T., Pfaff, H.: Fire severity and ecosytem responses following crown fires in California shrublands. Ecol. Appl. **18**, 1530–1546 (2008)
22. Park, H., Choi, J., Park, N., Choi, S.: Sharpening the VNIR and SWIR bands of Sentinel-2A imagery through modified selected and synthesized band schemes. Remote Sens. **9**, 80 (2017)
23. Fox, M.D., Fox, B.J.: The role of fire in the scleromorphic forests and shrublands of eastern Australia. In: Trabaud, L. (ed.) The Role of Fire in Ecological Systems, pp. 23–48. SPB Academic Publ. (1987)
24. Brown, J.K., Smith, J.K.: Wildland fire in ecosystems: effects of fire on flora. U.S. Department of Agriculture, Forest Service, Rocky Mountain Research Station, Ogden, UT. General technical report RMRS-GTR-42, vol. 2, p. 257 (2000)
25. https://sentinel.esa.int/web/sentinel/missions/sentinel-2. Accessed 08 Jan 2022
26. https://developers.google.com/earth-engine/datasets/catalog/COPERNICUS_S2_SR#bands
27. Vanderhoof, M.K., Fairaux, N., Beal, Y.J.G., Hawbaker, T.J.: Validation of the USGS Landsat burned area essential climate variable (BAECV) across the conterminous United States. Remote Sens. Environ. **198**, 393–406 (2017)
28. Piano anti-incendio boschivo della regione Basilicata (2020). http://www.protezionecivilebasilicata.it/protcivbas/files/docs/10/65/39/DOCUMENT_FILE_106438.pdf. Accessed 08 Jan 2022
29. Tassi, A., Gigante, D., Modica, G., Di Martino, L., Vizzari, M.: Pixel- vs. object-based landsat 8 data classification in Google earth engine using random forest: the case study of Maiella National Park. Remote Sens. **13**, 2299 (2021)
30. Praticò, S., Solano, F., Di Fazio, S., Modica, G.: Machine learning classification of mediterranean forest habitats in Google earth engine based on seasonal Sentinel-2 time-series and input image composition optimisation. Remote Sens. **13**, 586 (2021)

The Influence of Potential Flood Hazard Areas for Urban Adaptation to Climate Change

Simone Corrado$^{(\boxtimes)}$ ⓘ, Luigi Santopietro ⓘ, Francesco Scorza ⓘ,
and Beniamino Murgante ⓘ

Laboratory of Urban and Regional System Engineering (LISUT), School of Engineering,
University of Basilicata, Potenza, Italy
simone.corrado@studenti.unibas.it

Abstract. In the modern context of climatic uncertainty, the management of the resilient transformation process of the city emerges as a priority challenge for urban planning to develop a multidisciplinary and inter-scalar approach, transferring knowledge between sectors and encouraging the whole system thinking. In detail, the present work investigates the interferences between the potential flood hazard areas and the emergency planning of the municipality of Potenza. Critical nodes have been identified in the infrastructure network as essential features for coordinating rescue interventions to protect the population in a disaster. Indeed, the plan must be flexible in all emergencies, including unexpected ones. Therefore, a new perspective of risk management aims to reduce sources of risk and strengthen the ability to reduce the variability of performance, and complement the risk analysis with frameworks and practices that support system-wide resilience.

Keywords: Urban planning · Spatial planning · Climate change · Adaptation plan

1 Introduction

The international scientific debate believes that the increase of extreme events is linked to climate change and, in the future, the more evident effects will occur in Mediterranean countries. A study has already shown that the average temperatures in the Mediterranean region have risen by 1.4 °C since the pre-industrial era, 0.4 °C more than the global average [1]. Moreover, climate change is likely to affect design flood estimation by impacting rainfall Intensity-Frequency-Duration relationship [2]. Hence, river flooding, especially the local ones, will remain a significant physical threat to many cities, increasing frequency, and magnitude. Therefore, these conditions require an urgency of intervention that can be translated into the need to elaborate new development models for cities and territories based on the knowledge of the indivisible co-evolutionary relationship between man and climate [3]. Mainly, the cities can act as drivers for this paradigm shift [4]. Given the amount of open-data provided, they can be regarded as a laboratory for better understanding relational dynamics and exporting knowledge for defining local and regional development strategies [5]. Indeed, radicalizing the concept of ecosystem proposed by Eugene P. [6], "Ecosystem is any unit that includes all of the organisms in a given area interacting with the physical environment so that a flow of energy leads to an

© The Author(s), under exclusive license to Springer Nature Switzerland AG 2022
F. Calabrò et al. (Eds.): NMP 2022, LNNS 482, pp. 2146–2152, 2022.
https://doi.org/10.1007/978-3-031-06825-6_206

exchange of materials between living and non-living parts of the system," it is possible to define the urban ecosystem as an open system and strongly dependent on external inputs of matter and energy.

Consequently, the city is potentially vulnerable to external factors. By its nature, the urban context is a complex system where infrastructural, economic, and social components are strongly interconnected and difficult to evaluate as single-one [7, 8]. Moreover, in highly urbanized settlements, there is a wide spatial concentration of people and exposed values that, even for slight variations of the balance among systems, can be subjected to high levels of direct and indirect damages [9]. Based on these assumptions, urban resilience can be defined as the adaptability to change, the capacity to absorb disturbances, and keeping high standards of the functionality of the infrastructure network, physical, information, and energy [10–12]. In this new perspective, urban planning performs a crucial role in conflict management thanks to its natural knowledge of the spatial plurality of urban dynamics and all the potential risks associated with them. In the broadest meaning, the urban planner can support this transition through regulations and more informed spatial design even in the face of interfering extreme events. The purpose of this paper is to explain the critical role of land use planning in improving the performance of critical infrastructure systems under conditions of climate uncertainty. In Sect. 2, the study area is presented, and the motivation behind the choice is briefly described. Section 3 explains the methodology used in the study, focusing on geocomputational tools and solutions offered by GIS software. The spatiality of the investigated problem needs, besides a quick visualization, to generate the systems known as Spatial Decision SupportSystem [13]. In addition, Sect. 4 and 5 discuss the primary outcomes, and the flexibility of the plan and its integration into a broader range of initiatives against climate change by prioritizing strategic planning over the operative dimension, ass this is the only way to steer decisions in a well-defined medium-long term direction.

2   Area of Study

As already mentioned, the work aims to evaluate the potential interferences between local flooding areas and the civil protection plan proposed by the municipality of Potenza. The settlement area of Potenza Municipality has been selected since several seismic events in the past afflicted the area. Indeed, the settlement is classified among the highest seismic hazard values in the national context [14]. The damages caused by the latest earthquake highlighted the necessity to examine the vulnerability of lifelines, i.e., both transport and service infrastructure systems. The analysis of network systems in the study of seismic risk has become a vital issue because of the role that they play during both emergency management and the restoration of pre-event conditions.

Moreover, urban areas have been afflicted by local flooding phenomena in recent years, even for moderate rainfall events. As urban roads drainage systems are combined with the sewage schemes designed by local authority regulation, these cannot cope with runoff from impervious surfaces in extreme rainfall events. The current Civil Protection Plan of the municipality was drafted following ministerial directives, and it was approved in Nov. 2013. The plan identified the initiatives to be undertaken to improve the living conditions of people evacuating from their homes and the areas of first or medium-term assistance in case of extreme events.

3   Methodology

The national norm, according to article 5, paragraph 5, of law n. 401/2001 [15] has stated the criteria for the eligibility of Operational Coordination Centers (DI.COMA.C.) and emergency areas. Among the various criteria, areas where meteoric waters are free to flow have to be promptly excluded because they are not correctly regulated. Also, for network systems, as described in the previous paragraph, the priority in case of an unexpected event is total or partial-service preservation. Consequently, the indirect vulnerability of the system was assessed. To accomplish this purpose, methodology, based on geo-algorithms of spatial overlay in the GIS environment, identified the potential interferences between the emergency areas defined by the plan and those with storm-water management issues. Meaningful and valuable results are obtained from a previous study conducted over the same survey area [16]. Specifically, a hierarchical classification of the urban territory in terms of urban water management, which reveals an area's suitability regarding its intrinsic characteristics, has allowed identifying the overlapping areas between the two informative layers [17].

Further analysis of structural weaknesses has been carried out considering the localization of all geomorphological depressions within the city boundaries. This has allowed us to identify the functional and physical vulnerability of the urban transport network. Moreover, nodes requiring priority and consistent interventions have been highlighted to minimize the flood risk in an emergency (see Fig. 1).

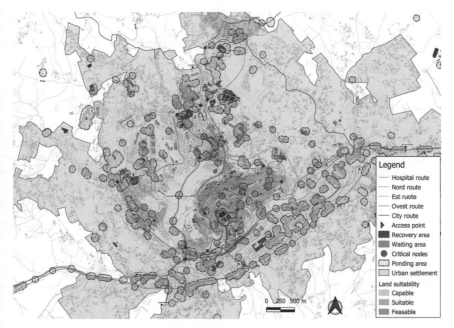

Fig. 1. Interference map between civil protection plan and potential local flooding sites.

4 Results and Discussion

The results derived from the spatial analysis showed a significant vulnerability in the case of simultaneous events. Both waiting and recovery areas have been identified that do not comply with the directives of the ministerial standard. The detailed results are described in the following Table 1.

Table 1. Results of the spatial analysis on out-of-standard plan choices.

Emergency area	Overall	Out-of-standard	Percentage
Waiting area	71	23	32%
Recovery area	29	9	31%
Critical node	13	5	38%

The table shows that approximately one-third of all identified areas fall within local flood-prone areas. Moreover, due to the steepness of the slopes, flooding at certain critical junctions can be significant [18]. A crucial node that requires further attention is access to the only hospital in the municipality (see Fig. 2). The spot is known to be flooded following every rainfall event as a ponding area falls along the main access road to the facility. Moreover, the road junction connects all strategic routes for civil protection purposes.

It seems clear that the risk of local flooding was not taken into account when the plan was drawn up due to the lack of studies on storm-water management in urban areas. But the awareness of both local decision-makers and citizens about sustainable and resilient water management is growing [19]. The concept is that the city is not only made up of buildings but also of people who, when involved, can improve the city's configuration and their knowledge of risks. This statement considers that a city is only as resilient as its citizens [20, 21]. Here, a sectorial approach to planning appears to be prevalent, characterized by the silo syndrome. This means that planning from a single diagnosis leads to a specific plan. It is taken for granted that the drafting of civil protection plans must start from an overall understanding of the context, beginning with the problems and then defining effective strategies. In this perspective, therefore, spatial planning is not a task to be performed by a single category of specialists trained in a single discipline. On the contrary, it presupposes a new strategic combination of skills of experts from a variety of disciplines.

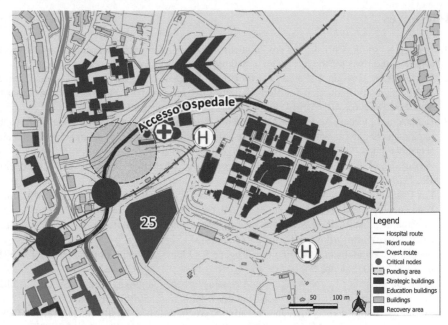

Fig. 2. The crucial access route to the only hospital is affected by local flooding.

5 Conclusion

The civil protection plan is a continually updated document that must take account of the evolution of the territorial layout. With even more short-term heavy rain events predicted, this paper supports examining the consequences of projected changes to climate, allowing better risk management of possible events and their interferences. The plan should be flexible in all emergencies, including unexpected ones. On this track, the flexibility of the plan could support the development of other "urban planning tools" facing climate change, such as Sustainable Energy and Climate Action Plans (see also [19, 20]) and implement the knowledge of the city built on a set of systems (such as green spaces, green infrastructures [21] waterproofed soils, energy system [22], active mobility [23–25], etc.). The city's preparedness for future risks depends on the quality of available risk scenarios and how these data are used to plan a more resilient city [26]. In this perspective, the proposed framework applied to the study area is necessary for decision-makers and the definition of the municipal climate adaptation plan. Future developments will be oriented to quantitatively assess the risk at critical infrastructure nodes by applying two-dimensional hydraulic modeling. In addition, further analyses will propose alternative routes to strategic facilities and implement sustainable urban drainage strategies, greening up the grey with the NBS solution, for improving the resilience of urban infrastructure [27–29].

References

1. Mrabet, R., et al.: Climate and environmental change in the Mediterranean Basin – current situation and risks for the future. First mediterranean assessment report, pp. 1–26 (2020)
2. Yilmaz, A.G., Hossain, I., Perera, B.J.C.: Effect of climate change and variability on extreme rainfall intensity-frequency-duration relationships: a case study of Melbourne. Hydrol. Earth Syst. Sci. **18**, 4065–4076 (2014). https://doi.org/10.5194/hess-18-4065-2014
3. Lorenzoni, I., Jordan, A., Hulme, M., Turner, R.K., O'riordan, T.: A co-evolutionary approach to climate change impact assessment: part I. Integrating socio-economic and climate change scenarios. https://doi.org/10.1016/S0959-3780(00)00012-1
4. Alberti, V., et al.: The future of cities: opportunities, challenges and the way forward (no. JRC116711). Joint Research Centre (2019)
5. Las Casas, G., Murgante, B., Scorza, F.: Regional local development strategies benefiting from open data and open tools and an outlook on the renewable energy sources contribution. In: Papa, R., Fistola, R. (eds.) Smart Energy in the Smart City. GET, pp. 275–290. Springer, Cham (2016). https://doi.org/10.1007/978-3-319-31157-9_14
6. Snehal, A.: Fundamentals of Ecology. Eugene P. Odum. Saunders, Philadelphia, xii + 384 pp. Illus. $6.50. Science (80-.) (1954)
7. Salat, S., Bourdic, L., Nowacki, C.: Assessing urban complexity. Int. J. Sustain. Build. Technol. Urban Dev. **1**, 160–167 (2010). https://doi.org/10.5390/SUSB.2010.1.2.160
8. Moroni, S.: Urban density after Jane Jacobs: the crucial role of diversity and emergence. City, Territ. Archit. **3**, 1–8 (2016). https://doi.org/10.1186/S40410-016-0041-1/METRICS
9. Manganelli, B., Pontrandolfi, P., Azzato, A., Murgante, B.: Urban residential land value analysis: the case of Potenza. In: Murgante, B., et al. (eds.) ICCSA 2013. LNCS, vol. 7974, pp. 304–314. Springer, Heidelberg (2013). https://doi.org/10.1007/978-3-642-39649-6_22
10. Scorza, F., Casas, G.L., Murgante, B.: Overcoming Interoperability weaknesses in e-government processes: organizing and sharing knowledge in regional development programs using ontologies. In: Lytras, M.D., Ordonez de Pablos, P., Ziderman, A., Roulstone, A., Maurer, H., Imber, J.B. (eds.) WSKS 2010. CCIS, vol. 112, pp. 243–253. Springer, Heidelberg (2010). https://doi.org/10.1007/978-3-642-16324-1_26
11. Scorza, F., Pilogallo, A., Saganeiti, L., Murgante, B.: Natura 2000 areas and sites of national interest (SNI): measuring (un)integration between naturalness preservation and environmental remediation policies. Sustainability **12**, 2928 (2020). https://doi.org/10.3390/su12072928
12. Redeker, C.: Strategies of urban flood integration - Zollhafen Mainz. In: Proceedings Conference on How Concept Resil. is Able to Improv. Urban Risk Manag. A Temporal a Spat. Anal., pp. 57–63 (2013). https://doi.org/10.1201/B12994-13
13. Murgante, B., Borruso, G., Lapucci, A.: Geocomputation and urban planning. Stud. Comput. Intell. **176**, 1–17 (2009). https://doi.org/10.1007/978-3-540-89930-3_1
14. Rota, M., Penna, A., Strobbia, C., Magenes, G.: Typological seismic risk maps for Italy. Earthq. Spectra **27**, 907–926 (2011). https://doi.org/10.1193/1.3609850
15. LEGGE 9 novembre 2001, n. 401 - Normattiva
16. Corrado, S., Giannini, B., Santopietro, L., Oliveto, G., Scorza, F.: Water management and municipal climate adaptation plans: a preliminary assessment for flood risks management at urban scale. In: Gervasi, O., et al. (eds.) ICCSA 2020. LNCS, vol. 12255, pp. 184–192. Springer, Cham (2020). https://doi.org/10.1007/978-3-030-58820-5_14
17. Saaty, T.L., Kearns, K.P.: The analytic hierarchy process. Anal. Plan. 19–62 (1985). https://doi.org/10.1016/B978-0-08-032599-6.50008-8
18. Corrado, S., Santopietro, L., Scorza, F.: Municipal climate adaptation plans: an assessment of the benefit of nature-based solutions for urban local flooding mitigation (2022). https://doi.org/10.31428/10317/10497

19. Scorza, F., Santopietro, L.: A systemic perspective for the sustainable energy and climate action plan (SECAP). Eur. Plan. Stud. 1–21 (2021). https://doi.org/10.1080/09654313.2021.1954603

20. Santopietro, L., Scorza, F., Rossi, A.: Small municipalities engaged in sustainable and climate responsive planning: evidences from UE-CoM. In: Gervasi, O., et al. (eds.) ICCSA 2021. LNCS, vol. 12957, pp. 615–620. Springer, Cham (2021). https://doi.org/10.1007/978-3-030-87013-3_47

21. Lai, S., Leone, F., Zoppi, C.: Implementing green infrastructures beyond protected areas. Sustainability 10, 3544 (2018). https://doi.org/10.3390/su10103544

22. Scorza, F.: Towards self energy-management and sustainable citizens' engagement in local energy efficiency agenda. Int. J. Agric. Environ. Inf. Syst. 7, 44–53 (2016). https://doi.org/10.4018/IJAEIS.2016010103

23. Scorza, F., Fortunato, G.: Cyclable cities: building feasible scenario through urban space morphology assessment. J. Urban Plan. Dev. 147, 05021039 (2021). https://doi.org/10.1061/(asce)up.1943-5444.0000713

24. Scorza, F., Fortunato, G., Carbone, R., Murgante, B., Pontrandolfi, P.: Increasing urban walkability through citizens' participation processes. Sustainability 13, 5835 (2021). https://doi.org/10.3390/su13115835

25. Fortunato, G., Scorza, F., Murgante, B.: Hybrid oriented sustainable urban development: a pattern of low-carbon access to schools in the city of Potenza. In: Gervasi, O., et al. (eds.) ICCSA 2020. LNCS, vol. 12255, pp. 193–205. Springer, Cham (2020). https://doi.org/10.1007/978-3-030-58820-5_15

26. Task Team on Preparedness and Resilience of the Inter-Agency Standing Committee (IASC). Emergency Response Preparedness (ERP): Risk Analysis and Monitoring, Minimum Preparedness, Advanced Preparedness and Contingency Planning. Draft for Field Testing. 55 (2015)

27. Pilogallo, A., Scorza, F.: Mapping regulation ecosystem services specialization in Italy. J. Urban Plan. Dev. 148 (2022). https://doi.org/10.1061/(ASCE)UP.1943-5444.0000801

28. Pilogallo, A., Saganeiti, L., Scorza, F., Las Casas, G.: Tourism attractiveness: main components for a spacial appraisal of major destinations according with ecosystem services approach. In: Gervasi, O., et al. (eds.) ICCSA 2018. LNCS, vol. 10964, pp. 712–724. Springer, Cham (2018). https://doi.org/10.1007/978-3-319-95174-4_54

29. Babí Almenar, J., et al.: Nexus between nature-based solutions, ecosystem services and urban challenges. Land Use Policy 100, 104898 (2021). https://doi.org/10.1016/J.LANDUSEPOL.2020.104898

Preliminary Results in the Use of WorldView-3 for the Detection of Cork Oak (*Quercus Suber L.*): A Case in Calabria (Italy)

Gaetano Messina[✉] , Giovanni Lumia , Salvatore Praticò ,
Salvatore Di Fazio , and Giuseppe Modica

Dipartimento di Agraria, Università degli Studi Mediterranea di Reggio Calabria, Località Feo di Vito, 89122 Reggio Calabria, Italy
gaetano.messina@unirc.it

Abstract. Cork oaks (*Quercus suber L.*) characterize many Mediterranean forest landscapes where they play important socio-economic and ecological functions for nature. This study, carried out in Mount Scrisi (Calabria Region, Italy) aims to map cork oak forests by using WorldView-3 (WV-3) high-resolution satellite image. For this aim, a supervised classification on WV-3's images was implemented to assess the potential performance of this sensor either in detecting the presence of cork oak woodlands and in distinguishing them from other spectrally similar tree species. Particular attention was paid to the distinction of cork oaks from olive (*Olea europaea* L.) and chestnut trees (*Castanea sativa,* Mill.). By exploiting the panchromatic image with 31 cm resolution and multispectral image bands through pansharpening, the objective was achieved obtaining a high accuracy in classification (OA = 0.88). Results confirm the usefulness of WV-3 in the applications of remote sensing (RS) on forestry for mapping species distribution and monitoring vegetation and environmental health.

Keywords: eCognition Developer · Geographic Object-Based Image Analysis (GEOBIA) · Segmentation · Image classification · Support Vector Machine (SVM)

1 Introduction

Cork oaks (*Quercus suber* L.) characterize many Mediterranean forest landscapes, where they play an important role in the socio-economic life of local communities and have a major ecological function for nature conservation [1–3]. In this framework, the use of remote sensed imagery for monitoring and mapping can provide helpful information for implementing further conservation actions [4, 5]. The mapping operation can be challenging, although performed with high-resolution imagery, due to the heterogeneity of the forest landscape [6]. In this case study supervised classification was exploited for mapping cork oak forest by distinguishing it from other tree species, in particular olive (*Olea europaea* L.) and chestnut trees (*Castanea sativa,* Mill.).

© The Author(s), under exclusive license to Springer Nature Switzerland AG 2022
F. Calabrò et al. (Eds.): NMP 2022, LNNS 482, pp. 2153–2162, 2022.
https://doi.org/10.1007/978-3-031-06825-6_207

2 Materials and Methods

2.1 Study Area

The study was carried out in Mount Scrisi, in an area which extends about 4.2 km^2 at $400 \div 500$ m a.s.l., in the Reggio Calabria province (Calabria Region, Italy) on the Tyrrhenian side of the Aspromonte (Fig. 1). The forest cover in this area is characterised by a thermo-mesophilous woodland dominated by cork oak. The climate is of sub-humid meso-Mediterranean type characterized by an average annual temperature between 14 and 16 °C and average annual rainfall between 700 and 1000 mm. The study area is recognized for the natural value of its cork oak forests, because of which it was designated as a Site of Community Interest under the Habitats Directive in 1995 [7]. Mount Scrisi got National legal reference of SAC (Special Areas of Conservation) designation in 2017-06 (DM 27/06/2017 - G.U. 166 del 18-07-2017).

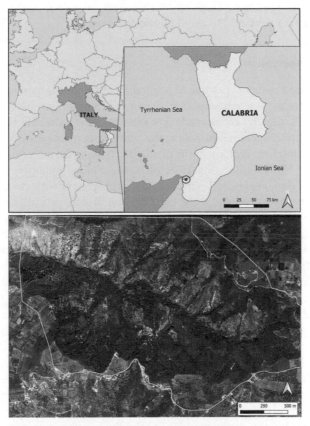

Fig. 1. The location of the study area, Mount Scrisi in the province of Reggio Calabria (Calabria Region, Italy) (top). True color WorldView-3 image of the study area (highlighted in yellow) (bottom).

2.2 WorldView-3 Data and Pre-processing

WorldView-3 (WV-3) is a multi-payload, high-resolution commercial satellite sensor launched in 2014 by DigitalGlobe [8]. WV-3 provides eight spectral bands with a spatial resolution of 1.24 m in the visible and infrared range of electromagnetic spectrum i.e., Coastal Blue, Blue, Green, Yellow, Red, Red edge, and two NIR bands, referred to in this work as NIR 1 and NIR 2, while the panchromatic band resolution is of 31 cm [9]. To exploit the advantages of both data in one image, by integrating the geometrical detail of the panchromatic image and the spectral information into each band of the multispectral image [10] pansharpen fusion method was performed in ERDAS Imagine environment [11]. In particular, the process was carried out by using the Hyperspectral Color Sharpening algorithm which allows combining WV-3's high-resolution panchromatic data with lower resolution multispectral data [12]. For resampling the multispectral image to the high-resolution image, the Cubic Convolution technique was chosen by using a 4 × 4 pixels moving window. The panchromatic and the multispectral images were separately orthorectified using as elevation reference the digital surface model (DSM) and subsequently co-registered to avoid alignment errors.

2.3 Image Segmentation

To extract the forest cover of cork oak, a geographical object-based image classification (GEOBIA) procedure was performed by using eCognition Developer 10.2 (Trimble GmbH, Munich, Germany) and according to the workflow showed in Fig. 2.

The classification was developed using the pansharpened image of WV-3. The image was firstly segmented into uniform multi-pixel objects using the multiresolution segmentation algorithm [13]. This algorithm is based on the relative homogeneity criterion and works by identifying single-pixel objects and merging them with closer objects [14]. Homogeneity criterion depends on the combination of the spectral and shape properties of the original image and the new objects generated by the merging process. The homogeneity criteria are governed by two parameters: shape and compactness. Shape is governed by a weight given to the shape of objects relative to color and can take a value between 0 and 0.9. A higher value results in a greater influence of shape over color. Compactness, which is derived from the product of width and length calculated over numbers of pixels, is the parameter that governs the importance of shape respect to the smoothness [15]. Finally, the size and dimension of the objects resulting from segmentation depends on the scale parameter. Segmentation was performed using the following layers (corresponding to the WV3's bands): Coastal Blue, Blue, Green, Yellow, Red, Red edge, NIR 1 and NIR 2. Launching the multiresolution segmentation, the following parameters were set: 110 for scale parameter, 0.1 for shape and 0.5 for compactness. Some trial-and-error tests were performed before choosing these parameters attributing different values to the segmentation parameters until the segmentation considered better (based on visual interpretation) was obtained [16]. After the segmentation was completed, the class "vegetation" was classified by using the Normalized Difference Vegetation Index (NDVI) [17]. This class includes all the vegetation in the image, both herbaceous and arboreal. The determination of the optimal NDVI value for the vegetation discrimination was performed by using the automatic thresholding algorithm which adapts the Otsu's method [18]. This algorithm works by implementing an iterative process in which all the possible thresholding values are tested until the optimal one is found, i.e., the one that maximizes the between-class variance of the image [19]. The objectives of this procedure is "to split" the image in two classes: "vegetation" and "non-vegetation". For this purpose, the multi-threshold segmentation algorithm was used. It splits the domain according to pixel values by creating image objects and classifying them basing on the optimal threshold value (NDVI's value in this case) [14].

"Vegetation" and "non-vegetation" classes' segmentation was separately "refined" reapplying the multiresolution segmentation by using the following parameters: 30 for scale parameter, 0.1 for shape and 0.5 for compactness.

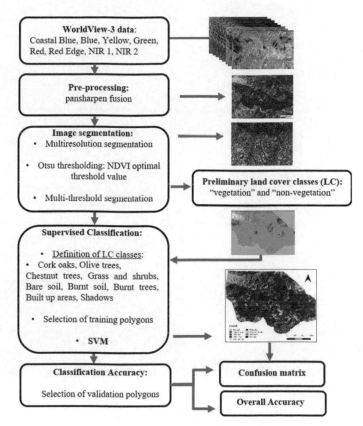

Fig. 2. Workflow of the methodology implemented.

2.4   Supervised Classification

A supervised classification of the imagery was implemented in eCognition by using Support Vector Machine (SVM) classification algorithm [20, 21]. SVM was chosen as it proved to be a more stable classifier in terms of accuracy as showed in [22]. Furthermore, SVM demonstrated its efficacy in several remote sensing applications in forest ecosystems [23–26]. The classification was implemented based on nine Land Cover (LC) classes: Bare soil, Built up areas, Burnt soil, Burnt trees, Chestnut trees, Cork Oaks, Grass and shrubs, Olive trees, and Shadows. The classes were chosen on the basis of some field surveys aimed at identifying the most common plant species present in the area. These were photographed and georeferenced as shown in Fig. 3. The selection of trainers (490 in total) was performed manually by on-screen interpretation while taking into consideration a good distribution among the LC classes and the different chromatic shades characterizing the study area.

Fig. 3. Google Maps screenshot showing the location of some of the photos taken during the study area field surveys (top). Photos of cork oaks taken during field surveys (bottom).

The supervised classification was performed by using SVM set with the following parameter: linear kernel-type with a model-type based on a C value (representing the size of misclassification allowed for non-separable training data) equal to 2. The training phase and the subsequent classification were separately performed on "non-vegetation" and "vegetation" image's parts. In particular, the software was firstly trained to classify only the classes not including vegetation or live vegetation: "Bare soil", "Built up areas", "Burnt soil", "Burnt trees", and "Shadows". Finally, the same thing was done on the part of the image first generally identified as "vegetation" classifying the following classes: "Chestnut trees", Cork Oaks", "Grass and shrubs", and "Olive trees".

2.5 Classification Accuracy

As a first step to carry out the classification accuracy validation a sample of 900 random points was created. Each point was labeled according to the defined LC classes by visual interpretation and field survey (ground truth). All polygons containing the sampling points were selected and for each of them the ground truths were compared with the classified LC class. The user's accuracy (UA - the ratio between the correctly classified polygons in a given class and all the classified polygons in that class), the producer's accuracy (PA - given by the ratio between the correctly classified polygons in a given class and the number of validation polygons for that class), and the Overall Accuracy (OA - the total percentage of correct classification) were computed [27] (Table 1).

3 Results and Discussions

For the map resulting from the supervised classification, showed in Fig. 4, the overall accuracy was 0.88 (Table 1).

Regarding UA and PA, the cork oak woodlands land cover class had high values as showed in Table 1 (0.90). The cork oak was correctly classified in most cases showing a distribution and a predominant presence compared to olive and chestnut in line with what was observed during the field surveys. The accuracy values for the classes Chestnut trees and Olive trees were mainly related to the difficulties encountered in discriminating the two species from the class Cork oaks. The explanation is traceable by observing the spectral signature of the three species that presents evident differences in their trend in the bands of the Red edge and NIR (Fig. 5). If compared, lower reflectance values characterize the signature of olive trees while higher values are found in the signature of chestnut trees. For this reason, there were no problems in the discrimination of the two species that were classified distinctly from each other. Instead, the spectral signature of cork oaks has a trend, in the bands indicated above, "intermediate" compared to that of olive and chestnut trees. In fact, in some cases, the cork oak has been confused with chestnut or olive.

Olive trees class shows a lower UA (0.60), is due to cases where olive trees have been misclassified as cork oaks (and Grass and shrubs in some cases). These errors can be explained by spectral characteristics and similar shape seen from the nadiral point of view of olive trees and grass and shrubs whose discriminating factor can be the height [3]. The use of features other than spectral features such as digital elevation models could improve the accuracy results of both the classes.

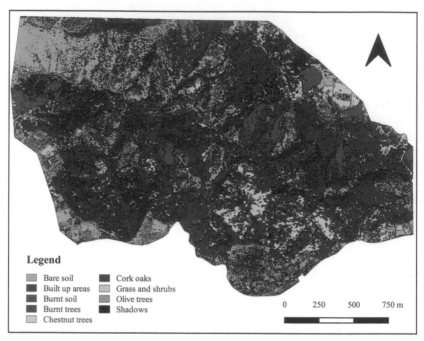

Fig. 4. Land use map of the study area obtained applying the implemented image object classification workflow.

Table 1. Confusion matrix for classification performed by using Support Vector Machine (SVM). The User's (UA) and Producer's Accuracy (PA) and the Overall Accuracy (OA) are reported.

Classes[a]	Ground truth									Tot	UA
	Bs	Ba	Brs	Brt	Cht	Co	Gr	Ot	Sh		
Bs	**144**	0	0	1	0	7	3	3	0	158	0.91
Ba	0	**9**	0	0	0	0	0	0	0	9	1
Brs	0	0	**44**	2	0	0	0	0	0	46	0.95
Brt	4	0	0	**28**	1	0	0	0	0	33	0.84
Cht	0	0	0	0	**146**	9	4	1	1	161	0.90
Co	2	0	0	0	9	**309**	7	4	12	343	0.90
Gr	2	0	0	0	2	0	**53**	0	0	57	0.92
Ot	0	0	0	0	0	13	8	**34**	1	56	0.60
Sh	1	0	0	1	1	2	0	0	**32**	37	0.86
Tot	153	9	44	32	159	340	75	42	46	**OA 0.88**	
PA	0.92	1	1	0.87	0.91	0.90	0.70	0.80	0.69		

[a]Bs = Bare soil; Ba = Built up areas; Brs = Burnt soil; Brt = Burnt trees; Cht = Chestnut trees; Co = Cork oaks; Gr = Grass and shrubs; Ot = Olive trees; Sh = Shadows

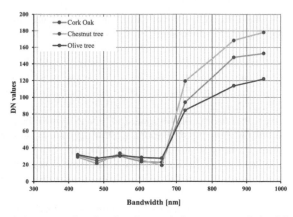

Fig. 5. The spectral signature of cork oak, olive and chestnut trees derived from the reflectance data of WorldView-3's image.

4   Conclusions and Further Research Perspectives

In this first step of the research activity, a supervised classification on WV-3's image was implemented to assess the potential performance of this sensor in detecting the presence of cork oak woodlands and distinguishing them from other vegetation, particularly from other spectrally similar tree species. Specific attention was paid to the distinction of cork oaks from olive and chestnut trees, widely present in the region under study. By exploiting the panchromatic image with 31 cm resolution and multispectral image bands the objective was achieved and they were obtained a high OA (88%) and high UA and PA for Cork oaks class, this despite of similarities with olive and chestnut trees in their spectral signature. Results confirm the usefulness of WV-3 in the applications of RS on environment and forestry for mapping species distribution and monitoring health by using proper vegetation indices [28]. The performed classification of cork oaks distribution suggests the possibility of improving the results obtained in terms of accuracy by using other data such as CHM thanks to which it is possible separating trees 'canopy from the grass, for example [29]. Furthermore, exploiting the presence of geometric patterns such as those determined by the planting scheme of the olive trees can help further distinct between the various species. Therefore, the direction of future research will involve the analysis of WV-3 images by considering other features besides spectral features for classification purposes.

References

1. Sousa, V.B., Leal, S., Quilhó, T., Pereira, H.: Characterization of cork oak (Quercus suber) wood anatomy. IAWA J **30**, 149–161 (2009). https://doi.org/10.1163/22941932-90000210
2. Modica, G., Pollino, M., Solano, F.: Sentinel-2 imagery for mapping cork oak (Quercus suber L.) distribution in Calabria (Italy): capabilities and quantitative estimation. In: Calabrò, F., Della Spina, L., Bevilacqua, C. (eds.) ISHT 2018. SIST, vol. 100, pp. 60–67. Springer, Cham (2019). https://doi.org/10.1007/978-3-319-92099-3_8

3. De Luca, G., et al.: Object-based land cover classification of cork oak woodlands using UAV imagery and Orfeo ToolBox. Remote Sens. **11**, 1238 (2019). https://doi.org/10.3390/rs1110 1238

4. Solano, F., Praticò, S., Piovesan, G., Chiarucci, A., Argentieri, A., Modica, G.: Characterizing historical transformation trajectories of the forest landscape in Rome's metropolitan area (Italy) for effective planning of sustainability goals. Land Degrad. Dev. **32**, 4708–4726 (2021). https://doi.org/10.1002/ldr.4072

5. Nolè, G., Lasaponara, R., Lanorte, A., Murgante, B.: Quantifying urban sprawl with spatial autocorrelation techniques using multi-temporal satellite data. Int. J. Agric. Environ. Inf. Syst. **5**, 19–37 (2014). https://doi.org/10.4018/IJAEIS.2014040102

6. Rogan, J., Miller, J.: Integrating GIS and remotely sensed data for mapping forest disturbance and change. In: Understanding Forest Disturbance and Spatial Pattern, pp. 133–171. CRC Press (2006)

7. Modica, G., et al.: Using Landsat 8 imagery in detecting cork oak (Quercus suber L.) woodlands: a case study in Calabria (Italy). J. Agric. Eng. **47**, 205–215 (2016). https://doi.org/10.4081/jae.2016.571

8. maxar. www.maxar.com. Accessed 12 Dec 2021

9. Ye, B., Tian, S., Ge, J., Sun, Y.: Assessment of WorldView-3 data for lithological mapping. Remote Sens. **9**, 1–19 (2017). https://doi.org/10.3390/rs9111132

10. Meng, X., Shen, H., Li, H., Zhang, L., Fu, R.: Review of the pansharpening methods for remote sensing images based on the idea of meta-analysis: practical discussion and challenges. Inf. Fusion **46**, 102–113 (2019). https://doi.org/10.1016/j.inffus.2018.05.006

11. Choudhury, M.A.M., Marcheggiani, E., Galli, A., Modica, G., Somers, B.: Mapping the urban atmospheric carbon stock by LiDAR and WorldView-3 data. Forests **12**, 692 (2021). https://doi.org/10.3390/f12060692

12. Padwick, C., Deskevich, M., Pacifici, F., Smallwood, S.: WorldView 2 pan-sharpening. In: ASPRS, Annual Conference, San Diego, USA (2010)

13. Baatz, M., Schäpe, A.: Multi-resolution segmentation: an optimization approach for high quality multi-scale. In: Beiträge zum, Agit XII Symp Salsburg, pp. 12–23 (2000). https://doi.org/10.1207/s15326888chc1304_3

14. Trimble Germany GmbH: Trimble Documentation eCognition Developer 10.1 Reference Book (2021)

15. El-naggar, A.M.: Determination of optimum segmentation parameter values for extracting building from remote sensing images. Alexandria Eng. J. **57**, 3089–3097 (2018). https://doi.org/10.1016/j.aej.2018.10.001

16. Messina, G., Praticò, S., Badagliacca, G., Di Fazio, S., Monti, M., Modica, G.: Monitoring onion crop "cipolla rossa di tropea calabria igp" growth and yield response to varying nitrogen fertilizer application rates using UAV imagery. Drones **5** (2021). https://doi.org/10.3390/drones5030061

17. Rouse, W., Haas, R.H., Deering, D.W.: Monitoring vegetation systems in the Great Plains with ERTS, NASA SP-351. In: Third ERTS-1 Symposium, vol. 1 (1974)

18. Otsu, N.: A threshold selection method from gray-level histograms. IEEE Trans. Syst. Man Cybern. **9**, 62–66 (1979). https://doi.org/10.1109/TSMC.1979.4310076

19. López-Granados, F., Torres-Sánchez, J., De Castro, A.-I., Serrano-Pérez, A., Mesas-Carrascosa, F.-J., Peña, J.-M.: Object-based early monitoring of a grass weed in a grass crop using high resolution UAV imagery. Agron. Sustain. Dev. **36**(4), 1–12 (2016). https://doi.org/10.1007/s13593-016-0405-7

20. Cortes, C., Vapnik, V.: Support-vector networks editor. Mach. Learn. **20**, 273–297 (1995). https://doi.org/10.1023/A:1022627411411

21. Vapnik, V.: Statistical Learning Theory. Wiley, New York (1998)

22. Modica, G., De Luca, G., Messina, G., Praticò, S.: Comparison and assessment of different object-based classifications using machine learning algorithms and UAVs multispectral imagery: a case study in a citrus orchard and an onion crop. Eur. J. Remote Sens. (2021). https://doi.org/10.1080/22797254.2021.1951623

23. Mountrakis, G., Im, J., Ogole, C.: Support vector machines in remote sensing: a review. ISPRS J. Photogramm. Remote Sens. **66**, 247–259 (2011). https://doi.org/10.1016/j.isprsjprs.2010.11.001

24. Adam, E., Mutanga, O., Odindi, J., Abdel-Rahman, E.M.: Land-use/cover classification in a heterogeneous coastal landscape using RapidEye imagery: evaluating the performance of random forest and support vector machines classifiers. Int. J. Remote Sens. **35**, 3440–3458 (2014). https://doi.org/10.1080/01431161.2014.903435

25. Hawryło, P., Bednarz, B., Wężyk, P., Szostak, M.: Estimating defoliation of Scots pine stands using machine learning methods and vegetation indices of Sentinel-2. Eur. J. Remote Sens. **51**, 194–204 (2018). https://doi.org/10.1080/22797254.2017.1417745

26. Wessel, M., Brandmeier, M., Tiede, D.: Evaluation of different machine learning algorithms for scalable classification of tree types and tree species based on Sentinel-2 data. Remote Sens. **10**, 1419 (2018). https://doi.org/10.3390/rs10091419

27. Congalton, R.G., Green, K.: Assessing the Accuracy of Remotely Sensed Data (2019)

28. Yu, L., Zhan, Z., Ren, L., Zong, S., Luo, Y., Huang, H.: Evaluating the potential of WorldView-3 data to classify different shoot damage ratios of Pinus yunnanensis. Forests **11**, 417 (2020). https://doi.org/10.3390/f11040417

29. Modica, G., Messina, G., De Luca, G., Fiozzo, V., Praticò, S.: Monitoring the vegetation vigor in heterogeneous citrus and olive orchards. A multiscale object-based approach to extract trees' crowns from UAV multispectral imagery. Comput. Electron. Agric. **175**, 105500 (2020). https://doi.org/10.1016/j.compag.2020.105500

Towards Quantifying Rural Environment Soil Erosion: RUSLE Model and Remote Sensing Based Approach in Basilicata (Southern Italy)

Valentina Santarsiero[1,2](✉) (iD), Gabriele Nolè[1], Antonio Lanorte[1], and Beniamino Murgante[2]

[1] IMAA-CNR, C.da Santa Loja, Zona Industriale, Tito Scalo, Italy
{Valentina.santarsiero,gabriele.nole, antonio.lanorte}@imaa.cnr.it
[2] School of Engineering, University of Basilicata, Viale dell'Ateneo Lucano 10, 85100 Potenza, Italy
{valentina.santarsiero,beniamino.murgante}@unibas.it

Abstract. Land degradation is a phenomenon that describes the degradation of soil quality, the causes are multiple, some dynamics related to agriculture have particularly influenced the process of degradation. Specifically, agricultural over-exploitation with unsustainable practices and the abandonment of agricultural land can cause pedological alterations of the most superficial layers of the soil that with intense rainfall and trigger the process of soil erosion. The aim of this work is to evaluate the dynamics and relationships between erosion processes and land cover changes. The approach employed is based on GIS and remote sensing. An initial analysis involved the application of the Revised Universal Soil Loss Equation (RUSLE) model to calculate soil erosion on an annual basis. The resulting data were then processed through a Getis-Ord local autocorrelation index to produce a persistent erosion map. The resulting dataset was compared to land cover classes of arable and brownfield areas. Finally, an attempt was made to quantify and test the relationships between erosion rate and the period of abandonment. Models and survey techniques have been applied in a rural area of Basilicata Region (South Italy) using exclusively a GIS Free and Open-Source Software and remote sensing approach.

Keywords: Soil erosion · Remote sensing · Abandoned arable land

1 Introduction

Land degradation is a complex phenomenon that describes the degradation of soil quality due to which agricultural land in particular is unproductive as a consequence of the loss of ability to produce crops and biomass caused by multiple factors that limit or inhibit the productive, regulatory and utilitarian functions as well as eco-system services that a natural soil can offer can offer [1–4]. The causes are diverse, including some agricultural

© The Author(s), under exclusive license to Springer Nature Switzerland AG 2022
F. Calabrò et al. (Eds.): NMP 2022, LNNS 482, pp. 2163–2172, 2022.
https://doi.org/10.1007/978-3-031-06825-6_208

practices that have particularly influenced the degradation process. In particular, agricultural overexploitation with unsustainable practices and land abandonment are causing ecological alterations that require contextual analysis to assess medium- and long-term effects [5–8].

The United Nations Convention to Combat Desertification (UNCCD) [2] has developed a methodology for qualitative assessment using a combined approach of the following sub-areas provides for the combined use of the following sub-indicators: land use and land cover change, loss of productivity, loss of ecosystem services, fragmentation, firers, etc. A first step to analyze land degradation is to highlight which territorial and phenomenological aspects are closely linked to this phenomenon. One of the most important indicators in the definition of the phenomenon is certainly the change in land cover followed by loss of productivity [2]. In the end land degradation is the reduction in the capability of the land to produce benefits from a particular land use under a specified form of land management. Soil degradation is one aspect of land degradation; others are degradation of vegetation or water resources [9]. Oldeman, recognized two categories of soil degradation: the first is soil degradation due to displacement of soil material such as soil erosion by wind and water; the second is related to degradation due to chemical process like loss of nutrients and organic matter, salinization, etc. [10].

Current climate change processes have been recognized as important factors in land degradation, e.g., global warming can affect seasonality and precipitation amounts [11–13]. Short and intense rainfall affects soils without vegetation cover, the resulting runoff removes the surface layer rich in organic matter from the soil. Arid, semi-arid and subhumid areas with sunny exposures are generally more at risk because they are often affected by this type of rainfall that triggers erosion processes [3, 4].

Agricultural overexploitation and abandonment of agricultural land are among the most cited causes to trigger phenomena of surface erosion, in particular some agricultural practices oriented towards mechanical processing and the use of chemicals can reduce soil quality and subsequent degradation. The abandonment of agricultural activities is also under the attention of the scientific community because the complete abandonment of soils, in addition to producing socio-economic and landscape impacts, can trigger a whole series of degradation phenomena. Soil erosion represent a most serious consequences of land degradation hazard in the Mediterranean Basin. It has an adverse influence on land degradation and agricultural productivity problems [14, 15]. Soil erosion consists in the detachment and transport of soil particles and also nutrients from the most superficial layers of the soil, decreasing the quality of the soil and reducing the productivity of the affected lands [16]. According to the Italian Institute for Environmental Protection and Research (ISPRA) about 10% of the national territory is at risk of soil degradation, and in particular in the Basilicata region the percentage of the territory affected by the phenomenon is about 25% [17]. The effects of soil erosion can be change in relation to territorial contexts and therefore depend on climatic factors, ecological, biological, pedological and topographical factors. Land abandonment, soil erosion and land degradation are closely linked and this connection is the subject of heated debates in the literature and needs in-depth methodological analysis and careful study of the dynamics and triggers that can often be divergent. The Basilicata Region (Southern Italy) is one of the Italian regions to have undergone over the years an intense process of exodus and demographic decline, which has led to intense agricultural abandonment with consequent land degradation and erosion processes [18]. In the literature, most of the studies of erosion of its are based on the use of RUSLE model, which takes into consideration several factors such as precipitation, soil erodibility factor, loss, to land cover and erosion control practices [19–21]. The present work concerned the estimation of the erosion rate by applying the RUSLE model in an area of Basilicata (Italy).

Starting from a study area located in Southern Italy (Basilicata Region) and affected by problems of agricultural abandonment, land degradation [3] and soil consumption [27], GIS and remote sensing techniques were used to perform a preliminary investigation to assess the relationship between land degradation and soil erosion. After an initial statistical survey, arable land and areas with post-crop vegetation potentially susceptible to land degradation were mapped based on erosion rates estimated using RUSLE methodology and grouped using autocorrelation techniques.

This work was based on the integration of remote sensing techniques and Geographic Information System (GIS) with the use of Free and Open-Source Software technologies and open datasets e freely available. The integration of the different datasets and the application of new models, represent an opportunity to study and monitor the evolution of the territory with a detailed temporal and spatial scale. The increasing efficiency of analysis models and the interoperability of the various data, represent a strength for the planner in terms of definition of plans and strategies consistent with real needs and environmental problems [22–24]. The purpose of this work is to quantify and identify areas at high risk of erosion and degradation, and then to compare erosion values with land cover classes, specifically in abandoned arable areas.

2 Materials and Methods

2.1 Study Area

The study area covered a portion of the Basilicata Region, made up of about 20 municipalities, located in the eastern part of the regional territory, bordering the Apulia Region. It is located in a territorial context particularly suited to agriculture and affected by the phenomenon of the abandonment of agricultural soils in recent decades, a territory, therefore, particularly vulnerable to soil erosion due to the combination of anthropogenic and natural factors. The climate is Mediterranean, with a pronounced bistagional regime characterized by hot, dry summers and wet, cold winters.

Agricultural areas of different types, such as arable land, permanent crops (olive groves, orchards, vineyards), permanent meadows, cover almost the entire study area (80% of the territory), the remainder is covered by natural areas (wooded areas, shrub and grassland) for about 18% [25] (Fig. 1, Table 1).

From a geomorphological point of view, it is possible to ideally divide the study area into two sub-areas: the western part characterized by a more complex and diversified morphology and the eastern part more heterogeneous. Geologically, the area considered falls almost entirely within the domain of the Bradanic Foredeep [26], which constitutes 43% of the entire Lucanian territory, and only in small part within the domain of the Apennine chain of hills [27].

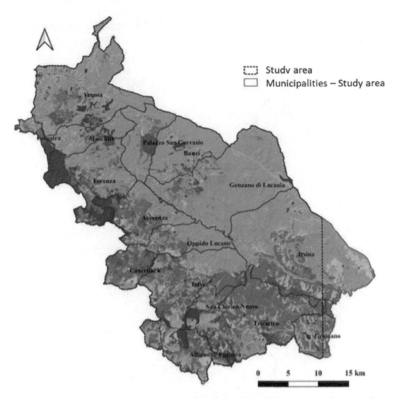

Fig. 1. Map of study area (dashed line) and related municipalities and land cover classes based on the 2013 Nature Map [25].

Table 1. Land cover based on the 2013 Nature Map expressed in square kilometers (km^2) and percentage (%)

Level Nature Map 2013	Km^2	Percentage %
Agricultural areas	1266.19	81.45
Artificial areas	12.44	0.80
Forest and semi-natural areas	2271.40	17.46
Water	3.77	0.24
Wetlands	0.79	0.05

The eastern part presents gentle morphologies, engraved by the valleys of the main water courses streams, and that are connected with regularity to the plains, to the marine terraces, the prevalence of clay outcrops creates a landscape at times gully.

The territory under analysis is affected by landslide phenomena of different nature, which are differ according to the lithology of the outcropping soil. In the study area,

858 landslides were surveyed up to 2014 by the IFFI Project (Inventory of Landslide Phenomena in Italy) [28], distributed as in Fig. 2.

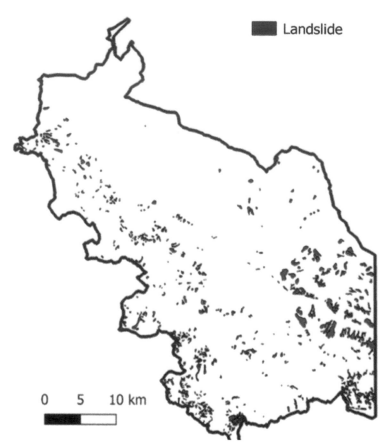

Fig. 2. Spatial distribution of landslides cataloged by the IFFI Project falling within the study area.

2.2 Methodology

The whole procedure has been realized with the open source software QGIS [29] and its plugins. The reference satellite data is represented by the Sentinel 2 medium-high resolution images provided by the Copernicus Mission [30] and the land cover classification was based on the 2013 Nature Map at scale 1:50.000 in a format freely available from the ISPRA website.

The first part of the work involved the implementation and application of useful models to estimate and map areas with high erosion rates.

Soil erosion was estimated using the RUSLE (Revised Universal Soil Loss Equation) [19], resampling all necessary parameters all necessary parameters to the spatial resolution of Sentinel 2A (10 m). The observation period of this work is from September

2019 to October 2020. The estimate of annual soil loss according to the RUSLE model is a function of five variables related to precipitation regime (R), soil characteristics (K), topography (LS), land crop cover and management (C), and conservation cropping practices (C), according to the following formula:

$$A = R * K * LS * C * P$$

The annual soil erosion assessment consisted of the following steps: geospatial data collection, estimation of soil loss with RUSLE for annual scenarios considering the changes recorded by K, LS and C factors.

The RUSLE values were first calculated monthly, through the implementation of specific model in QGIS to realize a batch processing in order to calculate the different parameters in a semi-automatic way. Then the monthly values were summed to have an annual RUSLE value.

The integration of remote sensing data and geostatistical analysis is an innovative approach for analysis and mapping based on factors influenced by the spatial and geographical component [4, 31]. The integration of remote sensing data and geostatistical analysis is an innovative approach for analysis and mapping based on factors influenced by the spatial and geographical component. By considering occurrences of a spatial variable (e.g., areas of persistent erosion), spatial autocorrelation measures the degree of dependence between events, considering their similarity and long-distance relationships [32, 33]. In particular, the issue of spatial autocorrelation is crucial to assess whether a particularly intense phenomenon in a specific area, implies the presence of the same in contiguous areas as well [34]. In the present work, autocorrelation indices were applied to erosion estimation maps derived from the application of the RUSLE model to produce an annual map of persistent erosion. Local indicators of spatial autocorrelation allowed us to locate clustered pixels by measuring the number of elements within the "fixed neighbor" file that are homogeneous. After several investigations, for this work we chose to use the local autocorrelation index Gi proposed by Ord and Getis [35]. Statistical interpretation of this index allows values to be grouped on the basis of a hot spot (pixel values above the mean) or cold spot (values below the mean).

The index was applied to the annual map of RUSLE values highlighting only pixels with positive autocorrelation. This allowed the development of a map highlighting areas with persistent erosion rates based on clusters of hot spots.

3 Results and Discussion

The purpose of this work was to relate erosion data to land cover in order to assess how this process may influence degradation phenomena and the relationship between the period of abandonment of agricultural activities with erosion rates.

Preliminary phase of the study was to identify on the basis of the Nature Map the abandoned agricultural areas and the areas under arable cultivation.

Next, a statistical investigation of the annual mean values of RUSLE was made with respect to cultivable areas and areas with post-crop vegetation (abandoned agricultural areas). The analysis on the annual values shows an erosion rate with the same order of magnitude in the two classes, but with a slightly higher value in the areas with post cultivation vegetation. In fact, the arable land areas show an erosion value of 22.92 (Mg-ha -1 -year -1), while the abandoned ones of (26.23 Mg-ha -1 -year -1). This small difference is extremely important to investigate because, generally, high values are noted in arable crops as they have long periods of the year with bare soil. The reason could be that areas with post-crop vegetation, have a cover type (expressed by the C-factor) and morphological context that could influence erosion.

Geostatistical operations carried out through the use of Getis & Ord.'s autocorrelation index, have made it possible to create a map of permanent erosion that takes into account the spatial and geographical relationship that may exist between contiguous areas [4]. These were compared to the Nature Map land cover classes to assess which classes were most affected in terms of surface area by permanent erosion. It emerged that, in percentage terms, the most representative classes are those related to arable land (22.48%) and areas with abandoned land (61.125) due to abandonment of agricultural activity. The results obtained from the statistical analyses were compared to the IFFI Project Landslide Map to assess which landslide type was affected by permanent erosion relative to the RUSLE data. The following histogram (Fig. 32) represents the result of the statistical analysis performed between the areas of the Landslide Map classified by activity status and type of movement compared with the Map of areas of permanent erosion identified to the spatial autocorrelation analysis performed by applying the Getis algorithm, what is inferred is that the areas identified as areas of permanent erosion are, for the majority, areas subject to surface and diffuse landslides, active and/or reactivated. The results obtained from the statistical analysis were compared with the Landslide Map of the IFFI Project, what inferred is that the areas identified as areas of permanent erosion are, for the majority, areas subject to surface and diffuse landslides, active and/or reactivated.

The map of permanently eroded areas was then compared to the areas affected by agricultural abandonment previously divided into decades (1990/2000, 2000/2010, 2010/2020). It was found that areas that were abandoned in the 1990s have RUSLE values of 22 (Mg-ha -1 -year -1) while those abandoned in the second and third decades have erosion values of 37 and 31. This first analysis shows that soils abandoned in the first decade present values of soil lost compared to the following decades, this probably because these areas have rinaturalized and stabilized over time. It is more difficult to discriminate and evaluate the abandonment from 2010 to the present, because these areas can be subject to vegetation rest. areas may be subject to vegetative rest and/or crop rotation, and therefore require further analysis. Subsequently, the average value of eroded soil for the agricultural areas, what can be deduced is that the average of the values of RUSLE in agricultural areas is about 19 (Mg-ha -1 -year -1), if the values of erosion for the arable land classes alone this amounts to about 17 (Mg-ha -1 -year -1) and about 33 (Mg-ha -1 -year -1) for extensive crops and complex agricultural systems.

4 Conclusion

The use of satellite data and GIS tools have provided useful data for estimating erosion risk, mapping and monitoring areas prone to degradation. These methods are mainly based on the use of indices obtained from the combination of different spectral bands, which emphasize and detect any change in the state of vegetation. Integration of soil erosion models (RUSLE model) with GIS and remote sensing have proven to be effective tools for mapping and quantifying areas and rates of soil erosion for the development of better conservation and monitoring plans for the land. In addition, the use of spatially explicit geostatistical surveys allows for a more accurate quantitative analysis of the various results obtained. In conclusion, it is possible to deduce from the analyses carried out that land cultivated with arable crops in the during the period considered present values of lost soil lower than that of the soils abandoned in the three decades analyzed.

References

1. Stringer, L.: Can the UN Convention to Combat Desertification guide sustainable use of the world's soils? Front. Ecol. Environ. **6**, 138–144 (2008). https://doi.org/10.1890/070060
2. A year in review: UNCCD (2017). https://www.unccd.int/news-events/2017-year-review. Accessed 27 Dec 2021
3. Santarsiero, V., et al.: Assessment of post fire soil erosion with ESA sentinel-2 data and RUSLE method in Apulia region (Southern Italy). In: Gervasi, O., et al. (eds.) ICCSA 2020. LNCS, vol. 12252, pp. 590–603. Springer, Cham (2020). https://doi.org/10.1007/978-3-030-58811-3_43
4. Cillis, G., et al.: Soil erosion and land degradation in rural environment: a preliminary GIS and remote-sensed approach. In: Gervasi, O., et al. (eds.) ICCSA 2021. LNCS, vol. 12954, pp. 682–694. Springer, Cham (2021). https://doi.org/10.1007/978-3-030-86979-3_48
5. Tucci, B., et al.: Assessment and monitoring of soil erosion risk and land degradation in arable land combining remote sensing methodologies and RUSLE factors. In: Gervasi, O., et al. (eds.) ICCSA 2021. LNCS, vol. 12954, pp. 704–716. Springer, Cham (2021). https://doi.org/10.1007/978-3-030-86979-3_50
6. Santarsiero, V., et al.: A remote sensing methodology to assess the abandoned arable land using NDVI index in Basilicata region. In: Gervasi, O., et al. (eds.) ICCSA 2021. LNCS, vol. 12954, pp. 695–703. Springer, Cham (2021). https://doi.org/10.1007/978-3-030-86979-3_49
7. Lanucara, S., Praticò, S., Modica, G.: Harmonization and interoperable sharing of multi-temporal geospatial data of rural landscapes. In: Calabrò, F., Della Spina, L., Bevilacqua, C. (eds.) ISHT 2018. SIST, vol. 100, pp. 51–59. Springer, Cham (2019). https://doi.org/10.1007/978-3-319-92099-3_7
8. Amato, F., Tonini, M., Murgante, B., Kanevski, M.: Fuzzy definition of Rural Urban Interface: an application based on land use change scenarios in Portugal. Environ. Model. Softw. **104**, 171–187 (2018). https://doi.org/10.1016/j.envsoft.2018.03.016
9. Blaikie, P., Brookfield, H. (eds.): Land Degradation and Society (2015). https://doi.org/10.4324/9781315685366
10. Oldeman, L.R.: Global extent of soil degradation. ISRIC BI-annual report, pp. 19–36 (1991)
11. Vicente-Serrano, S.M., et al.: Drought variability and land degradation in semiarid regions: assessment using remote sensing data and drought indices (1982–2011). Remote Sens. **7**, 4391–4423 (2015). https://doi.org/10.3390/RS70404391
12. Trenberth, K.E.: Changes in precipitation with climate change. Clim. Res. **47**, 123–138 (2011). https://doi.org/10.3354/CR00953
13. Barnosky, A.D., et al.: Approaching a state shift in Earth's biosphere (2012). https://doi.org/10.1038/nature11018
14. Zakerinejad, R., Maerker, M.: An integrated assessment of soil erosion dynamics with special emphasis on gully erosion in the Mazayjan basin, southwestern Iran. Nat. Hazards **79**(1), 25–50 (2015). https://doi.org/10.1007/s11069-015-1700-3

15. Gayen, A., Saha, S.: Application of weights-of-evidence (WoE) and evidential belief function (EBF) models for the delineation of soil erosion vulnerable zones: a study on Pathro river basin, Jharkhand, India. Model. Earth Syst. Environ. **3**(3), 1123–1139 (2017). https://doi.org/10.1007/s40808-017-0362-4

16. Nolè, G., et al.: Model of post fire erosion assessment using RUSLE method, GIS tools and ESA sentinel DATA. In: Gervasi, O., et al. (eds.) ICCSA 2020. LNCS, vol. 12253, pp. 505–516. Springer, Cham (2020). https://doi.org/10.1007/978-3-030-58814-4_36

17. Consumo di suolo, dinamiche territoriali e servizi ecosistemici Edizione 2021 Rapporto ISPRA SNPA (2021)

18. Statuto, D., Cillis, G., Picuno, P.: Using historical maps within a GIS to analyze two centuries of rural landscape changes in Southern Italy. Land **6**, 65 (2017). https://doi.org/10.3390/LAND6030065

19. Renard, K.G., Foster, G.R., Weesies, G.A., Porter, J.P.: RUSLE: revised universal soil loss equation. J. Soil Water Conserv. **46**, 30–33 (1991)

20. Predicting Soil Erosion by Water: A Guide to Conservation Planning with the ... - Kenneth G. Renard, Stati Uniti d'America. Department of Agriculture. Agricultural Research Service - Google Libri. https://books.google.it/books?hl=it&lr=&id=cQEUAAAAYAAJ&oi=fnd&pg=PR7&dq=Renard,+K.+G.+(1997).+Predicting+soil+erosion+by+water:+a+guide+to+conservation+planning+with+the+Revised+Universal+Soil+Loss+Equation+(RUSLE).+United+States+Government+Printing.&ots=HCNdubcwPi&sig=IbXeOKdSpCEr4-L8fZ0i2hJAwVQ&redir_esc=y#v=onepage&q&f=false. Accessed 27 Dec 2021

21. Terranova, O., Antronico, L., Coscarelli, R., Iaquinta, P.: Soil erosion risk scenarios in the Mediterranean environment using RUSLE and GIS: an application model for Calabria (Southern Italy). Geomorphology **112**, 228–245 (2009). https://doi.org/10.1016/J.GEOMORPH.2009.06.009

22. Murgante, B., Borruso, G., Lapucci, A. (eds.): Geocomputation, Sustainability and Environmental Planning, p. 269. Springer, Heidelberg (2011). https://doi.org/10.1007/978-3-642-19733-8

23. Las Casas, G., Scorza, F., Murgante, B.: New urban agenda and open challenges for urban and regional planning. In: Calabrò, F., Della Spina, L., Bevilacqua, C. (eds.) ISHT 2018. SIST, vol. 100, pp. 282–288. Springer, Cham (2019). https://doi.org/10.1007/978-3-319-92099-3_33

24. Murgante, B., Borruso, G., Balletto, G., Castiglia, P., Dettori, M.: Why Italy first? Health, geographical and planning aspects of the COVID-19 outbreak. Sustainability **12** (2020). https://doi.org/10.3390/su12125064

25. Carta della Natura—Italiano. https://www.isprambiente.gov.it/it/servizi/sistema-carta-della-natura. Accessed 28 Dec 2021

26. Doglioni, C.: Some remarks on the origin of foredeeps. Tectonophysics **228**, 1–20 (1993). https://doi.org/10.1016/0040-1951(93)90211-2

27. Tropeano, M., Sabato, L., Pieri, P.: Filling and cannibalization of a foredeep: the Bradanic Trough, Southern Italy. Geol. Soc. Lond. Spec. Publ. **191**, 55–79 (2002). https://doi.org/10.1144/GSL.SP.2002.191.01.05

28. IFFI - Inventario dei fenomeni franosi in Italia—Italiano. https://www.isprambiente.gov.it/it/progetti/cartella-progetti-in-corso/suolo-e-territorio-1/iffi-inventario-dei-fenomeni-franosi-in-italia. Accessed 28 Dec 2021

29. Benvenuto in QGIS! https://qgis.org/it/site/. Accessed 24 Nov 2021

30. Sentinel Hub. https://www.sentinel-hub.com/. Accessed 24 Nov 2021

31. Lanorte, A., Danese, M., Lasaponara, R., Murgante, B.: Multiscale mapping of burn area and severity using multisensor satellite data and spatial autocorrelation analysis. Int. J. Appl. Earth Obs. Geoinf. **20**, 42–51 (2012). https://doi.org/10.1016/j.jag.2011.09.005

32. Anselin, L.: Local Indicators of Spatial Association—LISA. Geogr. Anal. **27**, 93–115 (1995). https://doi.org/10.1111/j.1538-4632.1995.tb00338.x

33. The Interpretation of Statistical Maps on JSTOR. https://www.jstor.org/stable/2983777. Accessed 28 Dec 2021
34. Curran, P.J., Atkinson, P.M.: Geostatistics and remote sensing **22**, 61–78 (2016). https://doi.org/10.1177/030913339802200103
35. Ord, J.K., Getis, A.: Local spatial autocorrelation statistics: distributional issues and an application. Geogr. Anal. **27**, 286–306 (1995). https://doi.org/10.1111/J.1538-4632.1995.TB00912.X

A Simplified Design Procedure to Improve the Seismic Performance of RC Framed Buildings with Hysteretic Damped Braces

Eleonora Bruschi[1(✉)], Virginio Quaglini[1], and Paolo M. Calvi[2]

[1] Department of Architecture, Built Environment and Construction Engineering, Politecnico di Milano, Piazza Leonardo da Vinci 32, 20133 Milan, Italy
{eleonora.bruschi,virginio.quaglini}@polimi.it

[2] Department of Civil and Environmental Engineering, University of Washington, More Hall, Seattle, WA 98195, USA
pmc85@uw.edu

Abstract. The study presents an effective and easy to use design procedure for the seismic upgrade of frame structures equipped with hysteretic dampers, in order to attain, for a specific level of seismic intensity, a designated performance level. The frame-dampers structural system is idealized as an equivalent Single Degree of Freedom (SDOF) system, characterized through its secant stiffness and equivalent viscous damping. The global stiffness and strength of the equivalent SDOF system are distributed along the height of the frame according to a stiffness-proportionality criterion, which constrains the drifts of the braced frame to follow the first mode deformation of the main frame. To demonstrate the effectiveness of the design procedure, a case-study is chosen as paradigmatic of a category of existing buildings in Italy designed according to outdated codes. In order to assess the suitability of the design procedure, non-linear static analyses are performed on the upgraded building, showing a satisfactory agreement between the design target and numerical capacity curves. To have a deeper insight, also bidirectional non-linear dynamic analyses are performed in accordance with the provisions of the Eurocode 8.

Keywords: Seismic retrofit · Frame structures · Hysteretic damper

1 Introduction

Supplementary energy dissipation has been employed successfully both in new and retrofitted constructions in order to prevent structural damage, increase life-safety and achieve a desired level of performance [1, 2], appearing an appropriate and affordable solution to reduce the vulnerability of ordinary structures, such as residential, school and industrial buildings [3–5]. This technique consists in inserting special devices, called dampers, inside the structure, with the goal of achieving two effects, namely increase the structural stiffness, with consequent reduction of displacements, and dissipate much of the seismic energy, controlling both displacements and accelerations [6, 7].

© The Author(s), under exclusive license to Springer Nature Switzerland AG 2022
F. Calabrò et al. (Eds.): NMP 2022, LNNS 482, pp. 2173–2182, 2022.
https://doi.org/10.1007/978-3-031-06825-6_209

Even if this technique has been proved to be suitable for any dynamic excitation, independently of the frequency content (indeed it is effective also for structures resting on soft soils and for high-rise buildings) [8], the implementation of these systems for ordinary buildings is discouraged in practice. In fact, practitioners still have little confidence in using energy dissipation strategies due to unavailability of design procedures ready to be adopted and of seismic codes that properly address specific provisions [9]. Moreover, current codes disregard some potential issues investigated in recent studies such as the cyclic engagement of hysteretic dampers [10]. Several procedures have been proposed in literature for the design of supplementary energy dissipation systems (e.g., [11–15]), however, the common practice for the seismic upgrade of existing structures consists in determining the size and the suitable location of the dissipation units within the building starting from a trial configuration based on the engineer's expertise, and then verifying the retrofitted structure through dynamic or static analyses [16]. The properties and/or the number of the added dissipation units are iteratively changed until the target performance is reached; it is evident that this trial-and-error approach can be a laborious task.

The present study proposes an effective and easy to use procedure for the seismic upgrade of existing frame structures by means of hysteretic damped braces, consisting of two main parts: (i) a simple method based on the Capacity Spectrum Method [17, 18] to define the global properties of the damped brace system, described by means of an equivalent SDOF system; and (ii) a strategy to distribute the properties of the equivalent SDOF damped brace along the height and across each floor of the structure, which is derived from the literature [11] and based on the principle of distributing the stiffness and strength proportionally to the stiffness of each floor calculated via dynamic analysis.

The ease and effectiveness of the method is illustrated analyzing an existing 4-story frame building located in a medium/high seismic area and designed according to outdated standards [2], which needs to be retrofitted to comply with the performance requirements of the most recent Italian norm [19].

The procedure is applied to the case-study and the main results from non-linear static and dynamic analyses (NLSA and NLDA) are discussed. The effects of the distribution of the damper properties at the various floors are also highlighted by examining, as an alternative to the method recommended in the procedure, a second procedure available too in the literature [2].

2   Simplified Design Procedure of the Hysteretic Damped Braces

The design procedure is based on the Capacity Spectrum Method [17, 18] and consists of 5 steps [9]:

Step 1: definition of the capacity curve of the Main Frame
The capacity curve of the as-built structure ($V_F - d_F$) is determined via NLSAs following the prescriptions of the Italian Building Code (IBC) [19]: two lateral load distribution patterns are considered (i.e., a uniform pattern proportional to the floor masses m_i and a modal pattern proportional to the first mode eigenvector components ϕ_i, with i number of floors) are applied in both the positive and negative directions of each axis, considering 5% accidental eccentricity of the center of mass of each story, Fig. 1.

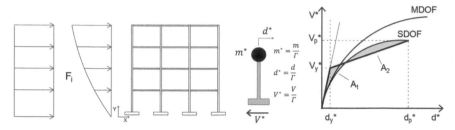

Fig. 1. Definition of the equivalent bilinear curve according to IBC [9]

The lowest $V_F - d_F$ curve is taken as the capacity curve of the main (unbraced) structure and then converted to the $V_F^* - d_F^*$ capacity curve of the equivalent SDOF system through the modal participation factor Γ according to the IBC [19].

Step 2: selection of the target displacement and definition of the equivalent bilinear capacity curve of the Main Frame
The target displacement d_p of the structure is selected depending on the required level of performance, according to the assumed design code, and applying the Eq. (1)

$$d_p = min \frac{\Delta_d \cdot h_i}{\delta_i} \tag{1}$$

where Δ_d is the target inter-story drift ratio, h_i is the height of the ith story, and δ_i is the difference between the first mode eigenvector components of the adjacent stories $(\phi_i - \phi_{i-1})$. The product $\Delta_d \cdot h_i$ represents the target drift of the ith story, if the inter-story height is uniform, i.e., $h_i = h_0$ for $i = 1, \ldots n$ (n = number of stories), then Eq. (1) turns into $d_p = \frac{\Delta_d \cdot h_0}{\delta_{max}}$, where $\delta_{max} = max(\phi_i - \phi_{i-1})$.

Once the target displacement d_p (and the corresponding base shear force of the main frame V_p^F) is assigned, the bilinear curve of the equivalent SDOF system is evaluated in accordance with IBC [19] and clause C.7.3.4.2 of the Commentary [20], by respecting three conditions, namely: (i) same initial stiffness as the initial stiffness of the MDOF capacity curve, (ii) crossing of the performance point (d_p^*, V_p^{*F}), and (iii) equivalence of areas A_1 and A_2 between the two curves, as shown in Fig. 1. This system is characterized by an equivalent secant stiffness $K_F^* = V_p^{*F}/d_p^*$ and an equivalent viscous damping ratio ξ_F (in percent) defined as [20, 21]:

$$\xi_F = \frac{\kappa_F \cdot 63.7 \cdot \left(V_y^{*F} d_p^* - V_p^{*F} d_y^{*F} \right)}{V_p^{*F} d_p^*} + 5 \tag{2}$$

where κ_F accounts for the energy dissipation capacity of the bare structure and can be taken as 1.0 for high damping capability, 0.66 for moderate damping capability, and 0.33 for low damping capability [17].

Step 3: check of the displacement for the relevant ξ_F
The equivalent bilinear capacity curve $V_F^* - d_F^*$ is converted into the capacity spectrum in the acceleration-displacement response spectrum (ADRS) space, where the spectral coordinates are defined as $S_a = V^{*F}/m^*$ (acceleration in m/s^2) and $S_d = d_F^*$

(displacement in m). The spectral displacement of the main frame for the considered seismic action is determined analytically. The secant stiffness of the capacity spectrum to the target displacement $K_F^* = V_p^{*F}/d_p^*$ is used to calculate the effective period of the main structure $T_F^* = 2\pi\sqrt{\frac{m^*}{K_F^*}}$ and hence the corresponding spectral displacement $S_d(T_F^*; \xi_F) = S_d(T_F^*; \xi = 5\%)\sqrt{\frac{10}{5+\xi_F}}$ where ξ_F is the equivalent viscous damping ratio of the unbraced structure defined by Eq. (2). In order to have a direct comparison to the capacity spectrum, the seismic action for the considered performance level is defined in terms of acceleration – displacement response spectra and plotted in the acceleration-displacement plane, with a set of coordinates defined by S_a and S_d. It should be noted that when the spectral values are plotted in ADRS format, the period is represented by straight lines radiating from the origin, see Fig. 2. The line with slope $\left(T_{eff}^1\right)^2$ identifies, where it crosses the 5% damped response spectrum, the spectral displacement d_e^1 of an elastic oscillator with period T_{eff}^1 corresponding to the effective period of the unbraced frame at maximum response d_p^*: if $S_d\left(T_{eff}^1; \xi_F\right) \le d_p^*$ the unbraced structure meets the performance requirement and the procedure ends (no retrofit is required); while, if $S_d\left(T_{eff}^1; \xi_F\right) > d_p^*$ the main frame alone is unable to meet the performance level and the damped brace system must be introduced.

Step 4: evaluation of the damped brace and the frame + damped brace capacity curve
In order to meet the target displacement d_p^*, additional damping must be supplied by the damped brace whose properties are evaluated through an iterative procedure. The ductility factor μ_{DB} of the damped brace is assigned as a design input depending on the employed damper technology. The damper yield strength V_y^{DB} is instead the unknown of the procedure, and, in case of devices characterized by an elastic-perfectly plastic behavior, V_y^{DB} coincides with V_p^{DB}. The equivalent viscous damping of the damped brace system ξ_{DB} is calculated according to Eq. (3) [21]:

$$\xi_{DB} = \frac{\kappa_{DB} \cdot 63.7 \cdot \left(V_y^{*DB} d_p^{*DB} - V_p^{*DB} d_y^{*DB}\right)}{V_p^{*DB} d_p^{*DB}} \tag{3}$$

where κ_{DB} can be selected based on experience and past applications on the employed damper technology or calibrated from experimental evidence.

The yield strength of the damped bracing system is hence evaluated through the energetic equivalence of Eq. (4)

$$\xi_{eff}^i \cdot \left(V_p^{*F} + V_{p,(i-1)}^{*DB}\right) = \xi_F \cdot V_p^{*F} + \xi_{DB} \cdot V_{p,i}^{*DB} \tag{4}$$

where i is the number of iterations and ξ_{eff}^i is the required effective viscous damping at the ith iteration, equal to $\xi_{eff}^i = 10 \cdot \left(\frac{d_e^i}{d_p^*}\right)^2 - 5(\%)$, where $d_e^i = S_d\left(T_{eff}^i; \xi = 5\%\right)$.

The process ends when the difference between d_p^* and $S_d\left(T_{eff}^i; \xi_{F+DB}^i\right)$ is sufficiently small, e.g., 0.05.

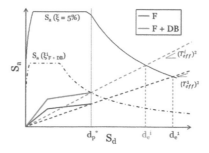

Fig. 2. Graphical iterative procedure for designing the damped brace system [9]

Step 5: evaluation of the Damped Brace distribution along the frame height

The distribution of the properties of the equivalent SDOF damped brace across the stories of the structure is performed on the basis of a proportionality criterion [11], which uses, as input parameters, the yield properties (V_y^{DB} and d_y^{DB}) of the equivalent SDOF damped brace and the components ϕ_i of the eigenvector associated to the first mode of vibration of the main frame. The damped braces are tuned in order to guarantee that the mode shape of the braced frame matches the first mode shape of the as-built structure. At each floor the properties of the braces equipped with hysteretic dampers are determined via Equations reported in Fig. 3, where N_{yi}^{DB} and K_i^{DB} represent the strength and stiffness of the single damped brace installed at the ith floor, respectively, and n_d is the total number of dampers per floor.

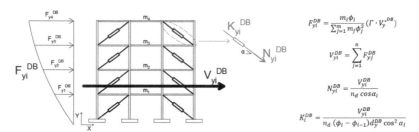

Fig. 3. Distribution of the properties of the equivalent SDOF damped brace across the stories of the structure [9]

3 Assessment of the Design Procedure

In order to prove the effectiveness of the proposed procedure, an existing 4-story reinforced concrete (RC) building is taken as case-study. This structure is located in Potenza (Italy), a medium/high seismic area with PGA of 2.45 m/s^2, and it is assumed to be founded on soil type B with topographic factor T_1 [2]. The main dimensions of the building are sketched in Fig. 4; information on materials, reinforcement, masses and loads are reported in reference [2].

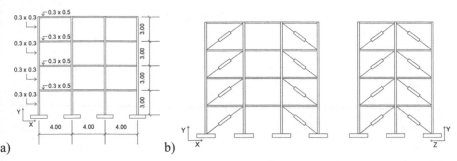

Fig. 4. Existing RC frame building in Potenza: (a) elevation view; (b) layout of steel hysteretic braces, installed in the perimetral frames [9]

A full 3D numerical model is formulated within the OpenSees framework [22], by using the *forceBeamColumn* element object [23] (in the form of the *beamWithHinges*) for beams and columns, as reported in reference [24] and applied in reference [9], and choosing a modelling approach consistent with the European design code [25].

The upgrade is carried out considering the seismic loads provided by IBC [19] for life-safety limit state (SLV), site of Potenza (Long 15° 48′ 20.1744″, Lat 40° 38′ 25.4688″), functional class cu = II, PGA = 2.45 m/s², soil type B and topographic factor T_1. Diagonal steel braces equipped with hysteretic dampers are inserted in the perimetral frames of each story, according to the layout shown in Fig. 4.

Since the building is designed in accordance with outdated codes, it is missing of seismic details; therefore, the hysteretic damper system is designed with the aim of keeping the main frame in the elastic range, avoiding structural damage. The target inter-story drift ratio is set to Δ_d = 0.005 m/m, corresponding to the target displacement d_p = 0.045 m of the MDOF structure (h_0 = 3.0 m, $\delta_{i,max}$ = 0.3306), and d_p^* = 0.036 m of the equivalent SDOF system, evaluated with a modal participation factor Γ = 1.27.

The design procedure is separately applied to both X and Z directions, choosing an equivalent damped brace system with a ductility factor μ_{DB} equal to 10 and κ_{DB} equal to 1, which corresponds to an equivalent viscous damping ratio ξ_{DB} = 57.3% [9].

The damped brace system is distributed along the height of the frame in accordance with the method illustrated in Sect. 2 (hereinafter called Method A). However, in order to highlight the effects of the damper distribution on the frame response, a second method described in literature [2], named Method B, has been investigated as well. By referring to the layout of Fig. 4, the resulting strength and stiffness of the dissipating braces calculated with either method are reported in Table 1, which shows that the Method A provides the highest values of strength and stiffness of the damping braces.

3.1 Numerical Investigation

In order to validate the procedure, NLSAs and NLDAs are performed on both as-built and retrofitted configurations. Figure 5 compares in the ADRS plane the capacity curves of the upgraded building: the design target is met in either building direction by the capacity curve of the structure upgraded according to Method A, while the target displacement is not achieved by the Method B counterpart.

Table 1. Properties of the damped braces [9].

	X-direction				Z-direction			
	Method A		Method B		Method A		Method B	
	K_i^{DB} [kN/mm]	N_{yi}^{DB} [kN]	K_i^{DB} [kN/mm]	N_{yi}^{DB} [kN]	K_i^{DB} [kN/mm]	N_{yi}^{DB} [kN]	K_i^{DB} [kN/mm]	N_{yi}^{DB} [kN]
1st	67.4	77.0	62.5	46.7	62.6	73.0	57.0	44.3
2nd	56.1	68.0	44.0	41.6	51.5	64.1	41.0	39.5
3rd	55.5	48.6	41.4	31.0	51.0	46.0	38.5	29.4
4th	52.3	22.3	38.0	16.5	47.6	21.1	34.3	15.6

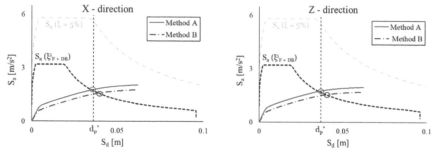

Fig. 5. Comparison of the capacity curves of the upgraded structure, with damper distribution according to either Method A (blue curve) or Method B (red curve)

Since NLSAs evaluate the response of the retrofitted structure in terms of global quantities only, inter-story drift ratios and shear forces at each story are assessed by applying bidirectional NLDAs in compliance with the IBC [19], considering two sets of seven artificial ground motions generated using the computer code SIMQKE [26].

Inter-story drift ratios Δ drastically decrease when the damped braces are introduced and, in particular, the damper distribution according to Method A produces at each floor maximum Δ values smaller than 0.5%, which is the design target drift ratio. In contrast, if Method B is adopted, Δ exceeds the specified limit at the first and second floors, where Δ is equal to 0.00514 and 0.00583 m/m respectively (Fig. 6).

As expected, the shear force V increases in the braced structure (Fig. 7), and the increase is higher for the stiffer distribution following Method A: e.g., at the third floor V grows by either 3.2% with Method B or 9.2% with Method A; however, the increment with Method A is always within 10%, which can be practically acceptable.

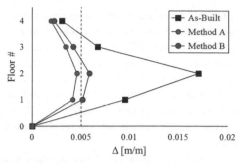

Fig. 6. Comparison of maximum inter-story drift ratio Δ obtained by bidirectional NLDAs with and w/o damped braces

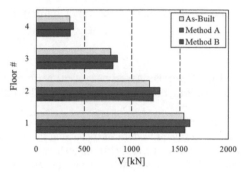

Fig. 7. Comparison of maximum shear forces at each floor obtained by bidirectional NLDAs with and w/o damped braces

4   Conclusions

A displacement-based design procedure based on the Capacity Spectrum Method [17, 18] has been developed for the seismic upgrade of frame structures via addition of hysteretic damped braces. The strength of the proposed method relies on its simplicity: only one NLSA of the as-built structure is performed at the beginning of the process, to determine the capacity curve of the main frame, and at each iteration the capacity curve of the upgraded frame is calculated by means of analytical equations; in this way, the iterative procedure can be implemented in a spreadsheet (such as Excel), and convergence is usually reached in few steps.

A case-study RC structure is considered to demonstrate the effectiveness of the suggested procedure. Once the overall properties of the damped brace system are determined, the distribution of the damped brace units at each floor has been evaluated in accordance with the method illustrated in Sect. 2, called Method A; a comparison with a second method available in literature [2], called Method B, has been investigated as well, in order to highlight the effects of the distribution method. Method A provides that the first mode of the braced and unbraced frames are the same, ensuring the simultaneous engagement of the dampers at the different floors, which is essential for the condensation of the MDOF structure to a SDOF system to be accurate. In contrast, method B, which assumes a distribution of the damped brace stiffness proportional to the stiffnesses of the different floors regardless the actual floor mass distribution may not lead to the same first modes for the braced and bare frames [9]. Indeed, the damped brace system designed according to Method A allowed to fulfill the performance requirements for the upgraded frame, providing that the maximum inter-story drift ratio was less than 0.5% at each floor, without an excessive increase in terms of shear demand. In contrast, for Method B the target performance was found not met at the first two floors, and also in terms of global results, the capacity curves on both directions did not meet the reference spectrum at the design displacement.

Owing to its ease, the procedure is aimed at enhancing the confidence of practitioners in using supplementary energy dissipation systems by providing a simple, fast, and handy procedure to tune the effective parameters of the damped braces. Further investigations are necessary to extend the method to irregular buildings, and include the constitutive behavior of steel hysteretic dampers characterized by a hardening ratio $r > 0$.

References

1. Di Cesare, A., Ponzo, F.C., Nigro, D.: Assessment of the performance of hysteretic energy dissipation bracing systems. Bull. Earthq. Eng. **12**(6), 2777–2796 (2014). https://doi.org/10.1007/s10518-014-9623-z
2. Di Cesare, A., Ponzo, F.C.: Seismic retrofit of reinforced concrete frame buildings with hysteretic bracing systems: design procedure and behaviour factor. Shock Vibr. **2017** (2017). Article ID: 2639361. https://doi.org/10.1155/2017/2639361
3. Aliakbari, F., Garivani, S., Aghakouchak, A.A.: An energy based method for seismic design of frame structures equipped with metallic yielding dampers considering uniform inter-story drift concept. Eng. Struct. **205** (2020). Article ID: 110114. https://doi.org/10.1016/j.engstruct.2019.110114
4. Garivani, S., Askariani, S.S., Aghakouchak, A.A.: Seismic design of structures with yielding dampers based on drift demands. Structures **28**, 1885–1899 (2020). https://doi.org/10.1016/j.istruc.2020.10.019
5. Quaglini, V., Pettorruso, C., Bruschi, E.: Experimental and numerical assessment of prestressed lead extrusion dampers. Int. J. Earthq. Eng. **XXXVIII**, 46–69 (2021)
6. Gandelli, E., Taras, A., Distl, J., Quaglini, V.: Seismic retrofit of hospitals by means of hysteretic braces: influence on acceleration-sensitive non-structural components. Front. Built Environ. **5** (2019). Article ID: 100. https://doi.org/10.3389/fbuil.2019.00100
7. Bartera, F., Giacchetti, R.: Steel dissipating braces for upgrading existing building frames. J. Constr. Steel Res. **60**, 751–769 (2004). https://doi.org/10.1016/S0143-974X(03)00141-X
8. Quaglini, V., Bruschi, E., Pettorruso, C.: Dimensionamento di dispositivi per la riabilitazione sismica di strutture intelaiate. Structural **237** (2021). Paper: 25, ISSN: 2282-3794. https://doi.org/10.12917/STRU237.25
9. Bruschi, E., Quaglini, V., Calvi, P.M.: A simplified design procedure for seismic upgrade of frame structures equipped with hysteretic dampers. Eng. Struct. **251**, 113504 (2022). https://doi.org/10.1016/j.engstruct.2021.113504
10. Gandelli, E., De Domenico, D., Quaglini, V.: Cyclic engagement of hysteretic steel dampers in braced buildings: a parametric investigation. Bull. Earthq. Eng. **19**(12), 5219–5251 (2021). https://doi.org/10.1007/s10518-021-01156-3
11. Mazza, F., Vulcano, A.: Displacement-based design procedure of damped braces for the seismic retrofitting of r.c. framed buildings. Bull. Earthq. Eng. **13**(7), 2121–2143 (2014). https://doi.org/10.1007/s10518-014-9709-7

12. Mazza, F., Vulcano, A.: Displacement-based seismic design of hysteretic damped braces for retrofitting in-elevation irregular RC framed buildings. Soil Dyn. Earthq. Eng. **69**, 115–124 (2015). https://doi.org/10.1016/j.soildyn.2014.10.029
13. Bergami, A.V., Nuti, C.: A design procedure of dissipative braces for seismic upgrading structures. Earthq. Struct. **4**(1), 85–105 (2013). https://doi.org/10.12989/eas.2013.4.1.085
14. Barbagallo, F., Bosco, M., Marino, E.M., Rossi, P.P., Stramondo, P.: A multi-performance design method for seismic upgrading of existing RC frames by BRBs. Earthq. Eng. Struct. Dyn. **46**, 1099–1119 (2017). https://doi.org/10.1002/eqe.2846
15. Nuzzo, I., Losanno, D., Caterino, N.: Seismic design and retrofit of frame structures with hysteretic dampers: a simplified displacement-based procedure. Bull. Earthq. Eng. **17**(5), 2787–2819 (2019). https://doi.org/10.1007/s10518-019-00558-8
16. Kim, J., Choi, H.: Displacement-Based Design of supplemental dampers for seismic retrofit of a framed structure. J. Struct. Eng. **132**(6), 873–883 (2006). https://doi.org/10.1061/(ASC E)0733-9445(2006)132:6(873)
17. ATC: Seismic evaluation and retrofit of concrete buildings. ATC-40, Applied Technology Council, Redwood City, California (1996)
18. Fajfar, P.: Capacity spectrum method based on inelastic demand spectra. Earthq. Eng. Struct. Dyn. **28**(9), 979–993 (1999)
19. CSLLPP (Consiglio Superiore dei Lavori Pubblici). D.M. 17 gennaio 2018 in materia di "norme tecniche per le costruzioni". Gazzetta ufficiale n.42 del 20 febbraio 2018, Supplemento ordinario n.8, Ministero delle Infrastrutture e dei trasporti, Roma (2018). (in Italian)
20. CSLLPP (Consiglio Superiore dei Lavori Pubblici). Circolare 21 gennaio 2019, n. 7 C.S.LL.PP. Istruzioni per l'applicazione dell'«Aggiornamento delle "Norme tecniche per le costruzioni"» di cui al decreto ministeriale 17 gennaio 2018, Roma (2019). (in Italian)
21. Dwairi, H.M., Kowalsky, M.J., Nau, J.M.: Equivalent damping in support of direct displacement-based design. J. Earthq. Eng. **11**(4), 512–530 (2007). https://doi.org/10.1080/ 13632460601033884
22. McKenna, F., Fenves, G.I., Scott, M.H.: Open system for earthquake engineering simulation. PEER report, Berkeley, CA (2000)
23. Scott, M.H., Fenves, G.L.: Plastic hinge integration methods for force-based beam-column elements. J. Struct. Eng. **132**(2), 244–252 (2006). https://doi.org/10.1061/(ASCE)0733-944 5(2006)132:2(244)
24. Bruschi, E., Calvi, P.M., Quaglini, V.: Concentrated plasticity modelling of RC frames in time-history analyses. Eng. Struct. **243**, 112716 (2021). https://doi.org/10.1016/j.engstruct. 2021.112716
25. CEN (European Committee for Standardization): Design of structures for earthquake resistance - part 3: assessment and retrofitting of buildings. EN 1998-3 Eurocode 8 (2005)
26. SIMQKE (SIMulation of earthQuaKE ground motions). https://gelfi.unibs.it/software/sim qke/simqke_gr.htm. Accessed July 2021

Saltwater and Alkali Resistance of Steel Reinforced Grout Composites with Stainless Steel

Sara Fares⬛, Rebecca Fugger⬛, Stefano De Santis[✉]⬛, and Gianmarco de Felice⬛

Department of Engineering, Roma Tre University, 00146 Rome, Italy
stefano.desantis@uniroma3.it

Abstract. The long-term performance of strengthening systems is of the utmost importance for the sustainable management of the building stock. Mortar-based composites have proved extremely effective for enhancing the ultimate strength of existing structures, but their durability has not been sufficiently investigated so far and still raises concern. This paper describes an experimental study on the saltwater and alkali resistance of steel reinforced grout composites with stainless steel textiles and lime mortars, named as SSRG, developed for applications to masonry structures. Bare textile and coupon specimens were aged in substitute ocean water and calcium hydroxide solutions for up to 5000 and 3000 h, respectively, and then subjected to direct tensile tests.

Keywords: Fabric reinforced cementitious matrix (FRCM) · Long-term behaviour · Mortar-based composites · Salt attack · Chloride attack · Stainless steel reinforced grout (SSRG)

1 Introduction

Mortar-based composites, made of high strength textiles externally bonded with inorganic matrices (cement, geopolymer or lime mortars), have assumed an important role within professional practice to repair and strengthen existing structures [1]. They include high-strength bidirectional textiles made of glass, carbon, aramid, basalt or polybenzoxazole (PBO) fibres, and are called fabric reinforced cementitious matrix (FRCM). When unidirectional textiles made of ultra-high-strength steel cords (UHTSS) are used, the systems are called steel reinforced grout (SRG).

These materials are an effective solution for improving the ultimate strength of existing structural members and have been successfully used on reinforced concrete [2, 3] and masonry [4, 5] structures. Their short-term effectiveness has been proven [6, 7], but, an improved knowledge has yet to be gained about their long-term behaviour. The importance of this issue is witnessed by the fact that those certification standards include compliance with several durability requirements [8–10].

For this reason, it is crucial to obtain a better understanding of the long-term performance of mortar-based composites when subjected to aggressive environmental conditions [11], in order to develop effective, durable, compatible and sustainable materials.

© The Author(s), under exclusive license to Springer Nature Switzerland AG 2022
F. Calabrò et al. (Eds.): NMP 2022, LNNS 482, pp. 2183–2191, 2022.
https://doi.org/10.1007/978-3-031-06825-6_210

Some studies have been carried out on brass-coated steel textiles [12], which showed that exposure to chemical attack and sodium chloride reduced mechanical properties. In this light of these outcomes, galvanized (zinc-coated) steel textiles were used, but their coating layer resulted not sufficient to fully protect the steel wires subject to salt attack in some investigations [13]. Stainless steel textiles could solve some of these problems, but their lower strength and stiffness and, above all, higher cost have made them less used than galvanized steels. This work describes an experimental study on the durability of a stainless steel reinforced grout (SSRG) system with natural hydraulic lime (NHL) mortar subjected to salt and alkaline attack.

2   Materials, Test Program and Experimental Setup

2.1   Material

The SSRG system used in this work comprised an AISI 316 [14] stainless steel textile (labelled as S) and a fibre-reinforced NHL mortar (L), classified as M15 according to [15]. The mechanical properties of the NHL mortar were obtained by means of flexural and compressive tests performed in accordance with [15, 16], on mortar specimens prepared in the laboratory after 28 days of curing. Compression tests (L-UC) gave an average strength $f_{cm} = 23.4$ N/mm^2 (coefficient of variation C.V. $= 18.4\%$, number of specimens tested N $= 36$), whereas the average flexural strength (L-IT) resulted $f_{tm} = 5.0$ N/mm^2 (C.V. $= 26.2\%$, N $= 18$) (Table 1).

Table 1. Results of compression and flexure test on mortar specimens.

Test code	N	f_{cm}, f_{tm} [N/mm^2]	C.V. [%]
L-UC	36	23.4	18.4
L-IT	18	5.0	26.2

As for the stainless steel textile, it consists of micro-cords arranged parallel to each other and held in place by small transverse stainless steel wires. The wires control the spacing of the cords, which is 6.30 mm, with each cord consisting of 19 filaments and having a cross-sectional area of 0.636 mm^2.

Bare textile and prismatic SSRG specimens were tested, both of which were 460 mm long and included 8 cords, with a cross-sectional area of 5.09 mm^2 [9, 10]. SSRG specimens were individually manufactured in plexiglass formworks and demoulded after 2 days. Specimens had 50 mm width and 10 mm thickness and included a layer of textile in the central plane. They were cured in water for 27 days and left for 7 days in the laboratory (18–20 °C temperature and 50–60% relative humidity) before testing.

Bare textile specimens, prior to testing, were provided with 50 mm × 80 mm × 3 mm aluminium tabs glued at the ends with a structural adhesive to ensure proper gripping. SSRG specimens were provided with two layers of GFRP wrap for a length of 80 mm, to promote a homogeneous load and prevent mortar splitting in the clamps [9, 10].

2.2 Test Program

The following tests were carried out in this work:

- Direct tensile tests on unaged bare textiles and SSRG specimens (labelled S-NC and SL-NC, respectively);
- Direct tensile tests on bare textile and SSRG specimens aged in substituted ocean water (labelled S-SWY and SL-SWY, respectively), prepared according to [17], at a temperature of 23 °C for 1000, 3000 and 5000 h. In the label, Y is the aging duration in thousands of hours.
- Direct tensile tests on bare textile and SSRG specimens aged in alkaline environment (S-AKY and SL-AKY) [18], at a temperature of 23 °C for 1000 and 3000 h. As before, in the label, Y is the aging duration in thousands of hours.

2.3 Experimental Setup

Tensile tests were performed under displacement control at 0.5 mm/min rate, in the direction parallel to the cords. A Material Testing System (MTS) universal load frame with a maximum capacity of 100 kN was used (see Fig. 1), which applied the load by a hydraulic actuator and recorded it by an integrated load cell at a sampling rate of 10 Hz.

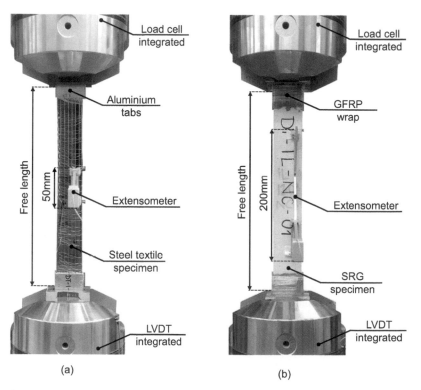

Fig. 1. Experimental setup of direct tensile tests on bare textile (a) and SRG (b) specimens.

The load was divided by the cross-sectional area of the textile to obtain the stress (σ). An MTS extensometer was placed in the centre of the specimens to measure the strain (ε), which had a base length of 50 mm for the bare textile and 200 mm for the SRG specimen (see Fig. 1). The strain, in the stress-strain response curves of the direct tensile tests, was obtained from the integrated LVDT data, corrected on the basis of that provided by the extensometer. This latter was removed during test execution at 80% of the expected ultimate load, to prevent its possible damage at cord rupture, as recommended by test standards [9, 10].

3   Test Results

3.1   Bare Textile Specimens

Figure 2 collects the stress-strain response curves of the direct tensile tests on bare textile specimens, performed after the completion of the ageing protocols, in which the curves of the individual tests are represented together with their average curve (dashed line).

The main results are listed in Table 2, consisting of the test code and number of the tested specimens (N), the mean value of the ultimate tensile stress (σ_f) and the yield stress (σ_y), defined as the stress corresponding to a residual strain at unloading of 0.1% [9], determined on the stress-strain curve, the strain (ε_f) corresponding to σ_f and the tensile modulus of elasticity (E_f), calculated as the secant stiffness between $\sigma_f/10$ and $\sigma_f/2$ [10]. The coefficient of variation (C.V.(σ_f), C.V.(σ_y)) and the percentage difference ($\Delta\sigma_f$, $\Delta\sigma_y$) with respect to the results of the reference unaged specimens are also listed.

The response curves of all the bare textile (see Fig. 2) showed an initial linear branch up to about $\sigma_f/2$, with E_f varying between 161.2 kN/mm^2 (S-SW5) and 177.6 kN/mm^2 (S-NC), followed by a gradual decrease in stiffness, until the progressive failure of the cords (see Fig. 3). The aged specimens did not show any deterioration of the tensile strength, in some cases the peak stress was even $1 \div 7\%$ higher than that of the unaged specimens ($\Delta\sigma_f > 0$), which was due to the unavoidable scatter of test results. Finally, the yield stress (σ_y) was 20% lower than σ_f, independently from the type of ageing.

Table 2. Tensile test results on bare textile specimens.

Test code	N	σ_f [N/mm^2]	C.V. (σ_f) [%]	$\Delta\sigma_f$ [%]	ε_f [%]	E_f [kN/mm^2]	σ_y [N/mm^2]	C.V. (σ_y) [%]	$\Delta\sigma_y$ [%]
S-NC	9	1597.0	2.3	–	1.15	177.6	1306.1	2.6	–
S-AK1	5	1637.2	2.9	2.5	1.40	173.1	1295.9	3.1	−0.8
S-AK3	5	1639.8	3.2	2.7	1.42	168.8	1325.4	2.5	1.5
S-SW1	5	1715.4	0.3	7.4	1.82	164.4	1371.5	0.4	5.0
S-SW3	5	1646.8	4.0	3.1	1.53	165.5	1378.9	1.4	5.6
S-SW5	5	1624.1	6.5	1.7	1.45	161.2	1337.1	6.7	2.4

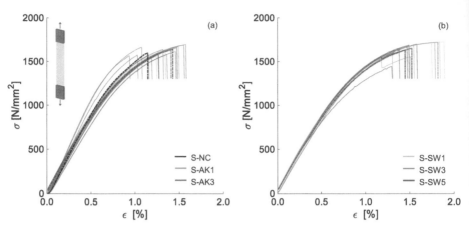

Fig. 2. Stress-strain response curves of bare textile specimens: unaged and aged in alkaline environment (a) and aged in salt water (b).

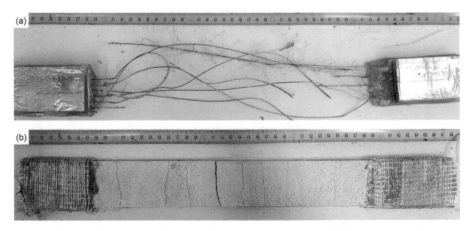

Fig. 3. Failure modes of bare textile specimens (a) and of SSRG specimens (b).

3.2 SSRG Specimens

Figure 4 collects the stress-strain response curves of the direct tensile tests on SSRG specimens, performed after the completion of the ageing protocols, in which the curves of the individual test are represented together with their average curve (dashed line). SSRG specimens showed an almost trilinear response curve with an initial branch for the un-cracked stage, a transition stage (crack development) and a third branch for the cracked specimen, followed by simultaneous failure of the steel cords (see Fig. 3).

Table 3 lists the test code and number of the tested specimens (N), the mean ultimate tensile stress (σ_u), the corresponding strain (ε_u) and the mean tensile modulus of elasticity of the un-cracked (E_I) and cracked (E_{III}) phases. The coefficient of variation (C.V.(σ_u)) and the percentage difference ($\Delta\sigma_u$) with respect to the results of the reference unaged specimens are also listed.

The unaged SSRG specimens (SL-NC) showed similar peak stress (σ_u = 1659.2 N/mm^2) and stiffness in the cracked phase (E_{III} = 183.2 kN/mm^2) as the bare textile, with a slight increase due to the matrix. The aged specimens showed no deterioration in tensile strength and a non-negligible decrease in tensile modulus of elasticity, which was attributed to the deterioration of the mortar matrix after ageing.

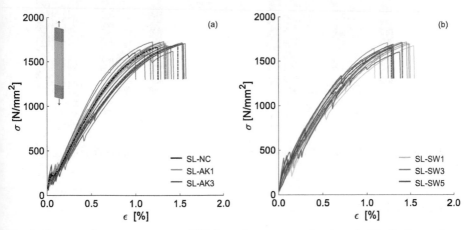

Fig. 4. Stress-strain response curves of SRG specimens: unaged and aged with alkaline environment (a), and aged in salt water (b).

Table 3. Tensile test results on SRG specimens.

Test code	N	σ_u [N/mm^2]	C.V. (σ_u) [%]	$\Delta\sigma_u$ [%]	ε_u [%]	E_I [kN/mm^2]	E_{III} [kN/mm^2]
SL-NC	9	1659.2	3.6	–	1.25	1178.3	183.2
SL-AK1	5	1691.0	0.9	1.9	1.33	864.6	141.7
SL-AK3	5	1687.9	0.8	1.7	1.48	970.7	128.5
SL-SW1	5	1636.6	1.8	−1.4	1.29	837.8	159.0
SL-SW3	5	1700.2	0.6	2.5	1.41	697.9	127.6
SL-SW5	5	1640.6	2.5	−1.1	1.28	807.1	150.5

4   Comparisons of Experimental Results

The average values of ultimate tensile stress (σ_u) of the bare textile (S) and SSRG specimens (SL) are compared in Fig. 5a and Fig. 5b, respectively. From Fig. 5a, it can be seen that the strength of the bare textile was not affected by the type of ageing performed. In some cases, as for the straight textiles, the percentage difference was even positive, meaning that there was an increase in peak and yield stresses. Moreover, for the SSRG specimens, it can be observed from Fig. 5b that the peak stress of the aged specimens was similar to that of the unaged specimens, which, in its turn, was equal to that of the bare textiles.

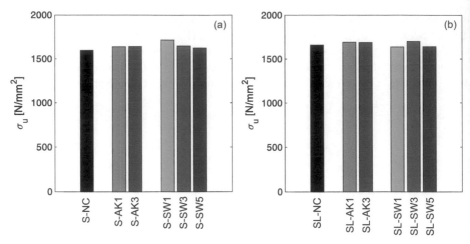

Fig. 5. Comparison of results of average ultimate tensile stress values of bare textile (a) and SRG specimens (b).

5   Conclusions

An experimental study was conducted on the salt water and alkali resistance of a stainless steel reinforced grout (SSRG) strengthening system with NHL mortar. Direct tensile tests on bare textile specimens gave a tensile strength of approximately 1600 N/mm^2 and a tensile modulus of elasticity of 180 kN/mm^2.

All the aged bare textile specimens showed no deterioration of tensile strength, independently from the aging process they underwent. Aged SSRG specimens showed approximately 5% higher peak strength and stiffness than the bare textile and the same tensile strength of the unaged specimens. The tensile modulus of elasticity in the cracked stage reduced by 10–30%.

The experimental research presented in this paper provided the first scientific evidence on the strength and durability performance of a stainless SRG system with lime-based matrix. Although further durability studies would be necessary, as there are many parameters involved, the SSRG system investigated in this research showed a good durability performance, which meets the requirements of the existing certification guidelines, combined with good mechanical properties, making it suitable for the effective and durable reinforcement of existing masonry structures. More in general, the outcomes of this study highlight the great potential of stainless steel reinforced grout systems for the life-span strengthening of the building stock, for which the highest cost of the material is balanced by the reduced efforts required by control and maintenance as well as by the highest confidence in the safety level of the strengthened structure.

Acknowledgements. This work was carried out within the Research Projects "ReLUIS", funded by the Italian Department of Civil Protection (2019–2021) and "RIPARA Integrated systems for the seismic protection of architectural heritage" funded by Regione Lazio (2021–2023). Kimia SpA (Perugia, Italy) is kindly acknowledged for providing strengthening materials. Funding is also acknowledged from the Italian Ministry of Education, University and Research (MIUR), in the frame of the Departments of Excellence Initiative, attributed to the Department of Engineering of Roma Tre University (2018–2022).

References

1. American Concrete Institute: Guide to design and construction of externally bonded fabric-reinforced cementitious matrix (FRCM) systems for repair and strengthening concrete and masonry structures. ACI 549.6 R-20. Farmington Hills, MI, US (2020)
2. Da Porto, F., Stievanin, E., Gabin, E., Valluzzi, M.R.: SRG application for structural strengthening of RC beams. Special Publ. **286**, 1–14 (2012)
3. Ascione, F., Lamberti, M., Napoli, A., Realfonzo, R.: Experimental bond behavior of Steel Reinforced Grout systems for strengthening concrete elements. Constr. Build. Mater. **232**, 117105 (2020). https://doi.org/10.1016/j.conbuildmat.2019.117105
4. Borri, A., Casadei, P., Castori, G., Hammond, J.: Strengthening of brick masonry arches with externally bonded steel reinforced composites. J. Compos. Constr. **13**(6), 468–475 (2009). https://doi.org/10.1061/(ASCE)CC.1943-5614.0000030
5. De Santis, S., De Canio, G., de Felice, G., Meriggi, P., Roselli, I.: Out-of-plane seismic retrofitting of masonry walls with Textile Reinforced Mortar composites. Bull. Earthq. Eng. **17**(11), 6265–6300 (2019). https://doi.org/10.1007/s10518-019-00701-5
6. de Felice, G., D'Antino, A., De Santis, S., Meriggi, P., Roscini, F.: Lessons learned on the tensile and bond behaviour of Fabric Reinforced Cementitious Matrix (FRCM) composites. Frontiers in Built Environment, 6 (2020). https://doi.org/10.3389/fbuil.2020.00005
7. Thermou, G.E., et al.: Bond behaviour of multi-ply steel reinforced grout composites. Constr. Build. Mater. **305**, 124750 (2021). https://doi.org/10.1016/j.conbuildmat.2021.124750
8. ICC Evaluation Service: Acceptance Criteria for Masonry and Concrete Strengthening Using Fabric-reinforced Cementitious Matrix (FRCM) and Steel Reinforced Grout (SRG) Composite Systems. AC434. ICC Evaluation Service (2016)
9. CSLLPP (Italian High Council of Public Works): Guidelines for the identification, the qualification and the acceptance of fibre-reinforced inorganic matrix composites (FRCM) for the structural consolidation of existing constructions (in Italian) (2018)
10. EOTA (European Organisation for Technical Assessment): Externally-Bonded Composite Systems with Inorganic Matrix for Strengthening of Concrete and Masonry Structures. EAD 340275-00-0104 (2020)
11. Franzoni, E., Gentilini, C., Santandrea, M., Zanotto, S., Carloni, C.: Durability of steel FRCM-masonry joints: effect of water and salt crystallization. Mater. Struct. **50**(4), 1–16 (2017). https://doi.org/10.1617/s11527-017-1070-2
12. Borri, A., Castori, G.: Indagini sperimentali sulla durabilità di materiali compositi in fibra d'acciaio. In: Proceedings of the 14th Conference ANIDIS, Bari (2011)
13. De Santis, S., Meriggi, P., de Felice, G.: Durability of steel reinforced grout composites. 1st edn. Brick and Block Masonry, pp. 357–363. CRC Press (2020). https://doi.org/10.1201/978 1003098508-48
14. ASTM International: Standard specification for chromium and chromium nickel stainless steel plate, sheet, and strip for pressure vessels and for general applications. ASTM A240/A240M-16. West Conshohocken, PA (2016)
15. CEN, European Committee for Standardization. EN 998-2. Specification for mortar for masonry - Part 2: Masonry mortar (2016)
16. CEN, European Committee for Standardization. EN 1015-11. Methods of test for mortar for masonry - Part 11: Determination of flexural and compressive strength of hardened mortar (2019)

17. ASTM International: Standard Practice for the Preparation of Substitute Ocean Water. ASTM D1141-98. West Conshohocken, PA (2013)
18. ASTM International: Standard Test Method for Alkali Resistance of Fiber Reinforced Polymer (FRP) Matrix Composite Bars used in Concrete Construction. ASTM D7705/D7705M-12. West Conshohocken, PA (2019)

Freeze/Thaw Effect on the Mechanical Properties of FRCM System

Salvatore Verre[1]($\boxtimes$) (ID) and Alessio Cascardi[2] (ID)

[1] Department of Civil Engineering, University of Calabria, 87037 Arcavacata di Rende, Italy
salvatore.verre@unical.it
[2] ITC-CNR, Construction Technologies Institute - Italian National Research Council, Bari, Italy

Abstract. Nowadays the use of FRPs (fibre reinforced polymers) and FRCMs (fabric reinforced cementitious mortar) is largely adopted in the field of the structural retrofitting. Nonetheless, the durability issue is a lightly investigated phenomenon, but it is important for the long-term effects on the effectiveness of external reinforcement. This paper presents the results about the main aspects related to the effects of the freeze/thaw cycles on the mechanical characteristics and on the stress-transfer mechanism between both the fabric and the matrix and the composite strip with respect to the concrete substrate. The investigated parameters were the number of layers and the FRCM-reinforcement type. In particular, the direct tensile, the single-lap and direct-shear test were all performed and compared.

Keywords: Fiber/matrix bond · Mechanical testing · Freeze/Thaw

1 Introduction

The FRPs (fiber reinforced polymers) have been firstly experienced, at the early'90, for the strengthening of existing structures such as masonry and reinforced concrete made, [1] and [2]. In the last years, an competitor system is fastly invading the market: the FRCM (fabric reinforced cementitious mortar); e.g. in [3, 4]. If the strengthening effectiveness is generally achieved the issue related to the bond strength is still open. In fact, the FRCM systems base their mechanical behavior and their effectiveness on physical interaction for the stress transfer to the substrate. Many phenomena are involved in the interaction between single fibers and matrix. Exposure to different degradation mechanisms may modify the stress transfer mechanism. The durability is a phenomenon little investigated, but it is important for the long-term effects on the effectiveness of external reinforcement on reinforced structures. Research reported in [5] analyzed the weathering cycle through deionized water in terms of flexural and bond capacity. The specimens were subjected for few days at the conditioning environment, and the study on the increase of the peak load and initial stiffness, with respect to the unconditioned specimens, was evaluated. A recent study focused on the development of a mathematical model able to predict the chosen mechanical parameters and damage index of brick wallets for a given number of freeze-thaw cycles through fragility curves for ductility based on the damage index.

© The Author(s), under exclusive license to Springer Nature Switzerland AG 2022
F. Calabrò et al. (Eds.): NMP 2022, LNNS 482, pp. 2192–2202, 2022.
https://doi.org/10.1007/978-3-031-06825-6_211

The research reported in [6, 7] studied the effect of different environmental conditioning on the FRCM specimens equipped with different types of fiber. The freeze–thaw cycle, in terms of flexural tensile and compressive strength on the mortar, induced a slight increment on the mechanical characteristics. In [7], the conditioned environment on the FRCM specimens in terms of mechanical characteristics (tensile test and bond pull-off) did not produce degradation. Finally, in [8], a review was reported on all tests conducted on FRCM systems subjected to freeze–thaw cycles, hydrothermal, and concentrated solution in order to evaluate the behavior and the mechanical characteristics in the long term. The principal aspects investigated in this experimental campaign were the effects of the freeze/thaw environment on the mechanical characteristics and the stress-transfer mechanism between the fiber and the matrix and the composite strip and the concrete substrate. The parameters investigated were the number of layers and the FRCM reinforcement type. In particular, the tests were performed through the main techniques as tensile test and the single-lap, direct-shear test.

2 Experimental Investigation

2.1 Mechanical Properties

The fabric mesh adopted in the FRCM system presents a development in mono-directional or bi-directional of fiber bundle and different fiber density. The strengthening system used in this study consists of fabric meshes embedded into cement-based mortar. Two types of fibers (see Fig. 1), Polyparaphenylenebenzobisoxazole (PBO) and Carbon (C), were used. PBO-fibers present the unbalanced (70–18 g/m^2) fiber bundles along two orthogonal directions, while the C-fibers present a monodirectional fabric mesh (182 g/m^2). In particular, the PBO-fiber bundles were spaced 10 mm and 20 mm in longitudinal and transversal bundles, respectively, while the C-fibers, the single fiber bundles, were spaced 10 mm from the center.

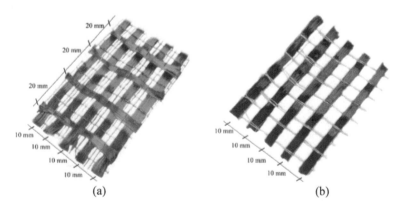

Fig. 1. Geometry of dry fibers: (a) PBO Fibers and (b) C Fibers.

In Table 1, the tensile strength, ultimate strain and elastic modulus were reported.

Table 1. Mechanical properties of the PBO and Carbon Fibers.

	PBO	Carbon (C)
Tensile strength [GPa] (CoV)	3.40 (0.04)	0.78 (0.03)
Ultimate strain [%] (CoV)	2.5 (0.08)	1.1 (0.27)
Elastic modulus [GPa] (CoV)	214 (0.04)	112 (0.07)

Tensile strength, ultimate strain, and elastic modulus of the fibers were measured in accordance with [9]. Four coupons of both fiber types were tested. The fiber strips, in accordance with [9], have a 500 mm length, moreover, at both specimen ends, two pairs of aluminum plates were attached by a thermosetting epoxy in order to guarantee a homogeneous constant pressure (see Fig. 2). At the center of the specimen an extensometer, to evaluate the fiber strain with a gauge length of 50 mm, was installed. For the evaluation of the stress, the area of a single yarn was evaluated according to Eq. 1:

$$A_f = 1000 \cdot p/\rho_f \cdot l \qquad (1)$$

where ρ_f is the density of fibers, l is the length of the yarn and p the weight. The tests, in displacement control at a rate of 0.5 mm/min, were conducted while both ends were clamped.

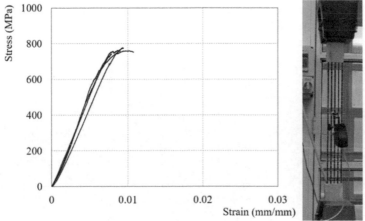

Fig. 2. Tensile test on dry C-fibers.

The mortar is a stabilized mono-component inorganic cementitious matrix as suggested by the manufacturer [10] for the fibers used. The mechanical properties (compressive and flexural strength) in terms of average values and the coefficient of variation (CoV) were summarized in Table 2. Three specimens were tested for each type of mortar considered. The load was applied with a rate equal to 100 N/s and 500 N/s for the flexural tensile and compressive test, respectively, in accordance to [11]. Lastly, the mortars used with PBO and C-fibers were called M-PBO and M-C, respectively.

Table 2. Mechanical properties of the matrix.

	M-PBO	M-C
Flexural tensile strength [MPa] (CoV)	7.08 (0.14)	7.77 (0.09)
Compressive strength [MPa] (CoV)	44.20 (0.08)	45.47 (0.05)

2.2 Specimen Preparation

All concrete joints had the rectangular cross section equal to 200×150 mm and a height of 400 mm. The compression and the tensile strength are equal to 26.1 MPa (CoV = 0.03) and 2.4 MPa (CoV = 0.05), respectively, in accordance with [9]. After the 28-day period of curing, the lateral surfaces with a width of 150 mm and 200 mm were sandblasted to remove cement wastes and improve the gripping with the reinforcement system, as suggested by [10]. The technique employed in the application of the composite strips for the single-lap direct shear (DS) and direct tensile (DT) specimens is similar and it is organized in three steps. The first step consists of an adhesive tape attachment of a formboard with a 3 mm thickness. A shape with a length and width of 260 mm (bonded length l) and 50 mm (width) was cut out of the formboard to accommodate the composite strips. Each composite strip was made individually and in particular, it was applied along the centerline of one face of the concrete prism for the DS specimen, while it was casted on the wood planer surface to ensure the planarity of the surface for the DT specimen (see Fig. 3a) and whose length was equal to 500 mm. Then, the mortar (called the internal layer) was applied by mold with light pressure (see Fig. 3b).

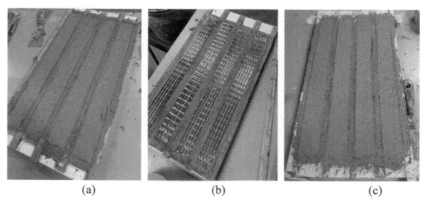

(a)   (b)   (c)

Fig. 3. Specimen preparation strengthened with one layer: (a) Apply the formboard and internal matrix layer, (b) and (c) apply external layer.

In the second step a fiber was placed on the internal layer. However, the second step is the most critical one, where the perfect alignment of the fibers and impregnation with the matrix must be ensured. Finally, in the third step the same operation as in step one are repeated (see Fig. 3c). The experimental campaign also includes composite strips equipped with two layers of fibers for the DS specimen. To make them, it is necessary, again, to perform the second and third steps. The composite strips equipped with two layers of fibers were composed of three layers of mortar and the second layer was called intermediate layer. In all applications of the composite strips, the bonded area was started at 20 mm off the edge of the top face of the concrete prism. Moreover, the top face was troweled smooth and then made planar. Finally, all specimens were stored and cured in laboratory for 28 days, while for the DT specimens while the demolding came about after 7 days.

2.3 Freeze/Thaw Protocol (FT)

All the specimens were conditioned in a chamber for one week at a relative humidity of 90% and a temperature of 38 °C ± 2 °C. Subsequently, the specimens were subjected to freeze-thaw cycles (20 in total). Each cycle consisted of 4 h at −18 °C ± 2 °C, then, 12 h at 90% humidity and a 38 °C ± 2 °C temperature, this procedure is in accordance with [9]. In Fig. 4 the environmental conditioning regime was reported.

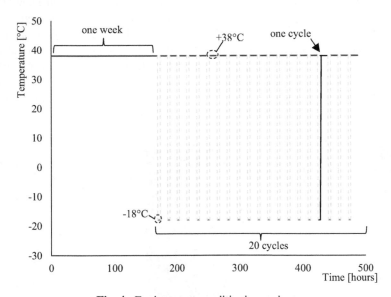

Fig. 4. Environment conditioning regime.

All specimens were named using the following notation DS (or DT) -PBO (or C) -U (or FT) -ηL -Z. DS and DT indicates the type of test, PBO and C indicates the type of fiber used; U and FT indicates the unconditioned specimen and Freeze/Thaw conditioning, respectively; ηL indicated the number of fiber layers and Z = specimen number.

2.4 Single Lap Direct Shear and Direct Tensile Test Protocol

Twenty-four FRCM-concrete joints specimen (control and subjected to conditioning) were tested by single-lap direct shear test set-up using the push pull configuration [12]. The test setup used (see Fig. 5a) is the classic push pull configuration. The steel frame was bolted by four steel bars to the machine used for the test. In addition, the shape of the steel frame was realized to restrain the movement of the concrete joint during the test and to allow the composite strip to be housed on the sides of the prism not involved in the test. The load was applied through a couple of aluminum tabs glued by thermosetting epoxy on the dry fiber at the top of the fibers (see Fig. 5a). For the composite strip equipped with two layers of fibers an additional aluminum tab was used to maintain the fibers parallel to the applied load and assure the load uniformly at the fiber bundles (see Fig. 5a). An L shape aluminum plate was attached to the dry fiber at the top of the concrete joint near the bonded area. The reaction of the latter was measured by two linear variable displacement transformers (LVDTs) attached at both sides of the composite strip, called LVDT a and b. The average value evaluated from the two LVDTs was called global slip s. All tests reported in this work were conducted under displacement control using a universal testing machine with a load cell of 100 kN and the rate adopted was 0.2 mm/min in accordance with [9]. In Fig. 5b the test setup was reported. Direct tensile tests carried out on twelve FRCM specimens (six control and six subjected to conditioning). FRCM specimens had a rectangular cross-section, while the thickness and length was equal to 10 mm (two layers of 5 mm) and 500 mm (considering the tabs), respectively. The gauge and the tabs used were of length 300 mm and 100 mm, respectively. Direct tensile tests were performed through clamped type grip under machine displacement control at a rate of 0.2 mm/min in accordance with [9]. The load was applied at the top the specimen while the other side was restrained at the end. In Fig. 5b the test setup was reported, in addition, strain measurements between the ends of the metal tab, were acquired by LVDT a and b, that were fixed lengthwise at the side of the specimen by aluminum supports (see Fig. 5b).

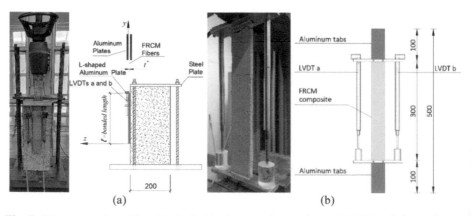

Fig. 5. Test setup adopted for: (a) single-lap shear test for specimen DS-FT-1 and (b) tensile test for FRCM specimen DT-FT-2 (Dimensions are in mm).

3   Results and Discussion

After FT treatment, all parts of the FRCM composites didn't show any crack patterns on the specimens' surface or at matrix to substrate interface. The mechanical properties of the mortar used in terms of flexural tensile and compressive strength was equal to 7.59 MPa (CoV = 4%) and 46.09 MPa (CoV = 3%), respectively, for M-PBO, while for the M-C was equal to 7.83 MPa (CoV = 5%) and 51.38 MPa (CoV = 2%), respectively. From the results, after the FT treatment, it should be noted that there was a slight increase for both types of mortar investigated in terms of flexural and compressive strength. In particular, the flexural strength evaluated showed an increase equal to 7% and 0.7%, while in terms of compressive strength it was equal to 1.4% and 16%, for M-PBO and M-C, respectively. The results of the DS tests presented in this study are reported in Table 3 in terms of peak load Peak Load P^* and average value of P^*_{avg} evaluated for each group of the specimens (unconditioned and conditioned). Moreover, in Table 3 a parameter Δ^* was summarized and it was evaluated using the following Eq. 2:

$$\Delta^* = \frac{P^*_{avg(conditoned)} - P^*_{avg(unconditioned)}}{P^*_{avg(unconditioned)}} 100\% \tag{2}$$

The failure observed for the specimens equipped with 1L and 2L was due to the debonding at the fiber–matrix interface. This type of failure is called telescopic failure in technical literature [12].

Figure 6 shows the comparison between the experimental curves of the unconditioned and conditioned specimens equipped with 1L and 2L. From the same figure, it should be noted that the conditioned environment increased the bond capacity with increments in terms of peak load for both strengthening systems. In addition, all conditioned specimens also showed an initial stiffness increment. A similar increase in terms of peak load of 7 and 8% for DS specimens equipped with 2L of PBO and C-FRCM, respectively. The major increases were observed in specimens strengthened with 1L of C-fiber with an increase of 14%. In Fig. 7 the stress-strain curves of DT specimens under tensile test were reported (unconditioned and conditioned) for DT specimens equipped with PBO-fiber and C-fiber, respectively. The experimental curves behavior was of a tri linear type. In particular, it can be divided in three main phases: the first linear phase represents the uncracked (I) stage, the second (II) is identified by crack development and the third (III) corresponds to the cracked stage (see Fig. 7). The values for the third stage are summarized in Table 4 in terms of stress (σ), strain (ε) and tensile modulus (E). Also, in the tensile tests it was possible to observe an increase in terms of stress of DT specimens with respect to unconditioned DT specimens. Comparing the experimental curves shown in Fig. 7, the global behavior in the third stage of the conditioned specimens shows similar stress values but a smaller strain range, in the case of PBO-fiber. While the C-fiber shows similar strain range but associated a greater stress value. In terms of the tensile modulus summarized in Table 4 and evaluated in the third stage, an increase of 6.8% and 26% was observed for the specimens with PBO-fiber and C-fiber, respectively.

Table 3. Experimental results of DS specimens.

	P^* [kN]	P^*_{avg} [kN]	Δ^* [%]
DS-PBO-U-1L-1	4.60	4.62	–
DS-PBO-U-1L-2	4.70		
DS-PBO-U-1L-3	4.56		
DS-PBO-U-2L-1	8.55	8.51	–
DS-PBO-U-2L-2	8.44		
DS-PBO-U-2L-3	8.54		
DS-C-U-1L-1	3.52	3.67	–
DS-C-U-1L-2	3.70		
DS-C-U-1L-3	3.79		
DS-C-U-2L-1	7.45	7.40	–
DS-C-U-2L-2	7.28		
DS-C-U-2L-3	7.47		
DS-PBO-FT-1L-1	4.61	4.74	3
DS-PBO-FT-1L-2	4.83		
DS-PBO-FT-1L-3	4.78		
DS-PBO-FT-2L-1	8.92	9.15	8
DS-PBO-FT-2L-2	9.16		
DS-PBO-FT-2L-3	9.36		
DS-C-FT-1L-1	4.19	4.19	14
DS-C-FT-1L-2	4.20		
DS-C-FT-1L-3	4.18		
DS-C-FT-2L-1	7.88	7.93	7
DS-C-FT-2L-2	8.02		
DS-C-FT-2L-3	7.88		

The increase of the capacity should be attributed to an extended curing time conditioned environment. Furthermore, the compatibility of the mortar with the substrate (concrete), high strength (mortar), and the low permeability of the materials (concrete and mortar) contributed to the increase of performance of the reinforcement [7].

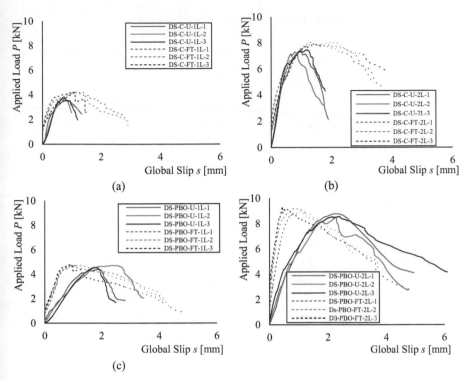

Fig. 6. Comparison between experimental curves (unconditioned and conditioned) of DS specimens equipped with: (a) 1L of PBO-fiber, (b) 2L of PBO-fiber, (c) 1L of C-fiber and (d) 2L of C-fiber.

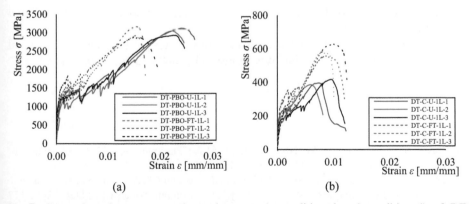

Fig. 7. Comparison between experimental curves (unconditioned and conditioned) of DT specimens equipped with: (a) 1L of PBO-fiber and (b) 1L of C-fiber.

Table 4. Experimental results of DT specimens.

	Strain [‰]	Stress [MPa]	Tensile modulus [GPa]
DT-PBO-U-1L-1	26.58	3111.30	108.87
DT-PBO-U-1L-2	24.55	3041.72	121.71
DT-PBO-U-1L-3	24.65	2931.57	119.88
DT-C-U-1L-1	11.94	396.15	46.20
DT-C-U-1L-2	10.04	384.42	47.05
DT-C-U-1L-3	12.35	554.29	47.43
DT-PBO-FT-1L-1	16.74	2920.58	122.65
DT-PBO-FT-1L-2	17.13	3155.39	126.62
DT-PBO-FT-1L-3	19.57	2899.08	125.03
DT-C-FT-1L-1	12.15	396.15	52.32
DT-C-FT-1L-2	8.09	581.72	68.73
DT-C-FT-1L-3	11.99	627.32	56.21

4 Conclusion

The paper presented the results of the experimental investigation on the effect of the freeze/Thaw environment on the different FRCM composite systems. Based on the results, the following conclusion can be made:

- The mechanical characteristics of mortar in terms of flexural tensile and compressive strength showed an increase between 0.7 and 16%;
- Freeze/Thaw environment did not affect the stiffness of FRCM composite in all stages;
- The DS specimens conditioned showed a gain in terms of both peak load and initial stiffness.
- The DT specimens conditioned showed an increase in terms of tensile modulus equal to 6.8% and 26% for the specimens with PBO-fiber and C-fiber, respectively.

References

1. Cascardi, A., Lerna, M., Micelli, F., Aiello, M.A.: Discontinuous FRP-confinement of masonry columns. Front. Built Environ. **5**, 147 (2020)
2. Micelli, F., Cascardi, A., Aiello, M.A.: Pre-Load Effect on CFRP- confinement of concrete columns: experimental and theoretical study. Curr. Comput. Aided Drug Des. **11**(2), 177 (2021)
3. Di Ludovico, M., Cascardi, A., Balsamo, A., Aiello, M.A.: Uniaxial experimental tests on full-scale limestone masonry columns confined with glass and basalt FRCM systems. J. Compos. Constr. **24**(5), 04020050 (2020)

4. Ombres, L., Mazzuca, P., Verre, S.: Effects of thermal conditioning at high temperatures on the response of concrete elements confined with a pbo-frcm composite system. J. Mater. Civ. Eng. **34**(1), 04021413 (2022)

5. Franzoni, E., Gentilini, C., Santandrea, M., Zanotto, S., Carloni, C.: Durability of steel FRCM-masonry joints: effect of water and salt crystallization. Mater. Struct. **50**, 201 (2017)

6. Donnini, J., Bompadre, F., Corinaldesi, V.: Tensile Behavior of a Glass FRCM system after different environmental exposures. Processes **8**, 1074 (2020)

7. Arboleda, D., Babaeidarabad, S., Hays, C.D.L., Nanni, A.: Durability of fabric reinforced cementitious matrix (FRCM) composites. In: 7th International Conference on FRP Composites in Civil Engineering in Proceedings of the CICE 2014. Vancouver, Canada (2014)

8. Al-Lami, K., D'Antino, T., Colombi, P.: Durability of fabric reinforced cementitious Matrix (FRCM) composites: a review. Appl. Sci. **10**, 1714 (2020)

9. National Research Council. Guide for the Design and Construction of Externally Bonded Fibre Reinforced Inorganic Matrix Systems for Strengthening Existing Structures. CNR-DT 215/2018, Rome, Italy (2020)

10. Available online: https://www.ruregold.com (2021)

11. UNI EN 1015-11:2007. Methods of Test for Masonry Units—Part 11

12. Verre, S.: Effect of different environments' conditioning on the debonding phenomenon in fiber-reinforced cementitious matrix-concrete joints. Materials **14**(24), 7566 (2021)

Structural Assessment of a Heritage Building in the UNESCO Site of Alberobello

Francesco Micelli[1]([✉]) [iD], Alessio Cascardi[2] [iD], and Salvatore Verre[3] [iD]

[1] Department of Innovation Engineering, University of Salento, 73100 Lecce, Italy
francesco.micelli@unisalento.it
[2] ITC-CNR, Construction Technologies Institute - Italian National Research Council, Rome, Italy
[3] University of Calabria, Arcavacata di Rende, 87036 Rende, Italy

Abstract. The cultural building Heritage is widespread all around the world and mainly consists of masonry structures. The UNESCO Map indicates about the 50% of this Heritage is placed in Italy, where seismic risk is recognized in the whole national territory. One of the most notorious sites is the city of Alberobello due to the presence of the well-known Trulli: the old city. With their cone-shaped roof made by stones, the Trulli are a unique example of architectural heritage. Moreover, the relatively new area is full of historical buildings that are an additional Heritage for the city. This study reports on the structural assessment of a large masonry structure faced in front of the Trulli zone, which is considered to be one of the first building of the old city. The building is the results of an aggregation of different units in which several masonry typologies met, at different ages. The historical and geometrical surveys were both performed in order to plan in-situ investigations aimed to reach an adequate level of confidence with respect to the mechanical properties of the masonry, according to the technical codes. A global analysis was performed by finite element method (FEM). Thus, the vulnerability assessment is herein reported and discussed for this complex building.

Keywords: Masonry · Heritage · Vulnerability analysis · Alberobello · Diagnosis

1 Introduction

Alberobello is a town belonging to the metropolitan area of Bari (in the south-east of Italy) where the famous Trulli site is located. It is one cultural site included in the World Heritage List by UNESCO, [1]. A Trullo is a long-lived example of spontaneous architecture (i.e. dwelling house). The historic center is entirely made up of these particular buildings, whose pyramidal shape makes them unique in the world. Here, between 1400 and 1500, a group of peasants build their homes using local materials, such as hard limestone easily found on site. At that time, the Kingdom of Naples established a tribute on each new construction. In order to avoid this tax payment, considered unfair and unpopular, they decided to build structures apparently devoid of the stability of ordinary homes and so low as to be easily hidden by the branches of trees, these structures were configured in what are now known as Trulli, [2]. After centuries, the "new" part

© The Author(s), under exclusive license to Springer Nature Switzerland AG 2022
F. Calabrò et al. (Eds.): NMP 2022, LNNS 482, pp. 2203–2212, 2022.
https://doi.org/10.1007/978-3-031-06825-6_212

of Alberobello was constructed just in front of the old one. Nowadays, the city center represents an additional value consisting of historic masonry building; which is poorly investigated. The present paper reports, in the next sections, a *step-by-step* procedure involving the mechanical and geometrical surveys, the in-situ investigations, the laboratory testing and the structural analysis aiming to assess the vulnerability index of a large and complex masonry building facing to the Trulli according to [3].

2   Description of the Building

The building, commonly called *"Palazzo dei Conti di Conversano"*, is located in the historic center of Alberobello. It was built in 1635. In a meticulous description of the building, dating back to 1822, numerous skirting boards are listed, used as mills, stables, oven, haystack and shops. In the upper part, the historical documents record thirteen rooms, among them communicating and adjacent to the family chapel. In the oratory tale, the Count placed a painting in 1636 depicting the Virgin of Loreto. The rooms on the main floor were accessed via a staircase, covered by a loggia, which housed the square. Here was the main door of the building, which disappeared due to some interventions dating back to the first half of the twentieth century. Following numerous extensions over the centuries and up to the current days, the building has a variable number of levels. A basement, a ground floor, a mezzanine level, a first floor and a roof can therefore be distinguished. A portion of the building also has a second-floor level with roofing. Figure 1a–d shows the relevant in plane views from a structural point of view with an evident geometrical irregularity both in plane and elevation. The evolutionary nature of the construction phases of the building highlights some peculiarities, such as the different quality of stones and the degree of the connection between the walls.

a)   b)   c)   d)

Fig. 1. Available geometry: a) underground, b) ground, c) mezzanine and d) first floor.

The building is founded directly on rock, sometimes by means of blocks of larger dimensions, and rarely, on one or two levels of square stones with an overall thickness greater than that of the walls, especially in the perimeter portions of the building, which appear to have been built in later times than the original internal core. The load-bearing structures are made up of walls of different types, according to the materials available at the time of the relative construction phases. Downstream of the observations and inspections carried out in-situ, it was considered appropriate to distinguish two macro categories of masonry, among those surveyed in the building in question:

- double face walls without transverse connection, with pseudo-regular blocks and thin joints (<10 mm);
- double face masonry masts in disorderly hard stone segments tied with mortar of poor mechanical properties according to non-aligned joints.

The first category of masonry characterizes the portion of the building that concerns the stallers and the perimeter walls on the ground and first floors. The second category mainly characterizes the internal core on the ground floor and first floor, where there are also arches made with squared and ordered blocks. Throughout the building there are niches and holes for flues that weaken the resistant males and justify the local crack pattern. During the structural inspections, no static instabilities due to possible seabed subsidence were found, confirming the fact that the building is based on a layer of hard limestone, not free of fractures, with the intrusion of vegetable soil. In the following paragraphs a punctual and quantitative evaluation of the properties inherent to the walls analyzed are provided.

The horizontal floors are almost exclusively made with barrel-type stone vaults, and not always half-round, pavilion, cross-shaped and in some cases sail or star-shaped, with the structural thickness of the supporting arches generally between 200 and 250 mm. From a preliminary analysis of the structure, following a visual inspection and subsequent plan metric survey, a set of structural criticalities emerges which are significant for the stability under static conditions and the response behavior under any seismic actions. With reference to this, the main features are summarized in the follow:

a) the building was built on a sloping surface;
b) severe irregularities in plan and elevation;
c) construction phases in different periods, with the consequent juxtaposition of existing masonry walls or absence of structural amortization between masonry facades built in subsequent periods;
d) prevalence of walls with disordered structure with partial or total absence of the respect of the rules of the art (e.g. aligned and regular mortar joints);
e) misalignment and correspondence between the vertical load-bearing elements with relevant misalignments between the various planes (see Fig. 2);
f) presence of disruptions in the wall facing on the first floor and barrel vaults on the mezzanine and ground floor.

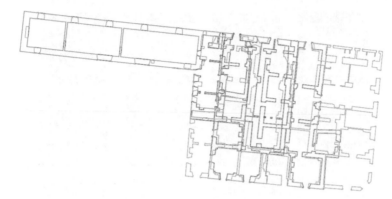

Fig. 2. Floors overlapping: ground (green), mezzanine (blue) and first (red).

The structure did not undergo seismic events of significant magnitude (>5) with a close epicenter according to the National Institute of Geophysics and Volcanology, [4]. Therefore, the important damage attested by the crack patterns is not attributable to seismic activity of the subsoil, but rather to the temporal evolution of the static problems in the long-term.

3   Description of the Building

The uneven nature of the building made it necessary to extract multiple stone segments, representative of rooms with different construction techniques, often associated with different periods of construction, in order to obtain an exhaustive mapping of the properties of the stone materials present building. The extracted stones were rectified and brought to a regular cubic dimension (with an edge size of 50 mm and 70 mm depending on the available dimensions). The weighing carried out in the laboratory, before and after the drying period in the oven, showed a high density of the segments. The specimens, after weighing, were subjected to mono-axial compression tests in the laboratory according to [5]. Endoscopic investigations consist of minimal invasive visual observations techniques that allow the inspection of portions inaccessible to direct observation. In the case under examination with endoscopy, it was possible to ascertain both the thickness of the external and internal leafs of the masonry "package", and the quality of the internal filling. From the investigations, the filling of the masonry walls turned out to be of good workmanship, with well-arranged split stone. The thickness of the external and internal leafs was found between 200 and 400 mm, with a mode value of 300 mm. The evaluation of the thickness of the segments constituting the vaults was also important, for which an almost constant value of 230 mm was recorded. Sclerometric tests were additionally performed. In order to deeply survey the masonry in the various regions, from a statistical point of view, a 500 mm × 500 mm grid frame with a thickness equal to 50 mm was made. Each series of measurements was made up of nine non-overlapping and equivalent tests. The selected 32 masonry textures are illustrated in Fig. 3. An average compressive strength of 19.25 MPa ± 19% was measured.

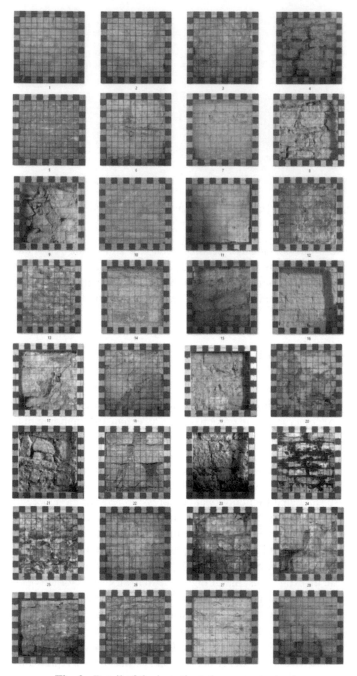

Fig. 3. Detail of the investigated masonry textures.

The masonry mechanical properties were computed by means of the Index of Quality Masonry (IQM) method reported in [6]. Rules of construction followed "to the letter" constitute the overall building measures which, if carried out during the construction of a wall, guarantee its reliability and ensure it is compact and structurally solid. By observing characteristic parameters of correct and efficient masonry construction, it is possible to establish a masonry quality index for each of the possible stress directions on a wall panel. An index of wall quality is therefore obtained for vertical loads, horizontal actions out of plane and finally for horizontal actions in the plane. The method considers the following parameters:

- High Quality Mortar/effective contact between elements. This condition, which is necessary for the uniform transmission of loads and the discharge of energy onto the ground, is obtained either by direct contact between square-cut blocks or through the mortar. Besides levelling the contact between the blocks, high quality mortar ensures the wall possesses cohesive resistance.
- Transversal Locking/presence of diatones. This condition prevents the wall dividing into various sections simply positioned one next to the other, and furthermore, allows to distribute the load along the entire width of the wall even when the load bearing is on the wall edge (e.g. a floor lying only on the internal part). This condition can be met thanks to "diatones", which are transverse blocks crossing the entire width of the wall.
- Form of Resistant Elements. The presence of blocks with a regular shape ensures the activation of the friction force, to which a large part of a wall's ability to resist horizontal actions are due. In fact, friction is principally activated through the effect of weight from masonry lying above the sliding surface, and is maximized for horizontal sliding. It can thus be seen that this is one of the conditions necessary in obtaining efficient locking between the wall elements.
- Dimension of the Elements. Resistant elements of large dimensions with respect to the width of the wall, ensure a high degree of stability to the wall, as do the diatones. Furthermore, thanks to their elevated dimensions, they are extremely heavy with good locking between them.
- Staggering of Vertical Joints/In Plane Locking. This condition allows for the application of a further resistance element in the wall: wedging between resistant elements (also known as "locking effect") which, together with friction, guarantees resistance to co-plane action.
- Horizontal Rows. This condition makes for good distribution of vertical load in that regular support is obtained. However, the horizontal aspect also assumes importance during seismic activity as it allows for oscillation around horizontal cylindrical hinges without damaging the masonry.
- High Quality Resistant Elements. Stone elements must be resistant and in good condition: intrinsically weak (e.g. mud bricks used in certain areas of the world) or heavily degraded elements invalidate other "to the letter" conditions.

After having evaluated the degree of respect of each of the above parameters, a proper formula is used to predict the different mechanical parameters. In such way, a 2.25 MPa $\pm 40\%$ and $0.037 \pm 30\%$ of the compressive and the shear strength was met, respectively.

These ranges were in accordance with the provision reported in the Italian Design Code [7] by selecting two different masonry typologies and respective mechanical parameters as reported in Table 1. The masonry distribution within the structure is illustrated in Fig. 4a and b. The same code imposes a Confident Factor (CF) according to the Level of Confidence (LC) with respect to the masonry strictures depending and the accuracy of the geometrical and mechanical investigations. It was reliably assumed a LC = 2 and a consequent CF = 1.2, [7]. Thus, all the evaluated mechanical strengths were divided by the CF.

Table 1. Mechanical parameters of the masonry typologies according to [7].

Label	Masonry typology [7]	Compressive strength (MPa)	Shear strength (MPa)	Young's modulus (MPa)	Shear modulus (MPa)	Weight density (kN/m^3)
		Min–Max	Min–Max	Min–Max	Min–Max	
A	Rough-hewn masonry with limited leaf and internal core	2.0–3.0	0.035–0.051	1020–1440	340–480	20
B	Tuff-based masonry	1.4–2.4	0.028–0.042	900–1260	300–420	16

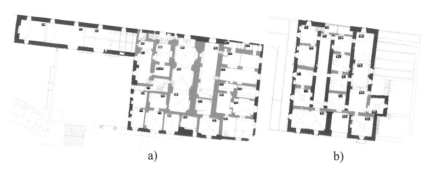

a)   b)

Fig. 4. Classification of the masonry typologies (green for A and red for B): a) ground and b) first floor.

4   Analysis

A global seismic analysis was carried out, with the aim of identifying the most evident deficiencies of the structure and the most suffering walls, as reported in this section. The first step in determining the seismic action is to establish the behavior factor [7] which quantifies the capacity of dissipation through internal damage. A value equal to 2.5 was computed for the case study. Thus, it was possible to obtain the design spectrum for the acceleration (see Fig. 5). In order to calculate the design seismic acceleration, the fundamental period of vibration of the building (T) must be determined. For example, the Italian code NTC08 provides a proper formula (see Eq. (1)), which can be used for buildings under 40 m height (H) and with a uniform distribution of the masses along the height. Consequently, the design spectrum acceleration (Sd) was assumed equal to 0.118g. Based on this result, the equivalent inertia forces were applied to the 2-D structures in order to preliminary verify the shear strength of the masonry walls (equivalent static analysis method). The check was satisfied in all cases.

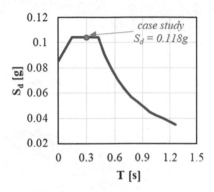

Fig. 5. Response spectrum.

$$T = C_1 \cdot H^{\frac{3}{4}} = 0.05 \cdot 11^{\frac{3}{4}} = 0.3 \text{ s} \tag{1}$$

The 3D behavior of the structures was evaluated through both the static and dynamic analyses and a FEM (finite elements model) in a MasterSap habitat, [12]. Numerical simulations are well-known tool in structural analysis of existing heritage; e.g. in [13–17]. Linear static analysis is the most common and traditional of the possible structural analyzes. It implies that the loads applied do not depend on time, or more exactly, very slowly between the initial instant of application t0 and the final instant of observation tf (quasi-static loads). By assuming also that the internal reaction force depends linearly on the displacements, through a matrix of constant stiffness K and that the external forces are constituted by loads independent of the displacement U, we obtain the classical equation of equilibrium for the almost static linear problems; i.e. KU = F. The maximum static deformation at the ultimate limits state was equal to about 6 mm, and relate to the wall misaligned on the first floor. For seismic combinations, obtained by the equivalent static analysis, the maximum lateral displacement was reduced, i.e. equal to about 4 mm, as shown in Fig. 6a. In seismic combination, on the other hand, the problems relating to the absence of transversal connection of the structural elements of the masonry walls of the stables and the second floor are highlighted, reaching a displacement of 70 mm (see Fig. 6b). In the static analysis, among the results, the maximum stresses (including ideal ones) and moments are reported, as well as the number of the element and the combination of relative load. The Fig. 7a and b show the envelopes of the normal stress and shear membrane tension values, respectively. The tension of greatest interest in masonry structures loaded with small eccentricities is undoubtedly the membrane action below, of compression with local peaks of 1.1 MPa and shear with values at the base around 0.06–0.07 MPa. The analysis shows how the bending stresses in the shell elements are sustainable, however close to the limit tolerable by the materials.

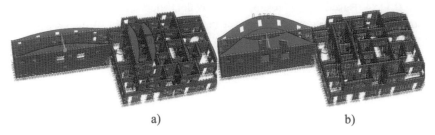

a)   b)

Fig. 6. Seismic deformations: a) 100%+X 30%+Y and b) 30%+X 100%+Y.

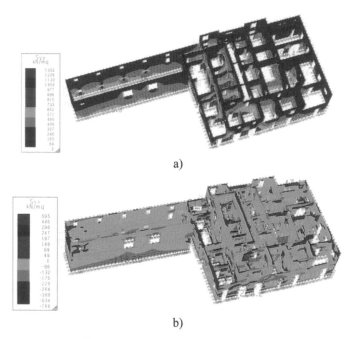

a)

b)

Fig. 7. Stress map: a) compression and b) shear.

5   Conclusions

A study was presented on a complex ancient building included in the UNESCO site of Alberobello, which was built in different ages, with different techniques and materials. The study illustrated the criteria, the operating methods, and the results of a structural vulnerability screening of the historic building called "Palazzo dei Conti di Conversano" located in Alberobello. A diagnostic plan was performed including visual inspections, measurements, materials testing in laboratory. Nondestructive testing was chosen in order to preserve the original building. Locally destructive tests were performed in order to detect the properties of building in non-visible regions or to measure the mechanical properties of materials. The first cognitive phase required numerous inspections and surveys of the masonry structure. On the basis of a first summary knowledge of the wall textures, of the structural architecture and of the relative system of rigidity, a campaign of locally destructive but non-invasive tests has been planned in order to measure the thicknesses of the bearing walls and vaults and the consistency of the wall. According to the results of the diagnostic plan analytical and numerical analyses were carried out. The results of the analyses. The study considered the scenario in static and seismic conditions, according to different levels of analysis. The results highlighted the vulnerabilities mostly due to some defects of construction and decay of mechanical properties of the ancient masonry.

References

1. UNESCO World Heritage List. https://whc.unesco.org/en/list/. Accessed Apr 2020
2. Marraffa, M.: I trulli di Alberobello: La storia della città attraverso i secoli. E. Cossidente (1967). (in Italian)
3. ICOMOS/Iscarsah Committee: Recommendations for the analysis, conservation and structural restoration of architectural heritage (2005). www.icomos.org
4. Istituto Nazionale di Geofisica e Vulcanologia (INGV). http://www.ingv.it/it/
5. UNI EN: Metodi di prova per pietre naturali—determinazione della resistenza a compressione, pp. 1–15. Ente Nazionale Italiano di Unificazione, Roma (2000)
6. Borri, A., Corradi, M., Castori, G., De Maria, A.: A method for the analysis and classification of historic masonry. Bull. Earthq. Eng. **13**(9), 2647–2665 (2015). https://doi.org/10.1007/s10 518-015-9731-4
7. Norme Tecniche per le Costruzione: Aggiornamento delle Norme tecniche per le costruzioni (No. 8). decreto 17-1-2018, Gazzetta Ufficiale 42, 20-02-2018, Ordinary Suppl (2018). (in Italian)
8. Heyman, J.: The Stone Skeleton: Structural Engineering of Masonry Architecture. Cambridge University Press, Cambridge (1997)
9. Heyman, J.: Spires and fan vaults. Int. J. Solids Struct. **3**(2), 243–257 (1967)
10. Heyman, J.: The safety of masonry arches. Int. J. Mech. Sci. **11**(4), 363–385 (1969)
11. Como, M.: Statics of Historic Masonry Constructions. Springer, Heidelberg (2013). https://doi.org/10.1007/978-3-642-30132-2
12. Lejeune, M.C.B.C.: Master Project Plans (2008)
13. Ombres, L., Verre, S.: Numerical modeling approaches of FRCMs/SRG confined masonry columns. Front. Built Environ. 5 (2019). https://doi.org/10.3389/fbuil.2019.00143
14. Ombres, L., Verre, S.: Analysis of the behavior of FRCM confined clay brick masonry columns. Fibers **8**(2) (2020). https://doi.org/10.3390/fib8020011
15. Micelli, F., Cascardi, A.: Structural assessment and seismic analysis of a 14th century masonry tower. Eng. Fail. Anal. **107**, 104198 (2020)
16. Frunzio, G., Monaco, M., Gesualdo, A.: 3D FEM analysis of a Roman arch bridge. In: Historical Constructions, pp. 591–598 (2001)
17. Baraldi, D., Reccia, E., Cecchi, A.: In plane loaded masonry walls: DEM and FEM/DEM models. a critical review. Meccanica **53**(7), 1613–1628 (2018). https://doi.org/10.1007/s11 012-017-0704-3

Mechanical Properties of Mortars for Structural Restoration of Historic Masonry Buildings

Maria Teresa Cristofaro$^{(\boxtimes)}$ ⓘ, Angelo D'Ambrisi ⓘ, Mario De Stefano ⓘ, and Marco Tanganelli ⓘ

Dipartimento di Architettura, Università degli studi di Firenze, Piazza Brunelleschi 6, 50121 Firenze, Italy

{mariateresa.cristofaro,angelo.dambrisi,mario.destefano,
marco.tanganelli}@unifi.it

Abstract. Mortars used for structural restoration of historic masonry buildings are made with the aim of ensuring the durability of masonry over time. This requirement is ensured if the mortars are compatible with historical materials in terms of physical, chemical and mechanical properties.

In this work an experimental study on the mechanical characterization of lime-based mortars used for the structural restoration of monumental historic buildings is presented. The experimental study was performed in laboratory on two distinct types of mortars: injection mortar and mortar for surface restoration. Compression and bending tests were performed on cubic and prismatic specimens. The results of laboratory tests were compared with those obtained from semi-destructive tests performed in-situ to obtain empirical formulations for the assessment of the in-situ strength.

Keywords: Mortars · Mechanical characterization · Structural restoration · Historic masonry buildings

1 Introduction

In the restoration intervention of historic buildings the choice of the mortar to use is very important. The repair mortars must be compatible with historical materials to guarantee the durability of the masonry in the long term. The compatibility criteria are defined on the basis of the original mortar characteristics, but often the performance of the mortar after its application on the masonry is not evaluated. International organizations, such as ICOMOS or ICCROM, recommend the use of materials similar in composition and properties to the original ones for the restoration works [1, 2]. In [3] a methodology that allows to characterize repair mortars in restoration projects was provided. The methodology was developed performing site and laboratory activities to obtain an effective characterization from a structural, physical, chemical, mineralogical and mechanical point of view.

The mortars for the restoration of historical masonry structures can be of different nature (aerial, hydraulic, cementitious and bastard). In the 20th century cement-based mortars were used in the restoration of the masonry structures. A strong repair mortar, as the cement-based mortar, is not advisable because masonry structures show the same

© The Author(s), under exclusive license to Springer Nature Switzerland AG 2022
F. Calabrò et al. (Eds.): NMP 2022, LNNS 482, pp. 2213–2222, 2022.
https://doi.org/10.1007/978-3-031-06825-6_213

degree of movement resulting from creep or thermal effects, therfore a repair mortar should be capable of accommodating movement [4, 5]. In [6–8] it has been shown that the historic masonry have suffered serious damage due to the incompatibility of the cement mortars with the old materials.

In recent years natural hydraulic lime (NHL) mortars have been widely used for the restoration of historic masonry structures. Their use has spread thanks to their compatibility with the historical materials constituting the masonries [9–14].

For the restoration of a historic or monumental building it is essential to guarantee not only the compatibility of the utilized mortar in terms of chemical and physical characteristics, but also an adequate level of mechanical strength [4, 6]. The UNI EN 459-1: 2015 standard [15] classifies natural hydraulic limes in different categories according to the compressive strength after 28 days: NHL5; NHL3.5; NHL2. The mortar mechanical properties can be determined through two types of tests: direct tests and indirect tests. Direct or destructive tests are tests that leads to destructive modifications and alter the tested materials, such as compression, shear and bending tests. Indirect or non-destructive tests are tests performed with methods that do not modify or alter the materials, such as penetrometric, ultrasonic and sclerometric tests.

The article presents the results of the first part of a more extensive experimental study on the mechanical characterization of natural hydraulic lime mortars for the restoration of historic buildings. The performed experimental campaign has the aim of comparing the results obtained with direct tests with those obtained with indirect tests and to propose a correlation law.

2 Materials

The tests were performed in the laboratory on two different types of natural hydraulic lime mortar: injection mortar and mortar for *scuci-cuci*.

2.1 Injection Mortar

The injection mortar is a premix based on natural hydraulic lime (NHL) compliant with the UNI EN 459-1: 2015 [15] standard, hydraulic binders, pozzolan, metakaolin, carbonate and very fine siliceous aggregates, fluidized to be easily injected. This mortar has a high mechanical strength, is permeable to water vapor and is free of any type of organic polymer. The injection mortar is a high strength mortar specific for the consolidation of masonry, according to the EN 998-2: 2001 standard [16], and is particularly suitable for the consolidation injections of masonry structures. The main characteristics of the injection mortar are reported in Table 1.

2.2 Mortar for *scuci-cuci*

The mortar for *scuci-cuci* is a transpiring mortar based on air lime, natural hydraulic lime (NHL) compliant with the UNI EN 459-1: 2015 standard [15], natural pozzolan, selected aggregates of natural stone, unground alluvial sands exempt from silt and/or marble powders and colored natural earths. This mortar is ideal for the restoration and reconstruction of the plasters of buildings of historical and monumental interest in accordance with the UNI 11488: 2021 [17] Cultural Heritage standard. The main characteristics of the mortar for *scuci-cuci* are reported in Table 1.

Table 1. Main characteristics of mortars.

	Injection mortar	Mortar for *scuci-cuci*
Max diameter of the aggregate		3.00 mm
Specific weight		~1500 kg/m^3
Granulometry	≤160 μ	
Mixing water	~32–34%	20–24%
Compressive strength	>UNI EN 196/1 > 10 N/mm^2	1.5 ÷ 5 N/mm^2
Flexural stength		0.5 a 1.5 N/mm^2
Elastic modulus	UNI 6556 13.000 ± 1.000 N/mm^2	

3 Experimental Tests

For each type of mortar, samples were prepared in accordance with the UNI EN 1015-11: 2019 standard [18]. 12 cubic samples 40 × 40 × 40 mm (code Ai Table 2), 12 cubic samples 50 × 50 × 50 mm (code Bi Table 2) and 12 prismatic samples 160 × 40 × 40 mm (code Ci Table 2) were prepared with the injection mortar. 12 cubic samples 40 × 40 × 40 mm (code Ac Table 2), 12 cubic samples 50 × 50 × 50 mm (code Bc Table 2) and 18 prismatic samples 160 × 40 × 40 mm (code Cc Table 2) were prepared with the mortar for *scuci-cuci* (Fig. 1).

The samples were subjected to three different tests: three points bending test; compression test; penetrometric test. Bending and compression tests were conducted using a 600 kN Instron hydraulic press. The bending test for the two types of mortar was performed at a constant loading rate considering the two extreme rate values 50 and 100 N/s provided by the UNI EN 1015-11: 2019 standard (Table 2). The compression test was performed considering two loading rate values: rate provided by the standard according to the class to which the mortar belongs; rate relative to the class lower than that to which it belongs in accordance with the UNI EN 1015-11: 2019 standard (Table 2) to evaluate any strength variations.

All cubic samples were subjected to compression test. The prismatic specimens were subjected to bending test (Fig. 2). Subsequently one of the two parts of each prismatic sample generated as a result of the bending test was subjected to a compression test while the other part was subjected to a penetrometric test (Fig. 3).

The penetrometric test was performed with a Gucci PNT-G penetrometer (Gucci and Barsotti [19]) which allows to determine the energy parameter P_g. From the energy parameter P_g it is then possible to estimate the mortar strength using a calibration curve. For each test 15 cavities must be executed evaluating the relative parameter P_g. The test is valid if at least 5 values of the parameter P_g are included within an interval ±25% of the mean value $P_{g,ave}$ of the parameters P_g.

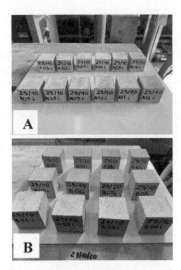

Fig. 1. Tested samples: a) cubic samples $40 \times 40 \times 40$ mm; b) cubic samples $50 \times 50 \times 50$ mm; c) prismatic samples $160 \times 40 \times 40$ mm.

Fig. 2. Flexural strength test.

Fig. 3. a) Compressive strength test – b) Penetrometric test.

4   Results and Discussion

The results of the experimental campaign are reported in the following. The results of the compression and bending tests are shown in Table 2. For each series of samples the average value of the specific weight SW_{ave}, the maximum load $F_{max,ave}$ and the strength f_{ave} are reported. The results show that when the loading rate and the sample size vary there are no significant variations in compressive and flexural strength for the same type of mortar.

The load-displacement curves of the bending tests for the two mortar types and for the two loading rates are reported in Fig. 4. Also in this case when the loading rate varies no significant changes in flexural strength are noticed.

Table 2. Results of compression and bending tests.

Code	N° samples	Section size [mm]	Test	Loading rate [N/s]	SW_{ave} [g/cm³]	$F_{max,ave}$ [N]	f_{ave} [N/mm²]
A_i	6	40 × 40 × 40	Compression	400	1.76	31002	19.4
A_i	6	40 × 40 × 40	Compression	200	1.75	31324	19.6
B_i	6	50 × 50 × 50	Compression	400	1.76	46844	18.7
B_i	6	50 × 50 × 50	Compression	200	1.75	44621	17.8
C_i	6	160 × 40 × 40	Bending	50	1.72	2303	5.4
C_i	6	160 × 40 × 40	Bending	100	1.73	2391	5.6
A_c	6	40 × 40 × 40	Compression	100	1.70	5123	3.2
A_c	6	40 × 40 × 40	Compression	50	1.70	5450	3.4
B_c	6	50 × 50 × 50	Compression	100	1.70	6975	2.8
B_c	6	50 × 50 × 50	Compression	50	1.69	7572	3.0
C_c	9	160 × 40 × 40	Bending	50	1.68	414	1.0
C_c	9	160 × 40 × 40	Bending	100	1.68	379	0.9

The results of the bending tests performed on the prismatic samples and of the compression and penetrometric tests performed subsequently on the parts of these samples obtained as a result of the bending test are reported in Tables 3 and 4. The symbol ′ refers to bending tests performed at a loading rate 50 N/s while the symbol * refers to bending tests performed at a loading rate 100 N/s. In Tables 3 and 4 the results relative to the flexural strength f_{bend}, the compressive strength f_{comp}, the average value of the parameters P_g relative to the 15 cavities $P_{g,ave}$, the average value of the 5 parameters P_g ranging between ±25% of the average value $P_{g,ave*}$ and the strength value f_m obtained from the manual of the utilized penetrometer are reported. From the correlation between f_{comp} and $P_{g,ave}$ an empirical relationship with linear law f_{form} was calibrated.

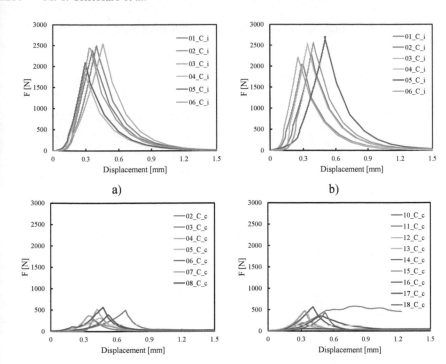

Fig. 4. Load-displacement curves of the bending tests. Injection mortar: a) loading rate 50 N/s; b) loading rate 100 N/s. Mortar for *scuci-cuci*: c) loading rate 50 N/s; d) loading rate 100 N/s.

In the case of the injection mortar the proposed formulation (Fig. 5a) provides strength values close to those obtained from the compression tests. The Gucci and Barsotti formulation [19] provides instead strength values that are more than 50% lower than those obtained from the compression tests (Table 3).

Also in the case of the mortar for *scuci-cuci* the proposed formulation (Fig. 5b) provides strength values close to those obtained from the compression tests and which fall within the range $1.5 \div 5$ N/mm^2 reported in the technical data sheet of the considered mortar. In this case the Gucci and Barsotti formulation [19] provides strength values close to those obtained from the compression tests.

The results relative to the correlation between f_{comp} and $P_{g,ave}$ reported in Fig. 5 show two different trends for the two considered mortars. For the injection mortar the results are concentrated around the linear regression straight line, which therefore presents small uncertainties in the strength estimation. In the case of the mortar for *scuci-cuci* the results are distributed at constant compressive strength in a wider range of $P_{g,ave}$ values showing a large dispersion.

Table 3. Results of the tests on injection mortar.

Code	f_{bend} [N/mm^2]	f_{comp} [N/mm^2]	$P_{g,ave}$	$P_{g,ave*}$	f_m [N/mm^2]	f_{form} [N/mm^2]
01_C_i′	5.57	16.27	970	970	6.44	13.78
02_C_i′	5.84	16.67	924	902	6.09	13.12
03_C_i′	5.98	17.06	875	861	5.87	12.43
04_C_i′	4.38	21.43	1852	872	5.93	12.87
05_C_i′	4.88	18.98	1265	981	6.50	13.93
06_C_i′	5.73	19.63	1338	1052	6.86	15.91
07_C_i*	5.95	16.18	906	1502	9.20	21.33
08_C_i*	6.32	14.13	981	1212	7.70	17.21
09_C_i*	6.01	17.33	1120	1708	11.31	24.26
10_C_i*	5.37	20.34	1502	1908	8.39	26.29
11_C_i*	4.78	18.58	1212	1345	8.22	17.96
12_C_i*	5.21	19.29	1708	1313	11.31	19.01

Table 4. Results of the tests on mortar for *scuci-cuci*.

Code	f_{bend} [N/mm^2]	f_{comp} [N/mm^2]	$P_{g,ave}$	$P_{g,ave*}$	f_m [N/mm^2]	f_{form} [N/mm^2]
01_C_c′	–	2.47	279	267	2.14	2.12
02_C_c′	1.15	2.26	245	229	1.84	1.86
03_C_c′	0.61	2.72	277	255	2.04	2.11
04_C_c′	1.23	3.24	298	269	2.15	2.27
05_C_c′	0.71	2.95	318	310	2.48	2.42
06_C_c′	0.91	2.64	373	366	2.93	2.83
07_C_c′	0.86	2.76	439	412	3.29	3.33
08_C_c′	1.31	2.66	422	401	3.21	3.21
09_C_c′	–	1.95	394	383	3.06	3.00
10_C_c*	0.33	2.72	392	347	2.78	2.98
11_C_c*	1.16	2.99	382	370	2.96	2.90
12_C_c*	0.47	1.94	470	461	3.69	3.57
13_C_c*	0.92	2.6	241	231	1.85	1.83
14_C_c*	1.32	2.97	332	322	2.57	2.52
15_C_c*	1.10	2.58	341	296	2.37	2.59

(continued)

Table 4. (*continued*)

Code	f_{bend} [N/mm^2]	f_{comp} [N/mm^2]	$P_{g,ave}$	$P_{g,ave}$*	f_m [N/mm^2]	f_{form} [N/mm^2]
16_C_c*	0.81	3.24	245	245	1.96	1.86
17_C_c*	–	3.31	310	302	2.41	2.36
18_C_c*	1.01	2.47	274	265	2.12	2.08

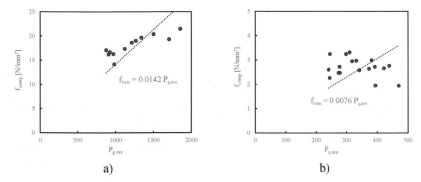

Fig. 5. f_{comp} vs $P_{g,ave}$ correlation: a) injection mortar; b) mortar for *scuci-cuci*.

5 Conclusions

Natural hydraulic lime mortars are used for the restoration of historic structures due to their compatibility with existing materials. In this work the first phase of a more extensive experimental campaign on the mechanical characterization of natural hydraulic lime mortars for the restoration and recovery of historic buildings was presented. The tests were performed in the laboratory on two different types of natural hydraulic lime mortar characterized by different mechanical properties: injection mortar and mortar for *scuci-cuci*.

The results of the compression and bending tests performed on the two types of mortar show that the loading rate and the sample size do not affect the mortar strength.

From the results of the compression tests and penetrometric tests performed on the two parts of the prismatic samples obtained as a result of the bending test two different formulations were calibrated for the two types of mortar. The proposed formulations provide strength values close to those obtained from the compression tests both in the case of the injection mortar and in the case of the mortar for *scuci-cuci*. The Gucci and Barsotti formulation provides instead strength values which in the case of the injection mortar are more than 50% lower than those obtained from the compression tests, while in the case of the mortar for *scuci-cuci* they are close to the experimental results.

In conclusion the performed experimental campaign has allowed to evaluate the correlation between direct tests (compression) and indirect tests (penetrometric) performed on low strength mortars (mortar for *scuci-cuci*). This correlation was instead not noticed for mortars with higher strength (injection mortar). In this work a simple evaluation methodology that can be used in-situ for estimating the mechanical characteristics of the applied restoration mortars and their stability.

The extension of this first experimental campaign to different types of mortar with the execution of direct (splitting test) and indirect tests (ultrasonic and sclerometric tests) and the application of accelerated aging procedures would allow to evaluate, with specific formulations for the various types of mortar, the mechanical characteristics and their stability over time.

References

1. Venice Charter: International Charter for the conservation and restoration of monuments and sites. Venice (1964). http://www.icomos.org/docs/venice_charter.html
2. Conclusions of the Symposium: Mortars, cements and grouts used in the conservation of historic buildings, Rome (1990). Mater. Struct. **23**, 235
3. Schueremans, L., Cizer, Ö., Janssens, E., Serré, G., Van Balen, K.: Characterization of repair mortars for the assessment of their compatibility in restoration projects: research and practice. Constr. Build. Mater. **25**, 4338–4350 (2011)
4. Mosquera, M.J., Benitez, D., Perry, S.H.: Pore structure in mortars applied on restoration. Effect on properties relevant to decay of granite buildings. Cem. Concr. Res. **32**, 1883–1888 (2002)
5. Hendry, A.W.: Masonry walls: materials and construction. Constr. Build. Mater. **15**, 323–330 (2001)
6. Degryse, P., Elsen, J., Waelkens, M.: Study of ancient mortars from Salassos (Turkey) in view of their conservation. Cem. Concr. Res. **32**, 1457–1563 (2002)
7. Moropoulou, A., Cakmak, A.S., Biscontin, G., Bakolas, A., Zendri, E.: Advanced Byzantine cement based composites resisting earthquake stresses: the crushed brick/lime mortars of Justinian's Hagia Sophia. Constr. Build. Mater. **16**, 543–552 (2002)
8. Rodriguez-Navarro, C., Hansen, E., Ginell, W.S.: Calcium hydroxide crystal evolution upon aging of lime putty. J. Am. Ceram. Soc. **81**(11), 3032–3034 (1998)
9. Faria, P., Silva, V.: Natural hydraulic lime mortars: influence of the aggregates. In: Hughes, J.J., Válek, J., Groot, C.J.W.P. (eds.) Historic Mortars, pp. 185–199. Springer, Cham (2019). https://doi.org/10.1007/978-3-319-91606-4_14
10. Vyšvařil, M., Žižlavský, T., Zimmermann, Š., Bayer, P.: Effect of aggregate type on properties of natural hydraulic lime-based mortars. In: Materials Science Forum, vol. 908, pp. 35–39. Trans Tech Publications (2017)
11. Isebaert, A., et al.: Pore-related properties of natural hydraulic lime mortars: an experimental study. Mater. Struct. **49**(7), 2767–2780 (2015). https://doi.org/10.1617/s11527-015-0684-5
12. Amenta, M., Karatasios, I., Maravelaki-Kalaitzaki, P., Kilikoglou, V.: The role of aggregate characteristics on the performance optimization of high hydraulicity restoration mortars. Constr. Build. Mater. **153**, 527–534 (2017)
13. Silva, B.A., Pinto, A.F., Gomes, A.: Natural hydraulic lime versus cement for blended lime mortars for restoration works. Constr. Build. Mater. **94**, 346–360 (2015)
14. Apostolopouloua, M., Armaghanib, D.J., Bakolasa, A., Douvikac, M.G., Moropouloua, A., Asterisc, P.G.: Compressive strength of natural hydraulic lime mortars using soft computing techniques. Procedia Struct. Integrity **17**, 914–923 (2019)
15. UNI EN 459-1:2015. Building lime. Part 1: definitions, specifications and conformity criteria, Brussels (2015)

16. EN 998-2:2001. Specification for mortar masonry: part. 2. Masonry mortar (2001)
17. UNI 11488:2021. Conservazione del patrimonio culturale - Linee guida per la classificazione, la definizione della composizione e la valutazione delle caratteristiche prestazionali delle malte da restauro (2021)
18. UNI EN 1015-11:2019. Method of test for mortar masonry. Part 11: determination of flexural and compressive strength of mortar (2019)
19. Gucci, N., Barsotti, R.: A non-destructive technique for the determination of mortar load capacityin situ. Mater. Struct. **28**(5), 276–283 (1995). https://doi.org/10.1007/BF02473262

A New Compatible and Sustainable FRLM Composite for the Seismic and Energetic Upgrade of Historic Buildings

Valerio Alecci[1], Mario De Stefano[1], Antonino Maria Marra[2], Fabrizio Pittau[3], Dora Pugliese[1(✉)], Rosa Romano[1], and Gianfranco Stipo[1]

[1] Department of Architecture, DIDA, University of Florence, Florence, Italy
dora.pugliese@unifi.it
[2] Department of Civil and Environmental Engineering, DICEA, University of Florence, Florence, Italy
[3] Department of Civil, Environmental and Architectural Engineering, University of Cagliari, Cagliari, Italy

Abstract. The historic building heritage's energy and structural requalification is a crucial topic usually investigated by separate approaches and procedures. This paper covers an integrated assessment of a new composite material to reduce the seismic vulnerability of historic masonry buildings while complying with the requirements of sustainable conservation, reduction of emissions, and energy saving. Firstly, research and selection of thermal mortars to be used as the matrix of the innovative composite material were carried out; then, the mechanical properties were investigated through compressive and bending tests. Furthermore, the mortar with better mechanical properties was used to assemble a composite specimen under the direct tensile test. Finally, dynamic thermohydrometric simulations by WUFI ® Pro software and EnergyPlus software were performed to check energetic contributions when applied on typically arranged masonry panels of existing masonry buildings.

Keywords: Innovative composite materials · Structural safety · Energy performances · Historical buildings renovation

1 Introduction

A large part of the historical heritage consists of masonry buildings, which are highly energy-intensive and vulnerable to seismic actions. In this scenario, an efficient solution for their requalification is the adoption of design strategies adequately aimed at obtaining an upgrade of both structural safety – after a proper investigation of the mechanical properties - and energy performance – after an accurate energy audit -, operating through non-invasive interventions on the building envelope.

Many studies are available in the literature on assessing historic masonry structures, reducing their seismic vulnerability [1–4], and increasing their thermal performance [5].

© The Author(s), under exclusive license to Springer Nature Switzerland AG 2022
F. Calabrò et al. (Eds.): NMP 2022, LNNS 482, pp. 2223–2232, 2022.
https://doi.org/10.1007/978-3-031-06825-6_214

However, the modus operandi is usually oriented to reach the two goals while following separate approaches.

In addition, only a few research works are available on an integrated approach to evaluate new methods of seismic and energy requalification of existing masonry building [6, 7]: a comparison between thermal performance and seismic capacity of different interventions on masonry structures, in terms of economic (€/m^2) and ecological (kgCO$_2$/m^2) costs is proposed in [8].

On the other hand, the scientific community is investigating innovative composites based on sustainable materials, improving the mechanical strength of masonry walls according to conservation criteria. Some innovative composite materials, known as FRCM (Fiber Reinforced Cementitious Matrix) composites, demonstrate to be promising for reducing the seismic vulnerability of masonry structures [9, 10]. FRCM composites have good behavior at high temperatures, a low installation cost, complete removability after use, and a good compatibility with the masonry substrate. In the last years, FRCM is being considered a good solution for the strengthening of existing masonry structures and preferable to FRP (Fiber Reinforced Polymer) composites, especially for applications on historical substrates, even if they are still under investigation [11, 12], and some details need to be deeply assessed mainly in terms of failure modes depending on the mortar matrix grade [13–15].

In this context, the current challenge for architects and engineers is the design of FRCM composites made of natural components, obtained through the reuse of recycled raw materials [16] or obtained from waste processing [17], as a starting point to reduce the environmental impact of the construction sector. The concept of environmental impact is to be understood by observing the "From Cradle to Cradle" approach, following the Life Cycle Assessment (LCA) method, in order to study the influence that the production process has on factors related to climate change [18]. Thanks to this intense interest in materials' environmental impact and life cycle, the scientific community addresses the design of innovative composite materials obtained by combining natural and sustainable mortars and fibres [19].

This work aims to investigate the performances of a new FRCM composite material in reducing the seismic vulnerability of masonry structures while being perfectly in line with the principles of energy efficiency indicated by the European standards for the energy requalification of historic buildings. The paper is organized into three parts. Firstly, after identifying a natural mortar matrix with potentially good mechanical and energy properties among the products present in literature and on the national and international market, three-point bending and compression tests were conducted on thermal plaster specimens carried out a vast experimental campaign. Then, in this first step, direct tensile tests on basalt fabrics were performed, and mortars with the best mechanical properties were selected. In the second phase, the mechanically tested thermal mortars were then evaluated in thermohygrometric properties through dynamic simulations with the WUFI ® Pro software and energy savings, using numerical simulation at the building scale with EnergyPlus software. Finally, in the third step, the thermal mortar having the best mechanical and thermodynamic properties was assembled with basalt fabric and tested in the laboratory under direct tension.

2 Mechanical Properties of the Constituent Materials

2.1 Mechanical Properties of Mortar Matrices

The first phase of the research work concerned a selection of commercial products or mortar mixtures described in the literature, according to the following parameters: composition, typological class according to the UNI Standards [20], aggregate size, minimum and maximum thickness for application on masonry substrate, thermal conductivity, recycled content. Then, a more in-deeply selection was performed according to the criterion of natural, eco-sustainable and environmentally friendly components. As a result, composition and thermal conductivity values λ [W/mK] of 11 selected matrices (labeled as INT.01–INT.11) are reported in Table 1.

Table 1. Composition of selected thermal mortars.

Specimen	Binder	Aggregates	Thermal conductivity [W/mK]
INT.01	Natural hydraulic lime NHL 3.5	Mixed	0,077
INT.02	Natural hydraulic lime NHL 3.5	Minerals	0,048
INT.03	Natural hydraulic lime NHL	Vegetables	0,050
INT.04	Natural hydraulic lime	Mixed	0,086
INT.05	Natural hydraulic lime NHL 3.5	Vegetables	0,064
INT.06	Natural hydraulic lime NHL 3.5	Mixed	0,048
INT.07	Slaked lime	Mixed	0,104
INT.08	Natural hydraulic lime NHL 3.5	Minerals	0,076
INT.09	Natural hydraulic lime NHL	Minerals	0,067
INT.10	Natural hydraulic lime NHL and pumice	Mixed	0,370
INT.11	Natural hydraulic lime NHL 5	Minerals	0,057

The experimental campaign was carried out at the Laboratory of Materials and Structures of the University of Florence. Three mortar specimens $40 \cdot 40 \cdot 160$ [mm^3] in size for each type of mortar were obtained using special standardized metal molds and tested, after curing for 28 days in a controlled temperature room, under three-point bending tests, according to UNI EN 1015-11 [21]. Then, axial compression tests on the two stumps obtained after the failure of each specimen due to bending test were performed (see Fig. 1).

Displacement control tests were carried out using an Instron-Satec universal hydraulic machine with a 600 [kN] hydraulic actuator. Results of the three-point bending, and compression tests are summarized in Table 2 in terms of compressive strength (f_c), compressive elastic Modulus (E_c), and flexural strength (f_b). Elastic Modulus was evaluated between 30% and 70% of the maximum load [22]. It can be noted that thermal mortars INT.06, INT.01 and INT.10 showed the highest compressive strength, flexural strength, and Modulus of elasticity, respectively.

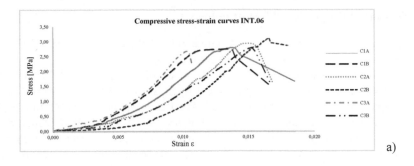

a)

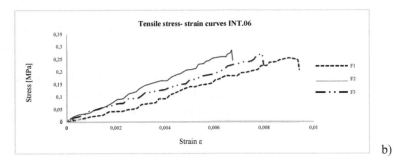

b)

Fig. 1. Compressive stress-strain curves of INT.06 a) and flexural stress-strain b).

Table 2. Mechanical properties of matrices.

Specimen	Compressive strength f_c [MPa]	Compressive Young Modulus E_c [MPa]	Flexural strength f_t [MPa]
INT.01	2,11	336,67	0,65
INT.02	0,40	153,59	0,08
INT.03	0,29	21,23	0,12
INT.04	0,73	143,68	0,02
INT.05	2,39	280,21	0,11
INT.06	2,86	319,86	0,29
INT.07	0,11	4,00	0,10
INT.08	1,23	187,77	0,02
INT.09	1,15	237,61	0,03
INT.10	1,87	470,81	0,07
INT.11	0,76	76,00	0,12

2.2   Mechanical Properties of Basalt Fabric

The basalt fabric fibre tested (Kerakoll Company) is a balanced biaxial net 17·17 [mm] with an equivalent thickness tf of 0.032 [mm] (Fig. 3).

Direct uniaxial tensile tests investigated the mechanical properties of the basalt fabric, performed under controlled displacement by an Instron-SATEC universal testing machine equipped with a 600 [kN] load cell. The load was applied with a rate of 0.25 [mm/min]. The tests were carried out according to the D3039/D3039M Standard Test Method for Tensile Properties of Polymer Matrix Composite Materials [20]. Properly designed aluminum tabs 120·75·0,8 [mm^3] in size were fixed at the two ends of the specimen by the Sikadur 31-CF bicomponent adhesive to ensure uniform transfer of the load to the longitudinal multifilaments of the fabric. Global displacements were acquired by an LVDT displacement transducer integrated into the universal testing machine. Local displacements were acquired through 50 [mm] strain gauges applied to the half-span of the fabric specimen. Different specimens of basalt fabric were assembled varying the number of multifilaments: in particular, four basalt fabric specimens consisting of one (1GS), two (2GS), three (3GS), and four (4GS) longitudinal multifilaments were prepared and tested: in Fig. 2, the stress-strain curves are plotted. The test results in Table 3 describe the tensile strength f_{tf}, Young's tensile Modulus E_{tf}, and ultimate strain ε_{tf}.

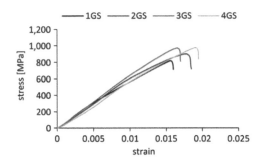

Fig. 2. Stress-strain curves of the basalt fabric specimens 1GS, 2GS, 3GS, 4GS.

The tensile strength f_{tf} was obtained by the following relationship:

$$f_{tf} = F_{max}/A_f$$

in which F_{max} is the maximum load value and $A_f = n \cdot b_f \cdot t_f$ is the equivalent cross-sectional area of the fibre, where n is the number of longitudinal fibre bundles, b_f is the pitch between the bundles of fibres and t_f is the equivalent thickness. Young's tensile Modulus E_{tf} was defined in the first straight line of the stress-strain curve.

Table 3. Mechanical properties of basalt fabric.

Specimen	Tensile strength f_{tf} [MPa]	Young's tensile Modulus E_{tf} [MPa]	Ultimate strain ε_{tf}
GS	865	60487	0,017

3 Hygrothermal Performances of the Matrices

With the objective to investigate the one-dimensional transient hygrothermal behavior of multilayer building components using the thermal plasters analyzed, a series of dynamic hydrothermal simulations were carried out with WUFI Pro 6.5.2 (Wärme-Und Feucht transport Instationär Heat and Moisture Transiency) according to UNI EN 15026:2008 [23].

Six types of thermal mortar were selected from those with the highest value of compressive strength f_c (MPa) and with the low thermal conductivity λ [W/mK] declared by products datasheets, referring to results from the experimental campaign reported in Table 2.

Three specific technical solutions for building masonry were considered for simulation selected from the list provided by UNI/TR 11552 [24], in particular:

M1) One-and-a-half-brick wall.
M2) Stone masonry.
M3) Sack masonry with weakly bonded filling.

The analysis was performed for the Florence climatic reference conditions, and they were carried out covering a 10-years' time so that the wall could achieve a dynamic equilibrium with the environment.

As a hypothesis of building seismic and energy renovation, the application of the thermal plaster was assumed on both sides of the wall considering 60 [mm] thick at the exterior and 40 [mm] thick at the interior of the wall. The analysis was performed for all the products applied to the three masonry solutions (M1, M2, M3) and aims to investigate the hydrothermal behaviour of different scenarios varying the thermal plaster adopted to replace the traditional plaster.

The water content and humidity inside the building components are two parameters that must be checked since the water inside the materials can change their characteristics. For all the investigated solutions, the water content variations in the wall depended only on seasonal variations. However, for all the investigated solutions, the variations of the water content in the wall depend only on seasonal variations. The products INT.01, INT.05 and INT.06 showed a better ability to be passed through the humid air, and a lower percentage of water content inside the wall with maximum values of 10–11 [kg/m^3].

In particular, the products INT.01 and INT.06 represent the ideal configuration for all three wall types located in Florence, and for this reason, they have been selected for the dynamic energy simulation at the building scale, using EnergyPlus software. The

Table 4. Performed different scenarios using dynamic energy simulation at the building scale.

Solution	Specimen	Thermal transmittance U [W/m²K]	Thickness [mm]
M1	Traditional plaster	1,34	20 [mm] int./20 [mm] ext.
	INT.06	0,449	40 [mm] int./60 [mm] ext.
	INT.01	0,514	40 [mm] int./60 [mm] ext.
M2	Traditional plaster	2,33	20 [mm] int./20 [mm] ext.
	INT.06	0,542	40 [mm] int./60 [mm] ext.
	INT.01	0,640	40 [mm] int./60 [mm] ext.
M3	Traditional plaster	1,02	20 [mm] int./20 [mm] ext.
	INT.06	0,449	40 [mm] int./60 [mm] ext.
	INT.01	0,517	40 [mm] int./60 [mm] ext.

model proposed for the analysis consists of a single room identical to the Labimed Test cell at the University of Florence [25], an outdoor laboratory test for energy performance assessment of building facades characterized by an insulated wooden envelope with opaque walls with a thermal transmittance of 0.29 [W/m²K] and a removable wall to test building envelope prototypes. The aim was to verify the potential energy savings achievable considering the mortars with better mechanical and thermohygrometric properties.

Table 4 shows the different scenarios evaluated using dynamic energy simulations for the three technical solutions (M1, M2, M3), with INT.06 and INT.01, considering their application on both sides of the test wall. The results from simulations were compared with the data values obtained from the case of a traditional mortar (20 [mm] thick at the exterior and 20 [mm] thick at the interior).

The results on the energy performances of the single wall show that there are no significant differences between INT.01 and INT.06; in the winter season, the insulating performance of the test wall increases in both cases, while in the summer season, the annual heat gains through the envelope decrease up to 77% compared to the case with traditional mortar (Fig. 3).

Finally, the simulations have shown different results between M1), M2) and M3) technical solutions in indoor comfort. The simulations have shown that improving heat transmission U value of the wall M1) in both the cases (INT.01 and INT.06) was not sufficient to improve the indoor comfort during the whole year: the simulations performed during the coldest day and the hottest day of the year show no significant changes in terms of zone mean radiant temperature. However, considering M2) and M3), the zone means radiant temperature values show a better profile during the winter than traditional mortar.

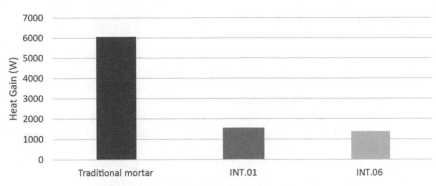

Fig. 3. Surface inside face conduction heat gain rate

4   Mechanical Properties of Composite Material

The basalt fabric was embedded in the INT.06, and three specimens labeled T-01, T-02 and T-03 (500·65·10 [mm³] in size) were tested under direct tension after curing for 28 days at room temperature. The tests were performed in displacement control, with a rate of 0.2 [mm/min], using an Instron-Satec universal machine equipped with a 600 [kN] load cell. The local displacements were captured using a proper 50 [mm] strain gauge positioned in the middle of each specimen.

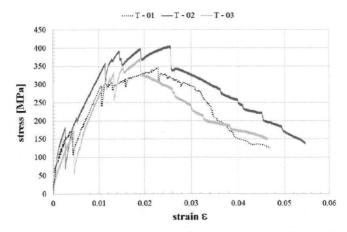

Fig. 4. Stress-strain curves of tensile tests on the three specimens of composite material

Figure 4 shows the mechanical behaviour of the tested specimens. As visible, the formation of the first crack in the matrix occurred at stress equal to 350 [MPa], and it caused a sudden reduction of the load-carrying capability; then, the loading branch highlighted a variation of its slope. When the peak load was reached, a multifilament of the basalt fabric failed. The post-peak softening branch was characterized by the increase of the existing cracks and by the formation of new cracks, and specimens highlighted large displacements without a high increase of applied load.

5   Conclusions

The restoration of the historical masonry building heritage is a highly topical issue. In fact, several intervention strategies have recently been adopted to promote a sustainable redevelopment to improve mechanical and energy performances of historical buildings by using green products with low environmental impact.

In this context, the selection and the analysis of many thermal mortars represent a first step aimed to find a natural and sustainable product to be used as matrix of a new composite material for seismic and energetic retrofitting. The results reported in this paper show that only INT.01 and INT.06 mortars have good mechanical and energy performance. In particular, the INT.06 product represents the ideal matrix from both points of view. Next step will involve the seismic and thermal check of masonry panels strengthened with such a composite and compared to un-strengthened ones, while enlarging the list of mortar mixtures to be evaluated as potential composite matrices.

This study helps to increase the knowledge of innovative technological systems for reducing the energy consumption of historic buildings and improving their mechanical performance and seismic resistance. The research contributes to the dissemination of proper energy and structural requalification practices, impacting the reduction of emission in the atmosphere, with a consequent environmental benefit for our society, as well as to the promotion of innovative technological solutions capable of improving the industrial leadership of the companies in the construction sectors.

References

1. Alecci, V., De Stefano, M., Luciano, R., Marra, A., Stipo, G.: Numerical investigation on the use of flat-jack test for detecting masonry deformability. J. Test. Eval. **49**(1), 537–549 (2020)
2. Alecci, V., Ayala, A., De Stefano, M., Marra, A., Nudo, R., Stipo, G.: Influence of the masonry wall thickness on the outcomes of double flat-jack test: experimental and numerical investigation. Constr. Build. Mater. **285**, 122912 (2021)
3. Karantoni, F., Tsionis, G., Lyrantzaki, F., Fardis, M.N.: Seismic fragility of regular masonry buildings for in-plane and out-of-plane failure. Earthquakes Struct. **6**(6), 689–713 (2014)
4. De Falco, A., Giresini, L., Sassu, M.: Temporary preventive seismic reinforcements on historic churches: numerical modeling of San Frediano in Pisa. In: Applied Mechanics and Materials (2013)
5. Litti, G., Khoshdel, S., Audenaert, A., Braet, J.: Hygrothermal performance evaluation of traditional brick masonry in historic buildings. Energy Build. **105**, 393–411 (2015)
6. Giresini, L., Casapulla, C., Croce, P.: Environmental and economic impact of retrofitting techniques to prevent out-of-plane failure modes of unreinforced masonry buildings. Sustainability **13**, 11383 (2021)
7. Giresini, L., Stochino, F., Sassu, M.: Economic vs environmental isocost and isoperformance curves for the seismic and energy improvement of buildings considering Life Cycle Assessment. Eng. Struct. **233**, 111923 (2021)
8. Mistretta, F., Stochino, F., Sassu, M.: Structural and thermal retrofitting of masonry walls: an integrated cost-analysis approach for the Italian context. Build. Environ. **155**, 127–136 (2019)
9. Prota, A., Marcari, G., Fabbrocino, G., Manfredi, G., Aldea, C.: Experimental in-plane behavior of tuff masonry strengthened with cementitious matrix-grid composites. J. Compos. Constr. **10**(3), 223–233 (2006)

10. Alecci, V., et al.: Experimental investigation on masonry arches strengthened with PBO-FRCM composite. Compos. Part B Eng. **100**, 228–239 (2016)
11. Sneed, L.H., D'Antino, T., Carloni, C., Pellegrino, C.: A comparison of the bond behavior of PBO-FRCM composites determined by double-lap and single-lap shear tests. Cem. Concr. Compos. **64**, 37–48 (2015)
12. Barducci, S., Alecci, V., De Stefano, M., Misseri, G., Rovero, L., Gianfranco, S.: Experimental and analytical investigations on bond behavior of basalt-FRCM systems. J. Compos. Constr. **24** (2020)
13. D'Ambrisi, A., Feo, L., Focacci, F.: Experimental and analytical investigation on bond between Carbon-FRCM materials and masonry. Compos. Part B Eng. **46**, 15–20 (2013)
14. D'Antino, T., Carloni, C., Sneed, L., Pellegrino, C.: Matrix–fiber bond behavior in PBO FRCM composites: a fracture mechanics approach. Eng. Fract. Mech. **117**, 94–111 (2014)
15. Alecci, V., De Stefano, M., Luciano, R., Rovero, L., Stipo, G.: Experimental investigation on bond behavior of cement-matrix–based composites for strengthening of masonry structures. J. Compos. Constr. **20** (2016)
16. Carbonaro, C., Tedesco, S., Thiebat, F., Fantucci, S., Serra, V., Dutto, M.: Development of vegetal based thermal plasters with low environmental impact: optimization process through an integrated approach. Energy Procedia **78**, 967–972 (2015)
17. Fořt, J., Čáchová, M., Vejmelková, E., Černý, R.: Mechanical and hygric properties of lime plasters modified by biomass fly ash (2018)
18. Napolano, L., Menna, C., Asprone, D., Prota, A., Manfredi, G.: LCA-based study on structural retrofit options for masonry buildings. Int. J. Life Cycle Assess. **20**(1), 23–35 (2014). https://doi.org/10.1007/s11367-014-0807-1
19. Monaco, M., Aurilio, M., Tafuro, A., Guadagnuolo, M.: Sustainable mortars for application in the cultural heritage field. Materials **14**(3), 5982021 (2021)
20. UNI EN 998-1: Specifiche per malte per opere murarie - Parte 1: Malte per intonaci interni ed sterni (2016)
21. UNI EN 1015 – 11: Metodi di prova per malte per opere murarie. Parte 11: determinazione della resistenza a flessione e a compressione della malta indurita
22. ASTM D3039/D3039M-17: Standard Test Method for Tensile Properties of Polymer Matrix Composite Materials. ASTM International, West Conshohocken, 15 October 2017
23. UNI EN 15026:2008: Hygrothermal performance of building components and building elements - assessment of moisture transfer by numerical simulation (2008)
24. UNI/TR 11552: Abaco delle strutture costituenti l'involucro opaco degli edifici (2014)
25. Romano, R., Gallo, P., Donato, A.: Smart materials for adaptive façade systems. The case study of SELFIE components. In: Littlewood, J., Howlett, R.J., Jain, L.C. (eds.) SUSTAINABILITY IN ENERGY AND BUILDINGS 2020. SIST, vol. 203, pp. 285–296. Springer, Singapore (2021). https://doi.org/10.1007/978-981-15-8783-2_24

Experimental Investigation on the Effectiveness of Masonry Columns Confinement Using Lime-Based Composite Material

Valerio Alecci[1] , Mario De Stefano[1] , Stefano Galassi[1(✉)] , Raymundo Magos[2], and Gianfranco Stipo[1]

[1] Department of Architecture, University of Florence, Firenze, Italy
{valerio.alecci,mario.destefano,stefano.galassi,
gianfranco.stipo}@unifi.it
[2] School of Architecture, Anáhuac University, Cancun, Quintana Roo, Mexico
raymundo.magos@anahuac.mx

Abstract. In this paper, the former results of an experimental campaign on the confinement effect of masonry square columns with FRCM composites wraps made of a textile embedded in a natural lime-based matrix are presented. The experimental results confirm the effectiveness of such a composite for increasing both the strength and ductility of axially loaded members, despite its low mechanical properties. These results are compared to the numerical ones obtained applying predictive formulas taken from the current literature. However, prediction of the strength increase provided by formulas taken from the literature and from the Italian guideline CNR-DT 215/2018 do not seem capable to fit the results obtained experimentally. The reason of this can be attributed to the very low mechanical properties of the natural matrix used to form the composite. Therefore, considering that the use of a natural matrix, i.e. a sustainable material compatible with the masonry member, is a fundamental requirement for strengthening masonry columns of buildings belonging to the architectural heritage, the authors believe that a subsequent effort should be made from the scientific community involved in this field to derive formulas capable of taking into account even composites made with very weak matrices.

Keywords: Masonry columns · FRCM composites · Confinement · Experimental campaign · Analytical predictions

1 Introduction

A wide part of the world historical heritage is constituted by masonry buildings. Their state of conservation and their response under seismic actions through in-situ investigations as well as analytical-numerical models [1–6] has always been the focus of many researches and studies. Historic masonry buildings are usually characterized by typical elements with both structural and decorative functions at the same time, such as arches,

© The Author(s), under exclusive license to Springer Nature Switzerland AG 2022

F. Calabrò et al. (Eds.): NMP 2022, LNNS 482, pp. 2233–2247, 2022.
https://doi.org/10.1007/978-3-031-06825-6_215

vaults, columns. In case of seismic events or due to load increases, these masonry elements show extreme vulnerability [7, 8]: this phenomenon is often fostered by the natural deterioration of the original material properties. These architectural elements enrich the masonry built, helping to enhance their aesthetics and typological character.

To preserve these buildings, appropriate retrofitting interventions are necessary.

In this ambit, an important topic is represented by the strengthening interventions on masonry columns.

In order to strengthen such masonry members, traditional interventions cover the use of iron bands or other metal devices to reduce their transversal deformation. In the past, metal hoops were applied and heated with subsequent cooling process in order to produce a pre-stress state and resulting in an active reinforcement. In the last two decades, fiber-reinforced polymer (FRP) composites were successively used, as passive intervention, to increase the load-bearing capacity of both masonry and reinforced concrete columns. The crucial problem concerning the precise evaluation of the effective collaboration of the composite material wrap and the calculation of the maximum strength of reinforced columns were largely experimentally and analytically investigated [9–11].

At the same time, great attention was devoted to the definition of a theoretical model for the mechanical behaviour of FRP reinforced columns. Campione and Miraglia [12] proposed a stress-strain relationship for a confined member applying the formula proposed in Mander et al. (1988) [13] to predict the strength of a concrete compressed column confined by means of steel hoops. Form and coefficients of the equation were variously adapted in order to fit experimental results obtained on FRP wrapped specimens and wide comparative studies on compressed members confined with FRP are available in the literature.

In the last decade, FRP composites were progressively substituted by more compatible and sustainable composites made of a fabric embedded in a cement-based mortar (FRCM, Fabric Reinforced Cementitious Matrix). Today, FRCMs are usually preferred to FRP composites, especially for strengthening historical and monumental masonry constructions and some issues involving the bond performances of such a composite are under experimental and numerical investigations [14, 15].

In fact, the scientific community raised the issue of the low physical-chemical compatibility of FRP composites with the masonry material, mainly in the case of historical and artistic buildings for which the sustainability of the intervention and the compatibility with the old substrate is strictly required. Concerning the strengthening of compressed masonry members, the use of FRCM composites is still in an early stage and only few experimental data and analytical studies are available in the literature [16–18]. In particular, load-carrying capacity, ductility property, failure mode of masonry columns strengthened with FRCM composites are still "open issues" that require a great effort of investigation from the scientific community.

Formulas in use to calculate the strength of masonry members confined with FRCM wraps are basically the same already used for FRP confinement, where contribution of the mechanical properties of mortar matrix is specifically considered. These formulas refer to an old expression used for confinement by means of steel reinforcement and that dates back to 1929 (Richart et al. 1929 [19]). It was proposed as a result of an experimental campaign aimed at evaluating the strength of axially loaded concrete cylinders confined with a steel spiral (Richart et al. 1928 [20]):

$$f_{cc} = f_{co} + k'f' \tag{1}$$

where f_{cc} is the strength of the confined member, f_{co} is the strength of the unconfined member, f' is the lateral uniform confining pressure and k' is a coefficient that Richart proposed equal to 4.1.

In subsequent years, other researchers used Richart's formulation for concrete columns reinforced with FRP wraps, but by reinterpreting f' parameter with the meaning of the confinement pressure corresponding to the ultimate strength of the reinforcement.

After performing a wide experimental campaign, some researchers (Campione and Miraglia 2003 [12]) found that a value of k' equal to 4.1 in Eq. 1 was not able to perfectly reproduce the strength of compressed elements wrapped with FRP; for this reason, they proposed a different value of the coefficient k'.

Today, in the literature numerous formulations can be found, where different values of k' parameter are adopted by researchers in order to fit the outcomes of their experimental campaigns: it follows that the problem of the correct evaluation of the confinement effect produced by composite materials is currently partially unsolved and a further effort, by the scientific community, is necessary in order to provide more targeted and reliable formulas.

2   Axial Strength Prediction of Masonry Columns Confined by FRCM Composites

In the last years, strengthening of masonry columns using FRCM composites is replacing the one based on FRP composites. The reasons depend on the greater compatibility between the inorganic matrices and the masonry substrate with respect to the epoxy matrices as well as the more breathability allowed to masonry which reduces the formation of moisture spot and the full reversibility of the strengthening intervention [21].

However, it is worth noting that the application of FRCM composites to masonry columns is very sensitive to the matrix type used to form the composite, both in terms of strength and ductility. In fact, it is well known that, for a chosen reinforcement textile, the use of a cement-based matrix, whose mechanical properties are higher than those of the masonry column, provides a very high strength increase, comparable to the one obtained using FRP composites. Nevertheless, the high increase of strength corresponds to a brittle behaviour of the masonry column, which fails due to the core disintegration which is kept together only by the strengthening wrap. Conversely, the use of a matrix characterized by mechanical properties not very different from those of the masonry member, such as a natural lime mortar, provides only a low strength increase but allows a much more ductile behaviour of the columns themselves. In fact, in this case, failure occurs with the spreading of vertical cracks while the masonry core preserves much more intact with a residual strength. Therefore, the choice of the matrix used to form the composite must be properly addressed for a targeted and correct strengthening intervention and this issue becomes more important in case of interventions on historical and/or monumental buildings.

The possible use of composites assembled with different types of inorganic matrices have led researchers to formulate different analytical models to predict the strength of a confined column, based on different coefficients calibrated to fit results of laboratory tests, often from themselves performed. Currently, the many formulations available, in the literature, for FRCM confined members are strongly affected by these coefficients and are basically derived from those previously proposed for FRP confined masonry columns. Two general expressions can be found in the literature: the first one has been derived from the study of confined concrete columns and is reported in Eq. 1; the second one is reported in the following Eq. 2:

$$f_{cc} = f_{co} \left[1 + k' \left(\frac{f_l}{f_{co}} \right)^{\alpha} \right] \tag{2}$$

Structure of Eq. 1 expresses that the strength of a confined concrete column is given by the strength of the column devoid of the reinforcement plus a rate depending on the effect of the uniformly distributed confinement pressure (in the case of a circular member) increased by the coefficient k' that incorporates the material and typology of the composite.

Structure of Eq. 2, instead, provides k' with a different meaning, that is as the coefficient which increases the axial strength of the strengthened column. The structure of this equation is currently the most preferred by the researches involved in this field.

Some authors even proposed an alternative formulation of Eq. 2 to predict the ultimate strength of a FRCM confined masonry column. This proposal is based on the reasoning that, if the confining composite material is a low volumetric fraction compared to the column size, its effect is negligible and the strengthening is ineffective, such as reported in [22]:

$$\begin{aligned} f_{cc} &= f_{co} & \text{if } \tfrac{f_l}{f_{co}} \leq 0.99 \\ f_{cc} &= f_{co} \left[0.88 + 1.324 \left(\tfrac{f_l}{f_{co}} \right) \right] & \text{if } \tfrac{f_l}{f_{co}} \geq 0.99 \end{aligned} \tag{3}$$

However, the main difference among the various formulations of Eq. 2, as proposed by the many authors, refers to the value of coefficients k'. Some authors propose constant values for k' like, for instance, Murgo and Mazzotti (2019) [17] who propose $k' = 1.53$. Other authors, such as Cascardi et al. (2018) [18], propose a coefficient k' as a function of the characteristic compression strength f_{mat} of the composite matrix, as follows:

$$f_{cc} = f_{co} \left[1 + k' \left(\frac{f_l}{f_{co}} \right)^{0.5} \right] \quad where \quad k' = 6\rho_{mat} \frac{f_{mat}}{f_{co}}, \; \rho_{mat} = \frac{4t_{mat}}{D} \tag{4}$$

where D indicates the diagonal of the rectangular section or the diameter of the circular section of the column.

Balsamo et al. 2018 (Eq. 5) [23] and the Italian guideline CNR-DT 215/2018 [24] (Eq. 6) compute coefficient k' based on the density g_m of masonry:

$$f_{cc} = f_{co} \left[1 + k' \left(\frac{f_l}{f_{co}} \right) \right] \quad where \quad k' = \left(\frac{g_m}{1000} \right)^{0.662} \tag{5}$$

$$f_{cc} = f_{co} \left[1 + k' \left(\frac{f_l}{f_{co}} \right)^{0.5} \right] \quad where \quad k' = \left(\frac{g_m}{1000} \right) \tag{6}$$

Currently, the reference code for confining intervention of masonry columns using a FRCM wrapping is the Italian guidelines CNR-DT 215/2018 [24]. These Guidelines recommend confining a masonry column using a continuous wrapping made of composite material with fibers aligned orthogonally to the geometrical axis of the column and embedded within an inorganic matrix in such a way as to counteract the transversal expansion.

According to these Guidelines, the ultimate axial strength f_{mcd} of the confined column is given by Eq. 2, which is re-written as in Eq. 7:

$$f_{mcd} = f_{md} \left[1 + k' \left(\frac{f_{l,eff}}{f_{md}} \right)^{\alpha_1} \right] \tag{7}$$

where f_{mcd} is the axial compressive strength of the confined column, f_{md} that of the unconfined column, $f_{l,eff}$ is the effective lateral confining pressure, k' is the coefficient of strength increase and α_1 is an exponent which is set equal to 0.5 if experimental data are not available.

The coefficient k' is computed as a function of the density of the composite matrix, according to Eq. 8:

$$k' = \alpha_2 \left(\frac{g_m}{1000} \right)^{\alpha_3} \tag{8}$$

where α_2 and α_3 are set equal to 1 in the absence of experimental data which can justify the adoption of different values.

Since only in the case of a circular column the lateral confining pressure f_l acts uniformly along the cross section and is entirely effective to counteract the lateral expansion, in the case of a square or rectangular column only a rate of it can be considered as actually effective because the stress distribution inside the cross section of the column follows the so called "arch effect". According to this theory, there is a stress concentration at the corners of the cross section while at the mid-sides the section the stress is practically zero. For this reason, the edges of the cross section are recommended to be rounded and the effective lateral confining pressure is introduced in Eq. 7 in the place of f_l and computed according to Eq. 9.

$$f_{l,eff} = k_H f_l \tag{9}$$

Coefficient k_H is denoted as coefficient of horizontal efficiency and depends exactly on the shape of the cross section of the column. In the case of a circular column, $k_H = 1$, while in the case of a rectangular column of sides B and H and diagonal D it is obtained through Eq. 10:

$$k_H = 1 - \frac{B^2 + H^2}{3BH} \tag{10}$$

Finally, to compute the lateral confining pressure, the Italian Guidelines provide Eq. 11:

$$f_l = \frac{2n_f \ t_f \ E_f \ \epsilon_{ud,rid}}{D} \tag{11}$$

where n_f and t_f are the number of layers and the equivalent thickness of the reinforcement, respectively, and $\epsilon_{ud,rid}$ is the strain of the FRCM composite, computed as follows:

$$\epsilon_{ud,rid} = \min\left(k_{mat} \cdot \eta_a \cdot \frac{\epsilon_{uf}}{\gamma_m}; 0.004\right) \tag{12}$$

0.004 being the conventional value of the strain for which the masonry core of the column is considered to be disintegrated and is held together only by the wrapping textile; η_a is the environmental conversion factor set equal to 1 and γ_m is the partial material safety factor set equal to 1.5 for Ultimate Limit States. To compute coefficient k_{mat} in Eq. 12, the following formulas are provided:

$$k_{mat} = \alpha_4 \cdot \left(\rho_{mat} \cdot \frac{f_{c,mat}}{f_{md}}\right)^2 \tag{13}$$

$$\rho_{mat} = \frac{4t_{mat}}{D} \tag{14}$$

where α_4 is set equal to 1.81 if experimental data are not available, $f_{c,mat}$ is the characteristic compression strength of the matrix and t_{mat} the overall thickness of the composite.

3   Laboratory Tests

An experimental campaign was carried out at the Materials and Structures Testing Laboratory of the University of Florence. The experimental program was organized into two main phases. The first phase concerned the mechanical characterization of all the constituent materials (of the masonry specimens and of the composite material). In the second phase, six full-scale square columns, $250 \times 250 \times 500$ mm^3 in size, were built assembling bricks and cement-lime mortar. After 28 days of curing under constant levels of temperature and humidity, three of the six columns were strengthened with a composite material made of a natural lime-based matrix and a balanced polyparaphenylenebenzo-bisoxazole (PBO) mesh. All the columns were then subjected to a uniaxial compression test to evaluate the effectiveness of the confinement provided by the Fiber Reinforced Lime Mortar composite.

3.1   Mechanical Characterization of Bricks and Cement-Lime Mortar Joints

Bricks and cement-lime mortar used to form the columns were mechanically characterized according to the standard UNI EN 772-1:2015 [25] and UNI EN 1015-11 [26], respectively, and the average values of corresponding results are reported in Table 1. The bricks used for the construction of the columns, $250 \times 120 \times 55$ mm^3 in size, were provided by San Marco Terreal Company. The bricks were tested under three-point bending tests, while uniaxial compressive tests were performed on cubic specimens of $50 \times 50 \times 50$ mm^3 in size, obtained by cutting the original bricks. Specimens of the cement-lime mortar used for the masonry assemblage, $160 \times 40 \times 40$ mm^3 in size, were subjected to three-point bending tests and then, compressive tests were performed on the stumps obtained after the bending test failure.

Table 1. Result of three-point bending tests and uniaxial compression tests on bricks and cement-lime mortar specimens.

Material	Compression strength [MPa] (standard deviation; CoV)	Flexural tensile strength [MPa] (standard deviation; CoV)
Brick	17.10 (0.84; 0.05)	5.60 (0.58; 0.10)
Cement-lime mortar joint	2.92 (0.18; 0.06)	0.44 (0.07; 0.16)

3.2 Mechanical Characterization of FRLM Composites

The composite material was obtained by embedding a PBO fabric in a natural lime mortar matrix prepared in the laboratory.

The lime-based mortar was mixed according to the following proportions: 1-part lime, 1-part water, 3-parts river sand. The mortar is fully natural, and it was mixed without any other additives. First, the mortar specimens, $160 \times 40 \times 40$ mm^3 in size, were subjected to three-point bending test and then the two stumps produced by the rupture of each prism were subjected to uniaxial compression test. Results of tests are shown in Table 2. It is important to note that strength of the natural lime mortar resulted lower than 1 MPa.

Table 2. Result of three-point bending tests and uniaxial compression tests on lime-based mortar.

Material	Compression strength $f_{c,mat}$ [MPa] (standard deviation; CoV)	Flexural tensile strength f_t [MPa](standard deviation; CoV)
Lime-based mortar	0.79 (0.12; 0.15)	0.64 (0.02; 0.03)

The PBO fabric, with the equivalent thickness of 0.014 mm provided by the manufacturer, is a bidirectional net with a weight of 22 g/mm^3 both in the warp and in the weft direction. The weaving of the PBO mesh provides sufficient spaces for the passage of the inorganic matrix (Fig. 1a) and, therefore, it improves the adhesion between the matrix and the PBO fabric. Tensile tests were performed which provided the tensile strength

$\sigma_t = 3,300$ MPa, the Young Modulus $E_f = 270,000$ MPa and the ultimate strain $\varepsilon_{uf} = 0.0149$.

The stress-strain diagrams (Fig. 1b) provided by the tensile test show a rather linear-elastic behavior until the peak load was attained and a brittle failure occurred.

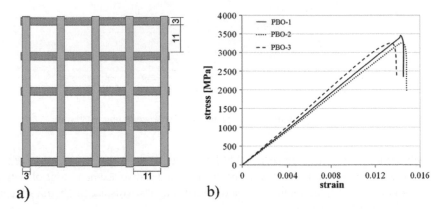

Fig. 1. a) PBO-fiber textile, geometry (mm); b) Stress-strain diagrams of the tensile test of PBO-fiber textile.

3.3   Mechanical Characterization of Columns

After the mechanical characterization of the constituent materials of both the masonry columns and the FRLM composite, six full-scale masonry columns were assembled as shown in Fig. 2a. Each specimen is a square column built with 16 bricks, arranged in pairs in a staggered way in a sequence of 8 rows and 10 mm thick mortar joints. After 28 days of curing, according to the CNR-DT 215/2018 [24], the corners of the of the three columns to be strengthened with FRLM composite were rounded in order to avoid localized stress concentrations at the edges and premature failure. Because the higher the radius of curvature of the corners the higher the load-bearing capacity of the column [27] and since the radius is recommended to be at least 20 mm in the above-mentioned standard [24], in the specific case a radius of curvature of 30 mm was adopted to round off the edges of the specimens (Fig. 2b).

Then, column faces were cleaned and smoothed in order to eliminate surface defects and wetted in order to prevent the specimens from absorbing the mortar mixing water during the application of the composite material. Subsequently, a first layer of matrix was applied on the four peripheral sides of the columns for a thickness of about 5 mm (Fig. 3a). Once the first layer was completed, the PBO mesh was quickly applied to the matrix around the column with an overlapping length of the net equal to 300 mm, as recommended in [24]. Finally, a second 5 mm thick layer of matrix was applied to cover the textile (Fig. 3b). In order to avoid a direct load on the composite wrap during the test, a gap of 20 mm was considered between the top of the column and the beginning of the wrap itself.

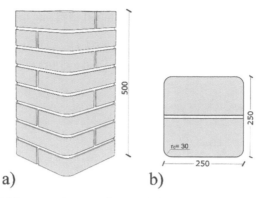

Fig. 2. a) Masonry columns to be tested; b) Corner rounding of 30 mm.

Each column was placed under a 3,000 kN universal press and it was subjected to a uniaxial compression test. Tests were performed under displacement control with a rate of 0.4 mm/min. Displacements were acquired through two displacement transducers placed at the top of the specimens.

Fig. 3. Columns strengthened with the FRLM composite; a) execution of the first layer of matrix and positioning of the textile; b) specimen completed with the second layer of matrix.

4   Experimental Results

The results of the uniaxial compression tests on the masonry columns, both unconfined and confined with the FRLM composite, are summarized in Table 3, where the peak load (F_{max}), the ultimate compression stress (σ) and the ductility factor (μ), computed as the ratio between the displacement corresponding to the ultimate load (conventionally assumed to be 80% of the peak load) and the displacement measured at the end of the elastic field [28], are reported.

Table 3. Results of uniaxial compression tests on the unstrengthened columns (NR-C1, NR-C2, NR-C3) and on the strengthened ones (LR-C1, LR-C2, LR-C3).

ID specimen	F_{max} [kN]	σ_C [MPa]	ε_u
NR-C1	556.35	8.90	1.02
NR-C2	529.65	8.47	1.14
NR-C3	512.04	8.19	1.15
Average	532.68	8.52	1.10
LR-C1	643.75	10.30	1.86
LR-C2	583.36	9.33	2.11
LR-C3	706.34	11.30	1.52
Average	594.43	10.31	1.83

The three unstrengthened columns, labelled as NR-C1, NR-C2, NR-C3 in Table 3, showed very similar behaviour during the tests and the same type of failure. The specimens highlighted vertical cracks triggered in the upper or lower sides and then quickly expanded up to the middle of the specimens (Fig. 4). Load-displacement diagrams showed a linear elastic behaviour up to the peak load without a softening phase while a brittle failure occurred (Fig. 5a).

Fig. 4. a, b) Reference failure mode of unreinforced masonry columns (NR-C1 specimen); c, d) Reference failure mode of the strengthened specimens (LR-C1).

The masonry columns strengthened with the FRLM composite were tested according to the same procedure followed for the unstrengthened ones. A first crack occurred from the top towards the middle of the specimen and, as the load increased, the surface of the lime matrix highlighted many cracks due to the beginning of the sliding between the fabric and the matrix. A first detachment occurred between the outermost layer of the matrix and the fabric, and then the sliding continued up to the complete detachment from the masonry substrate due to the failure of the column inner core. At the end of the test, the fabric remained mostly undamaged even if completely detached by the matrix layers (Fig. 4c) while the inner core of the specimen appeared largely damaged (Fig. 4d).

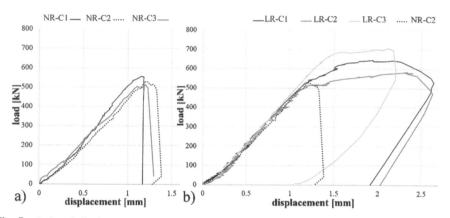

Fig. 5. a) Load-displacement diagram provided by the uniaxial compression tests of the unstrengthened masonry columns; b) Comparison between the load-displacement diagrams of the three strengthened specimens (continuous curves) and the reference NR-C2 unstrengthened specimen (dotted curve).

Load-displacement diagrams were compared to the curve of the NR-C2 unstrengthened column, taken as a reference (Fig. 5b). The comparison highlights that the behaviour of the unstengthened and the strengthened columns is rather similar in the ascending branch confirming that wrap scarcely collaborates in this phase becoming effective only after that the value of the maximum load of the unstrengthened column was reached. A large "post-elastic" branch can be appreciated corresponding to the phase of sliding between the fabric and the surrounding matrix. The FRLM wrap demonstrated to be able to provide an increase in load-bearing capacity and ductility. As well known, columns typically show a brittle behaviour and an increase of ductility is largely desired in seismic area.

Results in Table 3 show the effectiveness of the FRLM composite both in terms of strength and ductility increase which resulted to be 21% and 66%, respectively, higher than those detected on the unstrengthened columns.

Despite the used FRLM composite is made of a completely natural lime mortar of low mechanical properties, experimental results highlighted that it can be successfully used for interventions aimed at increasing the load-bearing capacity and ductility of axially compressed members. Such a composite could be used, in particular, in case of historical masonry buildings thanks to the natural mortar used as matrix and the fully compatibility with the historical substrate.

5 Analytical Predictions

The effectiveness of the FRLM composite has been analytically assessed using the predictive models described in Sect. 2. Numerical results have been compared to the experimental ones.

In particular, Eqs. (3)–(7) demonstrated that they are not able to correctly predict the strength of the confined columns as resulted by the laboratory tests. In fact, application of all formulas available in the literature leads to neglect any contribution of the FRLM composite wrap in increasing the strength of the masonry confined member. More precisely, all formulas provided a strength of the confined member basically coincident with that of the unconfined specimen, ranging from 8.52 to 8.55 MPa. The formula of Italian Guideline CNR-DT 215/2018 [24] estimated only a very minimal increase of strength (about 0.3%), so not relevant if compared to the actual 21% experimentally obtained. Even if the predictive formulas available in the literature are basically conservative, a similar underestimation leads to an erroneous evaluation of ineffectiveness of such a composite in strengthening masonry columns. Considering that in the literature [29] it is proved that the available predictive formulas, comprising the CNR one, are able to correctly estimate the load-bearing capacity increase of the wrapped columns, the discrepancy between analytical and experimental results obtained in this study can be mainly due to the very low mechanical properties of the natural mortar used as matrix of the FRLM composite with respect to the properties of the mortar used to bond the masonry units of the column, that directly influences the final value of the effective confinement pressure (Eq. 11).

It is clear that a fully natural mortar matrix cannot have high mechanical properties but, as experimentally determined, it can be anyway able of increasing both strength and ductility of axially compressed members. In particular, FRLM composite can be very useful – in terms of mechanical and compatibility requirements - for strengthening masonry columns of buildings belonging to the historical heritage and, for this reason, new experimental campaigns are in progress at the Laboratory of the University of Florence in order to evaluate the variation of the prediction capability of the existing formulas when the mortar grade is opportunely varied.

6 Conclusions

In this paper the effectiveness of FRLM composites for confining masonry square columns was experimentally assessed. Six full-scale columns were built at the Laboratory of Materials and Structures of the University of Florence and tested under axial compression. A natural and compatible lime-based mortar matrix was used to assembly the composite. Despite the use of a natural matrix with very low mechanical properties, strengthened columns highlighted an increase of strength and ductility of 21% and 66%, respectively. Predictive formulas from the literature and from the Italian Guideline CNR-DT 215/2018 do not perfectly fitted the experimental outcomes and did not confirm the strength increase of the confined columns as experimentally obtained.

Considering this, although wider experimental campaigns are needed for more general considerations, it can be concluded that:

- FRLM composites made of a natural matrix with low mechanical properties can be effective for increasing the strength and the ductility of masonry columns: this second aspect is crucial in case of historical building located in seismic areas;
- As a consequence of the previous point, a FRLM composite demonstrated to be a good strengthening solution in the case of interventions on columns of historical buildings where compatibility between new and original materials is a specific requirement;
- The equations currently present in the literature demonstrated to be able to fit the experimental outcomes of many researcher's investigations, where mortar used to form the composite had good mechanical properties, largely higher than those of the mortar joints of the column.
- As a result of the present investigation, the same equations present in the literature were not able to correctly predict the effectiveness of the used composite in increasing the strength of the confined columns and this result, probably, depends on the low mechanical properties of mortar that directly influences formula of confinement pressure;
- A further effort should be made by researchers also to take into account the contribution of the composite wrap in terms of ductility increase on confined member, this being a crucial aspect in seismic areas.

References

1. Alecci, V., Ayala, G., De Stefano, M., Marra, A.M., Nudo, R., Stipo, G.: Influence of the masonry wall thickness on the outcomes of double flat-jack test: experimental and numerical investigation. Constr. Build. Mater. **285**, 122912 (2021)
2. Alecci, V., De Stefano, M., Luciano, R., Marra, A.M., Stipo, G.: Numerical investigation on the use of flat-jack test for detecting masonry deformability. J. Test Eval. **49**(1), 537–549 (2020). https://doi.org/10.1520/JTE20190781
3. Giresini, L., Stochino, F., Sassu, M.: Economic vs environmental isocost and isoperformance curves for the seismic and energy improvement of buildings considering Life Cycle Assessment. Eng. Struct. **233**, 111923 (2021). https://doi.org/10.1016/j.engstruct.2021.111923
4. Galeotti, C., Gusella, F., Orlando, M., Spinelli, P.: On the seismic response of steel storage pallet racks with selective addition of bolted joints. Struct. **34**, 3808–3817 (2021). https://doi.org/10.1016/j.istruc.2021.10.001
5. Gusella, F., Orlando, M.: Analysis of the disipative behavior of steel beams for braces in three-point bending. Eng. Struct. **244**, 112717 (2021). https://doi.org/10.1016/j.engstruct.2021.112717
6. Beconcini, M.L., Cioni, P., Croce, P., Formichi, P., Landi, F., Mochi, C.: Non-linear static analysis of masonry buildings under seismic actions. In: Callaos, N., et al. (eds.) 12th International Multi-conference on Society, Cybernetics and Informatics (IMSCI 2018), International Institute of Informatics and Systemics (IIIS), pp. 126–131. Orlando, Florida, USA (2020)
7. De Falco, A., Giresini, L., Sassu, M.: Temporary preventive seismic reinforcements on historic churches: numerical modeling of San Frediano in Pisa. Appl. Mech. Mater. **352**, 1393–1396 (2013). https://doi.org/10.4028/www.scientific.net/AMM.351-352.1393
8. Croce, P., et al.: Influence of mechanical parameters on non-linear static analysis of masonry buildings: a relevant case-study. Procedia Struct. Integr. **11**, 331–338 (2018). https://doi.org/10.1016/j.prostr.2018.11.043

9. Borri, A., Grazini, A.: Masonry columns strengthening with FRP materials. In: L. Ceriolo, V. Zerbo (eds.) 2nd Congress on Mechanics of Masonry Structures Strengthened with FRP-materials: Modeling, Testing, Design, Control, pp.193–202. Cortina Ed., Padova (2004)

10. Corradi, M., Grazini, A., Borri, A.: Confinement of brick masonry with CFRP materials. Compos. Sci. Technol. **67**(9), 1772–1783 (2007). https://doi.org/10.1016/j.compscitech.2006.11.002

11. Krevaikas, T.D., Triantafillou, T.C.: Masonry confinement with fiber-reinforced polymers. J. Compos. Constr. **9**(2), 128–135 (2005). https://doi.org/10.1061/(ASCE)1090-0268(2005)9:2(128)

12. Campione, G., Miraglia, N.: Strength and strain capacities of concrete compression members reinforced with FRP. Cement Concrete Comp. **25**(1), 31–41 (2003). https://doi.org/10.1016/S0958-9465(01)00048-8

13. Mander, J.B., Priestley, M.J.N., Park, R.: Theoretical stress-strain model for confined concrete. J. Struct. Eng. ASCE **114**(8), 1804–1826 (1988). https://doi.org/10.1061/(ASCE)0733-9445(1988)114:8(1804)

14. Alecci, V., Barducci, S., De Stefano, M., Galassi, S., Luciano, R., Rovero, L., Stipo, G.: Reliability of test set-ups and influence of mortar composition on the FRCM-to-brick bond response. J. Test Eval. ASTM **49**(6), 4476–4495 (2021). https://doi.org/10.1520/JTE20200656

15. Barducci, S., Alecci, V., De Stefano, M., Misseri, G., Rovero, L., Stipo, G.: Experimental and analytical investigations on bond behavior of Basalt-FRCM systems. J. Compos. Constr. **24**(1) (2019). https://doi.org/10.1061/(ASCE)CC.1943-5614.0000985

16. Minafò, G., La Mendola, L.: Experimental investigation on the effect of mortar grade on the compressive behavior of FRCM confined masonry columns. Compos. Part B Eng. **146**, 1–12 (2018). https://doi.org/10.1016/j.compositesb.2018.03.033

17. Murgo, F.S., Mazzotti, C.: Masonry columns strengthened with FRCM systems: numerical and experimental evaluation. Constr. Build. Mater. **202**, 208–222 (2019). https://doi.org/10.1016/j.conbuildmat.2018.12.211

18. Cascardi, A., Miceli, F., Aiello, M.: FRCM-confined masonry columns: experimental investigation on the effect of the inorganic matrix properties. Constr. Build. Mater. **186**, 811–825 (2018). https://doi.org/10.1016/j.conbuildmat.2018.08.020

19. Richart, F.E., Brandtzaeg, A., Brown, R.L.: The failure of plain and spirally reinforced concrete in compression. Engineering experimental station. Bulletin **190**, 3–72 (1929)

20. Richart, F.E.; Brandtzaeg, A.; Brown, R.L.: A study of the failure of concrete under combined compressive stresses. Engineering experimental station Bulletin **185**, 3–102 (1928)

21. Zampieri, P.: Horizontal capacity of single-span masonry bridges with intrados FRCM strengthening. Compos. Struct. **244**, 112238 (2020). https://doi.org/10.1016/j.compstruct.2020.112238

22. Krevaikas, T.D.: Experimental study on carbon fiber textile reinforced mortar systems as a means for confinement of masonry columns. Constr. Build. Mater. **208**(2), 722–733 (2019). https://doi.org/10.1016/j.conbuildmat.2019.03.033

23. Balsamo, A., Cascardi, A., Di Ludovico, M., Aiello, M.A., Morandini, G.: Analytical study on the effectiveness of the FRCM-confinement of masonry columns. In: 7th Euro-American Congress on Construction Pathology, Rehabilitation Technology and Heritage Management (Rehabend 2018), pp. 1–9 (2018)

24. CNR-DT 215/2018, Istruzioni per la Progettazione, l'Esecuzione ed il Controllo di Interventi di Consolidamento Statico mediante l'utilizzo di Compositi Fibrorinforzati a Matrice Inorganica. Commissione di studio per la predisposizione e l'analisi di norme tecniche relative alle costruzioni, Roma (2018)

25. UNI EN 772-1: Metodi di prova per elementi per muratura - Parte 1: Determinazione della resistenza a compressione (2015)

26. UNI EN 1015-11: Metodi di prova per malte per opere murarie - Determinazione della resistenza a flessione e a compressione della malta indurita (2007)
27. Sneed, L.H., Baietti, G., Fraioli, G., Carloni, C.: Compressive behavior of brick masonry columns confined with steel-reinforced grout jackets. J. Compos. Constr. **23**(5), 04019037 (2019). https://doi.org/10.1061/(ASCE)CC.1943-5614.0000963
28. Alecci, V., Focacci, F., Rovero, L., Stipo, G., De Stefano, M.: Extrados strengthening of brick masonry arches with PBO–FRCM composites: experimental and analytical investigations. Compos. Struct. **149**, 184–196 (2016). https://doi.org/10.1016/j.compstruct.2016.04.030
29. Aiello, M.A., et al.: Masonry columns confined with fabric reinforced cementitious matrix (FRCM) systems: a round robin test. Constr. Build. Mater. **298**, 123816 (2021). https://doi.org/10.1016/j.conbuildmat.2021.123816

Preliminary Italian Maps of the Expected Annual Losses of Residential Code-Conforming Buildings

Eugenio Chioccarelli[1]([✉]), Adriana Pacifico[2], and Iunio Iervolino[2]

[1] DICEAM, Dipartimento di Ingegneria Civile, dell'Energia, dell'Ambiente e dei Materiali, Università degli Studi Mediterranea di Reggio Calabria, Reggio Calabria, Italy
eugenio.chioccarelli@unirc.it

[2] DiSt, Dipartimento di Strutture per l'Ingegneria e l'Architettura, Università degli Studi di Napoli Federico II, Naples, Italy

Abstract. The expected annual loss (*EAL*) is a metric often used to quantify the risk, including that related to seismic behavior of structures. It allows to combine the annual probability that the considered structure experiences any damage level, with the expected value of the consequences of such damages. In Italy, to promote seismic risk mitigation, a tax relief for retrofitting costs of existing buildings has been introduced. One of the two criteria adopted to demonstrate the improvement of the seismic performance of the structure after the retrofit is the comparison of (a simplified assessment of) the *EAL* before and after the retrofitting. In this context, it is interesting to quantify the *EAL* of code-conforming residential buildings of different structural typologies and supposed located in each Italian municipality. This kind of study descends from a recent Italian research project named RINTC – *Rischio implicito delle strutture progettate secondo le NTC* – that aims at the evaluation of the seismic reliability inherent to design according to the current Italian building code. Selecting some of the structures analyzed in the RINTC project, national maps of *EAL* for four code-conforming buildings are computed. Such structures are three- and six-storey reinforced concrete (RC) frame, infilled, buildings, and two- and three-storey unreinforced masonry (URM) buildings. The *EAL*s are obtained combing the results of probabilistic seismic hazard analyses, performed in accordance with an authoritative source model for Italy, with the fragility functions derived for two performance (i.e., damage) levels, namely *usability-preventing damage* and *global collapse*. Results, following those of the RINTC project, generally show that: (i) *EAL*s vary with the site and the considered structural typology; (ii) three-storey RC and two-storey URM are generally associated to the lowest *EAL* while the largest are often related to three-storey URM.

Keywords: Failure rate · Fragility functions · Probabilistic seismic hazard analysis · Seismic risk

© The Author(s), under exclusive license to Springer Nature Switzerland AG 2022
F. Calabrò et al. (Eds.): NMP 2022, LNNS 482, pp. 2248–2257, 2022.
https://doi.org/10.1007/978-3-031-06825-6_216

1 Introduction

In 2017 an Italian law enabled significant tax relief (up to the 85% of the total investment) for costs of structural seismic retrofitting of existing buildings [1, 2]. The amount of tax relief depends on the effectiveness of the seismic behavior improvement measured as a function of two parameters, computed with and without the retrofitting: (i) the ratio between the peak ground acceleration (*PGA*) causing the structure to exceed the *life-safety* limit state (as defined in [3]) and the design *PGA* adopted for the same limit state of a new-design building at the same site, (ii) the expected annual loss (*EAL*) computed combining the exceedance rates of six limit states, LSs, related to earthquakes, with the reconstruction costs associated to the exceedance of each LS. The LSs are defined for a range of damage states that goes from the onset of nonstructural damage to the conventional complete collapse of the building, namely: (i) onset and (ii) limitation of nonstructural damage, (iii) limitation of structural and nonstructural damage, (iv) life-safety, (v) collapse prevention, (vi) complete reconstruction. To the exceedance of each LS, a percentage of the complete reconstruction cost is associated (such costs vary between 0% to 100%). The mentioned law [1] provides a reference value of the *EAL* computed for a residential building conforming to the Italian code [3, 4], hereafter NTC08. Such a value is 1.13% assuming that each limit state is exceeded when the corresponding design seismic action (expressed in terms of *PGA*) is exceeded.

On the other hand, a large Italian research project, named *Rischio Implicito delle Strutture progettate secondo le NTC* or RINTC [5, 6], showed that, although the design seismic actions for a given limit state (i.e., design ground motion intensity) have the same exceedance return period over the country, the structural reliability of code-conforming buildings, measured as the failure rate with respect to two ad-hoc defined performance levels, that is, *usability-preventing damage*, UPD, and *global collapse*, GC, (although having similar names, they do not correspond with those of [1]) significantly changes across the Italian sites and structural typologies. The analyzed structures in the RINTC project were, among others, residential reinforced concrete (RC) buildings of three and six storeys (with infillings), and unreinforced masonry (URM) structures with two and three storeys. The structures were supposedly located in three Italian sites, Milan, Naples, and L'Aquila, chosen to be representative of different levels of design seismic hazard (low, medium, and high, respectively) according the probabilistic assessment for the country (e.g., [7]). Referring to each site, the structures were first designed according with NTC08 [3] (or its 2018 update [4]); then, the failure rates were computed with respect to UPD and GC.

The aim of the work described herein is to provide preliminary maps of conventional *EAL*s computed for each of the listed structure ideally located in each Italian municipality. To this end, the *EAL* is computed accounting for the seismic hazard of the sites evaluated by means of probabilistic seismic hazard analysis (PSHA) including the soil conditions, the seismic vulnerability of the structures with respect to UPD and GC, and the reconstruction costs defined in analogy with those indicated in [1]. However, due to several differences and limitations in the procedure (e.g., different number and definition of the considered performance/damage levels), the results of this study are not comparable with those obtained applying [1]; yet the maps shown here must be considered for comparison among different structural typologies over the country. It should also be noted that in a previous study, maps of failure rates of code-conforming buildings were developed at the municipality scale [8]. Such a paper shares most of the input models with the present work, yet results discuss seismic risk in a different perspective.

The remainder of this paper is structured such that the methodology for the computation of the *EAL* is presented first. Then, all the input models required for *EAL* assessment are described. Thus, results in terms of maps of *EAL* per structural typology are discussed. Some final remarks close the paper.

2 Methodology

PSHA allows computing the rate of earthquakes (i.e., mainshocks of seismic sequences) causing the ground motion intensity measure (*IM*) to exceed a threshold (*im*) at a site of interest characterized by a known soil class (θ), that is $\lambda_{im|\theta}$ [9]. The plot of $\lambda_{im|\theta}$ versus the possible *im* values is the so-called *hazard curve*. The probability that a building of a given structural typology reaches or exceeds a performance level (*PL*), that is, fails, given a known value of the *IM*, is defined as the fragility function, $P[PL \geq pl|IM = im]$. The latter can be used, together with the hazard curve, to compute the rate of mainshocks causing the failure of the building belonging to the considered structural typology and located on a soil class θ, $\lambda_{pl|\theta}$, Eq. (1):

$$\lambda_{pl|\theta} = \int_{im} P[PL \geq pl|IM = im] \cdot \left| d\lambda_{im|\theta} \right| \tag{1}$$

In the equation, $\left| d\lambda_{im|\theta} \right|$ is the absolute value of the differential of the hazard curve and it is assumed that the fragility is not dependent on the soil condition of the site [10].

Equation (1) can also be (approximately) applied to a given area (e.g., municipality) providing the rate of mainshocks causing a generic building of the structural typology to fail, indicated as λ_{pl}. This requires knowing the probability that the building is located on each possible soil condition, θ_i, that is, $P[\theta_i]$. Thus the total probability theorem results in Eq. (2) in which *i* varies between one and the number of soil classes considered by the ground motion prediction equation (GMPE) adopted for PSHA:

$$\lambda_{pl} = \sum_i \left\{ \int_{im} P[PL \geq pl|IM = im] \cdot \left| d\lambda_{im|\theta_i} \right| \right\} \cdot P[\theta_i] \tag{2}$$

Referring to the two performance levels defined in the RINTC project, that is, UPD and GC, Eq. (2) can be applied twice, computing λ_{UPD} and λ_{GC}. Thus, the *EAL* of the considered structural typology can be computed via Eq. (3), in which the annual probability of reaching or exceeding a generic performance level in one year is approximated by its annual rate and the expected value of the reconstruction cost given the performance level is $E[C_{UPD}]$ and $E[C_{GC}]$ for UPD and GC, respectively:

$$EAL = (\lambda_{UPD} - \lambda_{GC}) \cdot E[C_{UPD}] + \lambda_{GC} \cdot E[C_{GC}] \tag{3}$$

3 Input Data for the *EAL* Computation

3.1 Probabilistic Seismic Hazard

The probabilistic hazard assessment at the basis of the design seismic actions in the current Italian building code considers thirty-six seismic source zones for the country (except Sardinia Island) as described in [11], and adopts a logic-tree constituted by sixteen branches [7]. Among them, the branch named 921 is the one adopted here. The seismic parameters of each seismic source zone are defined via the mean annual number of mainshocks per magnitude bins, that is, the so-called *activity rates* (e.g. [12]), and the GMPE is that of [13]. The *IM*s for which PSHA is carried out for the purposes of this study are the *PGA*, and the pseudo-spectral accelerations associated to two spectral periods (T) that are of interest for the considered structural typologies (to follow), that is $T = \{0.15 \text{ s}, \ 0.5 \text{ s}\}$. Hazard analyses were performed, for each municipality, via the REASSESS software [14]. The hazard curves in terms of *PGA* on rock soil conditions, for all municipalities, are reported in Fig. 1a. In the same figure, the exceedance rate corresponding to a return period (T_r) equal to 475 years (yr) is presented together with the hazard curves computed for the sites of Milan, Naples, and L'Aquila; these sites (identified in Fig. 1b) are those considered in the RINTC project.

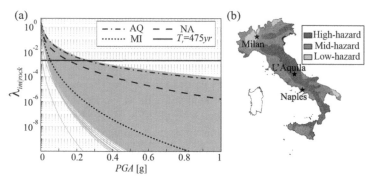

Fig. 1. (a) *PGA* hazard curves on rock computed via PSHA for all the Italian municipalities, (b) hazard classification according to PGA_{475} of each municipality.

To compute *EAL* maps, it is necessary to introduce some criteria to associate the RINTC structures to the Italian municipalities, other than those in which they were originally ideally located. To this aim, the values of *PGA* on rock corresponding to $T_r = 475yr$, that is PGA_{475}, computed for Naples (i.e., 0.15 g) and Milan (0.05 g) are (arbitrary) taken as bounds to define three hazard classes of the Italian municipalities. In other words, the sites characterized by a PGA_{475} lower than that associated to Milan were defined as low-hazard, sites with PGA_{475} lower than Naples (and larger than Milan) were considered mid-hazard and sites with PGA_{475} larger than Naples were high-hazard. The resulting classification is represented in Fig. 1b: the municipalities in low-hazard class are about 16% of the total, whereas those in mid- and high-hazard are about 47% and 36%, respectively (according to the source model, Sardinia is outside the definition range of the GMPE and is not considered hereafter).

Following such a classification, it is assumed that the RINTC buildings designed in L'Aquila, Naples, and Milan (see Sect. 3.3) are representative of buildings designed in any municipality belonging to the high-, mid- and low-hazard class, respectively.

3.2 Local Soil Classes at a Municipality Scale

According to Eq. (2), soil class probability is required for each considered municipality. Such an information can be retrieved profiting of the work of [12] that provides a database of local soil characterizations for a grid of about one million points covering the whole Italian territory. For each point, the soil class (from A to D) according to NTC08 is defined. Since the adopted GMPE accounts for the soil conditions referring to three categories, that is *rock*, *stiff* and *soft*, the soil classes from [12] were converted into these categories. More specifically, soil conditions that, according to NTC08, are identified as A correspond to rock, whereas soil conditions B correspond to stiff soil and soil conditions C and D correspond to the soft soil class of the GMPE.

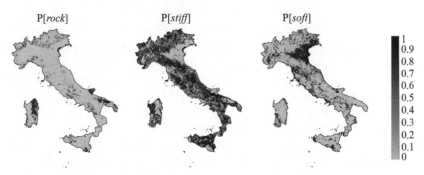

Fig. 2. Soil class probabilities in urbanized areas of Italian municipalities (adapted from [8]).

To quantify the probability that the building of a given municipality is located on a specific soil class, soil data can be combined with the data provided by the Italian *Istituto Nazionale di Statistica* (ISTAT) that identify the urbanized areas of each municipality (https://www.istat.it/it/archivio/222527, last accessed 21/12/2021). The computed percentages of soil categories in each urbanized area are treated as probabilities and are reported in Fig. 2. The largest percentages are associated to stiff soil in most of the municipalities; soft soil covers a non-negligible number of urbanized areas and is predominant in the north-eastern municipalities and along the coasts; finally, rock soil is a significant fraction only in a few areas.

3.3 Fragility Functions of Code-Conforming (RINTC) Buildings

In the RINTC project, a large set of residential buildings was designed, according to [3, 4], to be ideally located in a few Italian sites: L'Aquila, Naples, and Milan. Each structure was designed referring to *damage* and *life-safety* limit states as defined in NTC08. On the other hand, for reliability assessment, two different performance levels were defined:

UPD and GC. The former is reached if one of following conditions occurs (with some adjustments accounting for the peculiarities of specific structural typologies [10, 15]): (i) light damage in 50% of the main non-structural elements; (ii) at least one of the non-structural elements reaches a severe damage level leading to significant interruption of use; (iii) attainment of 95% of the maximum base-shear of the structure. GC, generally, corresponds to the deformation/displacement capacity associated to a 50% post-peak drop of the total base shear of the building.

Four different structural typologies are considered hereafter; they are characterized by the construction material and the number of storeys. The RC structures are moment resisting, infilled-frame buildings of three (RC3) and six storeys (RC6) [16], whereas the URM structures are two- and three-storey buildings (URM2 and URM3, respectively) made of perforated clay units with mortar joints [17]. All the structures are characterized by regularity in plan and elevation.

Lognormal fragility functions for each performance level and structure were defined in [18] as:

$$P[PL \geq pl|im] = \Phi\left[\frac{\ln(im) - \mu}{\sigma}\right] \tag{4}$$

where $\{\mu, \sigma\}$ are parameters. The adopted IM is the largest (between the two horizontal components) 5% damped pseudo-spectral acceleration at a vibration period close to that of the first mode of each model, indicated as $Sa(T)$: fragility functions for RC3, URM2 and URM3 refer to $Sa(0.15\ s)$, whereas $Sa(0.5\ s)$ is the chosen IM for RC6. The parameters of the fragility functions for both performance levels and for all the analyzed structures are provided by [8] and represented in Fig. 3.

Because the considered masonry structures are characterized by different architectural configurations, when URM buildings are of concern, Eq. (3) provides the expected annual loss per site, structural typology, and architectural configuration. Thus, the expected annual loss per site and structural typology can be computed by a weighted sum the EAL computed for each architectural configuration of the same typology. The weight associated to each architectural configuration is based, via expert judgement, on the representativeness of the architectural configuration with respect to the actual building portfolio (the sum of the weights for all the alternative architectural configurations of the same typology equals to one). The values of the weights are provided by [8].

3.4 Percentage Reconstruction Costs

The definition of the reconstruction cost for each performance level is a nontrivial task. It is herein assumed that the reconstruction costs of UPD and GC are the 50% and 100% of the reconstruction cost of the structure, respectively; thus, according to symbols in Eq. (3), $E[C_{UPD}] = 50\%$ and $E[C_{GC}] = 100\%$. These costs are those defined in [1] for life-safety (according to NTC08) and *complete reconstruction* limit states. However, the effect on results of the chosen reconstruction costs may be not negligible, as discussed in the following section.

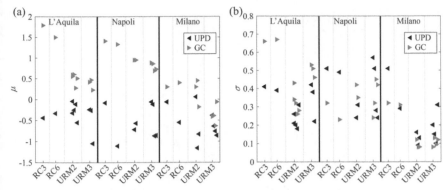

Fig. 3. Parameters of fragility functions: multiple values of μ and σ for URM2 and URM3 correspond to the different architectural configurations (parameters of RC6 refer to a different intensity measure expressing the fragility function than the other structural typologies).

4   Results and Discussions

As useful premise of this section, it should be noted that maps of the GC rates per structural typology, consistently computed with respect to this study, were reported in [19] showing that: (i) failure rates are different among different sites and structural typologies, (ii) in most of the municipalities, RC3 has the lowest failure rate that increases considering RC6 and URM2 and reaches the maximum for URM3. In this section, the analysis in terms of *EAL* allows to underline some additional issues.

The *EAL* computed for RC3, RC6, URM2, and URM3, supposed located in each Italian municipality (consistent with the hazard classes defined in Sect. 3.1), are shown in Fig. 4 from (a) to (d), respectively. It is to note first that, acknowledging the approach of [10], failure rates lower than 1E−05 are substituted by 1E−05 to avoid large extrapolations of hazard and fragility models. Thus, in some cases, the value of 1E−05 is associated to both λ_{UPD} and λ_{GC} resulting in an *EAL* equal to 1E−03%. This happens mostly for RC3 buildings located in northern or southern-east Italy (mild grey sites in the figure). Moreover, as shown in the figure, *EAL*s are different over the country and the considered structural typologies; this is in line with results of the RINTC project as well as the results of [8, 19].

The relative comparison of the maps in Fig. 4 shows that RC3 and URM2 are the buildings to which the lowest *EAL* is associated. RC3 buildings provide the lowest values in sites classified as low- and mid-hazard, whereas for high-hazard sites URM2 shows the lowest *EAL*s (i.e., best seismic performances). RC6 is generally comparable with RC3 over the country (in a few high-hazard sites RC6 has lower *EAL* than RC3). The largest *EAL*s are associated to URM3 in about 75% of the Italian municipalities; in the remaining sites, mostly belonging to the low- and mid-hazard classes, the largest *EAL*s are associated to RC6 and URM2. The mean value of the *EAL* computed for each map of the figure is 0.067%, 0.067%, 0.055%, and 0.113% for RC3, RC6, URM2 and URM3, respectively. The minimum value is equal to 1E-03 in each map. Finally, the maximum values are 0.38%, 0.33%, 0.21% and 0.58% for RC3, RC6, URM2 and URM3, respectively.

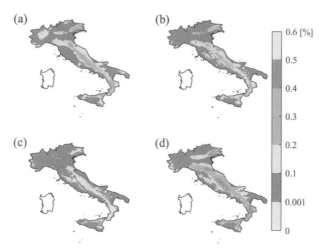

Fig. 4. *EAL* (percentage of the reconstruction cost) for (a) RC3, (b) RC6, (c) URM2 and (d) URM3.

The discussed comparison among the maps is partially in contrast with the conclusions of [19] being here URM2 (not RC3) the structure with the lowest *EAL* for several municipalities. This is motivated by the fact that, given the site, the values of the UPD rate for each structural typology do not have the same order as observed in the case of GC; i.e., referring to UPD, the failure rate for URM2 may be lower than RC3 and RC6, the failure rate for RC6 may be lower than RC3, and the URM3 failure rate may be lower than RC3 and RC6. This issue also means that, if different costs were associated to the considered performance levels, the results of maps' comparison may change.

5   Conclusions

This study evaluated the *EAL* associated to new code-conforming structures supposedly located in each Italian municipality. The considered structures were two- and three-storey unreinforced masonry and three- and six-storey reinforced concrete buildings that were designed, modelled, and analyzed in the RINTC project for three Italian sites (i.e., L'Aquila, Naples, and Milan). They were adopted to represent code-conforming buildings of the municipalities in high-, mid- and low-seismic hazard classes. These classes were identified according to the value of the PGA_{475} associated, via PSHA, to each (whole) Italian municipality. To compute the *EAL*, hazard curves resulting from PSHA (and accounting for soil conditions) were combined with the fragility functions of each structural typology considering two structural performances intended herein as damage levels: usability-preventing damage (UPD) and global collapse (GC). It was assumed that the repair costs associated to them is 50% and 100% of the total reconstruction cost of the building, respectively. Results, in terms of maps of *EAL*s per building typology, showed a clear dependency on the considered site and structural typology, confirming the findings of previous studies analyzing the seismic performance of code-conforming buildings with respect to a complementary point of view. It is also shown that RC3 is associated to the lowest *EAL* in low- and mid-hazard classes whereas URM2 has the lowest *EAL* in the high-hazard class; results for RC6 are comparable to RC3. In 75% of the sites, URM3 is the case with the highest *EAL*. However, in analyzing these preliminary results, their conventional nature and the effect of several limiting assumptions should be always considered.

Acknowledgements. The study presented in this article was developed within the activities of the ReLUIS-DPC 2019–2021 research programs, funded by the Presidenza del Consiglio dei Ministri—Dipartimento della Protezione Civile (DPC). Note that the opinions and conclusions presented by the authors do not necessarily reflect those of the funding entity.

References

1. Decreto Ministeriale: Infrastrutture e Trasp. 28/02/2017 n.58 (2017). (in Italian)
2. Decreto Ministeriale: Infrastrutture e Trasp. 06/08/2020 n.329 (2020). (in Italian)
3. CS.LL.PP.: Decreto Ministeriale 14 gennaio 2008: Norme tecniche per le costruzioni, Gazzetta Ufficiale della Repubblica Italiana, n. 29, 4 febbraio, Suppl. Ordinario n. 30. Ist. Polig. e Zecca dello Stato S.p.a., Rome (2008). (in Italian)
4. C.S.LL.PP. Decreto Ministeriale: Norme tecniche per le costruzioni, Gazzetta Ufficiale della Repubblica Italiana, n. 42, 20 febbraio, Suppl. Ordinario n. 8. Ist. Polig. e Zecca dello Stato S.p.a., Rome (2018). (in Italian).
5. RINTC-Workgroup: Results of the 2015–2018.The implicit risk of code-conforming structures in Italy. ReLUIS Report, Rete Dei Lab Univ Di Ing Sismica (ReLUIS), Naples, Italy (2018)
6. Iervolino, I., Dolce, M.: Foreword to the Special Issue for the RINTC (The implicit seismic risk of code-conforming structures) Project. J. Earthq. Eng. **22**, 1–4 (2018). https://doi.org/10.1080/13632469.2018.1543697. (in Italian).
7. Stucchi, M., Meletti, C., Montaldo, V., Crowley, H., Calvi, G.M., Boschi, E.: Seismic hazard assessment (2003–2009) for the Italian building code. Bull. Seismol. Soc. Am. **101**(4), 1885–1911 (2011). https://doi.org/10.1785/0120100130
8. Pacifico, A., Chioccarelli, E., Iervolino, I.: Residential code-conforming structural seismic risk maps for Italy. Soil Dyn. Earthq. Eng. **153** (2022). https://doi.org/10.1016/j.soildyn.2021.107104
9. Cornell, C.A.: Engineering seismic risk analysis. Bull. Seismol. Soc. Am. **58**(5), 1583–606 (1968). https://doi.org/10.1785/BSSA0580051583
10. Iervolino, I., Spillatura, A., Bazzurro, P.: Seismic reliability of code-conforming Italian buildings. J. Earthq. Eng. **22**(S2), 5–27 (2018). https://doi.org/10.1080/13632469.2018.1540372
11. Meletti, C., Galadini, F., Valensise, G., Stucchi, M., Basili, R., Barba, S., et al.: A seismic source zone model for the seismic hazard assessment of the Italian territory. Tectonophysics **450**(1-4), 85–108 (2008). https://doi.org/10.1016/j.tecto.2008.01.003
12. Iervolino, I., Chioccarelli, E., Giorgio, M.: Aftershocks' effect on structural design actions in Italy. Bull. Seismol. Soc. Am. **108**(4), 2209–2220 (2018). https://doi.org/10.1785/0120170339
13. Ambraseys, N.N., Simpson, K.A., Bommer, J.J.: Prediction of horizontal response spectra in Europe. Earthq. Eng. Struct. Dyn. **25**, 371–400 (1996)
14. Chioccarelli, E., Cito, P., Iervolino, I., Giorgio, M.: REASSESS V2.0: software for single- and multi-site probabilistic seismic hazard analysis. Bull. Earthq. Eng. **17**, 1769–1793 (2018). https://doi.org/10.1007/s10518-018-00531-x

15. Suzuki, A., Iervolino, I.: Seismic fragility of code-conforming Italian buildings based on SDoF approximation. J. Earthq. Eng. **25**(14), 2873–2907 (2019). https://doi.org/10.1080/136 32469.2019.1657989

16. Ricci, P., Manfredi, V., Noto, F., Terrenzi, M., Petrone, C., Celano, F., et al.: Modeling and seismic response analysis of Italian code-conforming reinforced concrete buildings. J. Earthq. Eng. **22**, 105–139 (2018). https://doi.org/10.1080/13632469.2018.1527733

17. Cattari, S., Camilletti, D., Lagomarsino, S., Bracchi, S., Rota, M., Penna, A.: Masonry Italian code-conforming buildings. Part 2: nonlinear modelling and time-history analysis. J. Earthq. Eng. **22**(2), (2918). https://doi.org/10.1080/13632469.2018.1541030

18. Iervolino, I., Baraschino, R., Cardone, D., Della Corte, G., Lagomarsino, S., Penna, A., et al.: Seismic fragility of Italian code-conforming buildings by multi-stripe dynamic analysis of three-dimensional structural models. Submitted for pubblication

19. Chioccarelli, E., Pacifico, A., Iervolino, I.: Italian seismic risk maps based on code-compliant design. In: 31st European Safety and Reliability Conference – ESREL, Angers France (2021). https://doi.org/10.3850/978-981-18-2016-8_678-cd

Structural Safety Assessment of Existing Bridge Decks: Numerical Analysis Assisted by Field Test Results

Dario De Domenico[(✉)] [iD], Davide Messina [iD], and Antonino Recupero [iD]

Department of Engineering, University of Messina, 98166 Messina, Italy
{dario.dedomenico,davide.messina,antonino.recupero}@unime.it

Abstract. The resilience of a metropolitan area relies on the structural safety of bridge structures that must be protected against any type of hazard. Indeed, bridges represent essential elements of the transportation system and play a crucial role in emergency situations for ensuring rescue activities. Many Italian bridges are approaching their natural service life and are affected by ageing and material deterioration (especially, corrosion of steel) due to the lack of periodical maintenance plans. Therefore, it is necessary to develop efficient and reliable procedures for the structural safety assessment of existing bridges. This work proposes an experimental-numerical methodology, illustrated in the context of a case study involving the prestressed concrete deck of the Longano viaduct (Sicily, Italy), which combines field test results (free vibration tests and static load tests) with numerical simulation. A nonlinear finite element model is calibrated based on field test results and used to explore the load-bearing capacity of the bridge deck at ultimate limit states under different hypothetical corrosion scenarios of the prestressing strands. The proposed methodology provides preliminary information on the structural health of existing bridge decks (under both serviceability and ultimate conditions) prior to performing extensive material test campaigns and can be applied to other similar bridges to rapidly detect vulnerable portions of a large infrastructure network.

Keywords: Bridges · Structural safety · Load-bearing capacity · Bridge deck · Nonlinear finite element analysis

1 Introduction

More than 80% of existing bridges in the Italian bridge stock were designed and built before the 1980s, following design regulations that are different from those in force today. Most of them (around 90%) were realized with a girder scheme with prestressed concrete (PC) beams and transversal diaphragms and an overlying slab made in reinforced concrete (RC). These structures are quite vulnerable to material deterioration phenomena that might have taken place over their structural lifetime, especially corrosion of steel. In many cases, durability aspects were not carefully considered at the design stage and most importantly, not regarded as a crucial performance requirement

© The Author(s), under exclusive license to Springer Nature Switzerland AG 2022
F. Calabrò et al. (Eds.): NMP 2022, LNNS 482, pp. 2258–2267, 2022.
https://doi.org/10.1007/978-3-031-06825-6_217

during the service life of the structure. Consequently, a scarce maintenance activity was performed over the years, and in most cases, there has been little attention paid to periodical management plans for safety checks, local repair/retrofitting actions, whenever needed. The occurrence of recent, unacceptably frequent cases of collapse of existing bridges in Italy [1, 2] has emphasized the presence of an outdated infrastructure system in need of a rational, widespread, and reliable quality control and management system.

The structural safety of existing bridges is a key condition for guaranteeing the resilience of a metropolitan area because bridges and viaducts play an essential role in emergency scenarios to ensure rescue activities. This fact is even more important in view of the large number of bridges in Italy, with approximately one bridge every two kilometers of the infrastructure network [3]. Bearing in mind the inherent degradation phenomena of existing bridges, most of which are approaching or even exceeding their natural service life, it is of utmost importance to systematically assess their structural health conditions through reliable, standardized, and comprehensive procedures that simultaneously consider serviceability and ultimate limit states. These procedures should aim at identifying critical portions of a large infrastructure network in order to give the managing bodies of the road network a clear understanding of the number, extent and type of repairing/retrofitting action to implement, e.g., localized, specific interventions via composite systems [4] or global interventions to improve the seismic bridge response via energy dissipation systems [5, 6], accompanied by clear prioritization schemes [7]. In this regard, a recent research theme is concerned with the development of efficient procedures for the structural safety assessment of existing bridges.

Along this research line, this contribution presents a combined experimental-numerical methodology, illustrated in the context of a case study represented by the Longano viaduct (Sicily, Italy), that is aimed at evaluating the structural health conditions of existing PC bridge decks. The procedure combines field test results (free vibration tests and static load tests) with numerical simulation via a finite element model (FEM). Linear analysis is performed for calibration purposes based on the field test results, thus addressing the serviceability conditions, and nonlinear analysis is carried out to investigate the ultimate limit states. Different hypothetical corrosion scenarios of the prestressing strands are incorporated in the nonlinear FEM, within a concentrated plasticity approach using plastic flexural hinges along the PC girders of the bridge deck whose moment-curvature relationships depend upon the corrosion degree. The proposed method provides preliminary information on the structural health of existing bridge decks (under both serviceability and ultimate conditions) prior to performing extensive material test campaigns and can be applied to other similar bridges to rapidly detect vulnerable portions of a large infrastructure network.

2   Presentation of the Case Study: The Longano Viaduct

The Longano viaduct [8] was designed and constructed in the first 1970s in the municipality of Barcellona (Sicily Italy) and belongs to the A20 highway between the provinces of Messina and Palermo. In this contribution, the attention is paid to the structures of the bridge deck. The viaduct is composed of two separate carriageways (in the two opposite directions), and its deck includes 3 spans of length 29.00 m (external spans) and 30.00 m (central span) with a simply supported scheme. Each span is composed of 4 I-shaped longitudinal girders (spacing 2.75 m), 5 transversal diaphragms and an overlying RC slab 20 cm thick and 11 m wide. The pre-tensioned girders have 42 steel strands with 0.6" diameter, 4 of which located in the top flange, 2 in the central part of the section (web), 26 in the bottom flange and the remaining 10 inclined from the supports (positioned near the top flange for the negative moments) to the mid-span (positioned within the bottom flange for the positive moments) by means of intermediate deviator blocks. A sketch of the plan section of the deck along with two representative girder sections is illustrated in Fig. 1. Based on the specifications retrieved from the original drawings of the bridge, concrete cylindrical strength >35 MPa and ultimate tensile strength of prestressing strands >1667 MPa are assumed in the analysis.

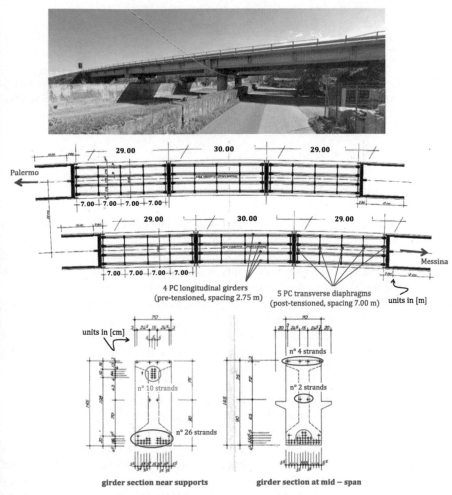

Fig. 1. Photograph and plan section (top) and two representative girder sections (bottom) of the Longano viaduct.

Visual inspection of the Longano viaduct revealed a state of diffused corrosion of some portions of the girders and soffit of the slab as shown in Fig. 2. The general bridge deterioration phenomena are justified by the proximity to the sea (the viaduct is located 1 km in front of the Tyrrhenian Sea) and the related high concentration of chloride ions from the surrounding marine environment. The concrete cover was found to be entirely removed in wide zones of the girder bottom face. The inspection did not reveal any damage phenomenon in the transverse diaphragms. Considering the deterioration state observed during visual inspection, a nonlinear numerical analysis procedure is proposed to investigate the structural safety of the viaduct for different hypothetical corrosion scenarios as described in the next sections.

Fig. 2. Photographs taken from visual inspection on the Longano viaduct.

3   Field Tests: Free Vibration Tests and Static Load Tests

The global structural health conditions of the Longano viaduct were investigated through field tests, namely free vibration tests and static load tests.

Free vibration tests were performed for identification of natural frequencies and mode shapes. Six vertical-axis accelerometers were deployed along the span of the viaduct (1000 mv/g), and free vibrations were triggered by two artificial excitations, namely hammer pulse load on a plate, and heavy truck (345 kN) step load passing through a 12 cm step as depicted in Fig. 3.

The acceleration time histories were post-processed through a Butterworth filter of order 6 (selecting an interval of interest as [3–40 Hz]) and then processed in the frequency domain to identify natural frequencies and mode shapes through the frequency domain decomposition (FDD) technique [9]. Although the two types of excitations induce vibrations that are, of course, of different magnitudes, they were comparatively analyzed to excite the largest number of vibration modes of the deck. In particular, it was observed that the hammer pulse excited a broader range of frequencies (including some higher-order modes) whereas the truck step load mainly excited the lower frequencies modes. Different positions of the loads were considered in the field test (at midspan and at the two quarters of the span). Based on the experimental outcomes, the first frequency was identified in the range (4.40–4.50) Hz, the second frequency in the range (4.85–4.95) Hz, the third frequency in the range (13.1–13.5) Hz and the fourth frequency 15.4 Hz. Based on the mode shapes obtained by the FDD, the first and third mode are of flexural type, while the second and fourth of torsional type. These results were quite in line with the expectations, i.e., with the numerical predictions obtained by the FEM using the linear elastic properties of the bridge deck, as clarified later.

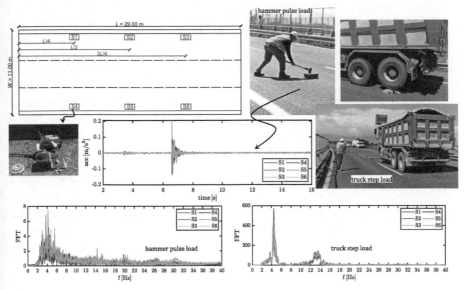

Fig. 3. Free vibration tests on the Longano viaduct triggered by two artificial excitation types.

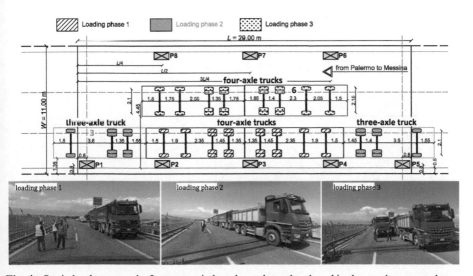

Fig. 4. Static load tests on the Longano viaduct through trucks placed in three subsequent phases.

Additionally, static load tests were performed to investigate the deflection response under the maximum serviceability loads prescribed by the Italian Technical Code

NTC2018. Traffic load models (TLMs) of the NTC2018 involve distributed and tandem (concentrated) loads that are conventional loads. In reality, in the static load tests the loads were applied by means of three-axle and four-axle heavy trucks filled with coarse gravel so as to reach a gross weight of 340 kN and 400 kN, respectively. Position of the trucks was designed to produce comparable deflection response in the bridge deck to that produced by the code compliant TLMs. The loads were introduced in three gradual phases to carefully check the occurrence of any cracking sign or unexpectedly large deflection. The deflections were monitored in each loading phase through eight digital surveyor's levels located in the bridge deck as shown in Fig. 4. After loading phase 3, the loads were then removed, and it was checked that residual deflection in any point of the bridge did not exceed the 15% threshold of the maximum values as recommended in the NTC2018 provisions. The maximum deflection (at mid span) was around 17 mm, in line with the expectations.

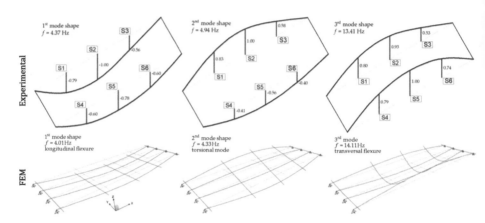

Fig. 5. Validation of FEM by comparison with experimental results from dynamic tests.

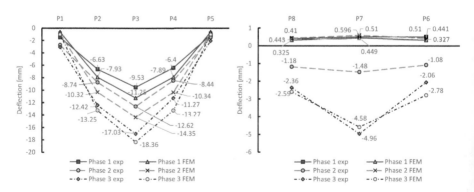

Fig. 6. Validation of FEM by comparison with experimental results from static load tests.

Both dynamic and static tests were experimentally simulated through a FEM developed in SAP2000, using beam elements (with 6 degrees of freedom per node) and adopting consistent sectional properties. The FEM was validated by comparison between numerical natural frequencies (calculated from modal analysis) and those obtained from experimental free vibration tests, as shown in Fig. 5. In general, the FEM was able to reproduce the first three modes of vibration with good accuracy, and the discrepancies between the natural frequencies were in the order of 10%. Similarly, deflection response of the static load tests was simulated through a linear static analysis in the FEM, by using load arrangements consistent with the heavy truck loads. Comparison between experimental and numerical deflections for the three loading phases of the static tests is shown in Fig. 6, from which it can be observed that a reasonable agreement was obtained. These results confirmed the validity of the developed FEM.

4   Ultimate Limit State Analysis of the Bridge Deck

Numerical static nonlinear analysis was carried out to investigate the structural behavior of the bridge deck under ultimate loading conditions. Starting from the load configuration of the loading phase 3 of the static load tests (maximum serviceability loads), the loads were increased monotonically until reaching the collapse of the deck. Considering the deterioration state of the bridge noted during the visual inspection, it is interesting to investigate the load-bearing capacity of the deck by incorporating the corrosion of the prestressing strands. In the literature, different models were developed to incorporate the effects of corrosion in the numerical analysis [10, 11].

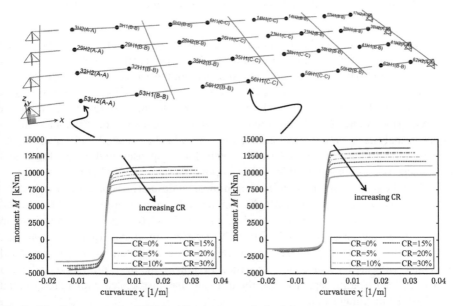

Fig. 7. Nonlinear numerical FEM with concentrated plasticity approach and moment-curvature relationships accounting for corrosion of prestressing strands for various CRs.

Due to the absence of experimental tests aimed to quantify the exact corrosion degree, a status of uniform corrosion was assumed for all the prestressing strands through six hypothetical corrosion scenarios in terms of corrosion rate (CR) or mass loss in percentage, namely 0% (uncorroded configuration), 5%, 10%, 15%, 20% and 30% (strong corrosion configuration). Sectional analysis principles were used to derive the moment-curvature relationships of the PC girders in the various corrosion scenarios, by reducing the area of resisting section of the strands according to the value of CR. A concentrated plasticity approach is adopted by assuming flexural plastic hinges at the two terminals of each segment of PC girders between two transverse diaphragms, as illustrated in Fig. 7. A plastic hinge length equal to the depth of the cross section was assumed.

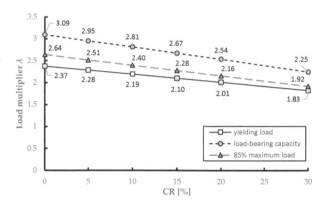

Fig. 8. Reduction of the load-bearing capacity with the CR of the prestressing strands.

Displacement-controlled nonlinear static analysis is performed in the FEM and the "vertical pushover" curves were extracted for each analysis (corrosion scenario) in order to identify the ultimate (collapse) load. In this manner, the evolution of the load-bearing capacity of the bridge deck depending on the corrosion degree of its prestressing strands was calculated. The results were elaborated in terms of dimensionless load multiplier $\lambda = R_{max}/R_0$, where R_{max} is the total maximum vertical reaction calculated in the nonlinear analysis and R_0 is the reference maximum load calculated in the static linear analysis under the maximum serviceability loads (loading phase 3 of the static load tests) in the uncorroded scenario. In this way, it is possible to quantify: 1) the load increase allowed by the stress redistribution in the nonlinear static analysis compared to the static linear analysis, which represents a rough estimate of the safety margin of the bridge deck in comparison to the maximum serviceability loads (NTC2018 loads); 2) the variation of the load-bearing capacity with the CR of the prestressing strands. Relevant results are illustrated in Fig. 8 for three characteristic values of the load, namely the yielding load (associated with the formation of the first plastic hinge), the maximum load (indicated as the deck load-bearing capacity) and the ultimate load (associated with a drop of load to 85% of the maximum value).

By inspection of the results in Fig. 8, it was noted that the yielding load multiplier in the uncorroded case (CR = 0) is 2.37, which indicates that the bridge deck has a good safety margin with respect to the formation of the first plastic hinge, that is, the loads should be increased of approximately two times and a half those of the loading phase 3 of the static load tests to exceed the conventional elastic limit of the bridge deck. This is an indirect confirmation of the structural safety of the bridge in its service conditions. On the other hand, the presence of corrosion, as expected, led to a significant decrease of all the three multipliers and, more specifically, the reduction trend of the load-bearing capacity (dashed line with circle markers in Fig. 8) is well described by the following regression (linear decaying) formula:

$$F_{corroded} = F_{uncorroded} - 0.028\text{CR} \ [\%] \tag{1}$$

that is valid for bridges having a structural scheme (simply supported), as well as geometric and mechanical characteristics similar to those of the considered case study. Equation (1) can be used to calculate a first estimate of the corrosion-induced degradation of the load-bearing capacity of existing bridges. As an example, for CR = 10% this formula would predict a considerable decrease of the load-bearing capacity of almost 30% compared to the uncorroded configuration. This formula should be used in combination with available experimental data on the actual corrosion degree of prestressing strands (whenever available) to give the managing bodies of the road network a first estimate of the influence of corrosion on the load-bearing capacity of existing bridges.

5   Conclusions

The resilience of a metropolitan area is strictly related to the structural safety of its infrastructure network. A methodology combining numerical analysis assisted by field tests results and involving both serviceability and ultimate loading conditions has been presented in this work for a simplified estimate of the safety conditions of existing bridges. The procedure can give a preliminary outlook of the bridge structural health conditions prior to performing comprehensive experimental tests on materials and structural elements, the latter being certainly useful for a more precise level of investigation in a future step. In the proposed approach, the response under serviceability loads (field test results) has been used to calibrate a numerical FEM of the bridge that has been subsequently employed for performing nonlinear static analysis to investigate the safety margin of the bridge at ultimate limit states. A simplified technique to incorporate the effects of corrosion of prestressing strands on the evaluation of the ultimate load-bearing capacity has been presented, which makes use of flexural plastic hinges (concentrated plasticity approach) whose moment-rotation relationships have been calibrated depending upon the corrosion rate. The proposed procedure, here described in the context of a representative case study, can be a useful tool for an expeditious load-bearing capacity assessment of other similar existing bridges.

References

1. Colajanni, P., Recupero, A., Ricciardi, G., Spinella, N.: Failure by corrosion in PC bridges: a case history of a viaduct in Italy. Int. J. Struct. Integr. **7**, 181–193 (2016)
2. Morgese, M., Ansari, F., Domaneschi, M., Cimellaro, G.P.: Post-collapse analysis of Morandi's Polcevera viaduct in Genoa Italy. J. Civ. Struct. Heal. Monit. **10**(1), 69–85 (2019). https://doi.org/10.1007/s13349-019-00370-7

3. Miluccio, G., Losanno, D., Parisi, F., Cosenza, E.: Traffic-load fragility models for prestressed concrete girder decks of existing Italian highway bridges. Eng. Struct. **249**, 113367 (2021)
4. Herbrand, M., Adam, V., Classen, M., Kueres, D., Hegger, J.: Strengthening of existing bridge structures for shear and bending with carbon textile-reinforced mortar. Materials **10**, 1099 (2017)
5. Moslehi Tabar, A., De Domenico, D., Dindari, H.: Seismic rehabilitation of steel arch bridges using nonlinear viscous dampers: application to a case study. Pract. Period. Struct. Des. Constr. **26**, 04021012 (2021)
6. Pucinotti, R., Fiordaliso, G.: Multi-span steel-concrete bridges with anti-seismic devices: a case study. Front. Built. Environ. **5**, 72 (2019)
7. Di Prisco, M.: Critical infrastructures in Italy: state of the art, case studies, rational approaches to select the intervention priorities. In: Proceedings of the fib Symposium 2019: Concrete - Innovations in Materials, Design and Structures, Krakow, Poland (2019)
8. De Domenico, D., Messina, D., Recupero, A.: A combined experimental-numerical framework for assessing the load-bearing capacity of existing PC bridge decks accounting for corrosion of prestressing strands. Materials **14**(17), 4914 (2021)
9. Brincker, R., Zhang, L., Andersen, P.: Modal identification of output-only systems using frequency domain decomposition. Smart Mater. Struct. **10**, 441 (2001)
10. Di Sarno, L., Pugliese, F.: Critical review of models for the assessment of the degradation of reinforced concrete structures exposed to corrosion. In: Proceedings of the Conference SECED 2019, Earthquake Risk and Engineering towards a Resilient World, Greenwich, London (2019)
11. Coronelli, D., Gambarova, P.: Structural assessment of corroded reinforced concrete beams: Modeling guidelines. J. Struct. Eng. **130**, 1214–1224 (2004)

Safety Management of Existing Bridges: A Case Study

Antonino Fotia[1] ⓘ, Maria Rosa Alvaro[2], Francesco Oliveto[2],
and Raffaele Pucinotti[1](✉) ⓘ

[1] Università Mediterranea di Reggio Calabria, Reggio Calabria, Italy
{antonino.fotia,raffaele.pucinotti}@unirc.it
[2] Ingegnere Strutturista Presso STACEC S.R.L., Bovalino, Italy
stace@stacec.com

Abstract. With the publication of the new guidelines on Risk Classification and Risk Management and Safety Assessment and Monitoring of Existing Bridges, a new phase in the management of bridges has begun that suggests the criteria of surveillance and monitoring. All these activities have led to the need to establish proper control procedures and intervention priority.

In this paper, a case study of an existing bridge in concrete reinforced is presented. The methodology includes the following steps: (i) use of aerial vehicles (Uavs) for image acquisition; (ii) creation of a three-dimensional model of the bridge containing information on the geometric and degradation of the concrete and reinforcements, (iii) elaboration through soft-computing techniques for defect detection; (iv) creating an "evolutionary" database, able to update the degradation in subsequent UAV inspections; (v) implementation of defects in a structural software.

The structural analyses were carried out using commercial software (Fata Next by Stacec Srl) capable of simulating different degradation models, with fiber beam elements with Forced-Based formulation in large displacements.

Keywords: Risk management · UAV · Arch bridges · Degradation · Crack identifications · Steel corrosion

1 Introduction

Continuous safety monitoring and maintenance of infrastructure are essential issues. Effects such as fatigue load, thermal expansion, contraction and external load reduce performance of bridges over time. As bridges become obsolete, requirement for inspection and maintenance increases. If maintenance works is not carried out, then their costs in the near future will increase non-linearly. Therefore, many countries have established a bridge maintenance plan [1–3]. In Italy, news guidelines on Risk Classification and Risk Management and Safety Assessment and Monitoring of Existing Bridges, were published [4]. Deterioration directly reflects the condition of the structures and are considered as one of the important parameters for monitoring structural health. Conventional Deterioration detections is carried out by human visual inspection; however, this method

© The Author(s), under exclusive license to Springer Nature Switzerland AG 2022
F. Calabrò et al. (Eds.): NMP 2022, LNNS 482, pp. 2268–2277, 2022.
https://doi.org/10.1007/978-3-031-06825-6_218

has limitations such as performance is highly linked to the inspector's experience, time consumption and limited accessible areas. Some approaches include the use of a camera or inspection based on unmanned aerial vehicles (UAVs) that allows to capture the maintenance state and therefore the analysis will be done on pictures.

In this paper, a methodology of identifying deterioration and crack by processing images acquired by a commercial UAV on a reinforced concrete bridge with a high level of degradation, is presented. Specifically, the structure and challenge are provided to automatically monitor and identify the appearance of deterioration on the bridge using an automated system based on UAV. A 3D model was created by photogrammetric techniques. Frame acquisitions points were designed on the trajectory in a way to make repeatable the acquisition in a different time and therefore easily comparable to obtain a timeline of the structure's evolution (and consequently of the degradation framework) [5]. Moreover, the images acquired by UAV are updated in a database to compare each other and establish the deterioration evolution during the time. It was entrusted to a deep learning processing for image classification and deterioration size estimation. The degradation tracing occurs thanks a masks application on the acquired images, thus defining the investigation area and consequently avoiding the information redundancy.

Convolutional neural networks (R-CNNs) were used for image processing, which significantly reduced the computational cost of degradation's detecting. The position of the degradation on the structure is defined as a function of the acquisition position of the drone. Finally, the length and thickness of the slot are calculated using various image processing techniques and validated through field testing.

Finally, the results obtained were inserted into a structural model and processed with the FATA NEXT software [5], going to compare the results obtained between the different models to which the level of degradation and crack were assigned.

2   Methodology

The proposed system is developed in 4 main phases, according to the workflow (Fig. 1) [6]:

– Preliminary

- inspection and identification of areas to be examined: step necessary to assess the presence of any obstacles that the drone may encounter, and definition of points of interest;

– Flight

- Design of the flight and acquisition plan: step necessary to define the flight path (way points), the spatial intervals of acquisition (choice of image acquisition positions), the inclination and rotation angle of the camera, travel speed, duration of the mission;
- Flight and image acquisition: the drone in an automated way makes a flight along the designed trajectory and acquires the images necessary for 3D modeling, the generation of orthophotos, the detection of degradation;

- Processing

  - 3D model Reconstruction of 3D and orthophoto of high resolution ensemble;
  - Detection of deterioration through CNN;
  - Degraded image processing assignment;
  - Visualization of deterioration on orthomosaic model;
  - Structural analysis of the infrastructure for different degradation scenarios;

- Viewing results

  - Creation of a historical database of points of interest: cracking framework;
  - Querying the layer overlay database.

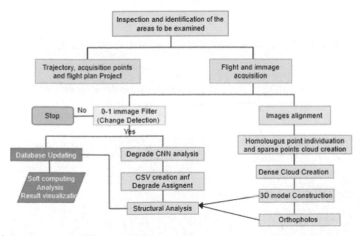

Fig. 1. Workflow proposed methodology.

3 Case Study

To test the methodology, it was decided to identify an area with a low level of traffic, and with all the necessary elements to be examined within the experiment.

The operations were tested in a road with low traffic in the territory of the city of Prunella (RC), South Italy, on a low density and traffic area. It is an interurban road infrastructure that connects the city center with a hamlet that develops along the bank of a river (Fig. 2).

Figure 3 shows the plan of the bridge deck consisting of 2 main lateral girders of 0.4 × 1.0 m, two internal girders of 0.3 × 0.65 mm and one central girders of 0.4 × 0.65 m. The deck is completed by transoms (transversal diaphragms) arranged at a distance of 2.90 m with a cross section of 0.4 × 0.75 m, while the superior arches has a cross section of 0.6 × 1.0 m.

Fig. 2. Case study: road and bridge in Reggio Calabria South Italy.

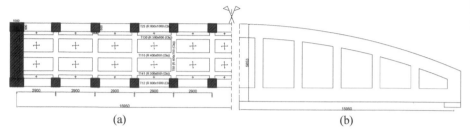

(a)   (b)

Fig. 3. Geometric properties of bridge: a) deck plan; b) elevation.

Figure 4 shows the images of the degradation of the deck beams where it is possible see as the deterioration it is present only on the deck girders.

Fig. 4. Actual degradation of the bridge

3.1 Preliminary Activities - Step 1

To determine the UAV trajectory it is necessary preliminarily to provides a first inspection to avoid interference with any obstacles during data acquisitions, to arrange take-off and/or landing point and data transmission, to verify the possible presence of overflight of areas subject to vehicular or pedestrian traffic and therefore plan an installation of "precautionary" parachutes for the mitigation of the hazard.

3.2   Flight - Step 2

Once the space of the trajectory and the points of interest for the inspection (areas to be acquired) were determined, through a flight planner the information was transposed and then the trajectory was designed and the take-off-landing/data transmission points were determined, and the image acquisition points (Fig. 5). In this way there is a repeatability of the flight and images (with the same camera used) easily comparable, especially in terms of variations in the structure (presence of degradation, improvement interventions, etc.)

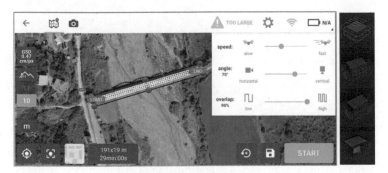

Fig. 5. Flight plan.

3.3   Processing – Step 3

Once the data and information necessary for the analysis of the bridge were acquired, the various phases of data processing were carried out.

3.3.1   3D Modeling

For 3D modeling the Agisoft Metashape software was used.

The elaboration process is composed of different phases:

1. Photo alignment: a phase that identifies the tie points with common characteristics in order to be adequately superimposed in the various photos;
2. Build dense cloud: this phase allows to build a dense cloud using dense image matching algorithms;
3. Build mesh: it consists in generating a polygonal model based on the newly created dense cloud;

Build texture: this phase allows to generate a two-dimensional raster image in a way to apply a photorealistic color to the mesh. The final result is visible in the Fig. 6.

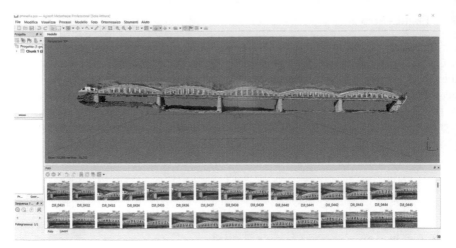

Fig. 6. 3D model.

3.3.2 Detection and Analysis Crack

In most studies, image processing follows the following stages: image acquisition; (2) pre-processing; (3) processing; and (4) identification of deterioration [7]. Recently, deterioration detection methods based on deep learning have been proposed [8, 9]. Features are extracted from training deterioration images from a convolutional neural network (CNN). In this document, deterioration detection has been defined as performing both the classification and localization of an image.

CNN has disadvantages such as a high computational cost and a long operating time. To overcome these drawbacks, the region proposal algorithm was proposed to quickly scan the significant region in the image. In particular, in this paper the R-CNN method combined regional proposals with rich features extracted from CNN and was applied to detect deterioration. R-CNN's steps consisted of extracting the region proposal from the input image using selective search, extracting features using CNN after cropping, and detecting the image after the bounding box regression.

The detected degradation was calculated using masks. The camera captured the image including the homologous points, then the mask was detected using the Region of Interest (ROI) algorithm.

3.3.3 Structural Analysis

Once the acquisitions on the decay have been processed and the 3D model of the bridge has been built, the structural analysis is carried out. Therefore, the 3D model of the bridge is automatically acquired within the FATA-Next finite element structural software, capable of automatically acquiring all the information on the degradation of the bridge by means of a CSV file. The CSV file, can be subsequently updated, each time that the phases described above highlight a change in the degradation conditions detected.

Degradation. 4 different levels of degradation were considered and inserted by CSV file in the software. In detail [5], degradation was considered by the introduction of indexes of cracking and corrosion [9]. If a uniform corrosion is considered, the dam-age indexes listed below can be taken into account [11–13].

– Reduction of the section of the reinforcing bars by the following equation:

$$A_s(\delta) = A_{s0}\big[1 - \delta_{As,u}(\delta)\big] \tag{1}$$

in which:
– $A_s(\delta)$ is the initial area of the reinforcing bars;
– $\delta_{As,u}(\delta) = \begin{bmatrix} 4\delta(1-\delta) & if & 0 \le \delta \le 0.5 \\ 1 & if & \delta > 0.5 \end{bmatrix}$, is the damage index of the reinforcing bars;
– $\delta = x(t)/d_0$, is the ratio between the thickness of the corrosion zone and the original diameter of the bar;
– Reduction of the resistance and ultimate deformations of the reinforcing bars and surrounding concrete:

$$f_{sy} = f_{sy0}\big[1 - \delta_{f_s}\big] \tag{2}$$

$$f_{su} = f_{su0}\big[1 - \delta_{f_s}\big] \tag{3}$$

$$\varepsilon_{su} = \varepsilon_{su0}\big[1 - \delta_{\varepsilon_s}\big] \tag{4}$$

where $\delta_{f_s} = 0.5\delta_{As,u}(\delta)$, while,

$$\delta_{\varepsilon_s} = \begin{bmatrix} 0 & if & \delta_{As} \le 0.016 \\ 1 - 0.1521\delta_{As}^{-0.4583} & if & \delta_{As} > 0.016 \end{bmatrix} \tag{5}$$

δ_{ε_s} is the damage index that taken into account of the reinforcing bars strain;
The degradation of compressive strength, for cracked concrete elements, is modeled by the following equation:

$$f_{c,rid} = \frac{f_c}{1 + K \cdot \varepsilon_{c,t}/\varepsilon_{c,0}} \tag{6}$$

where: K is a coefficient related to the roughness and diameter of the bars, as specified in [12]; f_c is the compressive strength of concrete corresponding to $\varepsilon_{c,0}$;

$$\varepsilon_{c,0} = 0.0017 + 0.0010 \cdot \left(\frac{f_{cm}}{70}\right) \tag{7}$$

$$\varepsilon_{c,t} = n_b \cdot w_m/b_i \tag{8}$$

in which, b_i is the depth of the considered portion of section, while w_m is the mean width of the cracks and n_b is the number of reinforcing bars.
In the Eq. (7), f_{cm} is the mean resistance of concrete (in MPa).
In the case of localized corrosion, others models can be considered [12, 13].

The Analysis Results

In this section, a short discussion of the results carried out by pushdown analysis, is presented. Generally, the pushdown analysis method consists of analyzing a structure under increasing gravity loads. Usually this state is reached after substantial changes in the geometric configuration of the structural system [14]. The load corresponding to this condition is defined as the failure load.

In this case, after the application of the gravity loads, a uniform pushdown analysis is accomplished, and consequently, the traffic loads are increased proportionally until the bridge collapses.

The capacity of the bridge can be expressed in terms of the load multiplier (LM) defined as the ratio of failure traffic load to the nominal traffic load:

$$LM = \frac{Q_{ftl}}{Q_{ntl}} = \frac{Failure\ traffic\ Load}{Nominal\ traffic\ Load} \tag{9}$$

In the paper three different scenarios were considered and compared:

1. bridge in absence of deterioration;
2. deterioration present only on the deck girders (actual scenario);
3. deterioration present on the deck girders and hangers (one of the possible scenarios hypothesized for the type of bridges under examination);

Therefore, based on the considerations made in the previous section, where the degradation was considered by the introduction of indexes of cracking and corrosion, in the Fig. 7 the Load Multiplier Vs Vertical displacement curves are compared. It is possible to verify that in both cases, of non-degraded bridge and degradation only in the deck girders, the curves present a similar trend. In fact, in both cases, a first sudden load loss is evident due to the rupture of 3 hangers in tension on the right side of the bridge; A 10% reduction of the loaded multiplier can be observed between the loads of "non degraded bridge" and of "degradation on deck girders".

A second sudden load loss is due to rupture of 3 hangers in tension on the left side of the bridge.

This shows that the degradation of the deck girders does not generally causes a high reduction in the bridge bearing capacity.

The hangers degradation's case results in a significant reduction of the load multiplier; in this case a progressive collapse can be observed, in fact, the hangers reach the collapse, in tension, one by one.

3.3.4 View the Results - Step 4

The proposed system is also composed of a system for displaying the results that allows to compare the evolution over time of the variations that have occurred on the bridge. Given the repeatability of the flight plan, the images acquired from a given established position will frame the same areas and therefore will be easily comparable; a filter (0–1) will therefore determine the addition of the image itself in the database whenever a change is identified compared to the previous temporal acquisition [15].

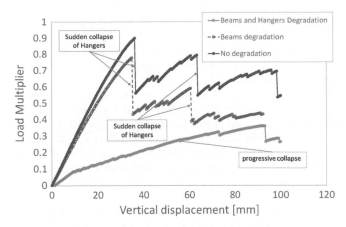

Fig. 7. Load multiplier vs vertical displacement

The degradation variations, depending on the acquisition point, will be associated with a particular element depending on the gripping point (waypoint determined in the preliminary phase) and recorded on a CSV file. The acquisition data and the results of the analysis can therefore be made available within a platform that can be queried by the user, making the results accessible to end users for priority intervention considerations.

4 Conclusions

The Italian infrastructure network ancientness and the absence of a reference database, in order to plan the maintenance interventions that the infrastructure network needs is one of the most widespread problems in our country. As bridges become obsolete, requirement for inspection and maintenance increases. In this paper, accordingly with the new guidelines on Risk Classification and Risk Management and Safety Assessment and Monitoring of Existing Bridges, a new methodology is proposed to establish proper control procedure and intervention priority.

With reference to a case study, an experimental and automated methodology capable of acquiring the geometric information and the degradation state of the Prunella bridge was presented.

I dati acquisiti da drone e processati con le tecniche di soft computing sono stati implementati all'interno del software commerciale FATA NEXT capace di simulare la riduzione della capacità del moltiplicatore di collasso sulla base di differenti scenari di degrado.

The data acquired by drone and processed by soft computing techniques, was implemented within the FATA NEXT structural software capable to simulate the load collapse multiplier reduction based on different degradation scenarios.

This information available on the visualization platform, helps and allows the Authority to determine the priority of maintenance interventions.

Acknowledgements. The authors thank the Mediterranea Technology Center srl enterprise for the cooperation during surveys phases and images elaboration.

References

1. Michael, J.: Ryall, Bridge Management, Elsevier Ltd., London (2009)
2. Wan, C., et al.: Development of a bridge management system based on the building information modeling technology. Sustainability **11** (2019)
3. Sousa, H., Rozsas, A., Slobbe, A., Courage, W.: A novel pro-active approach towards SHM-based bridge management supported by FE analysis and Bayesian methods. Struct. Infrastruct. Eng. Mainten. Manag. Life-Cycle Des. Perform. **16**, 233–246 (2020)
4. Ministero delle Infrastrutture e dei Trasporti, Consiglio Superiore dei Lavori Pubblici: linee guida sulla Classificazione e Gestione del Rischio e la Valutazione della Sicurezza ed il Monitoraggio dei Ponti Esistenti, Allegate al parere del Consiglio Superiore dei Lavori Pubblici n.88/2019, espresso in modalità "agile" a distanza dall'Assemblea Generale in data 17 April 2020
5. Barrile, V., Fotia, A.: A proposal of a 3D segmentation tool for HBIM management. Appl. Geomat. **14**, 197–209 (2021). https://doi.org/10.1007/s12518-021-00373-4
6. Stacec s.r.l, FATA-Next, v.2021.12.4, Bovalino (2021)
7. Barrile, V., Fotia, A., Leonardi, G., Pucinotti, R.: Geomatics and soft computing techniques for infrastructural monitoring. Sustainability **12**(4) (2020). https://doi.org/10.3390/su12041606
8. Peng, X., Zhong, X., Zhao, C., Frank Chen, Y., Chen, Zhang, T.: The feasibility assessment study of bridge crack width recognition in images based on special inspection UAV. Hindawi Adv. Civil Eng. **2020**
9. Barrile, V., Bernardo, E., Fotia, A., Candela, G., Bilotta, G.: Road safety: road degradation survey through images by UAV. WSEAS Trans. Environ. Dev. **16**, 649 (2020)
10. Barrile, V., Bernardo, E., Candela, G., Bilotta, G., Modafferi, A., Fotia, A.: Road infrastructure heritage: from scan to infraBIM. WSEAS Trans. Environ. Dev. **16**, 633–642 (2020)
11. Pucinotti, R., Tripodo, M.: The Fiumarella bridge: concrete characterisation and deterioration assessment by nondestructive testing. Int. J. Microstruct. Mater. Prop. **4**(1), 128–139 (2009)
12. Felitti, M., Oliveto, F., Stacec, S.r.l.: Influenza del degrado localizzato per corrosione delle armature sulla vulnerabilità sismica delle strutture. https://www.ingenio-web.it/28876-inf luenza-del-degrado-localizzato-per-corrosione-delle-armature-sulla-vulnerabilita-sismica-delle-strutture. Accessed 11 Oct 2020
13. Coronelli, D., Gambarova, P.: Structural assessment of corroded reinforced concrete beams: modeling guidelines. ASCE J. Struct. Eng. **130**, 1214–1224 (2004)
14. Rodriguez, J., Ortega, L.M., Casal, J.: Load carrying capacity of concrete structures with corroded reinforcement. Constr. Build. Mater. **11**(4), 239–248 (1997)
15. Khandelwal, K., El-Tawil, S.: Pushdown resistance as a measure of robustness in progressive collapse analysis. Eng. Struct. **33**(9), 2653–2661 (2011)
16. Barrile, V., Fotia, A., Bernardo, E., Bilotta, G.: Road Cadastre an innovative system to update information, from big data elaboration. In: Gervasi, O., et al. (eds.) ICCSA 2020. LNCS, vol. 12252, pp. 709–720. Springer, Cham (2020). https://doi.org/10.1007/978-3-030-58811-3_51

Seismic Retrofit of Concrete Reinforced Existing Buildings by Insertion of Steel Exoskeleton: A Case Study

R. Buda[1] , M. R. Alvaro[2], and R. Pucinotti[1][(✉)]

[1] Mediterranean University of Reggio Calabria, Reggio Calabria, Italy
{rocco.buda,raffaele.pucinotti}@unirc.it
[2] Structural Engineer at STACEC s.r.l., Bovalino, RC, Italy
stacec@stacec.com

Abstract. Buildings built in Italy after the 1950s represent about 50% of the existing building stock. Currently they have structural and energy deficiencies, compared to news standards, and their maintenance is considered, in many cases, unsustainable.

In recent years, several studies of these buildings have been carried out, but many of this have addressed individual problems relating to structural, energy and architectural deficiencies, without a real interdisciplinary coordination. For this reason, a study has been undertaken which aims to address these issues in an integrated way through a multidisciplinary approach in the retrofit of existing reinforced concrete buildings that deals with the aspects relating to structural strengthening, energy requalification and architectural renovation. In this paper only the first step of this study is presented related the structural aspects; in detail, a pushover analysis was carried out in order to define the performance point of the combined system (concrete reinforced building and steel exoskeleton) and the results of the analysis are presented and discussed.

Keywords: Life cycle · Sustainable and reversible repair; dry solutions · Eco-friendly materials · Reversible devices

1 Introduction

The need to promote the renewal and sustainable redevelopment of the built heritage is directing research towards performing and effective solutions, which no longer contemplate the classic demolition and reconstruction technique of existing buildings, but rather promote the enhancement of their architecture.

In recent decades, however, there has been a growing sensitivity to the issue of sustainable redevelopment almost always associated with strategies for reducing energy consumption. In this context, the problems related to the structural deficiencies of existing buildings are often neglected or placed in the background, although in some cases they are serious and evident.

Structural reinforcement techniques are very important, so much so that any type of intervention aimed solely at energy and architectural redevelopment can be

© The Author(s), under exclusive license to Springer Nature Switzerland AG 2022
F. Calabrò et al. (Eds.): NMP 2022, LNNS 482, pp. 2278–2288, 2022.
https://doi.org/10.1007/978-3-031-06825-6_219

defined inadequate; evidence of this, are all the damages caused by the recent Italian's earthquakes.

The tendency used to deal with individual problems was to operate in a sectoral way without involvement and coordination between the different disciplines, also neglecting the urban context [1–4].

Today, it is possible to do responses to the problems of obsolescence, vulnerability and resilience by a multidisciplinary approach, using innovative and integrated strategies allow to stretch the useful life cycle of a building. The application of this strategy through a multidisciplinary approach still remains a research and experimentation activity in progress within a framework that is not fully formalized.

An example is the application of external steel exoskeletons, this solution involves the formation of a second technological layer based on the existing and designed to overcome structural, energy and architectural deficiencies. From the structural point of view, the previous solution operates exclusively from the outside of the structure and together with the existing structure, contrasts seismic actions and, where necessary, also contributing to the absorption of accidental vertical loads, making the existing structure and the external steel exoskeleton work in parallel.

In the architectural field, the proposed solution is configured as an "adaptive double skin" [5], since it is designed to be able to adapt to any formal or technical changes.

This solution offers numerous advantages, as in addition to helping to increase the quality of life and well-being of the inhabitants, it also allows the building to remain in operation throughout the duration of the intervention [6, 7].

This work is part of this contest aimed at deepening the structural, energy and architectural problems of reinforced concrete existing buildings. In the paper, for the sake of brevity, only the structural aspects of this ongoing study, which concerns the structural reinforcement through a strategy of stiffening are discussed.

2   Case Study

The case study refers to a reinforced concrete building located near the Mediterranean University of Reggio Calabria.

2.1   The Existing Building

The building considered in the paper is a six storey residential building, built in the'70 of the last century, located in a seismic area. The building was designed in accordance with the anti-seismic rules in force at the time of its construction and of the handbooks commonly used [8–10] and built between 1968–1971. Live loads of 2 kN/m^2 are applied on all habitable stories, as well as 4 kN/m^2 on balconies and staircases, according to the 1967 Italian C.N.R. instructions [11].

It has an L-shaped plant (Fig. 1) with max dimensions of 28.0 m × 22.0 m. Two plus one plus three moment resisting frames having respectively 5, 3 and 2 bays, arranged along the x-direction, and three plus three moment resisting frames having respectively 2 and 5 bays, arranged along the y-direction make up the seismic-resistant structures.

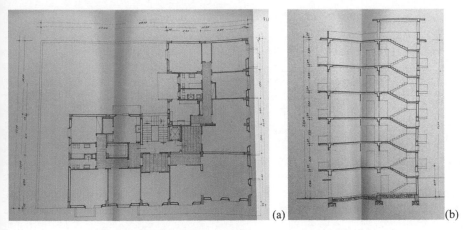

(a)   (b)

Fig. 1. Building: a) Horizontal section; b) vertical section

The first level has a height of 3.6 m, while the inter-storey height of other five storeys is 3.4 m.

The identification of the materials used for the examined building is obtained from the Technical Design Calculation Report and from the Test report. In detail, the Table 1 shows the types of employed materials (column 2), the yield strength and the ultimate strength of the steel (column 3).

Table 1. Materials used in the building construction.

Material	Type	Yield strength ultimate strength [MPa]	Design strength (allowable strength) [MPa]
Cement	R730	/ /	/
Concrete	Rbk250	/ 25	7.0
Steel	Aq50	≥270 500–600	160

Finally, column 4 shows the design strength of concrete (a medium quality type, called R_{bk} 250) and the design strength of steel, called Aq50 (column 4).

The building is an example of the typical construction practice of the seventies, with beams and pillars, forming a closed mesh both horizontally and vertically. The foundations are superficial and consisting of orthogonal reinforced concrete beams.

All beams and columns have rectangular cross-sections with the dimensions reported in the Table 2; shear reinforcement for the beams consists of transversal Ø6 stirrups with spacing equal to 15 cm, while the columns transversal stirrups are Ø6 with spacing between 20 cm and 30 cm.

Table 2. Geometric property of the structural elements.

Element [cm]	Foundation	1st floor	2th floor	3th floor	4th floor	5th floor	6th floor
Beams	50 × 120	30 × 80	30 × 70	30 × 70	30 × 60	30 × 60	30 × 60
Columns	/	40 × 95	27 × 90	27 × 85	27 × 75	27 × 65	27 × 55
Column 13	/	40 × 165	27 × 165	27 × 165	27 × 165	27 × 165	27 × 165

2.2   Retrofit Design Strategy

Generally, two different strategies can be performed to retrofit RC-Buildings: (i) increasing the displacement capacity or (ii) reducing the displacement demand [12, 13].

In this paper the retrofitting strategy is focused on reducing seismic-induced displacement demand into the capacity limits of the existing structure by the insertion of a steel exoskeleton.

In detail, strengthening strategy by the insertion of a suitable steel exoskeleton was performed to reducing seismic demand on the existing structure. Seismic performance of the building was evaluated by non-linear pushover analyses performed with distributed plasticity approach models, according to the N2 method [6, 12, 14, 15].

Analyses were performed along the two main directions of the building by two load patterns:

(i)   the first proportional to a constant distribution of accelerations;
(ii) the second proportional to the main first vibration mode in the considered direction.
      Simplified methodology was implemented according to the N2 method [7, 12–14].

All the analyses were carried out by using the FATA-Next Structural Analysis Program [16].

If Δ_{tar}^* denotes the displacement capacity of the existing structure, the stiffness of the retrofitted structure, in the hypothesis that the equivalent mass and the modal participation factor of the existing construction and the retrofitted ones remains the same, is given by the following equation:

$$k_d = \frac{m^* \cdot S_{el,ADRS}(\Delta_{tar}^*)}{\Delta_{tar}^*} \tag{1}$$

in which:

- $S_{el,ADRS}(\Delta_{tar}^*)$ is the elastic spectral acceleration corresponding to Δ_{tar}^*;
- m^* is the equivalent mass of SDOF system.

Furthermore, Eq. (1) is valid under the hypothesis that the yielding displacement of the exoskeleton corresponds to the yielding displacement of the existing structure [17].

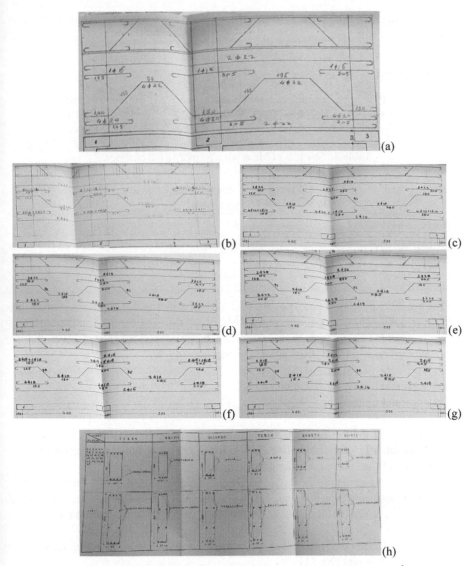

Fig. 2. Longitudinal reinforcements: a) Foundation beams; b) 1st floor beams; c) 2th floor beams; d) 3th floor beams; e) 4th floor beams; f) 5th floor beams; g) 6th floor beams; h) Columns

In the hypothesis that existing structure and the strengthening steel exoskeleton works in parallel, global lateral stiffness of the exoskeleton is obtained by the following equation:

$$\Delta k = k_d - k_{Ex} \tag{2}$$

In the Eq. (2) k_{Ex} represents the lateral stiffness of the existing structure, than, the global lateral stiffness of the exoskeleton is distributed locally to each level where each local system is composed of both the existing structure and the steel exoskeleton working in parallel [7, 14, 22].

2.3 Applications

In this section, the strengthening strategy described above is applied to case study.
In the analyses the materials strength shows in Table 3 were assumed.

Table 3. Strength of Materials assumed in the analyses.

Material	Mean yield strength f_{ym} [MPa]	Mean ultimate strength f_{um} [MPa]	Mean strength f_{cm} [MPa]
Concrete C20/25	/	/	20
Steel rebars (Aq50)	372*	547*	/

* Verderame et al. [18]

The models were developed using the FATA-Next Software [15] with reference to the eight (overlooking the accidental eccentricity between the mass centroid and the stiffness one) or sixteen load combinations (considering the accidental eccentricity between the mass). Moreover, two loads distribution are considered:

1. Proportional to first vibration mode;
2. proportional to masses.

In the Fig. 3a the pushover curves with the corresponding bilinear equivalent are presented at the life-safety limit state (SLV), for the existing building, overlooking the combinations of the seismic components.

The results, in terms of Capacity and Demand, are reported in the Table 4. In this case, only a critical situation is found, which corresponds to the pushover analysis carried out in the positive verse of the X direction for load distribution proportional to the first vibration mode.

In the Fig. 3b the combinations of the seismic components is taken into account by the following equations [20, 21]:

$$\pm(X) \pm 0.3(Y) \tag{3}$$

$$\pm(Y) \pm 0.3(X) \tag{4}$$

Table 5 detail the results of pushover analyses considering the combinations of the seismic components as indicated in the Eqs. 3 and 4. As it is easy to verify, in this case, the ratio C/D (between Capacity and Demand) is less than unity for 10 out of 16 cases examined.

The results of the sixteen load combinations provided by the software analysis are summarized in the Table 6, in the case of retrofitting by the insertion of steel exoskeleton.

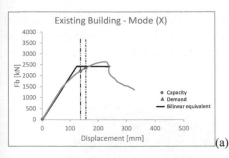
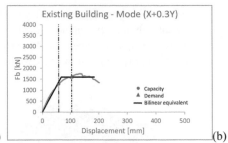

(a)   (b)

Fig. 3. Pushover results existing building: a) overlooking the accidental eccentricity; b) considering the accidental eccentricity.

Table 4. Results of Pushover analysis pre-intervention.

Pushover	F_{max} [kN]	Γ	F^*_{max} [kN]	α_u/α_1	$S_e(T^*)$ [g]	q_*	C [cm]	D [cm]	$\zeta_E = \frac{C}{D}$
Mode (+X)	3785	1.49	2536	0.961	0.334	1.2	8.65	9.69	**0.89**
Mode (−X)	3771	1.49	2526	1.06	0.336	1.3	10.82	9.64	1.12
Mode (+Y)	4749	1.43	3319	0.968	0.361	1.2	10.64	8.97	1.18
Mode (−Y)	4913	1.43	3434	0.999	0.352	1.2	11.20	9.21	1.21
Mass (+X)	6307	1.49	4226	0.793	0.398	1.0	11.38	8.15	1.39
Mass (−X)	6136	1.49	4111	1.49	0.40	1.0	8.34	8.10	1.03
Mass (+Y)	7547	1.43	5275	1.75	0.437	1.0	12.64	7.41	1.70
Mass (−Y)	7548	1.43	5276	1.55	0.428	1.0	12.06	7.56	1.59

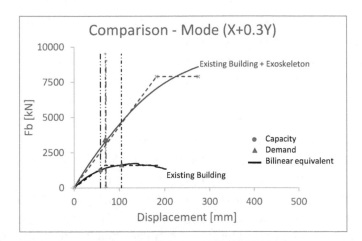

Fig. 4. Pushover results: bare building Vs existing building + steel exoskeleton.

As it is possible to observe, the safety index (i.e. ratio between Demand and Capacity) is always greater than unity, and no critical situation is found, confirming the effectiveness of the proposed project intervention. In particular, the seismic safety index value passed from 0.56 to 1.00 after insertion of exoskeletons.

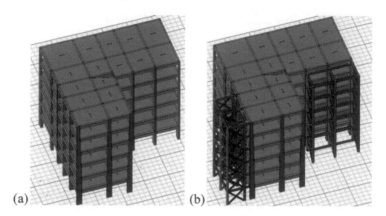

(a)   (b)

Fig. 5. Building Models; a) pre-intervention; b) and c) post-intervention.

Table 5. Results of Pushover analysis post-intervention.

Pushover	F_{max} [kN]	Γ	F^*_{max} [kN]	α_u/α_1	$S_e(T^*)$ [g]	q_*	C [cm]	D [cm]	$\zeta_E = \frac{C}{D}$
Mode+X+0.3Y	2317	1.60	1447	1.37	0.309	1.8	5.86	10.49	**0.56**
Mode+X−0.3Y	3202	1.60	2001	2.09	0.348	1.4	5.61	9.31	**0.60**
Mode−X+0.3Y	3210	1.60	2005	2.05	0.346	1.4	7.01	9.36	**0.75**
Mode−X−0.3Y	2326	1.60	1453	1.50	0.306	1.8	7.11	10.59	**0.67**
Mode+Y+0.3X	2837	1.51	1880	1.00	0.323	1.8	8.19	10.03	**0.82**
Mode+Y−0.3X	5408	1.51	3584	1.65	0.423	1.2	9.64	7.67	1.26
Mode−Y+0.3X	5852	1.51	3878	1.61	0.418	1.1	9.18	7.75	1.18
Mode−Y−0.3X	2895	1.51	1918	1.14	0.324	1.8	8.03	10.01	**0.80**
Mass+X+0.3y	4836	1.60	3022	1.87	0.409	1.1	8.09	8.11	**0.99**
Mass+X−0.3Y	4887	1.60	3054	1.94	0.409	1.1	7.84	7.92	**0.99**
Mass−X+0.3Y	4918	1.60	3073	1.90	0.409	1.1	6.61	7.93	**0.84**
Mass−X−0.3Y	4725	1.60	2952	1.95	0.409	1.1	7.99	8.10	**0.99**
Mass+Y+0.3X	6439	1.51	4267	1.63	0.435	1.1	11.12	7.45	1.49
Mass+Y−0.3X	7000	1.51	4639	1.37	0.477	1.1	10.25	6.79	1.51
Mass−Y+0.3X	7373	1.51	4886	1.32	0.469	1.0	12.00	6.90	1.74
Mass−Y−0.3X	6659	1.51	4413	1.71	0.430	1.0	8.69	7.53	1.15

Figure 4 show as the retrofit produced the following effects: (i) an increase of the bare building overall lateral stiffness; (ii) an improvements of the seismic safety indice; (iii) an increase of the capacity also due to the reduction of the eccetricity between the centre of mass (CM) and centre of stiffness (CS) in the design stage of steel exoskeleton In conclusion, the analyses proved that the interventions carried out through exoskeletons on the case study have allowed to achieve safety indices greater than or equal to one.

Table 6. Results of Pushover analysis post-intervention.

Pushover	F_{max} [kN]	Γ	F^*_{max} [kN]	α_u/α_1	$S_e(T^*)$ [g]	q_*	C [cm]	D [cm]	$\zeta_E = \frac{C}{D}$
Mode+X+0.3Y	4073	1.21	3371	1.87	0.456	1.2	6.91	6.91	1.00
Mode+X−0.3Y	6691	1.21	5538	1.76	0.583	1.0	6.91	5.56	1.24
Mode−X+0.3Y	6694	1.21	5540	1.69	0.584	1.0	6.91	5.56	1.24
Mode−X−0.3Y	4074	1.21	3372	1.71	0.456	1.2	6.91	6.91	1.00
Mode+Y+0.3X	4622	1.02	4527	2.01	0.384	1.2	8.18	8.18	1.00
Mode+Y−0.3X	9469	1.02	9275	1.70	0.549	1.0	8.18	5.90	1.38
Mode−Y+0.3X	9452	1.02	9259	1.68	0.544	1.0	8.18	5.96	1.37
Mode−Y−0.3X	4614	1.02	4520	1.75	0.38	1.2	8.18	8.18	1.00
Mass+X+0.3Y	7632	1.21	6317	1.66	0.644	1.0	6.28	5.03	1.24
Mass+X−0.3Y	9905	1.21	8198	1.82	0.703	1.0	6.23	4.61	1.35
Mass−X+0.3Y	9870	1.21	8169	1.16	0.703	1.0	6.91	4.61	1.49
Mass−X−0.3Y	8227	1.21	6810	1.23	0.643	1.0	6.61	5.04	1.31
Mass+Y+0.3X	9950	1.02	9746	1.77	0.561	1.0	7.33	5.78	1.26
Mass+Y−0.3X	13771	1.02	13489	1.97	0.66	1.0	6.54	4.92	1.33
Mass−Y+0.3X	13766	1.02	13484	1.36	0.654	1.0	8.18	4.96	1.64
Mass−Y−0.3X	9916	1.02	9713	1.46	0.556	1.0	8.18	5.84	1.40

3   Conclusions

A study was undertaken which aims to address the structural, architectural and energy deficiencies of existing buildings in an integrated way through a multidisciplinary approach that deals with the aspects relating to structural strengthening, energy requalification and architectural renovation. In this paper the first step of this multidisciplinary approach consisting in the retrofit of existing reinforced concrete buildings was presented. In detail, a pushover analysis was carried out to the case study of a six storey RC building in order to define the performance point of the combined system (bare building and steel exoskeleton). The results proved as the specific insertion of steel exoskeletons improve significantly the seismic response of RC existing buildings. The use of exoskeletons retrofit technique present the advantage that it also allows the building to remain in operation throughout the duration of the intervention.

References

1. Sugano, S.: State of the arts in aseismic strengthening of existing reinforced concrete buildings in Japan, Takenaka Technical Research Report, No. 25 (1981)
2. Japan Building Disaster Prevention Association (JBDPA): Guideline for Seismic Retrofit of Existing Reinforced Concrete Buildings, Translated by Building Research Institute (2001)
3. Bellini, O., Marini, A., Passoni, C.: Adaptive exoskeleton systems for the resilience of the built environment. J. Technol. Archit. Environ. **15**, 71–80 (2018)
4. Caverzan, A., Lamperti Tornaghi, M., Negro, P.: Taxonomy of the redevelopment methods for non-listed architecture: from façade refurbishment to the exoskeleton system, In: JRC, Conference and Workshop Reports, Proceedings of Safesust Workshop, Ispra, November 26–27, 2016
5. PRIN 2009: Ministero dell'Istruzione, dell'Università e della Ricerca; Coordinatore Scientifico Montuori M.; Nuove pratiche progettuali per la riqualificazione sostenibile di complessi di habitat sociale in Italia
6. Formisano, A., Massimilla, A., Di Lorenzo, G., Landolfo, R.: Seismic retrofit of gravity load designed RC buildings using external steel concentric bracing systems. Eng. Fail. Anal. **111** (2020)
7. Formisano, A., Di Lorenzo, G., Colacurcio, E., Di Filippo, A., Massimilla, A., Landolfo R.: Steel orthogonal exoskeletons for seismic retrofit of existing reinforced concrete and prestressed concrete buildings: design criteria and applications; Costruzioni Metalliche Nov-Dic **2020** (2020)
8. Regio Decreto-Legge 16 Novembre 1939-XVIII: n.2229 (suppl.Ord. alla Gazzetta Ufficiale n.92 del 18 aprile 1940; Norme per l'esecuzione delle opere in concglomerato cementizio semplice od armato
9. Ministero dei Lavori Pubblici: Consiglio Superiore, Circolare n.1472 del 23 maggio 1957; Armatura delle Strutture in cemento armato
10. Legge 25 novembre 1962, n. 1684, (Gazzetta Ufficiale 22 dicembre 1684, n. 326); Provvedimenti per l'edilizia, con particolari prescrizioni per le zone sismiche
11. Bollettino Ufficiale C.N.R., Anno 1 n.3 31 Maggio 1967; Ipotesi di Carico sulle Costruzioni
12. Tondini, N., Zanon, G., Pucinotti, R., Di Filippo, R., Bursi, O.S.: Seismic performance and fragility functions of a 3D steel-concrete composite structure made of high-strength steel. Eng. Struct. **174**, 373–383 (2018)
13. Faella, C., Martinelli, E., Nigro, E.: Seismic assessment and retrofitting of R.C. existing buildings, In: Proceedings of the 13th World Conference on Earthquake Engineering, Vanvouver, B.C., Canada, 1–6 August 2004, paper n. 84
14. Fajfar, P.: Capacity spectrum method based on inelastic demand, spectra earthquake. Eng. Struct. Dyn. **28**, 979–993 (1999)
15. Federal Emergency Management Agency (FEMA): Techniques for the Seismic Rehabilitation of Existing Buildings, FEMA 547/2006
16. Stacec s.r.l, FATA-Next, v.2021.12.4, Bovalino, RC, Italy
17. Ponzo, F.C., Di Cesare, A., Arleo, G., Totaro, P.: Protezione sismica di edifici esistenti con controventi dissipativi di tipo isteretico: aspetti progettuali ed esecutivi, Progettazione sismica, 1-2010, Eucentre Press (2010)
18. Verderame, G.M., Ricci, P., Esposito, M., Sansiviero, F.C.: Le caratteristiche meccaniche degli acciai impiegati nelle strutture in c.a. realizzate dal 1950 al 1980. Atti del XXVI Convegno Nazionale AICAP "Le prospettive di sviluppo delle opere in calcestruzzo strutturale nel terzo millennio", Padova, 19–21 maggio (2011)
19. Ministero delle Infrastrutture e dei Trasporti: D.M. 17 gennaio 2018, Norme Tecniche per le Costruzioni

20. C.S.LL.PP Circolare 21 gennaion 2019 n.7, Istruzioni per l'applicazione dell'«Aggiornamento delle "Norme tecniche per le costruzioni"» di cui al Decreto Ministeriale 17 gennaio (2018)
21. Fukuyama, H., Sugano, S.: Japanese seismic rehabilitation of concrete buildings after the Hyogoken-Nanbu earthquake. Cement Concr. Compos. **22**, 59–79 (2000)
22. Dolce, M., Manfredi, G.: Linee Guida per la Riparazione e il Rafforzamento di Elementi Strutturali, Tamponature e Partizioni, Dipartimento della Protezione Civile, Rete dei Laboratori Universitari di Ingegneria Sismica (ReLUIS) (2011)

Impact of Graphene-Based Additives on Bituminous Mixtures: A Preliminary Assessment

Filippo Giammaria Praticò[1] , Eliana Zappia[2]([✉]) , and Giuseppe Colicchio[1]

[1] Department of Information Engineering, Infrastructures, and Sustainable Energy (DIIES),
University Mediterranea of Reggio Calabria, Reggio, Italy
[2] Department of Civil, Energy, Environmental and Materials Engineering (DICEAM),
University Mediterranea of Reggio Calabria, Reggio, Italy
elianazappia92@gmail.com

Abstract. Asphalt pavements are subject to deterioration due to numerous factors such as traffic, temperature, rains, climatic change, and aging. This affects the expected life and the corresponding life cycle costing. Despite the fact that bituminous mixtures containing graphene can yield higher performance, there are still many uncertainties. These latter refer to graphene cost (as a function of type and market) and to the variability of results that can be obtained based on several factors, including percentage, type of graphene, process, components, and type of mixture. Based on the above, the objectives of the study presented in this paper have been confined into the preliminary assessment of the impact of graphene-based additives on bituminous mixtures. In the light of the above, analyses and tests were carried out on both the bitumen and the bituminous mixtures. Using a percentage of graphene-based additives of 10% by weight of the bitumen, an increase of mixture resistance was obtained.

Keywords: Graphene · Asphalt binder · Bituminous mixtures

1 Introduction

The graphene is a single layer of atoms of carbon arranged in a two-dimensional and honeycomb lattice nanostructure. A single honeycomb has six atoms (hexagon, see Fig. 1). Each atom of carbon has 6 electrons. Carbon has two electron shells, with the first holding two electrons and the second holding four electrons. These four electrons form sigma and pi bonds (see Fig. 1): 1) Three of the four outer-shell electrons of each atom occupy three sp2 hybrid orbitals that are shared with the three nearest atoms, forming σ-bonds. 2) The remaining outer-shell electron occupies a pz orbital that is oriented perpendicularly to the plane and these orbitals are responsible for most of graphene properties (thermal, mechanical, and electrical properties [1, 2]).

In 2004, Geim and colleagues at the University of Manchester first isolated monolayer samples from graphite [3]. Despite strong interest, graphene performance and costs depend on the number of layers present and on the overall quality of the crystal lattice.

© The Author(s), under exclusive license to Springer Nature Switzerland AG 2022
F. Calabrò et al. (Eds.): NMP 2022, LNNS 482, pp. 2289–2298, 2022.
https://doi.org/10.1007/978-3-031-06825-6_220

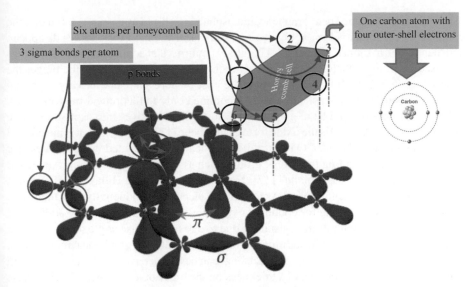

Fig. 1. Graphene honeycomb lattice (Adapted from By Ponor - Own work, CC BY-SA 4.0, [31, 32]).

The properties of bituminous binders are fundamental to the performance of the asphalt mix, despite the fact that the percentage of binder is only 5%–7%. To enhance asphalt performance, researchers have used many additives in the asphalt mix (including rubber waste [4]) and have set up different aggregate formulations [5]. The premature degradation of bitumen can be mitigated by adding carbon-based additives to reduce damage caused by increasing traffic loads [6] and climate changing conditions [7–10]. The characteristics that can be improved by adding nanomaterials such as graphene include rutting resistance, fatigue life and thermal cracking [11, 12]. Usually, graphene is added in a certain percentage by weight (e.g., from 0.5% to 10%, [13–15]). Some of the fundamental physical properties of graphene that influence the performance are the purity (wt%), the number of layers, the specific surface area (m^2/g), and the electrical conductivity (S/m) [1, 2]. It should be observed that pure graphene is a single-layer material, but graphene may also be a multi-layer material [16]. Specifically, the cost decreases as the number of layers increases. Conversely, the purity decreases as the number of layers increases. The base asphalt is chosen for its basic properties like penetration index, softening point, ductility, and viscosity. All the properties are tested using various methods. Based on Li's study [13], the first step is to heat the base asphalt (in this case A-70 Asphalt) to a flowing state (~135 °C) in an iron chamber. Next, a certain quantity of graphene (for example 0.5%, 1%, and 1.5% by weight) is added to the asphalt, which is then stirred by a high-shear mixer at 1000 rpm. Finally, they found that the incorporation of graphene gives the asphalt binder excellent anti-ageing property, increase in viscosity, and a better deformation resistance, which is due to the cross-linking network structure of graphene-asphalt [13]. The Yang's study [14] starts putting the asphalt into an oven at 120 °C for preheating. After that, it is rapidly transferred to a constant temperature silicon oil bath at 145 °C for heating to make it reach the melting state. Then, graphene and asphalt are stirred at low speed. Finally, the graphene-Asphalt mix is kept at high temperature (150 °C) until the bubbles disappear, and then cooled to room temperature. This study confirms that adding graphene promotes the viscosity of all asphalt and that the higher the content of graphene is, the bigger the viscosity is [13, 14]. However, note that the cost of pure graphene is very high and the use of a single-layer graphene could be confined only to theoretical purposes [16]. Other compounds derived from graphene are studied as additives into asphalt mix like graphene Oxide (GO). GO is the oxide form of graphene [17], where oxygen is introduced through chemical oxidation. GO is described as "heavily oxygenated, with the presence of many oxygen-containing functional groups such as epoxide, hydroxyl, carbonyl, and carboxyl groups on its basal plane" [17]. However, based on the Hidayah's [18] study, the oxygen functionalization on GO reduces the electrical conductivity and therefore, GO becomes less preferable for conductive polymer-based composites. The addition of GO can improve asphalt rutting resistance, due to a reduced phase angle of the complex modulus of asphalt [19–22]. In addition to graphene and GO, the graphene nanoplatelets, GNPs, consist of stacks of graphene sheets that can be characterized as "nano-discs with a diameter of sub-micrometer and a thickness on the order of a nanometer" [23]. These are identical to those found in the walls of carbon nanotubes, CNTs, but in a planar form [24]. They are 6–8 nm thick with a bulk density of 0.03–0.1 g/cm^3, oxygen content of less than 1%, and carbon content of 99.5% [16]. They are reported to have improved mechanical properties due to their unique size and morphology [16]. When GNPs are added at 2–5wt% to plastics or resins they make these materials electrically or thermally conductive and less permeable to gasses, while simultaneously improving mechanical properties like strength, stiffness, or surface toughness [25]. Based on Le's study [24], for large-scale applications, the material cost is $3–$4 per pound (where 1 lb corresponds to 0.45 kg), comparable to some existing asphalt modifier such as the styrene butadiene styrene [24], and it is significantly lower than the cost of multi-wall CNTs [16, 24–26].

2 Objectives and Task

Based on the content of the section above, the main objectives of the study presented in this paper are to analyse how the presence of graphene-based materials in the asphalt mix enhances asphalt properties.

In order to reach the objectives above, tests both on the bitumen and on the specimens (modified bitumen plus aggregates) were carried out.

The following tasks were carried out:

1. Analysis of the literature (cf. Sect. 1)
2. Design of experiments (cf. Sect. 1 and 2)
3. Production of samples (cf. Sect. 3)
4. Non-destructive tests (cf. Sect. 4)
5. Destructive tests (S_M, f_M, Q_M) (cf. Sect. 4)
6. Result analysis (cf. Sect. 4)

3 Materials and Methods

Tables 1 and 2 describe materials and methods. The materials used for the research include aggregates, a 50/70 penetration grade bitumen, HESAGON® SGG, and HESAGON® TGG (both patented). TGG and SGG are composed of a certain percentage of graphene (of different quality, supplied by NANOTECH EOOD), Polyethylene (PE), Polypropylene (PP) and a resin (polyvinyl Butyral, PVB). The SGG and TGG content was 10% by the total weight of the bitumen. The bitumen percentage by mix weight (B50/70) was approximately 3.20%, 5.60%, and 7.20%, while the corresponding dust ratio was about 2.2–2.4, 1.2–1.4, and 0.9–1.1.

Table 1. Samples and their gradation.

Specimen	Pb (%)	Psand (%)	DP
AC6o_33_N50_10%SGG	3.19	28	2.4
AC6o_35_N50_10%SGG	5.64	25	1.3
AC6o_37_N50_10%SGG	7.15	24	1.0
AC6o_34_N50_10%TGG	3.19	25	2.3
AC6o_36_N50_10%TGG	5.64	22	1.4
AC6o_38_N50_10%TGG	7.13	25	0.9
AC6o_27_ Without Additives	3.20	24	2.2
AC6o_28_ Without Additives	5.70	24	1.2
AC6o_29_ Without Additives	7.20	26	1.1

Symbols Psand = P2 mm-P0.063; DP = Dust Proportion = P0.063/Pb; Pb = Percentage of bitumen (by weight of mixture)

Table 2 summarises the standards used to carry out the tests (overall design of experiments).

Table 2. Main parameters and standards.

	Parameter	Standard
Volumetric properties	Gradation, G	UNI EN 933-1
	Pb [dim. less] (%)	UNI EN 12697-1
	Dust Proportion, DP [dim. less] (%)	UNI EN 933-1
	N [dim. less]	UNI EN 12697-31
	AV_N [dim. less] (%)	UNI EN 1097-6
	Thickness, t [mm]	UNI EN 12697-36

(*continued*)

Table 2. (*continued*)

	Parameter	Standard
	Diameter, D [mm]	UNI EN 12697-31
	Gmm [dim. less] (%)	UNI EN 12697-8
	Gmb_{DIM} [dim. less] (%)	UNI EN 12697-8
	Gmb_{COR} [dim. less] (%)	ASTM D 6752-03
	AV_{DIM} [dim. less] (%)	UNI EN 12697-8
	AV_{COR} [dim. less] (%)	ASTM D 6752-03
	VMA [dim. less] (%)	UNI EN 1097-6
	VFA [dim. less] (%)	UNI EN 1097-6
Surf. prop	a_{0_i} (with i = 1, 2, 3) [dim. less]	EN 134772-1
	r [Ns/m^4]	ISO 9053-2
Mechanical properties	MS [kg], MF [mm], MQ [kg/mm]	UNI EN 12697-34
	Ki (with i = 1, ..., 6) [dB re. to 1 N/m = dB calculated using 1 N/m as reference dynamic stiffness]	EN 29052-1 and ISO 7626-5 [27]

Symbols Pb: Percentage of bitumen. N: Maximum number of gyratory rotations, Nmax. AV_N: Air voids content at a given N. G_{mm}: theoretical maximum specific gravity. G_{mbDIM}, G_{mbcor}: Bulk specific gravity derived trough the dimensional method or the Corelok one. AV_{DIM} and AV_{COR}: corresponding Air Void content. VMA: Voids in mineral aggregate. VFA: Voids filled with asphalt. a_{0i}: Average of the sound absorption coefficient (a0) calculated in a given frequency range. r: Air flow resistivity (rayls per metre or $Pa \cdot s \cdot m^{-2}$). MS, MF, MQ: Marshall stability, Marshall flow, and Marshall quotient; Ki: Averages of the dynamic stiffness (K) in a given frequency range.

Production and tests were carried out as follows:

- Before producing the specimens, the B50/70 was tested. Note that the bitumen roughly corresponds to PG (64–22) based on [28].
- Graphene-based additives (10%wt) and B50/70 were previously mixed.
- Afterwards, specimens were produced. For each specimen, the blended aggregates were heated and dried at 160 °C for 8 h. The heated aggregate was placed in a heated container and mixed thoroughly. A crater (hole) was formed in the aggregate and the 50/70 penetration grade bitumen with graphene heated to 160 °C was added.
- Aggregate and bitumen were mixed thoroughly until the aggregates were well coated and were putted in a mold. A filter paper was placed at the bottom. Then the specimen was compacted to 50 gyrations (Nmax) using a Superpave Gyratory Compactor.
- The specimens were tested for volumetric properties, composition, and mechanics.
- One of the fundamental parameters of asphalt mix design is the bulk specific gravity (G_{mb}), equal to a mass (mass of aggregate and of the asphalt binder) divided by a

volume (of the aggregate, of the binder and of the air voids in a compacted specimen) and divided by the unit mass of water [29, 30].

- The Corelok test is performed by "the water displacement method using compacted specimens vacuum-sealed in special puncture-resistant bags" [29], according to the standard ASTM D 6752-03. First, the specimens were weighed at room temperature. Once the bags to use were calibrated, the specimens were inserted inside. These were sealed by removing the air inside through the Corelok vacuum-sealing device. The weight of the specimens sealed in water was then measured. We proceeded by opening the bag and weighing the specimens again in water. Finally, the values of the bulk specific gravity, G_{mb} was derived (ASTM D 6752-03).
- The Marshall Test was performed conditioning the specimens at 60 °C for 30 min. Then, the test started. It was conducted at a loading rate of 50.8 mm/min. Marshall stability, Marshall flow and Marshall quotient were derived. Marshall stability, S_M, represents the peak resistance load obtained during a constant rate of deformation [29]. Marshall flow, f_M, is the corresponding measure of the deformation (elastic plus plastic) of the specimen determined during the stability test [29]. The Marshall quotient, Q_M, is the ratio of S_M and f_M. The test was performed using a Marshall compression Tester. The results, in terms of Marshall stability, S_M, Marshall flow, f_M, and Marshall quotient, Q_M, are given below.

4   Results and Conclusions

As a part of the overall testing program, the following preliminary tests were carried out (see Table 3 and Figs. 2 and 3):

- Test on bitumen. Tests such as penetration, softening point, ductility, elastic recovery, and dynamic viscosity of the bitumen have been carried out. Results are shown in the following table (see Table 3).
- Tests on aggregates.
- Tests on bituminous mixtures.

Table 3. B50/70 test results and standards.

Test	B 50/70	Specifications	Standard
Penetration [dmm]	61	50–70	UNI EN 1426
Softening point [°C]	48.9	46–54	UNI EN 1427
Ductility [cm]	103	>80	UNI EN 13589
Elastic recovery, ER	0.1	>0.8 (*)	UNI EN 13398
Generalized ER	0.09	NA	UNI EN 13398
η at 60 °C	201000	>145000 (**)	EN 13702–2- AASHTO T316 - ASTM D4402

(*continued*)

Table 3. (*continued*)

Test	B 50/70	Specifications	Standard
η at 110 °C	1818		EN 13702–2 - AASHTO T316 - ASTM D4402
η at 135 °C	415	>295 (**)	EN 13702–2 - AASHTO T316 - ASTM D4402
η at 160 °C	152	20–100 (***)	EN 13702–2 - AASHTO T316 - ASTM D4402

Symbols Generalized ER: distance divided by ductility; η: Dynamic viscosity, [mPa·s = cP]. (*): for modified binders. NA: not available. (**): Scheda tecnica bitume stradale 50/70 ENI. (***): I quaderni tecnici ANAS Volume V.

The main volumetric and mechanical parameters are shown in Figs. 2 and 3, where aggregate gradation, bulk specific gravity, air void content, Marshall stability, Marshall flow, and Marshall quotient are represented as a function of the mixture type (with 10% of SGG, with 10% of TGG, or without additives).

Based on averages, note that:

- The bulk specific gravity increases for mixtures with additives. The corresponding porosity (AV, air void content) is higher for the specimens without additives while it is lower for the specimens with SGG and TGG. Further investigations are here needed.
- The application of graphene modified asphalt binder considerably improves the Marshall stability of the asphalt mix, even at the relatively low concentration of 0.32% by mixture weight. It could be assumed that it is due to the composition of SGG and TGG which contain graphene, plastic materials (PE and PP) and one resin (PB).
- The results in terms of percentages for Marshall stability and Marshall quotient are shown in Figs. 2 and 3. Specimens containing SGG show a Marshall stability that is twice the one of the specimens without additives.
- Similarly, the Marshall quotient appears to be higher for SGG samples than for the ones without additives.
- By referring to the use of TGG, this type of additive resulted in an increase in Marshall stability and Marshall quotient of about 30% compared to samples without additive.
- Further studies and experimental investigations are needed. The experiments detailed above are in progress and additional results are expected.

The authors would like to thank all who sustained them with this research, especially the NITEL (Consorzio Nazionale Interuniversitario per i Trasporti e la Logistica; project INFRAGRAPH). Furthermore, it is important to underline that this research addresses some of the objectives of the on-going Italian project IASNAF (Innovative Asphalts with Natural Fibers; where the DIIES department of the University Mediterranea of Reggio Calabria is one of the partners, and the ANAS S.p.a. is the project financier), and the European projects (i) LIFE "E-VIA" (ENV/IT/000201; into the LIFE2018 programme); and (ii) LIFE20 "SNEAK" (ENV/IT/000181; into the LIFE2020 programme) where

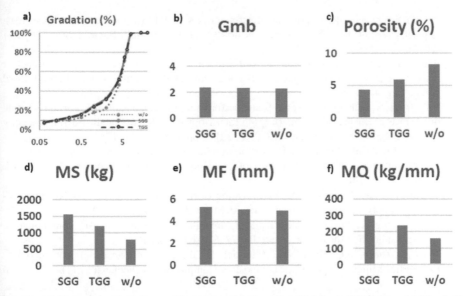

Fig. 2. (a) Gradation; (b) Bulk specific gravity; (c) Porosity; (d) Marshall Stability; (e) Marshall Flow; (f) Marshall Quotient.

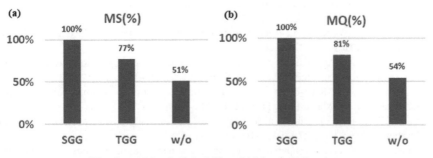

Fig. 3. (a) Marshall Stability; (b) Marshall Quotient.

the DIIES department of the University Mediterranea of Reggio Calabria is one of the partners, and the European Commission is the project financier.

References

1. Choi, W., Lahiri, I., Seelaboyina, R., Kang, Y.S.: Synthesis of graphene and its applications: a review. Crit. Rev. Solid State Mater. Sci. **35**(1), 52–71 (2010)
2. Si, C., Sun, Z., Liu, F.: Strain engineering of graphene: a review. Nanoscale **8**, 3207–3217 (2016)
3. Gerstner, E.: Nobel Prize 2010: Andre Geim & Konstantin Novoselov. Nat. Phys. **6**(11), 836 (2010)
4. Praticò, F.G., Moro, A., Noto, S., Colicchio, G.: Three-year investigation on hot and cold mixes with rubber, In: MAIREPAV 2016, 8th International Conference on Maintenance and Rehabilitation of Pavements (2016)

5. Praticò, F.G., Vaiana, R.: Improving infrastructure sustainability in suburban and urban areas: Is porous asphalt the right answer? And how? WIT Transp. Built Environ. **8**(4), 799–817 (2012)
6. Praticó, F.G., Moro, A., Ammendola, R.: Factors affecting variance and bias of non-nuclear density gauges for porous European mixes and dense-graded friction courses. Baltic J. Road Bridge Eng. **4**, 99–107 (2009)
7. Erlingsson, S., Ullberg, J.: Responses and performance of flexible pavements in cold climate due to heavy vehicle loading. In: The 10th International Conference on the Bearing Capacity of Roads, Railways and Airfields (BCRRA 2017), pp. 451–457, (2017)
8. Wen, Y.: Effects of selective distribution of alumina micro-particles on rheological, mechanical and thermal conductive properties of asphalt/SBS/alumina composites. Compos. Sci. Technol. **197** (2020)
9. Tayfur, S., Ozen, H., Aksoy, A.: Investigation of rutting performance of asphalt mixtures containing polymer modifiers. Constr. Build. Mater. **21**(2), 328–337 (2007)
10. Liu, K., Deng, L., Zheng, J., Jiang, K.: Moisture induced damage of various asphalt binders. Cailiao Yanjiu Xuebao/Chin. J. Mater. Res. **7**, 600–628 (2016)
11. Wu, C.L., Zhang, M.Q., Rong, M.Z., Friedrich, K.: Tensile performance improvement of low nanoparticles filled-polypropylene composites. Compos. Sci. Technol. **62**(10,11), 1327–1340 (2002)
12. Yao, H., et al.: Rheological properties and chemical analysis of nanoclay and carbon microfiber modified asphalt with Fourier transform infrared spectroscopy. Construct. Build. Mater. **38**, 327–337 (2013)
13. Li, X., et al.: Properties and modification mechanism of asphalt with graphene as modifier. Construct. Build. Mater. **14**, 3677 (2021)
14. Yang, L., Zhou, D., Kang, Y.: Rheological properties of graphene modified asphalt binders. Nanomaterials **10**(11), 2197 (2020)
15. Huang, G., et al.: Applications of Lambert-Beer law in the preparation and performance evaluation of graphene modified asphalt. Construct. Build. Mater. **273** (2021)
16. Wu, S., Tahri, O.: State-of-art carbon and graphene family nanomaterials for asphalt modification. Road Mater. Pave. Des. **22**(4), 735–756 (2021)
17. Du, J., Cheng, H.M.: The fabrication, properties, and uses of graphene/polymer composites. Macromol. Chem. Phys. **213**, 1060–1077 (2012)
18. Hidayah, N.M.S., et al.: Comparison on graphite, graphene oxide and reduced graphene oxide: synthesis and characterization. In: AIP Conference Proceedings, vol. 1892 (2017)
19. Liu, K., Zhang, K., Shi, X.: Performance evaluation and modification mechanism analysis of asphalt binders modified by graphene oxide. Construct. Build. Mater. **163**, 880–889 (2018)
20. Moreno-Navarro, F., Sol-Sánchez, M., Gámiz, F., Rubio-Gámez, M.C.: Mechanical and thermal properties of graphene modified asphalt binders. Construct. Build. Mater. **180**, 265–274 (2018)
21. Wu, S., Zhao, Z., Li, Y., Pang, L., Amirkhanian, S., Riara, M.: Evaluation of aging resistance of Graphene oxide modified asphalt. Appl. Sci. **7**(7), 702 (2017)
22. Zeng, W., Wu, S., Pang, L., Sun, Y., Chen, Z.: The utilization of graphene oxide in traditional construction materials. Asphalt Mater. (Basel). **10**, 48 (2017)
23. Fusco, R., Moretti, L., Fiore, N., D'andrea, A.: Behavior evaluation of bituminous mixtures reinforced with nano-sized additives: a review. Sustainability **12**, 8044 (2020)
24. Le, J., Marasteanu, M., Turos, M.: Graphene nanoplatelet (GNP) reinforced asphalt mixtures: a novel multifunctional pavement material, IDEA Program Final Report, NCHRP 173 (2016)
25. Cheap Tubes (2021). https://www.cheaptubes.com/. Accessed 26 Dec 2021
26. Le, J.L., Marasteanu, M.O., Turos, M.: Mechanical and compaction properties of graphite nanoplatelet-modified asphalt binders and mixtures. Road Mater. Pave. Des. **13**(3), 772 (2020)

27. Praticò, F.G., Fedele, R., Pellicano, G.: Pavement FRFs and noise: a theoretical and experimental investigation. Construct. Build. Mater. **140**, 274–281 (2021)
28. Montepara, A., Giuliani, F., Merusi, F.: L'analisi reologica dei bitumi stradali secondo i nuovi indirizzi europei: determinazione sperimentale della Zero Shear Viscosity, 52–06 (2006)
29. Abedali, A.H.: Asphalt Mix Design Methods, pp. 1–197. Asphalt Institute (2014)
30. Praticò, F.G., Moro, A., Ammendola, R.: Modeling HMA bulk specific gravities: a theoretical and experimental investigation. Int. J. Pave. Res. Technol. **2** (2009)
31. Wikimedia commons. https://commons.wikimedia.org/w/index.php?curid=92461297. Accessed 21 Dec 2021
32. Pixabay: https://pixabay.com/illustrations/carbon-atom-atoms-organic-4426054/. Accessed 21 Dec 2021

Cultural Heritage: Conservation, Enhancement and Management

Local Development Through the Connection Between Roots Tourism, Local Food and Wine

Sonia Ferrari[1]([✉]), Tiziana Nicotera[1], Anna lo Presti[2], and Ana Marìa Biasone[3]

[1] University of Calabria, Campus di Arcavacata, 87036 Rende, CS, Italy
sonia.ferrari@unical.it
[2] University of Torino, Lungo Dora Siena 100, 10153 Torino, Italy
[3] University of Mar del Plata, Dean Fune 3250, Mar del Plata, Argentina

Abstract. Today, food and wine tourism is an increasingly relevant form of *special interest tourism*. Tourists are interested in tasting food where it is produced but also in visiting the places of its cultivation and transformation. Food reflects local culture and identity and satisfies tourists' increasing search for *authenticity*. In this context, roots tourists, strongly linked to their land of origin, show a great interest in local traditional agri-food products. These products reflect the culture and lifestyle of their origins, even if food is not the main trip motivation. Once they return home, these tourists are eager to support their homeland and reconnect to it, so they consume these products and promote them among friends and relatives. The main aim of this research is to study if a real virtuous circle is established and the *place attachment* of this segment of visitors helps promote local productions together with the same region. A mixed methodological approach was adopted to render the research comprehensive, through a qualitative and a quantitative survey. The research's results show that roots tourists have strong ties with their homeland and they like to taste Italian food. Besides, they desire to help develop the economy of their motherland, recommending it as a tourist destination but also promoting its typical local food products. It represents a very effective way to favor the selling of these products in foreign countries and their export.

Keywords: Roots tourism · Food and wine tourism · Place attachment · Gastronomy tourism · Local food products · Typical products

1 Introduction

Tourism and local food and wine are strongly intertwined sectors (Hall et al. 2003). Gastronomy tourism is a type of special interest tourism expressed as the travels of tourists to experience local tastes (Stephen et al. 2008). It is a growing tourist segment, which is becoming more important every day (Tsai 2016). In fact, today's tourists want to taste and savour local typical food during their travels and cuisine is increasingly becoming a central tourism resource. For travellers, knowing and consuming local food means understanding the lifestyle and identity of the visited destinations better (UNWTO 2012). In addition, tourists are interested in tasting food where it is produced but also in visiting the places of its cultivation and transformation and food is every day a more

© The Author(s), under exclusive license to Springer Nature Switzerland AG 2022
F. Calabrò et al. (Eds.): NMP 2022, LNNS 482, pp. 2301–2312, 2022.
https://doi.org/10.1007/978-3-031-06825-6_221

significant element of the tourist experience (Quan and Wang 2004; Shalini and Duggal 2015). It satisfies tourists' increasing search for *authenticity*.

Frequently, once back home, they want to maintain contact with the visited destination and its community, especially if they have nurtured a form of *place attachment* towards it. If so, they return from that place and buy and consume the food that they have known during their holidays, promoting it among family and friends (Hidalgo and Hernández 2001; Jorgensen and Stedman 2006; Low and Altman 1992). In fact, place attachment implies a strong link with a territory and the desire to help its local community and business to develop. It is frequent when a tourist returns to the same place many times and when she/he owns a second home, especially if it is a family inheritance.

Place attachment is one of the main characters of roots tourism. It is a tourist segment generated by emigrants and their descendants who want to reconnect with their roots visiting their land of origin during holidays. For these tourists, the attachment to their family roots' village or region is evident and strong. It gives rise to a great interest in local traditional agri-food products that reflect the culture and lifestyle of their origins. However, food and wine are not the main trip motivation, but one of roots tourists' multiple ones. At the same time, once they return home, these tourists are eager to support their homeland and reconnect to it, so they consume these products and promote them among friends and relatives. In this way, a real virtuous circle is established and the *place attachment* of this segment of visitors amplifies it, promoting local production together with the same region.

This confirms the strong sustainability of this form of tourism and shows how important it can be to favour the development of inland areas that are often isolated and affected by depopulation and unemployment.

2 Literary Review

2.1 Food Tourism

Local food is a product with a brief supply chain in a limited geographic area (Autio et al. 2013; Czarnecki et al. 2021), but it also refers to food products with peculiarities that are linked to a territorial area (Fonte 2008). In fact, agri-food products, particularly local traditional ones, are among the goods that are best linked to their places of origin, not only for their raw materials but also because they hold local traditions and processing techniques. As a consequence, they reflect the identity, history and the culture of these territories (Bessière 1998; Giampiccoli et al. 2020; Zhang et al. 2019).

Nowadays, tourists and consumers are more and more interested in local food for many different reasons: health benefits, authenticity, and/or a strong link with an area (Winter 2003). Every day, they become more environmentally conscious, seek authenticity and are willing to pay more for it, particularly for products whose images or brands are linked to specific places (Melewar and Skinner 2018; Moulard et al. 2015). Consequently, labels of origin and place brands are becoming even more important tools in the marketing strategies of places, tourist destinations, products, and services, particularly as differentiation tools in a global market (Flack 1997; Papadopoulos 2018).

2.2   Place Attachment and Tourists' Behaviour

Many tourists develop a strong attachment to the destinations they visit. *Place attachment* is a concept born in environmental psychology. It is an affective bond between people and specific places, that is the consequence of positive experiences and satisfaction (Hidalgo and Hernández 2001; Jorgensen and Stedman 2006). It is based on affection (emotions, feelings), cognition (thoughts, knowledge, beliefs) and practice (action, behaviour) (Low and Altman 1992, p. 4). It is frequent when a tourist returns to the same place many times and when she/he owns a second home, especially if it is a family inheritance. Place attachment plays an important role in tourist experiences (Io and Wan 2018), increasing loyalty to a destination and the desire to revisit it (Chen and Phou 2013; George and George 2004; Prayag and Ryan 2012). Besides, it pushes to develop a brand, or, in tourism, to make efforts to help local business, promoting the destination and its typical products (Hosany et al. 2017; Prayag and Ryan 2012; Stedman 2006, Tsai 2016). Moreover, tourism is one of the factors that most influence local goods' exports, especially regarding food products (Fischer and Gil-Alana 2009; Madaleno et al. 2018; Mynttinen et al. 2015; Reis and Varela 2012; Webster 2002).

2.3   How Roots Tourists Promote Their Land of Origin

Often, emigrants and roots tourists feel place attachment to their homeland (Duval 2004; Ferrari and Nicotera 2018). These tourists, who are emigrants, and their descendants who want to reconnect with their roots, visiting their land of origin during holidays, show different degrees of home attachment, sometimes as *nostalgia* (Hollinshead 2004; Hui 2011; Li and McKercher 2016). In fact, roots tourism was born of diasporas, wars and migratory phenomena, and it includes people who travel because they feel a sense of nostalgia for the motherland and a desire to maintain ties with their family history (Coles and Timothy 2004; Huang et al. 2018; Sim and Leith 2013). They have strong spiritual and emotional involvement with their country of origin (Duval 2004; Lew and Wong 2004; Pelliccia 2012, 2018; Stephenson 2002; Timothy 2008).

Place attachment in emigrants is evident and they frequently pass it on to their children and grandchildren, who become tourists of second, third or even fourth generation roots (Ferrari et al. 2021). Through their families, these people have received the love, interest, and desire to learn about the history, culture and traditions of their ancestors, giving rise to a form of sustainable tourism in socio-cultural terms (Ferrari et al. 2021). For the community in the motherland, these visitors are often important, not only because of the demand that they generate, but also because they can encourage place and local production promotion. In particular, they can help improve place image by providing testimonials and giving life to a positive word-of-mouth (Newland and Taylor 2010). Furthermore, their demand frequently increases sales of products made in their family's place of origin by promoting them, and obtaining other important results such as fostering exports, stimulating local production, increasing community pride, rising sustainability and enhancing local attractiveness (Du Rand et al. 2003; Ferrari and Nicotera 2018). These tourists are interested in nostalgic products (Holbrook 1993) and become real ambassadors of the so-called *Made in Italy*.

2.4 Aim and Methodology

The main aim of this study is to investigate if roots tourists show a relevant interest in cuisine and food experiences during holidays in their regions of origin, if this reflects their place attachment and if their attitudes reveal the desire to support the international promotion, trade and export of agri-food products from their roots' countries.

With respect to the methodology, after an in-depth study regarding the state-of-the-art related to the academic research on the issue analysed, various types of surveys were conducted through different modes. Two methodologies were adopted to render the research comprehensive. In fact, a mixed approach in the research develops the strengths of the various adopted methods, and as a consequence, enables understanding and intuition that would otherwise be impossible to achieve. This approach is particularly suitable for the analysis of complex social issues, such as tourism-related themes (Creswell 2015; Morse 2016; Yin 2006).

In the first, exploratory phase of research, a qualitative methodology seems the most suitable because it is particularly interactive and in-depth. A qualitative survey was conducted in Italy with in-depth interviews and focus groups on a purposive sample of stakeholders and privileged witnesses of the food processing, trade and export industries, and the tourism sector. A total of 30 subjects were interviewed or participated in focus groups. The results have been processed through a thematic content analysis.

In the second phase a quantitative survey with focus on Argentina was carried on. In fact, this is one of the countries that historically was the destination of Italian emigrants, and its Italian community is the first one of Italians abroad. The target population consisted of all of the emigrants from Italy in Argentina and their descendants who live in this country today. The survey was conducted by a structured self-compiled questionnaire available online in English, Italian and Spanish. The questionnaire was designed to answer a series of research questions regarding the link of the respondents with their homeland of origin.

3 Findings

3.1 The Qualitative Survey

Qualitative research often precedes quantitative research. This is what was initially done by the authors, to investigate multiple aspects of roots tourism, including the interest in food and wine from various points of view -supply and demand- (Ferrari et al. 2021). In some cases, a following qualitative survey is also necessary to focus on a more specific theme that emerges from the mixed approach. In this case the object of investigation is represented by the relationship that binds visitors coming from another country to the food of their native land (culinary traditions, food and wine products purchased and consumed, and the promotional role of these tourists towards their compatriots).

To this end, between December 2021 and the first half of January 2022, two activities were carried out: in-depth interviews (for a total number of 18) aimed at different categories of stakeholders, e.g. roots tourists, companies of the agri-food sector, tour operators, and professionals who deal with events on Italian cuisine around the world;

and a focus group with 12 participants made up of producers of food and wine products from Calabria (a region of southern Italy) and experts in the sector[1].

The interviewees unanimously declare that there is a strong link between the Italian origins and the purchasing and consumption preferences for *Made in Italy* products in general, but above all for food products, both in the foreign countries where the roots tourists reside and during their holidays in the mother country. The purchase and consumption of these products are considered as a way to connect to one's identity heritage and to express the pride of belonging, to be shared with one's network of personal contacts.

One of the participants in the focus group says that cooking workshops are highly appreciated by tourists in general. For roots tourists, especially first-generation tourists, the experience is enriched by a sensory aspect: the memories they associate with food and recipes, especially the recollection of flavours and smells.

According to some participants, it is not easy to find local regional Italian products abroad and this is a limitation to the strengthening of the affective bond. However, these highly sought-after, exclusive and artisanal products could be used by Italian chefs in a creative way, helping to promote Italian excellence. This is one of the reasons why the Confederazione degli Italiani nel Mondo (Confederation of Italians Worldwide) is launching a project linked to the promotion of niche Italian products from small and medium-sized enterprises.

Some producers participating in the focus group export their products (which in some cases are niche goods) on e-commerce platforms (such as in the case of licorice root), others do not export their products because they are easily perishable commodities (for example, fresh cheeses or potatoes), or because of other problems in the distribution channels. However, they sell large quantities directly at their points of sale to Calabrian emigrants during their travels. Some manufacturers offer experiential activities. For example, one of them declares that it is possible to find the roots through the ancestors' work, the visit to the cultivation fields and the knowledge of how the farms of the past have changed.

A group of young producers have enhanced the ancient recipes of a typical traditional dessert, developing a relevant demand from abroad and especially by the immigrant communities of the municipality in which the company is based. Other companies have instead taken advantage of product innovation, being attractive to visitors and tourists through a new product linked to the territory (such as *clementine* juices), combining it with a museum dedicated to this typical citrus fruit.

Italians abroad could also contribute to countering the phenomenon of Italian Sounding. This phenomenon, in addition to obviously damaging Italian companies, irritates Italians abroad, especially for the unfair appropriation of a cultural identity. This is seen

---

[1] For their invaluable suggestions, we would like to thank: Accademia delle Tradizioni Enogastronomiche della Calabria (Valerio Caparelli), Fondazione ITS Iridea (Giorgio Durante), Consorzio di Tutela della Patata della Sila IGP (Pietro Tarasi), Santa Maria 120 (Umberto Ranieri), Molino Madre (Vittoria Caputo), Fattoria Biò (Mario Grillo), Romano Liquirizia (Vincenzo Romano), Medi Mais Calabria (Pierluigi Gallo), Caseificio Carbone (Davide Carbone), Maccheroni Chef Academy (Corrado Rossi), Dulcis in Fiore (Giovanni Piccolo), Azienda Agricola Spes (Marinella Meligeni).

as a lack of respect for the origin of the products, the provenance of the raw materials, the production techniques, without having any knowledge of the places' history. Italian emigrants and their descendants intend to carry out a work of "education", information, and raising of awareness about true quality Italian food. According to some interviewees, it is important to overcome the stereotype of the "Italian" product to make room for specific territorial productions. In this direction, it is suggested to better specify the places of origin of the goods on product labels, giving information about the territories and thus also favouring tourism. Others believe, instead, that the focus must be on the traditional Italian product and that it could be otherwise difficult to establish the products on foreign markets through restaurant chains and other BtoB activities.

Food and wine does not seem to be the main travel motivation of these tourists, but it is one of the main interests during their holiday. Travel is a way to appreciate food and wine even more, to get to know it live and to physically associate it with the places of origin. The journey becomes a stimulus to buy and consume products from tourists' regions of origin, not only during the stay, even visiting the production places, but also when they return home to emotionally repeat the experience. According to interviewees' answers, since it is difficult to compete abroad against low-cost Italian products of poor quality, the best way to promote the products is to taste them on site.

The suggestions put forward by interviewees to encourage the combination of "roots tourism - local food and wine" are: design of travel packages that can be called "enogastronomy of the roots"; creation of related events both in the country of origin and in the country of residence; design of "heritage" routes linked to physical places; creation of a unique brand to represent local regional excellences, making several products and producers travel in coordination; focus on tradition and innovation to make products more interesting and easier to export; and products presentation through storytelling.

The interviewees believe that, compared to the past, the possibility of promoting food and wine products has improved thanks to the advance of logistics systems and because of the fact that the COVID-19 pandemic got people accustomed to buying online. Ultimately, as one interviewee declares, "you cannot export a territory but the products that the territory produces and that represent it".

Even if it is not possible to generalize the results of these investigations, they constitute the basis for future structured research.

3.2   Italian Natives in Argentina

In the quantitative survey we were interested in answering a series of research questions pertaining to the connection of the respondents with their land of origin, the type of vacation experienced in Italy and their impressions, in an attempt to also investigate their level of interest in food experiences during the holidays in the country of origin and in local and typical agri-food products as well as their desire to promote them once back home. 1,545 subjects answered the questionnaire completely.

There are two ways to assess the bond that native inhabitants have with their countries of origin. It is possible to ask them directly how closely connected to their country of origin they feel. Otherwise, the attachment can be evaluated on the basis of their behaviour, asking them how much they perform the customs and traditions of their country of origin.

In this research both alternatives were examined. The bond declared by respondents is very strong, although it clearly weakens with each passing generation. An overwhelming majority of the respondents of all emigrants' generations said that they felt a *strong* or *very strong* connection with Italy, and, although we can assume that participation in the survey was higher among those who are more affectionate, the responses undoubtedly indicate the presence of a very close bond (see Table 1).

Table 1. Level (Scale 1–10 reclassified as follows: Very weak (1–2), Weak (3–4), Neither weak nor strong (5–6), Strong (7–8), Very strong (9–10)) of emotional connection with Italy stated by generation (percentage values). Source: direct survey

Generation					
	First	*Second*	*Third*	*Fourth*	*Other*
Very weak	3,1	1,9	1,6	3,4	10,9
Weak	0,6	0,7	1,1	4,5	2,2
Neither weak nor strong	0,6	2,7	6,8	8,9	8,7
Strong	4,9	12,3	24,6	32,2	23,9
Very strong	90,7	82,4	65,9	51,0	54,3
Total	100,0	100,0	100,0	100,0	100,0

Attachment to one's origins can also be inferred from the keeping of customs and habits of the land of origin. Respondents were asked to say to what extent the use of the Italian language or dialect of origin, religious traditions, civil festivities or culinary traditions are observed in their everyday life. It results that, on average, gastronomic traditions are the most respected. And this applies to all generations. Particularly, culinary traditions are those most preserved by the most recent generations (see Fig. 1).

In a way, the feeling of connection with Italy is mainly kept through the use of the Italian language and the maintenance of culinary traditions. In fact, these are the two variables most strongly correlated with the degree of attachment to Italy declared by the respondents. That is, those who said they felt very attached to Italy were also those who indicated the highest scores in these two variables.

Finally, experiencing local food is also one of the activities that have been indicated as the most frequent by those who have made a roots trip to their home regions (see Fig. 2).

It is evident that for roots tourists, as for many other tourists, local food represents a notable aspect of their travel experience. Presumably the appreciation of local food is even one of the elements that have a great effect on the propensity of native Italians and their descendants to promote and recommend their region of origin, both as a tourist destination and for its typical products.

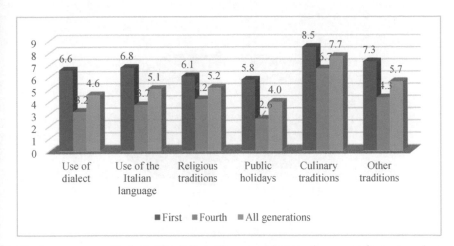

Fig. 1. Average score (Scale 1–10) attributed by respondents to the respect for customs by type of tradition. Comparison between first and fourth generation. Source: direct survey.

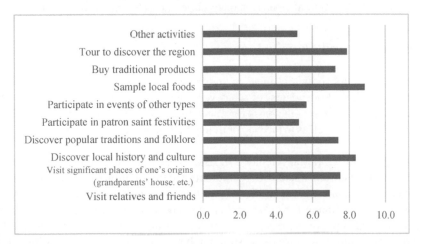

Fig. 2. Intensity (Scale 1–10) of the experience lived during the vacation (average score). Source: direct survey

The propensity for such promotion is generally very high. In fact, 68.2% of the respondents stated that they do this *often* or *very often* (see Fig. 3). This applies to all generations[2] and particularly to those who have already made a trip back to their homeland.

---

[2] From 87% of first generation to about 45% for the fourth generation respondents.

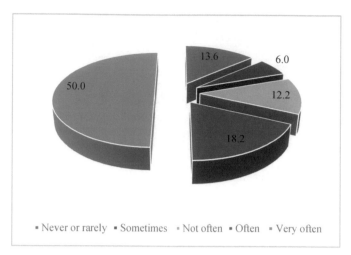

Fig. 3. Distribution of respondents by propensity (Scale 1–10 reclassifies as follows: never or rarely (1–2), sometimes (3–4), not often (5–6), often (7–8), very often (9–10)) to promote the region of origin and/or its local production (percentage values) Source: direct survey

4 Discussion and Conclusions

The research results show that there is undoubtedly a strong emotional link between emigrants and their descendants, of whatever generation, to their land of origin. This place attachment represents the basis for many different interests connected to Italy and to Italian lifestyle, as well as to the way of living, traditions and habits of specific regional areas.

Among the most important elements of Italian lifestyle, famous and loved all over the world, are food, wine and cuisine. For emigrants and their families they represent strong ties with the homeland, with memories of past times and with distant family and friends. They want to taste Italian food during their holidays but also when they are at home. Roots tourists also prove the desire to help develop the economy of their motherland of origin, promoting it as a tourist destination but also giving rise to a positive word-of-mouth about Italian and regional typical local food products. The latter represents a very effective way to favour the selling of these goods to other roots tourists during their holidays, but above all, the selling of these products in foreign countries and their export.

All this demonstrates how important it is for operators and, most importantly, political decision-makers, to be aware of the potential of this market segment and of the need to strengthen the relations with the communities of their fellow countrymen abroad. This is an area of research and governance planning so far neglected, which requires further studies and specific targeted investments before the latest generations of emigrants definitively move away from their motherland of origin, leaving it behind.

References

Autio, M., Collins, R., Wahlen, S., Anttila, M.: Consuming nostalgia? The appreciation of authenticity in local food production. Int. J. Consum. Stud. **37**(5), 564–568 (2013). https://doi.org/10.1111/ijcs.12029

Bessière, J.: Local development and heritage: traditional food and cuisine as tourist attractions in rural areas. Soc. Rural. **38**(1), 21–34 (1998). https://doi.org/10.1111/1467-9523.00061

Coles, T.E., Timothy, D.J.: Tourism, Diasporas and Space. Routledge, London (2004)

Creswell, J.W.: A Concise Introduction to Mixed Methods Research. SAGE Publications, Thousand Oaks (2014)

Czarnecki, A., Sireni, M., Dacko, M.: Second-home owners as consumers of local food. Int. J. Consum. Stud. **45**(2), 175–187 (2021). https://doi.org/10.1111/ijcs.12610

Du Rand, G.E.D., Heath, E., Alberts, N.: The role of local and regional food in destination marketing: a South African situation analysis. J. Travel Tour. Mark. **14**(3–4), 97–112 (2003). https://doi.org/10.1300/J073v14n03_06

Duval, D.: Conceptualizing return visits: a transnational perspective. In: Coles, T., Timothy, D.J. (eds.) Tourism, Diasporas, and Space, pp. 50–61. Routledge, Abingdon, Oxon (2004)

Ferrari, S., Nicotera, T.: First Report on Roots Tourism in Italy. EGEA, Milano (2021)

Ferrari, S., e Nicotera, T.: Il turismo delle origini in Calabria: indagine sulla domanda e sull'offerta per lo sviluppo di strategie di destination marketing. In: Regione Calabria, Quindicesimo rapporto sul turismo, pp. 119–169. Sistema Informativo Turistico, Regione Calabria (2018)

Ferrari, S., Hernández-Maskivker, G., Nicotera, T.: Social-cultural sustainability of roots tourism in Calabria, Italy: a tourist perspective. J. Vacat. Mark. (2021). https://doi.org/10.1177/135676 67211020493

Fischer, C., Gil-Alana, L.A.: The nature of the relationship between international tourism and international trade: the case of German imports of Spanish wine. Appl. Econ. **41**(11), 1345–1359 (2009). https://doi.org/10.1080/00036840601019349

Flack, W.: American microbreweries and neolocalism: 'Ale-ing' for a sense of place. J. Cult. Geogr. **16**(2), 37–53 (1997). https://doi.org/10.1080/08873639709478336

Fonte, M.: Knowledge, food and place: a way of producing, a way of knowing. Sociol. Rural. **48**(3), 200–222 (2008). https://doi.org/10.1111/j.1467-9523.2008.00462.x

George, B.P., George, B.: Past visits and the intention to revisit a destination: place attachment as the mediator and novelty seeking as the moderator. J. Tour. Stud. **15**(2), 37–50 (2012)

Giampiccoli, A., Mnguni, M., Dłużewska, A.: Local food, community-based tourism and well-being: connecting tourists and hosts. Czasopismo Geograficzne **91**(1–2), 249–268 (2020)

Hall, C.M., Sharples, L., Mitchell, R., Macionis, N., Cambourne, B. (eds.): Food Tourism Around the World: Development, Management and Markets. Butterworth-Heinemann, Oxford (2003)

Hidalgo, M.C., Hernandez, B.: Place attachment: conceptual and empirical questions. J. Environ. Psychol. **21**(3), 273–281 (2001). https://doi.org/10.1006/jevp.2001.0221

Holbrook, M.B.: Nostalgia and consumption preferences: some emerging patterns of consumer tastes. J. Consum. Res. **20**(2), 245–256 (1993). https://doi.org/10.1086/209346

Hollinshead, K.: Tourism and third space populations: the restless motion of diaspora peoples. In: Coles, T., Timothy, D. (eds.) Tourism, Diaspora and Space, pp. 33–50. Routledge, London (2004)

Hosany, S., Prayag, G., Van Der Veen, R., Huang, S., Deesilatham, S.: Mediating effects of place attachment and satisfaction on the relationship between tourists' emotions and intention to recommend. J. Travel Res. **56**(8), 1079–1093 (2017). https://doi.org/10.1177/004728751 667808

Huang, W.J., Hung, K., Chen, C.C.: Attachment to the home country or hometown? Examining diaspora tourism across migrant generations. Tour. Manage. **68**, 52–65 (2018). https://doi.org/10.1016/j.tourman.2018.02.019

Hui, A.: Placing nostalgia: the process of returning and remaking home. In: Davidson, T.K., Park, O., Shields, R. (eds.) Ecologies of Affect: Placing Nostalgia, Desire and Hope, pp. 65–84. Wilfrid Laurier University Press, Waterloo (2011)

Io, M.U., Wan, P.Y.K.: Relationships between tourism experiences and place attachment in the context of Casino Resorts. J. Qual. Assur. Hosp. Tour. **19**(1), 45–65 (2018). https://doi.org/10.1080/1528008X.2017.1314801

Jorgensen, B.S., Stedman, R.C.: A comparative analysis of predictors of sense of place dimensions: attachment to, dependence on, and identification with lakeshore properties. J. Environ. Manage. **79**(3), 316–327 (2006). https://doi.org/10.1016/j.jenvman.2005.08.003

Lew, A.A., Wong, A.: 13 Sojourners, guanxi and clan associations. In: Timothy, D., Coles, T. (eds.) Tourism, Diasporas and Space, pp. 202–221. Routledge, London (2004)

Li, T.E., McKercher, B.: Effects of place attachment on home return travel: a spatial perspective. Tour. Geogr. **18**(4), 359–376 (2016). https://doi.org/10.1080/14616688.2016.1196238

Low, S.M., Altman, I.: Place attachment. Springer, Boston, MA (1992)

Madaleno, A., Eusébio, C., Varum, C.: Purchase of local food products during trips by international visitors. Int. J. Tour. Res. **20**(1), 115–125 (2018). https://doi.org/10.1002/jtr.2167

Melewar, T.C., Skinner, H.: Territorial brand management: beer, authenticity, and sense of place. J. Bus. Res. (2018). https://doi.org/10.1016/j.jbusres.2018.03.038

Morse, J.M.: Mixed Method Design: Principles and Procedures, Vol. 4. Routledge, Abingdon (2016)

Moulard, J., Babin, B.J., Griffin, M.: How aspects of a wine's place affect consumers' authenticity perceptions and purchase intentions: the role of country of origin and technical terroir. Int. J. Wine Bus. Res. **27**(1), 61–78 (2015). https://doi.org/10.1108/IJWBR-01-2014-0002

Mynttinen, S., Logren, J., Särkkä-Tirkkonen, M., Rautiainen, T.: Perceptions of food and its locality among Russian tourists in the South Savo region of Finland. Tour. Manage. **48**, 455–466 (2015). https://doi.org/10.1016/j.tourman.2014.12.010

Newland, K., Taylor, C.: Heritage Tourism and Nostalgia Trade: A Diaspora Niche in the Development Landscape. Migration Policy Institute, Washington, DC (2010)

Papadopoulos, N.: Country, product-country, country-of-origin, brand origin, or place image? In: Ingenhoff, D., White, C., Buhmann, A., Kiousis, S. (eds.) Bridging Disciplinary Perspectives of Country Image Reputation, Brand, and Identity: Reputation, Brand, and Identity, pp. 11–31. Routledge, New York (2018)

Pelliccia, A.: Ulysses Undecided. Greek Student Mobility in Italy. Aracne editrice, Rome (2012)

Pelliccia, A.: In the family home: roots tourism among Greek second generation in Italy. Curr. Issue Tour. **21**(18), 2108–2123 (2018). https://doi.org/10.1080/13683500.2016.1237480

Prayag, G., Ryan, C.: Antecedents of tourists' loyalty to Mauritius: the role and influence of destination image, place attachment, personal involvement, and satisfaction. J. Travel Res. **51**(3), 342–356 (2012) https://doi.org/10.1177/0047287511410321

Quan, S., Wang, N.: Towards a structural model of the tourist experience: an illustration from food experiences in tourism. Tour. Manag. **25**(3), 297–305 (2012). https://doi.org/10.1016/S0261-5177(03)00130-4

Reis, J.G., Varela, G.: Can tourism encourage better export performance and diversification in Nepal. Econo. Premise Note Ser World Bank **127** (2012)

Rubinstein, R.I., Parmelee, P.A.: Attachment to place and the representation of the life course by the elderly. In: Altman, I., Low, S.M. (eds) Place Attachment. Human Behavior and Environment (Advances in Theory and Research), vol. 12. Springer, Boston (1992). https://doi.org/10.1007/978-1-4684-8753-4_7

Shalini, D., Duggal, S.: A review on food tourism quality and its associated forms around the world, Afr. J. Hosp. Tour. Leisure **4**(2), 1–12 (2015)

Sim, D., Leith, M.: Diaspora tourists and the Scottish homecoming 2009. J. Herit. Tour. **8**(4), 259–274 (2013). https://doi.org/10.1080/1743873X.2012.758124

Stedman, R.C.: Understanding place attachment among second home owners. Am. Behav. Sci. **50**(2), 187–205 (2006). https://doi.org/10.1177/0002764206290633

Stephen, L., Smith, J., Xiao, H.: Culinary tourism supply chains: a preliminary examination. J. Travel Res. **46**, 289–299 (2008). https://doi.org/10.1177/0047287506303981

Stephenson, M.L.: Travelling to the ancestral homelands: the aspirations and experiences of a UK Caribbean community. Curr. Issue Tour. **5**(5), 378–425 (2002). https://doi.org/10.1080/136835 00208667932

Timothy, D.J.: Genealogical mobility: tourism and the search for a personal past. In: Guelke, J.K. (ed.). Geography and Genealogy: Locating Personal Pasts, pp. 115–135. Ashgate Publishing Ltd., Farnham (2008).

Tsai, C.: T: Memorable tourist experiences and place attachment when consuming local food. Int. J. Tour. Res. **18**(6), 536–548 (2016). https://doi.org/10.1002/jtr.2070

UNWTO (United Nation World Tourism Organization): Global Report on Food Tourism. World Tourism Organization, Madrid (2012)

Webster, C.: The National Health Service: A Political History. Oxford University Press (2002)

Winter, M.: Embeddedness, the new food economy and defensive localism. J. Rural Stud. **19**(1), 23–32 (2003). https://doi.org/10.1016/S0743-0167(02)00053-0

Yin, R.K.: Mixed methods research: are the methods genuinely integrated or merely parallel. Res. Sch. **13**(1), 41–47 (2006)

Zhang, T., Chen, J., Hu, B.: Authenticity, quality, and loyalty: local food and sustainable tourism experience. Sustainability **11**(3437), 1–18 (2019). https://doi.org/10.3390/su11123437

Focus on the Role and Point of View of Municipal Administrations in the Apulia Region on the Phenomenon of Roots Tourism Through a Factor Analysis

Nicolaia Iaffaldano[1]( ) , Angela Maria D'Uggento[2] ,
and Vito Roberto Santamato[3]

[1] Ionian Department of Law, Economics and Environment, University of Bari Aldo Moro, Via
Lago Maggiore Ang. Via Ancona, 74121 Taranto, Italy
nicolaia.iaffaldano@uniba.it

[2] Department of Economics and Finance, University of Bari Aldo Moro, Largo Abbazia Santa
Scolastica, 53, 70124 Bari, Italy

[3] Department of Economics, Management and Business Law, University of Bari Aldo Moro,
Largo Abbazia Santa Scolastica, 53, 70124 Bari, Italy

Abstract. The paper deals with a very important topic for Italian southern regions
and therefore for Apulia and for smaller destinations like the internal and peripheral
areas that could become tourist destinations if they were properly organized to host
roots tourists.

In the past few centuries Italian southern regions were characterized by
poverty, misery and unemployment that obliged many people to emigrate to
distant lands in search of a better life. However, many of them never forgot
their places of origin. The municipalities of the Daunia Apennine foothills have
been heavily affected by past and current migratory phenomena. There, the phenomenon of depopulation must be curbed through the local territory and resources
development.

This research stems from the awareness that in Italy roots tourism has not yet
been studied in depth and even from the operational point of view little has been
done, i.e., there are few initiatives created for this tourism segment.

It was soon noted that in order to successfully plan the tourist offer for roots
tourists it was necessary to know this segment in depth, because it has specificities:
roots tourists' tastes, preferences and behaviours during their holidays are different
from other kinds of tourists.

In the light of this, it was necessary to understand what local administrators
think of the phenomenon of roots tourism, what interest they have developed in
this regard, the initiatives they have taken and how public administrators should
be made more aware of this segment.

Keywords: Roots tourists · Migration · Ancestral tourism · Genealogical
tourism · Municipal administrations · Socio-economic impacts · Word of mouth
effect · Attractiveness of the ancestors' land · Apulia region · Factor analysis

Although the paper is the result of a joint effort, Nicolaia Iaffaldano wrote paragraphs 2, 3.1 and
4; Angela Maria D'Uggento paragraph 3.1 and 3.2; Vito Roberto Santamato paragraph 1.

© The Author(s), under exclusive license to Springer Nature Switzerland AG 2022
F. Calabrò et al. (Eds.): NMP 2022, LNNS 482, pp. 2313–2324, 2022.
https://doi.org/10.1007/978-3-031-06825-6_222

1 Introduction

Apulia region is also part of the great history of Italian emigration between the mid-nineteenth and the twentieth century. There is no doubt that the agrarian crisis of 1887, due to the commercial break with France and to the consequent wine exports block in the transalpine market, which was the main economic resource of the region, triggered a crisis in the Apulian society. That crisis caused a long period of difficulty both for the agricultural economy and for the banking institutions that had granted numerous loans to small and medium-sized artisanal and commercial companies. It created an unprecedented condition of poverty and caused hunger, misery and unemployment in the population. The persistence of this condition led many Apulians to emigrate to distant lands in search of a better life (Bonerba 2010: 51).

The tourism of the origins, also known as roots tourism, genealogical tourism or return tourism, is an important form of tourism for our country, marked by very consistent migratory flows, not only in the past but also at the present time. Nowadays, however, the flows affect people having much higher cultural and competence levels than in the past.

From 2006 to 2020, Italian mobility increased by 76.6%, jumping, in absolute value, from just over 3.1 million registered in the A.I.R.E. (Registry of Italians Residing Abroad) to almost 5.5 million (Migrantes 2020). According to the F.I.E.I. (Italian Emigration-Immigration Federation) (www.fiei.org) the number of descendants of Italian emigrants in the world is estimated between 60 and 80 million. They represent a huge potential in terms of tourism demand, if we consider that today there are just over 60 million residents in Italy. The Apulia region strongly affected by past and current migratory phenomena will be the territorial context analyzed in the paper. In 2020 Apulia has been the seventh region of departure of the Italians residing abroad, equal to 367.996 (Migrantes 2020). A survey involving Apulian municipal administrators was conducted in order to understand the state of the art of several Apulian cities and the possible initiatives to be implemented to satisfy the needs and desires of roots tourists. The aim of the survey is to verify whether the phenomenon of roots tourism has been assimilated by municipal administrations in all its potential, how much attention they devote to it and whether they consider the tourist segment in question as a priority target or not (Iaffaldano et al. 2021: 118–130). In other studies, it has been pointed out how important is for Italy to focus on roots tourism, adopting appropriate strategies. Together with the national government, the local government should also develop policies and tools aimed at fostering the development of roots tourism, as well as implementing educational, exchange and reception programs for emigrants' communities and their descendants (Ferrari and Nicotera 2020). However, in Italy the interest in the migratory phenomenon is still modest and further studies are needed on roots tourism in order to define and implement marketing strategies aimed at developing this tourism segment.

2 Research Methodology and Aims

The aim of the research is to learn more about the attitudes and levels of awareness of local administrators with reference to the phenomenon of roots tourism. The survey aims to detecting what mayors and their delegates (in general, councillors for tourism of the Apulian municipalities) know or perceive with reference to roots tourism, its potential and effects to promote socio-economic development.

The web survey was carried out from the end of February to May 2021, targeting 257 administrators of Apulian cities in order to obtain their opinion on whether and to what extent roots tourists are attracted from ancestors' land and what they know about the preferences and behaviour of this type of visitors, as far as a positive impact of roots tourism on their municipalities is concerned.

The municipalities have been divided into 5 dimensional classes on the basis of the resident population (Emanuele 2011), namely: class a: 'very small municipalities' (population that does not exceed 5.000 residents), class b: 'small towns' (from 5.001 to 15.000 residents), class c: the so-called 'municipalities of the belt' (from 15.001 to 50.000 residents), class d: the 'medium urban centres' (from 50.001 to 100.000 residents), class e: the 'large city' (over 100.000 residents).

134 completed questionnaires were collected, with a response rate of 52.1%. This percentage is sufficient for the survey sample to be considered significant. The subjects who filled out the questionnaire were mayors or their delegates, usually tourism councillors. The questionnaire consisted of 28 questions, all closed-ended questions, except 3 open-ended questions, but only some of them were used in this research. In particular, the following research questions were examined:

– administrators' awareness and management of roots tourism flows in their municipalities,
– administrators' knowledge of the preferences and behaviours of this type of tourists,
– administrators' awareness of the potential and positive impact of roots tourism in their municipalities.

This paper introduces the main results of a quantitative survey carried out in all the municipalities of Apulia region. Exploratory and factor analyses and also the processing of the research results were conducted using SPSS rel. 25 software. Factor analysis was performed to reduce the many observed variables to a few underlying variables.

3   Main Results and Discussion

3.1   Some Results from the Exploratory Analysis

Exploratory data analysis was useful to understand some characteristics of the respondents. In particular, Fig. 1 shows the great importance that local administrators of smaller cities place on the contribution of roots tourism to overall tourist flows, compared to those of the largest cities in percentage values. Their positive perception of roots tourism is also confirmed by the higher values that respondents assigned to the fact that small cities are considered suitable tourist destinations (Fig. 2) and that 'word of mouth' can be one of the most effective means to spread and increase the image and reputation of roots tourists' native cities, i.e. the word of mouth activated by these types of tourists when they return to their place of residence. The answers with the highest rating are given by the administrators of the smaller municipalities where the risk of depopulation is greater.

This is a very significant aspect that shows how they are aware of the importance of word of mouth in terms of tourism promotion, (Fig. 3). Finally, the administrators of small cities consider these tourist flows as a crucial factor for socio-economic development of local territories (Fig. 4).

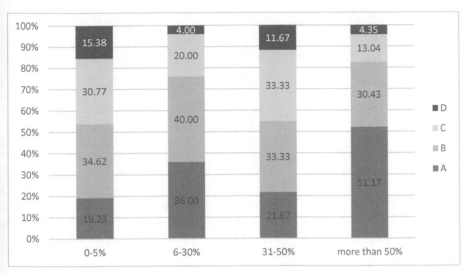

Fig. 1. According to your opinion, what percentage of the total annual tourism flows represents the roots tourism in your municipality?

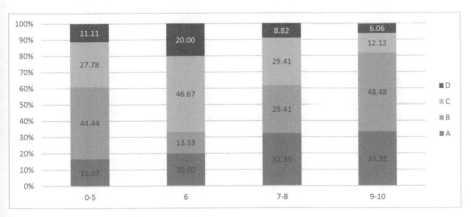

Fig. 2. On a scale of 1 (very little) to 10 (very much), to what extent can your municipality be considered a roots tourism destination?

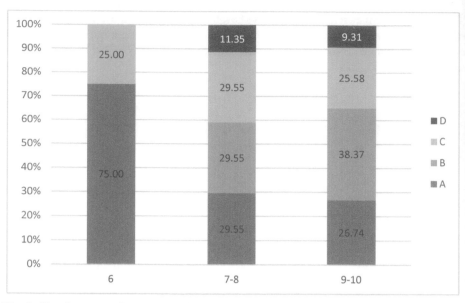

Fig. 3. To what extent from 1 (very little) to 10 (very much) do you think roots tourists could promote Apulia as a destination by word of mouth?

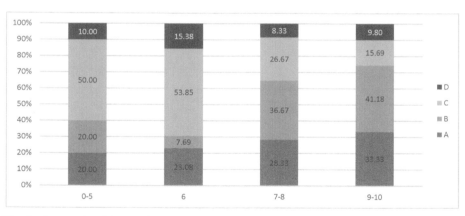

Fig. 4. On a scale of 1 (very little) to 10 (very much), to what extent do you think these tourist flows could be an important factor in the development of a local community?

3.2   A Multivariate Approach Through Factor Analysis

In order to reduce the number of the most interesting variables and to search for underlying (latent) variables represented by the observed variables (manifest variables), a factor analysis, using the extraction method principal axis factorization, was performed.

The values of the Kaiser-Meyer-Olkin Measure of Sampling Adequacy[1] (equal to 0.801) and the Bartlett's Test of Sphericity[2] (p = 0.000) confirm that factor analysis should be performed for our dataset.

To determine the number of factors to extract, we considered both theory and literature (de Lillo et al. 2007), but we also performed different analyses in which we extracted many factors and considered the most interpretable results.

Although the scree plot (Fig. 5) suggested a larger number of eigenvalues higher than 1, we decided to extract only 5 factors just to comment.

Starting with the sixth factor, in fact, each successive factor accounted for a progressively smaller proportion of the total variance. The variance explained by the selection of the 7 factors, which emerged from the scree plot, was 64.6, so the gain would have been only 12 points, but with a more complex pattern (Table 1).

After the first extraction, a Varimax rotation was performed and 5 components were considered to perform well in the analysis, explaining 52.2% of the variability (Table 1). Varimax rotation attempts to maximize the variance of each factor so that the total amount of explained variance is redistributed among the five extracted factors.

The values in Table 2 indicate the proportion of the variance of each variable/question that can be explained by the selected five factors. Variables with high values are well represented in the common factor space, while those with low values are not. These are the reproduced variances from the extracted factors and they are the same values on the diagonal of the reproduced correlation matrix.

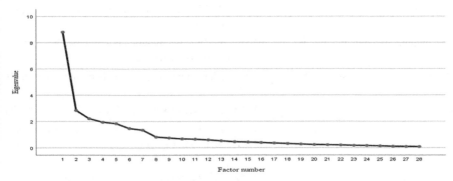

Fig. 5. Scree plot

---

[1] This measure varies between 0 and 1, and values closer to 1 are better. A value of 0.6 is a suggested minimum.

[2] It tests the null hypothesis that the correlation matrix is an identity matrix. An identity matrix is matrix in which all of the diagonal elements are 1 and all off diagonal elements are 0 and it is expected to reject this null hypothesis.

Table 1. Total variance explained by extracted factors.

Factor	Extraction			Rotation		
	Total	% of variance	% of cumulative variance	Total	% of variance	% of cumulative variance
1	8.466	30.235	30.235	3.604	12.871	12.871
2	2.505	8.946	39.181	3.589	12.817	25.688
3	1.911	6.823	46.005	3.322	11.863	37.551
4	1.567	5.595	51.600	2.222	7.936	45.487
5	1.519	5.425	57.024	1.888	6.743	52.230
6	1.104	3.943	60.968	1.836	6.558	58.788
7	1.022	3.652	64.619	1.633	5.831	64.619

Extraction method: principal axis factoring.

Table 2. Communalities.

3 Roots tourism destination	0.510	20 Interacting with local people	0.279
5 Flow percentage of roots tourists	0.512	21 Tourists satisfaction	0.303
11 Lower seasonality	0.415	22 Roots tourists play an important role in the promotion of the origin places through word-of-mouth, called native land 'ambassadors'	0.290
12 Interested in buying a house	0.666	26 Local development factor	0.626
16.1 Meeting relatives and friends	0.372	27.1 Economic availability of roots tourists	0.633
16.2 Finding ancestors' documents	0.525	27.2 Activating word-of-mouth in the place where they reside	0.591
16.3 Researching the history of the family	0.718	27.3 Typical local products promotion	0.538
16.4 Learning Italian language	0.568	27.4 Network among them	0.623
17.1 Visiting cultural/ historical/ religious sites	0.574	27.5 Interested in making investments, such us in tourist intermediation activities, or in export	0.826
17.2 Participating in traditional events	0.652	27.6 Interested in buying a property	0.821
18.1 Participating in religious events	0.480	27.7 Low season presences	0.390
18.2 Participating in other events	0.528	27.8 Longer average stays	0.465
18.3 Participating in tours to explore the region	0.535	27.9 Bringing ideas and projects	0.578
19.1 Local food tasting	0.734	27.10 Roots tourists are repositories of ancient traditions	0.441
19.2 Purchasing typical local products	0.722		

Table 3 shows the rotated factor loadings, which represent both the weights of the variables for each factor and the correlation between the variables and the factors. Since values represent correlations, they can range from −1 to +1. To make the table clearer, it was decided not to show correlations that are less than 0.40, since low correlations are unlikely to be meaningful for our analysis.

These factors are the results we are most interested in and they can be interpreted considering the aims of the research.

The first factor could be called "perceived development factors for roots tourism" because we found high loadings corresponding to the variables: 12 "interested in buying a house", 27.1 "economic availability of roots tourists", 27.4 "network among them", 27.5 "interested in making investments, such us in tourist intermediation activities, or in export", 27.6 "interested in buying a property", 27.9 "bringing ideas and projects".

The second factor could represent "the attachment to the place of origin, the interest in nostalgic products", local traditional goods that remind the roots tourists' country of origin, because it is mainly characterized by elements such as 17–18 "visiting cultural sites, participating in traditional and religious events", 18.2 "participating in other events", 18.3 "participating in tours to explore the region", 19.1 "local food tasting", 19.2 "purchasing typical local products", and it helps to attract new tourists through a positive word of mouth when they tell others about their experiences of the journey home.

The third factor has to do with "stimulating and satisfying the demand for identity research and strengthening cultural roots": they decide to travel to their ancestors' land to regain possession of their identity heritage stimulated by an ancestral link with their native land, given the high scores for 16.2 "finding ancestors' documents" conducting genealogical research, 16.3 "researching the history of the family", 16.4 "learning Italian language". According to some authors roots tourism is a niche in the cultural tourism segment (Novelli 2005).

The fourth factor deals with the various motivations of journey home of emigrants: 11 "the lower seasonality", 16.1 "meeting relatives and friends", and the more irrelevant 18.1 "participating in religious events".

The last factor seems to be mainly related to the perception of the administrators about the extent of the phenomenon: 3 "roots tourism destination" and 5 "flow percentage of roots tourists", so that we could call it "awareness of local administrators about roots tourists flows".

Moreover, it is interesting to interpret the identified factors considering at the same time some characteristics of the administrators' communities (Emanuele 2011).

As it can be seen in Fig. 6, the factor dealing with economic development opportunities mainly refers to administrators of small cities (marked A and B), while the second factor, "the attachment to the place of origin" does not seem to have a specific reference.

Factor 1, therefore, characterizes the perception of the administrators of the Apulian smaller municipalities at greater risk of depopulation: they think that roots tourism is a driving force for the local economy and brings wealth to the local community that would otherwise be destined for isolation (Ferrari and Nicotera 2021).

Table 3. Rotated component matrix.

Variables	Factor				
	1	2	3	4	5
3 Roots tourism destination					.684
5 Flow percentage of roots tourists					.675
11 Lower seasonality				.636	
12 Interested in buying a house	.592		.451		
16.1 Meeting relatives and friends				.571	
16.2 Finding ancestors' documents			.699		
16.3 Researching the history of the family			.770		
16.4 Learning Italian language			.661		
17.1 Visiting cultural/historical/religious sites		.512	.506		
17.2 Participating in traditional events		.668			
18.1 Participating in religious events		.421		.473	
18.2 Participating in other events		.638			
18.3 Participating in tours to explore the region		.614			
19.1 Local food tasting		.759			
19.2 Purchasing typical local products		.787			
20 Interacting with local people					
21 Tourists satisfaction					
22 Roots tourists are called native land 'ambassadors'		.430			
26 Local development factor	.496				.416
27.1 Economic availability of roots tourists	.624				
27.2 Activating word-of-mouth where they reside	.483	.520			
27.3 Typical local products promotion	.458	.464			
27.4 Network among them	.605	.451			
27.5 Interested in make investments	.898				
27.6 Interested in buying a property	.856				
27.7 Low season presences	.517				
27.8 Longer average stays	.495				
27.9 Bringing ideas and projects	.699				

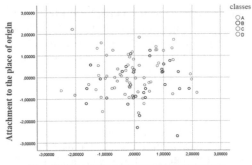

Fig. 6. Scatter plot of factors 1 and 2 by type/dimension of cities.

4   Conclusion

An ad hoc questionnaire was created to explore many aspects, and factor analysis allowed us to reduce the numerous variables connected to the most interesting latent factors. Five factors were extracted in order to explain the crucial dimensions of our research on the roots tourism.

The first factor emphasizes the role of roots tourism as an important driver of development through the injection of cash flows and new vitality in local economy. The interest in buying a house in Apulia is evident according to the answers given by municipal administrators: it perhaps reflects a mentality still linked to the possession of a property, considered as a status symbol and a form of close link with the land of origin; a mentality strongly felt by the emigrants of the first and second generation (Wagner 2014).

According to the mayors of Apulian municipalities, emigrants and their descendants residing abroad show an interest in making investments in the land of origin, such as journey home intermediation activity by designing and specializing travel itineraries for their fellow countrymen who emigrated abroad.

The second factor represents the ancestral and emotional attachment to the place of origin: local nostalgic products remind tourists of their country of origin and its amazing traditions and its highly emotional local authenticities (Cohen 1988; Corsale and Vuytsyk 2015; Hall and Müller 2004; Newland and Taylor 2010; Taylor 2001; Wang 1999; Williams et al. 1992). This emotional attachment helps attract new tourists through a positive word of mouth that happens when roots tourists tell other people about their experiences of their journey home (Basu 2005; Newland and Taylor 2010).

The need of emigrants for a deep knowledge of roots is clear when they search for a past social and cultural identity and travel to the land of their ancestors to regain possession of a tangible identity heritage and when they try to find ancestral documents in genealogical research (Yakel 2004). Moreover, by learning the Italian language, they build a bridge between the past and the present.

The fourth factor deals with the preferred travel periods and the various motivations of emigrants' journey home (Duval 2003), which is basically taken to meet relatives and friends and to participate in religious events.

The fifth factor is mainly related to the perception of the administrators about the quantitative extent of the phenomenon. Local administrators are aware of the economic importance of roots tourist flows, especially in smaller towns, where roots tourism is considered as an opportunity for the valorisation and revival of abandoned and marginal areas.

References

Basu, P.: Roots-tourism as return movement: semantics and the Scottish diaspora. In: Harper, M. (ed.) Emigrant Homecomings: The Return Movement of Emigrants. 1600–2000, pp. 131–150. Manchester University Press, Manchester (2005)

Bonerba, P.: Puglia, terra di migranti. In: Migrantes, F. (ed.) Rapporto italiani nel mondo 2010, pp. 51–61, Edizioni Idos, Roma (2010)

Cohen, E.: Authenticity and commoditization in tourism. Ann. Tour. Res. **15**(3), 371–386 (1988). https://doi.org/10.1016/0160-7383(88)90028-X

Corsale, A., Vuytsyk, O.: Long-distance attachments and implications for tourism development: the case of the Western Ukrainian diaspora. Tour. Plan. Dev. **13**(1), 88–110 (2015). https://doi.org/10.1080/21568316.2015.1074099

de Lillo, A., Argentin, G., Lucchini, M., Sarti, S., Terraneo, M.: Analisi multivariata per le scienze sociali. Pearson, Italia (2007)

Duval, D.T.: When hosts become guests: return visits and diasporic identities in a Commonwealth Eastern Caribbean community. Curr. Issue Tour. **6**(4), 267–308 (2003). https://doi.org/10.1080/13683500308667957

Emanuele, V.: Riscoprire il territorio: dimensione demografica dei comuni e comportamento elettorale in Italia. Meridiana **70**, 115–148 (2011)

Ferrari, S., Nicotera, T.: Roots tourism: Viaggio emozionale alla scoperta delle proprie origini. Turistica Italian J. Tour. **27**(4), 33–49 (2018)

Ferrari, S., Nicotera, T.: Il turismo delle radici in Italia: dai flussi migratori ai flussi turistici. Focus sulla Calabria. In: CNR, IRSS (eds.) XXIII 2018b/2019 Rapporto sul Turismo Italiano, pp. 577–594, Rogiosi Editore, Napoli (2020)

Ferrari, S., Nicotera, T.: Primo rapporto sul turismo delle radici in Italia. Dai flussi migratori ai flussi turistici: strategie di destination marketing per il 'richiamo' in patria delle comunità di italiani nel mondo. Egea, Milano (2021)

Migrantes, F.: Rapporto italiani nel mondo 2020. TAU Editrice, Todi, PG (2020)

Hall, C.M., Müller, D.K. (eds.): Tourism, Mobility and Second Homes: Between Elite Landscape and Common Ground. Channel View Publications, Clevedon (2004)

Huang, W.J., Ramshaw, G., Norman, W.C.: Homecoming or tourism? Diaspora tourism experience of second-generation immigrants. Tour. Geogr. **18**(1), 59–79 (2016). https://doi.org/10.1080/14616688.2015.1116597

Iaffaldano, N., Santamato, V.R., Ferrari, S., Nicotera, T.: Il turismo delle radici nel Mezzogiorno d'Italia: il ruolo delle amministrazioni comunali. In: Ferrari, S., Nicotera, T. (eds.) Primo rapporto sul turismo delle radici in Italia. Dai flussi migratori ai flussi turistici: strategie di destination marketing per il 'richiamo' in patria delle comunità di italiani nel mondo, pp. 118–130, Egea, Milano (2021)

MacCannell, D.: Staged authenticity: arrangements of social space in tourist settings. Am. J. Sociol. **79**(3), 589–603 (1973). https://doi.org/10.1086/225585

Newland, K., Taylor, C.: Heritage Tourism and Nostalgia Trade: A Diaspora Niche in the Development Landscape. Migration Policy Institute, Washington, DC (2010)

Novelli, M.: Nich Tourism. Contemporary Issues, Trend and Cases. Elsevier, New York (2005)

Taylor, J.P.: Authenticity and sincerity in tourism. Ann. Tour. Res. **28**(1), 7–26 (2001). https://doi.org/10.1016/S0160-7383(00)00004-9

Wagner, L.: Trouble at home: diasporic second homes as leisure space across generations. Ann. Leis. Res. **17**(1), 71–85 (2014). https://doi.org/10.1080/11745398.2013.869659

Wang, N.: Rethinking authenticity in tourism experience. Ann. Tour. Res. **26**(2), 349–370 (1999). https://doi.org/10.1016/S0160-7383(98)00103-0

Williams, D.R., Patterson, M.E., Roggenbuck, J.W., Watson, A.E.: Beyond the commodity metaphor: examining emotional and symbolic attachment to place. Leis. Sci. **14**(1), 29–46 (1992). https://doi.org/10.1080/01490409209513155

Yakel, E.: Seeking information, seeking connections, seeking meanings: genealogists and family historians. Inf. Res. **10**(1), 1–14 (2004)

The Management Models of a Tourist Destination in Italy

Angela Viglianisi$^{(\boxtimes)}$ and Francesco Calabrò⬤

Mediterranea University, Via dell'Università 25, 89124 Reggio Calabria, Italy
angela.viglianisi@unirc.it

Abstract. The paper reviews the evolution of key tourism destination concepts, with the aim to emphasize the extent of changes that occurred in understanding the term 'destination' over the past decades. A special emphasis is placed on the concept of the organization of Italian tourist destinations. Starting point of the analysis is the understanding of the development process of networks in tourism. Attention was paid to the relationship between the Italian approaches to tourism management (Tourism District, Tourist Local System SLOT, Destination Management (DM)). The governance of a complex system such as that of tourism requires a strong legislative anchoring and a daily practice which, based on what is established by law, is able to give answers to the different stakeholders of a territory and is able to indicate a development vision.

Keywords: Destination governance · Regional Tourist Organisations · Tourism policy

1 Introduction

In the last 10 years, the international specialized literature has coined the concept of "Tourist Destination"[1]. The Italian tourism system has traditionally shown a disconnected approach to the real market dynamics. Today, tourism organizations are going through a period of change because sector policies have shifted from the central government to regional offices. The conditions facilitating destination competitiveness are

---

[1] European Commission (2000, p. 149) defines a tourist destination as: "an area which is separately identified and promoted to tourists as a place to visit, and within which the tourist product is co-ordinated by one or more identifiable authorities or organisations". The concept of destination management has been developed by the original contribution of Laws (1995), Pechlaner and Weiermair (2000), Franch (2002) and contribution of AIEST (International Association of Scientific Experts in Tourism, www.aiest.org), among others, that analyse tourist systems as a unique group of actors localized in a common place. See the contribution of Capone in this same volume for any deepening.

This is the result of the joint work of the authors. Scientific responsibility is equally attributable, the abstract and Sects. 2, 3, 4, and 5 were written by A. Viglianisi while Sects. 1 and 6 were written by F. Calabrò.

© The Author(s), under exclusive license to Springer Nature Switzerland AG 2022
F. Calabrò et al. (Eds.): NMP 2022, LNNS 482, pp. 2325–2334, 2022.
https://doi.org/10.1007/978-3-031-06825-6_223

associated with the characterization of resources, making a distinction between inherited, created and support resources, destination management and cyclical conditions [1].

The great added value of these models is the aggregation of a set of factors in key areas of assessment of destination competitiveness, which, given their scope, allow comparable assessment at a global level, even if the importance attached to each of these dimensions may vary according to the culture of each country [2].

It is formulating a new paradigm that is "to make system", "to make network", in the tourism sector to respond adequately to the demands of the global tourist and take their rightful place in the international scenario of destinations [3].

This work aims to present a framework of indicators and is sufficiently objective to collect accurate and reliable data, being, at the same time, comprehensive to include the economic, social, cultural and environmental dimensions of tourism, given that studies of this nature are scarce in the literature [4, 5].

In order to theoretically frame what are the Tourism Organization Systems in Italy it seems useful to try to formulate a classification of the different typologies of the same, through the analysis of the main sources on the subject.

2   The Policies Alternative to Destination Management

Over the last decade, the role of destination governance has led researchers and practitioners to direct their attention towards analyzing the relationships that are established between government, businesses and the local community [6]. The literature provides several definitions of tourism context such as *S.L.O.T.* (Rispoli and Tamma 1995), *Destination* (Tamma 2000; Franch 2002; Brunetti 2002; Pechlaner 2002; Martini 2005), *Turist District* (Pencarelli and Forlani 2003; Sainaghi 2004; Della Lucia, Franch and Martini 2007; Cerquetti, Forlani, Montella and Pencarelli 2007; Franch, 2010), *territory-life system* (Golinelli 2000, 2002, 2008; Nigro, Trunfio 2003; Mastrobernardino 2004, Trunfio and Liguori 2006), *DMS* (Sheldon 1997; Ritchie 1993; Buhalis 1994; Sheehan and Ritchie 1997, 2005; Wang 2008). Each of these definitions is characterized by the strategic-management vision that emphasizes the coordination and integration between the various attractiveness factors.

In the literature, four ideal types of tourism system can be identified (see Fig. 1):

- *Market cluster*: it is a system where players, although located in the same area and belonging to the same tourism filière, do not establish co-operation relationships and do not recognise a unitary governance body. The system evolves following paths determined by exogenous factors, such as market dynamics and individual choices basically made as a result of mutual adjustments within a competitive approach.
- *District*: it is a system where, like in the market cluster, there is no unitary governance, but differently players try and establish long-lasting co-operation relationships and decision-making processes are jointly implemented. Evolutionary pathways imply coevolution of the various players and individual choices are based on a multi-lateral adjustment (partnership like).
- *Tourism local system*: it is a system characterised by close relationships among players and the existence of a governance body capable of orienting development

paths. Decision-making processes are then guided by a key player whose choices are amplified by the close interdependence among all organisations involved.
– *Constellation*: it is a system with a governance body having strong powers and acting as a core of the relationship network; then, while relationships among the various players are mere market interdependences, relationships with the key player are characterized by hierarchy. Decision-making processes are then guided by the key player, which determines the evolutionary paths the organisations will have to adjust to.

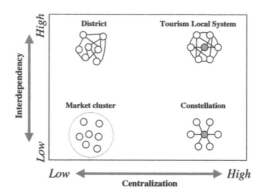

Fig. 1. Possible configurations of the system

In fact, in Italy we have three different typologies of model destination: the Tourist Local Systems (SLOT), the Tourism Districts (TD) and the Destination Management Organization (DMO).

2.1 The Systemic Approaches to Tourist Local Systems - SLOT

The definition of SLOT, Italian acronym: *Sistema Locale di Offerta Turistica* (low 135 of 29 march 2001), is: "a bundle of activities and factors of attractiveness situated in a specific place (site, locality, destination) can provide a well-constructed and integrated tourism offer, that represents a distinctive system of tourism hospitality enhancing local resources and culture [7, p. 41].

SLOT is characterized by the four following fundamentals [7, p. 41]:

– The system: a bundle of integrated activities rooted in the territory and requiring coordination and involvement of all stakeholders;
– The local: the reference being made to a specific area that determines the fundamental and peculiar characteristics representing a local attraction;
– The tourist offer: the aim of the system is to provide a wide range of tourism products;
– The local system of offer: it defines the area as a system open to relations with the external world.

The structure of SLOTs is based on two fundamental aspects: 1) the identification and convergence of different objectives towards one common aim that is shared by those involved, and on planning strategies that can develop the product; 2) the solid professionalism and sense of responsibility of the companies involved [8].

In this context, the tourism product has become systemic product and local system of tourism, it is a winning strategy for the development of a destination, which will make the time that it is not only a tourist attraction, but that also offers services, facilities and infrastructure that will enable the tourist an unforgettable positive experience.

SLOTs are organisms that create a strong link with the territory (to act local, think global), which shall take effective and innovative marketing strategies and communication, which must convince the public and private actors (companies) to abandon self-referential logic for yourself in a strategic organization and management activity, to do "real" system and to introduce elements of creativity and innovation to increase their competitiveness [9, 10]. The single operator (public or private, profit or non-profit) must be aware that it is increasingly difficult to acquire and enhance their ability to compete on their own, both in quantity and in quality, and in the presence of an application that claims with increasing force, personalization, creativity, responsibility, authenticity, quality/price ratio, infrastructure, environmental sustainability, and so on [11].

The basic elements of the new paradigm are essentially two.

1. To create a large number of interactions and a strong collaboration between all actors that can contribute to the process of construction, communication and sales of tourism products.
2. The use of the collective experiences and knowledge locally and quickly gained to realize the business opportunities that present themselves on a global level.

Determination is necessary to proceed to the construction of a network system in which the interest moves from simple and casual collaboration and/or economic transaction to a broader partnership to competition system or territory.

In Italy, Friuli Venezia Giulia, Trentino Alto Adige, Toscana, Piemonte e Valle d'Aosta are an example of SLOTs best practices.

2.2 The Tourist Districts - TD

The Tourist Districts are legal instruments (low 106 of 12 July 2011) that represent an experience of fruitful collaboration between administrations and private individuals.

There are only few contributions on tourism districts; one of them is a research carried out by ACI-CENSIS (2001) analyzing Italian territory: "a new model for an offer that draws its origin from natural vocations (sea, art, mountains, etc.), and that goes beyond the traditional reading by "points" (tourism municipalities), "lines" (for example the Romagna coasts, or the Amalfi Coast) or "systems" (for example: Val Gardena, Val Pusteria, etc.)".

The Tourist District allows the possibility to aim of requalifying and relaunching the tourist offer at national and international level by increasing the production of user services, and by increasing the development of the areas concerned by ensuring guarantees and legal certainties for the district businesses in relations with the administrations.

In fact:

1. They represent an instrument for valorising the territories which can, without doubt, constitute a facilitation for access to regional, national and community contributions for businesses members; allowed by the possibility of filing applications and initiating administrative procedures through a single collective procedure.
2 They allow companies that participate in tax breaks.
3. They constitute zero bureaucracy areas.

The districts are intended as a timed project of territorial economic development, deriving from the planning of public and private entities [12]. Their birth aims to enhance the autonomy, freedom and responsibility of individuals and groups with which power must confront [13].

2.3   The Destination Management Organization - DMO

Destination Management (regional laws) consists of the integrated management of those processes necessary in establishing an exchange between a destination and its visiting tourists.

Therefore, on the one hand, it involves the management of services offered and tourist attraction factors, while, on the other hand, managing demand, dependent on tourist flow and customer satisfaction [14]. Considering the various models, we can note the following principle functions for the DMO at the regional level:

– to maximize the long-term strategy in cooperation with other local organizations;
– to represent the regional interests and the regional tourism industry on the national level;
– to maximize profitability of local enterprises and maximize multiplier effects;
– to develop an homogenous and coherent region image;
– to optimize tourism impacts by ensuring a sustainable balance between economic benefits and socio-cultural and environmental costs.

The development of an area is strictly related to the DMO's capability of developing meta-management functions to manage and promote tourism activities. In a new approach to local and regional policies DMO play a crucial role in organizing interactions among the actors of the territorial network. However, to have local systems functioning effectively, a number of conditions are required: universally recognized and clear guidelines; effective and efficient planning and organizational systems; flexible and accessible institutional and organizational bodies; sufficient and stable financial resources, consistent with initiatives to be implemented.

The DMO plays a particularly critical and vital role in efforts to ensure that the expectations of stakeholders (both internal and external) are satisfied to the greatest extent possible [15].

Today around the world the critical role played by the DMO is recognized like fundamentally for enhancing of the tourism on all different levels or type of destination:

with-out the effective leadership and coordination of an efficacy DMO, a destination is ill-equipped to be either competitive or sustainable [16].

DMO represent a projectual methodology that is necessary for integrating and coordinating the efforts of public and private organisations towards a systematic development of the tourism product.

In this new context, the role of the DMOs is no longer simply limited to contributing to developing new tourism initiatives, but now includes the management of human and internal resources in various phases of evolution on a regional level in order to establish a relationship between the infrastructures present in the region and the existing market.

Therefore, it is imperative for DMOs to use legislative and management tools during planning and management of destinations in order to ensure that the benefits of tourism activity are shared fairly between all stakeholders and the sustainable practices safeguard the regeneration of resources utilized for the production of tourism [16].

A DMO will be will be the nerve center of the destination that holds crucial information pertaining to the market [17], it brings about a total management system and helps in coordinating and controlling the flow of tourists, tackles present trends and challenges and be that platform for all stakeholders to come in contact with the potential tourists.

3 The Common Components of the SLOT - TD - DMS Models

The Common Components of the three model are:

- All interventions of a public and/or private nature that aim to convert the place into a "destination" or to improve it. They may concentrate on specific areas (areas of large cities) or on more than one place.
- There is a need to analyse the "tourist system" in order to ensure that the different activities of the various operators working in the sector complement each other naturally and are correlated.
- The identification of a person or a direction room to guide the operation and the relative institutional structure.
- A strategic coalition in which the cooperation between autonomous companies makes the systems' activities far superior to the operative possibilities of the individual companies working alone. They aim to improve competitive advantages and to create a sustainable value for all those taking part. Operators working independently do not experience the benefits deriving from collaboration and the sharing of resources and new skills.
- They are integrated networks, a complex relational fabric which needs someone with planning and coordination abilities to oversee the optimal functioning of the system.
- They are not created and do not develop from a spontaneous union of private companies, but on the initiative of public and private subjects involved in the local product: public bodies (local bodies) and all other types of operators that have relations with tourism companies.
- Public subjects aim at creating wide-ranging economic-social benefits, while privates desire profit, in a vision that involves skills and resources that are useful for creating a long-lasting competitive offer.

4   The Difference Between the SLOT - TD - DMS Models

It is evident, however, that SLOTs and DMS are different to TDs in three main ways:

- the necessary presence of an Institution and a strategic leading company not only at the moment of the system's creation but also during its overall ad hoc management (normally this role does not coincide with a local company working in the sector);
- the open character of the system, in that it must interact with companies from other areas and countries;
- a local tourism product system is not created by a spontaneous grouping of companies in a given territory but is rather the fruit of public or mixed initiatives.

SLOTs are related to a network of companies rather than to a district. It is a form of union of companies and a network of relations for which the unit of analysis is not an individual company but a group of companies concentrated in an area directed by a strategic leader [7, p. 89].

If the theoretical-operational paradigm of the tourist district (DT) is based on a local system specializing in tourist activities according to the model of the industrial district [17], in the local tourist offer system (SLOTs) combine combination types on a territorial basis and public-private [18].

In current conceptualizations of destination management, destination management organizations (DMOs) are required to act as network managers. Destination Management Systems (DMSs) are online systems that support the activities of a DMO, for a given destination.

The DMS contemplates a streamline of types and functions, as well as particular emphasis on the role of the brand in the consolidation of the destination.

5   The Italian Experience of the Tourism Management Model

Puglia is among the tourist regions of the southern Mediterranean the one that has grown the most since 2000. In fact, one of the most important best practice in Italy is "Destinazione Puglia". It is a public sector agency for tourism promotion in the Puglia region.

The main key characteristics of this Destination Management System can be divided into two groups:

- *destination marketing strategies*: the promotion of a territorial image, brand management and internet web site management;
- *development of local resources*: support for companies in the sector, the enhancement of local resources and the integration of the skills provided by participating companies.

In any case it is essential that a destination's DMO be functional from both the strategic and the operational perspective; the leadership and coordination roles that a DMO must perform are the essence of ongoing, long-term success [15].

As for other regions, also in Puglia, the challenge is represented by the need to overcome a framework of institutional and operational fragmentation, ensuring the transition towards a governance of the activities connected to tourism marketing, the promotion of the territory, and hospitality services.

The purpose of this project is to promote Puglia territory through the creation of synergies between different players of the tourism industry [19], such as hospitality and food/beverage companies, local craftsmen, Pro Loco associations, political institutions, opinion leaders and stakeholders [20].

"Destinazione Puglia" is a highly innovative architecture of the Puglia tourism system and therefore the transformation process is very complex and unprecedented. This is a bottom up approach to DMO strategy. In fact, it is a question of favoring the overcoming of competitive logic, the achievement of agreements between the actors, public and private, local and national, to create a shared vision and stimulate the participation of these subjects in the overall development plan of the destination.

The purpose of "Destinazione Puglia" is to increase the organization and planning for tour groups who arrive in Puglia, paying particular attention to those projects, which regard above all organized tours during the low season.

Thanks to this platform, "Destinazione Puglia", is capable to engage local actors holding diverse resources and start a process of value co-creation. The strategy utilized to inform the local stakeholders about the existence and the potentials of the destination model proposed by "Destinazione Puglia" is to create events, focus group and roadshow presentations.

6 Conclusions

With respect to the various tourism system configurations, in the literature there is a debate on the superiority of one configuration over the other ones (meant as capability of supporting the sustainable development of a specific territorial area).

Actually, till date there is no study capable of producing univocal empirical evidences of such superiority. Conversely, it seems more meaningful to assume that the ability of a tourism system to generate competitive products is not only a function of its configuration, but also of the consistency between the configuration itself and a number of contingent factors related to the territorial area and the competitive system [21, 22].

This paper does not intend to provide any final answer; its aim is to participate in the current debate suggesting an interpretation model according to which the study of a territorial system requires its breaking down in various levels, each of them characterized by its own specificity in terms of problems, interpretation keys, evolutionary dynamics, functioning and governance approaches. Understanding the differences among the different levels of analysis is the only way to recognise the contribution provided by the various theories and to plan concrete and sustainable development processes. In summary, the good "tourism" regarded as an integral and central part of a territorial system becomes more efficient, financially independent, the more it is able to adopt a systemic organization and management [23].

The different typologies of destination could assume a strategic role for tourism in the very difficult historical moment following the pandemic.

The fundamental presupposition for successfully implementing a transversal policy of tourism is the coordination of public organisations (those in charge of planning), the establishment of alliances that share the same values, and cooperation between the public and private sectors [24–26].

DMSs have the winning strategy for the development of a tourist destination sustainable, as might be the South of Italy and the rest of the country (Italy) that has considerable tourist potential, which, however, is deficient in systemic forms for the management of product and territory, allowing the integration of a concrete product and territory in such a way that the product become global, but retaining its own identity and tradition [27].

References

1. Della Spina, L., Giorno, C., Galati Casmiro, R.: An integrated decision support system to define the best scenario for the adaptive sustainable re-use of cultural heritage in Southern Italy. In: Bevilacqua, C., Calabrò, F., Della Spina, L. (eds.) NMP 2020. SIST, vol. 177, pp. 251–267. Springer, Cham (2020). https://doi.org/10.1007/978-3-030-52869-0_22
2. Ruiz, G.: A relação entre o planejamento urbano e a competitividade dos destinos turísticos. Revista Brasileira de Pesquisa em Turismo 7, 260–280 (2013)
3. Cassalia, G., Ventura, C.: Un piano culturale integrato per la città di reggio calabria: la cultura come base per lo sviluppo locale dei territori. LaborEst, 10, 29–34 (2015). http://dx.medra.org/https://doi.org/10.19254/LaborEst.10.05 http://pkp.unirc.it/ojs/index.php/LaborEst/article/view/185. Accessed 21 Oct 2021
4. Dwyer, et al.: Inter-industry effects of tourism growth: implications for destination managers. Tour. Econ. 9, 117–132 (2003)
5. Park, J.: Developing a tourism destination monitoring system: a case of the Hawaii tourism dashboard. Asia Pacific J. Tour. Res. 14, 39–57 (2009)
6. Del Chiappa, G., Presenza, A.: Tourist destination and network's analysis approach. An empirical study on Costa Smeralda-Gallura. In: Proceedings of the 2011 Athens Tourism Symposium, 2–3 February, Athens, Greece (2011)
7. Rispoli, M., Tamma, M.: SLOT – Sistema Locale d'Offerta Turis tica. Risposte strategiche alla complessità. Rispoli M. e M. Tamma M. (eds.). Giappichelli, Torino(1995)
8. Manente, M., Furlan, M.C., Scaramuzzi, I.: Training strategies and new business ideas for the development of a local system of tourism supply: the role of partnerships. Libreria editrice Cafoscarina (1998)
9. Della Spina, L., Lorè, I., Scrivo, R., Viglianisi, A.: An integrated assessment approach as a decision support system for urban planning and urban regeneration policies. Buildings 7(4), 85 (2017). https://doi.org/10.3390/buildings7040085
10. Della Spina, L., Ventura, C., Viglianisi, A.: A Multicriteria assessment model for selecting strategic projects in urban areas. In: Gervasi, O., et al. (eds.) ICCSA 2016. LNCS, vol. 9788, pp. 414–427. Springer, Cham (2016). https://doi.org/10.1007/978-3-319-42111-7_32
11. Calabrò, F.: Integrated programming for the enhancement of minor historical centres. The SOSTEC model for the verification of the economic feasibility for the enhancement of unused public buildings|La programmazione integrata per la valorizzazione dei centri storici minori. Il Modello SOSTEC per la verifica della fattibilità economica per la valorizzazione degli immobili pubblici inutilizzati. ArcHistoR 13(7), 1509–1523 (2020)
12. Sangalli, F.: Le organizzazioni del sistema turistico, Milano (2007)
13. Duret, P.: Sussidiarietà e auto amministrazione dei privati, Padova (2004B)
14. Goelder, C.R., Ritchie, J.R.B.: Tourism: Principles, Practices, Philosophies. Wiley, New York (2003)
15. Ritchie, J.R.B., Crouch, G.I.: The Competitive Destination: A Sustainable Tourism Perspective. CABI Publishing, Wallingford (2003)

16. Buhalis: Marketing the competitive destination in the future. Tour. Manag. **21**, 97–116 (2000)

17. Della Spina, L., Giorno, C., Galati Casmiro, R.: Bottom-up processes for culture-led urban regeneration scenarios. In: Misra, S., et al. (eds.) ICCSA 2019. LNCS, vol. 11622, pp. 93–107. Springer, Cham (2019). https://doi.org/10.1007/978-3-030-24305-0_8

18. Calabrò, F., Cassalia, G., Lorè, I.: The economic feasibility for valorization of cultural heritage. the restoration project of the reformed fathers' convent in Francavilla Angitola: The Zibìb territorial wine cellar. In: Bevilacqua, C., Calabrò, F., Della Spina, L. (eds.) NMP 2020. SIST, vol. 178, pp. 1105–1115. Springer, Cham (2021). https://doi.org/10.1007/978-3-030-48279-4_103

19. Viglianisi, A.: Application of evaluation tools in support of decision making participation. The case study of Reggio Calabria, a Metropolitan City. Procedia Soc. Behav. Sci. **223**, 277–284 (2016). https://doi.org/10.1016/j.sbspro.2016.05.367

20. Agenzia REgionale del Turismo PUGLIAPROMOZIONE: https://www.agenziapugliapromo zione.it/portal/home. Accessed 21 Nov 2021

21. Calabrò, F., Iannone, L., Pellicanò, R.: The historical and environmental heritage for the attractiveness of cities. The case of the Umbertine Forts of Pentimele in Reggio Calabria, Italy. In: Bevilacqua, C., Calabrò, F., Della Spina, L. (eds.) NMP 2020. SIST, vol. 178, pp. 1990–2004. Springer, Cham (2021). https://doi.org/10.1007/978-3-030-48279-4_188

22. Tramontana, C., Calabrò, F., Cassalia, G., Rizzuto, M.C.: Economic sustainability in the management of archaeological sites: The Case of Bova Marina (Reggio Calabria, Italy). In: Calabrò, F., Della Spina, L., Bevilacqua, C. (eds.) ISHT 2018. SIST, vol. 101, pp. 288–297. Springer, Cham (2019). https://doi.org/10.1007/978-3-319-92102-0_31

23. Viglianisi, A., Rugolo, A.: IL DMS: un nuovo appoccio al turismo di Reggio Calabria. *Labor-Est*, *0*(21), 32–38 (2021). http://dx.medra.org/https://doi.org/10.19254/LaborEst.21.0 http:// pkp.unirc.it/ojs/index.php/LaborEst/article/view/756. Accessed 21 Oct 2021

24. Della Spina, L.: Cultural heritage: a hybrid framework for ranking adaptive reuse strategies. Buildings **11**, 132 (2021). https://doi.org/10.3390/buildings11030132

25. Della Spina, L.: Strategic planning and decision making: a case study for the integrated management of cultural heritage assets in Southern Italy. In: Bevilacqua, C., Calabrò, F., Della Spina, L. (eds.) NMP 2020. SIST, vol. 178, pp. 1116–1130. Springer, Cham (2021). https://doi.org/10.1007/978-3-030-48279-4_104

26. Della Spina, L.: A multi-level integrated approach to designing complex urban scenarios in support of strategic planning and urban regeneration. In: Calabrò, F., Della Spina, L., Bevilacqua, C. (eds.) ISHT 2018. SIST, vol. 100, pp. 226–237. Springer, Cham (2019). https:// doi.org/10.1007/978-3-319-92099-3_27

27. Franch, M.: Destination Management. Governare il turismo tra locale e globale. Giappichelli G. editore, Torino (2002)

Cultural Tourism in Historic Towns and Villages as Driver of Sustainable and Resilient Development

Paolo Motta[1,2,3,4](✉)

[1] ICOMOS SDG -Sustainable Development GoalsWG, Paris, France
[2] CIVVIH-ICTC-ISCES -ICOMOS Scientific Committees, Paris, France
[3] IASQ- International Association Social Quality, The Hague, The Netherlands
p.motta@socialquality.org, mottapa2@gmail.com
[4] EURISPES -BRICS Laboratory, Rome, Italy

Keywords: Cultural tourism · Heritage · Tangible and intangible patrimony · Social wellness · Circular and green economy

1 Summary Description

The Covid 19 pandemic has certainly caused changes that are destined to remain in many sectors, processes already underway for some years, among these the impacts on the entire tourism sector are most significant. Mass tourism of large numbers is unlikely to return as it was before the pandemic, due to a lower economic budget of average users in the coming years, to the logistical difficulties of long-range travel, and to a substantial exhaustion of the all-inclusive and low-cost model. In this renewed scenario, cultural tourism, also including the environmental one relating to cultural landscapes, assumes a particular role, also taking into account a growing demand from increasingly consistent flows to visit and learn about destinations different from traditional cities. art and territories still not well known. There is the opportunity to further enhance all aspects of cultural heritage, both the material one of monuments and sites, and the specific intangible one for each site and local community, overwhelmed by models of globalized tourism in recent decades, starting with the revival of traditions, music, handicrafts, gastronomy, etc. intangible assets to be preserved and re-evaluated. similarly, an action to safeguard and recover tangible assets, represented by the built heritage, monuments and minor sites. The session purpose is to animate the scientific debate on Cultural tourism as a fundamental element of the integrated and sustainable development of smaller towns, villages and rural agglomerates and their surrounding territories, supported by public authorities and private operators, collaboration with local communities.

2 Culture, Heritage, Tourism

After so many months of continuing emergency you can reasonable to affirm that some of the mass tourism models of the last few years are destined to be resized for a long time

© The Author(s), under exclusive license to Springer Nature Switzerland AG 2022
F. Calabrò et al. (Eds.): NMP 2022, LNNS 482, pp. 2335–2346, 2022.
https://doi.org/10.1007/978-3-031-06825-6_224

and that new forms of tourism, more sustainable and compatible with the environment are having a rapid increase. Among these IT cultural tourism no longer predominantly focused in some destinations such as "art cities", but extended to minor centers, villages and territories, rich in cultural, capital and natural resources so far not adequately enhanced. These represent a catalyst element for national tourism and proximity to the rediscovery of territories and areas also near neighbors but often forgotten where culture, heritage (tangible and intangible) and sustainable tourism are specific and non-replicable assets.

Culture : the COVID19 pandemic has been spreading around thousand of information, not only with e-learning scopes, also a wide platform of users, of different ages, accesses to cultural sites promoting museums, concerts, events, virtual tours, that in the mayor of cases never did before. Is hopefully expectable that for a relevant number this will become a habit also after the forced home confinement and a consequent increased interest for culture, whatever will be the specific sector. UNESCO and ICOMOS can play a leading role in facilitating this process and define guidelines, protocols, rules and tools to the operators and authorities at every level to continue to feed this process, and find common initiatives.

Heritage: this is the opportunity to further enhance all the heritage values, both the intangible ones, that are specific for each site and local community, that have been faded by globalized life of way patterns and this forced reflection pause is revaluating them from: music, handicrafts, gastronomy, and so on; the tangible ones, among the built patrimony , not only of monuments and sites of global value, but also the smaller and local ones, are a fundamental parts for a sustainable development. Rehabilitation, reuse and maintenance of heritage urban housings and monuments can also be a strong and permanent driver for local small enterprises and artisans. Both tangible and intangible patrimony and are assets worth to be carefully kept and enhanced.

Tourism: changes will be significant for the entire tourism sector as no massive places and gathering will be allowed for quite a long time affecting cruisers, resort villages, events, huge hotels, shopping and food arcades, and the average income reduction for a big quota of the tourist flows will force to revise the expense priorities, among them not being holidays or pleasure travelling. Probably we will return to a closer range tourism by individual means as private car, searching for natural and open air destinies, and to the discover of the often neglected sites and attraction that are nearby to each one residence. Rural and cultural tourism and countryside lodging should be one of the first sectors restarting, together with local gastronomy and handicrafts. All without the intermediation of multinational operators.

Therefore the safeguarding, enhancement and integration between these elements, taking into account the individual specificities of the territories and local communities, can and are fundamental tools for the durable and sustainable socio-economic development of the minor centers, villages and villages with asset value , present throughout our country and in many other similar, European and extra-European, contexts.

3   Tourism Main Issues

3.1   Cultural Tourism and Territorial Revival

The process of valorization of the heritage sites are certainly long and complex, require financial and human resources, integrated knowledge (including not just engineers and/or architects, but also economists, sociologists, environment experts etc.), creativity, intervention methods as well as exemplary interventions, detailed and global visions, organizations and specific instruments: technical and management agencies, municipal information centers to help citizens and investors, information systems, etc. all combined with the objective of providing solutions compatible with local characters and identities. Concepts of identity, customs and traditions, local history and sense of community, must be the basis of this process; identity goes far beyond the urban dimension of any single historic and heritage site, spreading in a larger spatial context, extended to surrounding territory, and aimed to revitalize the close relationship between urban and rural areas, that contribute to the meaning to each human settlement.

The aim is not to revive schemes of the past, but rather to reconnect the patrimony to its territorial reality through a system of physical connections anchored to the identified characteristics of heritage and environmental value. In the past, urban settlements interacted directly with agricultural land and the boundaries between the built and rural areas were not strictly defined, urban centers were part of this wider context with close relationship. Today is no longer that way and globally many small or medium-sized urban settlements and towns have been included in larger metropolitan areas losing their specific values. This minor historic urban tissue, once recovered its identity values, can become a real attraction pole of social relationships, the genetic blueprint of a territorial plan anchored to a system of connections, made of significant elements of the landscape, of morphological configurations, of roads network, of peculiar agricultural production, and strengthen the historical role of community spaces and landmarks as propulsive factor of a renewed identity and historic towns/ villages revival.

At least two conditions must be implemented to reach these permanent and resilient objectives, as:

– Local Information
  Collaborative action can transform vertical relations into horizontal ones, and top-down decisions into fully deliberated ones, fostering true community development. The need to support their arguments in a public arena to defend their own interests obliges the actors to research into all the elements in favour of their contentions, thus learning to know better their territorial resources and strengthening the internal bonds in the community. To be used all organizational methods which can contribute to weave a close community fabric fostering processes of social inclusion, generating values and cohesion among different social groups.
– Participation Role
  The participation process can be set up for the purposes of problem solving or problem setting to explore an uncertain situation, becoming an agent in the process of creating an "operative consensus" targeting a particular action. The participation process is thus considered as a resource, as an agent for breaking away from traditional frames,

as something that can subvert the usual ways of looking at and living in a territory, causing some aspects that were not previously attributed any value, perhaps because they were taken for granted, to emerge in all their importance.

3.2 Cultural Landscape Preservation

For a resilient and integrated development of heritage towns and villages in addition to the built patrimony a great importance is assumed by the environmental protection and landscape safeguard as important elements of the entire concept of heritage; the term "cultural landscape" embraces a diversity of manifestations of the interaction between humankind and its natural environment. Cultural landscapes often reflect specific techniques of sustainable land-use, considering the characteristics and limits of the natural environment they are established in, and a specific spiritual relation to nature.

Definition of the World Heritage Convention, cultural landscapes are cultural properties that represent the *"combined works of nature and man.... They are illustrative of the evolution of human society and settlement over time, under the influence of the physical constraints and/or opportunities presented by their natural environment and of successive social, economic, and cultural forces, both external and internal. Others, associated in the minds of the communities with powerful beliefs and artistic and traditional customs, embody an exceptional spiritual relationship of people with nature* [1].

Some towns and rural settlements still reflect among their tangible and intangible patrimony, specific techniques of land use that guarantee and sustain biological diversity. Others skills, preserved in communities with powerful artistic and traditional customs, embody an exceptional spiritual relationship of people with nature. To reveal and sustain the great diversity of the interactions between humans and their environment, to protect living traditional cultures and preserve the traces of those disappeared, these sites, called *"cultural landscapes"*, have been inscribed on the World Heritage List [2]. Protection of cultural landscapes can contribute to modern techniques of sustainable land-use and can maintain or enhance natural values in the landscape. The protection of traditional cultural landscapes is therefore helpful in maintaining biological diversity. Many recent ecological disasters result from the failure of maintaining populations in small towns and rural areas and consequent land abandon and ecosystem modification, becoming an issue of political strategies at the national level. So environmental evaluation is a central aspect of any cultural territorial plans to assure a good quality environment as a prior requirement to improve the attraction of minor historic centres. Natural ecosystems cannot be understood, conserved and managed without recognizing the human cultures that shape them, since both are mutually reinforcing and interdependent. Together, cultural diversity and biological diversity hold the key to ensure resilience in social and ecological systems.

3.3 Tourism Impact on SDG's

In UN Habitat- Agenda 2030 within the 17 Sustainable Development Goals (SDGs) sustainable tourism is firmly positioned, giving a contribute directly or indirectly to all of the goals. In particular, it has been included as targets in Goal 8: *"By 2030, devise and implement policies to promote sustainable tourism that creates jobs and promotes local culture and products"* [3] and in Goal 12: *"Develop and implement tools to monitor sustainable development impacts for sustainable tourism which creates jobs, promotes local culture and products"* respectively [4]. Tourism as actually developed in large scale has a relevant global influence on the climate change: increased mobility (as flight trips), hosting, energy supply, etc. provoke more atmospheric and water pollution, endanger environmental balance, modify the traditional landscape, so alternative models as cultural tourism must be favored. At the same time, being among the fastest growing economic sectors in the world, tourism is recognized as a vital contributor to job and wealth creation, environmental protection, cultural preservation and poverty alleviation. A well-designed and managed tourism can help preserve the natural and cultural heritage assets upon which it depends, empower host communities, generate trade opportunities, and foster peace and intercultural understanding. Therefore, the harnessing of cultural tourism's positive contribution to sustainable development and the mitigation of the sector's potential adverse effects are theoretically in line with the UN Habitat 2030 Agenda [5]. But at the same time there are negative impacts on the SDG's in terms of rising environmental pollution as for example: cruise ships in ports, touristic buses, water consumption for amenities, etc., but most dangerous are those affecting the intangible values and traditions of local communities from massive flows of disrespectful tourism .

3.4 Social Cohesion and Wellness

Pandemic has highlighted the value of social cohesion in smaller communities and the need to have a new overall vision to guarantee the well-being of the person as the main subject and, in particular for what concerns urban themes, substantially reviewing the economic representative paradigms in force with At the center the production that did not take into account social and environmental values. Hence the ambitious perspective for a substantial change of the territorial settlement model with a return to minor centers and villages which, now have new potential thanks to the advent of new technologies. If a similar design was of wide breath, cultural tourism could be one of the main development engines in the coming years, involving many complementary sectors between them and contributing substantially, reducing existing inequalities, to a widespread "fair and sustainable well-being". The objective is among those of the Italian BES "Benessere Equo Sostenibile" initiatives [6]: *"The need to undertake the path towards a new integral development model with conviction, which includes all the size of well-being"*…for the well-being and social cohesion of citizens, reducing the current territorial gap between urban and rural areas, and at the same time for greater safeguarding the tangible and intangible heritage and the environment that, given the specificities of our country, is a reasonably reachable result.

3.5 Circular and Green Economy

Another element that contributes to a renewed settlement model with the enhancement of the villages minor centers is represented by the Green Economy within the recent EU strategies, with the dissemination of the territory of services for monitoring and maintenance of innumerable products that will need structures widely distributed in each the urban settlements, with the consequent impacts on logistics and distribution networks. The need for the reuse and recycling of many members of everyday life will also be a valid contribution to environmental protection, reducing waste and residues often abandoned and harmful also for general health and harmful element for potential tourist-cultural attractivity. In fact tourism is among the main Priority Themes regarding green economy and resource efficiency adopted by the OECD, UNEP, EU, etc...

4 Cultural Tourism Improvement

4.1 The Actual Scenario

The rising tourism destinies and tours are often offered as "cultural", also if this seems often merely a label, that doesn't respond to what has in the past was considered a very serious approach to the ancient culture and heritage. The "Grand Tour" in the XIX century represented a real knowledge value, that inspired a great number of artists. Then there were travelers or voyagers, not generally non cultural prepared tourist. We assist nowadays, as the main heritage and historic cities are almost collapsing under the mass tourism flows, to the rise of alternative trails and routes, oriented to discover less known sites and rural heritage specificities of specific areas. Single organized tourist or we can call still visitors, discover and live rural areas, unknown for them merging and participating in the host community daily life, similarly do some innovative tour organization with limited groups of participants and carefully prepared agendas and visits. But the sector is attracting more and more traditional operators.

In the last decades cultural tourism has been increasing progressively and recently booming, representing a relevant quota of travelers and visitors interested in the cultural , artistic and historical attractions that each country presents. Therefore is necessary to enhance and implement the existing assets through rehabilitation and renovation projects, the creation of the necessary side services needed as hosting, transportation, information and so on....flows of this category of tourist are great and request attentive planning for the best use of the heritage site assuring at the same time the adequate safeguard.

Is needed an overall approach for the conservation and sustainable use of the cultural heritage. Is to be underlined that has to be considered not only to tangible, as monuments and buildings, but also to the intangible elements that characterize each historic site and define its quality of life. Most recently is included also urban landscape considering the urban areas as part of an "unicum" with their surrounding territory. This is an intervention sector for public authorities convinced that a sustainable and durable territorial development can be achieved only through integrated approach to such a complex theme. We consider that our experiences, not only studied in theory but also applied can give a contribution to the cultural heritage and landscape theme, focusing on those elements already acquired and capable to enhance the existing levels of protection, rehabilitation and re use of the existing heritage patrimony. This process can be carried out together with the private sector as capable to return relevant revenues if properly planned and managed: successful examples of public-private cooperation in this field are many and the models can be replied elsewhere. Every initiative for cultural tourism improvement has to pursue the goal to rebuild this ancient landscape searching for the rules that have guided the past and that could be yet applied in the future. It's not just a restoration process but also a sort of identity rediscovering of territories and people's habit, something that could rebuild the right conditions to put in value rural and urban heritage landscape. Being now accepted cultural heritage as a key factor for the development of territorial and urban integrated plans, regarding minor historical centers and their surrounding landscape design, is necessary that the multi-disciplinary approach must include the entire context within a global strategy shared by all competent authorities and actors. In this direction are implemented many of the ongoing programs and plans in several countries. In Europe, Asia and Africa many cultural heritage sites are result of ancient urbanization, as the crossroads of many civilizations that have taken place over them; minor historical centres of their territories are integral part of ancient landscapes, and their evolution processes are still tied to the surrounding agricultural and natural landscape. These towns are often the attraction poles of clearly defined geographical areas with specific features; maybe, if considered as single places, are not enough attractive for cultural tourism but potentially capable to become attractive tourism destinies within a network system composed by minor heritage sites and / or natural elements existing in the same territory. Is then interesting to focus attention on minor cities and towns well connected between them, located within a medium distance range from larger urban centres and major lines of communication. They can be considered minor not only for demographic reason, but also for economic ones with reduced resources; this alleged minority, where heritage assets can become a relevant economic resource and the implementation or enhancement of cultural tourism attractiveness, represents instead a factor worth to deepen for the improvement of the permanent economic returns, being also somehow a way of self-sustainability. A new approach that can be defined as *"Cultural System"* [7] a complex territorial and urban planning strategy that enhances the heritage and cultural patrimony in the context of local geographies and peculiarities. It is not just a combination of neighbouring assets, but a real system characterised by specific components, values and close connections deeply involving the cultural and heritage background. Is then necessary to implement networks of actors capable to impact on the public action for substantial changes through a broader adoption of integrated approaches and practices, accompanied by creative cultural territorial initiative. This strategy aims to guarantee long term conservation and permanent management of cultural heritage and landscape, taking in account the strong relationship between environment safeguard, spatial planning and socio-economical development through the identification of a specific operational tool as the "Cultural Tourism Master Plan".

4.2   Intervention Guidelines

Only by an integrated approach and permanent dialogue among institutional levels, all actors and local communities on single initiaitives and defined projects, can be reached the consensus and shared objectives needed to safeguard and sustainably develop "cultural tourism master plans" whose main guidelines can be resumed and must include:

Integrated Approach

- Comprehensive, cross-thematic and cross-sector policies in comparison to sector approaches. This is essential as cultural tourism is interacting with a variety of other field of actions
- Relevant sector policies, concepts and actions for the safeguarding and development of the cultural heritage are coordinated and oriented towards a common vision and objectives.
- Heritage is recognised as a cross-cutting and integrating theme as it is a unique feature and can be an important asset for the development of the area.

Smart Development

- It is not enough to raise public awareness of culture heritage significance; its importance as an economic asset and their usefulness to society and the individual also has to be underlined.
- A widespread asset like urban cultural heritage needs widespread protection and smart development and widespread consensus.
- Bring the relevant stakeholders together, to develop sustainable solutionsand bring them in line with the requirements to safeguard the cultural heritage assets, promoting a feeling of participation of the process.

Participation and Cooperation

- Involvement of relevant stakeholders in the development and implementation of the integrated strategy for the safeguarding and development of the cultural heritage assets, developing a joint vision, objectives and actions in favour of cultural tourism.
- Direct dialogue with and among the stakeholders– as local experts and as parties concerned – to coordinate their demands and orient them with the safeguard of the cultural heritage and actions for compatible tourism.
- Early integration of the national, regional, local responsible authorities and the community, to improve shared choices and the chances of successful processes.

Management and Implementation

- Implementation and compliance of policies and actions in support of the safeguard of the cultural heritage assets demands not only actions but also procedures and structures for their effective implementation, management and monitoring.
- Improvement of instruments governing the heritage sector and the territorial management as well legislative protection rules from massive tourism flows;
- Implementation should carried out by specific bodies with competences in fields ranging from agriculture to tourism, from environment and nature to public works, etc. with a general vision on conservation of the cultural heritage patrimony.
- Shared and coordinated objectives to provide guidance and a clear framework to the relevant stakeholders and management bodies on how to act for safeguard the cultural heritage.

4.3 Cultural Tourism Master-Plan

As previously stated in other paper (7) a Cultural Tourism Master Plan for the improvement of networks of minor heritage centres and their landscape systems will redefine the limits of the cities, their rural surroundings, the relationships between the centres that have shaped the territories, their physical morphology. Often these formal rules have been forgotten, buried by buildings put inside a context without any relation with the past and tradition of places. So, every plan has to pursue the goal to rebuild this ancient landscape searching for the rules that have guided the past and that could be yet applied in the future. It's not just a restoration process but also a sort of identity rediscovering of territories and people's habit, something that could rebuild the right conditions to put in value rural and urban heritage landscape. The goals and the objectives should be accompanied by a description of the basic characteristics of the intervention. Such characteristics refer to the key planning interventions, the financial plan and the organisational structures. The strategies of cultural territorial plans should be applied to each plan by the combination of various objectives:

- In free areas, the focus should be in water supply and sewage networks, as well as in the waste disposal, the improvement of the road network, and the creation of schools, health facilities, city hall office extensions, communication and internet networks, etc.
- n built areas, the focus should be in the rehabilitation of buildings, in physically strong cooperation with the owners, the inhabitants, and all private actors.
- In the minor historic centres, the main aim is the conservation, preservation and the improvement of the still inhabited traditional, built-up environment. Traditional business should be maintained as well as activities introduced to renew the local economy.

Important assets that can be improved by Cultural Tourism Master Plans are represented by the different cultural heritage sites & landscapes existing in any homogeneous territory, considering also that in the fast-growing urban areas, often the few remnants of tangible heritage in previous years, when attention was not yet developed, have been demolished. In many cultural districts, already identified or to be discovered and that need some further implementation or publicity, is then possible to create "cultural circuits" as an integrated touristic system to be organized not only enhancing the sites themselves, but providing all the complementary facilities, infrastructures and services that nowadays tourism requires. So a detailed survey of all the material and data available in the different tourism, historic and archaeology departments and agencies, is the first step to a first evaluation of the potential existing cultural and environmental landscapes.

5 Conclusions

5.1 General Considerations

The previous considerations and indications are based on the scenarios valid up to a few weeks ago, given that the latest development of the global pandemic will have an impact on the entire tourism sector that has not yet been fully assessed. Certainly there will be short-term effects in some specific sectors that have shown the greatest drawbacks in this period, among them those that involve a concentration of large tourist masses in limited spaces and places, also for the psychological aspects of potential users, such as example cruise ships, large resorts, theme parks, etc.

A decline in these sectors is plausible and a huge reduction in tourist reservations for the next season is already underway, accompanied by a parallel general decrease in flights to traditional tourist destinations, which brings great difficulties for many low-cost airlines, which is assumed by some to fail. But even before the crisis caused by the pandemic, signs of a crisis in the group holiday sector came from the failure of large international operators, including as examples the British Thomas Cook, the Germans Wave Reisen and H&H Touristik, who closed hotels, resorts and their own airlines. These tour operators each managed flows which in some cases amounted to over seven million a year, a significant percentage of the total. And even the urban B&B sector, in very strong global growth in the last years, is in crisis due to the growing limitations applied by many tourist destination cities in various countries, with the objective to regulate the rental sector and the lack of annual rentals for residents, will undergo a gradual downsizing in the next years, meanwhile can be expected a rise of demand in the rural accommodation and private house hosting. To these and other problems are adding rising campaigns by residents of tourism destiny "art cities" against the massive presence of tourists, whose flows are invading their roads changing their habits, rhythms and the entire social fabric, with the modification of trade, services, with a progressive change of owners and operators, from local to international groups.

5.2 Updated Forecasts

The forecasts made at the beginning of the pandemic, we now go beyond two years, for a recovery of the entire tourism sector after a pause that was estimated would have affected a couple of seasons, and in this sense supported by optimistic statements from operators and agencies, starting with the UN World Travel Organization (UNWTO), they have been completely denied. In fact, the traditional methods of mass tourism, starting with cruises, low-cost flights, all-inclusive packages, mega holiday villas, recorded drops in 2020 that were globally 80%, and in 2021 there was only one slight recovery in these sectors (8). Instead, individual and proximity tourism has not only governed but increased the number of presences, based on small structures such as B & B's, widespread hotels, agri-tourism, rural holiday homes, a sector in which Italy had long been equipped before the pandemic. Hence a lower negative impact on the whole of the national tourism sector, with limited drops, for example about half compared to Spain. However, still today, the emergency continues and without certainty about the future duration, a good part of the actors and authorities. agencies and private operators continue to think of a return to the status quo before

Covid 19, while the requests and methods of a growing share of citizens towards more sustainable and less standardized forms of tourism are changing, I personally believe to remain. Furthermore, since it has never been mentioned so far, the foreseeable decrease in personal and family economic resources must be considered, especially in the industrialized countries from which the greatest tourist flows come, due to lower income from work, end of benefits, higher tax burden, costs of growing travel, etc. all elements that will affect future tourism methods. The search for holidays in the open air, in contact with nature and the discovery of minor attractions, even near the usual residence and individually reachable, of minor reception facilities, give and will give a boost to tourism towards smaller centers, villages and rural settlements. In this perspective, environmental and cultural tourism becomes one of the main engines of the permanent and sustainable development of local communities, which will be direct beneficiaries without the various and costly intermediaries of mass tourism. As mentioned above, the closure and bankruptcy of large tourist agencies and numerous low-cost flight companies have had a dramatic and sudden impact on some tourist destinations, which had been organized through large international operators. The same is true for mega hotels and hospitality structures, now avoided by most of the users, who have had to eliminate many jobs, not only seasonal, with the consequent impact on local economies.

5.3 Recommendations

For these reasons, on which much have to be investigated, emerges that, following this pandemic crisis, is necessary to implement a profound revision of the actual tourist models, with the aim of identifying innovative methods, more related to compatible uses of the territories and shared by the inhabitants, which will have to become the main subjects as proposers, operators and final beneficiaries. Residents will also play a fundamental role for the protection of both natural environment and patrimony, of the intangible traditions and peculiarities of each territory, which are the locally specific and durable assets to be protected, being the indispensable elements for a sustainable and sustainable development.

Therefore must be reviewed the figure of the traditional tourist orienting it to become more a "traveler" or "visitor", curious about other realities and social contexts, interested to broaden his experience and transform holidays into a real enrichment opportunity, beyond the so-called cultural tourism "hit and run". At the same time, the envisaged direct management of the resources and economic returns, without the existing intermediation chain of external actors, will contribute significantly to assure a overall local development, not only economic but also improving permanent social quality and cohesion.

To reach this profound change that will require surely a long way, all the tourism players must be involved, to change the business model without loose the relevant economic returns represented by this sector on local economies. As happened for urban rehabilitation processes where building enterprises found new opportunities alternative to the new real estate projects. The actual crisis will modify many of the existing schemes and hopefully lead to develop new vision and approaches also in the future tourism strategies. Is therefore necessary and urgent that academia and NGO, very active in this forced isolation with declarations, webinars, calls, etc. extend the debate out of their own closed clubs but promote all together a huge campaign on all the media and social networks, to spread information, knowledge, strategies and objectives to the wider number possible of audience, to rise their awareness and participation feeling on the mentioned topics, with the aim to create a bottom-up movement that supports new initiatives within the SDG s of Agenda 2030 and the climate change global engagements.

Within UNESCO/ ICOMOS, ICTC and other scientific committees must become one of the promoters leading this initiative that has to be transmitted as soon as possible to the authorities, politicians and other big players that have to take in account such a rising demand for structural changes in the development model, that should be effectively sustainable also under the social and environmental aspects. UNESCO and ICOMOS can play a leading role in facilitating this process and define guidelines, protocols, rules and tools to the operators and authorities at every level to continue to feed this process, and find common initiatives.

A "statement" or similar declaration, supported not only by the Academia and NGO's, but shared with the public opinion in every country can be a useful contribute for defining a new vision of the overall development with a holistic approach, leaving out the actual consumerist inadequate models, based on social cohesion and climate change fight. This should be a single document, shared by all the various ICOMOS national and international committees/working groups, avoiding overlapping, to be submitted to the General Secretariat and to the other agencies interested on the tourism issues.

References

1. UNESCO-World Heritage Convention https://whc.unesco.org/archive/convention-en.pdf (1972)
2. UNESCO -World Heritage List. https://whc.unesco.org/en/culturallandscape/
3. UN Sustainable Development Goals. https://sustainabledevelopment.un.org/content/docume nts/21441EGMSDG_8_Concept_Note_15_Feb_2019.pdf
4. UN Sustainable Development Goals (2019). https://sustainabledevelopment.un.org/content/ documents/21252030
5. UN Habitat 2030 Agenda (2016). https://unhabitat.org/sites/default/files/2019/05/nua-english. pdf
6. Benessere.Equo Sostenibile. https://www.governo.it/sites/governo.it/files/BenessereItalia_rep ortannuale_ (2020)
7. MOTTA.P. - Some Reflections on Heritage & Cultural Tourism (2020). https://www.academia. edu/42798183/
8. UNWTO-COVID-19 AND TOURISM. https://www.unwto.org/covid-19-and-tourism- (2021)

Economic Feasibility of a Project for the Reuse of the Old Hospital of Nicotera as a Center for Eating Disorders and for the Enhancement of the Mediterranean Diet

Luca Santucci, Giuseppina Cassalia, and Francesco Calabrò[✉]

PAU Department, Mediterranea University, Via dell'Università, 25, 89124 Reggio Calabria, Italy
francesco.calabro@unirc.it

Abstract. The Mediterranean Diet is recognized as a healthy diet by studies conducted by Ancel Keys on various population samples, including that of Nicotera, in Calabria. In 2010, the Mediterranean Diet was inscribed by UNESCO on the World Intangible Heritage List. Since then, however, this recognition has not been sufficient to produce significant effects in terms of enhancing the territory. The study illustrated in this article aims to verify the economic feasibility of a project which, through the creation of a Center for Eating Disorders and a Botanical Garden dedicated to the Mediterranean Diet, contributes to triggering the development processes of the territory centered on the enhancement of this cultural heritage. The project combines health services, such as the Center for Eating Disorders, with cultural services for tourists and for educational purposes, such as the Botanical Garden.

Keywords: Management models · Economic feasibility · Recovery of the building stock · Mediterranean diet · Cultural tourism

1 Introduction

Over time, the Mediterranean Diet has been taken into consideration under two different profiles: first as a diet and, after, as a lifestyle.

This article illustrates a study conducted to verify the economic feasibility of a project to reuse some unused public buildings, located in Nicotera (VV, Italy), for services connected with the Mediterranean Diet [1].

1.1 The Mediterranean Diet as an Alimentary Regimen

The introduction of the concept of the Mediterranean Diet was thanks to Ancel Keys, an American doctor interested in the diffusion and pathogenesis of myocardial infarction.

The work is the result of the shared commitment of the authors. However, paragraphs 1 and 2 can be attributed to Giuseppina Cassalia; to Francesco Calabrò paragraph 3; to Luca Santucci paragraph 4. Paragraph 5 was written jointly by the authors.

© The Author(s), under exclusive license to Springer Nature Switzerland AG 2022
F. Calabrò et al. (Eds.): NMP 2022, LNNS 482, pp. 2347–2360, 2022.
https://doi.org/10.1007/978-3-031-06825-6_225

In 1952 Ancel Keys, in collaboration with his colleagues Gino Bergami and Paul Dudley White (Eisenhower's cardiologist and personal doctor) decided to start a study on eating behaviors [2, 3].

The study was conducted in Italy, in Nicotera, Crevalcore (in Emilia Romagna) and Montegiorgio (in the Marche), and was also extended to six other countries (United States, Finland, Netherlands, Yugoslavia, Greece, and Japan) [4].

With the results of their work, generally known as the "Seven Countries Study", Keys, Bergami and White have revolutionized modern physiology and generated health, social, economic, cultural, ecological nature implications in the approach to the problems of human nutrition and its interactions with the living environment of different populations; they contributed, in fact, to defining a modern food model, which for the first time was studied in a systemic way and, as research has shown, helps to reduce the risk of different diseases arising.

In addition to myocardial infarction, successive studies demonstrated that greater adherence to this diet can determine a reduction in the risk of gastric adenocarcinoma onset and antineoplastic effects on cancer leukemic cells, colon cancer cells and breast cancer [5, 6].

This is due to nutritional choices, based on the availability of agricultural, sheep farming and fishing derivates, present in the territories of the Mediterranean bioregion.

In the 1950s, thanks to the "Seven Countries Study", Nicotera was the home base of the team of scientists from different countries, pioneers of a research that has gradually assumed more and more importance among the international scientific community, and more recently for the increasingly cited and evident correlations with the environmental and economic sustainability of the planet and the sustainability of the social and health systems of developed countries.

1.2   The Mediterranean Diet Like a Lifestyle

Mediterranean Diet was inscribed in 2010 on the Representative List of the Intangible Cultural Heritage of Humanity, involving Italy, Spain, Greece, Morocco, and from 2013 also Cyprus, Croatia, Portugal [7, 8].

The recognition of this outstanding universal value identifies in the Mediterranean lifestyle a more balanced interaction between nature and humanity, according to the original etymology of the word "diet" that is lifestyle.

As stated by UNESCO, the Diet refers to all the practices, representations, expressions, knowledge, skills and cultural spaces with which people of the Mediterranean have created and re-created, over the centuries, a synthesis between the cultural and the social organization environment [9].

From this standpoint, it reveals a universe of abilities, rituals, symbols, and traditions related to food production, harvesting, fishery, zootechnics, conservation, processing, and cooking.

Therefore, Mediterranean Diet appears as an element that goes far beyond the simplistic sense linked to nutritional properties – as it generally appears instead - bringing itself values that in Italy must be necessarily extended to the whole Mediterranean area and not to specific areas (such as the town of Pollica in Cilento's area), emphasizing the individual specificities [10].

Not limited to nutritional values, then, attention should be given instead and especially to the ways and contexts where these foods are produced and consumed, or rather the cultural landscape that characterizes this lifestyle [11, 12].

For example, the UNI Reference Practice 25:2016 and the objectives of the UNESCO Chair on Mediterranean Diet, established at the Federico II University of Naples, go in this direction [13, 14].

2   An Enhancement Project of Mediterranean Diet to Local Development

Although Nicotera was one of the places where the research was developed, the immense heritage constituted by the Mediterranean Diet has never been valorized.

In fact, given the international character of the study and the extreme diffusion throughout the Mediterranean of this diet, the Mediterranean Diet could be an instrument able to produce positive effects in all of Calabria's territory, at least [15].

The Mediterranean Diet, in fact, constitutes a Cultural Heritage capable of triggering development processes on a cultural basis, as highlighted for example in the National Recovery and Resilience Plan - PNRR (Mission 1: Digitization, Innovation, Competitiveness, Culture and Tourism; Component M1C3: Tourism and Culture; Area M1.C3.2 Regeneration of small cultural sites, cultural, religious and rural heritage; Investment 2.2: Protection and enhancement of architecture and rural landscape) and by the National Research Program - PNR 2021–2027 (Research area 2: Humanistic culture, creativity, social transformations, society of inclusion; Intervention area 4: Creativity, design and Made in Italy; Sect. 5: Territories and enhancement of Made in Italy) [16, 17].

The Calabria Region, through the Regional Law of 7 November 2017, n. 40 "Enhancement of the Italian Mediterranean diet of reference of Nicotera" [18] has tried to start some form of enhancement, at the moment with very unsatisfactory results.

The project illustrated in this article is finalized to creating good spaces for the enhancement of both dimensions that characterize the Mediterranean Diet: health and cultural.

It is expected, in fact, the recovery of three currently unused property, in which to localize functions directly connected with the Mediterranean Diet [19, 20]:

- The so-called "Old hospital";
- Land confiscated from organized crime;
- The former modern cinema;

Inside the old hospital it is planned to build a center for eating disorders, where a series of specialists in disciplines related to this type of pathology (nutritionists, endocrinologists, etc.) will be able to offer advanced services for the treatment of increasingly widespread disorders, especially in the young population.

Inside the confiscated land, instead, it is planned to build a botanical garden that can contain most of the native Calabrian cultivars, underlie of the Mediterranean diet. The botanical garden will offer support not only for the purposes of further scientific research but also as a tool for the conservation of biodiversity and as a tourist attraction, where you can learn about the characteristic of this great heritage, but also taste its flavors [21–24].

Finally, inside the former cinema, a conference room should be built, to accomodate scientific and educational events or conferences, from the medical and informative point of view.

3 The Public-Private Partnership for the Creation and Management of Services of Public Interest

A project with aims as described above produces significant effects not only on Nicotera but throughout the surrounding territory, and therefore assumes public values.

On the other hand, the difficulties of public entities in providing services efficiently are known, particularly if not strictly connected with their mission: in this case, since the promoter of the project would be the Municipality, certainly the services envisaged by the project do not fall within the scope of those strictly inherent with a municipal administration.

For these reasons, the use of public-private partnership for the management phase is conceivable, without excluding the possibility of using private capital, also for the purpose of carrying out the initial investment.

Before proceeding with the next phases of the project, however, especially in view of the need to involve private stakeholders in the management and, if possible, also in the investment, it is necessary to verify the economic feasibility, also to avoid the widespread phenomenon of non-use of the property once recovered [25, 26].

This article, therefore, deepens the issue of the economic feasibility of the project to recover the so-called "old hospital" of Nicotera and some related properties, in order to realize services connected with the Mediterranean Diet.

3.1 Profitability and Public-Private Partnership Forms

In order to understand whether private parties may be interested in some form of partnership, it is necessary to first examine whether equilibrium conditions can be applied in the budget, and, where appropriate, if there is a management surplus sufficient to ensure sufficient profitability for a possible private capital investment. From this verification will depend in general whether the project is feasible or not, but also what type of private entity can be involved as a partner.

In relation to the capacity of the asset in question to generate revenue, in theory it can be hypothesized five (six) different conditions of profitability (Fig. 1) [27]:

The Choice of the Evaluation Technique
Each of the profitability bands shown in scheme 2, corresponds to different techniques to verify the feasibility and economic sustainability of the projects, in relation to the different purposes of the evaluation.

Band A   (High and medium-high profitability): In this case, the purpose of the economic feasibility assessment is to verify whether the incoming cash flows generated

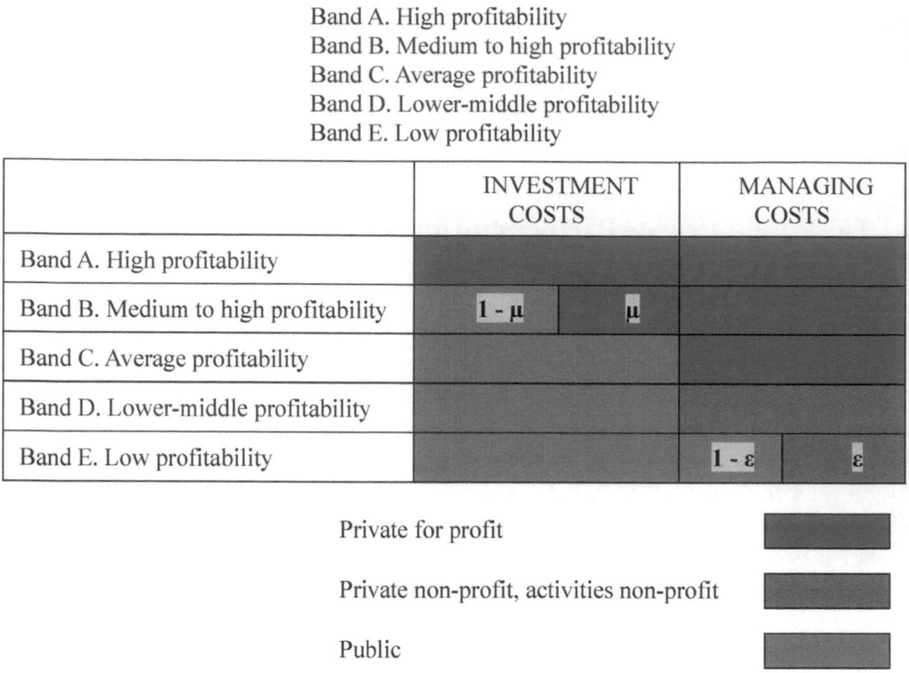

Band A. High profitability
Band B. Medium to high profitability
Band C. Average profitability
Band D. Lower-middle profitability
Band E. Low profitability

	INVESTMENT COSTS		MANAGING COSTS	
Band A. High profitability				
Band B. Medium to high profitability	$1 - \mu$	μ		
Band C. Average profitability				
Band D. Lower-middle profitability				
Band E. Low profitability			$1 - \varepsilon$	ε

Private for profit

Private non-profit, activities non-profit

Public

Fig. 1. Distribution of investment and managing costs between public and private entities

over time by the project, in addition to covering the entire operating costs, are also able to adequately remunerate the share of risk capital invested In this case, the financial analysis must be developed for a reasonable period of time, equal to the life cycle of the project; the technique to be used is the Discounted Cash Flow Analysis – DCFA.

Band B   (Average, medium-low and low profitability): in this case, the purpose of the assessment is to verify the economic sustainability related to a period of time equal to the life cycle of the project. The annual incoming cash flows generated by the project will therefore have to cover the related operating costs. The most appropriate valuation technique, in this case, is the Cash Flow Analysis - CFA: it differs from DCFA first of all for the time horizon, which in this case is equal to 1 and refers to the year when fully operational; further differences concern some of the items that are considered by the two techniques.

Band C   (Insufficient or nil profitability): the assessment must provide the public decision-maker with the elements to understand the social utility of the project. The most historically used technique is the Cost Benefit Analysis – CBA: the assessment will have to verify if the benefits, direct and indirect, internal and external, generated by the project is above its costs and therefore the community benefits from its implementation.

Purposes and evaluation techniques

Purpose	Technique
Verification of the profitability of an investment (feasibility)	Discounted Cash Flow Analysis - DCFA
Verification of the management balance of a project (economic sustainability)	Cash Flow Analysis - CFA
Verification of the public convenience of a project	Cost-benefit - ACB

3.2   The SostEc Model to Verify the Economic Sustainability of Management Models

For the purpose of verifying the economic feasibility of a project, it is possible to use the SOSTEC model, developed to provide the necessary information to identify the most appropriate form of public-private partnership to the case at hand [28, 29].

The Financial Economic Plan for the Feasibility Assessment of a Project
The SostEc Model identifies in the Economic and Financial Plan (PEF) the fundamental tool to express a judgment of convenience regarding the economic feasibility/sustainability of a project to be implemented in public-private partnership.

It includes estimated investment and management revenues and costs that must support the private individual and can be divided into 4 phases:

Phase 1. Estimation of investment costs
Phase 2. Estimation of revenues
Phase 3. Estimation of management costs
Phase 4. Verification of feasibility and/or economic sustainability

Scenarios Envisaged in the Case Study and Valuation Techniques
As explained above two hypotheses must be verified:

1. If, in the fully operational year, the operating revenue exceeds the costs;
2. If there is an operating surplus such as to ensure sufficient profitability for a possible investment of private capital.

The verification of the two hypotheses, although in both cases is based on the comparison between costs and revenues, in reality must refer to two very different time frames.

The first hypothesis must be verified through a Cash Flows Analysis that relates exclusively to the year when fully operational.

The second hypothesis, also, must be verified over a longer period of time, equal to the useful life of the investment: for this purpose it is possible to use the Discounted Cash Flows Analysis [30, 31].

4   The Financial Business Plan of the Study Case

In the present case, considering the level of definition of the project, that is of, feasibility, the estimate of the investment costs was obviously developed through a synthetic-comparative procedure, taking as a reference the parametric values for the Construction Cost reported in the typological price list of the Calabria Region [32].

Once estimated the cost of costruction, it has been possible to calculate the amount of the other sums available to the promoter, so arriving at the estimate of the cost of production (Tables 1, 2, 3, 4, 5 and 6)

Table 1. Cost of construction

A) Estimated sums for auction-based works Botanical Garden (Cost of Construction)			
	Unit construction cost	Physical quantity (m²)	Total
Green surfaces [2]	€ 150,00	5142,00	€ 771.300,00
Floor surfaces [2]	€ 95,00	1085,00	€ 103.075,00
Restaurant [2]	€ 1.232,00	140,00	€ 172.480,00
		A) Total sums for auction works	€ 1.046.855,00

(2) Typological price list 2017 Calabria region

Table 2. Cost of construction

A) Estimated sums for auction-based works Botanical Garden (Cost of Construction)			
	Unit construction cost	Physical quantity (m²)	Total
Old hospital renovation [1]	€ 1.700,00	869,25	€ 1.477.725,00
Ex cinema renovation [1]	€ 1.700,00	381,80	€ 649.060,00
Construction botanical garden [2]		6367,00	€ 1.046.855,00
		A) Total sums for auction works	€ 3.173.640,00
Of which: Amount for carrying out the work			€ 3.046.694,40
Amount for implementation of security plans (not subject to reduction) (4%)			€ 126.945,60

(1) Scale of cost s of construction and renovation of building artifacts, decision of adoption of council of the Order of Engineers of the province of Grosseto on date 24 July 2019
(1) Scale of cost s of construction and renovation of building artifacts, decision of adoption of council of the Order of Architects of the province of Messina on date 24 November 2019
(2) Typological price list 2017 Calabria region

4.1   Estimation of Revenues

On the basis of a study on the diffusion of pathologies in the nutrition and on tourist flows in Nicotera and in the province of Vibo Valentia, users have been estimated both for the Center for Eating Disorders and for the Botanical Garden [33].

Table 3. Sums available to the Promoter

B. Estimate of the sums available to the Promoter	
B 1. Economic work of the project and excluded from the contract	€ 0,00
B 2. Surveys, assessments and investigations	€ 6.132,00
B 3. Connections to public services (excluding VAT)	€ 2.000,00
B 4. Unexpected events (VAT included) (2%)	€ 63.472,80
B 5. Land or property acquisition	€ 0,00
B 6. Provision for price adjustments pursuant to art. 133, paragraph 3 of Legislative Decree 163/2006	Repealed
B 7. Incentive costs pursuant to art. 92, paragraph 5 of Legislative Decree 163/2006	€ no
B 8. Technical expenses related to design, construction management, daily assistance and accounting (7.3%)	€ 231.675,72
B 9. Technical expenses related to the coordination of safety in the design and execution phase (2%)	€ 63.472,80
B 10. Expenses for consultancy and support activities	€ 0,00
B 11. Any expenses for selection boards	€ 16.000,00
B 12. Expenses for advertising and, where applicable, for artistic works (VAT included)	€ 415,80
B 13. Expenses for laboratory investigations, technical checks and tests (1%)	€ 31.736,40
B.a) Total sums available to the Promoter excluding VAT	**€ 337.261,35**
B 14. VAT (if not recoverable) and any other taxes – 10% (B3 - B4 - B11 - B12 - B13)	€ 11.362,50
B 14. VAT (if not recoverable) and any other taxes - 22% (B2 - B8 – B9)	€ 66.281,67
B.b) Total sums available to the Promoter including VAT	**€ 414.905,52**

(1) www.gazzettaufficiale.it

Table 4. Cost of production

Summary I - Investment for the functional recovery of buildings (production cost)	
A) Sums of auction-based works	€ 3.173.640,00
B) Sums available to the Promoter	€ 337.261,35
Ia - Total investment for the recovery and re-functionalization of the properties excluding VAT	**€ 3.510.901,35**
Ib - Rounded total excluding VAT	**€ 3.511.000,00**
VAT (if not recoverable) and any other taxes	€ 390.808,67
Not rounded Total including VAT	**€ 3.901.710,02**
Rounded total including VAT	**€ 3.902.000,00**

Table 5. Investment for the usability of the properties

Summary II - Investment for the usability of the properties	
C.1) Sums for furniture	€ 129.372,81
C.2) Sums for hardware and software equipment	€ 47.753,73
II.a Total investment for the usability of the properties excluding VAT	€ 177.126,54
VAT (if not recoverable)	€ 38.967,84
II.b Total investment for the usability of the properties including VAT	€ 216.094,38

Table 6. Total cost of investment

Summary - investment estimate including VAT	
Ib - Investment for functional recovery of properties	€ 3.892.000,00
IIa - Investment for the usability of the properties	€ 216.094,38
IIIa - Investment in communication and marketing	€ 9.003,60
total investment estimate including VAT	€ 4.117.097,98

The unit prices of the services provided were estimated by comparison with the market prices currently charged for similar services in some cases assumed as best practices.

As a result, these two transactions enabled the amount of revenues to be estimated in a year when fully operational (Table 7).

Table 7. Revenues

C.2a) Estimated annual revenues when fully operational	Unit price	Quantity	Revenue
1) Revenues from specialist visits	€ 120,00	4500,00	€ 540.000,00
2) Revenues from overnight stays	€ 100,00	1500,00	€ 150.000,00
3) Average price for clinical analyzes	€ 50,00	800,00	€ 40.000,00
5) Average ticket price Botanical Garden	€ 6,00	10000,00	€ 60.000,00
6) Catering revenues	€ 15,00	5000,00	€ 75.000,00
Total 1) revenue from sales			€ 865.000,00

4.2 Estimation of Management Costs

Three different approaches were used in terms of operating costs. With regard to human resources, the organization chart of the necessary figures was drawn up preliminary, the annual cost of which was derived from the National Collective Labour Agreements. As regards the estimated cost for the purchase of consumer products, in particular for catering services, it was obtained as a rate of estimated revenue, while for the other annual operating costs reference was made to the best practices taken as reference (Table 8).

Table 8. Management costs

Summary - Annual management costs	
Annual total cost of human resources	€ 514.017,00
Annual total running costs	€ 55.190,94
Annual total management costs for consumable products when fully operational	€ 60.000,00
Annual total management costs for consumable products when fully operational	**€ 629.207,94**

4.3 Economic Feasibility of the Project

The estimation of investment and management costs makes it possible to verify the economic feasibility of the project, both in terms of operating equilibrium of the year at full capacity and the ability to remunerate any capital invested by private partners.

The first check carried out concerns the balance sheet of the year at full capacity, through the use of the Cash Flow Analysis, assuming that the investment is fully borne by the public entity (Table 9). The type of operator envisaged is non-profit, the outcome is positive.

As regards the verification of the capacity of the project to remunerate any private capital invested, two scenarios have been envisaged, both in terms of public-private co-financing: in the first scenery the public contribution to the investment is 50%, while the other 50% would be borne by private individuals; in the second scenery, with a public contribution of 80%, private investment would be just 20%.

The first scenario leads to a strongly loss-making NPV, while in the second scenario, despite the presence of a positive NPV, has a IRR of 6.47%, not very encouraging if compared to the riskiness of a highly innovative investment (Table 10, 11).

Table 9. Cash flow analysis Scenario 1

PROJECT INCOME STATEMENT (Senario 1)	
A) Value of production:	
1) revenues from sales and services envisaged by the project (Total 1 - Table C.2a)	€ 865.000,00
5) other revenues and income (Total 5 - Table C.2.2a)	
Total A) Value of production	**€ 865.000,00**
B) Production costs:	
6) for raw materials, ancillary materials, consumables and goods (Total Table C.3.3b)	€ 60.000,00
7) for services (utilities; repairs; cleaning; other ordinary maintenance services) (Total Table C.3.3b)	€ 55.190,94
8) for the enjoyment of assets (rent, leasing, etc.)	€ 54.251,00
9) for personnel: a) wages and salaries; b) social security contributions; c) severance pay; d) pensions and similar; e) other costs; (Table C.3.3a);	€ 514.017,00
10c) depreciation of furniture for eating disorders center (10%)	€ 4.500,58
10.1c) depreciation of specific equipment for eating disorders center (12.5%)	€ 5.324,13
10.2c) depreciation of office machines 20%)	€ 2.713,11
10.3c) amortization of generic equipment (25%)	€ 1.575,00
10.4c) amortization of catering furniture (10%)	€ 3.146,31
10.5c) depreciation of kitchen equipment (12%)	€ 2.337,12
10.6c) amortization of conference room furnishings (armchairs, etc.) 15.5%	€ 4.785,63
10.7c) depreciation of projection machinery and sound system (19%)	€ 2.919,73
10.6) Total depreciation	**€ 27.301,59**
11) provisions for risks; (2%)	€ 17.300,00
12) provisions for extraordinary maintenance (life cycle of the property years): (40)	€ 89.713,64
13) various management charges	€ 25.950,00
Total B	**€ 843.724,17**
Difference between value and costs of production (A - B) -> Operating Income (RO) or (MON)	**€ 21.275,83**

Table 10. Scenario 1

Project Scenarios					
	Types of assessment	Public co-financing	Estimated annual revenues when fully operational	Operating income before taxes in the year when fully operational	Type of managing entity
Scenario 1	CFA sustainability	100,00%	€ 865.000,00	€ 21.275,83	No profit

Table 11. Comparison between Scenario 1 and Scenario 2

				Discount rate	NPV	IRR	
Scenario 2	DCFA Profitability	50%	€ 865.000,00	5%	-€ 951.390,52	-	No profit
Scenario 3	DCFA Profitability	80%	€ 865.000,00	5%	€ 167.849,46	6,47%	No profit

5   Conclusions

The technical and economic feasibility project conducted shows that the intervention would bring multiple advantages on several fronts. First of all a new service to the community throughout the region with the creation of a center for eating disorders, the increase of job opportunities, the enhancement of the architectural heritage through the reuse and commissioning of the Old Hospital and the Former Cinema, the enhancement of the native flora through the creation of a botanical garden and a consequent enhancement of the tourist offer [34].

On the basis of the Cash Flow Analysis, both simple and discounted, conducted for the different scenarios, the sustainability of the intervention has emerged if the initial investment costs, for the recovery and the recasting of the buildings, are entirely the responsibility of the public entity. On the other hand, the involvement of a private party in the initial investment costs is unfavorable, as can also be seen from the last summary table. Estimated revenues can easily cover operating costs, including depreciation for furniture and provisions for extraordinary maintenance, potentially ensuring an "infinite life cycle" but fail to adequately remunerate any private capital invested.

References

1. Calabrò, F., Cassalia, G., Tramontana, C.: The Mediterranean diet as cultural landscape value: proposing a model towards the inner areas development process, Procedia Soc. Behav. Sci. **23**, 568–575 (2016)

2. De Lorenzo, A., Fidanza, F.: Dieta mediterranea italiana di riferimento. La dieta di Nicotera nel 1960, EMSI, Roma (2010)
3. Mancini, M.: Come è nato il concetto di Dieta Mediterranea? Neuromed, Napoli (2019)
4. Barbalace, P.: The Nicotera diet in the "Seven country study". (La dieta di Nicotera nello Studio dei Sette Paesi - 1957–60), "s.l.", EdiBios, Cosenza (2006)
5. Buckland, G., et al.: Adherence to a Mediterranean diet and risk of gastric adenocarcinoma within the European Prospective Investigation into Cancer and Nutrition (EPIC) cohort study. Am. J. Clin. Nutr. **91**(2), 381–390 (2010)
6. Casaburi, I., et al.: Potential of olive oil phenols as chemopreventive and therapeutic agents against cancer: a review of in vitro studies. Mol. Nutr. Food Res. **57**(1), 71–83 (2013)
7. UNESCO: Convention for the Safeguarding of the Intangible Cultural Heritage. In: Adopted by the General Conference at its 32nd Session, Paris (2003)
8. UNESCO: Intergovernmental Committee for the safeguarding of the Intangible Cultural Heritage. In: 8th Session, Baku (2013)
9. ICOMOS: International Cultural Tourism Charter. Managing Tourism at Places of Heritage Significance. In: Adopted by ICOMOS at the 12th General Assembly in Mexico (1999)
10. Council of Europe: Landscape Convention (as amended by the 2016 Protocol), European Treaty Series - No. 176, Florence, Italy (2000)
11. Council of Europe: Framework Convention on the Value of Cultural Heritage for Society, Council of Europe Treaty Series - No. 199, Faro, Portugal (2003)
12. Di Fazio, S., Modica, G.: Trasformazione del paesaggio, sistemi insediativi e borghi rurali. ArcHistoR **7**(13) (2020). https://doi.org/10.14633/AHR232
13. UNI Reference Practice: Mediterranean diet UNESCO intangible cultural heritage of humanity - Guidelines for promoting a lifestyle and cultural approach for sustainable development, p. 25:(2016)
14. UNESCO Chair on Mediterranean Diet, Federico II University of Naples: https://www.unescochairnapoli.it/targets/
15. Spampinato, G., Malerba, A., Calabrò, F., Bernardo, C., Musarella, C.: Cork oak forest spatial valuation toward post carbon city by CO_2 sequestration. In: Bevilacqua, C., Calabrò, F., Della Spina, L. (eds.) NMP 2020. SIST, vol. 178, pp. 1321–1331. Springer, Cham (2021). https://doi.org/10.1007/978-3-030-48279-4_123
16. Piano Nazionale di Ripresa e Resilienza - PNRR #NextGenerationItalia (2021)
17. Ministero dell'Università e della Ricerca: Programma Nazionale per la Ricerca – PNR 2021-2027, Testo approvato dal Comitato interministeriale per la programmazione economica con Delibera 15 dicembre 2020, n. 74
18. Regione Calabria: Legge 7 novembre 2017, n. 40 Valorizzazione Dieta mediterranea italiana di riferimento di Nicotera. BURC n. 109 (2017)
19. Calabrò, F., Mafrici, F., Meduri, T.: The valuation of unused public buildings in support of policies for the inner areas. the application of SostEc model in a case study in Condofuri (Reggio Calabria, Italy). In: Bevilacqua, C., Calabrò, F., Della Spina, L. (eds.) NMP 2020. SIST, vol. 178, pp. 566–579. Springer, Cham (2021). https://doi.org/10.1007/978-3-030-48279-4_54
20. Mallamace, S., Calabrò, F., Meduri, T., Tramontana, C.: Unused real estate and enhancement of historic centers: legislative instruments and procedural ideas. In: Calabrò, F., Della Spina, L., Bevilacqua, C. (eds.) ISHT 2018. SIST, vol. 101, pp. 464–474. Springer, Cham (2019). https://doi.org/10.1007/978-3-319-92102-0_49
21. Iuliano, L., Grotteria, M., et al.: Le piante delle tradizioni calabresi, Cosenza, ARSAA (2001)
22. Fideghelli, C.: Atlante dei fruttiferi autoctoni Italiani, Vol.1, CREA, Roma (2016)
23. Angotti, M., Bianco, G.. et al.: Varietà locali di fruttiferi in Calabria. Atlante della biodiversità, ARSAA, Cosenza (2007)
24. Blasi, C. e Biondi, E.: La flora in Italia. Flora, vegetazione, conservazione del paesaggio e tutela della biodiversità, Roma, Sapienza Università Editrice (2017)

25. Tajani, F., Morano, P., Locurcio, M., D'Addabbo, N.: Property valuations in times of crisis: artificial neural networks and evolutionary algorithms in comparison. In: Gervasi, O., et al. (eds.) ICCSA 2015. LNCS, vol. 9157, pp. 194–209. Springer, Cham (2015). https://doi.org/10.1007/978-3-319-21470-2_14

26. De Mare, G., Di Piazza, F.: The role of public-private partnerships in school building projects. In: Gervasi, O., et al. (eds.) ICCSA 2015. LNCS, vol. 9156, pp. 624–634. Springer, Cham (2015). https://doi.org/10.1007/978-3-319-21407-8_44

27. Calabrò, F.: Local communities and management of cultural heritage of the inner areas. an application of break-even analysis. In: Gervasi, O., et al. (eds.) ICCSA 2017. LNCS, vol. 10406, pp. 516–531. Springer, Cham (2017). https://doi.org/10.1007/978-3-319-62398-6_37

28. Calabrò, F.: Integrated programming for the enhancement of minor historical centres. The SOSTEC model for the verification of the economic feasibility for the enhancement of unused public buildings | La programmazione integrata per la valorizzazione dei centri storici minori. Il Modello SOSTEC per la verifica della fattibilità economica per la valorizzazione degli immobili pubblici inutilizzati. ArcHistoR **13**(7), 1509–1523. https://doi.org/10.14633/AHR280

29. Calabrò, F., Cassalia, G., Lorè, I.: The economic feasibility for valorization of cultural heritage. the restoration project of the reformed fathers' convent in Francavilla Angitola: the Zibìb territorial wine cellar. In: Bevilacqua, C., Calabrò, F., Della Spina, L. (eds.) NMP 2020. SIST, vol. 178, pp. 1105–1115. Springer, Cham (2021). https://doi.org/10.1007/978-3-030-48279-4_103

30. Nesticò, A., Moffa, R.: Economic analysis and operational research tools for estimating productivity levels in off-site construction [Analisi economiche e strumenti di Ricerca Operativa per la stima dei livelli di produttività nell'edilizia off-site], Valori e Valutazioni n. 20, pp. 107–128, ISSN: 2036-2404. DEI Tipografia del Genio Civile, Roma (2018)

31. Tajani, F., Morano, P., Di Liddo, F., Locurcio, M.: Un'interpretazione innovativa dei criteri di valutazione della dcfa nel Partenariato Pubblico-Privato per la valorizzazione del patrimonio immobiliare pubblico. LaborEst **16**, 53–57 (2018). http://dx.medra.org/https://doi.org/10.19254/LaborEst.16.09

32. Spampinato, G., Massimo, D.E., Musarella, C.M., De Paola, P., Malerba, A., Musolino, M.: Carbon sequestration by cork oak forests and raw material to built up post carbon city. In: Calabrò, F., Della Spina, L., Bevilacqua, C. (eds.) ISHT 2018. SIST, vol. 101, pp. 663–671. Springer, Cham (2019). https://doi.org/10.1007/978-3-319-92102-0_72

33. Ministero dei Beni e delle Attività Culturali e del Turismo: Piano Strategico di sviluppo del turismo 2017–2022. Italia Paese per Viaggiatori (2016)

34. Calabrò, F., Iannone, L., Pellicanò, R.: The historical and environmental heritage for the attractiveness of cities. The case of the Umbertine Forts of Pentimele in Reggio Calabria, Italy. In: Bevilacqua, C., Calabrò, F., Della Spina, L. (eds.) NMP 2020. SIST, vol. 178, pp. 1990–2004. Springer, Cham (2021). https://doi.org/10.1007/978-3-030-48279-4_188

The ICOMOS Draft International Charter for Cultural Heritage Tourism (2021): Reinforcing Cultural Heritage Protection and Community Resilience Through Responsible and Sustainable Tourism Management. New Approaches to Global Policies, Challenges and Issues Concerning Cultural Heritage Preservation and Enjoyment Within Tourism

Celia Martínez Yáñez[✉]

Art History Department, University of Granada, Granada, Spain
celiamarya@ugr.es

Abstract. The ICOMOS International Scientific Committee on Cultural Tourism decided to review the ICOMOS International Cultural Tourism Charter - Managing Tourism at Places of Heritage Significance 1999 at its Annual Meeting in Florence in 2017. The main objective of the revision was to update this document to the developments experienced by heritage theory and tourism practice during the past twenty years. In October and November 2021 the Scientific Council and Advisory Committee of the organisation have approved the resulting ICOMOS Draft International Charter for Cultural Heritage Tourism (2021): Reinforcing cultural heritage protection and community resilience through responsible and sustainable tourism management in view of its adoption at the ICOMOS General Assembly in 2022. The main aim of this paper is to reflect on the long process for drafting and approving this Charter and on its new conceptual approaches. Both show the emerging trends affecting cultural heritage globally and allow delving on the main issues and subjects that will characterize the future of cultural heritage tourism management and its links with today's environmental, social and cultural challenges.

Keywords: Cultural heritage · Cultural tourism management · Resilience, adaptation, and mitigation · ICOMOS Standard Setting Texts · COVID-19

1 Introduction

This paper aims to disseminate and expand the debate on the *[Draft] ICOMOS International Charter for Cultural Heritage Tourism 2021: Reinforcing cultural heritage*

C. Martínez Yáñez—Vice-president of the ICOMOS International Scientific Committee on Cultural Tourism & Coordinator of the ICOMOS Draft Charter for Cultural Heritage Tourism (2021).

© The Author(s), under exclusive license to Springer Nature Switzerland AG 2022
F. Calabrò et al. (Eds.): NMP 2022, LNNS 482, pp. 2361–2370, 2022.
https://doi.org/10.1007/978-3-031-06825-6_226

protection and community resilience through responsible and sustainable tourism management summarising its approval process and new conceptual approaches to cultural heritage conservation within tourism[1]. To do so we will delve into the reasons for the update of the previous ICOMOS International Charter on Cultural Tourism (1999), on the drafting evolution of the new one, on the impact that the COVID-19 has had on it, and on the main ideas behind its principles. We will use the Draft ICOMOS International Charter for Cultural Tourism 2021 itself, and our experience as its coordinator, as the main basis of the paper since, due to the novelty of the document, there are no analysis about it yet.

Having the symposium and its ICOMOS Heritage Days' ethos and subjects in mind, the paper will particularly reflect on three subjects:

– The [Draft] ICOMOS International Charter for Cultural Heritage Tourism 2021 approach to the social, economic, cultural, and environmental issues that are currently and increasingly re-framing and re-orienting cultural tourism and heritage preservation: Sustainable development strategies and the SDGs, climate change adaptation and mitigation solutions, and the stakeholder-centred approach to heritage practice.
– The challenges for cultural heritage planning and management posed by global tourism uncertainty and recovery (between overtourism and tourism disruption), with a particular attention to heritage destinations' carrying capacity and communities resilience and adaptation.
– How can the [Draft] ICOMOS International Charter for Cultural Heritage Tourism 2021 future implementation contribute to transform the unsustainable and unfair aspects of the tricky relationships between tourism, cultural heritage preservation and communities wellbeing and to enhance the positive ones.

2   Why a New ICOMOS Charter on Cultural Tourism?

2.1   The Goals of the Review

The ICOMOS International Scientific Committee on Cultural Tourism (ICTC) decided to review the ICOMOS International Cultural Tourism Charter - Managing Tourism at Places of Heritage Significance 1999 to update it to current tourism evolution and problematics in its 2017 Annual Meeting and Scientific Symposium in Florence, which was devoted to that subject. To address pertinent issues and progress on the review a Charter Review Taskforce, coordinated by the author, was appointed.

The main reason for the review was the unprecedented growth experienced by tourism in the previous twenty years and the need to avoid ill planed and unmanaged tourism and mass tourism negative impacts on cultural heritage and communities. A very important

---

[1] The Draft Charter has been drafted by the ICOMOS International Committee on Cultural Tourism through a task force composed of the following members: Celia Martínez (Coordinator), Fergus Maclaren (President of the ICTC), Cecilie Smith-Christensen, Margaret Gowen, Jim Donovan, Ian Kelly, Sue Millar, Sofía Fonseca, Tomeu Deyá, Ananya Bhattacharya and Carlos Alberto Hiriart.

ethos of the review was also the necessity to stress and support the requirements of cultural heritage conservation and protection within tourism, requirements that are implicitly understood but generally poorly respected and developed within this activity. On the one hand, since 1999, cultural heritage has become a "must" for any kind of tourism and leisure activity incrementing visitation until unexpected levels. On the other hand, by 2017, heritage conservation, visitors and sustainability concepts have been extended, and citizen and communities' demands and active involvement on heritage have constantly grown, all of which was demanding a general update of the document. There were a clear consensus that four main conditions were required to reach improvements concerning these aspects: 1) Pursuing a stronger cooperation among all actors concerned in tourism 2) Persuading the tourism industry to contribute to heritage preservation; 3) Fostering the identification and implementation of cultural heritage and communities' carrying capacity indicators able to inform the management of visitors and destinations; 4) Enhancing capacity building concerning the requirements for cultural heritage protection, valorisation and participatory governance among all tourism stakeholders, decision makers and communities.

With these guiding ideas in mind, the four main objectives of the first two years of the revision were:

- To update the 1999 International Cultural Tourism Charter to include current trends in heritage identification, protection, dissemination and sustainable, participative and balanced development within its principles, enhancing the following: Equitable gender participation and benefit in the cultural tourism domain; Reinforcing support for all communities affected by/participating in cultural tourism; Climate action; Carrying capacity several dimensions and application to the management of visitors and destinations; Professionalization of heritage dissemination and interpretation within the tourism field; Conservation awareness among the tourism sector.
- To analyse and study the alignment of the resulting document with ICOMOS and other relevant organizations' principles in this field: ICOMOS standard setting texts (ICOMOS s.d), UN Sustainable Development Goals and the Agenda 2030 (UN 2015), UNWTO doctrine and recommendations (UNWTO 2019), UNESCO Sustainable Tourism Programme and its future Visitor Management Assessment Tool (VMAT) (UNESCO 2022), among others.
- To raise the understanding and level of awareness about the several degrees of tourism's impact (positive and negative) on heritage properties and to encourage ICOMOS International Scientific Committees involvement in this process, as most of their thematic areas are impacted in some way or another by tourism.
- To support and empower ICOMOS members in efforts where tourism is a concern, reach a wider ICOMOS integration and alignment in the resulting document, and achieve a better understanding of current tourism problematics within ICOMOS and beyond.
- To produce an updated Charter able to confront the several complex and multilayered facets of heritage preservation and visitor management and to enhance ICOMOS/ICTC credibility and relevance in a tourism and conservation scenario that has changed dramatically this century (and that was about to drastically change again).

The reasons and aims of the review and the progress made to update the 1999 International Cultural Tourism Charter were presented at the Scientific Council and Advisory Committee meetings during ICOMOS General Assemblies of 2017 in New Deli, 2018 in Buenos Aires, and 2019 in Marrakech. In the later, the first version of the review launched a debate on whether the result was a major or minor change of the original document. The ICOMOS Board decided that the proposed draft was a major change and that the reasons for the review were strong enough to draft a new charter according to present time problematics and characteristics of tourism, while maintaining the 1999 Charter as an historic and pioneer standard setting text of its time, still relevant in many of its principles. It was also decided that the document would be circulated and adopted according to article 10 of the Rules of Procedure of the International Council on Monuments and Sites (ICOMOS 2019: 8–9).

2.2   The Impact of the COVID-19: The Draft of a New Document

The Charter Review Taskforce circulated the revised draft among ICTC membership in November 2019 introducing its input in the second draft. We were ready for the first round of consultation with ICOMOS International Scientific Committees, National Committees, Working Groups and membership when the COVID-19 deeply impacted our lives, work, and heritage conservation practice worldwide.

The ICT Charter Review Taskforce decided not to circulate this second draft of the revision until the text would be updated to the new situation, taking into account COVID-19's huge global impact on tourism: restrictions on travel, the new tourism national and international trends, i.e. the unprecedented fall of international tourism arrivals, the rise of domestic and local tourism, citizens and tourists preference for rural, natural and open-spaces, new tourismophobia and its targets, the decline in tourism revenues and the subsequent effect on funding for conservation and job losses. COVID-19 also affected several key subjects of the Charter that had to be re-thought: mass tourism, cultural heritage visitation conditions, carrying capacity, limits of acceptable change, site interpretation and presentation, climate change, cultural tourism's positive and negative effects on heritage sites and communities, communities' involvement and profit from tourism, resilience and capacity of adaptation, etc. In addition, the first draft had a strong focus on mass tourism's negative impacts on cultural heritage conservation and significance, and on the need to better manage tourism, but COVID-19 unexpectedly changed the situation: it was thought to be the end of mass tourism in certain countries and it forced to re-designing the flows and management of visitors according to safety measures. The impacts of COVID-19 also led us to include cultural heritage sector respond to COVID-19 and the recommendations that ICOMOS and other organizations, such as UNESCO and the UNWTO, were developing to confront these challenges.

All in all, the shift of approach and the need to carefully regard the new priorities compelled us to delay the consultation and approval process for a year and to establish a Drafting Committee in charge of writing a new draft (see note n.1), circulating the document for comments and introducing the input received in a new draft. The second draft of the Charter was distributed among the ICOMOS International Scientific Committees, National Committees, Working Groups and membership, as well as among some partner organizations, during April and May 2021. The third and final draft Charter has been substantially improved thanks to the input received from ICOMOS Italy, ICOMOS Finland, ICOMOS Australia, ICOMOS Norway, ICOMOS Venezuela, ICOMOS France and ICOMOS Spain; from the ICOMOS ISCs on Vernacular Architecture (CIAV), Archaeological Heritage Management (ICAHM), and on Training (CIF); from 12 individual ICOMOS members coming from all the regions of the world and from many other ISCs; and from 25 ICTC members. It has also been enriched by the input of several partner organizations: the IUCN TAPAS Group, the World Monuments Fund, the Cátedra UNESCO de Turismo Cultural Untref-Aamnba (Argentina), the Romualdo del Bianco Foundation, and representatives of the UNESCO World Heritage Sustainable Tourism Programme, whose involvement show the importance of this Charter[2].

Finally, the [Draft] *ICOMOS International Charter for Cultural Heritage Tourism (2021): Reinforcing cultural heritage protection and community resilience through responsible and sustainable tourism management*, including an annexure of useful references and international recommendations, was distributed by the International Secretariat across ICOMOS and approved by the ICOMOS Scientific Council and Advisory Committee in October and November 2021 and by the ICOMOS Board in March 2022 in view of its adoption at the ICOMOS General Assembly in 2022.

3   The Draft ICOMOS International Charter for Cultural Heritage Tourism (2021): Reinforcing Cultural Heritage Protection and Community Resilience Through Responsible and Sustainable Tourism Management

3.1   Structure and Overview of the Charter

The [Draft] ICOMOS International Charter for Cultural Heritage Tourism (2021) (The Draft Charter from now on) consists of a preamble, three objectives, a section targeting its audience, a background and seven principles. It aims to be universally applicable, taking into account existing and future regional guidelines. To ensure language inclusivity, it is available in English, French and Spanish, with translations into Italian, Chinese Mandarin, Arabic, Portuguese, Danish and Norwegian in progress.

The preamble defines cultural heritage tourism as "all tourism activities in heritage places and destinations, including the diversity and interdependence of their tangible, intangible, cultural, natural, past and contemporary dimensions". It recognizes heritage as a common resource, "understanding that the governance and enjoyment of these commons are shared rights and responsibilities". It also addresses the intensified tourism use of cultural heritage places and destinations and increasing concerns about the degradation of cultural heritage along with social, ethical, cultural, environmental and economic rights issues associated with tourism (ICOMOS ICTC 2021: 1).

These questions are further developed in the background, which highlights that "when responsibly planned, developed and managed through participatory governance,

---

[2] The compilation of the input received was assisted by Lorenza Stanziano (ICOMOS Spain) through the ICTC Emerging Professional Mentorship Initiative (EPMI).

tourism can provide direct, indirect and induced benefits across all dimensions of sustainability. However, unmanaged growth in tourism has transformed many places throughout the world, leaving tourism-dependent communities significantly altered and less resilient" (Ibid: 3). This is mostly due to the phenomenon of mass tourism and 'overtourism', which, together with the widespread promotion, marketing and use of cultural heritage, have caused a pervasive congestion and unacceptable degradation of tangible and intangible heritage, with associated social, cultural and economic impacts. These include gentrification and the rapid and insensitive commodification, commercialization and overuse of local culture and heritage, which "have placed irreplaceable assets, communities and cultural integrity at risk and resulted in negative and disruptive effects". Among the worst of these effects stand out the restrictions on rights of use, access to and enjoyment of cultural heritage by local people and visitors alike (Ibid).

The context within which these matters have been considered in the Draft Charter includes "the climate emergency, environmental degradation, conflicts, disasters, the disruptive effects of the COVID-19 pandemic, mass tourism, digital transformation and technological developments" (Ibid). These issues show that there is a need and opportunity to "recalibrate the perpetual economic growth-based approach to tourism, recognizing and mitigating its unsustainable aspects" (Ibid). This is why the Draft Charter is formulated within the Sustainable Development Goals and the UN's 2030 Agenda, the Climate action, and the right based approaches to participatory governance and cultural heritage management.

The Draft Charter aims to provide guidance for governments, decision makers within international, national and local government agencies, organizations, institutions and administrations, tour operators, tourism businesses, destination managers and marketing organizations, site management authorities, land-use planners, heritage and tourism practitioners, professionals, civil society and visitors, all of which share the responsibility to conduct cultural heritage tourism responsibly. It also aims to be a reference for educators, academics, researchers and students engaged with cultural heritage and tourism. It applies to the management of all cultural heritage properties and to the entire spectrum of their protection, conservation, interpretation, presentation and dissemination activities, since all are connected with, and influenced by, public use and visitation (Ibid: 2).

The main objective of the Draft Charter is to place the protection of cultural heritage and community rights at the heart of cultural heritage tourism policy and projects, by providing principles that will inform responsible tourism planning and management for cultural heritage protection, community resilience and adaptation. To do so it aims to align the work of cultural heritage and tourism stakeholders in the pursuit of positive transformative change, offering principles for regenerative tourism destination management that is conscious of heritage values, as well as their vulnerability and potential. It seeks the fair, ethical and equitable distribution of tourism benefits to and within host communities, contributing towards poverty alleviation and promoting the ethical governance of cultural heritage and tourism. To achieve these crucial aspirations, the seven principles of the Charter provide a framework for guidance on this subject that is not present in other documents concerning cultural heritage or tourism, calling for their integration into all aspects of cultural heritage tourism (Ibid).

We will deal with these principles, their conceptual innovations and their links with the global policies and problems affecting cultural heritage, tourism and life quality in the following section.

3.2   The Principles of the Charter: Global Policies and Problems Affecting Cultural Heritage, Tourism and Communities Wellbeing

- Principle 1: Place cultural heritage protection and conservation at the centre of responsible cultural tourism planning and management;
- Principle 2: Manage tourism at cultural heritage places through management plans informed by monitoring, carrying capacity and other planning instruments;
- Principle 3: Enhance public awareness and visitor experience through sensitive interpretation and presentation of cultural heritage;
- Principle 4: Recognize and reinforce the rights of communities, Indigenous Peoples and traditional owners by including access and engagement in participatory governance of the cultural and natural heritage commons used in tourism;
- Principle 5: Raise awareness and reinforce cooperation for cultural heritage conservation among all stakeholders involved in tourism;
- Principle 6: Increase the resilience of communities and cultural heritage through capacity development, risk assessment, strategic planning and adaptive management;
- Principle 7: Integrate climate action and sustainability measures in the management of cultural tourism and cultural heritage.

Principles 1, 2 and 5 stress that awareness and understanding of long-term protection and conservation requirements of heritage places is necessary in tourism planning and management. Cultural heritage is a significant resource for tourism and plays a major role in the attraction of travel, but its fragility and conservation requirements are insufficiently recognized. The aim of these principles is therefore to provoke a shift from the current focus on heritage tourist promotion and perpetual growth towards heritage tourism destination planning and visitor management through the carrying capacity and limits of acceptable change approaches. The most important innovations in this regard come from: 1) The call for the coordination between tourism planning and cultural heritage management across all levels of governance in order to identify, assess and avoid the adverse impacts of tourism on heritage fabric, integrity and authenticity. Heritage and Environmental Impact Assessments must inform the planning and development of tourism, going beyond the legal boundaries of cultural heritage properties to include infrastructure projects and management plans that might affect the integrity, authenticity, aesthetic, social and cultural dimensions of heritage places, including their settings, natural and cultural landscapes, host communities, biodiversity characteristics and the broader visual context. 2) The proposal to reinforce cooperation for heritage conservation within the tourism industry and other stakeholders, encouraging investors and enterprises to dedicate part of their revenues to this aim and to the responsible enhancement and promotion of less visited heritage places. Tourism' revenues must provide benefits to local communities and be collected and allocated in a transparent, fair, equitable and accountable manner making visitors aware of their contribution to cultural heritage funding and maintenance. 3) The identification of the several dimensions of carrying capacity (physical, ecological, social, cultural and economic) and their monitoring through ad hoc indicators. The monitoring of these indicators has to guide destination planning, the flows and management of visitors and the set up of site specific actions to limit group sizes, time group access, restrict entry, close sensitive areas, restrict or increase opening hours, zone compatible activities, require advance bookings, regulate traffic and/or undertake other forms of supervision.

Principle 3 proposes several measures to conduct the interpretation and presentation of heritage sites professionally and within an appropriate certification framework, bearing in mind that they are key to increasing site awareness and resolving possible conflicts and needs. Interpretation and presentation have to be inclusive and based on "interdisciplinary research, including the most up-to-date science and the knowledge of local peoples and communities" (Ibid: 8). They also have to address "conservation and community rights, issues and challenges, so that visitors and tourism operators are made aware that they must be respectful and responsible when visiting and promoting heritage" (Ibid). Interpretation methods should not detract from the authenticity of the place, but they can use appropriate, stimulating and contemporary forms of education and training, networks, social media and technologies, including augmented reality and virtual reconstructions based on scientific research (Ibid).

Principles 4 and 6 go beyond the 1999 International Cultural Tourism Charter focus on communities' participation to claim for the participatory governance of heritage commons. This fosters a shift of approach to this subject, which is consistent with ICOMOS current priority to emphasize right and people centred approaches to cultural heritage (ICOMOS 2017) and based on the principle of free, prior and informed consent' (United Nations Declaration on the Rights of Indigenous Peoples 2007). This means the inclusion of all types of communities in cultural tourism decision-making, taking into account the diverse and often opposing interests of experts, professionals, host communities and tourists, as well as a wide range of local, economic and political actors. As a result, capacity building in the field of cultural heritage and responsible tourism for all these kind of communities and actors needs to be urgently strengthened so that their participatory management is real, fair, balanced and able to increase and defend heritage preservation. Capacity building should also aim to "increase the ability of communities to foresee and reduce risks and to make informed decisions concerning cultural heritage management and tourist use of resources to minimize the negative societal and economic impacts of disruption or intensification of use" (Ibid: 11). This is particularly important considering "disruptions affecting tourism, ongoing systemic and pervasive global problems and emergent risks, all of which demand to enhance the resilience, adaptive and transformative capacities of communities to deal with future challenges related to climate change, loss of biodiversity and/or calamities that affect cultural heritage" (Ibid). In addition, tourism planning has to engage with and apply extensively heritage impact assessment (HIAs), environmental impact assessment (EIAs) and other relevant risk assessments, as well as mitigation and reduction measures which involve all stakeholders. An special consideration should also be paid to design models for assessing climate change impacts on cultural heritage, "which will become increasingly important in the future" (Ibid).

Finally, and conscious that climate emergency is an existential threat to the planet and the civilization "that jeopardizes cultural and natural heritage and threatens the livelihoods and wellbeing of people across the world" (Ibid: 12), Principle 7 focuses on reducing climate change impacts of tourism and to support the implementation of the 2030 United Nations Sustainable Development Goals dealing with tourism and cultural and natural heritage. This is a "shared responsibility of governments, tour operators, tourism businesses, destination managers and marketing organizations, site management authorities, land-use planners, heritage and tourism professionals, civil society and visitors". Consequently "tourism and visitor management must contribute to effective carbon and greenhouse gas reduction, waste management, reuse, recycling, energy and water conservation, green transport and infrastructures" (Ibid). Other interesting and still scarcely implemented proposal is to including messages about climate impacts on cultural heritage preservation within the presentation and interpretation of heritage places open to the public, making use of innovative technologies if appropriate (Ibid).

4   Conclusions

The main conceptual shifts and contributions of the [Draft] ICOMOS International Charter for Cultural Heritage Tourism 2021 are based on an aspirational desire of placing cultural heritage at the center of tourism activity, regarding it from the perspective of heritage commons and responsible management. Also on the reorientation from sustainable development towards responsible tourism management and collective and individual wellbeing, all of which depends highly of the humanity's ability to urgently stop climate change. These changes imply going beyond the purely economic dimension of development to base the assessment of cultural heritage tourism success in its effects on life quality, cultural heritage preservation and environmental awareness.

The Draft Charter demands the participatory governance of cultural heritage used by tourism considering ICOMOS itself as a community that defends a more humanistic, fair, solidary and sensible approach to tourism. Overall, it calls and provide principles "for transformative change towards a regenerative development paradigm that recalibrates the perpetual economic growth-based approach to cultural tourism" (Ibid: 4). As the background of the 2021 Charter notes "Any cultural tourism strategy must accept that cultural heritage protection, social responsibility and 'sustainability' are not merely options or brand attributes, but rather necessary commitments and, in fact, a competitiveness asset. In order to remain successful and sustainable in the long term, cultural tourism proponents must put this commitment into practice and become a force that supports community resilience, responsible consumption and production, human rights, gender equality, climate action, and environmental and cultural heritage conservation" (Ibid: 3).

COVID-19 has had a great influence on the Draft Charter, emphasising these conceptual changes that were on progress before the tourism collapse and accelerating the need to deal with them. As stressed by the principle 6 of the Charter, the massive decline in tourist activities due to the Covid 19 pandemic has exposed the vulnerability of many heritage places and the communities hosting cultural tourism, demonstrating that tourism must actively contribute to recovery, resilience, heritage conservation and life quality, and that heritage places and host communities must consider adaptation options (Ibid: 10). In fact, the pandemic suggested the advent of a new world and a regenerative and responsible tourism, which was contradicted by the first postCOVID-19 recovery and increasing numbers of visitors in the always same and unchanged most famous cultural destinations, and in rural and natural landscapes too. This apparent return to mass visits to the always crowded heritage sites, together with the restrictions, obstacles and uncertainty that the new variants could impose again, show that the disease and/or other disasters could continue affecting tourism, the ideas behind this document and its implementation.

We hope that, once is approved, the Charter can be operationalised globally and supported by regional guidelines and that its adoption and implementation will contribute to addressing the global issues that are affecting and will continue to affect cultural heritage, the environment, people and tourism in the years to come.

References

ICOMOS: Charters and other doctrinal texts (s.d). https://www.icomos.org/en/what-we-do/involvement-in-international-conventions/standards. Accessed 6 Jan 2022

ICOMOS: ICOMOS International Cultural Tourism Charter (Managing Tourism at Places of Heritage Significance). Adopted by ICOMOS at the 12th General Assembly in Mexico, October 1999. http://www.icomos.org/charters/tourism_e.pdf. Accessed 6 Jan 2022

ICOMOS: Case Studies Carried Out Within the 'Our Common Dignity Initiative 2011–2016: Rights-Based Approaches in World Heritage'. Rapports Techniques. ICOMOS Norway, Oslo (2017)

ICOMOS: Rules of Procedure of the International Council on Monuments and Sites (2019). https://www.icomos.org/images/DOCUMENTS/Secretariat/2018/Rules_of_Procedure/ICOMOS_RulesOfProcedure_EN_20191122_amended.pdf. Accessed 6 Jan 2022

ICOMOS ICTC: [Draft] ICOMOS International Charter for Cultural Heritage Tourism (2021): Reinforcing cultural heritage protection and community resilience through responsible and sustainable tourism management. https://www.icomosictc.org/. Accessed 6 Jan 2022

UN: United Nations Declaration on the Rights of Indigenous Peoples (UNDRIP) (2007). https://www.un.org/development/desa/indigenouspeoples/declaration-on-the-rights-of-indigenous peoples.html. Accessed 6 Jan 2022

UN: Transforming our World: The 2030 Agenda for Sustainable Development (2015). https://sdgs.un.org/publications/transforming-our-world-2030-agenda-sustainable-development-17981. Accessed 6 Jan 2022

UN/FAO: Free, Prior and Informed Consent (FPIC) – An Indigenous Peoples' right and a good practice for local communities. Manual for project practitioners (2016). https://www.fao.org/3/i6190e/i6190e.pdf. Accessed 6 Jan 2022

UNESCO: World Heritage and Sustainable Tourism Programme (s.d). https://whc.unesco.org/en/tourism/. Accessed 6 Jan 2022

UNWTO: Compilation of UNWTO Declarations, 1980 – 2018. UNWTO, Madrid (2019). https://doi.org/10.18111/9789284419326

The Haenyeo Community: Local Expert Facilitator for Tangible-Intangible Cultural Heritage and Its Economic Contributions to the Jeju Society, Korea

Hee Sook Lee-Niinioja[⊠]

Helsinki, Finland
leeheesook@hotmail.com

Abstract. "Culture of Jeju Haenyeo (women divers)" in Korea was inscribed in 2016 on the UNESCO Representative List of the Intangible Cultural Heritage of Humanity. Jeju Island is a volcanic area off the southern coast of Korea, and a women diving community collects seafood under the sea. Each haenyeo has her mental sea map and commands local knowledge on the winds and tides without oxygen masks. The knowledge is acquired through repetitive diving experience, transmitted to generations. Moreover, representing the island's character and spirit, the culture of Jeju haenyeo has enhanced women's status and environmental sustainability. They imprint an indispensable part of Jeju identity since most families have facilitated haenyeo. Questions arise on haenyeo's contributions to the Jeju society through the tangible place (bulteok) and the intangible tradition (Jamsugut ritual) as a cultural space. It questions further the impact of the current natural (climate change) and social (pandemic) circumstances on haenyeo's life: how are these factors integrated into them, regardless of Jeju Island as a tourist place worldwide. Among many possible solutions, strengthening Haenyeo Association as a sustainable community seems the best choice. This paper discusses Jeju haenyeo in cultural identity, heritage, collective memories and community bearers in the Jeju society.

Keywords: Jeju Haenyeo · Tangible Bulteok · Intangible Jamsugut · Haenyeo association · Cultural identity · UNESCO ICH

1 Culture of Jeju Haenyeo (Women Divers)

1.1 History

Jeju Island (Fig. 1) populates about 600,000, and women haenyeo earn a living from the sea. The haenyeo's first appearance was not sure due to the lack of historical documents, but their diving-associated vocation presumably existed before the foundation of the Tamna Kingdom in the most southern part of the Korean peninsular 2,000 years ago.

Until the mid-17 century, men divers also collected seafood before transferring to women's primary task. In the Joseon era, Lee Gun's *Jeju Topography* (1629) wrote

© The Author(s), under exclusive license to Springer Nature Switzerland AG 2022
F. Calabrò et al. (Eds.): NMP 2022, LNNS 482, pp. 2371–2382, 2022.
https://doi.org/10.1007/978-3-031-06825-6_227

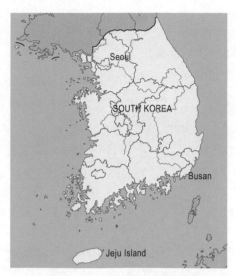

Fig. 1. Map of Korea with Jeju Island. Image: NordNordWest, edited by the Author.

hacnyeo's gathering abalones. During his governorship (1694–96), Lee Ik Tak recorded in *Jiyoungro*k his ordering dried abalones from jamnyo (haenyeo). Moreover, on inspecting the Jeju coastline, Lee Hyun Sang, the Jeju governor (1702–03), published *Tamna Sullyeokido* (1702, Fig. 2) as the earliest surviving drawings. One of the 42 portrays Jeju haenyeo.

1.2 Haenyeo

As additional material for a deeper understanding of the UNESCO ICH inscription, the Korea Cultural Heritage Foundation has published a book, "Haenyeo" (2016). An average haenyeo holds her breath for one minute, diving 10 m underwater to gather marine products without oxygen masks. When a haenyeo exhales after each dive, she makes a sound. A haenyeo works for six-seven hours a day in summer and four-five hours in winter. She dives about 90 days a year.

A haenyeo develops her map of reefs and the habitat for shellfish, local knowledge on the winds and tides of the sea (Fig. 3). Three categorizations (sanggun, jungun, hagun) find in the haenyeo's capacity levels, and the less experienced haenyeo consult with the most skilled sanggun about the weather forecast instead of the official forecasts and guidance. However, haenyeo communities need to perform a shamanistic ritual for the sea goddess (Jamsugut), including prayers for safety at sea and an abundant catch before diving. Songs rowing out to the sea is part of Jeju people's cultural identity, while haenyeo, jumping into the wild sea depending on a small buoy (tewak), symbolizes the spirit. It is no wonder why the Jeju Provincial Government designated them to represent the character of Jeju Island.

Geographically, the land's volcanic soil is not fertile enough for large-scale agriculture; thus, haenyeo should be the primary labourer in their families, assisting the

Fig. 2. A painting of *Jamnyeo (Haenyeo)* doing *muljil* near Yongduam (Yi Hyeong-sang, *Tamna Sullyeokido*). Image: Scanned from Culture of Jeju Haenyeo, p.39.

advancement of women's status by the public appreciation of women's work in a male-centred Confucian culture in Korea. The independent and strong haenyeo is considered a symbol of gender equality. The eco-friendly searching method of haenyeo's diving work encourages sustainability, although the wish for a big catch is counterbalanced by an individual's limited ability to stay underwater. The culture of Jeju haenyeo exemplifies living in agreement with nature. They have a deep respect for each other: they are competitors in gathering products yet trusting companions in their dangerous sea work. They ascertain their group work to monitor the safety of nearby haenyeo during the dive. For the senior, the haenyeo associations designate a specific place as "the sea for grandmothers." The income of the collective work from the designated parts of the sea is for community projects, such as building primary schools for community children, testifying to solidarity and harmony among the haenyeo and their communities.

Fig. 3. Haenyeo, female diver diving and collecting seafood under the sea for living. Image: (https://www.flickr.com/people/79121222@N00)

Haenyeo demonstrates human creativity in natural resources, improving her diving skills based on her physical condition. Local knowledge on marine ecology has been transmitted-accumulated to the communities. Moreover, a worldview has shaped the traditional way of life by fusing indigenous beliefs and values. Shamanism is the oldest belief. Good and evil spirits inhabit the world, and the shaman communicates with them through rituals involving incantation, music and dance.

In a word, Jeju haenyeo's culture refers to conducting muljil (diving underwater to catch seafood) and daily living. The tangible heritage contains bulteok (a resting place), tools, vessels, costumes, and Haeshindang (the shrine for the god). The intangible property is songs, tales and sayings, rituals of Yeongdeunggut and Jamsugut, and fishing knowledge.

1.3 UNESCO ICH Inscription and Its Contributions

The UNESCO ICH 2003 Convention guides four key definitions. First, intangible cultural heritage should be traditional, contemporary, and living simultaneously. It represents traditions and contemporary rural and urban practices in which diverse cultural groups participate. Second, intangible cultural heritage should be inclusive. People can share cultural expressions similar to those practiced by others beyond locations, such as the neighbouring village or a city on the opposite side or migrating settlers in a different region. It has been passed in generations and evolved in response to their environments, endowing identity and continuity. It connects people toward their future. Third, intangible cultural heritage should be representative. It is not a simple cultural product for its exclusivity or exceptional value, but it flourishes within communities and depends on those whose knowledge of traditions, skills, and customs are transferred to the rest of the community. Fourth, intangible cultural heritage should be community-based. It is only recognized by the communities, groups, or individuals that create, maintain, and transmit it (https://ich.unesco.org/en/what-is-intangible-heritage-00003).

"Culture of Jeju Haenyeo" was inscribed on the UNESCO Intangible Cultural Heritage of Humanity 2016, expecting the following contributions to Jeju people and outsiders. (1) Koreans have become conscious of the importance of local lifestyle and information as integral parts of the nation's cultural heritage. (2) The inscription facilitates the international visibility of intangible cultural heritage on local knowledge and practices. It raises alertness of women's work as intangible cultural heritage and supports sustainable growth. (3) Haenyeo learns their diving work as hard labour and a cultural heritage entitled to worldwide credit. (4) Communities are encouraged to value local ways of natural resources. (5) Respect for cultural diversity on indigenous, folk, and local knowledge and practices will encourage dialogue among communities worldwide (https://ich.unesco.org/en/RL/culture-of-jeju-haenyeo-women-divers-01068).

2 Cultural Identity, Heritage, and Collective Memories

Culture consists of behavioural patterns constituting the characteristic attainment of human groups. Acquired-transferred by symbols, it has traditional ideas and their attached values. Culture systems can be regarded as productions of action or conditioning influences on further action. Culture as a visual language includes its embodiments in artefacts (Geertz 1973). Moreover, cultural identity has been formulated through cultural heritage, whose legacy confers monuments, objects, and traditions or living expressions. Tangibility is built heritage and other physical products of human creativity with cultural significance; intangibility concerns the practices, representations, expressions, knowledge, object, and cultural spaces. Transferred-recreated through generations, it promotes a sense of identity, continuity, and respect for cultural diversity.

Identities in culture bear a heritage character and the combined element of the locality. A relationship between culture, place identity and participants' representation maintains local identity as a combination of historical, social, economic and political processes. Cultural identity underlines the similarities shared (Pritchard and Morgan 2001). Identity arises from a place with a history and posits without a fixed essence. This identity generated in the past goes under place, time, history and culture (Hall 2003).

Remembrance and commemoration of the past are the present's nucleus. It ties to identity and is an innate part of the heritage process as people remember the past (Walker 1996). Without memory, a sense of self, identity, culture and heritage is lost (Lowenthal 1985), and what is remembered is defined by "the assumed identity" (Connerton 1989).

In this regard, collective memory functions as a framework for studying societal remembrance, proposing a possibility of construction, sharing, and passing on by communities. Each collective memory relies on specific groups described by space and time. The group builds the memory, and the individuals do remember. In other words, individual memory is understood through a group context; collective memory develops further as people keep their history. Symbols, architecture, and literature are references for binding people to past generations and influencing their memory. A ritual is a succession of activities following a characteristic community tradition with sacral symbolism (Halbwachs 1925/1992).

Cultural identity, heritage and collective memories are intermingled and fused to mark the individuality-community of Jeju haenyeo in distinction.

3   A Tangible Cultural Space: Bulteok

Despite its original meaning as a site for bonfires, a bulteok (Fig. 4), an outdoor dressing site at coastlines, represents the Jeju haenyeo community. As haenyeo wearing conventional diving suits in cotton fabric conduct muljil for long hours, a bonfire is necessary near their workplace to warm up between their muljil operations. In certain villages, haenyeo sailed a boat to a far workplace and would make the bonfire onboard. However, an old-fashioned bulteok was replaced by a modern one with house showers and heating when haeneyo started to wear rubber diving suits from the mid-1970s.

A bulteok is a rounded, stone-walled space protected from the wind, where haenyeo light fires using firewood prepared to warm up and rest in small groups. They talk about their lives and prepare for their dive. It is also a place to educate each other on muljil and the sea ecology, town information and events. As a complex cultural space, the bulteok has different names depending on the material to build it. If stone walls surround it, it calls "doldam bulteok."

The convenient position of doldam bultoek is near the sea and in protected places from the strong wind. The rocky ground is too hard for the haenyeo to walk barefoot in their diving suits and to carry their tewak buoy and firewood. Bulteok should be within walking distance from the muljil workplace for haenyeo's carrying their mangsari full of seafood.

The preserved bulteok is a one-sided, straight, stone wall or a square with an open side, similar to the Korean alphabet "digeut." In the case of a round-shaped bulteok, the opening is close to the sea for haenyeo's easy entering and exiting from the site.

Fig. 4. Pyongdaeri bulteok. Image: Visit Jeju. https://m.visitjeju.net/kr/detail/view?contentsid= CNTS_000000000021512

A squared bulteok is a little curved on one side covered with stone walls that haenyeo detoured along with them. The height of the stone wall is a person's height to look inside when haenyeo rest. They would only see their heads from the outside.

Since 1970, haenyeo's adopting rubber diving suits and modernized house heater has changed their working hours. They could work five hours in the sea and spend less time at the bulteok, resulting in the old-fashioned bulteok's functional loss. Although its physical disappearance has transferred to haenyeo's collective memories, the doldam bulteok sites still exist in many fishing villages, renovated and used by some haenyeo.

As a bulteok facilitated the firewood storage and a joint chat among haenyeo, it adheres to rules and regulations. The seats were distributed according to their skill level. For sanggun (the best), it was called "sanggundeok"; for junggun (the middle) "junggundeok"; for hagun (the lowest) "hagundeok". Around bulteok, haenyeo shifted to their wetsuits, got their equipment ready while talking, and ran towards the sea with tewak around their waists. After about 40 min' collecting labour, they changed into dry clothes in the bulteok site and lit a fire to warm up. When they were warming up, they only wore wetsuits. They used a thick ttudegi outfit to thaw their frozen bodies in the wintertime. When the tide was good, they needed many old cotton diving suits. In reality, they owned a few and would dry them by the fire before changing into a different one (Byun et al 2015, Han 2013).

In a word, Bulteok has been a place of knowledge transmission from experienced divers to new haenyeo since ancient times. Throughout winter, the bulteok was critical for their warmth and shelter from the wind. It was the resting place where haenyeo could converse in other seasons sharing their stories of diving and superstition.

4 An Intangible Cultural Memory: Jamsugut

Haenyeo regards the sea as nature with respect and gratitude. The sea as their livelihood has been expressed in folk beliefs. Jamsugut (Fig. 5) is a shamanistic ritual in Jeju coastal villages and is held between January and March of the lunar calendar, but the date can differ by the village. As a grouped ritual for all the villagers, Jamsugut is a festival by haenyeo for their safety and good fishing for the year, compared to Yeongdeung by the whole society.

The ordinary ritual steps of Jamsugut can be seen in Donggimneyong-ri, where the performance takes all day. At dawn, the participants bring the sacrifices to the altar, set all pillars, and begin the ritual. Besides wishing for no accidents and a bountiful fishing harvest, the ritual prays for the excellent relationship of the community. Jamsugut has two characteristics. First, the village men and the members of the Fishing Village Cooperative join in praying with them, prepared and managed by each village's haenyeo association. Second, haenyeo displays a firm bond as a community with passion and seriousness to hold the ritual. Due to the haenyeo association's operating Jamsugut, all haenyeo stay in the headquarters and pray for wellness throughout the day. Other villagers, including community leaders, visit them with gift money and liquor donations and take a bow to haenyeo. It is the day for the villagers to pray for safety, a good fishing harvest, and a big festival to strengthen the bond of all the villagers and enjoy a meal together. It is "Haenyeo's Day".

Fig. 5. Jeju Island shamanic ritual to Yeongdeung, god of wind, UNESCO Intangible Cultural Heritage 2009. South Korea's Cultural Heritage Administration. Image: Wikimedia.

In practice, with seven days left before the ritual day, five-six leaders of the haenyeo association put all preparations as the ritual host. Every house is decorated with a horizontal straw rope to block bad luck or death in the family. The leader haenyeo hardly leave their homes and care about food. On the eve of the ritual day, the sacrifices are prepared at the president house of the haenyeo association. About four rice bags make different types of rice cake and are dedicated to Jamsugut. For the ritual expenses, haenyeo raises money and receives donations from other villagers. In addition, every time they collect seafood, they save some percentage of their profit in advance and use it to continue the following year's ritual instead of raising funds every year.

5   The Haenyeo Association: Local Expert Facilitator for the Living Community

Jeju haenyeo are the socioeconomic agents whose vocation involves collecting shellfish and contributing to their community development. Each village creates a self-sustaining organization, "the haenyeo association," classified as the respective Fishing Village Cooperatives. The organizations (1) assist haenyeo with the Fishing Village Cooperatives, (2) control-secure their duties and rights, (3) manage the fishing grounds, (4) regulate the diving rights, (5) coordinate the diving period, (6) represent the collective opinion, and (7) advocate the interests.

In principle, the cooperative nature of diving work is through consensus. The haenyeo association in each village is a citizens' assembly and a community for economic production. Members discuss and decide the dates to start and end the diving season at the meetings, including working hours and the minimum size of the catch and forbidding technology for excessive fishing. When community events such as weddings or funerals occur, the association determines their suspending work from the sea. The general meeting is held every year but is more irregular, convening haenyeo to near bulteok when preparing or finishing their muljil.

In Jeju haenyeo, muljil demonstrates a communal trait. It is an unwritten rule to share work with each haenyeo facing similar characteristics. The favourable time and hours for haenyeo's diving is decided on as a group. Although it is an individual's choice of the muljil operation and no one puts pressure on them, its task has a communal element due to its operative nature. Haenyeo can seek practical benefits from the association, maintain the order in their communities, and handle problems in the sea when acting on the agreed collective ideas.

The haenyeo association also looks after each diver's family events, the health, and the shared issues among the same village members. They collectively secure each member's economic interests and distribute duties to raise profits of their village communities. For example, they conserve the marine environment and resources to prevent water pollution and avoid overfishing against resource depletion. Haenyeo volunteers and supports the villagers' unification in profit-making businesses, including village festivals.

The haenyeo community combines a powerful bond of autonomy and independence. Its management depends on the justified authority and the decentralized power system. The association adopts a horizontal consensus among its members; thus, resolutions should come from all members' agreements. Decision-making is through democratic consensus and group dialogues; thus, the resolutions of the haenyeo association bears a powerful binding force, which each haenyeo must follow.

The association designates an area for over 60-year-old haenyeo only within a region near the coast to do muljil. It also allocates to the weak and sick haenyeo, who can no longer fish at the most shallow and safe sea. Above all, the practical function of the association regulates the overall procedures of diving without an oxygen tank. Other tasks include managing fisheries by getting rid of useless seaweed, supervising the sea during the banned period, and responding to fighting over the fishing domain.

Additional agendas are: manage the joint fund for the association, the village, or the shared facilities; support the villager-led ritual ceremony; share the rights to fish with the village newcomers; deal with the woman's fishing right, who left the village through marriage but later returned; punish those who fail to participate or violate the regulations on collective work; regulate the bulteok management; divide-use the fishery rotationally; designate the non-fishing days; determine haenyeo's release from her task by an accident and other members' dealing it.

6 Conclusion

It is hoped to combine the author's empirical experiences in Jeju Island, haenyeo's collective memory through interviews, UNESCO ICH inscription texts, and other research sources to approach this paper topic. Unfortunately, the strict regulations of the pandemic situation worldwide in these years have prevented the author's visit to Korea. Instead, this unexpected circumstance prompted the author to focus on written materials around haenyeo, keeping questions of the impact of climate change and the pandemic across Jeju Island.

Two articles can shed on light these questions. Rahman's "Haenyeo Divers Threatened by Climate Change" (June 14, 2021) informs that UNESCO ICH inscription 2016 attracts much tourism to Jeju, particularly haenyeo's working life due to their connection to the ocean. However, climate change's transforming the ocean could end this traditional practice. Seas off Korea's southern coast have warmed approximately 1.2 degrees Celsius, creating new species' arrival and the declination of the existing underwater habitat. The disappearance of large seaweed causes marine life to perish. Haenyeo, without oxygen tanks under the erosive environment, is compelled to dive deeper to catch their muljil, frightening their work capability.

One diver says: "With the rise in water temperatures, the diversity of species has dropped markedly. In the past, there was a lot of gulfweed and brown algae, which served as hideouts and spawning grounds for many fish. But brown algae have disappeared and been replaced by hard corals." The year 1960 was the culmination for Jeju haenyeo with 26,000 divers compared to the current 4,500 divers. Some women do not train their daughters on diving skills but choose to educate or send them to work in the city. Accordingly, environmental degradation has worried of the haenyeo's life, and the vanishing divers facilitate the disappearance of inherited generations' knowledge.

Another article, "Wisdom is needed to overcome the dilemma" in *Dream Jeju* (2020 summer), reveals how Jeju Island has sustained in nature and underlined solidarity actions facing COVID19. The Chairman of the Jeju Special Self-Governing Province Council concerns the global pandemic as a continuing inexperienced circumstance at length. Jeju citizens are suffering, and the medical sacrifices are at the limit. He hopes that sincere tourists visit Jeju as a driving force for the local economy's revitalization, while Jeju must do thorough quarantine activities to prevent the spread of the pandemic to the local community as follows:

> In July, the beautiful sea adds blue, and the green forest gives off a refreshing scent. As usual this holiday season, Jeju has been an island of rest for tourists who visit here with residents … Now we must gather wisdom to overcome this situation. Jeju is an island - an isolated space. Due to this nature, Jeju might be relatively safe from the spread of the virus. Jeju must further strengthen its quarantine measures. Residents and tourists must observe them.

Bearing these critical issues in mind, Jeju haenyeo's cultural identity, heritage, collective memories, and duties-agenda handling of the haenyeo association testify to a living heritage regardless of unforeseen natural or social circumstances. As manifested in the paper title, all the elements performed by individuals or communities play as custodians to care for Jeju haenyeo's tangible and intangible cultural heritage, facing the future.

References

Byun, K.H., Kang, E.J., Yoo, C.G., Kim, K.H.: Spatial transformation and functions of Bulteok as Space for Haenyeo on Jeju Island Korea. J. Asian Architect. Build. Eng. **14**(3), 533–540 (2015). https://doi.org/10.3130/jaabe.14.533

Connerton, P.: How Societies Remember. Cambridge University Press, Cambridge (1989)

Geertz, C.: The Interpretation of Cultures. Basic Books, New York (1973)

Halbwachs, M.: On Collective Memory. Chicago University Press, Chicago (1925/1992)

Hall. S.: Representation: Cultural Representations and Signifying Practices. SAGE Publications, London; The Open University, Thousand Oaks (2003)

Han, R.H.: Jeju Haenyeo community - Bulteok. Open Jejusi **81**, 12–13 (2013)

Korea Cultural Heritage Foundation: Culture of Jeju Haenyeo (Women Divers). K-Heritage. **4**, (2016)

Lowenthal, D.: The Past is a Foreign Country. Cambridge University Press, Cambridge (1985)

Pritchard, A., Morgan, N.J.: Culture, identity and tourism representation: marketing Cymru or Wales? Tour. Manage. **22**(2), 167–179 (2001). https://doi.org/10.1016/S0261-5177(00)00047-9

Rahman, I.: Haenyeo Divers Threatened by Climate Change, 14 June 2021. https://www.ajplus.net/stories/haenyeo-divers-threatened-by-climate-change. Accessed 20 Feb 2020

The Jeju Special Self-Governing Province Council: Wisdom is needed to overcome the dilemma. Dream Jeju, Summer 2020

UNESCO: What is Intangible Cultural Heritage? https://ich.unesco.org/en/what-is-intangible-heritage-00003. Accessed 20 Feb 2021

UNESCO ICH Inscription 2016 (Republic of Korea). Culture of Jeju Haenyeo (women divers). https://ich.unesco.org/en/RL/culture-of-jeju-haenyeo-women-divers-01068. Accessed 13 Dec 2021

Walker, B.: Dancing to History's Tune. Queen's University Belfast, Institute of Irish Studies (1996)

"Experts and Professionals": Intangible Cultural Heritage Custodians and Natural Resource Management in Poland and Canada

Agnieszka Pawłowska-Mainville[✉]

University of Northern British Columbia, 3333 University Way, Prince George, BC V2N 4Z9, Canada
Agnes.Pawlowska-Mainville@unbc.ca

Abstract. Intangible cultural heritage custodians play a main role in activities that deal with safeguarding of own cultural elements, including natural resources. These experts are knowledgeable about resource-use and sustainability, and make decisions based on their cultural values and familiarity of the area. However, these community experts rarely receive the same recognition and acknowledgement as 'professional' experts. Two case studies, one of Polish beekeepers and another of Anishinaabeg (Indigenous) harvesters in Canada, show how intangible cultural heritage can be used to strengthen local decision-making and reinforce sustainable livelihoods. Because intangible cultural heritage empowers communities, the framework may not only assist local experts in obtaining the recognition they deserve, but can also foster a better working relationship between States and communities in the area of natural resource management.

Keywords: Intangible cultural heritage · Land-based cultures · Sustainable livelihoods · Indigenous · Beekeepers

1 Introduction

Support for local community self-determination has been growing in popularity in recent years. Communities are reviving own governance structures and leadership to attain more influential decision-making power in the discourse of public interest, especially in heritage management. The 2003 UNESCO *Convention for the Safeguarding of Intangible Cultural Heritage* (ICH) has been instrumental in changing the lens of 'expertise' towards ICH-custodians. Since ICH centers on a ground-up approach, communities have been relying on the concept to establish measures at attaining protection of their heritage places and spaces, with many groups taking on the process of ecological sustainability, themselves.

"It's been a long time coming" tells me Abel Bruce, an Anishinaabeg Elder, who was involved in establishing the Indigenous-led Pimachiowin Aki UNESCO World Heritage Site [1]. The Pimachiowin Aki world heritage site nomination was designated by the UN organization in 2020, but the effort to protect their traditional territory took the collaborating communities over a decade. Recognition of their traditional territory, their resource

© The Author(s), under exclusive license to Springer Nature Switzerland AG 2022

F. Calabrò et al. (Eds.): NMP 2022, LNNS 482, pp. 2383–2392, 2022.
https://doi.org/10.1007/978-3-031-06825-6_228

management system, and their right to be self-determining were reluctantly recognized by State governments in Canada along with the UNESCO site – and became a precedent-setting story to other Indigenous and local groups. On the other side of the world too, a group of arboreal beekeepers in Poland, known collectively as Bractwo Bartne (the Beekeeping Brotherhood), have likewise been struggling to have their voice heard. The organization came to life over a decade ago, however, the history of tree-beekeeping by a community-based collective has been in existence since time immemorial. This article will present two distinct case studies of land-based ICH practices, one with an Indigenous community in Canada, and the other, with a group of beekeepers in Poland, to show how local knowledge experts are using the ICH framework as a platform to have their voices heard, and their livelihoods safeguarded.

A brief note on methodology. A large part of this work arises from long-term rela-tionships of trust with members of these communities. As a non-Indigenous scholar, I have been examining the Asatiwisipe Anishinaabeg cultural and natural resource stew-ardship as a form of community self-determination on the Pimachiowin Aki UNESCO World Heritage Site in Canada for over fifteen years. My relationship with the Polish bee-keepers began as part of my research on the Intergovernmental Science-Policy Plat-form on Biodiversity and Ecosystem Services (IPBES) Values assessment, where I led a group of apiarists, journalists, community organizations, and activist on two IPBES Indigenous and Local Knowledge Dialogues. Establishing relationships is important in community-based [2] and Indigenous research [3], and while this article builds on aca-demic scholarship related to ICH, it also heavily relies on the oral histories and ecological knowledge of the Polish beekeepers and Anishinaabeg harvesters.

2   The Case Studies

2.1   Poland: Cultural Apiarists and Tree-Beekeepers

Honeybees across the globe are undergoing significant stress [4] and efforts are being made to protect pollinators from disappearance [5]. In eastern Europe, cultural beekeep-ers have been instrumental in ensuring the survivability of the European black honeybees (*Apis mellifera mellifera*) indigenous to the region [6]. Based on a special relationship with the bees, cultural apiarists and tree-beekeepers are models of governance where the keepers work in tandem with the 'bee family' to create socio-environmental rules and norms guided by the bees, and sustained by cultural heritage.

In Poland, beekeeping is an activity that is passed down from generation to gen-eration, and tree-beekeeping ('bartnictwo' in Polish) is a unique form of apiary where the beekeeper builds homes for bees based on bee habits and behaviours. The ancient practice is based on cultural ecological knowledge and on the belief that bees are forest creatures who live near the sky and man should not interfere with the development of the bee family [6, 7]. The tree-beekeepers hollow specific areas in the trunks of living trees (called tree-hives) or put up hives in logs up on the tree or on the ground (called log-hives) (Fig. 1). The shape of hives is a direct imitation of natural beehives and based on knowledge of bee behavior; the shape also aids in the extraction of bee products while also protecting the hide and the bee colony from animals [8]. The beekeepers visit the hives only a few times per year; they climb up the tall pine tree by means of the traditional tree-climbing 'leziwo' method so as not to injure the tree. The tree-beekeeper collects honey only in the fall, after the feast of the Nativity of the Blessed Virgin Mary which occurs on September 8th. Beekeepers can collect only small amounts of honey so that the bees can have food for the winter.

Fig. 1. *Log-hive in a tree.* The hive can be 3–4 m high from the ground. Source: Bartnictwo.com. Taken from http://cudaregionu.fundacja-hereditas.pl/2021/11/10/bractwo-bartne/ on 2022/02/26.

Tree-beekeeping is considered a noble occupation because, according to local tenets, only a righteous and honest man can be a beekeeper [8]. As part of the efforts to revitalize the practice, the Polish and Belarusian members affiliated with Bractwo Bartne nominated tree-beekeeping to the UNESCO's Representative List of the Intangible Cultural Heritage of Humanity. The element was recognized by UNESCO in 2020, augmenting the region's heritage to be globally recognized.

Tree-beekeeping is special in the sense that it is a different approach to beekeeping: here, the keeper must give up control over nature and make a hive that a bee family will like – or they will not come [7, 8]. This knowledge of bees extends to forest management practices, specifically, to the idea that, if honeybees are to be protected, forests require species diversity rather than monocrops [9]. Cultural apiarists advocate for more sustainable methods of resource management because ecological sustainability ensures their livelihood. Until the discovery of sugar beets, "civilizations relied on honey for the sweet taste and we know that as early as 2600 years ago in Egypt, where people had clay hives, humans had that close relationship with bees", tells me Andrzej Pazura, one of Poland's leading tree-beekeepers [10]. For generations then, traditional knowledge holders are guided by cultural protocols and local values, believing that bees take care of the humans by providing then with honey, wax, pollination, but also that humans must care for the bees. Thus, while polinator contributions to humankind's well-being can be readily affirmed, many beekeepers also believe that a mutual and reciprocal relationship with these important pollinators, is a requirement for a healthy world.

ICH-custodians obtain their knowledge from their ancestors; they may not be scientists, biologists or politicians or have any formal higher-education. Their expertise however, provides a unique lens with which decisions about forests, pesticide-use, and nature-based livelihoods can be made. For example, customary *lex loci* used by Slavonic peoples of the region included tree-marking symbols called 'klejmo' [eng.: clay-moo] (Fig. 2). These symbols were a method to determine the type of tree that a keeper has selected for his tree-hive, a sign informing others in the community that the tree has been claimed [11]. Since, the tree needs to be adequately healthy, thick and tall enough to both protects the bees and withstand strong winds, this legal system illustrates intricate knowledge of forest ecology and disease management. By regulating exclusivity and collective (and often, familial) ownership, this adaptive, heritage-sourced resource management system continues today.

The expertise garnered over years of lived-experience, however, tends to be challenged by State-licensed experts. When local communities disagree with political and economic decision-making, a clash of 'experts' ensues. This is currently the case in Eastern Poland, where the Bractwo Bartne is engaged in a conflict with State Forests over logging of primeval forest areas. In an attempt to safeguard their livelihood and sustain large trees for hives, Bractwo Bartne is requesting to preserve large tracts of forested areas by pushing for a Forest Stewardship Council (FSC) certification. In response, the municipality is claiming that the brotherhood is acting against 'collective regional responsibility' [12]. By requesting 'official administrative documentation', the municipality is questioning the brotherhood's role as a 'community-based expert group' [12].

Consequently, the brotherhood is using the ICH framework to push for the acknowledgement of their living heritage as one that is unequivocally bound to the trees. Arguing

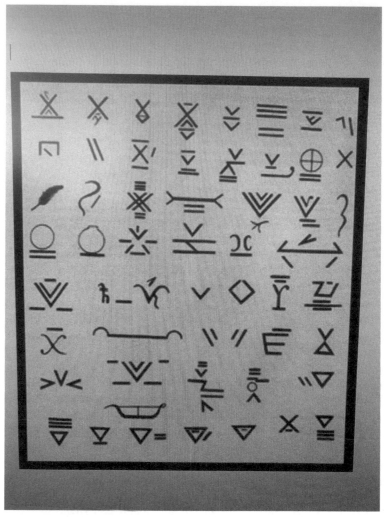

Fig. 2. *Tree-hive symbols from Belarus.* The symbols were used by Slavonic people throughout Eastern Europe for exclusivity and identification. Source: https://pl.wikipedia.org/wiki/Znami%C4%99_bartne, last accessed: 2022/02/26.

that their inherited skills, traditional methods and tools, numerous social practices as well as culinary and medicinal traditions, guide collective resource-governance decisions, a beekeeper's livelihood is directly tied to the forest's ecology. The brotherhood's UNESCO ICH designation raises the thorny question of who is the expert in forest ecology: the keeper of forests or the feller of trees. The Polish debate tells us that the ICH framework can propel customary knowledge systems to have more negotiating power over ecological decision-making, and that integration between policy sectors and the scales of bee-keeper decision-making can be important drivers of environmental change, especially in the realms of institutional influence.

2.2   Canada: Asatiwisipe Anishinaabeg (Indigenous) Harvesters

Indigenous cultural heritages in Canada involve a dazzling variety of elements ranging from specific social practices and cultural expressions, to oral traditions and cultural landscapes. The boundaries between intangible and tangible heritage are fluid and vary from community to community across this large continent. Devaluation of certain practices and territoriality by colonial policies of land dispossession, cultural genocide, forced relocations, and involuntary enfranchisement for the past few hundred years, has led many Indigenous communities to reinterpret the recognition of their rights through court cases, social movements, and local community acts of resistance [13]. In recognition of such difficulties, the UNESCO 2003 *Convention for the Safeguarding of Intangible Cultural Heritage* refers particularly to Indigenous peoples who "play an important role in the production, safeguarding, maintenance and recreation of the[ir] intangible cultural heritage" [14]. While the document was not written with Indigenous people in mind, it does recognize Indigenous communities as requiring special consideration.

Unlike Poland, Canada's ICH policy work is occurring outside of the convention. In fact, not only has Canada not ratified the 2003 UNESCO Convention, but the State does not seem very interested in the discourse [15]. This leaves communities in Canada, including Indigenous people, to carry out own ICH safeguarding. For example, the First Nations involved in the Indigenous-led Pimachiowin Aki World Heritage Site relied on the concept to push their UNESCO nomination forward in 2016; Heritage Saskatchewan, a provincial organization, has received ISO certification to engage the ICH framework. Likewise, the First Peoples Cultural Council, an Indigenous cultural organization, published a report in 2021 about ICH being important to Canada's reconciliation process with Indigenous people [16]. And, while the Province of Quebec has instilled the principle into its official policy in 2016, I leaned on the ICH concept to assist Inninuwak Elders to protect their trapline territories from development in 2013 and 2014 [17]. These examples show that, although there is very little ICH work done at the national level in Canada, communities and groups are relying on the framework to promote and document their living heritage.

The Asatiwisipe Anishinaabeg are an Indigenous group with whom I have worked with for over a decade. Like many Indigenous people in Canada, the Anishinaabeg view themselves as true experts of their territories and they are determined to protect their lands and resources in the way that that is customary to their Anishinaabeg culture (Fig. 3). The trappers I work with follow traditions and protocols that align with the cultural tradition called *ji-ganawendamang Gidakiiminaan,* 'keeping the land/we take care of our land' in Anishinaabemowin, the Ojibwe language. *Ji-ganawendamang Gidakiiminaan* arises out of Anishinaabe laws, customs, and a certain way of doing things or performativity. Maintained for generations in tandem with oral traditions consisting of stories and songs, this cultural tradition is manifested by.

harvesting sites, habitation and processing sites, traplines, travel routes, named places, ceremonial sites, and sacred places such as pictographs associated with powerful spirit beings. These attributes are dispersed widely across a large landscape and concentrated along waterways, which are an essential source of livelihood resources and a means of transportation. Anishinaabe customary governance

Fig. 3. *The Asatiwisipe Anishinaabeg cultural landscape.* Only an expert knows that water and abundant fish lay beneath the trees on this winter photo. Source: ©Pawlowska-Mainville.

and oral traditions ensure continuity of the cultural tradition across generations [18].

As a "compelling example of the inseparability of an indigenous culture and its local environment" [18], *ji-ganawendamang Gidakiiminaan* illustrates the local community's extensive ties to the land and vast knowledge of the area. Because, it is so comprehensive and inclusive of diverse customs and approaches, *ji-ganawendamang Gidakiiminaan* directly shapes people's interactions with the land and with each other. This means that expert resource-users rely on cultural traditions to make decision about their territory.

Many of these ICH custodians have never attended high school, and much of their expertise is gained from time spent harvesting: trapping, hunting, fishing, and gathering. These experts have been taught by their Elders to monitor the wildlife in the area, to know how much fish to catch from a river or lake, and how many furbearers to take so the animal populations remain sustainable. Because the knowledge about the landscape is gained by going out on the land, many Anishinaabeg harvesters have the best and most detailed information of black bear and beaver locations, of moose kill-sites, and of places where sacred stories are told. They also know the best time to hunt geese, where the best pickerel fishing sites are, and where the most recent fire occurrences took place. Having spent most of their life as stewards of their territory, these expert knowledge-holders can make the best decisions about their lands and resources.

Rather than negotiate for acceptance in colonial spaces, the Indigenous Nations involved in the Pimachiowin Aki world heritage site have shown the world a different civilizational discourse. In place of a colonial ladder that charts civilization from hunter-gatherers, to farmers, to industrialized modernity, *ji-ganawendamang Gidaki-iminaan* offers the possibility of overlapping and intersecting civilizations, honouring non-European heritage in the present, not the past [19].

3   Valuation of Experts and Professionals

Although the focus of ICH is on the intangible, cultural heritage is indivisible from its counterpart, tangible heritage. The two are inherently connected as no artifact, building, bone, instrument, etc., has any meaning without the values assigned to it. This duality is especially important for land-users like beekeepers and Indigenous harvesters who are dependent on the land and the resources for their livelihood. To thrive, these ICH custodians need the tangible and intangible elements of their heritage; and, in order to be healthy, the tangible environment too, requires caretaking by these land-users. In that sense, intangible heritage gives life to tangible (natural) heritage through knowledge systems, skills, and relationships of people with the local environment.

Land and resource management is the main point of contention between cultural custodians who hold inter-generational knowledge about the resources, and the formally-appointed professionals. While ICH experts argue for more ecological approaches towards resource management by challenging economically-driven decision-making, 'professionals' with certifications and administrative powers often end up being decision-makers about lands that affect local community livelihoods. This includes conflicts over logging, economic development initiatives, or even industrial pollination or trapping laws [19]. In situations when competing systems of 'expertise' become visible, community members can rely on the ICH framework to attain a 'title' that recognizes and affirms their cultural knowledge, skills, and expertise. This is particularly important when formalized professionals (State-run administration, scientists, or representatives of political bodies) operate in parallel with power structures legitimized by customary law and local traditions.

Given the fact that boreal forest areas have historically been perceived to be *terra nullius,* meaning 'empty land' in Latin (the doctrine was also a foundation of colonialism in the Americas), ICH serves as a powerful force in acknowledging historically marginalized voices and identities. Recognizing local experts in natural resource management challenges the idea that natural spaces are 'empty' and thus can be administered, developed, or settled for 'progress' by the State. The ICH framework permits the strengthening - and perhaps even outright visibility, of local governance to outsiders. By recognizing stewards of the local landscape, ICH helps place community ICH custodians as the 'face' of their cultural landscapes. ICH, then, not only offers a more authoritative voice in conflicts that threaten their livelihoods, but in Settler-States like Canada, the framework helps forward reconciliation with Indigenous territoriality [14].

Therefore, based on their intricate and long-term knowledge, the Polish and Anishinaabeg ICH custodians, significantly contribute to long-term data analysis and management of natural resources. Already, their customary governance systems have served for that purpose since time immemorial, and as an element of their cultural heritage, these systems (which have generally been subdued by colonial governments), can once again drive cultural, ecological, and economic sustainability of an area. From my experience with these experts, the intangible cultural heritage framework as articulated by the 2003 UNESCO Convention has already assisted local land-users and cultural resource experts in attaining a higher level of self-determination than previous policy approaches. The Convention's interpretation of heritage as alive and requiring community-engagement directly challenges the "authorized heritage discourse," [16] and more community-based research is needed to better reinforce implementation of the ICH discourse in natural spaces. Ultimately, an ICH-based understanding of 'expertise' over natural landscapes is another mechanism to assist cultural custodians to implicitly or explicitly leverage points towards embarking on environmentally sustainable and socially fair future pathways.

4 Conclusion

The community-centered understanding of heritage is what makes the UNESCO *Convention for the Safeguarding of Intangible Cultural Heritage* unique: communities choose their ICH, decide its meaning, value, extent of descriptions, and the steps for protection. This means that, rather than the top-down approach associated with 'professional' expertise, community experts play a main role in activities that deal with safeguarding of own cultural and natural heritage. Local experts may not have the professional designations of 'real' experts, however, custodians like Polish tree-beekeepers and Anishinaabeg harvesters in Canada represent a wealth of knowledge and history that can contribute to sustainable resource management and local economic development. By putting faith in, and partnering with 'traditional experts', government-sanctioned professionals can bring viability to cultural heritage professions. Supporting land administration governed by 'ordinary' people does not signify that governments and professionals working within natural resource management are forced to acquiesce power. Rather, supporting ICH and custodians helps provide a platform for individual community experts to finally get the 'professional' recognition they deserve.

References

1. Bruce, A.: Personal communication, 10 Mar 2019
2. McKinnon, S. (ed.) Community Based Participatory Research: Methods UBC Press and Canadian Centre for Policy Alternatives Practice and Transformative Changes. Vancouver (2018)
3. Tuhiwai-Smith, L.: Decolonizing Methodologies. Led Books, London & New York (1999)
4. Oldroyd, B.P.: What's killing American honey bees? PLoS Biol. 5(6), e168 (2007). https://doi.org/10.1371/journal.pbio.0050168
5. VanEngelsdorp, D., Hayes, J., Underwood, R.M., Pettis, J.S.: A survey of honey bee colony losses in the United States 2008. J. Apic. Res. 49, 7–14 (2010). https://doi.org/10.3896/IBRA.1.49.1.03
6. Pawlowska-Mainville, A., et al. : Tree-beekeeping and Apiary in Poland, Report in Response to the IPBES-ILK Dialogue, Methodological Values Assessment. IPBES (2021)

7. Wróblewski, R.: Krótka historia pszczelarstwa polskiego [Short History of Polish Apiary]. Sądecki Bartnik, Stroże (2015)
8. UNESCO: Tree-Beekeeping ICH. https://ich.unesco.org/en/RL/tree-beekeeping-culture-01573. Accessed 26 Feb 2022
9. Jacques, P.: Monocropping cultures into ruin: the loss of food varieties and cultural diversity. Sustainability 4, 2970–2997 (2012). https://doi.org/10.3390/su4112970
10. Pazura, A.: Personal communication, 16 Feb 2022
11. IPBES Dialogue Meeting. Prince George,12 Feb 2021
12. Bartoszewicz, Z.: Czy w Augustowie Mieszkancy beda mieli wplyw na las? 6 December 2021. https://augustow.org/2021/czy-w-augustowie-mieszkancy-beda-mieli-wplyw-na-las. Accessed 19 Feb 2022
13. Pawlowska-Mainville, A.: Asserting declarations: supporting indigenous customary governance in Canada through intangible cultural heritage. Patrimonio: Economía cultural y educación para la paz (MEC-EDUPAZ). 1(19), 346–381 (2021)
14. UNESCO: Text of the Convention for the Safeguarding of the Intangible Cultural Heritage. http://www.unesco.org/culture/ich/en/convention. Accessed 13 May 2021
15. Pawlowska-Mainville, A.: 'Stored in Bones': Safeguarding Indigenous Intangible Cultural Heritages. University of Manitoba Press. Accepted for Publication, Manitoba (2022)
16. Aird, K., Fox, G.: Indigenous living heritage in Canada. In: First Peoples' Cultural Council and Fox Cultural Research, April 2020. IndigenousLivingHeritageCanada.pdf Accessed 15 May 2021
17. Pawlowska-Mainville, A.: Aski Atchimowina and Intangible Cultural Heritage. In: Expert Report and Presentation to the Clean Environment Commission of Manitoba on the Keeyask Generating Station on behalf of the Concerned Fox Lake Grassroots Citizens, Participant, CEC Manitoba (2014)
18. Aki, P.: Nomination for the inscription of Pimachiowin Aki on the World Heritage List. Pimachiowin Aki World Heritage Site Project (2016)
19. Smith, L.: Discourses of heritage: implications for archaeological community practice. Nuevo Mundo Mundos Nuevos [En ligne], Questions du temps présent, mis en ligne le 05 octobre 2012. https://doi.org/10.4000/nuevomundo.64148. Accessed 14 May 2021

The Role of the Expert in Holistic Approach to ICH: Case Study in the Community of "Sites Remarquables du Goût (Outstanding Sites of Taste)" in France

Catherine Virassamy[✉]

ICOMOS, Paris, France
catherinevirassami@gmail.com

Abstract. A holistic approach to ICH is particularly visible in the work conducted with the "Fédération Nationale des Sites Remarquables du Goût (https://www.sites-remarquables-du-gout.fr/)" to be included into the UNESCO Intangible Cultural Heritage National Inventory. About twenty registered practices and research work conducted by the expert appointed by the committee, show the indissociable link between agricultural production, landscapes and know-how pertaining to the work carried on by the farming communities, tangible, and intangible cultural heritage. The recording of these skills stresses the point that they altogether belong to the local culture and history. We have sampled three sites within this research work; "Fishing of freshwater fish in the Dombes ponds", "Elaboration of Armagnac brandy", livestock breeding and "Salers cheese making", more than just practice, these are acknowledged processing, territories, architectures, types of food, ways of life and sustainable development respect, all of them fully decoded within the tangible and intangible dimension.

Keywords: Intangible heritage · Communities · Role of experts · Holistic approach · Tangible heritage · National Inventory · Agricultural production · Fishing skills · Livestock breeding · Food heritage · Rural heritage · Sustainable development

1 Genesis of the Approach

The concept of "Sites Remarquables du Goût (outstanding sites of taste)" (SRG) is the starting point of the study. What is a "Site Remarquable du Goût"? First of all: a tight association between a site and a product. Through this spontaneous association the shaping of the landscape, the territory and environment is tightly related to the farming community's choice of agricultural and food production throughout the years… Sea or freshwater or river products, land products, cheeses, brandies, wines…

© The Author(s), under exclusive license to Springer Nature Switzerland AG 2022
F. Calabrò et al. (Eds.): NMP 2022, LNNS 482, pp. 2393–2403, 2022.
https://doi.org/10.1007/978-3-031-06825-6_229

Products of outstanding sites of taste © Catherine Virassamy 2021

The definition of the SRG refers:

- firstly, to the notion of tangible heritage, the sites' production being linked to the local resources by farming communities: crops, livestock, agricultural and food production,
- secondly, to the notion of intangible cultural heritage (ICH) according to the 2003 Convention on the ICH by its goals - safeguarding, respect for and awareness of the communities and identified "practices, representations, expressions, knowledge and know-how…"

The community is composed of a network of local associations or groups of individuals. Its role is to protect the identity and quality of agricultural production. About sixty sites and groups in France animate the local network characterized by its 4 facets: agriculture, culture, environment, tourism.

The experts were to focus on the observation and the understanding of a combination of elements in order to emphasize the intangible cultural heritage carried on by the SRG community. Their task was also to define the positioning of the community through its practices beyond the value of the product and the territory, and to observe the modes of transmission and valorisation that have perpetuated its vitality.

About twenty research studies were conducted in about twenty sites, within the National ICH Inventory framework, resulting into multiple issues: economic for the sector, environmental through the protection and the respect of the biodiversity, touristic by perpetuating festivals, fairs, and markets, cultural through the museography, the discovery of the built and landscaped heritage, and educational by the animations being organized.

2   Position of the Expert Researcher

2.1   Accompanying the Community in a Research Project

The research is based on the expression, the awareness, the recognition, and the inspiration of a community in accordance with the Immaterial Cultural Heritage issues. The project objectives established by the expert with the community consist in;

- doing an inventory of the knowhow linked to the "Sites Remarquables du Goût" products, and their conservation;
- stressing the link between tangible and intangible cultural heritage and these communities;

- integrating the broader concept of sustainable development while respecting the ecosystem;
- contributing to the people well-being through quality food.

2.2 The Research Methodology

Based on a site and an associated product, the project aims to identify and describe the know-how/practices, the holders, the modes of valorisation and transmission that generate the products of the SRG in its multiple components. The research was carried out within and in cooperation with the SRG community, local associations representing institutions, professionals, and communities in a collective approach.

Its purpose was;

- to capture several sources of information.
- to define the essence of the know-how in its context, based on verbal data collected from a questionnaire from the field survey, visuals: photographic documentation, films of the interviews, visits, and documented data (publications, reports, official texts), to understand, verify, confirm, and illustrate the interviews and practices.
- to transcribe these data as directly as possible, highlighting the reality and authenticity of the practice, essential sources of the form.

This also means taking into account other practices that are complementary to those studied, such as handicrafts or cultural mediation in educational and touristic activities.

Defined, conducted, and organized by an architect as an expert, via a global and inclusive approach around ICH, the research was built up with a representative of the SRG community and the contribution of local associations. Throughout the surveys undertaken, the research was lead along the following steps: definition of the subject to be observed, raising the awareness of the local associations about the SRGs, implementation of the survey: collective meetings, interviews, visits, and practical workshops.

First, the following questions were raised in reference to the Convention for the Safeguarding of the Intangible Cultural Heritage, about the practices within the related communities:

Who are the people who have shaped landscapes, structured architectures, sustained, and processed the food products of high quality specific to a territory?

What are the particular knowledge and skills developed, perpetuated, and valued, associated with the sites (of taste) and products to be tasted that contribute to sustainable development, to the livelihood, the well-being of the populations and consumers of the territories?

Secondly, the criteria for the selection of the sites responding to the approach were defined as follows: to offer traditional products, specific local cultures and varieties, the result of an inseparable link between the sites and their productions, to be carried on by a living community, to share local, sustainable, and evolving know-how.

The identified skills are for example, linked to the products of the sea and the ponds, of the earth and the livestock. They have been perpetuated for generations by the local communities who lived from them, like transformed landscapes and architectures, and have been generated by local know-how. These perpetuated ancestral skills, practiced for their livelihood consist in different techniques of fishing, collective governance, gathering and harvesting, livestock breeding and pastures, wine distillation, food processing and preservation, confectionery etc.

2.3 The Survey Implementation

The local associations were contacted and informed about the National ICH Inventory by handing an interview guide describing the survey method and the questions, particularly the profiles of the operators to contact, i.e. the holders of know-how but also all the contributors to the safeguarding and revitalization of the practices in agreement with the Convention objectives (paragraph 3 of art. 2) that is ensuring the viability of the relevant intangible cultural heritage.

Each survey lasts two to three days and begins with the meeting of the local representatives and members of the community to bring together the key doers of the site: representatives of the communities, politicians and institutions, holders of well-known practices, trade unions, representatives of the training and educational systems, promotion, and tourism. This first cross-disciplinary meeting makes it possible to understand the aspects, the history of the skills and knowledge, the local issues, the local development and the synergies of the operators, major holders of essential know how, mostly combining tradition and innovation strategies.

This key phase makes it possible to bring out the notion of intangible cultural heritage and the "sense of identity and continuity" as indicated in the Convention.

The survey grid outlines the guideline for the interviews and visits, each lasting about an hour, with notes and photos taken on the spot. The individual and collective interviews allow us to systematically meet the holders of the know-how, farmers, producers, stock breeders, fishermen, watercress farmers, craftsmen, bakers, confectioners, cheese-makers, cheese maturers, processors, packaging and marketing professionals, restaurant owners, trainers, cultural mediators, communicators, historians, local scholars, and tourism professionals whose know-how perpetuates the authenticity, the quality and the particularity of the products concerned.

Observation and participation in the practice are crucial. Observation of places, practices, gestures, but also culinary workshops and tasting encapsulate sensations and what can't be worded such as the shop practice, the efficiency of the gesture, the sound of the stone, smells, flavours, and tastes. It is a question of observing the practices in their context along the entire process until the tasting of the final product.

The survey framework:

- To Identify and observe the sites: environment, geography, landscapes, cultivated areas, orchards, architecture, workshops, specific areas, etc.,
- To identify gestures, instruments, machines, and tools,
- To record life and territory stories
- To identify social and technical evolutions, innovations, and the inscription of the practice in its historical continuity,
- To taste the products, to differentiate their organoleptic qualities,
- To identify traditional and new recipes, traditional culinary practices,

- To identify the labels and other modes of recognition of this local cultural heritage,
- To identify the institutional and private sectors involved: agriculture-food, environment, agronomy, culture, architecture, tourism, processing, trade, distribution, consumption, crafts, catering, equipment, packaging, design etc.
- To understand the ecosystem in which the know-how is integrated,
- To link agricultural, architectural, environmental, and economic approaches.

So many elements linked to the productions and the pertaining know-how.

2.4 Collecting Testimonials: Examples of Verbatim

The testimonies reflect the gustative and nutritional quality linked to its environment, its appearance, or its smell… perpetuated and expressed by the custodians of the know-how.

The carp of the Dombes ponds has proven nutritional qualities and is one of the fish with the lowest fat content (2.5% lipids) after cod; it is rich in vitamin D and vitamin B12. Each Salers cheese in Auvergne has its own identity:

- *To the eye: its crust is golden, thick, blooming with red and orange spots.*
- *To the nose: AOP Salers gives off subtle aromas of mountain flowers.*
- *To the taste: it reveals a taste of the land, the taste of the farm.*

The golden Chasselas table grape from Moissac gives beautiful bunches of white grapes with golden tones, tasty, crunchy, and sweet pulp.

Only the "charcutier" experienced in smoked cured meats can judge the intensity of the coniferous and juniper tree sawdust smoke in the "tuyés" (Franche-Comté farm fireplaces to smoke the meat).

3 Examples of "Taste Skills"

3.1 Example of Pond Fishing in the Eastern France in the Dombes Ponds

The expert conducted his survey, accompanied by a representative of SRG, according to the procedure previously defined at a collective meeting with community representatives, individual interviews, and participation in pond fishing. The link between tangible and intangible heritage is obvious. Freshwater pond fishing in the French site of Dombes has been perpetuated by the local community for nearly 800 years. It is part of the management cycle of 1200 ponds where fishing is alternatively practiced with cereal farming, hence creating an exceptional ecosystem. This age-old custom represents a magnificent and rare expression of respect for the environment and passion for a territory, through the mutual aid and conviviality that characterizes the fishing of freshwater fish in the ponds of the Dombes. Today, 1200 tons of quality fish are produced annually, 60% of which are carp, the emblem of this remarkable site. The participants mentioned social meetings contributing to the transmission of the practice, festive events such as meals with fishermen, families, and friends on fishing days and also a Customary Law[1] ruling the use of the waters and becoming the law (Fig. 1).

---

[1] A. Rivoire Coutumes, usages et bibliographie des étangs de la dombes et de la bresse.[Commentaires et bibliographie par A.Truchelut] – Editions de Trévoux 1982.

Fig. 1. Dombes fishing © Catherine Virassamy 2016

3.2   The Case of Salers Cheese Making in Auvergne in France

Salers cheese is an example of traditional farmstead production with a "farmyard character"[2]. Salers is a cow breed, a very particular one: very resilient and motherly. The cow can be milked only when her calf is next to her. Salers flocks are moved into the Auvergne (central France) high pastures (800 to 1000 m) from April to mid-November. There they graze the rich grass and fragrant flora. In the past, cheese was only made in the shepherd's hut built from local stones and covered with slabs called "Lauzes". It is a specific stone hut called "Buron" where the shepherd stays with his herd during the summer. It is also used to allow cheese to mature in the cellars for at least one year. Today the "Burons" are transformed into housing or into eco-museum showing local know-how, Salers cows breeding, cheese making etc. There are hundreds 'Burons' - stone huts - in Auvergne, the symbols of the regional landscape. The age-old expertise of Salers cheese making has evolved with the young generations who have taken over the family farms. To make a traditional Salers cheese you must use raw, unheated milk, straight after milking in the mountain. Today cheese making uses the traditional processing either in the 'Buron' or in hight tech laboratories at the farm. It takes about 400 L to make a 35 kg round cheese. It embodies both healthy food and its sustainable process. Every year, the departure to the mountain is marked by a local feast where all the inhabitants accompany the shepherd. It is followed by a festival after the harvest, the " Cow and Cheese Festival" in August, the "Livestock fair" in September, and the autumn cattle show (Fig. 2).

Fig. 2. Salers cheese making © Catherine Virassamy 2016

---

[2] Paul Garouste, La laiterie dans le cantal, Imprimerie de Bonnet-Picut 1986.

3.3 The Case of Armagnac Brandy Elaboration in Gascogne in Southwestern France

The survey was carried out according to the same process, involving a collective meeting at the outset, individual interviews with women and men winegrowers, distillers, cooperative workers, coopers, oenologists, restaurant owners, historians, representatives of public authorities and professional unions, visits to farms and Eco museums, observation of practices such as distillation, and tasting. Armagnac brandy has been produced for 700 years from the distillation of white wine, whether it is made on the estate or ambulant still. Its production is linked to the activity of the historical towns of southwestern France, "the Bastides" or New Towns of the 13th century, whose urbanism is characterized by a system of perpendicular streets displayed around a central square, where the early forms of public and religious life were established, escaping from feudal power. With charts and public facilities, they are characterized by the commercial development, especially of wines (Fig. 3).

Fig. 3. Armagnac brandy elaboration © Catherine Virassamy 2021

The vitality of the practice is linked to the use of a particularly resistant local white grape the Baco variety[3], characterized by its finesse and floral notes. The smoothness of Armagnac and the richness of its aromas are the result of the continuous Armagnac distillation in one single heating of the wine in a column still, specific to Armagnac.

After seasoning for many years in oak barrels, the brandy "with forty virtues", invoked by the Prior of Eauze in a work kept in the Vatican, is smooth, elegant, and invigorating. It is distinguished by the famous "Armagnac millésimé", made from a single year's harvest, not having been blended with another vintage.

The inextricable links between viticulture, urbanism and architecture is recognized by labels and protection and the particularly active local life, also labelled "CittaSlow", towns where life is good. Armagnac festivals, brotherhoods and local associations celebrate in French or in Gascon language, the brandy that gives work and enlivens the local community.

---

[3] Duhart (Frédéric), «Les créations de François Baco (1898–2011). Naissance et destin d'une collection d'hybrides producteurs directs français», Territorios del vino, n° 7, 2011, 32 pages, en ligne: http://www.territoriosdelvino.fhuce.edu.uy.

Social practices and festive events contribute to the practice's transmission and continuity. "Armagnac en fête" (last weekend of October) has been held in the "Place Royale of Labastide d'Armagnac" little town for over twenty years, celebrating the end of the vine harvest and the beginning of the distillation. This event gathers the Armagnac market, around numerous Armagnac producers, associations of the Sites Remarquables du Goût from different regions of France and offers animations on heritage and know.

4 Results About Holistic Approach

4.1 Follow-up by an Expert Ambassador and Mediator

The survey about pond fishing led to the production of a video and photo reports published as part of an exhibition or on the Internet.

After all the surveys, the expert was officially recognized by the community as an "Ambassador of the Sites Remarquables du Goût".

He contributes to the awareness of the cultural dimension of agricultural know-how and to the resulting inspiration. The valorisation of the practice through his professional and private network, interventions in articles, conferences, related to the knowledge of taste and ICH are all part of his mission.

The research allowed the formulation of gestural or oral knowledge, of the role of community members, as well as its acknowledgement by the community. The practice is divulged in different forms but also in the occurrence of local production fairs and markets organized in the SRG sites via the expert or the community.

4.2 Follow-up on Sustainable Development

One of the main outcomes of the holistic approach is the link between the ICH and the respect of sustainable development. The ICH contributes to sustainable development through its link with the natural heritage and a community, the different domains of practices of the convention such as knowledge and practices concerning nature and the universe, traditional craftsmanship, social practices, and rituals and festive events. The studied practices are carried out through these three domains linked in an ecosystem, in a territory and in a community.

The following ICH practices can be included in this framework: sustainable fishing, livestock breeding, harvesting, and gathering, processing of local resources. They are linked to the environment in which these practices are carried out by preserving the biodiversity, by reducing greenhouse gas emissions, by protecting carbon sinks through the preservation and maintenance of the natural heritage.

Pond fishing is a reference in the respect of sustainable development. In respect and preservation of the local biodiversity: the alternative practice of filling and draining the ponds enriches the soil with organic matter for the crops, and the fish feed on the grain crops. This practice has contributed to the formation and preservation of a natural reserve of flora and thousands of birds. This Dombist ecosystem allows fishing, hunting and agricultural practices all over the territory. In reduction of greenhouse gases emissions processes: fishermen don't use electricity, chemical processes, or hydrocarbon energy: fishing is done on foot and by hand with a very large net. People also use pond's clay to make building bricks for local traditional habitat with low environmental impact. In sustainable water management: rainwater feeds the pond; fishing is done by draining the pond. In valorisation of eco-friendly knowledge: educational actions in elementary school, exhibition, interactive booklets, but also by environmental, landscape and architectural labels, are conducted by communities.

5 Review of the Data Collected

5.1 The Expression Identity - Quality

How is the notion of "quality" expressed? It is considered as the practice and the product identity. It is expressed through the testimonies of the operators of the community, at each level.

It is also expressed through farming practices that respect the environment: crops and livestock Dombes ponds, Auvergne pastures, methods of integrated agriculture or pastures rich in biodiversity, or in specific places, architectures, or containers, "tuyés", cheese chestnut tubs, or ancestral gestures, smells, textures, multiple materials indissociable from the practices.

5.2 The Expression of Visuals and Photographic Reports

The photographic report accompanies the interviews and visits, captures the environment, the sites, the landscapes, the architectures expressing this feeling of continuity, the history, the past and the present of the sites in many items: vegetation, materials, crafts, social life, constant and immutable gestures.

The visuals refer to the original communities: monks, founders of the ponds of les Dombes, peasants and builders of the "tuyés", shepherds who made the Salers cheese, and farmers. They also mention other holders of traditional crafts such as the "buronnier", the builder of stone shelters and slab roofs, the "tavillonneur" the maker of shingles constructions of the farms and "tuyés", and finally social and landscape organizations in perpetual mutation.

5.3 The Expression of Transmission

One of the requirements of the survey is to meet the holders of the know-how elaborated in a "traditional" context, farmers or craftsmen, the last custodians of the know-how, to describe this and the inbred skills, to relate workshop stories, and in a "contemporary" context in cooperatives or factories that perpetuate the know-how by using new mechanical/electrical tools for making equivalent products, while respecting the process and the logic of the practice. Research expresses the evolution of fluctuating and/or constant know-how.

The transmission is linked to the family circle, the apprenticeship masters, training given by agricultural schools and professional unions, and restaurant owners who reveal and cultivate the consumer's taste.

The other aspect of transmission concerns the framework of the practice, a delicate approach linked to the traditional repetition of gestures, the evolution or innovation which takes place with new operators and new frameworks, high tech laboratories, and drones that monitor the herds in the pastures.

These effective transmissions indicate the arrival of new generations into traditional communities, putting into practice innovative process and techniques in places designed to respect hygiene, safety, animal welfare and to reduce physical strain. However, we observe new types of farms that have been redesigned and adapted to both traditional and current practices, in line with the evolution of communities and social life. For example, cows are grouped in free-stall barns, equipped with a chip indicating the amount of food to be provided automatically.

The changes also concern the creativity in cooking, in processed products, in packaging, in the evolution of consumer's tastes.

5.4   The Expression of Mediation

The operators are an instrumental for the survey in connection with the expert by:

– their knowledge of the field, by the actions of
   their communication and educational actions they lead in the schools and the canteens,
– their activity in tourism and marketing operations and defined other field such as short circuits, fine and organic stores, shop markets, eco museums and public financing.

The newcomers, recently settled in rural areas who get involved in local life, contribute greatly to the development of local know-how and the territory, through innovative activities in terms of exploitation, reception, marketing, or manufacturing of derived products. New farmers, organic farmers, creators of gites, inns, restaurants, local businesses, trade of quality products manufactured in a sustainable development approach, creators of local partnerships between producers and processors such as collective approaches in short circuits, promote the animation of certain territories rich in local products of exception.

The recognition of territories, skills and products are supported by well identified local communities combining the respect of tradition, modernity, and sustainable development.

6   Conclusion

The transversal ICH approach implemented by the expert's research, highlights the elements in conformity with the Convention for the Safeguarding of the Intangible Cultural Heritage:

– The sense of identity and continuity is revealed in the sites, by the stories about ruptures and continuities in the respect of the tradition and the environment without overlooking the innovation and creativity in a permanent movement.

- The participatory approach of the communities through educational and cultural actions, including training in the sites and landscape examining, and tourism is due to the network's associations.
- The interaction between nature and history is expressed through the specificity of the communities, sites, products, skills and knowledge, tastes, and visuals rendering the cultural identity of a territory.
- The multidisciplinary approach engaged from the beginning of the research with operators from different fields, aims at including ICH into a systemic logic, which is confirmed by the nature of the SRGs, that are related to agriculture, culture, tourism, and the environment.
- The approach specifies the issues of ICH Inventory, as a vector of sustainable development through environmentally friendly farming practices, specific taste and nutritional qualities contributing to the well-being of people. Another issue is the expression and promotion of the food heritage in its cultural diversity, of people and knowledge.

Beyond the inventory form, beyond the observations and research carried out during the surveys, it is appropriate to question the effect induced by the survey, on the issue of the patrimonialization of the territory through the holders, on the awareness they raised about cultural value of their practices. What are the effects and projects induced by the recognition due to the inscription in the National Inventory of Intangible Cultural Heritage[4] supported by an holistic approach?

---

[4] National Inventory of Intangible Cultural Heritage: https://www.culture.gouv.fr/Thematiques/ Patrimoine-culturel-immateriel/Le-Patrimoine-culturel-immateriel/L-inventaire-national-du- Patrimoine-culturel-immateriel.

The Cultural Routes of the Council of Europe and the Journey of Ulysses: Shared Values and Good Practices

Roberta Alberotanza[1], Francesco Calabrò[1] (iD), Mariangela Monaca[2] (iD), and Carmela Tramontana[1](✉) (iD)

[1] Mediterranea University of Reggio Calabria, 89100 Reggio Calabria, Italy
carmen.tramontana@unirc.it
[2] University of Messina, 98122 Messina, Italy

Abstract. The Journey of Ulysses testifies to the existence of an ancient network of relationships between the territories bordering the Mediterranean: for this reason, it is here assumed as the guiding thread of the proposal for a cultural itinerary, which, similar to the certified Cultural Routes of the Council of Europe, encourage cultural exchanges between communities and, in such a way, mutual knowledge. This contribution, starting from the exposition of these inspiring principles, aims to identify in the journey of Ulysses the values assumed by the Council of Europe as its foundation, universal values such as democracy, the rule of law and the richness of dialogue and diversity among peoples. To strengthen the hypothetical candidacy for the Cultural Routes of the Council of EuropeProgramme of Ulysses's Journey, we make use of the analysis of certain best practices from the itineraries already certified, which with this chosen path provide common principles: the Mediterranean as an area of reference and themes. The ultimate aim of the whole proposal is therefore the establishment of a cultural path that, in line with the criteria and values established by the Council of Europe, favours knowledge, enhancement and the use of the material and intangible, cultural and natural identity of the places affected by the Itinerary.

Keywords: Cultural routes · Cooperation · Democracy · Intercultural dialogue · Cultural heritage

1 Introduction

The Journey of Ulysses testifies to the existence of an ancient network of relationships between the territories bordering the Mediterranean: for this reason, it is assumed here as the guiding thread of the proposal for a cultural itinerary, which, like the certified Cultural

The work is the result of the shared commitment of the authors. However, paragraph 1can be attributed to Francesco Calabrò; to Roberta Alberotanza paragraphs 2, 2.1, 2.2; to Mariangela Monaca paragraphs 4, 4.1; to Carmela Tramontana paragraphs 3, 3.1, 3.2, 3.3, 3.4. Conclusions were written jointly by the authors.

© The Author(s), under exclusive license to Springer Nature Switzerland AG 2022
F. Calabrò et al. (Eds.): NMP 2022, LNNS 482, pp. 2404–2416, 2022.
https://doi.org/10.1007/978-3-031-06825-6_230

Routes of the Council of Europe, encourages cultural exchanges between communities and, in this way, mutual knowledge.

Ulysses travels in the Mediterranean meeting numerous populations, with very different identities: he, embodying the values of Greek culture, becomes the emblem and vector of this civilisation that merges with others, thus laying the foundations for Western culture.

This contribution, starting from the exposition of the inspiring principles, aims to identify in the journey of Ulysses the values assumed by the Council of Europe as its foundation, those of universal values such as democracy, the rule of law and the richness of dialogue and of diversity among peoples, which are declined in the Homeric poem through the experiences of the protagonist, his attitude towards the populations he meets and the social and environmental context in which the events take place. To strengthen the hypothetical candidacy for Cultural Routes of the Council of Europe Programme of Ulysses' Journey, we make use of an analysis of some best practices among the itineraries already certified, which with the chosen path have common principles, the Mediterranean as an area of reference and themes. The ultimate aim of the whole proposal is therefore the establishment of a cultural path which, in coherence with the criteria and values established by the Council of Europe, favours the knowledge, enhancement and use of the material and intangible, cultural and natural identity, of the places affected by the Itinerary, according to an integrated approach, which following a multidisciplinary logic, takes the cultural, environmental and economic-social needs into account, combining resources, knowledge and skills [1, 2].

2   Inspiring Principles

A fundamental reference for the general objective of the idea is certainly the Council of Europe, which since 1949, has brought together 47 countries from Europe to Asia, under the mission of promoting democracy and protecting human rights and the rule of law in Europe. According to article 1 of its statute: "The Council of Europe aims to bring about a closer union between the Members in order to protect and promote the ideals and principles which are their common heritage and to foster their economic and social progress" [3].

These principles are taken up again in the European Cultural Convention of 1954 in order to strengthen, deepen and further develop a specific European culture, with culture and cultural heritage as the starting point. It establishes to pursue common goals and an action plan to achieve an integrated European society, celebrating universal values such as human rights and diversity, encouraging the implementation of measures to safeguard and encourage the development of each member's contribution to the common cultural heritage of Europe [4]. Again, on the European Union front, the Charter of Fundamental Rights, reaffirming the principles of democracy and the rule of law, seeks to promote balanced and sustainable development and ensures the free movement of persons, goods and services in compliance with diversity of cultures and traditions of European peoples which are fundamental in the light of the evolution of society, social progress and scientific and technological developments [5]. In this synthetic panorama, culture plays a fundamental role in promoting human rights, the practice of democracy and the rule of law, as the main and concrete tool for the implementation of these common values [6]. All this is implemented through a cultural policy with strong governance which, through transparency and participation, identity and dialogue, aims to respect and tolerate diversity in today's globalised world. A further fixed point will be the concept of conservation of cultural heritage, as an essential principle for building a peaceful and democratic society, for sustainable development processes and for the promotion of cultural diversity [7]. Another reference - last but not least - is the Council of Europe Landscape Convention, which, by promoting the protection, management and planning of European landscapes, promotes European cooperation by helping to develop a new culture of the territory. This vision of the territory is the basis of the transversal approach with which the proposed cultural itinerary of the Journey of Ulysses is pursued [8].

For these purposes, the Council of Europe makes use of a series of programmes, including that relating to Cultural Routes.

2.1   The Cultural Routes of the Council of Europe

The Cultural Routes Programme was launched by the Council of Europe in 1987 with the Santiago de Compostela Declaration [9].

The Cultural Routes of the Council of Europe, characterised as a certified European cultural path, represents an invitation to travel and to discover European cultural heritage, which can be achieved through the development of a network of people and places linked together by history and a common heritage, which promotes shared European culture and evocation. These itineraries, within the broad framework of cultural tourism, concretise and assume as their ultimate goal the founding values of the Council of Europe, demonstrating, through journeys in space and time, how the heritage of the various countries and cultures of Europe contribute to creating a shared and living cultural heritage, to be implemented - in general - as a tourism cooperation project. This cooperation network is aimed at promoting a path or a series of paths based thus on an historical journey, a cultural concept, a figure or a phenomenon of transnational and notable significance for the understanding and respect of common European values.

The main objectives of the Cultural Routes can therefore be summarised as follows:

– Promote awareness of a European cultural identity and European citizenship.
– Promote intercultural and inter-religious dialogue through a better understanding of European history.
– Safeguard heritage for the improvement of the living environment and social, economic and cultural development.
– Giving a place of honour to cultural tourism, with a view to sustainable development.

2.2   Certification of the "Cultural Route of the Council of Europe"

Reinforcing the potential of Cultural Routes is the Enlarged Partial Agreement on the Cultural Routes of the Council of Europe (EPA) established in 2010, which reaffirms the above concepts in terms of sustainable territorial development and social cohesion from a political and financial position, with particular attention to issues of symbolic importance for the discovery of lesser-known destinations [10].

Obtaining the "Cultural Route of the Council of Europe" certification is a guarantee of excellence for the network proposing this path: this certification can be granted to projects that deal with a theme that satisfies certain eligibility criteria and that involve priority actions in certain identified fields from Resolution CM/Res (2013) n. 67 [11]. The criteria concerning the themes are:

– the theme must be representative of European values and common to at least three countries of Europe;
– the theme must be researched and developed by groups of multidisciplinary experts from different regions of Europe so as to ensure that the activities and projects which illustrate it are based on consensus;
– the theme must be illustrative of European memory, history and heritage and contribute to an interpretation of the diversity of present-day Europe;
– the theme must lend itself to cultural and educational exchanges for young people and hence be in line with the Council of Europe's ideas and concerns in these fields;
– the theme must permit the development of initiatives and exemplary and innovative projects in the field of cultural tourism and sustainable cultural development;
– the theme must lend itself to the development of tourist products in partnership with tourist agencies and operators aimed at different publics, including school groups.

The priority fields are:

– Co-operation in research and development
– Enhancement of memory, history and European heritage
– Cultural and educational exchanges for young Europeans
– Contemporary cultural and artistic practices
– Cultural tourism and sustainable cultural development

The networks set up for the realisation of the project will have to:

– present a conceptual framework based on research carried out into the chosen theme and accepted by the different network partners;
– involve several Council of Europe member States (at least three) through all or part of their project(s);
– ensure that the projects proposed are financially and organisationally viable;
– have a legal status, either in the form of an association or a federation of associations;
– operate democratically.

Through this programme, the Council of Europe offers a model of transnational, cultural and tourist management that fosters - moreover - synergies between national, regional and local authorities and a series of associations and socio-economic actors.

To date, the CoE has certified 45 itineraries covering a number of different themes, from architecture and landscape to religious influences, from gastronomy and intangible heritage to the great masters of European art, music and literature. They differ in type:

– transnational projects (several countries are involved, of which at least 3 European countries)

– trans-regional projects (usually neighbouring countries are involved)
– regional projects (if the historical, cultural, artistic and social interest of the projects goes beyond the borders of the region or state)

and can be classified into:

– Linear Route, historical infrastructure
– Territorial Route, territorial contiguity
– Reticular pattern, virtual route

based on the theme and the type of network set up, Fig. 1.

Linear Route	Territorial Routes	Reticular pattern
Historical Infrastructure	*Territorial Contiguity*	'Virtual' Route
- Santiago de Compostela Pilgrim Routes - The Via Francigena - The Via Regia	- Iter Vitis - Roman Emperors and Danube Wine Route - Routes of the Olive Tree	- The Hansa - European Mozart Ways - European Route of Jewish Heritage

Fig. 1. Examples of the classification of cultural itineraries, based on the theme and type of network constituted. Reworked by CoE.

Ultimately, the itineraries offer multiple recreational and educational activities aimed at all citizens, both European and non-European, thus representing a key resource for responsible tourism and sustainable development of the territories.

3   The CoE Itineraries of Reference

The journey of Ulysses is set in the Mediterranean Sea area that is the background to all the adventures narrated in the Homeric poem, but it is also the physical element that unites populations belonging to different cultures and states today, once the protagonists of intense cultural and economic exchanges: therefore the Mediterranean is characterised as the geographical and cultural area - the core zone - in which the project will attempt to implement the principles.

Research on the Mediterranean as a cultural space will move towards the knowledge, conservation, management and enhancement of cultural heritage, both tangible and intangible, starting from the UNESCO sites present in the territories concerned.

The scientific comparison, animated in appropriate contexts, will deepen theoretical aspects and case studies with the aim of enhancing the "koinè" of the Mediterranean and verifying the possible existence, even today, of common traits in different territories and societies, or even, in current accounts of its multiple stratifications over a thousand years, of a cultural landscape of the Odyssey which can still be found.

In this regard, it is worth dwelling on some best practices among the Routes recognised by the CoE which - as for the itinerary of Ulysses - have the sea and the Mediterranean area as their reference element. Specifically, reference is made to the following Routes: Aeneas Route, Phoenicians' Route, Iter Vitis, Routes of the Olive Tree, [12], Fig. 2.

Cultural Route of reference			Certification year	Countries involved
AENEAS ROUTE *Associazione Rotta di Enea*	Cultural route of the Council of Europe Itinéraire culturel du Conseil de l'Europe	COUNCIL OF EUROPE CONSEIL DE L'EUROPE	2021	Albania, France, Greece, Italy, Tunisia, Turkey
LA ROTTA DEI FENICI	Cultural route of the Council of Europe Itinéraire culturel du Conseil de l'Europe	COUNCIL OF EUROPE CONSEIL DE L'EUROPE	2003	Albania, Belgium, Cyprus, Croatia, France, Greece, Italy, Lebanon, Malta, Palestine, Spain, Tunisia.
Iter Vitis *Les chemins de la vigne*	Cultural route of the Council of Europe Itinéraire culturel du Conseil de l'Europe	COUNCIL OF EUROPE CONSEIL DE L'EUROPE	2005	Armenia, Austria, Azerbaijan, Croatia, France, Georgia, Germany, Greece, Italy, Macedonia, Malta, Moldova, Portugal, Romania, Slovenia, Spain, Hungary
THE ROUTES OF THE OLIVE TREE	Cultural route of the Council of Europe Itinéraire culturel du Conseil de l'Europe	COUNCIL OF EUROPE CONSEIL DE L'EUROPE	2009	Albania, Algeria, Bosnia and Herzegovina, Cyprus, Croatia, Egypt, France, Jordan, Greece, Italy, Lebanon, Libya, Malta, Morocco, Portugal, Serbia, Syria, Slovenia, Spain, Tunisia, Turkey

Fig. 2. Best practice reference for the Journey of Ulysses

3.1 Aeneas Route

Firstly, the Route of Aeneas, which, recently awarded by the CoE, constitutes an essential term of comparison with Ulysses' journey, in terms of themes, reference area and project proposal [13].

The Route of Aeneas is an archaeological route that extends from the coasts of Turkey to the coasts of Lazio in Italy along a maritime and, in some points, land route. This Route is inspired by the legend of Aeneas as it was narrated by the Latin poet Virgil. Legendary "father" of Roman civilisation and a timeless source of inspiration for the artistic and cultural creation of European humanity over the centuries, the Trojan hero Aeneas remains a symbol of European identity. *"Establishing a mythological and historical link between Troy in Asia Minor and the foundation of Europe, the legend of Aeneas is characterised by a strong dimension of intercultural encounters between people and places. The figure of Aeneas thus embodies the values of dialogue and understanding between the peoples of the Mediterranean, of empathy and human solidarity, of coexistence and mutual enrichment, of respect for the other, of peace, of multiculturalism and of intercultural dialogue… the Aeneas Route aims to promote cross-border cultural cooperation and dialogue in Europe, along destinations that illustrate the richness of our shared European heritage"* in this way the values of the Council of Europe are declined and outlined in the reference documents, defining this path and the figure of Aeneas as key elements for an archaeological route that starts from Turkey and connects 5 European and Mediterranean countries through the legend of Aeneas. Over the centuries, Aeneas' narrative has become a common cultural heritage that unites different Mediterranean countries and civilisations and is the subject of countless paintings, mosaics, sculptures and works of art. Starting from the archaeological sites of Troy and Antandros (Turkey) and ending in Rome (Italy), the route brings together a series of rural landscapes, natural areas and archaeological sites, some of which are well known and inscribed on the UNESCO World Heritage list.

The route offers today's travellers the prospect of an adventurous discovery of a common European archaeological and cultural heritage. Travelling along the Route of Aeneas, visitors can explore off-the-beaten-track destinations through sustainable travel alternatives ranging from nautical, natural and landscape tourism to trekking and guided tours.

3.2 The Phoenicians' Route

Another reference is the Phoenicians' Route, certified in 2003, which *aims to stimulate intercultural dialogue in the Mediterranean by sharing the values of the Council of Europe, especially human rights and democracy. The route passes through various non-European countries - some of which are the theatre of conflicts - and contributes to promoting freedom of expression, equality, freedom of thought and religion and the protection of minorities. This network proposes a way of working together for peace and mutual respect in the Mediterranean* [14]. The Phoenicians' Route follows the great navigation routes which, from the 12th century BC, were used by the Phoenicians, great navigators and merchants, for trade and communication throughout the Mediterranean. Through these routes, the Phoenicians and other great civilisations of the Mediterranean contributed to the birth of a "koiné": a Mediterranean cultural community. The route covers 18 countries, many of which are located in North Africa and the Middle East, and strengthens the historical links between Mediterranean countries. These links are represented by a great heritage that originated with the ancient civilisations of the Mediterranean and which are found in various archaeological, ethnic, anthropological, cultural and naturalistic sites and also in the significant intangible heritage of the Mediterranean. The cities of the Mediterranean were the stopping places for travellers on the Phoenicians' Route, used to exchange artefacts, knowledge and experiences. In this sense, the travel experience along the Phoenicians' Route aims to show the traveller common routes, connecting countries of three continents and over 100 cities that originated from the ancient civilisations of the Mediterranean.

3.3 Iter Vitis

Winemaking has always been a symbol of European identity. The technical knowledge, indispensable for production, has contributed greatly over the centuries to the construction of European citizenship, which unites regions and populations, and of national identities. Many countries in the Mediterranean area share a common denominator: their cultural landscape. One of the main objectives of the route is the safeguarding of wine biodiversity, promoting its uniqueness in the globalised world [15]. These are the reasons with which this route enters by right among the Routes of the CoE. The culture of wine and winemaking and the viticultural landscape are an important part of European and Mediterranean gastronomy. The European rural landscape is an important heritage with a highly added value. Wineries, people and the technologies connected to them, are important components of European and Mediterranean culture, which was also handed down in the form of oral tradition. The quality of life in rural areas can also be considered a model for the future and a heritage to be protected. In this spirit, the traveller can discover remote lands, from the vineyards of the Caucasus to those of Western Europe, learn about cultivation techniques, winemaking, storage and transport, thus becoming familiar with the myths and symbols associated with this rich culture. Cultural and educational meetings are also organised in the villages crossed by the itineraries.

3.4 Routes of the Olive Tree

The cultivation of the olive tree has marked not only the landscape, but also the daily lives of Mediterranean peoples. Considered a mythical and sacred tree, the olive tree is associated with rituals and traditions and has influenced lifestyles, creating a specific ancient civilisation, the olive civilisation. The Routes of the Olive Tree retrace the footsteps of the Olive Civilisation, from Greece to the Euro-Mediterranean countries. The interaction between this tree and human civilisation has produced a lively and rich cultural heritage, founding the daily habits of Mediterranean peoples. It plays a fundamental role in aspects of food and wine, art and traditions and finally in the social development of the areas concerned. These words justify the values of the CoE in this project: *Routes of the Olive Tree are routes of intercultural discovery and dialogue concerning the olive tree, a universal symbol of peace. These itineraries are a bridge towards a new cooperation between remote areas, otherwise condemned to isolation, since they bring together all the operators involved in the use of the olive tree (artists, small producers and farmers, young entrepreneurs, etc.), threatened by today's crisis. In these difficult times, this is one way to defend the fundamental right to work* [16]. In this route, the traveller can touch the civilisation of the olive tree first hand and appreciate its landscape, products and traditions. There are various cultural itineraries that run through the countries of southern Europe and North Africa, from the Balkans to the Peloponnese in Greece, and as far as to Morocco. There are also routes by sea, to underline the important maritime connection between the port cities of the Mediterranean. Among the itineraries various activities related to the olive tree are exhibitions, concerts and tastings.

4 The 'Journey of Ulysses' Project

As mentioned, the Cultural Itinerary will touch the places of the Odyssey, explicitly cited or hypothesised by scholars through the numerous studies dedicated to the poem. Following the stages of Ulysses' journey, the common values of democracy, solidarity and intercultural dialogue that are the foundation of the CoE through cultural heritage (material and intangible) will be rediscovered and emphasised, together with the local communities first - and then with the users of the Itinerary and landscape that has been handed down to us. The project acts as a channel for intercultural dialogue and promotes better knowledge and understanding of the cultural identity of the territories while preserving and enhancing the cultural landscape inherent in them [17]. All this will take place while taking into account the UNESCO sites situated along the route, so that, and also thanks to them, a solid tourist network can be established in the name of sustainable growth for the territories, with the prospect of peace and prosperity for future generations.

4.1   Values of the Council of Europe Promoted by the Journey of Ulysses

It is certainly not the first time that we contrast ourselves with the Odyssey, with the values it transmits, with the meanings attributed to it, with the places hypothesised as scenarios for the poem. There were also many approaches used for its interpretation: from historical-anthropological ones to historical-philosophical ones, from the ethical-social to the psychological and psychoanalytical one.

For the purposes of the project. everything revolves around the figure of Ulysses and the territories he explores. That for Ulysses was not only the navigation between spaces and lands, but between identities that, when they met, meant a profound exchange of values, beliefs and ideals. An identity that arises from being part of an ethnos, the custodian of a language, a culture, a religion. Despite strong cultural and political fragmentation, the differing Greek communities were united by a uniform polythcism, considered, like the language, a concrete sign of common identity. Polytheism, despite its complexity, means acceptance. The Greek man was therefore already accustomed to the complexity of being different from himself, and had hospitality as his own value. Ulysses - a Greek - is the emblem of a hero who therefore represents not a community, nor Greece, but the whole of humanity. This is Ulysses, a universal hero. Ulysses is the man who, despite being Greek, lives without borders, becomes the prototype of modern man. Thanks to his courage and thirst for knowledge, he is able to weave networks of relationships between identity and globality. He becomes the emblem of a meeting, having inherited a great communicative ability directly from the god Hermes. But Ulysses is also the hero of nostos, of returning home: the Ulysses who returns to Ithaca is no longer the bold hero eager to grasp knowledge: the journey has made him different, a wiser man, able to to give to his people a new perspective of life; enlarged, enriched by those he met, by what he had learnt.

And it is precisely in this reading that the Ulysses' Journey Itinerary embodies one of the values proclaimed by the Council of Europe, the richness of diversity and encounter, the importance of dialogue and welcoming others through the knowledge of cultural heritage which, together with the implementation of responsible tourism, becomes the main tool for promoting intercultural dialogue, cooperation, mutual understanding and - ultimately - building peace [18, 19].

Yet there is more. By shifting the attention away from the protagonist, one discovers the world that surrounds him and which helps to give meaning to his adventures: nature is the background to the narrated events, the landscapes that surround the characters, the society in which they live. First of all there is the sea, the Mediterranean.

Much of the poem tells us of a Mediterranean civilisation in which people move by sea and through it entertain relationships, trade products of various kinds: cities arise mainly on the coasts, where there are natural landing places. The sea as a large connecting infrastructure, when other types did not exist or were much less efficient. Then, there is frastructure that also influences knowledge: seamen who know how to build boats, but also orientate with the stars and read the winds [20].

There is the natural world as a whole: the spontaneous vegetation of the Mediterranean islands, with its scents and colours that we can only imagine, looking at those actual ones; the forests where the timber for boats was obtained; sources of drinking water, which allowed life, before the aqueducts were built; the orography of the territories [21].

In this scenario we can see the structure of a complex society, the society of an Archaic Age, a society in transition which, as Codino and Calzecchi Onesti underline, experiences the first forms of assembly [22], which will then give rise to democracy, a further founding value of the CoE. And a society that feels the need for justice, for laws that regulate human relationships: when Ulysses decides to go and learn about the land of the Cyclops, he tells his companions that he wants to understand if the people who live there *"...they are violent, savage, without justice, or lovers of guests and have a pious mind towards the gods."* (L. IX, vv. 175–176) [23].

The following table, Table 1, attempts to outline the project, referring directly to the CoE criteria for the Evaluation of the Itinerary proposals:

Table 1. .

CoE criteria	'The journey of Ulysses' itinerary
Involves a theme representative of European values and common to at least three European countries	The theme of the Journey of Ulysses physically affects three European countries (Italy, Greece, Spain, Malta and Turkey) and embodies the values of democracy, encounters between cultures and hospitality in the figure of Ulysses, of his path and of the poem in general
Is the subject of transnational and multidisciplinary scientific research	The Homeric poem has always attracted the attention of scholars of all times, for the many ideas it offers: in the itinerary project the network that is established involves multidisciplinary skills ranging from historian to evaluator - for example - of all the countries touched by the itinerary. It also intends to implement, not only research activities related to the cultural and natural heritage in terms of knowledge, but also conservation and enhancement: following the references of the European Green New Deal to create a network of scientific cooperation that contrasts climate change for the ecological transition
Enhances European memory, history and heritage and contributes to the interpretation of today's diversity in Europe	Ulysses and his journey in the Mediterranean shows a cross-section of a past society and in the descriptions he records the places of the past, many still recognisable. The values transmitted by that society, democracy, justice, hospitality, respect for others, together with the search for beauty, are the foundations of the future Western world. Taking into account the relevance of the figure of Ulysses

(continued)

Table 1. (*continued*)

CoE criteria	'The journey of Ulysses' itinerary
Supports cultural and educational exchanges for young people	The project, in a nutshell, envisages and promotes cultural and educational exchanges among the youngest, in order to pass on these values over time
Develops exemplary and innovative projects in the fields of cultural tourism and sustainable cultural development	Cultural tourism has a place of honour in the Ulysses' journey project, as the only means of implementing its aims, which through a sustainable use of resources allow balanced growth of the territories involved due to the correct enhancement of its cultural heritage (material, intangible and naturalistic). Therefore, the projects that will be developed will fully embrace the network of places involved in order to make them usable in a sustainable and balanced way
Develops tourism products and services aimed at different groups	The itinerary, in addition to preparatory actions for cultural tourism, such as the recovery of historic centres, archaeological areas or the improvement of connecting infrastructures, the safety of naturalistic areas or the creation/consolidation of the economic network that governs everything, includes the creation of tourist services diversified by target and type, thanks to the multiplicity of ideas offered in the chosen theme. To this end, a diversified partnership will be used, capable of responding to all project requirements

5   Conclusions

The proposal to nominate the Journey of Ulysses to the Cultural Routes Programme of the Council of Europe is a difficult thing: the multiplicity of themes and ideas left by Homer in his poem highlights such a complexity that it can only be managed through an integrated and multidisciplinary approach that makes this complexity the real strength of the whole proposal.

In a now globalised world, offering the itinerary to future users offers the possibility to immerse themselves in the world of Ulysses both of today and yesterday, embodying his values, which are still current, will be an experience of growth, where the encounter with the other will be a wealth.

Succeeding in the creation of this network, where Ulysses himself is a network, would be a great goal, even if the Itinerary does not fully meet the requirements of the CoE which should certify its excellence: the realisation of the project would in any case trigger the implementation of virtuous processes, of cultural cooperation in which the nodes of the network will have to work in a democratic and participatory way with respect to its management, research and the implementation of activities.

The results of the research activities that will arise from the implementation of the idea are referred to in subsequent studies, with the general aim of giving value and visibility to those territories that have remained outside the mechanisms of economic, cultural and social growth and that have so much to offer in terms of of resources, where often the cultural and naturalistic heritage still remains intact or requires targeted measures to be rediscovered.

References

1. Cassalia, G., Tramontana, C., Ventura, C.: New networking perspectives towards mediterranean territorial cohesion: the multidimensional approach of cultural routes. Proc. Soc. Behav. Sci. **223**, 626–633 (2016). https://doi.org/10.1016/j.sbspro.2016.05.371
2. Calabrò, F.: Integrated programming for the enhancement of minor historical centres. The SOSTEC model for the verification of the economic feasibility for the enhancement of unused public buildings | La programmazione integrata per la valorizzazione dei centri storici minori. Il Modello SOSTEC per la verifica della fattibilità economica per la valorizzazione degli immobili pubblici inutilizzati. ArcHistoR **13**(7), 1509–1523 (2020). https://doi.org/10.14633/AHR280
3. Statute of the Council of Europe: London, 5.V.1949
4. European Cultural Convention: Paris 19 December 1954
5. Charter of Fundamental Rights of the European Union: Nizza, 7 December 2000
6. Mallamace, S., Calabrò, F., Meduri, T., Tramontana, C.: Unused real estate and enhancement of historic centers: legislative instruments and procedural ideas. In: Calabrò, F., Della Spina, L., Bevilacqua, C. (eds.) New Metropolitan Perspectives. SIST, vol. 101, pp. 464–474. Springer, Cham (2019). https://doi.org/10.1007/978-3-319-92102-0_49
7. Council of Europe Framework Convention on the Value of Cultural Heritage for Society, Faro, 27.X.2005
8. Council of Europe Landscape Convention, Florence, 20.X.2000
9. The Santiago de Compostela Declaration, 23 October 1987
10. Enlarged Partial Agreement on Cultural Routes (EPA), COE (2010)
11. Council of Europe: Resolution CM/Res, n. 67 (2013)
12. https://www.coe.int/it/web/cultural-routes/by-theme
13. https://www.coe.int/it/web/cultural-routes/aeneas-route-4
14. https://www.coe.int/it/web/cultural-routes/the-phoenicians-route
15. https://www.coe.int/it/web/cultural-routes/the-iter-vitis-route
16. https://www.coe.int/it/web/cultural-routes/the-routes-of-the-olive-tree
17. International Cultural Tourism Charter. Managing Tourism at Places of Heritage Significance, Adopted by ICOMOS at the 12th General Assembly in Mexico, October 1999
18. Campolo, D., Calabrò, F., Cassalia, G.: A cultural route on the trail of greek monasticism in Calabria. In: Calabrò, F., Della Spina, L., Bevilacqua, C. (eds.) New Metropolitan Perspectives. SIST, vol. 101, pp. 475–483. Springer, Cham (2019). https://doi.org/10.1007/978-3-319-92102-0_50
19. Calabrò, F.: Promoting peace through identity. Evaluation and participation in an enhancement experience of calabria's endogenous resources | Promuovere la pace attraverso le identità. Valutazione e partecipazione in un'esperienza di valorizzazione delle risorse endogene della Calabria. ArcHistoR, **12**(6), 84–93 (2019). https://doi.org/10.14633/AHR146
20. Lopez, L., Pérez, Y.: Turismo, patrimonio e cultura: verso un'educazione territoriale. Un caso di studio in Galizia (Spagna). LaborEst, **21**, 18–24 (2020). https://doi.org/10.19254/LaborEst.21.03
21. Solano, F., Praticò, S., Piovesan, G., Chiarucci, A., Argentieri, A., Modica, G.: Characterizing historical transformation trajectories of the forest landscape in Rome's metropolitan area (Italy) for effective planning of sustainability goals. Land Degradation Development, July, ldr.4072 (2021). https://doi.org/10.1002/ldr.4072

22. Spampinato, G., Malerba, A., Calabrò, F., Bernardo, C., Musarella, C.: Cork Oak forest spatial valuation toward post carbon city by CO2 sequestration. In: Bevilacqua, C., Calabrò, F., Della Spina, L. (eds.) New Metropolitan Perspectives. SIST, vol. 178, pp. 1321–1331. Springer, Cham (2021). https://doi.org/10.1007/978-3-030-48279-4_123
23. Codino, F.: Preface of Omero, Odissea. Translation of Rosa Calzecchi Onesti. ET Classici Einaudi (2014)

The Journey of Ulysses: A Cultural Itinerary Among the Shores of the Mediterranean for the Promotion of Dialogue and Sustainable Development as Tools for Peace and Territorial Growth

Roberta Alberotanza[1], Francesco Calabrò[1] (ID), Mariangela Monaca[2] (ID), and Carmela Tramontana[1(✉)] (ID)

[1] Mediterranea University of Reggio Calabria, 89100 Reggio Calabria, Italy
carmen.tramontana@unirc.it
[2] University of Messina, 98122 Messina, Italy

Abstract. Retracing the stages of the Odyssey journey, the project aims to build a Cultural Itinerary which, reconsiders the legacy left by the poem, fosters Dialogue between peoples and the Sustainable Development of territories. Through the knowledge of local places and identities, scientific research will promote new forms of education for the younger generation, with particular attention to cultural diversity: this will contribute to building a perspective of peace and prosperity, according to the principles of the Council of Europe and the Faro Convention - as regards the involvement of local communities and the role of cultural heritage for society - and the european green new deal, as regards sustainable development. the mediterranean area is the physical-geographic space in which the idea is located, in which to create (or rather recreate) the dense network of cultural and economic exchanges of the past, of which the homeric story is testimony in the widespread material and immaterial evidence of the territories that characterise it. the project will make use of this network to identify shared values by the different peoples of the mediterranean and build, as well as relations between territories, a network of subjects, capable of implementing strategies and policies aimed at achieving sustainable and inclusive growth, which identifies - for example - sustainable tourism as a possible relaunch for the areas involved, thus making them more competitive and attractive.

Keywords: Cultural heritage · Mediterranean · Sustainable development · Networks · Integrated approach

1 Introduction

The journey of Ulysses, narrated by Homer in the Odyssey, is one of the most fascinating stories ever, where myth mixes with reality, and events over time become a paradigm

---

The work is the result of the shared commitment of the authors. However, paragraph 1 can be attributed to Roberta Alberotanza; to Carmela Tramontana paragraphs 2, 3; to Mariangela Monaca paragraph 4; to Francesco Calabrò paragraphs 5, 6. Conclusions were written jointly by the authors.

© The Author(s), under exclusive license to Springer Nature Switzerland AG 2022
F. Calabrò et al. (Eds.): NMP 2022, LNNS 482, pp. 2417–2427, 2022.
https://doi.org/10.1007/978-3-031-06825-6_231

with strong symbolic values, still current today. This contribution proposes guidelines for the construction of a cultural itinerary which, through the knowledge of local places and identities, promotes scientific research and prospects new strategies of youth education, favours intercultural dialogue, sustainable development, collaborators.

in the building of a peaceful and dramatic society, in line with the values promoted by the Council of Europe, thus contributing to the prospect of peace and prosperity [1, 2].

The project, focusing its incipit on shared values, intends to promote the knowledge, enhancement and use of everything (identity, material and intangible, cultural and natural) heritage of which the places affected by the itinerary are keepers. this will happen starting from the unesco sites in their vicinity, through the establishment of a sustainable tourism network in order to trigger or consolidate virtuous processes of economic and social growth for the territories involved [3]. Properly planned and responsibly managed cultural tourism, by establishing forms of governance in which different cultures, rights holders and stakeholders participate, can be a powerful vehicle for cultural heritage conservation and sustainable development. Responsible tourism promotes and creates awareness of cultural heritage, provides opportunities for individual and community well-being and resilience, and creates respect for the diversity of other cultures. It can therefore contribute to intercultural dialogue and cooperation, mutual understanding and peace building [4, 5].

A further objective of the project is to create a lasting network of scientific cooperation, which in addition to cultural and natural heritage, is concerned with contrasting climate change, in the field of ecological transition, which finds a valid basis in the European Green New Deal, in order to increase the well-being of current communities and future generations. The project is inspired by a new growth strategy aimed at making society just and prosperous, with a modern, resource-efficient and competitive economy, in which economic growth will be dissociated from the use of resources [6]. According to these premises, the implementation of the idea requires an integrated approach, which according to a multidisciplinary logic, takes into account cultural, environmental and economic-social issues, combining resources, knowledge and skills.

2   Ulysses, Travel and the Network

The unifying element of the whole idea, identifying the perimeter of the project, is the Mediterranean and its shores: it is the background to all the adventures narrated in the Homeric poem, but it is also the physical element that unites populations today which belong to different Cultures and states, once protagonists of intense cultural and economic exchanges: therefore the Mediterranean will also become the geographical area in which the project will attempt to implement the principles of the Green New Deal and the Council of Europe, which through its program on cultural [7, 8], contributes to the consolidation of common values and roots in Europe and enhances lesser known territories, though rich in values.

Obviously, there can be no certainty in the location of the places mentioned in the Odyssey, except for a few cases such as, for example, Troy, Scylla and Ithaca. There are, in fact, numerous hypotheses relating to the other places mentioned. It is an extremely mobile and imaginative scenario, which expands from east to west, in a temporal space that stretches from the history of the first colonies to late antiquity, Fig. 1 [9–13].

Fig. 1. The Journey of Ulysses: 1. Troy, 2. Land of the Ciconians, 3. Land of the Lotus Eaters, 4. Land of the Cyclops, 5. Island of Aeolus, 6. Land of the Laestrygonians, 7. Circe's Island, 8. Pillars of Hercules, 9. Land of the Sirens, 10. Scylla and Charybdis, 11. Island of the Sun, 12. Ogygia, 13. Island of the Phaeacians, 14. Ithaca.

Whereas, crossing the border of met-history, in consideration to the purposes of the project, an inclusive logic will be followed, which takes into consideration all positions in order to increase the number of communities that will activate cultural exchanges, thus favouring a greater dissemination of the culture of Peace and Dialogue.

The places touched by Ulysses constitute a network, like the network served to Homer to tell a story. A network that, like that of the fishermen of Mare Nostrum crossed by the legendary Ulysses, smells of the sea, of "stories", of experiences, a network intertwined with ideas, thoughts, words, costumes and traditions.

As Ulysses' voyage was not just navigation between spaces and lands, but between identities that, when they met, staged a type of dance of values, beliefs and ideals, in which everyone brought himself, and made the self his own, and other, without ever losing his identity, driven by the desire to know the other and to "connect" with him. Despite the cultural limit of the border of Greece, Ulysses is the man who lives without borders, the prototype of modern man, torn between globalising tensions, identity defence and migration, all necessary for survival: on his journey each will bring themselves, will keep both identity and cultural facies, but they will be reshaped during the encounter with the other to give life to something new, a new social cultural economic identity in which everyone will be the protagonist of interfaces and exchanges.

Ulysses does not "weave a web" of relationships: he is himself the "web", the driving force of the dialectic between self and other, between identity and globality.

The project will make use of this network idea to identify shared values between the relationships left by the Homeric stories and the material evidence: a network of relationships, territories, subjects, exchanges. The countries initially involved are: Greece, Italy, Malta, Morocco, Spain, Tunisia, Turkey.

On the basis of the scientific evidence that will emerge during the development of the project and the political conditions that will occur over time, other possible countries to be involved are: Algeria, Croatia, France, Libya, and also those countries bordering the eastern Mediterranean and the Adriatic.

3 Design Lines

The Cultural Itinerary "The Journey of Ulysses" aspires to promote intercultural dialogue and sustainable development in the Mediterranean, through the establishment of transnational networks of cities in order to promote and improve their competitiveness [14].

In particular, the actions aimed at promoting sustainable development will concern the conservation and enhancement of natural and cultural heritage, tangible and intangible, present in the territories involved in the project, starting from the heritage already recognized and protected under different forms: Natural Protected Areas, sites and elements of cultural interest, starting from UNESCO sites, and, more generally, the landscapes and cultural landscapes.

The project will favour, in particular, scientific exchanges, as mentioned, concerning the conservation and enhancement of natural and cultural heritage, the fight against climate change and the active participation of citizens in such processes, in line with the principles established by the Faro Convention [15]. At the same time, initiatives will be developed aimed at developing the agri-food supply chains in the territories involved, in particular enhancing identity production and connected traditional knowledge [16].

The Cultural Itinerary will be usable from a tourist point of view thanks to the involvement of tour operators, but also by acting locally in synergy with the shipping companies, with companies operating in the field of accommodation, with tourist guides, also through the involvement of any trade organisations.

The key to the proposal is right here: in the concept of cultural tourism, in the role it plays both in promoting the culture of peace and the protection of heritage, and in the multiple opportunities for inclusive economic growth and sustainable development through the creation of places of work and the regeneration of rural and urban areas.

For these aspects, in addition to other references, the Muscat Declaration, drawn up on the occasion of the second UNWTO/UNESCO World Conference on Tourism and Culture in 2017, underlines the importance of the interaction between these sectors:

- making sustainable cultural tourism more "peace-sensitive" so that it can help strengthen global citizenship and encourage peace-related site visits, cross-border travel, exchange visits and religious tourism;
- promoting cultural experiences, cultural exchange and dialogue through innovative tourism models that facilitate guest-guest interaction, enhancing cultural diversity and heritage, involving and empowering local communities;
- raising awareness of the impact of sustainable cultural tourism on peaceful societies among international, national and local stakeholders and conducting further research on the role of tourism in building peace, reconciliation and security;

– stressing the importance of protecting historic heritage sites with cultural sharing and religious values while preserving and enhancing social diversity for the benefit of all its people, that of the wider region and the world [17].

For the aspects related to the growth of the territories, a further reference for the project lines is the Siem Reap Declaration on Tourism and Culture drawn up by UNWTO / UNESCO in 2015 in order to build a new partnership model capable of:

– Promoting and protecting cultural heritage through tourism activities that contribute to raising awareness of the values inherent in it through the experience of the tourism chain
– Connecting people and fostering sustainable development through cultural itineraries through cooperation across regional or national borders to encourage, facilitate and build governance models and certifications to guarantee the quality of cultural itineraries;
– Promoting fairness of governance structures, public/private partnerships and marketing activities along the entire cultural path;
– Consulting local communities and involving them as stakeholders in the formulation and management of tourism along cultural itineraries. [18]

Particular attention will be paid to young people: the values of the project will be conveyed through direct involvement in school project and the promotion of cultural exchanges between schools.

4   Yesterday's Itinerary for Today's World

What remains today of the world described by Homer? What traces, if any, are still visible? What lessons can be drawn from reading the poem, applicable for facing the challenges of contemporaneity?

We will therefore try to build a Cultural Itinerary that attempts to grasp the symbolic component of the poem, distinguishing first of all the material component from the immaterial one, but also looking for reciprocal influences, in particular how the material substrate conditions, in an almost decisive way, the evolution of the immaterial dimension.

In this reading of the territory of reference, in search of its cultural values, continuous references are certainly the UNESCO's Conventions, like the Convention Concerning the Protection of the World Cultural and Natural Heritage (1972), the Convention on the protection of the underwater cultural heritage (2001), the Convention for the Safeguarding of the Intangible Cultural Heritage (2003) and the Convention on the Protection and Promotion of the Diversity of Cultural Expressions (2005) [19–22].

The new NDICI-Global Europe instrument combines several previous EU external financing instruments. It aims to support the countries most in need to overcome long-term development challenges and will contribute to the achievement of the international commitments and objectives agreed by the Union: in this sense, it could be a valid tool for the realization of the project and/or individual actions that will affect the entire path [23].

Of the material component, as already mentioned, the most important element is the sea, in particular the Mediterranean: the project will try to highlight this element, favouring the knowledge and enhancement of certain aspects, natural or the result of ingenuity and manmade. In line with the aforementioned principles relating to the European Green New Deal, the 2030 Agenda, we will have, for example:

– As regards the natural aspects, those connected to life in the seas, that is to marine biology, but also to winds and currents, and to other dynamics that, in particular in recent decades, have endangered sea-life, starting from marine pollution; on this front, the cognitive aspects related to technologies for prevention and the elimination of pollutants will be studied. Then there are a whole series of other issues related to the natural dynamics in the sea, from the phenomena of coastal erosion to the energy that can be obtained from wave motion, up to forms of cultivation, that is, fish farming. Protected marine areas, sea museums scattered along the coasts, fish companies, but also research centers that are developing technologies for the protection of the marine environment and its sustainable use may be involved in this part of the project.

– As regards human interaction, the most important component is the art of navigation, from the construction of boats (today, from design) in all their components, to the port infrastructure system, to the system of necessary knowledge necessary to go to sea. In the development of this component, nautical clubs, sailing schools, fishermen, shipyards, shipping companies, cruise lines, but also universities in which there is teaching connected with all these issues will be fundamental partners. Another fundamental aspect of the "artificial" dimension are the cities: the cities that exist today along the coasts but above all the remains of those that existed when the poem was written, therefore the archaeological areas of the Magna Graecia age and all the underwater heritage, largely unknown or otherwise not usable.

The immaterial dimension of the project essentially focuses on six themes:

– Democracy
The assembly forms described in some passages of the Odyssey obviously do not reflect democratic systems such as those considered as such today, but still testify to the existence, even then, of forms of citizen participation (according to the citizenship rights of the time) in decision-making processes. Today, if on the one hand democratic systems are experiencing a profound crisis for various reasons, on the other hand different forms of participation are being tested, the most relevant of which, for the purposes of the project, will be that provided for by the Faro Convention: the citizens as protagonists, through patrimonial communities, of the conservation, enhancement and transmission to future generations of Cultural Heritage.

– Migratory flows
The different testimonies of forms of welcome towards foreigner present in the Odyssey can inspire new ways to face the challenges of increasingly intense migratory flows from the Southern hemisphere to richer countries.

– The power supply
Today's challenges regarding nutrition are complex: if on the one hand there is the problem of guaranteeing sufficient food for all (a challenge also highlighted by the

2030 Agenda in Objective 2 of the Sustainable Development Goals) [24] and, on the other, the need exists to guarantee the safety and quality of food. In the richest countries, then, it becomes a priority to fight the perverse effects of well-being and poor nutrition on ever larger sections of the population as much as the fight against food waste. In this sense, the Mediterranean Diet is a lifestyle, as well as a diet, able to reconcile those different needs [25].

– The play dimension

Sports, body care, but also music, dance and parties in general have always been indispensable expressions of the human soul: the project will explore its Mediterranean roots and their contemporary evolutions.

– Craft activities

The progressive mechanisation of production processes has led to a strong crisis in artisanal activities, increasingly unable to withstand the competition of industrialised systems. New communication technologies, together with increasing accessibility to innovation, can constitute relevant tools, able to restore competitiveness to the various craft sectors that draw their origins from antiquity, such as the production of wicker and ceramic containers.

The project will foster opportunities for exchange and knowledge and the birth of transnational relationships that support virtuous development processes.

– religiosity

The spiritual dimension is one of the constants of human life, in all its different manifestations and is widely represented in the poem: starting with Athena, who accompanies Ulysses throughout the journey, but also Poseidon who hinders him, Zeus who peeps out from time to time, like the other gods of Olympus. The project will promote forms of inter-religious dialogue and moments of Nondenominational spirituality of reflection on ethical and eirenic bases, with the aim of recovering the sense of the sacred in humanity and considering human values intangible and inviolable.

5 Subjects Involved

The project initially involves the following types of subjects:

– the universities, research centres and technological districts of the cities touched by Ulysses' trip or in any case of neighbouring cities, as regards research activities;
– the ICOMOS of the countries concerned, for specialist advice on the issues of heritage protection, conservation, enhancement and management, also through the construction of cultural itineraries and the organisation of sustainable forms of management and tourist use of the heritage, with the possible involvement of the thematic scientific committees concerned;
– the UNESCO clubs of the territories involved, under the supervision of the respective national federations, for territorial animation activities;
– high schools, starting from classical high schools, including all schools in which the Odyssey is part of the study program, for activities related to the world of youth;
– local authorities (municipalities, provinces/metropolitan cities, regions) and peripheral offices of state administrations (superintendencies, secretariats);

– Local Action Groups (LAGs) and, above all, FLAGs (Fisheries Local Action Group), for the involvement of local business systems.

The Promoting Committee of the project is composed as follows: ICOMOS Italy, Mediterranean University of Reggio Calabria and University of Messina; FICLU; MiC Secretariat of the Calabria Region; Superintendence of Archeology, Fine Arts and Landscape of Reggio Calabria - Vibo Valentia; Municipality of Scilla; FLAG of the Straits.

In the continuation of the activities it will also be possible to involve other types of subjects, such as the Ministries, the Italian Cultural Institutes, ICOM etc.

This form of diversified partnership, which includes - in the full spirit of the project – Institutions, universities, research centres, entrepreneurs and associations will be able to respond to all project needs and, at the same time, to the indications of the Faro Convention on community involvement.

6 Project Road Map

The project uses an integrated approach [26] can be divided into two levels, national and international, and three phases:

Phase 1: Preparatory activities
This phase has already started and is aimed at creating the necessary conditions for a profitable development of the project; in turn, it is divided into 4 sub-phases:
Phase 1.a Scientific knowledge: aimed at building the minimum knowledge base for project development;
Phase 1.b Territorial animation: currently managed by the FICLU and the FLAG of the Straits and includes:

– Involvement of UNESCO Clubs and Italian LAGs / FLAGs;
– Identification of schools associated with UNESCO and, subsequently, with ISESCO (for Islamic countries);
– Involvement of classical and scientific high schools;
– Involvement of other local actors (seafarers, tour operators, local historians, shipowners, citizens, other associations, etc.);
– Activation of cultural exchanges (European Voluntary Service - EVS; Erasmus + etc.).

Phase 1.c Construction of the partnership: Construction of the partnership is handled by all the parties who initially initiated the project and who constitute the aforementioned Promoting Committee.
Phase 1.d Institutional dialogue: includes both at a national and international level; it will be taken care of by all the partners and foresees:

– Relations with the Ministries and the Council of Europe
– Relations with the Italian Cultural Institutes of the countries involved

- Relations with the UNESCO National Commission, the Regions, local authorities
- Relations with the Superintendencies and Regional Secretariats
- Relations with local and regional administrations in Italy

Phase 2: CoE Candidacy Proposal

The conclusion of Phase 1 of the project will allow for the preparation of the application dossier for the Council of Europe program. The new phase will also allow the development of other projects (research; territorial animation; education in schools; cultural tourism paths, etc.), defining the governance of the project and identifying the proposing party for possible candidates, possibly with a legal personality different from that of the partner subjects. This aspect could take on a complex architecture, with the possibility that new entities will arise, different for each of the States involved and in turn linked by forms of partnership to be defined. In this phase, particular attention will be paid to the aspects related to Communication.

Phase 3: Implementation and management (start up)

The initial implementation and management phase includes, first of all, the organisation of all the services necessary for the use of the itinerary, including communication and marketing actions. To do this, it will be necessary to involve local actors entitled to provide individual services, the definition of production regulations and the signing of partnership agreements.

Even if an ad hoc subject is set up for the presentation of applications and any others for the implementation of projects, the partners will continue to have at least a monitoring, tutoring and accompanying function as long as the new subject is not sufficiently consolidated.

Trying to draw up a time schedule, taking into account all the surrounding situations, we could hypothesize:

Phase 1: Preparatory activities 6 months.
Phase 2: Applications 9 months.
Phase 3: Implementation and management (start up) 21 months.

7 Conclusions

The proposed itinerary fully embraces the metaphor of travel, not only a journey of man but a journey of his soul, which creates connections, encounters, generates dynamics of belonging: a sea journey that has the image of the network, as that of the fishermen of Mare Nostrum navigated by the mythical Ulysses.

The Journey of Ulysses, interpreted in these terms, with its multiple readings, becomes an itinerary that promotes Dialogue and Sustainable Development in the Mediterranean as instruments of Peace and growth for the territories: Ulysses becomes a metaphor for modern man who - yesterday as today - explores and welcomes diversity, recognising in it a mutual enrichment. His path instead becomes the symbol of a network, whose interconnected nodes cooperate for the same purpose. Culture, reflected in heritage and traditions as well as in contemporary art, languages, cuisine, music, crafts, museums and literature, shapes community identities and promotes respect and tolerance, cultural tourism can make a significant contribution to socio-economic development and the empowerment of local communities [27].

As is clear from the discussion, the proposed theme opens up multiple lines of research: is the journey of Ulysses to be considered a metaphor or an evocation of the myth of the Homeric hero? At the moment it is not possible to dare to answer the questions: they will certainly be clearer with the subsequent development of the research.

The results of the research activities that will arise from the implementation of the idea are referred to in subsequent studies, with the awareness of the extreme complexity of the theme: complexity understood not as an obstacle or difficulty, but a qualitative characteristic of an articulated structure of elements that each offer multiple ideas.

References

1. Statute of the Council of Europe, London, 5.V.1949
2. European Cultural Convention, Paris 19 December 1954
3. Campolo, D., Calabrò, F., Cassalia, G.: A cultural route on the trail of Greek monasticism in Calabria. In: Calabrò, F., Della Spina, L., Bevilacqua, C. (eds.) New Metropolitan Perspectives. SIST, vol. 101, pp. 475–483. Springer, Cham (2019). https://doi.org/10.1007/978-3-319-92102-0_50
4. International Cultural Tourism Charter. Managing Tourism at Places of Heritage Signifi-cance. In: ICOMOS at the 12th General Assembly in Mexico, October 1999
5. Calabrò, F.: Promoting peace through identity. evaluation and participation in an enhancement experience of calabria's endogenous resources I Promuovere la pace attraverso le identità. Valutazione e partecipazione in un'esperienza di valorizzazione delle risorse endogene della Calabria. ArcHistoR. **12**(6), 84–93 (2019). https://doi.org/10.14633/AHR146
6. The European Green Deal, Bruxelles, COM (2019)
7. Enlarged Partial Agreement on Cultural Routes (EPA), COE (2010)
8. Council of Europe: Resolution CM/Res, n. 67 (2013)
9. Braccesi, L.: Sulle rotte di Ulisse. L'invenzione della geografia omerica, Laterza, Bari (2010)
10. Luce, J., Stanford, W.B.: The Ulisse Quest, Phaidon, Londra (1974)
11. Armin Wolf - Hans-Helmut Wolf, Die wirkliche Reise des Odysseus: Zur Rekonstruktion Des Homerischen Weltbildes, Langer, München (1983)
12. Ballabriga, A.: Les fictions d'Homere. Presses Universitaires de France, Parigi, L'invention mythologique et cosmographique dans l'Odyssée (1998)
13. Geisthövel, W.: Unterwegs mit Odysseus durch das Mittelmeer, Artemis und Winkler, Düsseldorf 2007; inglese trans. Il Mediterraneo di Omero: A Travel Companion, Haus, Londra (2010)
14. Calabrò, F., Della Spina, L., Aragona, S.: The evaluation culture to build a network of competitive cities in the Mediterranean. In: Calabrò, New Metropolitan Perspectives - The Integrated Approach of Urban Sustainable Development, Transaction of Technical Publications (2014)
15. Council of Europe: Convenzione quadro sul valore del patrimonio culturale per la società, Faro, 3 ottobre 2005
16. Barrile, V., Malerba, A., Fotia, A., Calabrò, F., Bernardo, C., Musarella, C.: Quarries renaturation by planting cork Oaks and survey with UAV. In: Bevilacqua, C., Calabrò, F., Della Spina, L. (eds.) New Metropolitan Perspectives. SIST, vol. 178, pp. 1310–1320. Springer, Cham (2021). https://doi.org/10.1007/978-3-030-48279-4_122
17. Muscat Declaration on Tourism and Culture: Fostering Sustainable Development (2017)

18. Siem Reap Declaration on Tourism and Culture – Building a New Partnership Model (2015)
19. Convention Concerning the Protection of the World Cultural and Natural Heritage, Paris (1972)
20. Convention on the protection of the underwater cultural heritage, Paris (2001)
21. Convention for the Safeguarding of the Intangible Cultural Heritage, Paris (2003)
22. Convention on the Protection and Promotion of the Diversity of Cultural Expressions, Paris (2005)
23. Neighbourhood, Development and International Cooperation Instrument – Global Europe (NDICI – Global Europe)
24. United Nations: Agenda 2030. Sustainable Devolepment Goals. https://sdgs.un.org/goals
25. Moro, E.: La Dieta Mediterranea UNESCO. Un modello di sviluppo sostenibile tra mito e realtà. LaborEst **12**, 17–24. (2016) https://doi.org/10.19254/LaborEst.12.03
26. Calabrò, F.: Integrated programming for the enhancement of minor historical centres. The SOSTEC model for the verification of the economic feasibility for the enhancement of unused public buildings | La programmazione integrata per la valorizzazione dei centri storici minori. Il Modello SOSTEC per la verifica della fattibilità economica per la valorizzazione degli immobili pubblici inutilizzati. ArcHistoR. **13**(7), 1509–1523 (2020). https://doi.org/10.14633/AHR280
27. Tramontana, C., Calabrò, F., Cassalia, G., Rizzuto, M.C.: Economic sustainability in the management of archaeological sites: the case of Bova Marina (Reggio Calabria, Italy). In: Calabrò, F., Della Spina, L., Bevilacqua, C. (eds.) New Metropolitan Perspectives. SIST, vol. 101, pp. 288–297. Springer, Cham (2019). https://doi.org/10.1007/978-3-319-92102-0_31

In the Wake of the Homeric Periples: Escapes, Rejections, Landings and Emergencies in Italy from the Second Post-war Period Up to the Pandemic Times

Salvatore Speziale[(✉)]

University of Messina, 98168 Messina, Italy
salvatore.speziale@unime.it

Abstract. Since preomeric times, the Mediterranean routes have met the largest natural watershed in the Mediterranean, the Italian peninsula. The old and new Ulysses, fleeing from their Ithacas and looking for new homes, daily land on the Italian and European coasts, posing logistical, social, political and cultural problems that were crucial for the past and are equally crucial for the present and for the future.

The history of these flows in the time framework going from the second post-war period to the Covid pandemics times of these days can serve to better understand and deal with them. At the same time, the analysis of the italian reactions to the arrivals of immigrants and to the increasing presence of them in the territory, that include also a long chain of emergencies and delayed laws here sketched, must arouse new points of view and encourage new solutions that are more in step with the times. A different and deeper awareness of the migratory phenomenon can also bring to a revaluation of the positive and different contributions of migrants to the complex history of the Mediterranean.

Keywords: Migration · Emergency · Mediterranean

1 The New Ulysses: Mediterranean and Italy, Crossroads of Migrations

More than 280 million people today live outside their country of origin and in the last twenty years the number of those who have left their homeland has increased by about 50% [1]. It would be enough to consider these data to place the migratory phenomenon in the right geographical, demographic and historical perspective: a global factor of ever-increasing proportion with a very long-lasting process [2]. Actually, who considers the historical evolution of the planet, and particularly of the Mediterranean space, feel the richness of a millenary history made up of voyages, encounters and clashes that in the case of the *Mare nostrum* are essentially favored by the geographical proximity of African, Asian and European shores: from the land of the lotophages to that of the lestrigones, from the land of the Cicones to that of the nymph Calypso. Therefore, a wider acknowledgment of this history would favor a more correct vision of migrations and a more concrete definition of spaces, distances and borders for those who live around

© The Author(s), under exclusive license to Springer Nature Switzerland AG 2022
F. Calabrò et al. (Eds.): NMP 2022, LNNS 482, pp. 2428–2437, 2022.
https://doi.org/10.1007/978-3-031-06825-6_232

this sea as "frogs around a pond"[1] and who, for opportunities and/or by coercion, jump from one shore to the other, from North to South and vice versa. Italian and European migration to Africa, for example, deserves a space that historiography has not granted yet. Conversely, the history of African and Asian migration to Italy is certainly useful for identifying lines of continuity and turning points both in the actual consistency and in articulation of migration paths, both in the perception of the ordinary citizens and in the vision of politicians and legislators [3, 4].

2 From a Land of Emigration to a Land of Immigration (1945–1989)

The transition from a land of emigration to a land of immigration is hardly felt by the Italian population and quite misinterpreted by the political and administrative authorities. The first immigrations are the forgotten ones made by former subjects of the Italian colonies, Jews in transit from Central and Eastern Europe to Israel or other destinations and refugees from Istria and Dalmatia [5–9]. The last Italian refugees from Africa and the first African and Asian students and workers follow them, in the 1960s and 1970s. In the meanwhile, a new Italian emigration from the South to the North of Italy takes place reducing the flow towards North Europe and overseas countries. In the 1970s, therefore, the South of Europe and the South of Italy are seen, in Africa and the Near East, as a North of a South even further south. In Italy, on the contrary, there is a lack of organic and collective perception of the phenomenon except in a few politicians[2] and scholars [10]. The Italy of those years, consciously or not, is at the same time an old country of emigration and a young country of immigration. This ambiguous condition entails a series of problems, first of all the fact that immigration arises in connection with issues of exploitation, illegality and clandestinity and in the absence of a long-lasting and far-reaching strategy in dealing with them. To this it must be added the compliance of a political action aimed at the containment and rejection of refugees who do not fall within the cases contemplated by the Geneva Convention, in the absence of specific legislation that exceeds the limits of the "geographical" "reserve" and the "temporal reserve"[3] [3]. Therefore, in front of the two most important immigrations of the 1970s - that of Tunisians in Sicily and that of Yugoslavs in Friuli - different measures are adopted and different reactions at local and national level are observed. They are always influenced by concerns about job competition and identity threat and are, therefore, accompanied by forms of refusal [3].

---

[1] Platone: Fedone, 109 a-b.

[2] See the opening speech of the Prime Minister Aldo Moro at the Conferenza Nazionale dell'Emigrazione (1975). L'emigrazione italiana nelle prospettive degli anni Ottanta. In: Proceeding of the Conferenza Nazionale dell'Emigrazione (Rome 1975/02/24–1975/03/01), C.N.E., Roma (1975).

[3] The first reserve limits the reception of refugees to Europeans and, in particular, to dissidents from socialist regimes. The second one limits the reception to those persecuted for causes prior to 1951. They remained in force until the "Martelli Law" of 1990. They are obvious legacies of World War II and the opposing blocs.

The approval of the "Foschi Law" in 1986 - in the wake of the fears spread by the terrorist attack in Fiumicino Airport in December 1985 - rather than an organic law on immigration is a reorganization of previous circulars[4] [11]. Furthermore, while providing partial solutions to delicate issues, such as family reunification, and while opening up to an amnesty of illegal immigrants, it creates a recruitment system that is difficult to manage. Ambiguity, lack of planning and perspective continue to mark immigration policy while fear and disinformation begin to mark a widespread way of perceiving the phenomenon, which, from a quantitative and qualitative point of view, is very varied and in marked progression. If foreigners residing in Italy are almost 165,000 in 1969, they are 20,000 more in 1975, exceed 280,000 in 1981[5] and rise to about 365,000 in 1991[6].

3   A New Perception of the Migratory Phenomenon (1989–2000)

The fall of the Berlin Wall in 1989 opens Western Europe to new flows from the East while flows from Africa and Asia are strengthened and diversified due to the exacerbation of conflicts [12]. Everything is now projected on a media landscape in which the growing presence of immigrants stands out, placing the word immigration in close connection with the words irregularity-marginality-exploitation-underworld. The murder of Jerry Masslo, a Nigerian laborer who fled South Africa owing to the persecution of the apartheid regime and vainly seeking asylum in Italy because of the "geographical reserve", is certainly one of the causes of the turning point. The assassination, perpetrated in 1989, although preceded by other racially motivated murders, catalyzes a vast movement of opinion that takes the form of anti-racist demonstrations and promotes the demand for a new immigration law.

It is the law no. 39 of February 28th 1990 or "Martelli Law" to try to give answers to the demands expressed by society and mediated by the political forces of right and left. The "geographical reserve" for asylum seekers is abolished. The regularization of those who prove to be resident on December 31st 1989 is granted, conceding the new status to approximately 225,000 people. Various types of residence permits are prepared and the subordination of the regularization of the remaining immigrants to the employment contract is sanctioned. However, the integration policies entrusted to the Regions are vague and the problem of overcoming the amnesty with an annual programming of admissions remains. With these and other limits, the law bridges the gap with the other European States and allows Italian participation in the Schengen agreements and the 1990 Dublin Convention. The effects of these treatises will be felt only towards 2005, when the number of applicants will grow exponentially, with effects and unexpected reactions from the Italian side[7].

---

[4] Law no. 943 of December 30th 1986.

[5] Many irregular immigrants recorded in Ascoli's book miss in the official data of the Ministry of Interior.

[6] https://ebiblio.istat.it/digibib/Censimenti%20popolazione/Censimentipopolazioneresidente dal1861/RML0050288Pop_res_cens_1861_1991.pdf, last accessed 2021/11/18.

[7] The Schengen Agreements stipulate that border countries supervise the European border. The Dublin convention, on the other hand, establishes that asylum seekers must submit their application in the first EU country they arrive in. Both have come into force in Italy since 1997.

Another fact of considerable political, social and media importance is the immigration from Albania in 1991: the first great exodus to Italy produced by the crisis of the socialist countries and the opening of the eastern borders after the fall of the Berlin wall. Since March 1991, thousands of Albanians have reached Italy by makeshift means and, in August, about 20,000 people seize the Vlora freighter and head to Bari, without water or food. The hard line adopted by the Andreotti government, which denies the authorization to enter Italian ports to ships coming from Albania with migrants on board and threatens the indictment for aiding clandestine immigration for their respective captains, prevents the preparation of measures of hospitality. To avoid a humanitarian catastrophe, the landing takes place anyway but in the most confusing conditions.

Despite everything, immigration for political or economic reasons is still not part of a long-term strategy in the 1990s. Each new crisis, such as the dissolution of Yugoslavia and the war that follows, the war in Somalia, the wars in the Gulf, with the consequent arrivals of refugees, always takes public institutions by surprise compelling them to counter "continuous emergencies". Therefore, despite Italy being a country where immigration has long been a "structural" phenomenon, Italians continue to think and act in a "conjunctural" way and to see Italy as a country of emigration. As proof of this, the new and current citizenship law of 1992 (Law no. 91 of February 5th 1992) favors the maintenance and obtaining of citizenship of Italians abroad and restricts the acquisition possibilities for immigrants in Italy despite long years of regular residence[8] [13]. In addition, the 1990s see new legislative efforts to address immigration always as an "emergency". In 1995, there are two "urgent" decrees of the Dini government to reorganize immigration laws, in order to grant another amnesty[9] and to create temporary centers for the unidentified[10]. The "Turco-Napolitano Law", on the other hand, dates back to March 1998, therefore delayed on the entry into force of the Schengen Agreements and the Dublin Convention[11]. The law has three basic objectives: to control flows,

---

[8] It is granted only if requested by sons of foreigners born in Italy who have maintained their residence without interruption until the 18th year. Naturalization by residence is also difficult. The simplest way appears to be by marriage. A recent attempt to change access to citizenship on the basis of ius soli ends unsuccessfully during the Gentiloni government in 2017..

[9] It is the Decree no. 489 of November 18th 1995. Disposizioni urgenti in materia di politica dell'immigrazione e per la regolamentazione dell'ingresso e soggiorno nel territorio nazionale dei cittadini dei Paesi non appartenenti all'Unione europea. In: Gazzetta Ufficiale, Serie Generale, 270, November 18th 1995. It comes into effect on November 19th 199 and lapses due to failure to convert into law. Cfr. https://www.gazzettaufficiale.it/eli/id/1995/11/18/095G0539/sg, last accessed 2021/12/20. See also Colucci M.: op. cit., pp. 114–115.

[10] Law no. 563 of December 29th 1995. Conversione in legge del Decreto-Legge n. 451 del 30 dicembre 1995, recante disposizioni urgenti per l'ulteriore impiego del personale delle Forze armate in attività di controllo della frontiera marittima nella regione Puglia. In: Gazzetta Ufficiale, Serie Generale, 303, December 30th 1995. It comes into effect on December 31st 1995. https://www.gazzettaufficiale.it/atto/vediMenuHTML?atto.dataPubblicazioneGazzetta=1995-12-30&atto.codiceRedazionale=095G0603&tipoSerie=serie_generale&tipoVigenza=originario, last accessed 2021/11/20.

[11] Law no. 40 of March 6th 1998. Disciplina dell'immigrazione e norme sulla condizione dello straniero. https://www.camera.it/parlam/leggi/98040l.htm, last accessed 2019/08/20.

to facilitate integration and to simplify expulsions. For this last purpose, an annual planning of quotas is established, the card and the permit residence are introduced for social reasons and the expulsion procedures are simplified thanks to the Temporary Residence Centers (CPT). However, a new regularization or amnesty is also approved.

The 1990s, however, saw a progressive increase in the foreign presence in Italy, witnessed by the comparison of the 1991 and 2001 censuses, by the Caritas and Migrantes dossiers (the first was in 1991) and by trade unions data. If in 1991 the registered foreigners are, as mentioned, about 356,000, in 2001 their number almost quadruples, exceeding 1,330,000[12] [14, 15].

It should be added that since 1993 the natural balance of the Italian population has gone negative. The balance turns positive only by adding the births and deaths of resident immigrants. If this is seen together with the now structural aging of the Italian population, it can be deduced how, with a blocked immigration, the population would decrease more and more and the relationship between active and passive population would change in an increasingly irreversible way.

4   Recent Diasporas Between Shipwrecks, Landing and Immigration Policies (2001–2021)

If the Italian population continues to decrease in the first decade of 2000, the foreign one, on the other hand, has a notable increase. In fact, at the 2011 census, the foreign population reaches 4,570,317 units, more than tripling compared to 2001 and reaches 7.5% of the national population[13] [16]. The situation profoundly changes in the second decade. The number of resident immigrants increases very weakly and the exit of resident foreigners from Italy to other European countries is recorded: in 2021, the residents are 5,035,643 with a reduction of 5.1% compared to 2020[14]. It is a phenomenon that may underlie the beginning of a structural migratory outflow in clear contrast to what is felt by public opinion and in connection with the long economic crisis and the recent pandemic outbreak of Covid-19.

The configuration of foreign presences in Italy is also changing. One reason is given by the enlargement of Europe to the East that favors the immigration of people who are no longer non-EU citizens, visibly changing the percentage ratio between European immigrants and African and Asian ones since 2001. Another cause, in addition to that of the previous and recent conflicts in the Mediterranean, is given by the attack on the Twin Towers on September 11th 2001. The laws are influenced by the renewed fears for immigrants and for Islamic terrorists that burst into political campaigns. The "Bossi-Fini Law" (Law no. 189 of July 7th 2002), overcoming the "Turco-Napolitano Law" aims to «make the foreign presence more precarious and less protected by social and legal safeguards» [3]. However, the law contains the same contradictions as the previous ones. It binds immigration to the employment contract, the flow decree and the residence

---

[12] ISTAT: 14° censimento della popolazione e delle abitazioni 2001.

[13] http://www.caritasitaliana.it/materiali/Pubblicazioni/libri_2011/dossier_immigrazione2011/scheda.pdf, last accessed 2019/10/23. ISTAT: Censimento della popolazione e delle abitazioni, 2011; http://dati-censimentopopolazione.istat.it/Index.aspx# (last accessed 2019/10/23).

[14] One of the most treated issues is immigration and Covid pandemic.

contract, favors expulsions and sanctions the use of the Navy against the trafficking of illegal immigrants. But it also decrees the largest amnesty ever made (634,728 applications accepted against 217,000 of the "Turco-Napolitano Law", of 244,000 of the "Dini Decree", and of 215,000 of the "Martelli Law"). This contradiction should be seen as the umpteenth effect of a policy that, riding on the fears of the moment, does not plan the future around a structural and very long-lasting phenomenon. As proof of this, the gradual tightening of immigration continues and worsens following the international economic crisis that started in 2008–2009. The constant social alarm, revitalized by media in everyday life, places the now old security-immigration link increasingly in the foreground, producing legislative restrictions and increasing the distance between the actual migratory flow and the perception that one has of it. It is a vicious circle that produces notable and controversial results: the Berlusconi-Gaddafi agreement, issued in 2008, that establishes Libyan control of the migrants' flow to the Mediterranean and whose violations of human rights are soon denounced[15]; a law on public security, issued in 2009, that commits irregularity into a crime, extends the time required to obtain Italian citizenship following marriage, lengthens the maximum period of detention in the Identification and Expulsion Centers (CIE)[16].

The second decade, marked by the worsening of the economic crisis and the social crisis culminating during the pandemic outbreak, with the further precariousness of work and the growing economic and social insecurity, sees the exacerbation of positions towards immigrants increasingly pointed to as a plausible cause of loss safety and workplaces as well as identity risks. Furthermore, the strong destabilization in the Mediterranean produced by the so-called "Arab Springs" is linked to the internal facts of Europe, triggering new controversies on the right to asylum. Faced with the "wave" of 2011[17], the Italian government decrees a state of national emergency, grants a residence permit for humanitarian reasons, creates large reception centers, the ENA (North Africa Emergency), later called CAS (Extraordinary Reception Centers). The fall of

---

[15] Trattato di amicizia, partenariato e cooperazione tra la Repubblica italiana e la grande Giamahiria araba libica popolare socialista, Benghazi, August 30th 2008. The treaty is ratified by Italy on February 6th 2009 and from Libya on March 2nd 2009. It is particularly important the article no. 19.

[16] Law no. 94 of July 15th 2009. https://www.gazzettaufficiale.it/gunewsletter/dettaglio.jsp?service=1&datagu=2009-07-24&task=dettaglio&numgu=170&redaz=009G0096&tmstp=1248853260030 (website accessed on November 29th 2021). See also Colucci M.: op. cit., p. 153.

[17] According to the Department of Public Safety and the ISMU arrivals by sea in recent years: 2001: 20.143; 2002: 23.719; 2003: 14.331; 2004: 13.635; 2005: 22.939; 2006: 22.016; 2007: 20.455; 2008: 36.951; 2009: 9.573; 2010: 4.406; 2011: 64.261; 2012: 13.267; 2013: 42.925; 2014: 170.100; 2015: 153.842; 2016: 181.436; 2017: 119.369; 2018: 23.370; 2019: 5.852 (al 12/09/2019). http://www.interno.gov.it/sites/default/files/cruscotto_statistico_giornaliero_12-09-2019.pdf, last accessed 2019/09/05. http://www.libertaciviliimmigrazione.dlci.interno.gov.it/it/documentazione/statistica/cruscotto-statistico-giornaliero, last accessed 2021/12/19; https://www.ismu.org/dati-sulle-migrazioni/#1539609382198-bf55b132-c04376c9-dfb9, last accessed 2021/12/19. The first peak, short and sudden, takes place in 2011, the second, longer, begins in 2013, increases until 2016 and falls again in the second half of 2017.

almost all the political interlocutors of the southern shore of the Mediterranean in 2011 hastens the search for new political figures to whom to delegate the relocation of the anti-immigration border. Firstly, the agreement signed by Interior Minister Anna Maria Cancellieri of the Monti government and her counterpart Fawzi al-Taher Abdulali dated June 18th 2012, that reaffirms the agreements signed with Gaddafi with the vain emphasis on respect for human rights. Secondly, the memorandum drawn up by the Minister of the Interior Marco Minniti and signed by the Prime Minister Gentiloni and the Head of the Libyan national unity government Fayez al-Serraj dated February 2nd 2017[18]. The effect of these and other subsequent negotiations are ideally aimed at combating human trafficking, controlling borders and managing immigration. Practically, it is to block or delay the routes of illegal immigration on the south shore in the detention centers denounced for the violation of human rights.

In fact, after the increase in the years 2014–2016, landings have sharply fallen since the summer of 2017 precisely because of the Italy-Libya agreements. So this reduction takes place well before the advent of the first Conte government (June 1st 2018) that follows an electoral campaign in which the fear of immigrants, the security of the country and the closure of borders are dominant themes[19]. Despite the data, the government inaugurates a further tightening of anti-immigration measures based no longer on laws dedicated to migrants but on security. The "Security Decree"[20] and the "Security Decree Bis"[21], dated November 28th 2018 and June 14th 2019 respectively, have raised observations by the Head of State and negative assessments by some political forces, the Church and part of the public opinion. The critical points are a possible increase in mortality at sea and a plausible rise in illegal immigrants due to the augmented difficulty in obtaining a residence permit. According to various experts, the paradox of the "security decrees" seems to be that of responding to "exacerbated" security requests, producing further insecurity that leads to new and tougher security requests [17]. Therefore, these decrees would increase insecurity: at sea, for migrants still in the hands of smugglers; in rescuing, owing to the new punitive clauses; on land, given the difficulty for the migrants in finding hospitality due to the created climate of tension and in placing themselves in

---

[18] https://www.repubblica.it/esteri/2017/02/02/news/migranti_accordo_italia-libia_ecco_cosa_contiene_in_memorandum-157464439/, last accessed 2019/09/11. To what is established in the memorandum are added the restrictions on the right of asylum sanctioned by Law no. 46 of April 13th 2017.

[19] The XXVII Rapporto Immigrazione is rightly entitled: Un nuovo linguaggio per le migrazioni and highlights the exponential growth of news concerning immigration in recent years in connection with the issue of the threat to security and to public order. Caritas e Migrantes: XXVII Rapporto Immigrazione 2017–2018, op. cit., p. 2 (syntesis).

[20] It is converted into Law no. 132 of December 1st 2018. https://www.gazzettaufficiale.it/eli/gu/2018/12/03/281/sg/pdf, last accessed 2021/09/05.

[21] It is converted into Law no. 77 of August 8th 2019. https://www.gazzettaufficiale.it/eli/id/2019/06/14/19G00063/sg, last accessed 2021/09/05.

the sphere of legality despite the need for workforce[22] [18–25]. An umpteenth vicious circle, therefore, is already evident at the beginning of the second Conte government (September 5th 2019-February 13th 2021) when the Covid-19 pandemic upsets Italy, Europe and the whole world, further changing the picture of international migration and its perception [18]. Firstly, the progressive shifting of the political focus from the "migration emergency"; secondly, the smoothing of some critical points contained in the "Security decrees" subsequently incorporated in the Law no. 130 of October 21st 2020; lastly, the almost total absorption of political activity and media attention by the Covid-19 pandemic.

5   The New Ulysses from Africa, Asia and Europe to Italy, Between Past and Future

From a historical point of view, a much more detailed excursus on migrations than this, from the Second World War to the Pandemic times, divided into the aforementioned phases, would surely better highlight a series of factors that are not apparently evident and linked together in a short-term analysis. Beyond the evident connection of migration in Italy to global mobility and economic, political and climatic changes and beyond its clear insertion into a very long-lasting process, the complex stratification and diversification of successive migrations is to be emphasized. The long period sees, in fact, migrations of people of different languages, cultures, religions and origins driven by different motivations, usually compelling and dramatic, in different phases of national and international history. From Africa, for example, there are women and men from former colonies who have lived in Italy since the postwar period and whose painful and little known history is quite distinct from those of Somalis and Eritreans arriving in more recent times. It seems self-evident but, from the standardizing point of view of the present, unfortunately it is not.

The long term would also allow us to observe how some errors of assessment and action that arise at the beginning of immigration in post-war Italy are maintained and even worsened over time. It is the case of the late awareness by the inhabitants of the peninsula of the transformation taking place, from country of emigration to country of immigration. From this "original sin" the politics of "emergencies" which characterizes the way of dealing with the phenomenon would partly derive even when this, after the turning point between the 1980s and 1990s, becomes an established fact. It is also the case of the crystallized link between immigration and insecurity, irregularities, illegality,

---

[22] Among the numerous contributions on these topics see those of R. Bottazzo (https://www.meltingpot.org/Puglia-Ecco-come-il-decreto-sicurezza-fabbrica-insicurezza.html#.XXqJmCgzaUk, last accessed 2021/09/09); L. Borga (https://www.ilfoglio.it/sound-check/2018/11/12/news/piu-immigrati-irregolari-meno-sicurezza-223989/, last accessed 2021/09/09); P. Mele (http://www.rainews.it/dl/rainews/articoli/La-precarieta-e-l-insicurezza-sono-aumentate-con-il-decreto-sicurezza-Intervista-a-Chiara-Peri-39383b53-3cb1-4846-ae76-d3ef61555725.html, last accessed 2021/09/09); F. Marcelli (https://www.ilfattoquotidiano.it/2018/11/16/il-decreto-insicurezza-di-salvini-e-paradossale-e-cancella-la-dignita-umana/4764253/, last accessed 2021/09/10); L. Matarese (https://www.huffingtonpost.it/entry/chiamiamolo-decreto-insicurezza_it_5db572a7e4b079eb95a50b14, last accessed 2021/12/21).

precariousness and exploitation: a complex consequence of the legislator's ability to deal with migration only as a "conjunctural" fact and not as a "structural" process. Furthermore, one should consider the recent but decisive population changes, such as the negative birth/death balance: "the demographic winter" denounced also by Pope Francis on December 26th 2021. It should make people reflect on the future of the Italian and the European population in general and on the need to promote strong birth policies and the regularization and the social integration of immigrants, rather than rising fear of ethnic substitution. Lastly, more recent and perhaps temporary factors should be considered, such as the incipit of stagnation in the number of immigrants residing in Italy. Maybe this specific juncture is pushing people to reconsider the current worsened economic and social conditions of Italy, its decreased attractiveness and, on the other hand, its increased push-back strength and its hindering acceptance and inclusion, worsened by the current pandemic wave.

Just as perpetually emergency actions in the long run seem to show their intrinsic inexcusability, similarly some information politics pointing at reiterating some aspects of immigration to the detriment of others seem to show their effective weakness. However, they have gradually rewritten with effectiveness the language of and on migration in an alarmist key in a maybe no-returning process [19].

In a society in which the relationships between life and death are worsening, in which the gap between active and passive population is widening, in which a closed and defensive cultural model collides with an open and inclusive one, an effort of reflection is needed and not dictated only by the paths of the economy and the walls of politics. An effort that in an ethical discourse rewrites a new philosophy of migration and cohabitation looking at the long history of the Mediterranean rightly to draw from it ideas for a better future [20]. The troubled routes of Ulysses, with their shipwrecks, landings, encounters and returns could be perfect symbols of the ancestral struggles of the man in search of a home-land, symbols that could help peoples to welcome castaways, listen to their stories, learn how to help them, helping them to help us.

References

1. Caritas e Migrantes: XXX Rapporto Immigrazione. Verso un noi sempre più grande, Tau, Todi (2021)
2. Fazzi, P.: Globalizzazione e migrazioni. Breve storia dall'età moderna a oggi, FrancoAngeli, Milano (2014)
3. Colucci, M.: Storia dell'immigrazione straniera in Italia. Dal 1945 ai nostri giorni, Carocci, Roma (2018)
4. Speziale, S.: Tra Africa, Asia ed Europa: l'interminabile "emergenza" dell'Italia delle migrazioni dal secondo dopoguerra a oggi. La Chiesa nel Tempo. **35**(3), 11–28 (2019)
5. Morone, M.: L'italianità degli altri. Le migrazioni degli ex sudditi coloniali dall'Africa all'Italia. Altreitalie. **50**, 71–86 (2015)
6. Deplano, V.: La madrepatria è una terra straniera. Libici, eritrei e somali nell'Italia del dopoguerra (1945–1960), Le Monnier-Mondadori, Firenze (2017)
7. Ravagnan, M.: I campi Displaced Persons per profughi ebrei stranieri in Italia (1945–1950). Storia e Futuro. **50**, 1–13 (2019). file:///C:/Users/User/Downloads/blog_3514c35ac00dc2fcab1226ab4969fdbc.pdf. Accessed 05 Dec 2021

8. Crainz G.: Il dolore e l'esilio: l'Istria e le memorie divise d'Europa, Donzelli, Roma (2005)
9. Pupo, R., Il lungo esodo. Istria: le persecuzioni, le foibe, l'esilio, Rizzoli, Milano (2006)
10. Ascoli, U.: Movimenti migratori in Italia, p. 7, il Mulino, Bologna (1979)
11. Favaro, G., Tognetti Bordogna, M.: Politiche sociali e immigrati stranieri, Carocci, Roma (1989)
12. De Cesaris, V.: Il grande sbarco. L'Italia e la scoperta dell'immigrazione, Guerini e Associati, Milano (2018)
13. Panella, L.: La cittadinanza e le cittadinanze nel diritto internazionale. Editoriale Scientifica, Napoli (2008)
14. Gli stranieri in Italia: analisi dei dati censuari, p. 32. ISTAT, Roma (2005)
15. Ferruzza, A., et alii (eds): Stranieri in Italia: analisi dei dati censuari, ISTAT, Roma (2006)
16. Caritas e Migrantes: Dossier Statistico Immigrazione. XXI rapporto. Oltre la crisi, insieme, (synthesis) (2011)
17. Villa, M.: Sbarchi in Italia: il costo delle politiche di deterrenza. Ispionline, 1 Oct 2018. https://www.ispionline.it/it/pubblicazione/sbarchi-italia-il-costo-delle-politiche-di-det errenza-21326. Accessed 04 Sep 2019
18. Salvadego, L., Savino, M., Scotti, E. (eds): Migrazioni e vulnerabilità. La rotta del Mediterraneo centrale. In: Proceedings of the 2nd Doctoral Colloquium of the Accademia Diritto e migrazioni, Macerata, 5–6th December 2019, Giappichelli, Torino (2021)
19. Villa, M.: Migrazioni e comunicazione politica. Le elezioni regionali 2018 tra vecchi e nuovi media, FrancoAngeli, Milano (2019)
20. Di Cesare, D.: Stranieri residenti. Una filosofia della migrazione, Bollati Boringhieri, Torino (2017)
21. Bottazzo, R.: https://www.meltingpot.org/Puglia-Ecco-come-il-decreto-sicurezza-fabbrica-insicurezza.html#.XXqJmCgzaUk. Accessed 09 Sep 2021
22. Borga, L.: https://www.ilfoglio.it/sound-check/2018/11/12/news/piu-immigrati-irregolari-meno-sicurezza-223989/. Accessed 09 Sep 2021
23. Mele, P.: http://www.rainews.it/dl/rainews/articoli/La-precarieta-e-l-insicurezza-sono-aum entate-conil-decreto-sicurezza-Intervista-a-Chiara-Peri-39383b53-3cb1-4846-ae76-d3ef61 555725.html. Accessed 09 Sep 2021
24. Marcelli, F.: https://www.ilfattoquotidiano.it/2018/11/16/il-decreto-insicurezza-di-salvini-e-paradossale-e-cancella-la-dignita-umana/4764253/. Accessed 10 Sep 2021
25. Matarese, L.: https://www.huffingtonpost.it/entry/chiamiamolo-decreto-insicurezza_it_5db 572a7e4b079eb95a50b14. Accessed 21 Dec 2021

Proposal for the Enhancement of Archaeological Sites and Places of Spirituality, in the Territories of the Colonnie Magnogreche of the Province of Reggio Calabria

Maria Savrami[✉]

Byzantine and Christian Museum, Ministry of Culture, 10675 Athens, Greece
marianna_savrami@yahoo.gr

Abstract. The key-study is to propose a plan of reinforcing of the cultural heritage identifying 6 routes that could improve all types of resources present in the territory.

1 Route is related to the magno Greek colonies of the province of RC, more Crotone for its characteristics common to the territory of Reggio in particular the photo shows the mechanism of Antikythera which was found in Greece and represents the first calculator designed and built on the principles of Archimedes and Pythagoras (originally from Crotone). **2** Route deals with the analysis of the territory of the Greek linguistic minority differentiating the Greek territory, understood as the whole territory that in the past belonged to this culture, and the hellenophone island as the territory where the Greek idiom is still spoken today. **3** Route highlights 5 possible leads to the use of the environmental naturalistic assets. **4** Route enhances the resources of the Byzantine Greek Rite. **5** Route refers to religious festivals as a tourist attraction. **6** Route proposes the enhancement of the two typical and unique products present in the Calabrian territory: bergamot and Cedar, using the funds provided by the EC for the "cultural itineraries". (example Olive Road in Greece, Italy and other Mediterranean countries).

Conclusion: only by using the three essential elements of the cultural landscape (the good, man and the environment), as a unique reality, can we arrive at the sustainable development of an integrated system of valorization of the cultural environment.

Keywords: Byzantine Calabria · Byzantine Tradition · Byzantine Architecture

The development of the cultural Heritage presupposes the making of networks which allow a holistic utilization of the architectural, environmental and cultural environment of a region. One of the acts which are leading to the establishment of thematic networks is the reestablishment of the private assets of a cultural, historical, environmental and ecclesiastical character which can gather a general interest. It is therefore necessary to identify "networks" and "routes" which can create a system of areas, which can be a candidate for economical enforcement. The goal is certainly that the Calabrian cultural resources should no any longer constitute as outer parts, far away from the local life, but that the local populations should consider them as a new chance for development, for

© The Author(s), under exclusive license to Springer Nature Switzerland AG 2022
F. Calabrò et al. (Eds.): NMP 2022, LNNS 482, pp. 2438–2448, 2022.
https://doi.org/10.1007/978-3-031-06825-6_233

production of a new wealth. In the same time this should happen in a sustainable and long lasting process, with a close collaboration with the local tourism systems. For the region, in essence, the development of the cultural heritage must have as the main task the consolidating of the development for the tourism sector. This can be happened through the increasing number of tourists and the increasing average time of the tourists spend in the area. The general goal so is to enforce the promotion of the local cultural heritage in order to achieve the creation of cultural tourist routes that, taking as a great advantage the rich local history, the local identities (linguistic minorities) and testimonia. All these can create the conditions for a different and more interesting tourist offer, among its various components, focusing on a better quality.

1   Routes

Route we can call the geographical path which the visitor has to to follow and respect, and follows the purposes of the trip. Therefore, each route cannot be separated from a series of information and services which must be carefully designed and presented and also should be based on all the resources. It should be a close connection between the guiding theme and the places of the route, presented in the territory.

Specific directions included in a travel route must include:

- the date, the duration of the trip divided by days and nights;
- th scheduled stops, places to visit
- the types of transportation to be used
- the type of accommodation, reception services
- the costs which are included and not included in the participation fee

1.1   The Magnogreche Colonies in the Province of Reggio Calabria

Magna Graecia was a fertile ground for philosophy, arts, mathematics, sciences that were spread throughout the Mediterranean and became the foundations of the Western culture and philosophy. In archaeological areas, the focus should be in growing the network through making of events, including the most important areas but also and the smaller sites which are scattered throughout the territory creating places of tourist attraction. The cities to be involved are:

Colony of Rhegion: located in the current municipal territory of Reggio di Calabria;
Colony of Epizephyr Locri: located in the current territory municipal of Portigliola;
Colony of Kaulon: located in the current municipal territory of Monasterace;
Colony of Metauros: located in the current municipal territory of Gioia Tauro **Colonia Medma:** located in the current municipal territory of Rosarno;
Colony of Gerace: located in the current municipal territory of Gerace;
Bruzzano zeffirio colony: located in the current municipal territory of Bruzzano Zeffirio;

Colony of Kroton: located in the current municipal territory of Crotone. Only site not belonging to the province of Reggio Calabria, but that can be considered as such, because it has common origins and cultural testimonies almost the same as those of Reggio Calabria.

In all the previous areas or cities we mentioned, The cultural activities which can be presented are the following: classical theater performances, of the most prestigious names of national and international level, or cultural artistic, musical and literary events which can gain tourist flows even in time periods except the summer season. As an example, the organizing of national or international conferences dedicated to famous people of the world such as Pythagoras, for his "school" founded in Crotone, (Calabrian magnogreca land) or exhibitions on ancient technology during the time of Magna Graecia etc., such as the famous "mechanism of Antikithira. This ancient mechanism was found at a depth of 60 m. by sponge fishermen during a shipwreck near the island of Antikthyra They were valuables, which a Roman ship carried from Rhodes to Rome during the time of Julius Caesar in the 1st century B.C. Those events will increase the interest for a visit of the archaeological sites of the area in some of them fortunately have developed real activities of marketing of local products with good quality at the same time the increase of the number of visitors will becoming an motive for industries of the area to produce quality products linked to local tradition.

The general strategy is based on the key action-"theme park"-the form of "integrated" economic development (which provides, in other words, exchanges and co-operation between cultural services, tourism, craftsmen, enhancement of products, etc.).

1.2   The Greek Area and the Hellenophone Island

The route of the Grecanica Area is characterized by those territories recognized as the «holly center» of the hellenophone linguistic minority of Calabria. The southern slope of Aspromonte still preserves unchanged the traces of its ancient nature as a crossroads on the Mediterranean basin. This area played for many centuries the role of a real island and cultural stronghold. This happened for a number of reasons such as the protection of the connections through historical times due to the particularly impervious hinterland.

The town, although urbanized, is located largely within the boundaries of the Aspromonte National Park, is a peaceful and quiet natural environment. The Centers of Condofuri, Galliciano, stronghold of the Greek, Roghudi keep the most evident traces of the magno-Greek culture. The Greeks of Calabria as they still call them up to our days, is a subject of many studies and researches as well as a source of cultural exchange and initiatives to protect historical linguistic minorities. The population in the 1991 as was counted was 2000.

The hellenophone island extends today mainly along the valley of the Great River Amendolea, in the province of Reggio. The villages are located about 15 km from the coast, generally all exist on mountains providing a difficulty to access and surrounded by ravines, dominated by the south side of the Aspromonte. The Valley of Amendolea includes Amendolea, Galliciano, Roghudi, Chorio Di Roghudi, Roccaforte,and in addition the new migratory settlements of those of San Giorgio Etra. The hydro-geological instability and the territorial marginality, the strong emigration, the difficulties in human communication made the work of the public administration completely feasible. The almost hostile nature landscape have played a fundamental role in the progressive disappearance of the last Greeks of Calabria. The area includes the territory of the following municipalities plus other small towns (Condofuri Marina, Galliciano, Palizzi Marina, Pentedattilo) that are part of these municipalities: Melito Porto Salvo, San Lorenzo, Bagaladi, Roghudi, Roghudi Nuovo, stronghold of the Greek, Codofuri, Bova, Bova Marina, Brancaleone, Africo, Bagaladi, Palizzi, Galliciano, Samo, Motta San Giovanni, Montebello, Amendolea, Condofuri, Cardeto, Sant'agata Del Bianco, Staiti.

The Grek Area is one of those European areas that was not participated in the economic and industrial development and suffered a lot from the depopulation and emigration. The lack of accommodation and hotel facilities has been one of the main problems that in the past put an obstacle in the tourist growth of the inland areas. The GAL Area Grecanica with an important action of authorization and restoration of small rural centers, creating a training course for operators, supports the birth of Pucambu ("somewhere", in Greek of Calabria), an agency for the development of rural tourism, becomes a central interpoint between the tourist market and provides a further training for all operators.

1.3   The Natural Heritage

The third route includes five possible ways which all have as a starting point the Aspromonte Park. The Aspromonte National Park is a protected natural area located within the province of Reggio Calabria.The Aspromonte is a mountain mass that is part of the complex called Southern Alps or Calabrian Alps.

The morphology of the territory is rather tough, except for the top, whose profile is softened by the schistose nature. The complex has a conical-pyramidal shape, with a central relief, the Montalto, of 1956 m, from which several ridges branch out interspersed with deep valleys. Hydrography is characterized by rivers. These are short streams, with torrential regime, which flow on Stony GRETS, downstream also quite wide. The flow rate of the rivers is fed by the surface outflows of rainwater. Given the rainfall regime, with rains concentrated in the winter period, these streams flow impetuously in the winter period, forming suggestive waterfalls at high altitude. With the decrease in precipitation, in the spring, the flow gradually reduces until it is completely exhausted, so that, in the summer and autumn, the rivers are completely dry. In the Aspromonte National Park there are also numerous villages: Bova, Gerace, Mammola and San Giorgio Morge to, which boast an ancient history and deserve a particular visit as they arouse great interest. The ancient villages offer to the visitors a very compact historical, cultural and artistic interest rich in testimonies and finds from various eras (Magna Graecia, Roman, Medieval and Renaissance), preserving all the glory of the past. The traditional villages are formed by small houses, one attached to the other and narrow streets, small squares, ancient churches, Noble palaces and fascinating panoramic views. All over the year there are numerous religious festivals that still carry the history and traditions, of the era. There are also many cultural events and numerous festivals of local products during the year. The villages are a destination for many visitors and represent a real tourist resource of the park territory. Starting from these historic centers located within the Aspromonte Park, the five routes pass through valleys, high floors, waterfalls, rivers, protected areas, cliffs and characteristic Rocky forms.

1 Road: Montalto-Saba Sibilia - Fiumara Bonamico (Upper Part)- Zervo plans - Tabular Ridge - Aspromonte National Park;

2 Road: Monte Sant'elia-Costa Viola-Limina - Fiumara Amendolea-Torrente Casalnuovo-Aspromonte National Park;

3 Road: Stilaro Valley-Marmarico waterfalls-Monte Stella - Monte Consolino - Fiumara Di Gerace - Fiumara Amendolea - Aspromonte National Park;

4 Road on the Tyrrhenian Side: Strait of Messina-Scilla Cliff - Costa Viola-Monte S. Elia - Piana degli ulivi - Southern Dolomites-Aspromonte National Park.

5 Road Path on Theonico Slope: Pentadattilo Cliff - Fiumara Di Melito-Fiumara Dell'amendolea-Capo Spartivento - Bay of gelsomini - Capo Bruzzano - Fiumara Bonamico - Valley Of The Great Stones - Fiumara Condojanni - Fiumara Allaro, Fiumara Stilaro-Aspromonte National Park.

1.4   Byzantine Itinerary, of the Italian-Greek Monasteries

Also this route for its special character and for the some other unique characteristics that make it to be something spontaneous it can be divided into two special routes:
   a) places of spirituality b) the streets of Greek Orthodox worship.

Today, more than ever, the connection between high living standards and natural enviroment is very strong, is expressed mainly with the search of ourselves,but also through visits to places of art and culture, to the sources of mysticism spirituality, and nature. In Calabria there is a great heritage linked to spirituality, wich at the same time serves as a unique potential for the locals and visitors to develop their inner world. The strong link between spirituality, quality of life and this magnificent territory creates a new project.which overcomes the traditional concept of religious tourism (connected mostly monasteries, to sacred places without having any connection to the territory around and mainly connected with modest accommodation facilities). The new conception is that the places will offer a quality offer, linked to the sources of the sacred and spiritual life, but also to landscapes and places which will offer quality services.

1.4.1   The Italian-Greek Monks, Between the VIII and the XI Cent

Asceticism was the key note, but all the Saints explained the spirituality of the soul into an intesive hard working. In the formation of cenobi,many other monks were participated. (Sant'elia Speleota in Melicucca). They founded new monasteries (Sant'elia the younger in Seminara, San Nicodemo on Mount Kellerano, San Giovanni Theristis in Stilo) some of which became places of culturein later years, such as the Monastery of Seminara, which hosted Barlaam and Telesio. Their life was humble and their goal was to serve tthe others and to pray in the cenobium: the Italian-Greek ascetics often gathered around the first cave with other hermits, who in their turn used additional caves around. With such away the first monastery communities were created. The monks devoted themselves to work offering their labors for the sake of others, often becoming masters in agriculture and teaching how to draw sustenance from the resources of nature (St. Nicodemus and St. John Theristis, for example, introduced the practices of Mulberry and silkworm).

Today they proclaimed saints in the places where they lived and are recognized patrons of the territories such as Mammola (St. Nicodemus), Bova and Africo, which keeps the relics of St. Leo and his protection. From the Greek-Byzantine monastery of San Giovanni Theristis in Bivongi, the symbolic path of the approaching between the Orthodox Church and the Roman Church started there. Points of that route are the places dedicated to the celebration of the Greek-Byzantine Rite. In the territory of Reggio the embrace between the two churches becomes strong enough in all the sites where you can attend the celebration of a ritual that is not only the memory of a religious historical past, but a lively liturgy that serves the faith. the sacred monastery of St. John Theristis, since 1995 officially handed over by the commune of Bivongi to the Orthodox Archdiocese of Italy for a period of 99 years. This monastery is the first in Italy to have been founded by Athonite monks from directly from Mount Athos. it is, throughout Europe, the only Byzantine monastery in which liturgical services are regularly celebrated. We are in the territory that Paolo Orsi called "the Holy Land of basilianism in Calabria". In the background of the Stilaro Valley, the sounds of nature create a a great scenery of spirituality and keeps with a unique way the essence of the Greek Orthodox Rite. The monastery is dedicated to the Italian-Greek saint who lived as a monk in these mountains, dedicating himself to prayers and miracles for the poor people. The Basilian monks abandoned the monastery in 1662 and remained silent until 1994, when the monks of Mount Athos arrived there. From the Holy Mountain to the rest of the area reflects the energy that turns the light into faith. Light and prayer create the common path that leads to the sacred places of this route. The intensive religious faith which existed in Byzantine times led to the construction of churches and monasteries, and finds one of the greatest expressions in the jewel of the "Cattolica Di Stilo". Pazzano offers a historical-religious synthesis: the cave carved into the rock, which in the eighth century welcomed the Italian-Greek monks, today is a sanctuary, where the Latin faithful venerate the statue of Mary SS. of Monte Stella. Following the intersection of the Byzantine Rite with the Latin rite, we reach Gerace. The Small Church of San Giovanni Crisostomo or San Giovannello is the oldest Greek-Byzantine cult building (i Near the church, the important Norman-Byzantine building of the Cathedral. Around many churches, which make Gerace a masterpiece of religious architecture in the province of Reggio, as if conclude the embracement between the Latin Church and the Byzantine church, worship and asceticism,these are the foundations of the cave of Sane The Saint one of the three patrons of Gerace. His painted icon is kept in the Church of San Giovannello just before reaching Galliciano, the detour towards Staiti leads to the discovery of the red bricks that form the ruins of the Byzantine Abbey of Santa Maria dei Tridetti (X–XI cent).

1.5   The Celebration of Liturgical Rites

The route ends on the Amendolea Valley: the grecanica area where, in previous time, the celebration of liturgical rites allowed the preservation of the grecanico idiom, still used by the elderly.

The monks of Mount Athos are responsible for the Greek Orthodox chapel of Our Lady of Greece (located in Galliciano). From the silence of the sacred stones rise the notes from the prayers that awaken from the ruins of the four Byzantine churches (SS. Annunziata, Santa Caterina, San Sebastiano, San Nicola), which stand alone on the opposite side, in the shadow of the castle of Amendolea.

Seminara 14 Agosto Madonna Dei Poveri
Polsi 1–2 Settembre Madonna Della Montagna
Polsi 14 Settembre Festa Della Croce
Melicucca 10–11 Settembre S. Elia Speleota
Seminara Taureana 24 Luglio S. Elia Il Giovane, S. Filareto, S. Fantino
Mammola 12 Marzo, 12 Maggio, Ia Domenica Di Settembre S. Nicodemo
Bova 5–7 Maggio S. Leo
Africo 5 Maggio, 12 Maggio S. Leo
Bivongi 27 Febbraio S. Giovanni Theristis
Reggio Calabria Ii Sabato Di Settembre Madonna Della Consolazione
Palmi 16 Luglio Madonna Del Carmine
Bagnara 15 Agosto S. Maria E I Xii Apostoli
Bagnara Marinella 22 Settembre M. Ss. Di Portosalvo
Taureana 31 Luglio Madonna Dell'alto Mare
Polistena 10 Luglio Madonna Dell'itria
Gioia Tauro 12 Settembre Madonna Di Portosalvo
Sinopoli 8 Settembre Maria Ss Delle Grazie
Rosarno II Domenica Di Agosto Madonna Di Patmos
Gerace 23 Agosto Immacolata, S. Antonio, S. Veneranda, S. Junio
Gallico 22–25 Agosto M. Ss. Delle Grazie
Motta S. G. 15 Agosto M. Ss. Del Leandro
San Lorenzo 27 Luglio, 12/15 Agosto M. Ss. Della Cappella
Melito Portosalvo 25 Marzo M. Ss. Di Portosalvo
Bova Marina 7 Agosto M. Del Mare
Brancaleone 7 Agosto M. Del Carmine
Bombile Di Ardore 1–2–3 Maggio M. Della Grotta
LocriI Domenica Di Agosto Immacolata
Gioiosa IonicaIii Domenica Di Agosto M. Del Carmine
RoccellaI Domenica Di Luglio M. Delle Grazie
Monasterace 26 Giugno M. Di Portosalvo
Siderno 7 Settembre Maria Ss Di Portosalvo

1.6 The Road of Bergamot and Cedar

In Calabria, on the way to the places that bergamot and Cedar are cultivated, among the traces of Magna Grecia, a route is presented between two seas, unknown to the most of the people, in which the visitor can discover the flavors and colors of wild nature. It can be an interesting route that will lead us to know more about the daily life of Calabria linked to bergamot and cedar, the Riviera, and their link with history, Calabria through time was always a region which its economy was based on Agriculture, which consists from both Agriculture and livestock. The small towns that make the larger area created their own local gastronomies. They contain ingredients, cooked not only in different ways, but often could be recognised, from area to area, with different names, have made them characterized, making them particularly tasteful and finally are making from the gastronomic point of view Calabria is very rich and stimulating The presence of different populations, whose influence remains strongly in the cultural heritage and the landscape of the region itself, linked to a continous exchange of geological phaenomena, Olive oil, wine, pork, fish to dairy products, vegetables and fruits, bergamot, cedar, are the raw materials of extraordinary quality. In particular, the Riviera is a showcase of excellence characterized by some very precious flavors: like the D. O. C. Wine of Verbicaro, or chili pepper, with its Academy and the Festival, and the gastronomic-cultural festival "the Cedar in the Riviera", and the cultivation of bergamot. The quality of cultivated cedar which, and called "smooth Cedar of Diamond" (large cut and fragrant, intended largely for candying).

Although numerous attempts have been made in different citrus areas of the world (from the United States of America to Florida and California, to North Africa, South America, to the Ivory Coast) to obtain the acclimatization of this plant, ultimately, it can still be said that almost all of the world's production is concentrated in Calabria, in particular in the province of Reggio In this small strip of land that, along the coast and for a distance of about five kilometers, it opens at the entrance of the Strait of Messina (towards Scilla with the offshoots up to Rosarno) and extends to Roccella Jonika. The area cultivated with bergamot has now reduced to below 1500 hectares; in the Thirties, in the past it was 2400 hectares. The number of manufacturing companies is about 1500. In the province of Reggio Calabria, 45 processing industries operate. The data provided by the Consorzio del bergamot, for the 1995–96 vintage indicate in 16,000 tons the fruits of bergamot started to transform with a yield in essence equal to 8,120 tons. From a recent survey it is estimated that the employees of the monopoly of both products: cedar and bergamot and related activities are about 3,000 Bergamot has been protected by a D. O. P. mark since 2001. On September 24, 2011, the first International Congress was held, at the Hall of the Bergamot Museum. This Congress opened a discussion of a special interest for the possible uses of bergamot in the food and pharmaceutical industry.

The Cedar today in Italy is mainly cultivated and processed in Calabria, in the coastal strip of the upper Tyrrhenian cosentino that goes from Belvedere Marittimo to Tortora, called Riviera dei Cedri with Santa Maria del Cedro in the center, where this tree grows spontaneously. In the lowest part of Northwestern Calabria the mixture of various population of the Mediterranean created new cultural roots which last for a long time. They still find continuity to welcome the New without forgetting the old. On these lands Greeks and Romans, Byzantines, Lombards, Arabs, and then again Normans, Angevins and Aragonese brought elements of differentcultures, which were then influenced local cultures with beneficial effects. And this site should by itself be an open invitation to be visited, starting from Santa Maria del Cedro, seaside places and inland, covered all by rocky spurs or built near the sea.: they descend from mountains that frame a of the most beautiful coasts of Calabria, embellished with beautiful cliffs with imaginative names.

The road of bergamot and Cedar, could be an annual Intercultural Dialogue, a journey between Mediterranean countries. It could include cultural visits, events, activities aimed at strengthening the historical value of bergamot.

With bergamot and Cedar central axis, the goal will be to develop the dialectic of Mediterranean cultures, the documentation and preservation of the "culture of bergamot and cedar" and their international promotion, on behalf of the local economy, through the creation of common activities between cultural tourism and sustainable cultivation of Bergamot. Cedar and their products are part of Mediterranean cuisine, with beneficial effects on health.

Bergamot derivatives can be exploited as the main products of the Mediterranean rural economy through a network of Chambers of Commerce, Museums, Festivals, small and medium-sized privatei ndustries.The Media and private organizations, could also become a symbol of socio-economic and cultural development of Calabria.

- Educational activities and training programs that will relate not only to students, but also to representatives of associations from different sectors of the local economy.
- promotion activities for the use of bergamot and cedar products, with the aim of containing the use of bergamot in traditional cuisine.
- promotion and enhancement of bergamot derivatives
- creation of multimedia and information stands.
- We will also try to develop certified alternative tourism programs with high cultural standards, using bergamonte, bergamot as a characterizing element for the promotion and widespread of the local economy.

2 Conclusions

In order to promote the the whole territory and to create options, for a drastic development, we should organise a very strict and modest master plan that will cover the common goal of development of the territory and proceed to a very spesific approach." But to do this the Integrated Project must be:

- based on a cental dea for development, explained and shared not only by the institutions but by the entire territory;
- based to the collaboration of the 1st and 2nd degree local magistrates.
- the creation of complex project, consisting of specific directions that require a very good management with characteristics of new methods and organisation.
- the identification of unitary, organic, and integrated in order to achieve the effective achievement of the objectives in the set time.
- effective and timely monitoring;
- the ability to effectively manage the network of relationships with others institutional subjects with more or less significant roles;
- an evaluation and ante of the project.

If you use the three essential elements of the cultural landscape (the good, man and the environment itself) as a unique reality, then you define an integrated system of enhancement of the environment cultural territories of the magno-Greek colonies of the province of Reggio Calabria:

$$Heritage + Man + Territory = Unique\ Reality$$

Of course for the implementation of all these proposals:

a) additional infrastructure and facilities are required due to improve security, access and usability.

b) it is necessary to carry out complementary services and activities for the enhancement, with particular reference to the infrastructural and plant equipment necessary for the imlementation of cultural activities, research and training, hospitality and catering services. For the creation of areas and premises for compatible economical activities (craft shops, Natural shopping centers, production of multimedia content, etc.);

c) it is necessary to adopt Environmental Quality Systems and certification of the services offered;

d) it is necessary to create a coordinated image through: editorial Productions, signage, reception services, etc.

e) it is necessary to complete recovery and restoration interventions and activate necessary actions to build, starting from this cultural heritage, a regional system of cultural landmarks able to make Calabria more attractive for visitors-tourists and for tour operators to work for a sustainable tourism sector and the culture industry. The proposed strategy is to identify the realistic goals that can be started immediately available for valorization and fruition actions, focusing on the quality of the interventions.

References

1. Albanese, G.: Istituzione di paesaggi Protetti nel territorio del "Basso Tirreno Reggino" Costa Viola e Piana degli Ulivi. Larufa editore, Reggio Calabria (2001)
2. Ferri, A.: I Beni Culturali e Ambientali. Giuffrè, Milano (1995)
3. Ardi, A.: Parchi d'Europa, guida alle riserve a agli ambienti naturali. Mondadori, Milano (1994)
4. Barucci, P., Becheri, E.: L'industria turistica nel Mezzogiorno. Rapporto, Il Mulino (2006)
5. Borrello, C., Idotta, F., Romeo, R.: L'ecomuseo dell'area grecanica. Proposta per la creazione di un ecomuseo della cultura grecanica nella vallata dell'Amendolea. Istar, Reggio Calabria (1999)
6. Campolo, D.: Progetto Di Valorizzazione Di Un Centro Storico: Il Caso Pentedattilo. Aracne editrice, Roma (2006)
7. Campolo, D., Vecchio RuggerI, S.: Linee guida per il recupero del Borgo di Pentidattilo. Laruffa, Reggio Calabria (2008)
8. Cassalia, G.: I siti UNESCO: il ruolo del patrimonio immateriale identitario nei processi di sviluppo locale. Laruffa editore, Reggio Calabria (2009)
9. Condemi, F.: Gli ellenofoni del 2000 in Calabria. Edizioni Ellenofoni di Calabria, Reggio Calabria (1999)

10. Cuteri, F.: Guida alla Calabria Greca. Citta Calabria edizioni, Catanzaro (2011)
11. Della Spina, L.: Procedure di valutazione della qualita abitativa. Gangemi Editore, Roma (1999)
12. Malaspina, M.: "Se l'Aspromonte diventasse patrimonio percepito, Laruffa, Reggio Calabria (2009)
13. Malaspina, M., Mollica, E.: La formazione per le competenze necessarie al territorio: proposte per l'Aspromonte Grecanico. Laruffa editore, Reggio Calabria (2010)
14. Minuto, D.: Catalogo dei monasteri e dei luoghi di culto tra Reggio e Locri, Ed. di Storia e letteratura, SIT regione Calabria, Roma (1977)
15. Mollica, E.: L'investimento nelle risorse culturali. In: Quaderni PAU n. 2, Gangemi Editore, Reggio Calabria (1991)
16. Mollica, E.: Le Aree Interne della Calabria. Rubbettino, Soveria Mannelli Catanzaro (1996)
17. Mollica, E., Malaspina, M.: Programmare, valorizzare e accompagnare lo sviluppo locale. Laruffa Editore, Reggio Calabria (2012)
18. Mollica, E., Mirenda, G.: Insediamenti industriali ed artigianali in Calabria. P§M Edizioni, Roma (1990)
19. Nucera, E.: La Grecia di Calabria nell'alto medioevo. Citta del sole Edizioni, Reggio Calabria (2009)
20. Russo, F.: Storia dell'arcidiocesi di Reggio Calabria. Laurenziana, Napoli (1961)
21. Stellin, G., Rosato, P.: La valutazione economica dei beni ambientali. Metodologia e casi di studio. UTET Libreria, Torino (1998)
22. Stroppa, C.: Sviluppo del territorio e ruolo del turismo. Clue, Bologna (1976)
23. Teti, V.: Il senso dei luoghi. I paesi abbandonati di Calabria. Donzelli editore, Roma (2004)
24. Violi, F.: Demetra e Persefone: la palma di Bova, Quaderni Cultura greco calabra, IRSSEC Bova marina (2004)
25. Violi, F.: Tradizioni popolari greco-calabre, Apodiafàzzi. Reggio Calabria (2001)

Fortifications for the Control of the Early Medieval Valdemone

Fabio Todesco[✉]

Dipartimento di Ingegneria, Università di Messina, Messina, Italy
ftodesco@unime.it

Abstract. The contribution concerns the early medieval Sicilian fortifications built by the Byzantines to contrast the Islamic threat. After the conquest of the island by the Muslims, the sources testify to the rebellion of the Byzantine people between 962 and 965 A.D. and expressly mention the cities of Taormina, Mîqus, Rametta and Demenna as strongholds of the revolt. The geography of the rugged and mountainous territory of Valdemone, however, suggests the presence of other points of revenge and control of the coasts and the mouths of rivers, considered privileged axes for the penetration of the island.

The location of the cities of Mîqus, like that of Demenna, was never unequivocally identified, however, the consideration of the relative immutability of the orography of the territory and its ancient road system allows us to carry out some significant reflections on the still uncertain location of some sites mentioned by the sources.

Keywords: Sicilia · Valdemone · Mîqus

1 Sicily in the Early Middle Ages

1.1 Sicily in the Byzantine Age

The Islamic expansion began in the seventh century AD but the conquest of Sicily began in 827 AD with the landing in Mazara and the siege of the main cities and fortresses of the island.

The sources testify that already in the seventh century, with the coming of Constantine III in Siracusa, there was a marked increase in military contingents in the island (Cracco Ruggini 1980). In 634/36 the Muslims invaded Syria, followed by Egypt up to Tripoli; in 646 the general 'Amr conquered Alexandria and then a fleet led by Mu'away attacked Cyprus, Rhodes and Coo.In the following years also in Sicily Islamic incursions began (Amari 1880–81) that became more frequent forcing the Byzantines to reorganize the Thema of Sicily to try to contrast this war pressure. The Themi were administrative and military units that provided for the allocation to the peasants of agricultural funds transmissible to heirs in exchange for compulsory military service and pursued the purpose of motivating the army as well as contain military expenses. The consolidation of imperial power required the strengthening of military power and the control of the

© The Author(s), under exclusive license to Springer Nature Switzerland AG 2022
F. Calabrò et al. (Eds.): NMP 2022, LNNS 482, pp. 2449–2459, 2022.
https://doi.org/10.1007/978-3-031-06825-6_234

main junctions of the internal road network that guaranteed, if necessary, an effective defensive strategy based on the creation of a checkerboard consisting of strong places in which they were allocated military continents and points of alert between them visually connected that, thanks to an effective internal road network, they allowed, if necessary, the sudden intervention of a military contingent where there had been need. The exarchate reacted with an action of control of the seas and with the assignment of plots of land to the soldiers in an attempt to make them faithful by linking them to the possession of the land. The testimonies of the Islamic chroniclers translated by Michele Amari allow to formulate a hypothesis about the state of the Sicilian fortifications: *"Il paese fu ristorato dai Rum, i quali vi edificarono fortalizi e castelli, né lasciaron monte che non v'ergessero una rocca"* (An Nuwayri, BAS 1981) ed ancora: *"I Rum ristorarono ogni luogo dell'isola, munirono le castella ed i fortilizi e incominciarono a far girare ogni anno nella stagione propizia intorno alla Sicilia delle navi che la difendevano e talvolta, imbattendosi in mercanti musulmani, li catturavano"* (Ibn Al'Athir, BAS II, p. 363). The exaggeration of the two chroniclers, however, testifies to the greater effort made by the Byzantines in the defense of the island, probably consisting in the strengthening of the main strategic nodes such as the passes and the nerve centers for the government of the island. The new organization of the territory determined a network of road relations of spontaneous character that went along with the orography of the territory.

1.2   The Islamic Conquest of the Island

At the beginning of the ninth century the aghlabites of Qayrawân conveyed towards Sicily the insurgent thrust exercised by both religious groups and the military caste of Gund.

The conquest of Sicily took place from 827 AD when the Muslims led by Asad Ibn Al'Furat landed in Mazara and continued to the east its conquest. The defensive system implemented by the Byzantines was based on a checkerboard of fortified places located in sites naturally defended and visually connected to each other that referred to the fortress of Castrogiovanni (Enna), barycentric in the island, where it was stationed a military contingent capable of intervening where there had been a need.

The aghlabita dynasty began with the conquest of Palermo but only thirty years later there was the collapse of the defensive system with the fall of Castrogiovanni that determined the subsequent fall of other fortresses first in the Val di Mazara, then Val di Noto and finally Valdemone (Maurici 1992).

The penetration of the Muslims in the government of the island found a significant resistance in Valdemone, coinciding with the north-eastern part of the island that ended in 902 with the fall of the main city-fortress present. The relative tranquillity lasted hardly a sixty years since the cities of Valdemone rebelled and remained besieged until the definitive conquest happened in 965 with the fall of the four principal cities of Valdemone: Taormina, Mîqus, Demenna and Rometta.

One of the possibilities of understanding the system put in place by the Byzantines can not be separated from the consideration of the viability of the island along which goods were carried men and ideas.

1.3   The Road System in the Island

The road system in imperial times is documented by the Itineraries and the Tabula Peutingeriana (Pace 1935), in the Middle Ages by the Cosmography of the Anonymous of Ravenna (7th century) and by the Geography of Guidone (9th century). Between 1138 and 1154 Idrisi wrote, for order of Ruggero II, the nuzhat 'al mustaq that allows to fill the gaps between the period of writing of the Itineraries up to the redacted map of the baron Samuele Von Schmettau. Sicily has a complex of roads of remote antiquity on which provides us with more details the first map of the island, drawn up by Schmettau in 1719–21, before the reorganization of the roads on the island wanted by Charles III in 1774–77. (Bianchini 1841, p. 427).

By adopting the criterion, already applied by Columba (Pace 1935, p. 427) of comparing with the most recent cartography the ancient roads as well as with the orography of the territory, elements of significant importance can be deduced (Columba 1910) (Fig. 1).

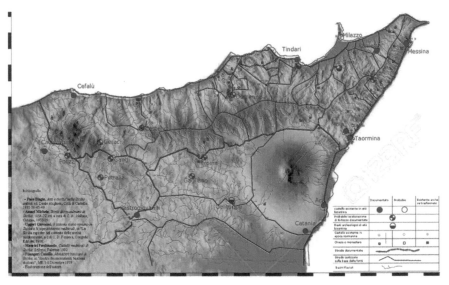

Fig. 1. Dislocation of the fortifications in Valdemone in the Byzantine age in relation to the road network and to the religious presence and to the strong places documented in the Norman age.

From the description of 'Al 'Muquaddasî (Muquaddasi, BAS, p.669) we know that the Valdemone area was defended by some important naturally fortified sites: Rimtah, Tabarmîn, Cefalù, Damannas, Y.n.f.s (Mîqus). Certainly among these main centers there were a series of minor aggregations, in which part of the population gathered as for example near the Rocca di Novara, south of Tindari in the site of Gioiosa Vecchia, in Roccabadia, Monte Castiddaci or in the Argimusco plain. Of the documented sites, only the first three are certainly locatable and possess very similar typological and functional characteristics: at Rametta as at Taormina or Cefalù, but also in fortresses outside Valdemone such as Platani on Monte della Giudecca near Cattolica Eraclea

(AG) or Castrogiovanni (Enna), a naturally defended plateau, integrated by walls, gave the possibility of overpowering and controlling the access routes.

Starting from the knowledge acquired from the comparison of cartographies and from the presence of centers of ancient origin, it is possible to hypothesize roadways that connected them and through which goods, knowledge and ideas were carried.

1.4   The Defensive Strategy

The defensive strategy implemented by the Byzantines referred to the fortresses of Cefalù, Enna and Butera, arranged along a north-south axis that crossed Sicily from the Tyrrhenian Sea to the Mediterranean. In Valdemone, the last territory to fall into the hands of the Muslims, the fortresses of Demenna, Taormina, Rometta and Mîqus' constituted a defensive chessboard which availed itself of strategically dislocated lookout points that allowed a defense lasting over seventy years. The finding of the visual connections among the inhabited sites in Byzantine age allows to formulate hypotheses on the possible localizzation of points of sighting functional to the control (Fig. 2).

Fig. 2. Visual relations between the forts inhabited in the Byzantine period.

In order to verify the visual connections between the Byzantine cities, some sections of the territory have been developed. This has allowed to verify the sites in which they are ascertained such visual connections and the sites in which the localization of the site presupposes the existence of a point of sighting that could transmit eventual signals (Fig. 3).

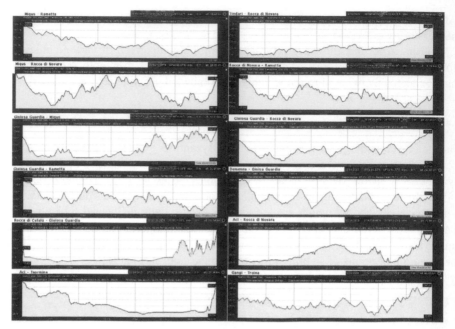

Fig. 3. Verification of the visual connections among some documented centers and points of sighting of Byzantine age. In many cases direct connections are not possible therefore it can be assumed the presence of numerous points of sighting localized on the highest peaks of the island.

While some fortresses cited by contemporary sources are still easily identifiable today because they insist in the same site (this is the case of Taormina, Rometta,…) in other cases the attempt to trace the sites cited by the sources is full of difficulties. In fact after this period the sources do not allow to identify the location of several fortresses. Thus the fortresses of *Qal'at Abd Mumin, Qal'et al. Armanin, Quastaliasah, Gabal abu Malik, Baqara, M.s.kan* are all toponyms which are not easy to identify (Amari 1883). In Valdemone, remain of uncertain location the fortresses of Mîqus' and Demenna, although the latter remains present in a series of place names that attest to its presence near the area between San Marco d'Alunzio and the Rocche del Crasto, in the territory of Galati mamertino. Archaeological finds (Filangeri 1978) testify the presence of a Paleocastro on Pizzo san Nicola even if it is appreciable a human frequentation of the whole area (Filangeri 1983).

Well documented is instead the history of Rometta whose defensive strategies, as in the Byzantine fortification practice, integrated with works of fortification the site already naturally defended.

The fortress of Rametta was located on a large plateau on a hill overlooking the river Saponara naturally defended by high steep walls whose only access routes were barred by walls of fence. The position of the city with respect to the mouth of the river, a privileged axis for penetration into the interior of the island, does not allow the control access routes constituted but allowed the visual connection with the castle of Saponara from where it was possible to control the main axis of the river and a tributary that allows you to reach the internal road network of the island.

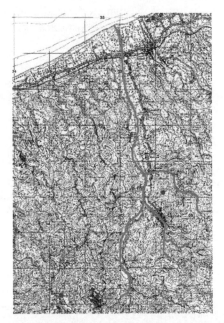

Fig. 4. Cartographic excerpt of the territory of Rametta. Note the location of the town that does not allow the vision of the mouth of the river. The control of transit is allowed by the site of the castle of Saponara which was a fundamental point of sighting for the fortress of Rametta.

Fig. 5. Particular of the mastio endowed with cistern of the castle of Saponara.

The city of Rametta referred to a system of visual references able to keep under control the mouth of the Saponara torrent, which has always been used as a percurrency towards the most internal areas (Fig. 4). It is a square-shaped tower located at the highest point of the hill (Fig. 5) at the foot of which is located the historic center of Saponara. The small fortress, still present at the confluence of the stream with one of the arms that feed it, is made with a poor building technique related to the availability of building material and, certainly, its first nucleus must have served as a lookout point. From here it is possible to have a visual relationship with the fortress of Rametta and to control at the same time the river basins of the fiumare Scarcella and of the torrent Cardà at their confluence with the torrent Saponara that constituted as many ways of penetration towards the viability of the ridge (Fig. 6).

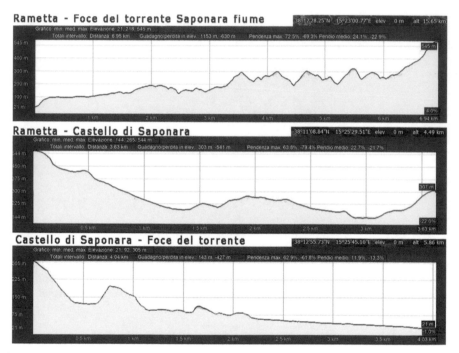

Fig. 6. Detail of the terrain sections that allow to verify the visual connection between two points. Particularly the first graph refers to the connection (not possible) between the mouth of the river (21 m.slm) and Rametta (545 m.slm); the second to the connection Rametta-castle of Saponara; the third shows the visibility of the mouth of the river from the castle of Saponara.

In the case of the fortress of Mîqus' the identification of the site is made possible by the news provided by Muslim chroniclers who report that the city was evacuated by the defenders in 902, after the fall of Taormina ['Ibn 'al Athir, I, p.424; 'An Nu-wayri, III, p.185]. Amari wrote that Mîqus' "*... torna forse a Mandanici o Fiumedinisi...*" concluding that this fortress was close to what was anciently called Monte Miconio (Amari 1933-39, p. 105), located east of Rometta and west-southwest of Messina. Seybold identified it with the castle whose ruins overhang, about 5 km south of Monte Scuderi, Fiumedinisi.

Relatively to the toponimo it is possible to formulate some observations: in the History of the Muslims of Sicily it is read that the arabian chroniclers called this earth *Biq.s, B.n.f.s, Tif.s, B.b's, Bn's*. Edrisi locates it «*...da Monteforte 15 miglia* (arabiche) *per mezzogiorno (...) tra Messina e Taormina, (...) si arriva per sentieri alpestri* una terra *Miq's, M.nîs...*», according to the various manuscripts.

The Nallino believed that «*...il luogo risponde tra il capo di Scaletta ed il Monte Scuderi, sia Artalia o Pozzolo sup. o Giampilieri ecc., castello par che non ne rimanesse neanco al tempo di Al-Edrisi...*». He concluded that the place under consideration could be identified with Mandanici and that the distance could be inaccurate in Edrisi's manuscript (Fig. 7).

Fig. 7. The fortress of Miqus seen from Piano Margi.

Fig. 8. The fortress of Miqus seen from Belvedere Castle.

Another reference to Mîqus' is found in the *Mu'gam 'al buldân* of Yaqût who writes «*... e nei monti di Micos delle miniere di vitriolo, di ferro e di piombo...*» confirming a location, for this site, rich in mineral deposits as is the case only in some points of the Peloritani Mountains (Yaqut, I, p.204). Considering the orography of the land, Monte Scuderi, where during some surveys some coins and medals referable to the VII-VIII century were found, offered optimal opportunities for human settlement, the best of the whole area, by those who needed to control the main nodes from which departed the road links that related between them the centers of western Sicily. From the summit of this mountain, moreover, it is possible to have a visual connection with the nearest "*strong places*" of the Byzantine territorial chessboard: Taormina to the south, Rometta and Mount Antennamare to the north, the latter certainly used at least as a point of sighting able to control at the same time the Tyrrhenian and the Ionic sides.

Even in the case of the site of Mount Scuderi, therefore, the defensive model assumed is similar to that of other Byzantine fortifications whose location is certain and where, from the summit plateau, it is not possible to control access to the way of penetration towards the interior of the territory constituted by the river Nisa. The point of lookout that controlled the access toward the whole valley, could be identified with the Belvedere Mountain to Fiumedinisi on which still today insist the ruderi of a small fortilizio that co-munque manifests phases chronologically more recent (Villari 1981). From this point of observation privileged, in whose underlying lowland they have been recovered traces of human presences from the age neolitica until that classical (Bacci 1978) it is possible to have under control the way of penetration towards the inside constituted from the greto of the Nisa and it is at the same time possible to establish a visual relationship with the Scuderi Mountain beyond that with the site of Taormina also with the castle of Force of Agrò and with the fortilizio of Head S. Alessio. The masonry visible today, are referable to various construction phases that span a few centuries, although the building that is observed today could insist on another pre-existing whose function hypothesized sighting place, thus confirming a continuity of human settlement in the site (Fig. 8).

The underground cistern covered with a pointed barrel vault is located close to the south wall of the castle and its similarity to the cistern located in the castle of Saponara is of significant importance.

The castle of Monte Belvedere has similar characteristics but is slightly larger and has a more articulated configuration of the spaces inside the enclosure. It is a fortress placed on the top of the mountain able to control the access to the valley formed by the Nisa river and which is in visual contact with the top of Mount Scuderi where it is hypothesized the city of Mîqus' was located. It consists of a tower leaning against the fence that surrounds the top of the mountain enclosing a quadrangular space. The observation of the masonry of the building shows several interventions, some of which are to be considered historical; the possibility of reading them is inhibited by the last intervention whose indiscriminate filling of the mortar joints does not allow to distinguish the physical superimpositions between the different portions of masonry; therefore, the hypotheses on the genesis of the fortress, in the absence of sources, can only be limited to observing its typology as well as its location, in order to formulate hypotheses based on the history of this interesting evidence. Also in this case, however, the abandonment of the site has been the main cause of the degradation. The cornerstones of the tower and the worked stones used in the door and window jambs were stolen (Fig. 9, 10).

The entire summit of the mountain on which the fortress is located is affected by a fracture that crosses it in an east-west direction determining a lowering of the masonry to the south of about 90 cm. The restoration of the building in the recent past, unfortunately, has not affected the stability of the slope if not for some limited interventions of stabilization necessary to ensure accessibility to the site for which it is considered still active instability in place.

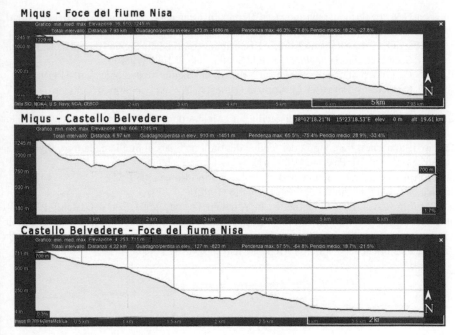

Fig. 9. Detail of the terrain sections that allow to verify the visual connection between two points. In particular the first graph refers to the connection (not possible) between the mouth of the river (18 m.slm) and Miqus (1229 m.slm); the second to the connection Miqus-Belvedere castle; the third shows the visibility of the mouth of the river from Belvedere castle.

Fig. 10. Belvedere castle seen from west

References

Amari, M.: Storia dei Musulmani di Sicilia, 1854–72, ed. a cura di C. A. Nallino, Catania (1933–39)

Amari, M., 'Muquaddasî, A.: 'Ahsan 'at taqasim, Biblioteca Arabo Sicula, vol. II (1880–81)

Amari, M., Nuwayrî, A., Biblioteca Arabo-Sicula: trad. it., Torino-Roma 1880–81, rist. Edizioni Dafni, Catania (1981)

Bacci, M.G.: Fiumedinisi 1978-79. In: BCA Sicilia, anno III, n1-2-3-4 (1982)

Colomba, G.M.: Per la topografia antica di Palermo. In: Centenario della nascita di Michele Amari, vol. II, Palermo, pp. 395 [1]–426 [32] (1910)

Cracco Ruggini, L.: La Sicilia tra Roma e Bisanzio, in "Storia della Sicilia" diretta da R. Romeo, vol.III, Società ed. Storia di Napoli e della Sicilia, Napoli 1980 (1980)

Filangeri, C.: I ruderi di un paleocastro sui Nebrodi, in «Sicilia Archeologica», n. 51 (1983)

Filangeri, C.: Ipotesi sul sito e sul territorio di Demenna, in "Archivio Storico Siciliano", serie IV, Palermo (1978)

Filangeri, C.: Monasteri basiliani di Sicilia, in "Mostra dei codici e dei monumenti basiliani siciliani", Messina 3–6 dic. 1979, Regione Siciliana Ass. BBCCAA e PI, Palermo (1980)

Martino de Spucches, F.S.: Storia dei feudi e dei titoli nobiliari di Sicilia dalla loro origine e ai nostri giorni, vol. III, Palermo (1924–41)

Athîr, I.: Biblioteca Arabo-Sicula, trad. it., vol. 2, Torino-Roma (1880–81)

Maurici, F.: Castelli medievali di Sicilia, Sellerio (1992)

Pace, B.: Arte e Civiltà nella Sicilia antica. In: Alighieri, D., (ed.), vol. I, Città di Castello (1935–49)

Villari, P.: Monte Giove e Fiumedinisi, Messina, 1981; Idem I giacimenti preistorici del Monte Belvedere e della Pianura Chiusa di Fiumedinisi (Messina). Successione delle culture nella Sicilia nord-orientale, in "Sicilia Archeologica", pp. 46–47 (1981)

The Church of San Sebastiano in the Valley of Pagliara (Messina) Formerly the "Priorato di Santa Maria di Billimeni in Terra di Savoca" in the Byzantine Mediterranean

Francesca Passalacqua[(⊠)]

Dipartimento di Ingegneria, Università degli Studi di Messina, Messina, Italy
francesca.passalacqua@unime.it

Abstract. The church of Santa Maria of Polimena, today dedicated to San Sebastiano, is located on the left bank of the river Pagliara, in the Ionian Valley.

Documentary sources on the building date back to the 16th century when it was one of the Basilian monasteries still active in the territory annexed to the *Archimandritato of Santissimo Salvatore of Messina,* but its foundation dates back to earlier times.

The present research intends to deepen the study of the existing church building, almost completely neglected by historiographic studies, through analysis of the monument and identification of documentary sources. The sources, in fact, acquire an important documentary value for this building since they testify to the existence of the building over a long period (16th–19th centuries).

The investigation is based on a cross-referenced study of bibliographic sources and archival documentation and allows us to clarify the steps of a complex and articulated transformation of the building from its greatest splendour to its abandonment.

The article is, first of all, structured on an analysis of the bibliographic and archival sources that document the presence of the settlement among the Basilian foundations in the Ionian Peloritan territory, where the presence of Byzantine settlements is notable.

A comparison between the results of the sources and a survey of the state of the sites led to an assessment of the transformation of the building from its foundation to the complete definition of the structure, also considering what no longer exists.

Evaluation of the state of the sites has led us to think about greater safeguarding for these settlements, considering that the church needed consolidation works, which began in the 1990s and have not yet been completed.

Keywords: Sicily · Italo-Greek monasticism · Architecture heritage

1   Introduction

The church of Santa Maria of Polimena is located in the Ionic Valdemone, on the left bank of the river Pagliara, also known as *Santa Maria of Bullumeni or Bollomerio*, today

© The Author(s), under exclusive license to Springer Nature Switzerland AG 2022
F. Calabrò et al. (Eds.): NMP 2022, LNNS 482, pp. 2460–2469, 2022.
https://doi.org/10.1007/978-3-031-06825-6_235

dedicated to San Sebastiano. In the twelfth century, the territory - the ancient *Barony of Savoca* - counted numerous Greek monasteries supervised by the abbot Saint Luca at the head of the *Archimandritato* of Messina.

The documentary sources concerning the building date back to not earlier than the sixteenth century, a period in which the "Priorato di Santa Maria di Billimeni della Terra di Savoca", between 1532 and 1555, was annexed among the last monasteries (about twenty) still active then in the territory to the *Archimandritato of Santissimo Salvatore*.

The sixteenth century *visitationes* and the pastoral visits until the late nineteenth century bear witness to the transformation of the church and the monastery. In 1557, the church building was lacking a high altar and the monastery, considered insufficient, needed to be enlarged and, in the *visitatio* of 1580, it is clear that the 'orders' of the *Archimandritato* were respected: the high altar and a side altar were realized and the monastery enlarged.

This period probably saw the monastery's greatest splendour; the church was rich in decorative elements and, on the high altar, there was a triptych composed of a painting of the Virgin Mary flanked by Saints Peter and Paul.

Today this painting is preserved at the Museum of Messina and only the central part remains, with the title of *Madonna del Gelsomino* attributed to Antonello de Saliba (1466–1535).

The church, in spite of the passing of the centuries and the abandonment, still rises imposingly with its top crenellation. The church needs consolidation interventions, which have been started but have not yet been concluded, for safeguarding the building. The interventions required confirm the succession of enlargements as being the cause of multiple instabilities.

2   The *Archimandritato* of Messina and the 'Barony of Savoca'

The church of Santa Maria of Polimena is located in the Ionic Valdemone, on the left bank of the River Pagliara, also known as *Santa Maria* of *Bullumeni* or *Bollomerio*, today dedicated to San Sebastiano (Figs. 1, 2 and 3).

In 1131, Roger II instituted the *Archimandritato* of Messina and, in October of the same year, the Latin bishop Ugo (bishop from 1127 to 1139) gave his approval by endowing it with Greek churches [1]. In the same year, Abbot Luca (12th century) was appointed first archimandrite of the monastery of *Santissimo Salvatore*, to which the monasteries of the Peloritanian territory and some of the Calabrian monasteries were subjected [2].

Abbot Luca, from Rossano, received numerous fiefs in Sicily and, in particular, the 'Barony of Savoca', which included a vast territory between the sea to the east and the Peloritani mountains, between the Agrò and Pagliara rivers, as far as the borders with Santa Lucia del Mela to the west, and counted numerous monasteries [3].

The *Archimandritato* of Messina, however, declined drastically in the following centuries compared to the Latin church, until 1538, when the *Archimandritato* seat was moved to the Abbey of San Pantaleone, south of Messina, since a military garrison was built on the ruins of the monastery of Santissimo Salvatore, on the peninsula of San Raineri, to defend the city [4]. In the last decades of the sixteenth century, only about a hundred monks lived in the remaining monasteries, among which the priory of Santa Maria di *Bullimeni* stands out.

Documentary sources concerning the building date back to the sixteenth century at the earliest, a period in which the '*Priory of Santa Maria di Billimeni della Terra di Savoca',* between 1532 and 1555, was one of the last monasteries (about twenty) still active in the territory of Messina.

Fig. 1. Pagliara (ME), the Church of Santa Maria Polimena, main facade and side view (Francesco Romagnolo, 2017)

Fig. 2. Pagliara (ME), the Church of Santa Maria in Polimena, view of the apse (Francesco Romagnolo 2017)

3   The Church of Santa Maria in Polimena in the Sixteenth Century

The first sources concerning the building date back to the first half of the sixteenth century during the Archimandrite of Abbot Annibale Spadafora of the monastery of *Santissimo Salvatore* (1532–1555). The act in which it is inferred that the '*Priorato di S. Maria di Billimeni in terra di Savoca*' is present among the active monasteries in the territory of Messina is preserved in the State Archive of Palermo, Conservatory Real Patrimonio, vol. 1305. The document also reports the displacement, for disposition of Emperor Charles V of 18 August 1539 and of Pope Paolo III of June 1542, of the seat of the *Archimandritato* to the monastery of Saint Pantaleone, in the village of Bordonaro, south of Messina, confirming therefore the existence of the church of Saint Maria di Polimena before 1542, whose prior was Abbot Matteo Xiria [5].

The *visitationes* of the sixteenth century and the pastoral visits of the seventeenth and eighteenth centuries recount the transformation of the church and the monastery. In the text of the first *visitatio*, on 19 December 1557, [6] the church building was found to lack the high altar and therefore it was ordered to enlarge the church and adapt it to liturgical worship and build an adequate location to place the Blessed Sacrament.

The monastery, considered insufficient for use and housing, needed to be enlarged in a southerly direction and in such a way that there were "two rooms per cells and a kitchen containing an adequate fireplace" [6].

In the *visitatio* on 13 March 1580 [7] it can be seen that the 'orders' of the previous *visitatio* were respected: the high altar, a side altar for the Blessed Sacrament and the monastery, which had already been enlarged, needed further works of enlargement and adaptation.

This period probably saw the monastery's greatest splendour; the church was rich in decorative elements and, on the high altar, there was a triptych composed of a painting of the Virgin Mary flanked by Saints Peter and Paul.

Today this painting is preserved at the Museum of Messina and only the central part remains, with the title of *Madonna del Gelsomino* attributed to Antonello de Saliba (1466–1535) [8].

This painting dates back to the early sixteenth century, a period when de Saliba painted several polyptychs of the Madonna and Child between Saints. The painting was described as follows in the visitatio of 1580: «*Visitvit altare maius ipsius ecclesiae in quo estat quattrum antiquum com imagine beatissimae virgin Mariae cum suis ianuis seu copertoriis, com quibus clauditur et aperitur in quibus videntur picte ab una parte imago Sanctii Petri ab alia Santi Pauli*» [9].

The following documentary traces refer to the first decades of the eighteenth century. In the pastoral visit of 1726 [10], the high altar is flanked by eight altars (the chapel of the Blessed Sacrament, the altar of St. Sebastian and the altars of St. Anthony Abbot, St. Barbara, St. Anthony of Padua, St. Charles Borromeo, St. Joseph, and St. Michael the Archangel).

From this, it seems quite clear that in the last decades of the sixteenth century and in the first decades of the seventeenth century the church underwent an important typological transformation. The probable single nave of the original building was enlarged and transformed into a three-nave structure, marked by strong pillars supporting round arches.

There are numerous examples of church buildings with characteristics comparable to the church of San Sebastiano in the Valdemone territory. The need to enlarge the place of worship led to the enlargement of the existing churches, extending or rotating their axis and, in some cases, the building was completely rebuilt. Simple and essential systems that repeated a well-defined model spread widely in this geographical area and, in general in Sicily, between the sixteenth and the late seventeenth century. The construction of a new building started from the transept, with the construction of the apses, semi-circular or quadrangular, at the end of the transept. Following this, the construction of the central nave began, almost always marked by a double series of columns (in the most important examples) or pillars that articulate the entire space [11] (Fig. 3).

The following pastoral visit of 1745 [12] also confirms the presence of the sacristy, in compliance with post-Tridentine norms, as well as changes to the main façade with new opening on the façade. From 1809, the church is indicated as "Chiesa Matrice di San Sebastiano" [13] and, unlike the church that continued to be restored and embellished, the monastery - of which only few remains are standing - was abandoned by the monks [14] (Fig. 4).

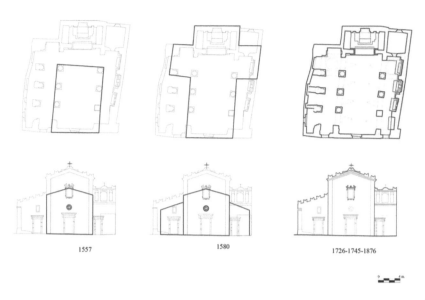

1557   1580   1726-1745-1876

Fig. 3. Pagliara (ME), the Church of Santa Maria di Polimena, reconstructive hypothesis 1557–1580–1876 (graphic design Natali Saldone, 2017)

The year 1876, engraved on the keystone of the triumphal arch, is witnesses of the further finishing works of the church: the lunetted vault of the hall of the main nave, as well as the stuccoes covering the pillars of the naves, modify the facies of the church embellished only by the limestone covering of the high altar (Fig. 5).

The historian Gaetano La Corte Cailler, who visited the church in 1903, as a testimony to its past, describes it as follows: «[…] E' una bella chiesa con tre porte sul prospetto che danno accesso a tre navate separate da pilastri con qualche stucco, decorazione e numerose cappelle […] la chiesa dev'essere antica, ma sulle due sepolture più antiche non si riscontrano che le date 1737 e 1739 […] Notevoli due quadretti ovali in cornice intagliata dorata, che si trovano in alto ai lati del grande arco in fondo alla navata centrale. A destra del terzo altare, è una tavola malandata esprimente la madonna dell'Idria con due Santi ai lati, opera dei principii del '500. Sotto si legge: PER DEVOZIONE D ANTONIO PATTI RESTAURATO 4 DICEMBRE 1876. Nel secondo altare a sinistra è una bella tela con la figura ben conservata, di S. Antonio, e sotto la firma IACOBUS IMPERATRIX PINAT 1641. All'altare maggiore è una tavola della Madonna sedente col bambino nudo ritto in piedi ed al quale ella porge dei fiori» [15].

Fig. 4. Pagliara (ME) the Church of Santa Maria di Polimena, view of the main nave (Giuseppe Spadaro, 20th century)

4   Evolution and Transformations of the Monastic Settlement

The church still stands imposingly with its crenellations at the top, despite the passage of time and neglect, which has made it necessary to carry out unfinished consolidation work to protect the building.

The relief of the church shows a plan with three naves separated by massive pillars covered with stucco decorations. However, the interior is uneven: the aisles are not parallel and both follow the course of the ground. The north aisle ended with the altar of St. Sebastian and the masonry widens out well beyond the path, following the contour of the ground; the south aisle, more regular, houses the recesses of the chapels in the perimeter masonry and opens out, at the end, onto the sacristy. The outer walls are flanked by the monastery rooms, of which there are few remains (Fig. 6).

The high altar, which concludes the nave, is described in these words in the 1970s: «poco oltre si innanza l'altare maggiore edificato con tecniche e materiali differenti [...]. La scalinata è in pietra: la parte centrale e le colonne dell'icona, tra le quali era collocato il quadro di Santa Maria Polimena di A. de Saliba, sono in marmo: le volute laterali dell'altare ed i sei gradini soprastanti sono in legno; lo stemma sormontato da corona, le figure angeliche e le altre decorazioni sono a stucchi; la parte più addossata alla parete, forse la più antica, è in pietra intagliata con motivi geometrici e floreali. Altre colonnine in marmo nero, ora trasportato nella chiesa di SS. Pietro e Paolo completavano l'altare» [16].

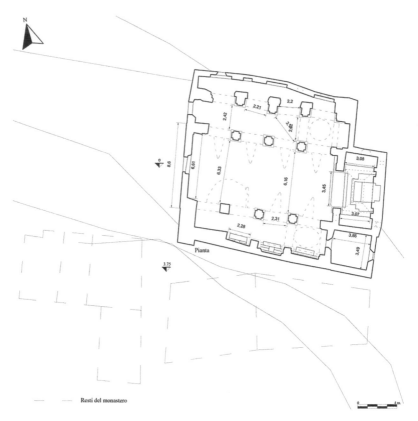

Fig. 5. Pagliara (ME). Church of Santa Maria di Polimena, church floor plan and the area of monastery (graphic design Natali Sardone 2017)

Consolidation and embellishment work dating back to 1876, safeguarding the monument by inserting chains in the naves and transept, modified the interior appearance with a substantial stucco work that partially altered the original appearance of the interior (Fig. 6).

The work, which remained incomplete, included the construction of the lunetted vault in the nave, the probable insertion of small domes in the side aisles (traces of which remain in the transept area) and the covering of the gross vault of the main apse. The stucco cladding would thus have masked the wooden roof, completely transforming the original facade of the church.

The relief of the building, compared with the reports of the sixteenth century and pastoral visits up to the late nineteeth century, suggest the transformations of the architectural structure.

From the reading of documentary sources and material analysis, it is possible to assume the existence of a primitive church, delimited to the current hall of the faithful. The first indications are that the church was enlarged with the construction of the apse and chapels next to the transept area. Further enlargements between the sixteenth and seventeenth centuries resulted in the construction of a total of eight altars, defining the present tripartite space. The exterior also shows the enlargements over the centuries: the main façade is characterised by a salient façade in which it seems quite clear that it was transformed from a simple gabled façade in which the entrances and the holes above were modified [13] as was the bell tower (of which there is no documentary evidence).

Restoration work carried out by the *Soprintendenza* of Messina in the 1990s confirmed the evolution of the church's layout as the cause of the structure's major disruption and collapse. There seems to be no continuity between the wall structure of the nave and the side walls. The latter have undergone at least two expansions, the downstream wall has been leaned against without an adequate connection to the outside, and the foundations have been built at different times, heights and terrain.

5   Conclusions and Future Prospects

The church of San Sebastiano, despite the passage of centuries and its abandonment, still stands imposingly with its battlements, next to the River Pagliara.

From reading the archival documents, analysing the architectural artefact, and the material investigations, it can be concluded that there probably already existed a primitive church in initially, perhaps, with a single nave.

Between the fifteenth and sixteenth centuries it was enlarged and provided with two altars. As the documents shows, the additional altars were built at later times. In those years, the church was embellished with important sacred vestments and work of art that testify the importance of the presence of the Basilian monks in the area.

Therefore, it is hoped that the restoration works started in the 1990s will be completed and the church will once again be used by citizens as an important monument of the territory.

The building can thus be included in an interesting itinerary of the settlements that belonged to the *Archimandritato of Santissimo Salvatore* of Messina in the Ionian Peloritan area.

Fig. 6. Pagliara (ME), Santa Maria of Polimena Church, main altar (Natali Sardone 2017)

References

1. Scaduto, M.: Il monaschesimo basiliano nella Sicilia orientale. Rinascita e decadenza sec. XI–XIV, Edizioni Storia e Letteratura, Roma (1947)
2. Fonseca, C.D.: Pontificati sede aptavit: la ricostruzione della chiesa vescovile di Messina (seco. XI–XII) in Messina il ritorno della memoria, Edizioni Novecento, Palermo, p. 36 (1994)
3. Calleri, S., Savoca segreta, I.S.C.R.E., Catania, p. 44 (1972)
4. Reina, G.: Itinerari italo-greci in Sicilia, Marsilio, Venice, p. 66 (2016)
5. Cascio, A.: Il priorato basiliano di Santa Maria di Polimena nella Terra di Savoca, Archivio Concetto Marchesi, Roccalumera, pp. 119–160 (2013)
6. Cascio, A.: Archivio di Stato di Palermo (ASP), Conservatoria Real Patrimonio. Opus Citaum. **1305**, 122–126
7. Cascio, A.: ASP, Conservatoria del Real Patrimonio. Opus Citaum. **1320**, 6, 125
8. Cascio, A.: Opus Citaum, p. 127
9. Pugliatti, T.: Pittura del Cinquecento in Sicilia. In: La Sicilia orientale, Napoli, Electa Napoli, 1993, pp. 15–27. Musolino G., Palazzo Ciampoli tra arte e storia, Rubbettino, Soveria Mannelli, p. 192 (2016)
10. Cascio, A.: Archimandritato Santissimo Salvatore, Messina, Pastoral visits. Opus Citaum, 116 (1762)
11. Passalacqua, F.: La basilica di Santa Maria Assunta di Randazzo (XIII-XIX secolo), Caracol, Palermo, p. 48 (2017)
12. Archivio Parrocchiale Chiesa Madre di Savoca, Pastoral visits (1745)
13. Archimandritato Santissimo Salvatore, Messina, Pastoral Visits (1809)
14. Archimandritato Santissimo Salvatore, Messina, Pastoral Visits 1818, 1822, 1825, 1833
15. Corte Cailler La, G.: Comune e provincia di Messina nella storia e nell'arte. In: Quaderni dell'Archivio Storico Comunale di Messina "Nitto Scaglione" e della Biblioteca Comunale di Messina Tommaso Cannizzaro Messina, vol. 1, p. 137 (2017)
16. Spadaro, G.: Chiesa di Santa Maria di Polimena. Scheda aggiuntiva relazione finale Soprintendenza per i beni architettonici, artistici e storici di Catania (1979)

On the Roads of Bruttium Between Italìa and Kalavrìa. Rediscovering Ancient and New Cultural Landscapes

Mauro Mormino[1]([envelope]) [iD], Mariangela Monaca[1] [iD], Francesco Calabrò[2] [iD], and Emanuele Castelli[1] [iD]

[1] Università degli Studi di Messina, Piazza Pugliatti, 1, 98122 Messina, Italy
mmormino@unime.it
[2] Università Mediterranea di Reggio Calabria, Via dell'Università, 25, 89124 Reggio Calabria, Italy

Abstract. The Bruttium: a land nestled between the ancient Italìa and Kalavrìa, whose history is fully interconnected with the wider context of the Byzantine Mediterranean Sea. This contribution aims to reanalyse events and dynamics of the diverse identities which have inhabited it (e.g. Pagans, Jews, Greek and Latin Christianity, Arabs) paying particular attention to their intertwining, to the shape of an inclusive cohabitation, inserting them in the wake of the Greco-Roman past but also projecting them into the future to reevaluate the landscape forms and its cultural heritage with an eye to the prospect of an economic and social recovery on firm cultural foundations.

Keywords: Cultural Landscape · Byzantine · Kalavrìa · Cultural heritage enhancement

1 Introduction

The enhancement of the Calabrian cultural heritage requires the preliminary understanding of its specific characteristics, which differentiate this Region from the neighboring ones [1].

Catastrophic events of various kinds (earthquakes, wars, etc.) have left few visible traces of the magnificent past of the Calabrian colonies of the Magna Graecia period: today the remains of Sibari, Crotone, Locri, Reggio, just to mention the most important, testify to a sumptuous past but difficult to understand for the layman, certainly less attractive than, for example, the present contemporary heritage in Campania or Sicily.

In the Calabrian territory, on the other hand, there still exists a heritage of great value and interest, potentially capable of making this region competitive with respect to other southern regions, if properly enhanced [2]: it is the heritage of the Byzantine age.

The work is the result of the shared commitment of the authors. However, paragraphs can be attributed to Mauro Mormino; to Mariangela Monaca paragraph 3; to Francesco Calabrò paragraph 1; to Emanuele Castelli paragraph 4. Conclusions were written jointly by the authors.

© The Author(s), under exclusive license to Springer Nature Switzerland AG 2022
F. Calabrò et al. (Eds.): NMP 2022, LNNS 482, pp. 2470–2484, 2022.
https://doi.org/10.1007/978-3-031-06825-6_236

It is not a question of individual monuments or historical centers, let alone characterized by considerable dimensions or particularly sumptuous materials and decorations (such as in Ravenna), but rather a system of nodes, born between the sixth and tenth centuries. d. C. from a military defense strategy, closely connected with a widespread system of contemporary religious settlements, which make the spiritual dimension one of the distinctive features of the Calabrian territory [3].

From the cognitive point of view, this heritage has been the subject of numerous studies, which will be briefly recalled later.

As will be seen, it is precisely in that period that this peculiar settlement system was born: it is made up of very few larger nodes on the coast, equipped with ports and easily defensible, and a constellation of small towns in the inland areas, strategically connected with the coastal watchtower system, against Saracen invasions; agriculture and crafts have flourished around these small towns for centuries, guaranteeing a comfortable survival for the populations [4, 5].

Even today this complex settlement system distinguishes the regional territory, despite the transformations that have taken place over the centuries, particularly profound during the twentieth century; a series of historical events, however, including, for example, the mechanization of production processes in agriculture and the displacement of the geopolitical center outside the Mediterranean, following the discovery of the Americas, have progressively marginalized and impoverished this territory [6].

Today the transmission to future generations of the cultural heritage, material and intangible, present in the small towns of the Calabrian inland areas is being repented by the progressive depopulation: it is therefore urgent to intervene for its enhancement before these testimonies are definitively lost and, with them, the most authentic spirit of this region [7–9].

This article, therefore, intends to offer an interpretation that looks at the whole of the Calabrian settlement system that arose between the sixth and tenth centuries. d. C., underlining its exceptional character, in order to contribute to its adequate enhancement by setting itself two main objectives:

- To briefly highlight the requests and logics, military and religious, which determined their genesis;
- To trace some possible trajectories for its enhancement.

2 Time of Change: From Bruttium to Kalavrìa

At the end of the Fifth century monasticism was quite widespread and had given rise to conflicts which the pontiff's epistle intends to remedy. The presence of a female monasticism is also confirmed by a report from the Life of Saint Phantinos the Elder according to which a female monastery stood next to the tomb of the saint in Taurianum. It is also necessary to remember the vitality of the monastery founded by Cassiodorus in Squillace between 555 and 560, the Vivarium; as well as the existence of a monastery in Tropea, one in Taurianum and, probably, one in Reggio, according to the testimony of pope Gregory the Great (590–604) [10].

The Sixth century constitutes, in hindsight, a fundamental period in the development of the Calabrian monastic tradition and in the political-administrative organization of the ancient Bruttium. In both cases, these are the consequences of the Gothic Wars (535–553) which also had an impact on the religious and social substratum of the lands returned to Roman-Byzantine imperial rule after the Gothic invasion of Italy in the first half of the century. When the Byzantine armies arrived in Calabria, leaded by the general Belisarius, the urban, strategic and administrative situation was not little complex: towns were scarce and only few of the bishopric seats were housed in an urban context.

Most of them were installed in rural sites: near to ancient aristocratic *villae*, at post stations along the main roads or coinciding with the official residences of the papal estate administrators.

During but especially after the Gothic Wars, Justinian walled ancient cities and founded other in new sites [11–15]. The first step was the reconstruction and redefinition of the main ports and harbours along the Tyrrhenian (e.g. Reggio, Vibona) and Ionian (Scolacium, Crotone) coasts in order to facilitate direct communication with the main administrative centres such as Rome and Constantinople and with local ones both towards Campania and Puglia.

The Justinian operation envisaged a new urbanization policy – the cities were the seat of civil, military and ecclesiastical authorities – and the creation of new city walls and fortifications around them. One of the most characteristic aspects of the period, an indication of the need to defend the territory more incisively from the risk of future invasions, was to provide the inhabited centres with an additional defensive system located in the upper part of the city (the so-called *praetorium* or *praitorion*) or on the hills and promontories located at the margin. Two interesting cases are constituted by the fortification undertaken by Belisarius on the area now occupied by the Aragonese castle of Reggio or by the construction of the acropolis of Scolacium on a promontory overlooking the Gulf of Catanzaro.

The latter is perhaps the most interesting.

In the case of Reggio the chosen area matched generally the layout of the city walls dating back to the Chalcidian colonization of the Eighth century B.C.; but in that of Scolacium a building operation *ex novo* was undertaken following construction methods referring to coeval Byzantine military architecture. However, the purpose of those interventions was also to create new spaces to protect the population and to house both civil and ecclesiastical public buildings rendering them more defensible in the event of a siege.

The Church cooperated in the fortification movement started by Justinian to ensure ancient urban episcopal seats, providing equally lands, resources and funds in the case of new foundations in the rural areas (e.g. Scolacium acropolis was built on monastic lands). The roman ecclesiastical authorities realized how important was to group the population or *rustici* inside walled towns (*castra*) to guard them – leaving unfortified villages (*vici*) – and force itinerant bishops to reside there permanently. This process, which caused a small internal migration of the local population, had a further dual purpose: to protect the rural territory by concentrating the forces for its defence and to make fiscal control more efficient [15].

However, at the end of the Sixth century the Justinian project was far from complete: the Lombard invasion of Southern Italy and the emergence of the competitive Duchy of Benevento (571) actually showed its weaknesses. Lombard forces easily occupied villages and town in the North area of the ancient Bruttium (known for its silver mines), overcoming the resistance of the unarmed population and the tenuous grasp of the local aristocracy.

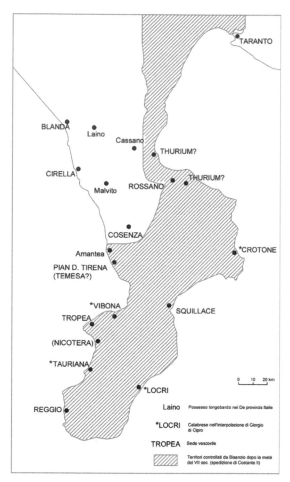

Fig. 1. Byzantine possessions in Calabria after the campaigns of Constant II (second half of the 7th century). Map of the authors taken from: G. Noyè, L'economie de la Calabre de la fin du VI au VIII siècle. Cahiers de recherches médiévales et humanists, 2014 – 2, n° 28

The Byzantine resistance moved South, convincing Constantinople to undertake a new defensive strategy.

By this time mountain peaks and rocky reliefs in the inland became the subjects of a new fortification system (the so-called *kastron*), housing military garrisons and local population, erected on critical sites and controlling larger spaces as in the case of Tiriolo: at the top of a mountain looking at the same time to the Tyrrhenian and the Ionian coasts. In fact, under Constans II (641–668) with the institution of the Byzantine Duchy of

Calabria the Isthmus of Catanzaro became a key area in the defence of the southern portion of the ancient Bruttium (Fig. 1) [15].

Despite attempts to recover the lost territories, in the middle of the Seventh century also the name Bruttium lost his previous meaning. Now identifying the lands under Lombard rule (also called *Brettia*) while the term Calabria (*Kalavrìa*) became to indicate the southern part – the Byzantine province – where Constantinople concentrated its military and administrative efforts in the following century [16].

The reign of Leo III (717–741) was a defining moment in the stabilisation of the Byzantine-Lombard borders on the Crati river after a successful military campaign recovering territories once lost and celebrated, for example, with the renaming of the city of Santa Severina as Nikopolis.

But the reconquest had two direct consequences of particular importance:

– Along the Lombard *limes* new *kastra* were built in the inland while some of them already set on mountains and rocky peaks re-founded or transferred on even more defensive and strategic sites to control even the seashores from the height. A reaction to the growing Arab threat coming from the sea. The presence of small villages and towns along the coast, the principal roads/pilgrimage routes and rivers persisted but carefully studied. Located in the vicinity of hills and not too far from fortified mountain settlements, they preserved the commercial trade or encouraging the cultivation of fields in the plains [15].
– The ancient papal estates were progressively transferred directly to the Byzantine imperial administration or distribute to the aristocratic elites and to regular soldiers as remuneration, while the bishopric seats – the ancient and the new ones – attached to the obedience of the Patriarchate of Constantinople and no more to Rome. A decision obviously opposed by the popes in the following centuries and that would be resolved only with the Norman conquest of Southern Italy [17–20]. It was the result of a re-Hellenization project, hosted by the emperor but also the indicator of profound cultural and social changes[1] (Figs. 2 and 3).

Speaking of socio-cultural aspects, religious one is certainly the most indicative to understand the choices made by Leo III. As regards the Greek monastic sphere proper, for example, a first migratory movement was hypothesized during the Sixth century that would have contributed to increasing the eastern monastic presence in Southern Italy [21–23]. Later, after a probable new insertion at the beginning of the Seventh century – coming from the Balkans and Peninsular Greece (Patras) due to the Avar-Slavic invasions – the subsequent fall of Jerusalem (637) and Alexandria (641) into Arab hands would have given force to new migratory movements of Eastern-rite Christians both towards

---

[1] What briefly explained in this section regarding the urban-architectural situation and changes of Byzantine Calabria (in an urban and rural environment), without claiming to be exhaustive, intends to identify some essential research points or *case studies* in a much broader subject that requires further investigations that will be conducted in the subsequent stages of the investigation. Attention that will also concern the historical-artistic data, through field surveys and the use of the most up-to-date bibliographic tools, e.g. Ricciardi L., Corpus della Pittura Monumentale Bizantina in Italia, II/Calabria. Rubbettino, Soveria Mannelli (2021).

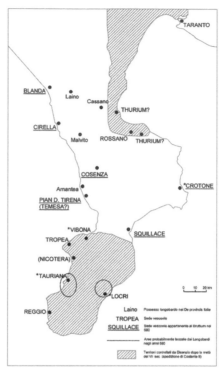

Fig. 2. Byzantine possessions in Calabria at the end of the 7th century. Map of the authors taken from: G. Noyè, L'economie de la Calabre de la fin du VI au VIII siècle. Cahiers de recherches médiévales et humanists, 2014 – 2, n° 28

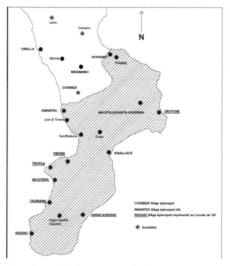

Fig. 3. Byzantine possessions in Calabria at the end of the 8th century. Map of the authors taken from: G. Noyè, L'economie de la Calabre de la fin du VI au VIII siècle. Cahiers de recherches médiévales et humanists, 2014 – 2, n° 28

the regions of Southern Anatolia that towards the West involving, obviously, also the regions of Southern Italy such as Sicily and Calabria which for centuries had hosted a conspicuous Greek presence. It is not easy to assess the actual numerical incidence of these migrations but it can be assumed that the social component of the exiles was at least varied, consisting of lay people from all walks of life and, obviously, the clergy: the traces of this important presence are however deduced from the local liturgical traditions [24, 25]. While, as regards Sicily and Calabria, it allowed the autochthonous Greek element to strengthen confirming a successful widespread re-Hellenization [26–28]. It can therefore be assumed that Italo-Greek monasticism – developed between the Seventh and the Eight centuries – shows a strong oriental influence and was able, above all thanks to the Greek language as a profoundly unifying linguistic element, to adapt to the reality of Southern Italy: in both hermitic and cenobitic forms. A main characteristic of Italo-Greek monasticism, in fact, was its adaptability, a consequence – we could say – of the geographical reality of places and political events. Alongside monastic settlements based on forms of life in common according to the classic formula of the *coenobium*, in the Italo-Greek area it is possible to observe the predilection for a life of solitude and prayer: the latter a preferred modality, which unites the local monasticism to the great Byzantine monastic traditions of Mount Athos or Mount Olympus of Bithynia [29–31].

About the ties between *Kalavrìa* and Mount Athos it seems appropriate to recall here the figure of a Calabrian monk named Nicephorus the Naked lived in the Tenth Century. The information about Nicephorus does not depend on a *bìos* expressly dedicated to him but from some evidence deducible from other hagiographical texts. His figure is important for two reasons: (1) it testifies the link between the South of Italy and the great Byzantine monastic centres; (2) attests to the growing fame of Mount Athos as a destination for Greek monks from the outskirts of the Empire. The information in our possession on Nicephorus comes from the two *bìoi* of Athanasius the Athonite (920 ca.-1000): the founder of the first monastery on Athos. The so-called *Vita A*, written by Athanasius of Panagiou around the first quarter of the Eleventh century and the *Vita B*, more recent but anonymous. Beside them there is a hypothetical original (lost) *bìos* that would have served as a model for the following ones [32, 33]. In any case, both surviving texts underline the presence of aspiring monks, coming from distant regions, attracted by the teachings of Athanasius the Athonite. Among them are men from Lombard Italy, Calabria and Amalfi. Faced with this generic information, both *Vita A* and *Vita B* present some passages about a famous disciple of Athanasius. This monk is Nicephorus the Naked, former companion of Saint Phantinos the Younger, who moved to Mount Athos. The *Vita B* describes him an ascetic, perhaps already a long-time resident on the Holy Mountain, going to Athanasius with some disciples asking him to guide them by becoming their abbot. But this is a paradoxical aspect. In fact, between the Ninth and Tenth centuries, Eastern Christianity had seen the growth of dialectical tensions between the proponents of hermitic life and supporters of the cenobitic option. Nicephorus, as an Italo-Greek monk, knew well and had practiced both forms of monastic life, it is no coincidence that his figure has a balancing role. Athanasius, in fact, respecting their eremitic impulse offers to Nicephorus and his disciples the possibility of continuing to practice it not in the wilderness but inside a monastic community [31].

Between the Ninth and the Tenth century, the *Kalavrìa* was an integral part of the Byzantine Empire and subject to direct control by Constantinople. The varied ethnic-cultural and linguistic *facies* which, as we have seen previously, were a characteristic of the region since the Late Antiquity now acquires – *mutatis mutandis* – different inserts, attesting the profound political, social and military changes that occurred between the Eighth and Ninth century. Alongside the Greek communities, which constitute one of the majority presences, there are in fact the Latin ones (linked mainly to the authority of the Church of Rome) together with the Jewish survivors and the Islamic presence. Minority groups, attested in the area as a consequence of the various military campaigns conducted by the rulers of the East and the West, are those composed of Franks, Armenians, Slavs and Bulgarians together with others ethnic minorities – coming from Central-Eastern Europe and the Mediterranean Basin – whose small number does not allow to trace a secure profile.

In any case, what can be understood in observing the general situation is the direct involvement of Calabria in the various expansionist projects of the three main political forces of the period ready to "tighten it" on several fronts: Byzantium in the East and in the territories of the North-East; the Frankish Empire in the North interested in expanding its political influence in Southern Italy and the Emirate of Sicily (together with its allies in North Africa) mainly along the coasts. The *Kalavrìa* was therefore a politically unstable land, a reality on which the *bìoi* of the main saints of the period – monks of Sicilian and Calabrian origin – offer an evident and useful cross-section for of historical reconstruction. Approaching the so-called Italo-Greek monastic hagiography means facing a vast and heterogeneous group of texts written in a period between the Ninth and Thirteenth centuries [34]. The protagonists of these reports are mainly monks, hermits, *higumenoi* founders of monasteries linked to the Church of Constantinople but not without ties to the Church of Rome and the papal throne. The exquisitely hagiographic nature of these texts, their narrative and rhetorical *topoi* and the encomiastic style must not, however, represent the prelude to a too hasty judgment on the quality of the information they pass on to us. By now abandoned the historiographic prejudice of the past for the modern historian the *bioi* of the Italo-Greek saints appear as historical sources capable of throwing an unprecedented light on some dark phases of political-religious history and on the daily life because, it should be remembered, Italo-Greek monasticism is an integral and indispensable phase in the history of Southern Italy.

With the beginning of the slow and inexorable Arab occupation in 827, Sicily began to acquire the features of a land hostile to the Christian monastic presence. If we turn our attention to the contemporary hagiographic sources, in fact, we find the peculiar signs of a migratory flow of the Greek island population towards centres and regions still firmly in the hands of the Empire or, in any case, far from the theatre of the Arab conquest. The destinations chosen by the Italo-Greek populations show a particular predilection for Greece and Constantinople, the latter not just a political centre but also a symbol of a religious and cultural identity to which these populations knew they were part. Observing the migratory phenomenon within the Byzantine Empire which originated in the Ninth century, it was not only Sicily that had to deal with a slow movement of population since, almost simultaneously and in order to face the Arab danger in the area as well, also the Greek inhabitants from the Aegean islands began to move towards peninsular Greece and, whenever possible, to the great cities of the Empire [35–37]. The political crisis that resulted from these events, however, caused equally the reaction of the Eastern Empire, both from a military and administrative point of view, which strengthened its presence in Southern Italy now considered a vital outpost for Byzantine political and military survival in the area. But Constantinople was not the only ideal destination for the Italo-Greek monks. With a process that will become more evident between the Tenth and the Eleventh centuries because, once again, to the ever-present Arab threat, even the city of Rome, if not a place of permanent residence, will become a place of pilgrimage being the closest apostolic sanctuary and, regardless of disputes with the Constantinopolitan patriarchate, the main religious seat in the Italian peninsula [38, 39] (Fig. 4).

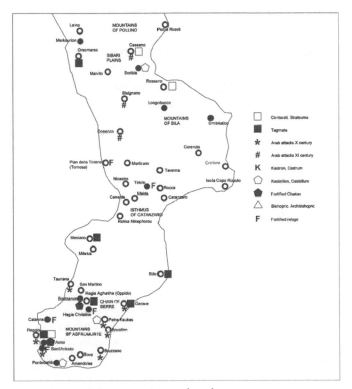

Fig. 4. Byzantine possessions in Calabria in the 9th-10th century. Map of the authors taken from: Noyè, G.: Byzantine Calabria. In: Cosentino, A.: A Companion to Byzantine Italy, Koninklijke Brill NV, 2021

The slow advance of the Islamic forces pushed the monastic population of Sicily towards the eastern Ionian coasts, finally moving to Calabrian lands beyond the Strait of Messina which has now become a final border to cross, searching for a lost material and spiritual security. The event becomes an important *topos* of the Italo-Greek hagiography as revealed in the *bìoi* of two monks between the Ninth and Tenth centuries:

– The *bìos* (BHG 850) of Elias the Younger (823–903).
– The *bìos* (BHG 1611) of Sabas the Younger († 990/995ca.).

The movement of monks, priests and laity unites the history of the two regions even more, and allows us to recognize the centrality of this area, in which the political clash between the great empires of the Mediterranean becomes palpable and evident but becoming also a clash between religious beliefs and cultural identities. To the waning of at least part of the Sicilian monastic presence corresponded the strengthening of the Greek element in Calabria defining the Byzantine religious component and contributing into making the region a chosen land for Greek monasticism - almost a new Thebaid [40, 41] - where these seeds found fertile ground because of a centuries-old local Christian traditions but also acquired new specific and Calabrian features.

3 The Tools for the Enhancement of Cultural Heritage

The enhancement of cultural heritage can be undertaken by subjects of a very different nature.

The Council of Europe Framework Convention on the value of cultural heritage for society, better known as the "Faro Convention", first of all identifies citizens and local communities as the primary subjects responsible for the conservation and enhancement of heritage [7].

The Italian legal system in turn identifies a plurality of subjects who, in different ways, can participate in this safeguard action, from the Municipalities to the Superintendencies, passing through the Regional Secretariats.

At an even higher level there are the initiatives and programs promoted by supranational organizations, such as:

– the European Union, which within its founding Treaty provides for actions to safeguard and develop the European cultural heritage (art. 3 TEU) [42];
– the Council of Europe, which has undertaken numerous initiatives in favor of cultural heritage in its various meanings, among which, in addition to the aforementioned Faro Convention, it is worth mentioning the European Landscape Convention and the creation of the European Institute of Cultural Routes (IEIC) [43, 44].

The first article of the European Landscape Convention gives this definition of "Landscape": "*a certain part of the territory, as perceived by the populations, whose character derives from the action of natural and/or human factors and their interrelations*", while article 11 establishes the Landscape Award of the Council of Europe.

Finally, but certainly not least, UNESCO, which through various Conventions provides for forms of safeguarding and enhancement of the cultural and natural heritage of world interest, including the 1972 Convention concerning the protection of the world's cultural and natural heritage, which this year celebrates its fiftieth anniversary [45].

The Convention, in particular, provides for 6 types of goods susceptible to protection, 3 of a cultural nature and 3 of a natural nature. In order for the World Heritage Committee, the deliberative body on the matter, to insert an asset in the World Heritage List, it is necessary that its exceptional and universal value be recognized, evaluated according to ten possible criteria, 6 of a cultural nature and 4 nature, reported in the Guidelines for the application of the World Heritage Convention [46]:

Cultural criteria:

i.   to represent a masterpiece of relative human genius;
ii.   to show an important interchange of human values over a period of time or in a cultural area of the world, in relation to developments in architecture or technology, monumental arts, urban planning or landscape design;
iii. to represent a unique or exceptional testimony of a cultural tradition or a living or vanishing civilization, or
iv.   to be an outstanding example of an architectural or technological or landscaping building or ensemble that illustrates a significant stage or stages in human history, or
v.   to represent an exceptional example of a traditional human settlement or land use that is representative of one or more cultures, especially if it has become vulnerable to the impact of irreversible change; or
vi.   to be directly or tangibly associated with living events or traditions, with ideas and beliefs, with artistic or literary works of universal value (the committee considers this criterion should justify inclusion in the list only in exceptional cases and together with other cultural criteria or natural);

Natural criteria:

vii. to represent outstanding examples of major stages in the history of the earth, including the presence of life, significant geological processes taking place for the development of the land shape or significant geomorphic or physiographic features, or
viii. to be an outstanding example of ecological and biological processes occurring in the development and evolution of terrestrial ecosystems, fresh, coastal and marine ecosystems, and plant and animal communities, or
ix.   to contain superlative natural phenomena or areas of exceptional natural beauty and aesthetic importance, or
x.   to contain the most important and significant habitats for the on-site conservation of biological diversity, including those containing threatened species of exceptional universal value from a scientific or conservation point of view.

4   The Enhancement of the Cultural Landscape of Byzantine Calabria

According to UNESCO [47, 48]: *"The term "cultural landscape" embraces a diversity of manifestations of the interaction between humankind and its natural environment. Cultural landscapes often reflect specific techniques of sustainable land-use, considering the characteristics and limits of the natural environment they are established in, and a specific spiritual relation to nature. Protection of cultural landscapes can contribute to modern techniques of sustainable land-use and can maintain or enhance natural values in the landscape. The continued existence of traditional forms of land-use supports biological diversity in many regions of the world. The protection of traditional cultural landscapes is therefore helpful in maintaining biological diversity.*

In 1992 the World Heritage Convention became the first international legal instrument to recognise and protect cultural landscapes. The Committee at its 16th session adopted guidelines concerning their inclusion in the World Heritage List.

The Committee acknowledged that cultural landscapes represent the "combined works of nature and of man" designated in Article 1 of the Convention. They are illustrative of the evolution of human society and settlement over time, under the influence of the physical constraints and/or opportunities presented by their natural environment and of successive social, economic and cultural forces, both external and internal."

From the foregoing paragraphs it is evident that the Calabrian settlement system born in the Byzantine age was born from the thought and action of man, but is closely connected with the natural characteristics of the territory.

It is the orographic conformation of Calabria, its position in the center of the Mediterraneo, the enormous development of its coasts, which inspire and allow the birth of that peculiar settlement system: it is, therefore, the result of the combined action of man and nature, as foreseen by the definition of Cultural Landscape given by the World Heritage Committee in its 16th assembly [49].

This settlement system, together with the conformation of the territory, in turn, determines the network of infrastructural connections, the methods of exploitation and cultivation of the land, the system of regimentation and use of water, in a very delicate balance handed down to us by past generations and today in serious danger [50].

5   Conclusions

This article essentially represents the starting point of a research that rests on the enormous amount of studies already conducted up to now on the cultural heritage of Byzantine Calabria, with the aim of identifying the most congruent forms of enhancement.

The continuation of the research will essentially move in two directions:

- the systematization of the information already available on the heritage in question in order to make the organic nature of the original design that gave rise to the specific Cultural Landscape of Byzantine Calabria more and more evident;
- the identification of the most appropriate forms of valorisation of this heritage, starting from the involvement of the communities, in the spirit of the Faro Convention, up to the verification of the feasibility of candidacies that recognize the exceptional and universal value of this heritage [51, 52].

As regards the first line of research, the activities will have an international horizon, also concerning territories other than Calabria, such as: Turkey, Greece, Cyprus, Albania and Macedonia, in order to verify the existence of similar recurring characters to the Calabrian ones, which can be united by common enhancement tools [53].

As regards the second line of research, however, the focus will be mainly on identifying forms of enhancement by different subjects, whose feasibility, sustainability and effectiveness will be carefully evaluated.

References

1. Calabrò, F., Cassalia, G.: Territorial cohesion: evaluating the urban-rural linkage through the lens of public investments. In: Bisello, A., Vettorato, D., Laconte, P., Costa, S. (eds.) SSPCR 2017. GET, pp. 573–587. Springer, Cham (2018). https://doi.org/10.1007/978-3-319-75774-2_39
2. ICOMOS: The Paris Declaration on heritage as a driver of development. In: 17th General Assembly ICOMOS (2011)
3. Campolo, D., Schiariti, C., Tramontana, C.: Alle origini dell'Umanesimo: un itinerario culturale per una rete competitive delle città del Mediterraneo. LaborEst 9(2014), 9–13 (2014)
4. Spampinato, G., Malerba, A., Calabrò, F., Bernardo, C., Musarella, C.: Cork Oak Forest spatial valuation toward post carbon city by CO_2 sequestration. In: Bevilacqua, C., Calabrò, F., Della Spina, L. (eds.) NMP 2020. SIST, vol. 178, pp. 1321–1331. Springer, Cham (2021). https://doi.org/10.1007/978-3-030-48279-4_123
5. Barrile, V., Malerba, A., Fotia, A., Calabrò, F., Bernardo, C., Musarella, C.: Quarries renaturation by planting cork oaks and survey with UAV. In: Bevilacqua, C., Calabrò, F., Della Spina, L. (eds.) NMP 2020. SIST, vol. 178, pp. 1310–1320. Springer, Cham (2021). https://doi.org/10.1007/978-3-030-48279-4_122
6. Valtieri, S.: Scambi culturali alle radici dell'Umanesimo. LaborEst 9(2014), 5–8 (2014)
7. UNESCO: Convention for the Safeguarding of the Intangible Cultural Heritage. In: General Conference at its 32nd session, Paris (2003)
8. Council of Europe: European Cultural Convention, European Treaty Series - No. 18, Paris, France (1954)
9. Council of Europe: Framework Convention on the Value of Cultural Heritage for Society, Council of Europe Treaty Series - No. 199, Faro, Portugal (2003)
10. Otranto, G.: La cristianizzazione della Calabria e la formazione delle diocesi. In: Leanza S. (ed.): Calabria cristiana, società, cultura nel territorio della Diocesi di Oppido Mamertina-Palmi. Atti del Convegno di Studi, Palmi-Cittanova 21–25 novembre 1994, pp. 47–48. Soveria Mannelli, Rubettino (1999)
11. Noyé, G.: Les villes des provinces d'Apulie-Calabre et de Bruttium-Lucaine du IVe au VIe siècle. In: Brogiolo G.P. (ed.): Early Medieval Towns in the Western Mediterranean, pp. 97–120. Mantua, Editrice S.A.P. (1996)
12. Noyé, G.: La città calabrese dal IV al VII secolo. In: Augenti A. (ed.), Le città italiane tra la tarda Antichità e l'alto Medioevo, Atti del Convegno – Ravenna 26–28 febbraio 2004, pp. 510–515. Firenze, All'Insegna del Giglio (2006)
13. Noyé, G.: L'espressione architettonica del potere: praetoria bizantini e palatia longobardi nell'Italia Meridionale. In: Martin J.M., Peters-Custot A,, Prigent A.(eds.): L'héritage byzantine en Italie (VIIIe-XIIe siècle), II, pp. 415–420. Rome, Publications de l'École française de Rome (2012)
14. Noyé, G.: The Still Byzantine Calabria, a case study. In: Gelichi S., Hodges (eds.): New Directions in Early Medieval European Archaeology. Spain and Italy Compared. Essays for Riccardo Francovich, pp. 221–266. Turnhout, Brepols (2015)
15. Noyé G.: Byzantine Calabria. In: Cosentino S. (ed.), A Companion to Byzantine Italy, pp. 436–444. Leiden-Boston, Brill (2021)

16. Minasi, G.: Le Chiese di Calabria dal quinto al duodecimo secolo: cenni storici, p. 111. Lanciano e Pinto, Napoli (1896)
17. Grumenl, V.: L'annexion de l'Illyricum oriental, de la Sicile et de la Calabre au Patriarchat de Constantinople. Recherches de science religieuse. **40**, 191–200 (1951–1952)
18. Anastos, M.V.: The transfer of Illyricum, Calabria and Sicily to the jurisdiction of the Patriarchate of Constantinople in 732–733. Studi bizantini e neoellenici **9**, 14–31 (1957)
19. Kreutz, B.M.: Before the Normans. Southern Italy in the Ninth and Tenth Centuries, pp. 116–136. University of Pennsylvania Press, Philadelphia (1991)
20. Brandes, W.: Finanzverwaltung in Krisenzeiten. Untersuchungen zur byzantinischen Administration im. 6–9. Jahrhundert, pp. 368–370. Löwenklau, Frankfurt am Main (2002)
21. Cappelli, B.: Il Monachesimo Basiliano ai confini Calabro-Lucani, Napoli 1963, pp. 15–16 (1963)
22. Ekonomou, A.J.: Byzantine Rome and the Greek Popes. In: Eastern Influences on Rome and the Papacy from Gregory the Great to Zacharias, A.D. 590–752, pp. 43–60. Lexington Books, Lanham (2007)
23. Herrin, J.: Margins and Metropolis. Authority across the Byzantine Empire, pp. 37–39. Princeton University Press, Princeton (2013)
24. Jacob, A.: Deux formules d'immixtion syro-palestiniennes et leur utilisation dans le rite byzan-tine d'Italie méridionale. Vetera Christianorum **13**, 29–64 (1976)
25. Parenti S.: A Oriente e Occidente di Costantinopoli. Temi e problemi liturgici di ieri e di oggi, pp. 149–215. Libreria Editrice Vaticana, Città del Vaticano (2010)
26. Guillou, A.: Grecs d'Italie du sud et de Sicile au moyen age: les moines. Mélange de l'École Française de Rome **75**(1), 70–110 (1963)
27. Borsari, S.: Il Monachesimo Bizantino nella Sicilia e nell'Italia Meridionale pre-normanna, pp. 32–46. Istituto Italiano per gli Studi Storici, Napoli (1963)
28. Mccormick, M.: The Imperial Edge: Italo-Byzantine Identity, Movement and Integration, A.D. 650–950. In: Ahrweiler H., Laiou A.E (eds.), Studies on the Internal Diaspora of the Byzantine Empire, pp. 35–45. Dumbarton Oaks Library and Collection, Washington D.C. (1998)
29. Flusin, B.: L'hagiographie monastique à Byzance au IXe et au Xe siècle. Modèles anciens et tendances contemporaines. Revue Bénédictine **103**, 31–50 (1993)
30. Morini, E.: L'eredità ascetica del monachesimo calabro-greco. Hellenika Menymata **6**, 31–47 (2002)
31. Morini, E.: Monastic interactions between Calabria and Mount Athos in the Middle Ages. In: Crostini B., Murzaku I.A. (eds.), Greek Monasticism in Southern Italy. The Life of Neilos in Context, pp. 85–88. Routledge, London (2017)
32. D'Ayala, V.L.: La Vita di Atanasio l'Athonita di Atanasio di Panaghiou, pp. 271–273. Città Nuova, Roma (2017)
33. Talbot, A.-M.: Life of Athanasios of Athos, version B. In: Greenfield, R.P.H., Talbot, A.M. (eds.) Holy Men of Mount Athos, pp. 264–267. Dumbarton Oaks Medieval Library and Collection, London (2016)
34. Re, M.: Italo-Greek Hagiography. In: Efthymiadis, S. (ed.) The Ashgate Research Companion to the Byzantine Hagiography, I, pp. 227–258. Routledge, Ashgate (2011)
35. Gabrielli, F.: Greeks and Arabs in the centra Mediterranean Area. Dumbarton Oaks Papers **18**, 59–65 (1964)
36. Christides, C.: The Conquest of Crete by the Arabs (ca. 824.). A Turning Point in the Struggle Between Byzantium and Islam, pp. 158–163. Akadēmia Athēnōn, Athens (1984)
37. Signes-Codoñer, J.: The Emperor Theophilos and the East. Court and Frontier in Byzantium During the Last Phase of Iconoclasm, pp. 206–2190. Routledge, London-New York (2014)
38. Russo, F.: La 'peregrinatio' dei santi italo-greci nelle tombe degli Apostoli Pietro e Paolo a Roma. Bollettino della Badia Greca di Grottaferrata **22**, 89–99 (1968)

39. Caruso, S.: Sulla cronologia del dies natalis di s. Vitale da Castronovo di Sicilia. In: Lucà, S., Perria, L. (eds.), ΟΠΩΡΑ. Studi in onore di mgr Paul Canart per il suo LXX compleanno, II, pp. 131–312, nt. 36. Bollettino della Badia Greca di Grottaferrata, Grottaferrata (1998)
40. Russo, F.: I monasteri greci della Calabria nel secolo XV. Supplemento al 'Liber Visitationis' di Atanasio Calceopulo del 1457-1458. Bollettino della Badia Greca di Grottaferrata. **16**, 117–134 (1962)
41. Giannini, P.: Il monachesimo basiliano in Italia. Bollettino della Badia Greca di Grottaferrata **41**, 5–18 (1987)
42. European Union: Consolidated version of the Treaty on European Union, Official Journal of the European Union C 326/13, Maastricht, Netherlands (1992)
43. Council of Europe: Landscape Convention (as amended by the 2016 Protocol), European Treaty Series - No. 176, Florence, Italy (2000)
44. Council of Europe: Resolution CM/Res(2013)66, confirming the establishment of the Enlarged Partial Agreement on Cultural Routes (EPA). In: Committee of Ministers on 18 December 2013, 1187bis meeting of the Ministers' Deputies (2013)
45. UNESCO: Convention concerning the protection of the World Cultural and Natural Heritage. In: General Conference at its Seventeenth Session, Paris (1972)
46. UNESCO World Heritage Centre: Intergovernmental Committee for the Protection of the World Cultural and Natural Heritage, Operational Guidelines for the Implementation of the World Heritage Convention, WHC.19/01 (2019)
47. Luengo, A., Rössler, M. (eds.): World Heritage Cultural Landscapes. Elche (2012)
48. Von Droste, B., Plachter, H., Rössler, M. (eds.): Cultural Landscapes of Universal Value. Components of a Global Strategy. Fischer Verlag, Jena (1995)
49. Campolo, D., Calabrò, F., Cassalia, G.: a cultural route on the trail of Greek monasticism in Calabria. In: Calabrò, F., Della Spina, L., Bevilacqua, C. (eds.) ISHT 2018. SIST, vol. 101, pp. 475–483. Springer, Cham (2019). https://doi.org/10.1007/978-3-319-92102-0_50
50. Calabrò, F.: Promoting peace through identity. evaluation and participation in an enhancement experience of calabria's endogenous resources I Promuovere la pace attraverso le identità. Valutazione e partecipazione in un'esperienza di valorizzazione delle risorse endogene della Calabria. ArcHistoR., **12**(6), 84–93 (2019). https://doi.org/10.14633/AHR146
51. Calabrò, F.: Integrated programming for the enhancement of minor historical centres. The SOSTEC model for the verification of the economic feasibility for the enhancement of unused public buildings I La programmazione integrata per la valorizzazione dei centri storici minori. Il Modello SOSTEC per la verifica della fattibilità economica per la valorizzazione degli immobili pubblici inutilizzati. ArcHistoR. **13**(7), 1509–1523 (2020). https://doi.org/10.14633/AHR280
52. Lorè, I., Meduri, T., Pellicanò, R.: Ipotesi di valorizzazione delle risorse identitarie circostanti il castello San Niceto nella provincia di Reggio Calabria. LaborEst **15**(2017), 13–20 (2017). https://doi.org/10.19254/LaborEst.15.02
53. Calabrò, F., Mafrici, F., Meduri, T.: The valuation of unused public buildings in support of policies for the inner areas. the application of SostEc model in a case study in Condofuri (Reggio Calabria, Italy). In: Bevilacqua, C., Calabrò, F., Della Spina, L. (eds.) NMP 2020. SIST, vol. 178, pp. 566–579. Springer, Cham (2021). https://doi.org/10.1007/978-3-030-48279-4_54

Sustainable Tourism and Its Role in Preserving Archaeological Sites

Dimah Wahhab Ajeena[✉]

University of Technology, Baghdad 10076, Iraq
ae.21.10@grad.uotechnology.edu.iq

Abstract. Heritage and archaeological sites face many challenges that lead to the deterioration of their physical condition, thus declines its historical value. These challenges may be due to environmental conditions, inhabitants themselves, the authorities' neglect of these sites' maintenance and/or the increase of tourism. The concept of sustainable tourism has emerged as one of the foremost global tourism patterns at the present time for its clean method in preserving nature sites and balancing urban heritage by preserving its identity and landmarks and marketing it in a sustainable manner for tourism which benefits all parties involved in the tourism process.

Being rich in many factors, Babylon Governorate is a great tourist attraction due to prior civilizations on its land and the Ancient City of Babylon, the many religious holy shrines, the abundance of its agricultural lands and the presence of the Euphrates River makes it eligible for all kinds of tourism especially archaeological tourism, if it is exploited in a sustainable way, it may be one of the most prominent destinations of local tourism.

This research aims to determine the extent of the ability of sustainable tourism to revive ancient city of Babylon for tourism, by building a theoretical framework to describe the relationship between sustainable tourism and archaeological sites to create a set of strategies and apply them to the current reality of the archaeological city using description and analysis to measure the extent of its efficiency then end the research with concluding that the small percentage of adopting sustainable tourism mechanisms due to a number of administrative, social and technical obstacles, prevents the tourism industry from flourishing and benefiting from this vital resource.

Keywords: Heritage · Archaeology · Sustainable tourism · Historic sites · Babylon

1 Introduction

The historical and archaeological sites represent the physical part of a nation's civilization and mission and give cities their distinctive identity, which makes it an important tourism destination. Therefore, interest in archaeological and historical sites has increased in an era when the tourism industry is considered one of the most important global industries, but it is endangered by some negative factors that affect these areas and lead to their

© The Author(s), under exclusive license to Springer Nature Switzerland AG 2022
F. Calabrò et al. (Eds.): NMP 2022, LNNS 482, pp. 2485–2495, 2022.
https://doi.org/10.1007/978-3-031-06825-6_237

deterioration. There is a need to balance the protection of historical areas and the development of tourism due to the close relationship between them, and from here emerged a global trend aimed at exploiting archaeological and historical sites for tourism while ensuring their sustainability within sustainable tourism trends that seek towards optimal investment of natural and urban resources on the one hand while preserving the historical environment of the tourist interface on the other hand, so this research reviews how sustainable tourism is linked to archeological sites and its ability to promote those sites that suffer from negative effects of traditional tourism by analyzing the case of the ancient city of Babylon as a case study.

2   Historic Buildings Definition

There are several definitions of historic buildings, most notably:

2.1   The National Trust for Historic Preservation [1] Definition of a Historic Site

It is a building that has been constructed for 50 years or more, or a building that was legally classified by government agencies or local as historical because it has a special value, as it expresses a particular architectural style or a site of important historic value.

2.2   The Wisconsin Historical Society Definition of a Historic Site

It is any building or structure registered in the National Register of Historic Sites and designated as a historic site or property according to a specific international or local law.
  And as for Arabic:

2.3   Charter of Preservation and Development of Urban Heritage in Arab Countries 2003 [6]

It is all that is built by man i.e., cities, villages, neighborhoods, buildings and gardens of archaeological, architectural, urban, economic, historical, scientific, cultural or functional value, and is classified into three types:

Heritage buildings. They are buildings of historical, archaeological, and artistic value, including decorations, fixed furniture and the environment associated with it.

Urban heritage areas. These include cities, villages and neighborhoods of historical value including all their components like urban fabric, public squares, roads, allies, infrastructure services, etc.

Urban heritage sites. These include buildings associated with a natural environment, either natural or man-made.
  This research targets urban heritage areas represented in the ancient city of Babylon (see Fig. 1).

Fig. 1. The ancient city of Babylon

3   Challenges Facing Historic Areas

Urban heritage faces many challenges resulting from these factors:

- Natural disasters [5]: such as earthquakes, floods, rising groundwater levels, etc.
- Human behavior such as the absence of cultural awareness of people, the lack of awareness of historical values of those areas, corrections and adjustments made at the expense of the nature of these areas out of ignorance of their value.
- Neglect of authorities such as the Ministry of Tourism and the lack of financial allocations to maintain these areas with a shortage of competent technicians to carry out conservation operations in these areas [12].
- Armed conflicts that damage buildings, such as wars, political disturbances, and successive violent waves [11].

4   Tourism; Definition and Types

4.1   There Are Many Definitions that Dealt with the Subject of Tourism as an Activity, as Follows

- The Swiss researcher Marcel Hunziker [14], defines tourism as the relations and phenomena that result from travel and the temporary residence of a person outside his usual place of residence, as long as this temporary residence will not turn into a permanent residence and is not linked to a profitable activity for this foreigner.

- Cambridge Dictionary, Retrieved
  It is the group of activities that serve tourists and contributes to providing accommodation.
- World Tourism Organization definition [9].
  It is a humane and social phenomenon based on the movement of individuals from their permanent places of residence to areas outside their community for a temporary period of not less than 24 h and not more than a whole year for any of the known purposes of tourism, except for study and work.

4.2 Tourism Has Many Types that Are Classified According to the Purpose of Tourism, This Research Will Only Mention the Following Related Type

Heritage tourism: It has several definitions, most notably:

The concept of heritage tourism, according to the National Fund of Preservation of Historic Heritage, refers to it as follows [15]:

- It is a travel that aims to experience activities and places and learn about artifacts and historical people's stories, their cultures, natural resources, etc.
- It is where tourists aspire to understand the civilizations that flourished and collapsed. This tourism is based on the visit of museums, statues, historical sites, buildings and archaeological remains, which represent visible historical memories of antiquity [3].

4.3 Definition of Sustainable Tourism

- As defined by the World Tourism Organization
  "It is the kind of tourism that meets the needs of tourists and host sites, protects and provides opportunities for the future. It represents the guiding rules in the field of resource management in a way that fulfills the requirements of economic, social and cultural issues, and achieves cultural integration, environmental factors, biodiversity and support of life systems".
- and in another definition
  It is the optimal use of tourist sites without compromising them, i.e., the entry of tourists in balanced numbers, their awareness and knowledge beforehand in the importance of the tourist areas and dealing with them in a friendly manner to prevent damage to both parties [13].

Tourism is a means of exchanging cultures between people, so it cannot in any way be separated from history, as the great demand for historical sites made this type of tourism constitute 10% of the global tourism movement [7], 8 so tourism and antiquities are closely related and cannot be ignored if we want to achieve sustainable development for both parties. While tourism may represent a threat to historical sites, at the same time, it is the main fuel for reviving, marketing and preserving them, so there is no contradiction between heritage tourism and antiquities if they are exploited in a sustainable manner. Sustainable tourism seeks to achieve sustainability according to three overlapping axes [10]:

Economic Sustainability. Creating new opportunities for investment and creating a diversity in the economic return and sources of income.

Environmental Sustainability. Preserving the ecological balance and historical areas in resource consumption.

Social Sustainability. Creating job opportunities, strengthening people's identity and openness to the world.

4.4 Importance of Sustainable Tourism

Sustainable tourism is clean tourism that supports the environment and natural resources with the least possible negative effects in the environment and the highest positive effects on the tourist area environmentally, socially, and financially, and it is a good place for spreading awareness in the area among tourists and residents equally [8].

From the foregoing we find that sustainable tourism is an effective tool in maintaining a balance between pleasing the tourists, obtaining their satisfaction, the preservation of natural, cultural, and environmental resources, the preservation of ecological balance and biodiversity, good management of resources, enrichment of historical areas, strengthening social aspects, creating job opportunities and diversification of income sources.

The question may arise on how can archaeological areas be revived by activating sustainable tourism methods?

Escalating from the previous presentation, a set of methods or strategies were deduced to improve the tourism of archaeological areas within clean sustainable tourism mechanisms to ensure the preservation and exploitation of these areas.

Optimally at the same time, as follows:

- Facilitating access to archaeological sites by creating an efficient system of public transportation using buses and public vehicles and linking them to the main roads of the city to facilitate transporting tourists and visitors to and from archaeological sites and reduce the number of personal vehicles.
- Giving more attention to these areas by the authorities, represented by the Ministry of Tourism and Antiquities and other local organizations, with the need to cooperate with the private sector to ensure sustainable tourism services in the long term represented in attracting investors and implementing advanced tourism programs that keep pace with international tourism which requires great economic, technical and artistic capabilities that can only be acquired with the cooperation of all parties.
- Enriching these areas with elements of attraction, both at the cultural level, by holding festivals, events, and art exhibitions, or by holding recreational activities or folklore celebrations that would encourage the visitor to extend their stay in area.
- Building hotels, tourist villages and housing units in shapes and designs that mimic the structural patterns of these cities, using similar materials, colors and patterns in harmony with the place, to simulates that era for the tourist, taking into account the use of environmentally friendly materials and the provision of thermal and environmental comfort and the use of sustainable treatments to provide electricity, water systems, wind energy and recycling as much as possible.

- Increasing outdoor spaces around archaeological areas and planting them with different types of trees, establishing reserves, zoos, and water bodies, if possible, to support biodiversity.
- Using efficient waste treatment systems such as recycling and other methods, by using recyclable bags and packages such as paper bags, and the prohibition of using plastic or toxic materials as required reconsidering the current laws and enacting new laws that pertain - at least - to these areas.
- The need for professional media to market monuments globally. Every successful business needs a huge media system which we lack in Iraq to promote an idea and convince others to try it. It can attract the Western world from the far reaches to the first civilization of mankind.
- Avoiding traditional tourism methods, bringing updates to tourism activities, and designing innovative programs that include commiserating with cultural and religious backgrounds, age and gender in order to provide the best services and programs to attract different levels of visitors and satisfy their needs and please different manners.
- The necessity of enhancing social awareness of the population about the importance of these areas, how to preserve and not sabotage them, by setting up awareness-raising campaigns or seminars, making them involved in the decision-making process, creating job opportunities such as traditional technical positions or tourism activities such as tourist guides, etc. Thus enhance their sense of belonging to the area and to their history and civilization. Citizens are the main circle in the tourism sustainability process, so no development is achieved separately from the population.
- Conducting comprehensive studies on the archaeological area and the surrounding region, such as site studies, soil investigations, survey works, etc., with the aim of revitalizing the building functionally, whether with the same previous function or any other, compatible with its original function, by making necessary modifications required to adopt the proposed function, within the so-called adaptive reuse, which aims to adopt a new appropriate function to a site or historical facility that enhances its cultural and economic values (Fig. 2).

Fig. 2. Wall relief from ancient Mesopotamia

5   The Ancient City of Babylon as a Case Study

The Gate of God, the Hanging Gardens City, one of the Seven Wonders of the World, the ancient city of Babylon, which goes back in history to more than four thousand years. It was the cradle of a civilization that affected the world till this day with its architecture and laws. The ancient city of Babylon is located on the Euphrates River, 85 km south of the capital Baghdad. The city of Hilla is an extension of the historical city of Babylon, whose ruins are located 8 km north of the city center of Hilla. Kish and Borsippa and other ruins of the ancient city are located within the region of the city of Hilla. Its most prominent features are: The Lion of Babylon, Hanging Gardens, Babylon's Walls, Ishtar Gate and many more (Fig. 3).

Fig. 3. The gate of the ancient city of Babylon (Ishtar Gate)

6   Babylon Governorate; Tourism Qualifications and Obstacles

6.1   Historical Qualifications [2]

The governorate of Babylon is characterized by the presence of many archaeological, religious and tourism sites, the most important of which are the ancient city of Babylon, Kish and Borsippa, and religious sites such as the holy shrine of Job; the Prophet of God and Imran (peace be upon him). Also, there are several tourism areas around the city, including Babylon Tourist Resort, Al-Mahanawiya Island, which is estimated at 800 acres, and is among investment projects and the dam of Al-Hindiya (Fig. 4).

Fig. 4. The maze of the city of Babylon

6.2   Geographical Qualifications

The city of Hilla is the center of the Babylon Governorate, and it is the main city of the middle Euphrates region, which include the governorates of Babylon, Qadisiyah, Karbala and Najaf. The city of Hilla is located about 100 km south of Baghdad, 41 km south-west of the religiously important city of Karbala and 8 km south-east of the historically important city of Babylon. In addition to the city being on a transit road to Najaf and Karbala, there are many religious sites and dozens of holy shrines in the city itself.

6.3   Economic Tourism Market Qualifications [4]

The interest in improved facilities, ease of access to natural attractions and the rise in global interest in the environment have given the tourism market an attractive global character. It is now one of the widest and fastest growing markets in different parts of the world with vast wild and rural areas where animal colonies can be seen in their natural habitats. From this perspective, the city of Hilla appears as the clear access point to the natural spaces of the Babylon Governorate. Small folk hotels can be immediately improved to accommodate those visitors, especially since the city has an abundance of agricultural lands, fields and orchards.

7 Obstacles

The governorate suffers from many obstacles, the most important of which is the neglect and marginalization of the tourism sector as a productive sector, the weak role of the private sector in the development process of tourism, the lack of professional confidence due to fluctuation of laws related to this sector, the weak role of supporting and complementing sectors of the tourism process, such as the industrial and agricultural sectors, the lack of expert and efficient staff in the field of tourism and hotels, the deterioration of infrastructure services, the adverse air and land transportation, the absence of professional media that draws attention to monuments, the political turmoil and successive wars had the greatest impact on the deterioration of the tourism-making process in Iraq in general.

Analyzing the ancient city of Babylon according to the above-mentioned criteria to measure the extent of its achievement of sustainable tourism factors, then adopting the measurement in Table 1 as an approximate relative measurement of the study case.

Table 1. Indicators of achieving sustainable tourism in the ancient city of Babylon

Achievement		Mechanism	Row
0%		Efficient Transmission System	1
5%	▶	Tourism Investment	2
10%	●	Festivals and Events	3
5%	▶	Building Sustainable Hotels and Housing Units	4
0%		Afforestation of Open Spaces	5
5%	▶	Update Programs and Events	6
5%	▶	Social Awareness	7
0%		Adaptive Use	8
0%		Waste Recycling Systems	9
0%		Professional Media	10

● Achieved, ◖ Averagely achieved, ○ Not achieved

According to the percentage (approximate ratios) that was adopted to sort the measurement table, it was found that the percentage of achieving sustainable tourism mechanisms in the ancient city of Babylon is approximately 30% as a result of the absence and lack of standards or basic methods in upgrading tourism in the governorate in general and in the monuments of Babylon in particular. Following are the conclusions by this research.

8 Conclusion

The archaeological and historical areas are the core of people's experiences and their distinctive feature, which requires an increase in social awareness in the value of these areas and their great economic, social and environmental benefits. Hence, sustainable tourism aims at developing these areas economically and culturally in a sustainable manner that preserves the continuity of its identity. By analyzing the status of the ancient city of Babylon as a case study, and despite its inclusion on the UNESCO World Heritage List in 2019 and its enormous potential for practicing various types of tourism as one of the oldest civilizations known to man in the world, however, it is still subjected to serious deterioration that may lead to the obliteration of its features and identity. The research has found that the percentage of adopting sustainable tourism mechanisms is small, no more than 30%, due to a number of administrative, social and technical obstacles, which prevents the tourism industry from flourishing and benefiting from this vital resource. Therefore, a set of measures was attained that would improve governorate's tourism status through:

- Enhancing the population's social awareness of the importance of their history through media, seminars, etc.,
- Establishing a system of efficient public transportation methods to facilitate access to archaeological sites.
- Activating the role of government and other authorities in different sectors, whether public, private, or mixed.
- Tourism is a huge industry that needs to combine all efforts and authorities, as it needs professional staff, advanced technologies, and massive capabilities.
- Marketing it culturally in media by holding parties and festivals and through social media. Building attractive hotels and housing units that mimic the ancient Babylonian era to captivate tourists (Fig. 5).

Fig. 5. A reconstructed gate of the ancient city of Babylon

- Using sustainable technology in construction or water and wind systems and recycling.
- Updating tourism programs and developing innovative activities to attract visitors.
- Taking care of outdoor spaces, afforestation, establishing reserves, etc.

- Preparing comprehensive and extensive studies on archaeological areas and the surrounding environment with the aim of reviving the worn out and damaged ones, perhaps by reusing them in a manner consistent with their surroundings and their original function to ensure their sustainability and continuity.

References

1. A privately funded organization that seeks to preserve America's historic regions. http://www.perservationnation.org
2. Al-Janabi, A., Majid, A.: Characteristics of the tourism movement in al-Hashemia district, University of Babylon/College of Education for Humanities- Babylon University. J. Human. **27**(2) (2019)
3. Al-Lahham, N.: Tourist Planning for Heritage Areas Using the Environmental Effects Assessment Technique
4. Al-Maamouri, H.: Tourism Development of The City of Hilla According to The Development of The Most Important Centers (Parallel Development). World Trade Organization (WTO)
5. Al-Mahari, S.: Preserving Historic Buildings, Buildings from the City of Muharraq
6. Charter of Preservation and Development of Urban Heritage in Arab Countries, the League of Arab States (2003)
7. Dodds, R.: Sustaining tourism. Sustainable tourism expert and professor at Ryerson University, Canada
8. Gharaibeh, K.: Professor in the Department of Basic Sciences at Al-Balqa University/Jordan, Ph.D. in Cities and Population Development from University of Baghdad in 1995, researcher in the affairs of ecotourism, sustainable development, historical geography, urban and environmental planning (1995)
9. The World Tourism Organization: It is the United Nations agency responsible for promoting responsible, sustainable, and accessible tourism for all (1975)
10. Mediterranean Stories Program, Peoples Collaboration across Borders, Cultural Heritage and Sustainable Tourism
11. Stovel, H.: Risk Preparedness: A Management Manual for World Cultural Heritage (1998)
12. Arab Republic of Egypt, Institute of National Planning, n.d. sustainable cultural heritage tourism applied to Historical Cairo
13. The Ninth Extraordinary Session of the Governing Council of the United Nations Environment Programme. Dubai (2005)
14. Walter, H.: The Department of Tourism, and Institute at the University of St. Gallen
15. What is Heritage Tourism. http://www.culturalheritagetourism.org

Sustainable Transportation for Healthy Tourist Environment: Erbil City-Iraq a Case Study

Shahad Ali Dawood$^{(\boxtimes)}$ and Wahda Sh. Al-Hinkawi

Architectural Engineering Department, University of Technology, Baghdad, Iraq
ae.21.03@grad.uotechnology.edu.iq

Abstract. This paper discusses one of the most significant concepts in urban development, which is Ecotourism. The concept emerged with social and cultural that contribute to improving several psychological aspects, whether at the individual level or the country level. The concept of Ecotourism has also occurred as a modern and practical alternative that preserves nature and contributes to its growth and thus achieving health sustainability of cities. This study clarifies the points of interest in the healthy environment of these places and identifies the possibilities to achieve them within the framework of urban design using sustainable transportation. The study also develops recommendations used in sustainable urban transport which can be applied in Erbil city in Iraq to create an appropriate spatial health environment in the tourist areas in terms of increasing the sustainability of these places and in attracting tourists to healthy environments, which in turn will contribute to the health of the world.

Keywords: Urban development · Sustainable transportation · Ecotourism

1 Introduction

The environment is considered as a framework in which humans live and inhabit and is measured in the extent of how safe and well this environment is, whether it is natural or created by a man whereas, Ecotourism is known as the travel to natural, environmentally friendly areas that maintain the well-being of the local population. Blangy and Wood [1] expressed Ecotourism as traveling to unpolluted natural areas to enjoy and admire the natural landscapes, whether plant or animal, as well as other cultural aspects that exist in the past and present. Weaver [2] defines Ecotourism as a form of tourism that enhances learning experiences. Ecotourism is a relatively recent concept that came to express a modern type of environmentally friendly tourism activity, which the individual engages in while preserving the natural and civilized innate heritage of the place in which he lives. In other words, how is the environment employed to represent a healthy ecological pattern of tourism that the individual can resort to economic and social benefits.

Ecotourism has been denoted by many terms such as soft tourism, responsible tourism, green tourism, alternative tourism, and the similarity that fuses these terms is the connection with nature and its benefits for humans [3]. The development of Ecotourism is designated as one of the countries' economic and social development goals

© The Author(s), under exclusive license to Springer Nature Switzerland AG 2022
F. Calabrò et al. (Eds.): NMP 2022, LNNS 482, pp. 2496–2504, 2022.
https://doi.org/10.1007/978-3-031-06825-6_238

because of its significant contribution to the economy of their cities. The United Nations considered the year 2002 as the International Year of Ecotourism and praised the focus on global efforts aimed at connecting sustainable tourism development with the preservation of natural areas. In addition, adopting a healthy environment with tourism is one of the best global tourism patterns to be followed at present due to its elements such as preserving the health of society, preserving natural areas from pollution, and trying to be environmentally sustainable for the tremendous advantage of all participating sectors [4]. Furthermore, Ecotourism decreases environmental pollution through saving in energy consumption, as well as contributing to the development of a healthy, social, and cultural lifestyle for the society members and thus contribute to the development of cities.

1.1   The Impact of Ecotourism on Individual Behavior and Mental Health

There are important inter-relationships between nature, tourism, and mental health [16]. The psychological aspect is one of the considerable health pillars that tourism aims for in which various forms of indoor or urban tourism may also generate happiness and recovery [17]. The goal of creating a healthy environment in the tourist areas is to develop a healthy human life and thus raise the community, which will therefore be the basis for a healthy society that leads to attracting tourists to the region. Studies have shown that human behavior is the focus of many researchers in the field of Ecotourism [5]. To create a healthy environment, according to George [6], the objectives of healthy tourism can coincide with the goals most achieved in Maslow's theory. The theory refers to physiological, safety, and social needs, self-realization, and how tourists are seeking these outcomes to eliminate physical and mental tension and to attain psychological rehabilitation. Saayman [7] identified seven psychological and social factors that motivate tourists to travel, such as escaping from the daily routine, self-discovery, and evaluation, relaxation, participating in recreational activities, gaining a certain social status, strengthening family bonds, in addition to increasing their social interaction.

1.2   Ecotourism Framework

Ecotourism is affected by several environmentally reliable factors by tourists as shown in Fig. 1, which can be summarized as:

(1)   The attractions must be mostly nature-based or other attractive tourist elements, such as archaeological and heritage areas, shrines, etc.
(2)   Focus on the tourist's interactions with these environmental landmarks.
(3)   Tourists' education, which is achieved through certain mechanisms related to tourism management of these landmarks, such as through educational presentation on screen or through professional tour guide staff.
(4)   Expertise and skilled tourism management that must follow the principles of tourism sustainability and conservation practices related to the environment, social, cultural, and economic [2].

Fig. 1. Ecotourism framework.

As depicted in Fig. 1, the main goal of Ecotourism is to achieve an environment that lures tourists by improving environmental quality and enhancing the conditions of the tourist areas [8]. This is achieved via fostering urbanism with Ecotourism for designing external spaces of these tourist areas [9]. There are several urban strategies to reach a healthy environment in urban design. For example, preparing plans for sustainable development, accommodating field expansion, preparing urban plans committed to growing visitors season, contributing to the expansion of green areas, and sustainability of urban infrastructures.

Although the concerted efforts by most countries to pursue a healthy environment for tourism, several negative aspects, however, have led to some drawbacks. This is due to unsustainable transportation in numerous ways, which consequently generate negative effects on health such as noise and pollution through the emission of harmful gases due to their association with long travel distances and their impact on the climate. As a result, Ecotourism has been linked to attaining a purely economic benefit regardless of the environment's enhancement [10].

One way of solving such an issue is via sustainable transportation. The sustainability of transport routes and the use of environmental transport, directly contributes to performing a balanced healthy environment, indicating that green or sustainable transport is achieved with well-studied and practically applied strategies [8] which will be discussed in the next section.

2   Sustainable Transportation

One of the main techniques for conducting environmental sustainability in transportation is by integrating planning of transportation with the development of a comprehensive urban design [11]. This can be outlined as:

(1) The process of coordination between the tourism sector and the public sector in the planning and implementation of transport systems, considering the health aspects. Transport-related policies should be linked with the environment, energy, and urban land usage in tourist areas.

(2) Organizing decisions related to the process of environmental transportation, informing tourists or local visitors of several environmental transportation options outside and inside the tourist areas, and clarifying the benefits of environmental improvement if these regulations were followed and economically supporting environmental methods.

(3) Planning transportation systems in which there are ways to encourage pedestrians and bicycles in urban tourist areas, as they are environmentally friendly and human health, in addition to providing alternatives to private cars with attractive and safe public transportation.

(4) Preserving tourist sites, reducing noise and noise pollution, by planning smart transport networks that limit this phenomenon, and building green transport networks.

(5) Benefiting from global experiences in the field of sustainable and environmental tourism transport and trying to apply them within the availability of local resources.

2.1 Eco-friendly Transportation Types

Ecotourism is changing rapidly, as heritage and nature destinations have become the most attractive element at present, especially during the current pandemic, and therefore, these areas are obliged to meet more strict environmental requirements, especially in the urban environment. For the sustainability and preservation of these areas, it is possible to use eco-friendly transportation means such as

(1) Bicycles are an old, environmentally friendly way to reach destinations. They are often used by the young age group. They do not have any harm to the environment. They are safe in terms of noise and carbon dioxide emissions. They are used as a fun way in tourist areas to explore, in addition to their real use.

(2) Low-fuel, hybrid, or electric scooters, which are motorcycles of a special type designed in several shapes and suitable for different weights and are safer and have a low impact on the environment.

(3) The use of electric/hybrid or solar-powered vehicles. Vehicles that do not require gas to operate have rapidly developed over the past years due to almost zero-fuel cost and preserving the environment. These cars are charged effortlessly using sockets at home or on the go whereas solar-powered cars are the one that uses photovoltaic energy from sunlight to charge the batteries that power the electric motors.

(4) Walking is the most straightforward mode of transportation that is environmentally friendly and supports individual health at the same time. It is often used for hiking in tourist areas, with beautiful transportation routes, i.e., places to achieve the pleasure of walking for individuals. The splendor lies in its attractiveness through safe walking represented by increasing occupancy rates and being present in these spaces to achieve safety for these places and this applies to tourist places [12].

(5) Participation in public transportation, whether through electric buses or metro networks, and tourist buses designed in an environmentally friendly manner.

2.2   Benefits of Electric Power Transportation

The environmental benefits are integrated in terms of transport strategies related to electric power [13]. The developed countries have started using these technologies. This experience can be transferred through the establishment or placement of electric or photovoltaic truck stations as points on tourist roads or during tourist areas and their practical application because of their effectiveness in creating an ecological balance and nature preservation. This has been applied in some tourist areas such as China, and many companies have also begun to support these energy-consuming vehicles that are environmentally friendly, such as Tesla company which manufactures revolutionary electrical cars. This is a kind of green energy with the low ejecting polluting gases and noise.

2.3   Examples of Using Ecofriendly Transportation

(1) **Photovoltaic Energy Transportation Strategy in China**: the transnational experience in China in ecological sites is an example of the use of environmentally friendly energy, as electric minibuses were used for tourist transport that derives their fuel from light energy, by solar panels on the roofs of these buses in the designated parking spots as shown in Fig. 2. In addition, solar or electric energy charging points on tourist roads are experimented with for charging these minibuses while moving on the road [12].

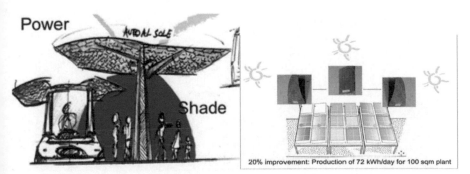

20% improvement: Production of 72 kWh/day for 100 sqm plant

Fig. 2. The electric transportation system in the terracotta warriors' archeological side of Xi'An, China [12].

The initiative implemented in China and the use of photovoltaic energy, which is an important source of clean energy, can operate many machines and vehicles that would preserve the environment by reducing emissions. As well as being economically appropriate, which in turn contributes to preserving historical tourism areas as well by reducing vibrations of transport buses which in the long term can harm the topography of the land with the continuity of movement through time.

Statistically, the annual fuel consumption for the minibuses compared to photovoltaic was measured with the Tonne Petroleum Equivalent (TPE) for 32 km per year. When calculating diesel fuel for daytime operation with air conditioning, except for night lighting, because tourists often go to these tourist areas during the day, a value of (2.8 to

4 TPE) is generated from the consumption energy equivalent to (32 to 46 MKh). While an electric minibus can operate at over the same covered distance consuming energy approximately around (4.5 MKh). As a result, the project involved developing innovative solar-energy roofs that support Ecotourism economically. The proposed system works on the sunlight self-tracking roofs which comprise the use of composite sheets containing properly oriented glass fibers and piezoelectric films in different segments. When these chips are electrically stimulated, they can vary according to the intensity and position of the stimulated plate. This electrical stimulation can be automatically triggered by the positioning of the solar panels' N-beam sensors in different areas of the PV roof.

(2) Using Solar Energy for Cable Car in Malaysia: the hybrid solar system has great potential as one of the renewable energy technologies for power generation. It is an effective solution for green alternative energy generation with ease of operation, maintenance, logistics, and cost problems, providing reliable 24-h supply at an effective cost in addition to nature conservation. The solar hybrid system installed in the Malaysian tourist area [14], the middle and upper stations of the Langkawi cable car in is shown in Fig. 3. It is the first tourist complex solar energy project that provides the project with reliable energy. The main objective of installing the system was to generate electricity while preserving nature by avoiding cutting down trees if a system was used, thus reducing carbon dioxide emissions and noise.

Fig. 3. Solar panels on top of top and middle cable car stations in Langkawi, Malaysia [14].

(3) The Pedestrian Movement Strategy Applied in Italy: Ten Italian cities released a detailed assessment highlighting the use of a sustainable local strategy to deal with the needs of modern urban mobility, after being implemented in three Italian cities such as Milan, Bologna, and Florence, which are famous for their low private car ownership rates, excellent public transport systems, and high coverage of riding lanes [15]. Bicycles, pedestrian zones, no-go zones, and innovative and effective mobility plans prepared for the COVID-19 emergency. In these cities, the local government developed, in a short time, effective programs thanks to its strong vision of a new green deal and for a sustainable future. The experience that was applied proved a change towards a sustainable pure environment that can be applied in the tourist areas to obtain a healthy environment.

As it expanded and applied in eight other cities in Italy, it formed a new vision for a sustainable system because of high environmental awareness. Bicycle paths were created, and the walking system encouraged better physical activity, in addition to the use of electric transport in public transport. This analytical framework provided, and the studies presented from these cities prove that with the focus on the central role in the sustainable environmental shift through urban planning and intelligence in encouraging the use of environmentally friendly transport by reducing public transport prices and providing bicycles, scooters, and electric vehicles.

3 Case Study on Erbil City in Iraq

It is possible to apply the strategies and case studies in Subsect. 2.3 in the northern resorts, as Ecotourism has not taken its lead role. In Erbil governorate, even though it has the elements to attract multiple and diverse natural tourism and is suitable for practicing Ecotourism activity in its various forms, with the lack of studies dealing with Ecotourism in Erbil city. Therefore, it is possible to apply the pedestrian movement paths that were applied in Italian resorts, utilizing bicycles and motorcycles, or use electric buses.

Fig. 4. Mount Korek cable cars in Erbil, Iraq.

There is also the possibility of applying sustainable energy forms in Mount Korek, as shown in Fig. 4 which has a cable car operating based on fuel (diesel) and replaces them with solar panels to operate the cable cars and transport tourists while taking advantage of the sunny summer climate to achieve better utilization of energy storage in the panels on the slopes of the mountains, where they will serve as the lens for the rays of the sun to obtain a renewable energy source in addition to reducing visual pollution and noise resulting from the sounds of generators in this location. Also, solar panels can generate power for the cafe and restaurants designated at the top station of car cable station on top of Mount Korek.

To implement urban transformations based on green principles, we must have smart urban management of the tourist place that would play a useful element in the health of the individual. Therefore, dealing with monitored transportation systems, stations, security experts, and the workforce with a simultaneous schedule and supervision strategy requires adequate planning. If a disturbance occurs at a station, for example, many mathematical algorithms map out new communication methods and alternatives to reduce the passengers who pass through that station so as not to waste transmission energy that would disturb a healthy ecosystem [18]. A defect in the social distancing system is currently being developed to reduce the spread of epidemics as an alternative to producing a healthy city from scratch. The basic policy of the application, which in turn contributed to the establishment of a healthy environment in cities [15] despite the absence of a fixed framework system that can be transferred from one experience to another. It is possible to rely on many planning alternatives and administrative policies that are subject to the officials of the tourist area, and which are applied according to well-studied approaches to be used in the development of the local healthy environment.

The possibility of applying them in the city of Erbil through the establishment of special lanes for cyclists and pedestrians and the provision of multiple stops with Providing environmentally friendly transportation with safe corridors would raise the healthy ecosystem of the region. It is clear that the city of Erbil is rich in these green roads, rich in a pure atmosphere, and has very stunning views, which will thus contribute to increasing the tourist attraction of the region.

4 Conclusion

The urban environmental transformation has been linked to the support of stakeholders occupying an influential role in creating a renewed healthy environment. The benefits of establishing an ecotourism system were explained as traveling to green resorts provides psychological comfort, relaxation, as well as enriching social interaction. The proposal suggested as the case in ARBIL city to use Solar panels, solar self-tracking devices, smart powered systems, and vehicles PV-powered materials which will be one of the elements in an integrated healthy ecosystem that contributes to the sustainability of tourist areas and contribute to the healthy investment of cities.

The possibility of manufacturing air filters that work on solar energy so that each panel is designed as a tree and works with the capacity of one tree inside the site where cultivation cannot be achieved, such as historical areas to preserve them from carbon dioxide emissions that cause their extinction which will be discussed in our next work.

References

1. Blangy, S., Wood, M.E.: Developing and implementing ecotourism guidelines for wildlands and neighboring communities, pp. 32–54. Ecotourism Society (1993)
2. Weaver, D.B.: Mass and urban ecotourism: new manifestions of an old concept. Tour. Recreat. Res. **30**(1), 19–26 (2005)
3. Slee, B., Farr, H., Snowdon, P.: The economic impact of alternative types of rural tourism. J. Agric. Econ. **48**(1–3), 179–192 (1997)
4. Bushell, R.: Healthy tourism. In The Routledge Handbook of Health Tourism , pp. 119–130. Routledge (2016)

5. Dyer, P., Gursoy, D., Sharma, B., Carter, J.: Structural modeling of resident perceptions of tourism and associated development on the Sunshine Coast. Australia. Tour. Manag. **28**(2), 409–422 (2007)
6. George, R.: Marketing South African Tourism. Oxford University Press, USA (2004)
7. Saayman, M.: Marketing Tourism: Products & Destinations: Getting Back to Basics. Leisure C Publications (2006)
8. Buckley, R.: A framework for ecotourism. Ann. Tour. Res. **21**(3), 661–665 (1994)
9. Grenier, D., Kaae, B.C., Miller, M.L., Mobley, R.W.: Ecotourism, landscape architecture and urban planning. Landsc. Urban Plan. **25**(1–2), 1–16 (1993)
10. Mullins, P.: Tourism urbanization. Int. J. Urban Reg. Res. **15**(3), 326–342 (1991)
11. Schiller, P.L., Kenworthy, J.R.: An Introduction to Sustainable Transportation: Policy, Planning and Implementation. Routledge (2017)
12. Aversa, R., Petrescu, R.V., Apicella, A., Petrescu, F.I.: Modern transportation and photovoltaic energy for urban ecotourism. Transylv. Rev. Admin. Sci. Special **2017**, 5–20 (2017)
13. Wu, M., Wu, Y., Wang, M.: Energy and emission benefits of alternative transportation liquid fuels derived from switchgrass: a fuel life cycle assessment. Biotechnol. Prog. **22**(4), 1012–1024 (2006)
14. Ali, B., Sopian, K., Rahman, M.A., Othman, M.Y., Zaharim, A., Razali, A.M.: Hybrid photovoltaic diesel system in a cable car resort facility. Eur. J. Sci. Res. **26**(1), 13–19 (2009)
15. Barbarossa, L.: The post pandemic city: challenges and opportunities for a non-motorized urban environment. An overview of Italian. Sustainability **12**(17), 7172 (2020)
16. Buckley, R. C. "Therapeutic mental health effects perceived by outdoor tourists: A large-scale, multi-decade, qualitative analysis." Annals of Tourism Research (2019).
17. Buckley, R.: Nature tourism and mental health: parks, happiness, and causation. J. Sustain. Tour. **28**(9), 1409–1424 (2020)
18. Spencer, J., Nguyen, H.: A landscape planning agenda for global health security: Learning from the history of HIV/AIDS and pandemic influenza. Landsc. Urban Planning. **216**, 104242 (2021)

The Urban Blue Space as a Wellness Tourism Destination

Reema Hamza Yassin[✉] and Rawaa Fawzi Naom Abbawi

Department of Architectural Engineering, University of Technology, Baghdad, Iraq
ae.21.36@grad.uotechnology.edu.iq,
Rawaa.f.abbawi@uotechnology.edu.iq

Abstract. The term blue space encompasses all accessible surface water as a parallel concept to green space, not as a sub-division of it, it is defined as open-air natural or manmade environments that features water, reachable by individuals either adjacently (being in, on, or near water) or virtually (to have the vision, hearing, or sensing water); blue space has a lot of health-related qualities, these qualities are well studied in previous literature, but there is a gap in connecting it with the wellness tourism destination. The research aims to investigate the fitness of blue urban space as a wellness tourism destination. From reviewing the literature, research had extracted the main concept of blue space factors on the one hand and the components of wellness tourism on the other hand. The research supposes there is fitness between them, which was applied as a conceptual framework. The methodology of the research had used an analytical, conceptual framework to assess the availability of success factors in urban blue space Zawraa Park et al. in Baghdad, Iraq, and check for its fitness to make the park a wellness tourism destination. The research concluded that the blue space enhances the park's wellness and fits it to be a wellness tourism destination according to its place making spatial attributes social interaction services provision and physical well-being.

Keywords: Blue space · Water landscape · Therapeutic landscape · Water tourism · Wellness tourism · Post-COVID tourism · Blue space tourism

1 Introduction

Tourism, in general, is important in building and identifying a unique personality for the city which enhances its attractiveness, The city of Baghdad is distinguished by many religious, cultural, commercial and other tourist attractions, which contribute to enhancing its tourism capacity (Al-Hinkawi 2021) Nowadays a huge interest in well-ness and a healthy lifestyle increased after the pandemic era and a major part of it is wellness tourism. Wellness tourism can work besides cultural and heritage tourism, in which the original function of the buildings is displaced and invested in tourism, which connects the space's identity with its original function, activating the tourism in the city that leads to empowerment and development of society (Al-Saadawi 2021; Hussein 2021).

1.1 Wellness Tourism

Nowadays a huge interest in wellness and healthy lifestyle increased after the pandemic era and a major part of it is the wellness tourism. The World Health Organization (WHO) defines health as a state of well-being on the mental, physical and social aspects and

© The Author(s), under exclusive license to Springer Nature Switzerland AG 2022
F. Calabrò et al. (Eds.): NMP 2022, LNNS 482, pp. 2505–2515, 2022.
https://doi.org/10.1007/978-3-031-06825-6_239

health is not limited to diseases or disabilities absence (WHO 2018). Several researches are interestingly focusing on advocating health tourism by implementing two subsets, medical and wellness tourism. On the other hand, other researches consider curative tourism as a third subcategory of health tourism. Wellness tourism tourists are usually in search for rejuvenation and improvements in their overall health and well-being.

To differentiate between medical and wellness tourism it's stated that medical tourism deals with the tourists who are traveling to tourism destinations for the purpose of treatment (Charak 2019), while wellness tourism it relates to the adopting of a healthy lifestyle beside its being an asset to the economic and worldwide implications and is becoming one of the most important topics for research in tourism studies. It implies individual and group travel to specialized resorts and it is becoming a growing tourism specialty and traveling to places to maintain mental and physical health (Kazakov and Oyner 2019) from this definition of wellness tourism, it is understood that it is the type of tourism related to salutogenetic health: the branch of health that is concerned with reasons and circumstances to create and preserve health rather than to treat diseases [the branch of health that treat diseases is called pathological health] (Völker and Kistemann 2011) the tourist destination to obtain salutogenetic health would be the therapeutic landscape. After reviewing the previous literature research a knowledge gap was found to be that blue space specifically needs to be studied as a wellness tourism destination as previous studies focused on traditional therapeutic landscape and not the **water element**. This research aims to focus on the space that water surfaces create in the landscape, this space is called: Blue space. From The literature review, extraction of conceptual framework was made to achieve this aim.

The **objective** of this research is to test if blue space, as a type of therapeutic landscape, is a fit to become a wellness tourism destination, the methodology used was to use an analytical conceptual framework to check the availability of blue space concept attributes in a local water body Zawraa park et al. in Baghdad and then to test their fitness to make a wellness tourism destination.

1.2   Landscape and Well-Being

Therapeutic landscapes are environments that provide healing profits and they are defined as environments where the physical and man-made environments, social conditions, and human perceptions come together to create an atmosphere that is useful to healing. This definition cleared the significance of understanding the health-promoting qualities of a specific space physically and socially, but besides that, the more available subjective ways by that society may interpret and use that space (Bell et al. 2018). Therapeutic landscapes can be: (i) Institutions that provide traditional western medical treatments as well as alternative health therapies such as natural-based therapies. (ii) Constructed attractions, such as spa hotels, hospitals, community gardens, and so on, (iii) Vacation places with natural qualities that offer healing experiences as well as a sense of comfort and thankfulness to individuals looking for health and well-being (Majeed and Ramkissoon 2020).

Gesler in his book "healing places" categorize natural environment healing aspects into four aspects:

- Natural: belief in nature as a healer, beauty, aesthetic pleasure, remoteness, immersion in nature, specific element of nature

- Built: the sense of trust and security, Affects the senses, Pride in building history, symbolic power of design
- Symbolic: the creation of meaning, physical objects as symbols, the importance of rituals
- Social: Equality in social relations, Legitimization and marginalization, therapeutic community concept, social support

Many people believe that physical, mental, and spiritual healing can be obtained by having time outdoors where they can be exposed to undisturbed nature, and several natural healing power more than others and the most important one is water because interpreted by space users as a resource to a 'healing sense of place' (Gesler 2003).

1.3   Blue Space, Blue Health, and Wellness Tourism

As demonstrated earlier, water elements provide the most important natural element in natural healing environments. The United Nations World Tourism Organization (UNWTO) declared freshwater to be the natural resource of the tourism industry that's the most scarce and critical (UNWTO 2003). In reviewing the literature it was found that there is a research gap which is: there are up to date, very few studies implemented the study of therapeutic landscape as a wellness tourism destination and there is a research gap about studying the blue space specifically a wellness tourism destination as previous studies focused on traditional therapeutic landscape and not the water element. This research aims to focus on the space that water surfaces create in the landscape, this space is called: Blue space. The objective of the research is to study the relationship between wellness tourism and blue space, The word blue space encompasses all accessible surface water as a parallel concept to green space, not as a sub-division of it, it is defined as open-air natural or manmade environments, which precisely feature water and are reachable by individuals either adjacently (being in, on or near water) or virtually (to have the vision, hearing, or sensing water) (Volker and Kistemann 2011; Grellier et al. 2017). Blue space can be generally divided according to scale into:

- Mega scale: includes natural water surfaces connected to the ocean, such as seas, bays, gulfs, lagoons, or estuaries.
- Macroscale: includes flowing inland water surfaces like rivers, streams, or canals of different sizes, flow rates, turbulence, and transported sediments.
- Medium to micro-scale: Stagnant inland water bodies like lakes, ponds, pools, or basins of different sizes and turbidity.
- Microscale: includes other urban blue elements which are not water bodies, such as geysers or waterfalls. Fountains and other artificial water features are included in one of the last three categories depending on their appearance (Völker et al. 2016).

And another recent more comprehensive categorization for the blue space is the one mentioned in the book " Urban Blue Spaces" that categorizes the blue space according to blue health concept, which is a concept that describes health-related to accessibility to water bodies i.e. blue urban space, the book categorized it into six main categories:

- Constructed coastal spaces: Promenade, pier.

- Natural coastal spaces: Sandy beach, stony beaches, sand dunes, sea cliffs, salt marsh, estuary.
- Lakes and other still water bodies: Natural lake, artificial lake, reservoir, pond, wetland, fen, marsh, bog.
- Rivers, streams, and canals: Large river with man-made banks, large river with natural banks, medium-sized river with man-made banks, medium-sized river with natural banks, stream that has a mix of man-made or natural banks, urban canal, rural canal, waterfall or rapids.
- Docks, ports, and marinas: Dock, harbor, marina.
- Other blue infrastructure: Ornamental water feature or fountain, mineral spring, thermal spring, outdoor skating, curling or ice hockey rink, swimming pool (Bell et al. 2021)

1.4 'Exposure' to Urban Blue Space and Wellness Tourism

There are three types of exposure to Landscape and urban blue space according to Kenniger et al. (2013):

- Intentional: Intentional interactions are the interactions that happen with intention from the user of landscape, an example of that is wildlife viewing, gardening, or walking in park. For water that would be deliberately chosen direct exposure that could be in [e.g. bathing], on [e.g. boating] or by [e.g. Resting, cycling, walking alongside] the water.
- Indirect: Indirect interactions are the kind of exposure that don't need a person to be in a state of physical presence in nature, and covers activities such as observing virtual sceneries of nature, or watching it through a window, for water that would be as an example a view from a home/office/building's openings.
- Incidental: Incidental interactions happen when a person has a physical presence in nature, but the reciprocal action with nature is a by-product of another activity, such as passing by a water body whilst cycling to work. Incidental interactions are differentiated from indirect interactions by the necessity of the physical presence of natural elements (E.g. visual exposure during a commute; if the path is picked due to its proximity to water, this is 'intentional') (Keniger et al. 2013).

According to all that the research question was generated as:

What are the characteristics of blue space that help to create a salutogenetic health effect and contribute to making it a wellness tourism destination?

1.5   Blue Space Components and Attributes

Components of blue space environment include aquatic and terrestrial components, their environmental circumstances, qualities and impacts can individually and directly influence the users' exposure to blue space, for the better or the worse, affecting behavior and psychological aspects (Mishra et al. 2020).

Blue urban space is known as an area consisting from land and water and has at minimum one land-water edge (e. g. shoreline, riverbank), often secluded from the surroundings by at least one physical edge which has voids (pores) that serve for the flow of people, An indisputable determinant of urban blue space is its concrete and abstract relation with water, that greatly affects the characteristics of the space. The boundaries (edges) of urban blue space, that serve to define it as an urban interior, can change according to topographic conditions, functional layout, and surrounding urban structures (Breś and Krośnicka 2021).

2 Research Methodology

The approach used in this research was qualitative, The analytical framework was designed by researchers after a comprehensive literature review, the methodology was to select assessment factors of the blue space wellness quality and then to check for their availability in the chosen blue space sample; after that testing each factor if it was fitting as a wellness tourism destination component. The research hypothesis was that **blue space slautogenetic factors contribute to creating a wellness tourism destination**. The components of wellness tourism were stated as Social interaction, Physical fitness, and Individual interaction (emotional, spiritual, and mental). (Esfandiari and Choobchian 2020) the factors that are used to assess the blue space quality include Accessibility, Design quality, Health and well-being, Water connections, Physical activities (Bell et al. 2021) these factors alongside factors that determine the quality of space: experienced space, symbolic space, social space, and activity space from (Völker and Kistemann 2011) (Table 1) were used for assessing salutogenetic health effects of blue space.

To test the hypothesis the sample chosen was a local blue space in Al Zawraa park in Baghdad, Iraq. It was chosen based on its being the largest park in Baghdad, Iraq the capital city of Iraq, and is a destination for many tourists visiting Baghdad whether be it local from another governorate or a foreign tourist from outside Iraq. Data: spatial plans from google earth at the date Dec. 29. 2021, images were taken by the researcher at the date Dec. 29. 2021, at weather conditions partly clouded sky and 16/9 degree temperature.

The analytical procedures were combined and clarified in one table (Table 1) analytical framework test the fitness of blue space as a wellness tourism destination by applying

Fig. 1. An aerial view of Al-zawraa Park shows the relationship between land and water. (Resource: Google earth, Dec. 2021)

Table 1. Conceptual analytical Framework and fitness to tourism components

Main concept of blue space factors	indicators	Availability in Al-Zawraa blue space	The components of wellness tourism		
			Social Interaction	physical fitness,	Individual interaction (emotional, spiritual and mental)
Accessibility	• site visibility	Available	✔		✔
	• Pedestrian bicycle or car access	Available	✔	✔	✔
	• car parking	Available	✔		
	• inclusive access	Available	✔	✔	✔
Design quality	• on-site circulation	Available	✔	✔	✔
	• views and landmarks	Available	✔		✔
	• inclusion of cultural heritage values	Not Available			
	• site furniture fitting the context	Available	✔		✔
	• safety and security	Available	✔		
Facilities	• range of facilities	Available	✔	✔	✔
	• accessibility of facilities	Available	✔	✔	✔
	• amount of seating	Available	✔		✔
	• quality of nature	Available	✔		✔
	• degree of shelter or shade and lighting	Available		✔	✔
Health and well-being	• genius loci	Available	✔		✔
	• sense of being away	Available	✔	✔	✔
	• contact with nature	Available	✔	✔	✔
	• sensory stimulation	Available	✔	✔	✔
	• contemplation	Available	✔	✔	✔
Water connections	• land-water connectivity	Available	✔	✔	✔
Water connections	• water visibility	Available	✔	✔	✔
	• access from and to water	Not Available			
	• water safety equipment	Available	✔	✔	✔

(continued)

Table 1. (*continued*)

Physical activities					
Physical activities	• formal sports activities	Not Available			
	• informal sports	Available	✔	✔	✔
	• water sports	Available	✔	✔	✔
	• children's play and activity zoning	Available	✔	✔	✔
experienced space	• territoriality	Not Available			
	• the beautiful natural environment	Available	✔	✔	✔
	• identity/sense of place,	Available	✔		✔
	• removal from everyday stress	Available	✔	✔	✔
	• Place meaning.	Available	✔	✔	✔
symbolic space	• The origin of spiritual nature	Not Available			
	• Beliefs of healing power for the place	Not Available			
	• Spiritual space	Not Available			
social space	• Shared rituals	Available	✔	✔	✔
	• historical context	Not Available			
	• everyday activities	Available	✔	✔	✔
	• relative equality	Available	✔	✔	✔
	• social relations	Available	✔	✔	✔

them to Al-Zawraa park blue space and using them as a checklist to see the percentage of how much it fits to serve as a wellness tourism destination.

Analytic Conceptual Framework

The analytical tool (Table 1) used in this research was to assess the water body Zawraa park et al. to see if the attributes of blue space are available in it and then test each attribute if it serves the objectives of wellness tourism components (Figs. 1, 2, 3, 4, 5, 6, 7, 8, 9, 10, 11,12, 13, 14 and 15).

3 Results

From the studying of al Zawraa park blue space, it was found that the blue space attributes available at the park make it a successful blue space and match the criteria that were taken from the literature review about the urban blue space features as most of the features that constitute the concept of blue space were available on-site as demonstrated by pictures above and 7 out of 41 factors were unavailable only which is a small percentage in comparison to what's available i.e. 34 factors out of 41, a further step was to check if the available factors serve to fit as a component of wellness tourism destination and the

Fig. 2. Pedestrian bicycle or car access and inclusive access. (Source: taken by author, 2021)

Fig. 3. Panoramic view of site visibility and on-site circulation. (Source: taken by author, 2021)

Fig. 4. Inclusion of cultural heritage values of Islamic design elements (axis and fountains leading to the main water body), shade and lighting appear in the picture too. (Source: taken by author, 2021)

Fig. 5. Water features connected by canals to the main water body. (Source: taken by author, 2021)

Fig. 6. Sense of being away from contact with nature achieved by Provision of vegetation and birds. (Source: taken by author, 2021)

Fig. 7. Water artwork by artist Fadhil Al-Rubaiy serves as an entrance connected to the water canal providing sensory stimulation and a sense of place. (Source: taken by author, 2021)

Fig. 8. Availability of water sports. (Source: taken by author, 2021)

Fig. 9. Caution signs forbid fishing and swimming in the lake. (Source: taken by author, 2021)

Fig. 10. Availability of facilities and shops on-site. (Source: taken by author, 2021)

Fig. 11. Availability of maintenance services. (Source: taken by author, 2021)

Fig. 12. Availability of children's playground. (Source: taken by author, 2021)

Fig. 13. Availability of landmarks. (Source: taken by author, 2021)

Fig. 14. Zoo information map. (Source: taken by author, 2021)

Fig. 15. Al-Zawraa park information map. (Source: taken by author, 2021)

result was that 33 out of 34 were available in the interaction component section, 25 out of 34 were available in the physical component section, 32 out of 34 were available in the individual component section, these high numbers indicate that a successful blue space can serve as a destination to wellness tourism, which means that the hypothesis of the research, **blue space slautogenetic success factors make it a fit to be a wellness tourism destination,** is a valid hypothesis.

4 Conclusion

Blue space has a lot of health-related qualities, these qualities are well studied in literature but are not connected to studies of making it a wellness tourism destination, and that lead to the research problem which was that there was a knowledge gap in studying blue space as a wellness tourism destination. this research concludes that Blue urban spaces should be considered a destination to wellness tourism, especially in the post-COVID times as they can provide a lot of health benefits because of their ability to enhance well-being and salutogenetic health effect and at the same time they fit the components of wellness tourism destination requirements, the factors that should be available in the blue space to make it successful wellness tourism destination are: (1) According to placemaking: place well-being, water connections, and experienced space. (2) According to spatial attributes: experienced, symbolic, social, and active space. (3) According to social interaction: social space and accessibility. (4) According to services provision: Facilities. (5) According to physical well-being: design quality, physical activity. These factors were studied in detail and made a clear analysis of how they should be appropriated to serve the wellness tourism components, the factors that impact wellness tourism the most were found to be activity space, social space, health and well-being, and place well-being attributes as they impact all of the three components of wellness tourism: social interaction, physical fitness, and individual interaction. the recommendation to improve the role of blue spaces more in well-being tourism and to enhance the quality of urban blue space in the site of Al-Zawraa park is to include cultural heritage, providing access from and to water, enable formal sports activities, provide spiritual spaces, and include social shared rituals on the site which will help to increase its attractiveness as a wellness tourism destination.

References

Charak, N.S.: Role of spa resorts in promoting India as a preferred wellness tourism destination – a case of Himalayas. Int. J. Spa Wellness **2**(1), 53–62 (2019). https://doi.org/10.1080/24721735. 2019.1668672

World Health Organization (2018). https://www.who.int/about/who-weare/constitution#:~:text= Health%20is%20a%20state%20of,absence%20of%20disease%20or%20inrmity

Kazakov, S., Oyner, O.: Wellness Tourism: A Perspective Article, p. 1. Emerald Publishing Limited (2019). https://doi.org/10.1108/TR-05-2019-0154

Al-Hinkawi, W.S., Zedan, S.K.: Branding for cities: the case study of Bagh-dad. In: IOP Conference Series: Earth and Environmental Science, vol. 779, no. 1, p. 012037. IOP Publishing (June 2021)

Hussein, S.A., Al-Taee, M.D.: The management of heritage buildings in historical urban areas according to cost-benefit methods. In: IOP Conference Series: Earth and Environmental Science, vol. 779, no. 1, p. 012045. IOP Publishing (June 2021)

Al-Saadawi, B.A., Al-Hinkawi, W.S.: Role of urban malleability's mecha-nisms on sustaining cities. In: IOP Conference Series: Materials Science and Engineering, vol. 745, no. 1, p. 012165. IOP Publishing (March 2020)

Völker, S., Kistemann, T.: The impact of blue space on human health and well-being–salutogenetic health effects of inland surface waters: a review. Int. J. Hyg. Environ. Health 214(6), 449–460 (2011)

Wilbert, M.G.: Healing Places, p. 8, 9. Rowman and Littlefield (2003)

Bell, S.L., Foley, R., Houghton, F., Maddrell, A., Williams, A.M.: From therapeutic landscapes to healthy spaces, places, and practices: a scoping review. Soc. Sci. Med. 196, 123–130(p. 123,124) (2018)

Majeed, S., Ramkissoon, H.: Health, wellness, and place attachment during and post health pandemics. Front. Psychol. 11 (2020)

UNWTO: Djerba Declaration on Tourism and Climate Change, UNWTO (2003)

Grellier, J., et al.: BlueHealth: a study program protocol for mapping and quantifying the potential benefits to public health and well-being from Europe's blue spaces. BMJ Open 7(6), e016188 (2017)

Völker, S., Matros, J., Claßen, T.: Determining urban open spaces for health-related appropriations: a qualitative analysis on the significance of blue space. Environ. Earth Sci. 75(13), 1–18 (2016). https://doi.org/10.1007/s12665-016-5839-3

Bell, S., Fleming, L.E., Grellier, J., Kuhlmann, F., Nieuwenhuijsen, M.J., White, M.P. (eds.): Urban Blue Spaces: Planning and Design for Water. Routledge, Health and Well-Being (2021)

Keniger, L.E., Gaston, K.J., Irvine, K.N., Fuller, R.A.: What are the benefits of interacting with nature? Int. J. Environ. Res. Public Health 10(3), 913–935 (2013)

Esfandiari, H., Choobchian, S.: Designing a wellness-based tourism model for sustainable rural development (2020)

Mishra, H.S., Bell, S., Vassiljev, P., Kuhlmann, F., Niin, G., Grellier, J.: The development of a tool for assessing the environmental qualities of urban blue spaces. Urban Forestry Urban Green. 49, 126575 (2020)

Breś, J., Krośnicka, K.A.: Evolution of edges and porosity of urban blue spaces: a case study of Gdańsk. Urban Plan. 6(3), 90–104 (2021)

Sequere Pecuniam!

Christer Bengs[✉]

Aalto University, PO BOX 11000, 00076 Aalto, Finland
christer.bengs@aalto.fi

Abstract. Planning and architecture are disciplines geared towards finding best practices, and thus connected to innovative approaches. The multi-disciplinary nature of urban studies lends itself to a form of inference, abduction, which proceeds from clues and theoretical assumptions to explanations. Meaningful explanations require an understanding of the object under scrutiny as well as heuristic techniques for analysis. Any particular society in history is compressed into some essential narratives, which facilitate for understanding that society. Our concurrent predicament is to follow the money. We can, however, choose how to do it!

Keywords: Urban studies · Semiotics · Heuristics

Urban studies seem to be void of a generally accepted body of theory as well as common research methods. That may be considered a default, but it may also be harnessed for deliberately detecting a multitude of creative entries. According to tasks at hand, established fields of academia provide alert researchers of urbanism with a plethora of intellectual models to be exploited. This contribution sketches creative approaches to urbanism, including abduction, a heuristic model for conceptual analyses as well as a general method for scrutinizing narratives. The presentation also includes a general model for analysing built environment.

The aim of the discussion below is to pinpoint how to reach intelligible explanations by employing semiotic models. The underlying aim is to juxtapose the two dimensions of urban studies, the real world of causal relations with the imaginary world of meanings, that is, the true with the meaningful and reasonable. The intention is to indicate how production of significance in urban matters could be improved by employing suitable methods found in adjacent fields of study. Due to their general character, the presented methodological approaches are applicable in studying built environment as well as planning and design. They are thought to display an exercise in learning-by-doing, that is, the essence of producing meanings.

1 Matter, Mind, Inference

A fundamental distinction is made between matter, the world and all what there is in space and time, and mind, how we conceive the world. For animists of the past, such a split would have been unfathomable. If one thinks every part of the world is animated, the split between mind and matter does not make much sense! Plato bridged the gap between the age-old and newer ways by claiming that we possess innate knowledge, matching

© The Author(s), under exclusive license to Springer Nature Switzerland AG 2022
F. Calabrò et al. (Eds.): NMP 2022, LNNS 482, pp. 2516–2525, 2022.
https://doi.org/10.1007/978-3-031-06825-6_240

an allegedly real world of eternal and immutable ideas [1]. Aristotle denied the notion of innate knowledge by claiming that the essential form of everything is inherent in the thing itself. The nature of things is best understood by observing this world [2].

The main point of Plato when professing innate knowledge was mathematic. Concurrently, it is the foundation of science. How do matter and mind relate to invention and/or discovery? Is mind a subset of matter or vice versa, matter a subset of mind? Are they separate or overlapping? If mathematical relations are a constituent part of matter, do we invent them or discover them (see Fig. 1).

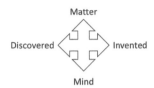

Matter

Discovered ⟨ ⟩ Invented

Mind

Fig. 1. Discovery and invention in relation to matter and mind

Plato thought that the world of matters was a representation of an ideal one, and art - being a representation of this world - a representation of a representation of the ideal. Beauty was to him a property that existed in an ideal form, independent of mind or matter [3]. This notion of classical architecture was destroyed only by nascent modernism [4]. According to Aristotle´s moderate realism, beauty is a property of matter (*fundamentum in re*) that the mind abstracts from graceful things.

Our existence is based on the ability of abstracting matter, and all abstractions involve a process of putting matters into proportion and proportions into perspective, in short, to relate. Cause (*causa*) refers to matter, while reason (*ratio*) refers to mind explaining matter. But how to distinguish cause from reason? Consider Samuel Butler´s (1835–1902) statement that the hen is a way for the egg to produce another egg. Producers of eggs may agree while chicken farmers may disagree.

Cause and effect - as different from premises and conclusion - are not part of logic, but they are logically relevant in the sense that valid inferring requires premises. The main forms of inference include three factors: theory, observations and explanations. Deduction indicates that observations are subsumed under a theory, ending with explanation. Induction implies observations and explanations in order to contribute to a better theoretical understanding. Abduction is run differently starting with observations, which are informed by plausible theories, ending up with explanations (see Fig. 2).

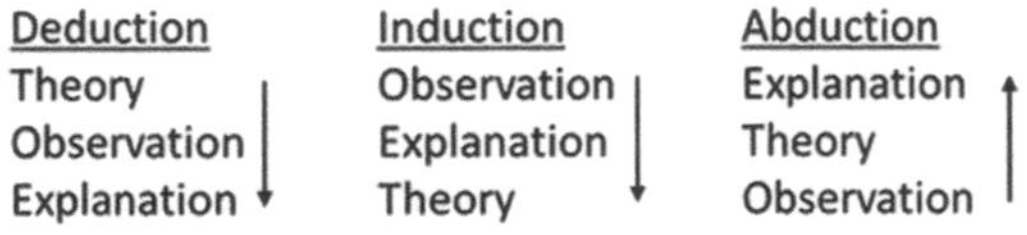

Deduction	Induction	Abduction
Theory	Observation	Explanation
Observation	Explanation	Theory
Explanation	Theory	Observation

Fig. 2. Forms of inference

In the following, we shall focus on abduction, which is a common way for inferring in everyday life, and, due to its heuristic nature, a major path for researchers to proceed.

We do not perpetually apply theories on everyday empirical findings. Neither do we extrapolate statistical probabilities when making everyday decisions. Human uniqueness is thought to spring from the capacity of thinking rationally. Matters are, however, seldom rationalised beforehand, but rather afterwards as part of justification [5].

2 Signs

Signs are categorised in communicative (symbols, icons) and noncommunicative (indices) ones. Symbols and icons are communicative in the sense that they take part in an exchange between intentional senders and receivers. When speaking, we communicate with other intentional beings (or with ourselves). Art works are seen as tokens of authors´ intentions. Indices lack this communicative aspect, because they relate to causality. They do not cover intentions, providing fake clues are not planted in order to cheat the investigator, like in screen plays. Abduction starts with an observation of an index (a clue), which alludes to causal relations. During the 19th century, this way of inferring was expanding in art history, psychoanalysis and criminology [6].

In practice, any matter conceived as an index may also be viewed as a symbol. A blossoming fruit tree is an index, manifesting spring and foreseen harvest, but for thoughtful persons it could mean anything, from the order of the world to peace of mind. By choosing the context, we determine the meaning. In this sense, we are free to invent meanings. Our capacity to do that is limited only by our capacity to discover or invent relations. Sometimes this predisposition is reviewed in testing intelligence.

Consider various historical ways of applying visual perspectives [7]. Urban history indicates at least four different ways of conceiving and shaping the physical environment with regard to whether we see (or not see) and are being seen (or not being seen). From the point of view of the planner or designer, the choice of perspective is deliberate, but what does it stand for as an index of social relations? Is the rule of modern society materialised in the logic of the Panopticon, a one-sided system for controlling the many by the few? Or is it a manifestation of the intention to improve functionality and safety? We can choose our point of view. Various explanations may enrich the conclusions, but they may blur the essential as well.

The double attention to communicative as well as non-communicative signs has always been there. However, the idea of communicative signs allegedly emerged only in late Antique and early Medieval times. Before that, people would have clung to indices when interpreting nature and environment [8]. Surely, ancient art generated meanings, but those may have been meanings that manifested causal relations.

Why so? In the ancient view of the world where everything was animated, matter would speak for itself. A reminiscence of this is found in the use of talismans and relicts. Here, art historians often go wrong as the meaning of magic objects is not to stand for something else, but to act in their own right. This view of the world is now much overlooked, but probably still among us, illustrated by how cause and reason are blurred.

Proper nouns are communicative, but sometimes also indexical. One example of this may be the bank called Banco di Santo Spirito. Founded in 1605, it existed under that name until 1992. The purpose of the bank - allegedly the first national bank in Europe - was to provide capital for construction, and for other commercial purposes as well [9,

10]. An intuitive interpretation of the name would embrace an allusion to the concept of trinity: Father, Son and the Holy Spirit. The Vatican, being the ultimate authority on these matters, would be the obvious point of reference. That was verbatim not the case as the name alluded to the Hospital of the Holy Spirit (founded in 1201), which was the owner and beneficiary of the bank.

A second way of interpreting the name is to stick to its essence, the Holy Spirit, which is an expression for a mediating and communicating force. The Nicene Creed of 325 CE emerged in a strongly centralised Empire, addressing the question of Father and Son – perhaps mimicking an elevated Emperor and the terrestrial Subdued - but said little about the Holy Spirit. This third image of God was consolidated by the Church Fathers by the end of the 4th century, at the time when the Christian faith was made exclusive state religion [11]. Thus, Holy Spirit alludes to centralised authority and alert guardiancy.

A third and thrilling parallel, is the invention of money-based economy in the 7th century BCE. This was a revolutionary event. For the first time in European history, there was a common yardstick for measuring the value of everything. The consequences for social life, politics and culture were dramatic, and the occurring money-based economy coincided with emerging city states (*polis*) and the birth of philosophy, freed from religious ballast [12]. The Roman concept of the Holy Spirit might have been an impossible tenet to conceive without a prehistory of a millennium of monetary economy. A bank named Santo Spirito sounds like a remote eco of the dawn of money.

Proper nouns are of course only one of many possible clues to follow. When discussing any matter, we may relate it to its properties or to the various contexts that it is part of. To do that, one has to provide the object with a somewhat diversified set of abstract characteristics.

3 Objects

Any object under scrutiny may be viewed as possessing a given composition. In addition, it is part of hierarchically occurring processes. When a particular process is upheld, it constitutes a structure of fairly stable relations. Changing processes cause structures to be modified and compositions to alter. This way of conceptualising has been fruitfully applied in studying ecosystems [13].

The model is useful in providing the opportunity to arrange matters under investigation in a hierarchical manner. It is very general, but flexible in the sense that it can be scaled down or up according to the extension of the studied object. The model also provides a possibility to investigate how systems of different orders affect one another by applying statistics or evaluations in order to define so called tipping points, which indicate the stage at which a series of small incidents becomes significant enough to cause a larger and more important change (Fig. 3).

The global level as pictured from outer space can indicate the structure of climatological and geological processes, vegetation and ice cover determining human habitats. At the regional level we can detect major types of land use and urban-rural relations in terms of transportation veins and land cover. At the level of urban agglomerations, urban-rural relations are featured more clearly in terms of main arteries, and decentralised structures with centres, subcentres and edges [14].

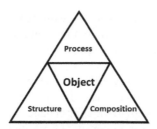

Fig. 3. Environmental model

The urban fabric is clearly distinguishable with respect to public, common and private land use. An experienced eye would detect the historical roots of any particular urban landscape, synthesised in the concept of urban morphology. Depending on functions, land use and building traditions, morphologies indicate division of public and private space, property structure, standard ways properties have been exploited, connections between private and public space, orientation of buildings and orientation of openings and windows [15]. The silhouette of North American cities seems to mimic the land value gradient, and this feature is increasingly present at the global level as well.

The overall environmental features indicate human measures at various levels, but it also reflects the economic processes behind, that is, how the environment is produced in any particular place. The concurrent driving force behind development are the foreseen revenues, which correspond to profitability of land use. Hence, concentration of building rights is not primarily a question of functional considerations. At the city level, crammed centres and high buildings do not necessarily save land, nor function well, but intensive exploitation is first and foremost a question of realising potential property values. The simple reason is that investors seek profits and high-rise is a standard response, despite the fact that dwellers and many sections of the economy (tourism, heritage) would suffer. In this rigmarole, different categories of property owners are treated unequally and lesser neighbours may have their properties devaluated by godfathers.

The big shift took place when city building was subsumed under speculation. The regulation of Paris under Haussman in the 1850s and 1860s is a classic example. There, the biggest body of Medieval buildings in Europe was torn down in the name of increased density and progress, while the factual effects meant reduced overall density and redistributed opportunities for speculation [16]. This idea was further attempted by Corbusier´s plan for Paris, which implied high-rise colossuses and a new round of total destruction [17]. Fortunately for Paris, the plan was not realised, but the logic of urban development was formulated.

The paradigmatic shifted, from seeing urban environment as a stable resource for sustainable life of the many, to viewing it as destruction in order to facilitate for speculation by the few. The switch included monopolisation of land ownership, construction and finance. The altered building and planning codes of Sweden is an instructive example. The law of 1874 (Byggnadsstadga) stipulated the foreseen shape of built-up areas and included mandatory plot division, maximum number of storeys and defined architectural form. The code of 1987 (Plan-och Bygglag) did not address the foreseen result, but only the process and listed stakeholders. Consequently, the already strongly monopolised building sector reached an even more dominant position. Municipalities were not allowed to make plans without projects being initiated - mostly by the biggest developers [18].

The paradigm shift of 19th century Paris was factually codified in Swedish legislation under the pretext of collaborative planning. The general public was brought to believe that further monopolisation would provide for enhanced public participation. The overall effect of constant rebuilding and intensified exploitation is perpetual destruction of historical structures. The heritage sector has reacted already since the time of Haussman to this havoc by focusing conservation measures on "monuments" [19]. Lost is thus the sense of urban morphology as a valuable asset, destroyed in the name of progress, ideologically backed by historicism.

4 Heuristics

Even in culture-minded Italy, many wonderful ancient cities have been tainted by out-of-place financial palaces at their centres. In a series of fusions, Banco di Santo Spirito ended up in UniCredit, now one of Europe's biggest banks. From the 1960s till now, the process has been contaminated with corruption charges, political scandals and murder charges, and a continuous restructuring of global finance as a corollary to global concentration of wealth. How come that the Holy See ended up in a moral swamp?

The situation is not new, however. In 1750, Pope Benedict XIV, known for his condemnation of usury, reorganised the bank and restricted its lending activities. However, in 1786 the bank was one of the first to issue paper money [20]. Nowadays we would call it "fiat money" that derives its solvency mainly from an issuer´s credibility. In 1923, the bank was made a joint stock company, and in 1935, the fascist government gained a controlling interest in the bank, which reflects the keen relations between politics and banking. Altering Samuel Butler's previously reiterated remark: for politicians, banking is a means for reproducing political power, for bankers, politics recreate financial assets!

Where to find moral credibility in society? In order to conceptualise further, we may engage in the use of the so-called semiotic square. This is a tool to use in analysing concepts. S1 indicates a positive concept (compassion), contrary to S2 (greed). S1-S2 is called the complex axis S. On the neutral axis -S underneath, we find the negations -S2 (not-greed) and -S1 (not-compassion), which contradict S2 respective S1. The lower level implies the upper level, which means the upper level is a part (hyponym) of the lower one (hypernym). The choice of S2 constitutes the creative act of the scheme [21] (Fig. 4).

Why to choose "compassion" and "greed" for further analysis? Compassion is a good equivalent for the essence of the Christian faith as expressed in the Sermon of the Mount. Greed seems to indicate the core of Adam Smith´s idea that personal profit is required to maximise economic yield, which will ultimately benefit all. The two views seem to contradict one another, and Pope Benedict XIV´s condemnation of usury does not contradict that notion.

What could not-compassion and not-greed stand for? As the -S level indicates a hypernym of the S level, it stands for a not-personified entity, and represents possible

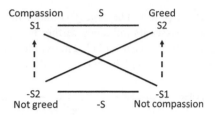

Fig. 4. The semiotic square

societal manifestations of the two negations. Let us consider two historical illustrations addressing not-compassion and not-greed.

Adam Smith's social idea of maximised utility boils down to greedy competition. According to Smith, a necessary precondition is the existence of free markets, unhampered by monopolies [22]. But control of landed resources, and land ownership in particular, always implies the presence of monopoly and thus market restrictions. Two and half centuries after Smith, age-old feudal structures are still present, and the Royal Family continues to be a major land holder in Britain.

In effect, land-use planning does not gear the system towards a more market-oriented system. Planning in Britain and elsewhere seems to be apt to establishing landed monopolies that transcend the economy [23]. Monopolisation is a signum of all sectors of the economy, in particular those that relate to land use. In advanced economies, the Smithian criterion has never been present and therefore it is doubtful whether his view of maximised overall utility - turning personal greed to overall not-greed - is valid at all. Not-compassion surely indicates an economic system based on greed, but till now the perpetual concentration of wealth suggests that utility serves the greedy few, not the greedy many.

Thomas More sketched in his Utopia of the early 16th century a society based on equal possibilities and standardised life forms among its citizens [24]. His idea was to eliminate the possibility of greed by creating a society of equal rights and opportunities. Greed would have been extinct since there would have been nothing to gain from being greedy, rather the contrary. More even included standardised housing and town planning into his regulations. His Utopia depicts a version of an ideologically liberal society, with strict rules for reaching maximum moral utility. The root of his views is to be found in the Christian gospels of love and compassion. More was decapitated by King Henry VIII, who killed the Utopian and usurped the fortunes of the Catholic church. More professed a society of non-greed, but became himself a victim of greed.

5   Narrative

Structural and semiotic traditions have concluded that stories comprise a standard set of 6 facets, called actants. The actant model could provide a tool for understanding the syntax and paradigms of changing worldviews, and thereby contribute to a better understanding of our predicament as urban professionals.

In any narrative, the subject yearns for an (often abstract) object. This is labelled the axis of desire. The subject is promoted by a helper, while an opponent acts against those two. This is called the axis of power. The object is under the influence of the sender who initiates the action, and passes its influence further to a receiver who profits from the action. This is the axis of transmission [21]. The whole setup seems to include the notion of action as the axis of power, and the context as the axis of transmission (Fig. 5).

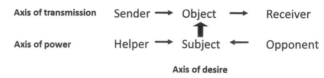

Fig. 5. The actant model

Consider four narratives that may be labelled Christianity, Utopia, Classical Economy and Uchronia:

(1) Traditional Christianity places the Devote as the subject in centre, striving for the Salvation of her/his soul. The sender is God, and Mankind is the receiver. Spirituality serves as the helper and Matter as the opponent.

(2) In More's Utopia, the subject is the Community, yearning for Fairness. The sender is Place as Utopia equals a place that is still to be found, and the Receiver is Mankind. The ultimately good is to be found in the temporal world, not in Christian eternity. The primacy of place reflects the early 16th century period of global exploration. Spirituality as helper is now backed up by mundane Rationality, but Matter is still the opponent. The ethical base is still Christian compassion, but discussed in terms of communal organisation and human rights in our temporal world.

(3) In Classical Economy as sketched by Adam Smith, the subject is the Entrepreneur who strives for enhanced Productivity by rationalising the production and thereby improving productivity of human labour. The sender is Time that requires being optimised, and the receiver is Mankind. The helper are unrestricted Markets and the opponent are the Monopolies.

(4) In Uchronia, the subject is the Investor, pursuing instant Profits. Now, the sender is the Future, where the ultimate good is located. When the time horizon shrinks, Future seems to fuse into Present. Being replaced by continuous present, the future as a common interest fades away. The receivers are those Fortunate who catch the day (*carpe diem*). A remarkable shift from the Classical Economy narrative is that helper and opponent have swapped places. Now Monopoly represents the most profitable condition for investments, and market Competition constitute an opponent to speculative profits.

The four narratives mirror a transformation of factual actants, which gradually turned the traditional world upside down. Perhaps surprisingly, a big shift seems to have occurred between Classical Economy and Uchronia. To understand the change, one must realise the shift in economic thought. In the classical tradition of Smith, Ricardo and Marx, the source of value was thought to be human labour.

The neoclassical revolution, starting in the late 19th century, switched the focus from labour to price. Any realised price was "right" as someone had been willing to pay it. The price of labour followed the same logic, and therefore the price-setting of labour should be free. Unemployment was allegedly caused by those, who did not want to sell their labour cheap enough [25]. Trade unions are seen as monopolies that try to manipulate labour markets to their own advantage. Despite antedating trade unions, employers' associations are seldom viewed in the same way.

The focus on prices indicates the profitability of monopolies, which can manipulate price levels by controlling supply. The most profitable monopolies are those where necessities (communication and transport, energy, water, food, housing, healthcare) are not subdued to competition. This explains why immobile natural monopolies are popular targets for privatisation. In the name of enhanced competition, public monopolies are privatised for the sake of establishing private monopolies.

6 Conclusions

A supreme medium for organising social life may be traced back to the dawn of monetary economy. The most powerful concurrent monopolies are financial institutions, big enough to privatise profits and socialise losses by grabbing public funds and shaking states when needed.

Academic research as well as institutions producing innovations are depending on the benevolence of funders. That does not exclude the option for us to try to understand our predicament. If evidence-based truth suggests that investors run the world, our investigations lead us to follow the money: Sequere pecuniam! - This state of affairs does not prevent us from taking a stand and formulate our own views. We may be deprived of meaningful matters, but not of meanings.

References

1. Richard, K.: Plato. In: Zalta, E.N. (ed.) The Stanford Encyclopedia of Philosophy (Fall 2017 Edition), https://plato.stanford.edu/archives/fall2017/entries/plato/
2. Marc, C.S., Reeve, C.D.C.: Aristotle's metaphysics. In: Zalta, E.N. (ed.). The Stanford Encyclopedia of Philosophy (Winter 2020 Edition). https://plato.stanford.edu/archives/win2020/entries/aristotle-metaphysics/
3. Nickolas, P.: Plato's aesthetics. In: Zalta, E.N. (ed.). The Stanford Encyclopedia of Philosophy (Fall 2020 Edition). https://plato.stanford.edu/archives/fall2020/entries/plato-aesthetics/
4. Hearn, F.: Ideas That Shaped Buildings. The MIT Press, Cambridge, Massachusetts (2003), see also: Hersey, G.L. Architecture and Geometry in the Age of the Baroque. University of Chicago Press, Chicago (2002)
5. Cheng, E.: The Art of Logic – How to Make Sense in a World That Doesn't. Profile Books, London (2019)
6. Ginzburg, C.: Treads and Traces: True, False, Fictive. University of California Press, Berkeley (2012)
7. Bengs, C.: Perspectives in perspective. seeing, visualising, ruling. In: Prezioso, M. (ed.) Capitale Umano e Valore Agguinta Territoriale. Geografia-Economico Politica, vol. 25 pp. 19–49, Aracne editrice, Roma (2018)
8. Kjorup, S.: Semiotik. Studentlitteratur, Lund (2004)
9. Banco di Santo Spirito. Wikipedia. https://en.wikipedia.org/wiki/Banco_di_Santo_Spirito. Accessed 15 Nov 2021

10. Institute for the works of the religion. Wikipedia. https://en.wikipedia.org/wiki/Institute_for_the_Works_of_Religion. Accessed 15 Nov 2021
11. State Church of the Roman empire. Wikipedia. https://en.wikipedia.org/wiki/State_church_of_the_Roman_Empire. Accessed 15 Nov 2021
12. Jaeger, W.W.: The Theology of the Early Greek Philosophers: The Gifford Lectures 1936. Wipf and Stock Publishers, Eugene Oregon (2003)
13. Peck, S.: Planning for Biodiversity: Issues and examples. Island Press (1998)
14. Bengs, C., Schmidt-Thomé, K.: Urban-rural relations in Europe. ESPON 1.1.2, Final report. Centre for Urban and Regional Studies, Helsinki University of Technology, vol. 277 p. + 9 annexes (2005). https://www.espon.eu/programme/projects/espon-2006/thematic-projects/urban-rural-relations-europe
15. Carmona, M., Heath, T., Oc, T., Tiesdell, S.: Public Places – Urban Spaces: The Dimensions of Urban Design. Elsevier, Oxford (2005)
16. Pinkney, D.H.: Napoleon III and the Rebuilding of Paris. Princeton University Press (1972)
17. Le Corbusier: The City of Tomorrow and its Planning. Dover Publications (1987)
18. Bengs, C.: The housing regime of Sweden: Concurrent challenges - Part A: Aims, effects and interpretations. Espoo: Aalto University publication series SCIENCE + TECHNOLOGY 14 (2015). http://urn.fi/URN:ISBN:978-952-60-6474-1
19. Bengs, C.: Pluralitas non est ponenda sine necessitate. In: Bevilacqua, C., et al. (eds.) NMP 2016, Procedia – Social and Behavioural Sciences, vol. 223, pp. 561–567 (2016)
20. Banco di Santo Spirito. Wikipedia. https://en.wikipedia.org/wiki/Banco_di_Santo_Spirito. Accessed 27 Feb 2022
21. Greimas, A.J.: Structural Semantics: An Attempt at a Model. University of Nebraska Press, Lincoln (1984)
22. Smith, A., Sutherland, K. (eds.): An Inquiry into the Nature and Causes of the Wealth of Nations. Oxford University Press, Oxford (2008)
23. Bengs, C., Rönkä, K.: Competition restrictions in housing production – a model for analysis. Econ. Model. 11(2), 125–133 (1994)
24. More, T.: Utopia. Penguin Classics, London (1978)
25. Mazzucato, M.: The Value of Everything. Penguin Random House, UK (2018)

Re-thinking the Resilience Paradigm in Cultural Heritage

Zachary M. Jones[✉] [ID]

Politecnico di Milano, Milan, Italy
zachary.jones@polimi.it

Abstract. Resilience has become an enduring and widespread concept in cultural heritage policy and practice. The term has even been ascribed as a defining attribute of persistent tangible and intangible heritage. Despite its diffused adoption and use, there has been limited reflection or investigation into the concept at a theoretical level. This missing research becomes even more apparent upon examining other disciplines that have evaluated and critiqued the concept more thoroughly – especially the social sciences. This paper takes a first step in examining some of these shortcomings within the cultural heritage field by presenting a trans-disciplinary literature review. The first section covers the handling of resilience within research and practice in cultural heritage. The second section covers some of the key critiques emerging within other disciplines. The third section examines emerging paradigms in cultural heritage and how these can support new reflections on resilience. The paper concludes by identifying and discussing three future research avenues that could further explore new conceptualizations beyond resilience within the cultural heritage sector.

Keywords: Resilience · Critical heritage studies · Conservation policy

1 Resilience in Cultural Heritage

The concept of resilience has by now diffused across disciplinary boundaries – at times labelled a buzzword, fuzzy concept and even so broad as to be considered useless [1]. It is difficult to simply define resilience as one of the main complaints against it is its wide range of usage and meanings. One definition proposed by Walker et al. [2] succinctly summarizes the main attributes as "the capacity of a system to experience shocks while retaining essentially the same function, structure, feedbacks, and therefore identity." This is perhaps the general definition intended in documents like the 2021 European heritage Green Paper which in fact provides no set definition of resilience, but repeatedly refers to resilience as something to enable, enhance and strengthen. Throughout the paper there is an implied assumption of what is intended by resilience and that it is an inherently positive concept. Meanwhile the Shelter Research Project goes to great lengths to define the concept of resilience as well as specific subsets of resilience including cultural, economic, environmental, governance and institutional, social. They define resilience as "The ability of an historic urban or territorial system-and all its social, cultural,

© The Author(s), under exclusive license to Springer Nature Switzerland AG 2022
F. Calabrò et al. (Eds.): NMP 2022, LNNS 482, pp. 2526–2534, 2022.
https://doi.org/10.1007/978-3-031-06825-6_241

economic, environmental dimensions across temporal and spatial scales to maintain or rapidly return to desired functions in the face of a disturbance, to adapt to change, and use it for a systemic transformation to still retain essentially the same function, structure and feedbacks, and therefore identity, that is, the capacity to adapt in order to maintain the same identity" [3]. The researchers recognize the potential to reach for a 'desired' state but the overall focus remains primarily on a return to an idealized pre-existing state. Their resilience approach is heavily defined by a stakeholder and participatory approach that sets it apart from many other resilience-based approaches within cultural heritage. Yet their work remains closely aligned with Disaster Risk Management and reflects the focus on returning to a pre-disaster scenario.

Various other interpretations or approaches of resilience might argue for a greater flexibility or adaptability to the concept, but its origins derive from the ability to absorb and overcome stresses that threaten existing conditions. Perhaps in part due to its wide usage, it has remained an enduring and popular concept capable of capturing the imagination, informing research and driving local community-led initiatives and international policy documents alike. In recent years, resilience has been found to be particularly well suited to the disciplines of cultural heritage and conservation, now used as a defining attribute of tangible and intangible heritage that has endured for centuries or millennia [4]. In the *2021 European Cultural Heritage Green Paper* by Europa Nostra and ICOMOS, resilience features throughout as a key element to be improved or enhanced, reflecting a steady growth of the concept in cultural heritage over the last 20 years in research as well as international documents and charters such as the *2007 Policy Document on the Impacts of Climate Change on World Heritage Properties* and *2015 UN Framework Convention on Climate Change* [5]. Much of this work has been framed in preparation for physical threats from climate change and derives from Disaster Risk Management (DRM) to prepare historic sites, landscapes, and practices against future disasters – fires, floods, earthquakes, temperature extremes, etc. Research has till now largely focused on developing various analyses, methodologies, and frameworks to identify vulnerabilities [6]. Despite this profusion, and in contrast to other disciplines, there has been limited reflection or critique of resilience at a theoretical level – generally assumed and accepted as a positive attribute to pursue. Problematically, this ignores the risks of either reinforcing embedded inequalities in the preservation of heritage or developing a kind of counterfeit resilience focused on the symptoms rather than the causes of threats.

For example, much of the current research focuses on developing tools to assist practitioners in identifying cultural heritage most at risk based on a defined set of criteria. For example, Rodríguez-Rosales et al. [7] assess sites in Spain and Cuba for their risk of coastal flooding through a vulnerability matrix that measures the possibility of erosion, pollution, weather fluctuations, extreme temperatures, fire, tourism, etc. Likewise, Wang [8] uses GIS technologies to map cultural sites at the most risk of flooding in Taiwan. Meanwhile, Marchezini et al. [9] discuss the need for a Cultural Heritage Articulated Warning System that would coordinate heritage sector actors with national defense and geological services to improve responses through early warning systems. The Cultural Heritage Risk Index [10] and the Climate Vulnerability Index (CVI) developed by the James Cook University in Australia are two other tools that assess various threats from climate change – including economic, social, and cultural vulnerabilities.

Evolving research has come to recognize the importance of the social aspect of resilience and involving the local communities within these matrices as well as in the decision-making processes themselves [11]. This expansion of thinking on resilience mirrors wider attempts to use Socio-Ecological Systems (SES) approaches to increase the concept's overall sustainability [12, 13] and to position it as a more proactive rather than merely reactive force [14]. While these contributions are very effective at responding to certain disasters, they are less prepared for wider-ranging social or economic shifts from mass migration to war, global economic crashes to pandemics. For example, the COVID-19 global pandemic has accelerated discussions supporting resilience approaches. Yet with such a strong focus on economic recovery, will the eventual response outcomes merely aim at returning to previous levels of mass tourism and pollution that threaten tangible and intangible heritage alike?

Within applied research there remains a clear focus on traditional conservation approaches. The Heritage Research Hub highlights the completed and ongoing cultural heritage research projects funded at the European level within the JPI CH, Horizon 2020 and Horizon EU schemes. Amongst the past and ongoing projects, four deal specifically with the concept of resilience: ARCH; HYPERION; RESCULT; SHELTER. Combined, these projects represent 13.7 million euros of funding promoting and advancing the resilience concept. All four projects share a common focus on DRM to primarily investigate the physical vulnerabilities of built heritage, developing new approaches, technologies and frameworks to better predict and respond to a range of climate-related disasters. Both the ARCH and SHELTER project do involve local communities to address aspects of social resilience in preparing and responding to disaster-related threats. The findings and outputs of these projects represent valuable resources for historic cities worldwide facing ever increasing emerging climate-related disasters. Current and upcoming Horizon EU Calls will extend this approach with several Research and Innovation Actions on cultural heritage devoted to the effects of climate change, sustainability and resilience. This ongoing research within cultural heritage and conservation is of course important and necessary to respond to the physical vulnerability of heritage sites and a range of other issues as they tackle immediate and emergency challenges. This research and work is vital, but there remains the need to investigate alternative approaches and new concepts that may be more appropriate in certain situations. If resilience continues to play a key role in guiding future funding, research, policy and practice, critical reflections on its conceptualization and use are urgently needed to ensure that harmful policies or practices are not needlessly duplicated and spread. This contribution aims at highlighting and to a degree problematizing the current usage of the resilience concept within the cultural heritage field. Drawing on critiques from other fields provides the opportunity to identify gaps within the existing discussion and propose future research tracks to address these issues.

2 Resilience Critiques in Other Fields

Outside of the cultural heritage sector, the critique of resilience frameworks is more extensive than currently present in the field. One initial issue pertains to the practicalities and useability of resilience-based research as most methodologies lack the depth to capture all possible aspects of resilience, often with social considerations left out [15]. Practitioners in the field may also struggle to collect the necessary data to gauge or determine resilience according to frameworks developed by researchers [16]. Yet the main critique from scholars in the social sciences regard the aim of resilience to return to a previously existing status quo – often with the assumption that this previously existing condition is an ideal or preferred one [17]. Such a position fails to recognize potential inefficiencies, inflexibility, or injustices that may exist within present Socio-Ecological Systems, particularly under capitalist and neoliberalist policies [18, 19]. Mackinnon and Derickson [20] argue that, within social theory, resilience tends towards conservative rather than progressive policies that fail to challenge unjust power structures, shift responsibilities from the government onto the shoulders of local communities and struggle to encompass issues transcending local, regional, or global scales. Tierney [21] posits that resilience inherently upholds neoliberalism and can actively prevent some social groups from improving their socio-economic conditions.

Other scholars instead propose that resilience-based strategies do in fact have the capability to affect transformative socio-economic change depending on the criteria used [1, 22]. Evolutionary Resilience has been proposed in urban and regional planning that conceptualizes the ability to bounce forward and enhance performance of complex Socio-Ecological Systems [23, 24]. Matin et al. [12] define equitable resilience by taking into account social vulnerability and differential access to power, knowledge, and resources within their approach. Yet others propose entirely new concepts to replace resilience such as robustness [25] or resourcefulness [20] to develop more precise policies and strategies that can both weather threats from future disasters as well as grow beyond current Socio-Ecological Systems when and where needed. While these new concepts propose stretching the boundaries of typical definitions of resilience, they don't reflect current understanding or approaches within cultural heritage and represent an important starting point from which to re-evaluate the concept.

3 Emerging Heritage Paradigms

In practice, we can observe a willingness on the part of cities and decision makers to seek out new ways to experiment with their built heritage and its engagement with cultural policy [26] as well as exploring wider territorial development schemes [27]. These practices reveal a rethinking of the role of cultural heritage that work in part towards implementing a Historic Urban Landscape approach [28]. However, there have been emerging calls from scholars for even more severe systematic change. Holtorf [29] suggests adopting a cultural resilience approach that goes beyond the perpetual 'heritage at risk' framework that can absorb disturbance and acknowledges the inevitable loss of physical heritage as we know it today. Seekamp and Jo [30] echo these sentiments by introducing a new interpretation of resilience which can accept loss and change rather than a strict conformity to preserve all threatened heritage sites at all cost. They propose an alternative heritage policy based on transformative continuity by establishing a new listing of World Heritage Sites in Climatic Transformation. These discussions mark an important point of departure in the thinking on resilience within cultural heritage, but as seen in the discourses from other disciplines, there remains significant space to analyze and critique resilience at a theoretical level.

Without necessarily questioning the resilience paradigm, the push for change and transformation within conservation and cultural heritage remains primarily focused on physical aspects. The discourse has been framed by the question of whether to conserve or not, exploring potential alternatives to caring and valuing heritage beyond conservation [31]. In particular, the Heritage Futures Research Programme investigates these approaches by embracing concepts of uncertainty and transformation that are often overlooked or ignored by cultural heritage policies [32]. The researchers argue that current practice is ultimately unsustainable and that by learning to let go of some sites or memorialize them through loss can build greater resilience – taking an opposing perspective from most established heritage actors. With a focus on physical conservation, many of the examples promoted present an idyllic image of the slow decay of sites [31], reminiscent of Winckelmann. However, these arguments may not apply to the many situations where built heritage is either intentionally demolished for developmental purposes or intangible heritage disappearing as a result of shifting socio-economic systems.

Future theoretical explorations into the resilience concept could go even further, leading to new heritage reflections and paradigms. For example, MacKinnon and Derickson's [20] stipulation that resilience can reinforce and prop up capitalist and neoliberal systems could explain instances where restrictive physical conservation regulations ultimately lead to the gentrification and privatization of heritage structures and spaces – such as in the example of the restoration of the Battersea Power station in London [33]. The long-awaited restoration was ultimately used as a tool by developers to deflect criticism against property speculation and development aims through the creation of new narratives and privatized spaces [34]. From a resilience-critical perspective, such actions could be reframed as merely keeping in place and supporting existing socio-ecological systems that privilege certain communities over others. Additional examples include instances of the complete loss of historic structures due to competing economic incentives. In Istanbul, historic structures have been intentionally destroyed to convert the areas into parking spaces which are deemed more profitable [35]. While scholars have already widely critiqued such practices, they have not yet been framed in relation to resilience discourses. In these examples the links with resilience may not be immediately evident, but they suggest the possibility that the intensive focus on the aspects of physical resilience of built heritage at the expense of social or cultural concern can lead to heritage becoming a burden rather than a resource [36]. Even efforts to support community involvement and public participation that have been heavily promoted for the last twenty years within the conservation movement [37–39] are limited in their focus and aims. Generally, these calls do not go as far as to question embedded decision-making powers or local agency in the face of national or global actors. A more reflective or critical view on resilience could reframe the goals of such initiatives to go beyond consensus building and work towards true systems of cooperative governance or co-cities [40]. By no means exhaustive, these few examples illustrate some of the possible issues to be explored through new reflections on resilience.

4 Discussion: Exploring New Research Avenues

By reflecting on current usage in research and practice within cultural heritage and considering some of the objections from other fields, this paper has taken a first step in pointing out the need for more in-depth theoretical reflections on the resilience concept. To conclude, this paper calls for new research avenues to be developed. Deriving from the review of existing literature and scholarship across sectors, three key issues emerge that future cultural heritage research can address:

1) Lack of any conceptual or theoretical level critique of the resilience concept represents a significant oversight leaving a blind spot in research, policy and practice. Namely, the role of existing policies and practice in upholding unequal and unjust Socio-Ecological Systems.
2) Differentiating between DRM and more widely branded 'resilience' approaches. This separation will support those efforts while shifting the rest of the field past a permanent disaster mindset that prioritizes physical conservation above all other considerations.
3) Little counter current research that questions the main upheld paradigms of the cultural heritage field. This theoretical reflection and critique will support and link emerging approaches questioning long-accepted convention and practice.

The first and main aim of this paper is to call for more in-depth investigation into the theoretical concept of resilience within cultural heritage in order to address some of the gaps identified in this paper. This call for new research should explore if the existing resilience concept within the cultural heritage field is too narrowly conceived and represents a limiting rather than a transformative force. Findings from this work may challenge some commonly held and accepted views within the discipline by starting from a point of critique rather than a blanket acceptance of resilience as is often the case. Such research would significantly add to the ongoing discussion on resilience and will ultimately provide a theoretical alternative on which to develop future approaches and policy. This call to critically analyze the resilience concept within cultural heritage, represents a necessary step in reassessing the use of the term within the discipline. This approach avoids the common pitfall of assuming resilience as a universal good [17] and recognizes the potential for undesirable or unequal conditions to exist within or be promoted by resilience [41] while also contributing to the widely agreed upon need for greater clarity [4] pertaining to the term resilience. Research should explore the congruity or divergence in how the concept is defined and measured across the cultural heritage sector to identify existing bias in how the concept is being activated as well as gaps in current practice. Such research would provide a more precise conceptual picture of resilience within the field.

This research could lead to the development of new theoretical frameworks and form the basis for new policies and approaches within cultural heritage, supporting emerging calls for change in the field. Other recent approaches have taken an incremental approach in an attempt to rehabilitate the concept of resilience [1] by making it adaptive [42] or community focused [11], yet without the critical analysis of the base concept of resilience these updated frameworks fail to collectively address a broader range of embedded issues. Disaster Risk Management approaches respond to very specific needs of specific contexts. There will of course remain a need for the ongoing applied work relating to DRM and cultural heritage, but the development of this new theoretical concept will be better prepared to handle a wider set of issues pertaining to both existing and future Socio-Ecological Systems. Unlinking more general resilience thinking and practice from DRM would generate theories more capable of responding to calls for greater transformative adaptability, either in response to changing Socio-Ecological systems or the ability to go beyond those systems when they become stagnant and restrictive. In contrast to existing resilience frameworks and approaches that primarily target sites of disasters, new 'post-resilience' theories will be more elastic and prepared for systematic changes like those posed by COVID-19 as well as other threats, natural evolutions, or societal and cultural changes.

Expanding research will bring new knowledge pertaining to the existing understanding and usage of the resilience concept in the cultural heritage field and to go beyond it. While some recent scholarship has explored possible post-neoliberal futures, there is a general agreement that one iteration cannot be predicted or defined but rather multiple variations may exist simultaneously [43]. The development of a new theoretical framework or frameworks can better position the cultural heritage field to be more robust and respond to a range of unexpected changes across the Socio-Ecological system. These new research avenues would ultimately aim at developing more forward-looking approaches based on continuous improvement rather than a fixed dedication to previous conditions.

References

1. Davidson, J.L., et al.: Interrogating resilience: toward a typology to improve its operationalization. Ecol. Soc. **21**(2), 15 (2016)
2. Walker, B., Gunderson, L., Kinzig, A., Folke, C., Carpenter, S., Schultz, L.: A handful of heuristics and some propositions for understanding resilience in social-ecological systems. Ecol. Soc. **11**(1) (2006)
3. Shelter Project: D.1.2. Building of best/next practices observatory. Sustainable Historic Environments hoListic reconstruction through Technological Enhancement and community-based Resilience (2019)
4. Conolly, J., Lane, P.J.: Vulnerability, risk, resilience: an introduction. World Archaeol. **50**, 547–553 (2018)
5. Fatorić, S., Seekamp, E.: Are cultural heritage and resources threatened by climate change? System literature review. Climatic Change **142**(1–2), 227–254 (2017)
6. Ramalhinho, A.R., Macedo, M.F.: Cultural heritage risk analysis models: an overview. Int. J. Conserv. Sci. **10**(1), 39–58 (2019)
7. Rodríguez-Rosales, B., et al.: Risk and vulnerability assessment in coastal environments applied to heritage buildings in Havana (Cuba) and Cadiz (Spain). Sci. Total Environ. **750**, 141617 (2021)
8. Wang, J.-J.: Flood risk maps to cultural heritage: measures and process. J. Cult. Herit. **16**(2), 210–220 (2015)
9. Marchezini, V., Iwama, A.Y., Pereira, D.C., da Conceição, R.S., Trajber, R., Olivato, D.: Designing a Cultural Heritage Articulated Warning System (CHAWS) strategy to improve disaster risk preparedness in Brazil. Disaster Prevent. Manage. Int. J. **29**(1), 65–85 (2020)

10. Forino, G., MacKee, J., von Meding, J.: A proposed assessment index for climate change-related risk for cultural heritage protection in Newcastle (Australia). Int. J. Disaster Risk Reduct. **19**, 235–248 (2016)

11. Fabbricatti, K., Boissenin, L., Citoni, M.: Heritage community resilience: towards new approaches for urban resilience and sustainability. City, Territory Architect. **7**(1), 1–20 (2020). https://doi.org/10.1186/s40410-020-00126-7

12. Matin, N., Forrester, J., Ensor, J.: What is equitable resilience? World Dev. **109**, 197–205 (2018)

13. Partelow, S.: Coevolving Ostrom's social–ecological systems (SES) framework and sustainability science: four key co-benefits. Sustain. Sci. **11**(3), 399–410 (2015). https://doi.org/10.1007/s11625-015-0351-3

14. Sesana, E., Bertolin, C., Loli, A., Gagnon, A.S., Hughes, J., Leissner, J.: Increasing the resilience of cultural heritage to climate change through the application of a learning strategy. In: Moropoulou, A., Korres, M., Georgopoulos, A., Spyrakos, C., Mouzakis, C. (eds.) TMM_CH 2018. CCIS, vol. 961, pp. 402–423. Springer, Cham (2019). https://doi.org/10.1007/978-3-030-12957-6_29

15. Stojanovic, T., et al.: The "social" aspect of social-ecological systems: a critique of analytical frameworks and findings from a multisite study of coastal sustainability. Ecol. Soc. **21**(3), 20 (2016)

16. Béné, C., Chowdhury, F.S., Rashid, M., Dhali, S.A., Jahan, F.: Squaring the circle: reconciling the need for rigor with the reality on the ground in resilience impact assessment. World Dev. **97**, 212–231 (2017)

17. Cretney, R.: Resilience for whom? Emerging critical geographies of socio-ecological resilience. Geogr. Compass **8**(9), 627–640 (2014)

18. Olsson, L., Jerneck, A., Thoren, H., Persson, J., O'Byrne, D.: Why resilience is unappealing to social science: theoretical and empirical investigations of the scientific use of resilience. Sci. Adv. **1**(4) (2015)

19. Tanner, T., Bahadur, A., Moench, M.: Challenges for Resilience Policy and Practice. Overseas Development Institute, London (2017)

20. MacKinnon, D., Derickson, K.D.: From resilience to resourcefulness: a critique of resilience policy and activism. Prog. Hum. Geogr. **37**(2), 253–270 (2013)

21. Tierney, K.: Resilience and the neoliberal project: discourses, critiques, practices—and Katrina. Am. Behav. Sci. **59**(10), 1327–1342 (2015)

22. Boschma, R.: Towards an evolutionary perspective on regional resilience. Reg. Stud. **49**(5), 733–751 (2015)

23. Rega, C., Bonifazi, A.: The rise of resilience in spatial planning: a journey through disciplinary boundaries and contested practices. Sustainability **12**(7277), 18 (2020)

24. Yamagata, Y., Sharifi, A. (eds.): Resilience-Oriented Urban Planning: Theoretical and Empirical Insights. Cham: Springer (2018)

25. Capano, G., Woo, J.J.: Resilience and robustness in policy design: a critical appraisal. Policy Sci. **50**(3), 399–426 (2017)

26. Jones, Z.M.: Cultural Mega-Events: Opportunities and Risks for Heritage Cities. Routledge, Abingdon (2020)

27. Tricarico, L., Jones, Z.M., Daldanise, G.: Platform spaces: when culture and the arts intersect territorial development and social innovation, a view from the Italian context. J. Urban Affairs 1–22 (2020)

28. Bandarin, F., Van Oers, R. (eds.): Reconnecting the City: The Historic Urban Landscape Approach and the Future of Urban Heritage. Wiley (2015)

29. Holtorf, C.: Embracing change: how cultural resilience is increased through cultural heritage. World Archaeol. **50**(4), 639–650 (2018)

30. Seekamp, E., Jo, E.: Resilience and transformation of heritage sites to accommodate for loss and learning in a changing climate. Clim. Change **162**(1), 41–55 (2020). https://doi.org/10.1007/s10584-020-02812-4

31. DeSilvey, C.: Curated Decay: Heritage Beyond Saving. University of Minnesota Press, Minneapolis (2017)

32. Harrison, R., et al.: Heritage Futures: Comparative Approaches to Natural and Cultural Heritage Practices. UCL press, London (2020)

33. Chandler, A., Pace, M.: The Production of Heritage: The Politicisation of Architectural Conservation. Routledge (2020)

34. Vijay, A.: Dissipating the political: battersea power station and the temporal aesthetics of development. Open Cultural Stud. **2**(1), 611–625 (2018)

35. Aygen, Z.: International Heritage and Historic Building Conservation: Saving the World's Past. Routledge, New York (2013)

36. de Clippele, M.-S.: Protecting cultural heritage: whose burden? In: Second All Art and Cultural Heritage Law Conference. Université de Genève (2016)

37. De Cesari, C., Dimova, R.: Heritage, gentrification, participation: remaking urban landscapes in the name of culture and historic preservation. Int. J. Herit. Stud. **25**(9), 863–869 (2019)

38. Neal, C.: Heritage and participation. In: Waterton, E., Watson, S. (eds.) The Palgrave Handbook of Contemporary Heritage Research, pp. 346–365. Palgrave Macmillan UK, London (2015). https://doi.org/10.1057/9781137293565_22

39. Pendlebury, J., Townshend, T.: The conservation of historic areas and public participation. J. Archit. Conserv. **5**(2), 72–87 (1999)

40. Iaione, C.: The CO-city: sharing, collaborating, cooperating, and commoning in the city. Am. J. Econ. Sociol. **75**(2), 415–455 (2016)

41. Walker, B., Salt, D.: Resilience Practice: Building Capacity to Absorb Disturbance and Maintain Function. Island press (2012)

42. Bui, H.T., Jones, T.E., Weaver, D.B., Le, A.: The adaptive resilience of living cultural heritage in a tourism destination. J. Sustain. Tour. **28**(7), 1022–1040 (2020)

43. Davies, W., Gane, N.: Post-neoliberalism? an introduction. Theory, Culture Soc. **26**, 3–28 (2021)

Subsidence Monitoring in the Duomo di Milano: Half a Century of Measuring Activities

Luigi Barazzetti[1(✉)], Mattia Previtali[1], and Fabio Roncoroni[2]

[1] Department of Architecture, Built Environment and Construction Engineering, Politecnico di Milano, Piazza Leonardo da Vinci 32, Milan, Italy
{luigi.barazzetti,mattia.previtali}@polimi.it
[2] Polo territoriale di Lecco, via Previati 1/c, Lecco, Italy
fabio.roncoroni@polimi.it

Abstract. Constant monitoring activities of vertical settlement in the Duomo di Milano started during the '60 when the rapid changes of the water table seriously affected the stability of the Cathedral. This paper describes the origin and evolution of the monitoring system, which is still operative thanks to continuous maintenance and technological advancements in measurement and processing methods. Nowadays, measurements are still acquired twice a year as a part of the traditional monitoring work, which is based on several other systems, including automatic and manual (mechanical) sensors and tools. A digital archive with more than 50 years of data is constantly updated, continuing a tradition that followed major restoration interventions and ordinary maintenance carried out by Veneranda Fabrica del Duomo di Milano (VFD). After more than half a century, the collaboration between VFD and Politecnico di Milano to monitor the Duomo is still active.

Keywords: Digital archive · Monitoring · Subsidence · Time series

1 Introduction

1.1 Overview

The origin and evolution of the monitoring system able to capture relative vertical movements of the pillars of the Duomo di Milano are strictly correlated to the restoration of the four central pillars of the tiburium, which was carried out between 1981 and 1984, after restoring 21 pillars, and ten additional years of experimental investigation (Fig. 1).

Carlo Ferrari da Passano (Architect of Veneranda Fabrica del Duomo di Milano, the institution founded in 1387 in charge of the completion and conservation of the Duomo) describes the causes of static instability of the Duomo, which [1] can be summarized as follows:

© The Author(s), under exclusive license to Springer Nature Switzerland AG 2022
F. Calabrò et al. (Eds.): NMP 2022, LNNS 482, pp. 2535–2544, 2022.
https://doi.org/10.1007/978-3-031-06825-6_242

- historical and constructive aspects, among which the size and constructive methods of foundations and columns, the construction of the round arches, the main spire, and the three smaller spires;
- "recent" causes, i.e., vibrations because of the vehicular traffic around the Duomo (on three sides, North, East, and South) and the subway line M1 (North), and extraction of water from the subsoil and consequent rapid variations in the level of the water table.

However, the worsening of static conditions was clear since the beginning of the 1960s, when Ferrari da Passano was hired as a technical director by the administration of the Fabrica (in 1961). He started continuous inspections and new monitoring measurements. A series of safety measures were taken to face changes in the static condition, including the construction of temporary reinforcement systems of the pillars. The numerous details are out of the scope of this manuscript, and the reader is referred to [1].

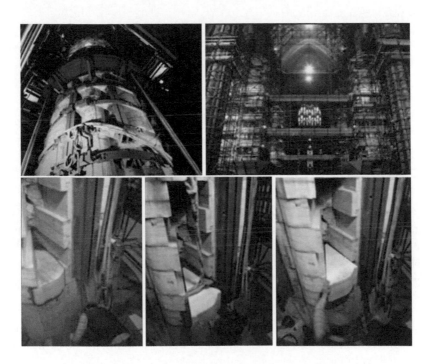

Fig. 1. Some images of the restoration of the pillars.

Among the causes previously described, the extraction of water from the subsoil and the consequent lowering of the water table changed a secular equilibrium, triggering subsidence processes and differential movements of the pillars. Shown in Fig. 2 is the variation of the water table level in Milan. As can be seen, rapid changes occurred in 1950–1970. A report by Croce [7] shows a constant water table level measured in the period 1900–1920 (provided by Ufficio Tecnico Servizio Acquedotto Comune di Milano). The level constantly changed in the period 1920–1957/58, notwithstanding it was relatively slow with an estimated average speed of about 0.2 m/year. The variation became rapid in 1957/58–1970 with a speed of about 1.5 m/year.

In 1972, it was decided to close the traffic around the cathedral and block water extraction activities from the ground, closing all the wells in the Cerchia dei Navigli. These measures were beneficial and led to the inversion of the water table level curves already starting from 1973.

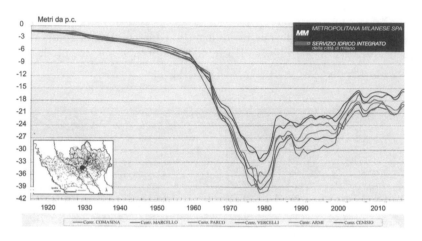

Fig. 2. Variation of the water table level in Milan (in meters), figure retrieved in [2].

After briefly reviewing the actual configuration of the monitoring system of the Cathedral of Milan, the paper will illustrate and discuss the origin and evolution of the historic system used to track vertical movements, which is still operative after more than half a century.

1.2   The Actual Monitoring System of the Duomo di Milano

Although this paper mainly illustrates and discusses the method used to monitor vertical differential movements, the (actual) whole system can be synthetically summarized as follows:

- monitoring of vertical displacement of pillars and structures around the Duomo (i.e., the system described in this paper);
- tilt monitoring system for 32 pillars based on the plumb-bob principle. Measurements with ±0.1 mm precision are repeated every six months;
- tilt monitoring system of the facade and the apse based on a vertical plummet on a micro-metric sled. Measurements with ±0.2 mm precision are repeated every six months;

- tilt monitoring system of the main spire based on the plumb-bob principle. The wire is connected on top of the main spire, just below the Madonnina. Measurements are taken at three different levels: two on the main spire and one inside the church;
- continuous tilt monitoring system with an automatic plumb-bob tool installed on the main spire;
- measurement of crack variations;
- convergence measurement the tiburium carried out with LVDTs with an additional manual comparator with a precision of ±0.01mm;
- high precision geometric leveling network around the main spire;
- differential GNSS based on three receivers to measure relative movements between the main spire and the scaffolding during restoration work: on top of the main spire, on the scaffolding, and on the terraces used as a reference (this system is no longer active at the moment for the progressive removal of the scaffolding);
- continuous static monitoring system based on bi-axial tiltmeters and vibrating wire extensometers, coupled with hygrometers and temperature sensors (inside the church);
- continuous dynamic monitoring system based on mono- and bi-axial seismometers (inside the church and on the main spire);
- a weather station installed on the highest accessible level of the main spire;
- manual tilt monitoring system for the bell tower of the church of San Gottardo, based on both the plumb-bob principle (inside) and the vertical plummet (outside);
- continuous tilt monitoring system with an automatic plumb-bob sensor installed inside the bell tower of the church of San Gottardo.

Therefore, the monitoring system of the cathedral includes several other sensors and tools, including automatic and manual sensors and digital or optical/mechanical devices. Monitoring data are provided in real-time or with a lower acquisition frequency. More details about these additional sensors and the achievable information are described in [3–6]. Moreover, other sensors installed in the past were replaced and are no longer active.

The following section describes the first system installed in '60, which is today still operative with several enhancements carried out over time.

2   Monitoring Vertical Displacements Induced by Subsidence

2.1   Origin of the System

Croce [7] describes the beginning of the measurements able to detect vertical displacements between a set of pillars of the Duomo. First measurements were carried out in 1961, revealing an average differential variation between pillars 28 (towards the facade, North-West corner) and 74 (one of the tiburium pillars) of 2.1 mm/year in the period June 1961–May 1970. The overall differential movement detected is 17.8 mm.

Figure 3 shows the initial scheme of the monitoring system, which was installed in collaboration between Veneranda Fabrica del Duomo di Milano and the Institute of Surveying, Photogrammetry, and Geophysics of Politecnico di Milano, the Polytechnic University of Milan. Prof. M. Cunietti was the coordinator of the measurement activities carried out using high-precision geometric leveling. Measurements continued under the supervision of Prof. A. Giussani, who collaborated with Cunietti for several years. The authors still acquired measurements twice a year after collaborating for about 20 years with Giussani. In other words, the monitoring process is also a multi-decade tradition relying on the transfer of expertise among specialists.

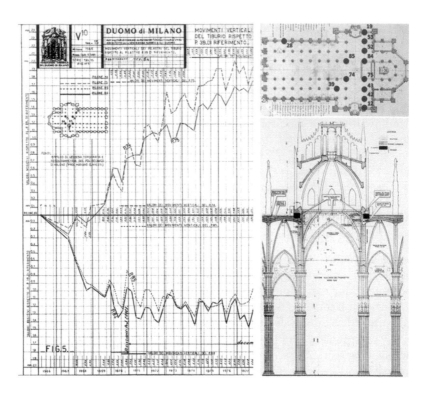

Fig. 3. Graph with some differential movements between the 4 pillars of the tiburium. Scheme with original monitoring points and sections along the transept.

First measurements were able to track the movement of the four pillars of the tiburium (n. 74-75-84-85), additional pillars forming a line along the transept (n. 12-42-43-52-53-19), pillar n.28 (towards the facade, N-W side), and pillar n.39, which was assumed as reference for all measurements. Such initial configuration was then extended and enhanced, adding benchmarks on several other pillars and the area outside the Duomo.

Shown in Fig. 3 is also part of a historic graph showing the relative vertical displacement between the four pillars of the tiburium, which highlights an overall variation between pillars 74–75 and 84–85 of about 3 mm in the period 1966–1977.

2.2   Modern Configuration of the System

The actual configuration of the monitoring system is the result of continuous technological advancements that led to the progressive change of instruments and data processing methods. However, all the changes carried out in different years considered the need for a continuous time series of data. In other words, measurements acquired today are still registered in the system of original measurements, offering a continuous archive with more than 50 years of data.

The configuration and location of measuring points are still based on the 1969 setup. Nowadays, the system includes an external leveling network on the structure around the Duomo (such as Palazzo Reale, Galleria Vittorio Emanuele II, the Church of San Gottardo, the main square, Piazza Fontana, and some streets like Delle Ore, Pattari, Agnello, Radegonda, and S. Raffaele).

Additional points are installed on the external structures of the Duomo, forming a closed loop that is connected to an internal leveling network made up of 59 benchmarks installed on the pillars. The general scheme is shown in Fig. 4.

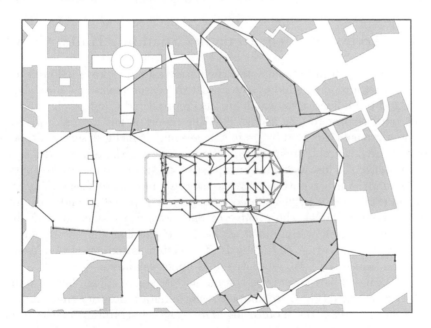

Fig. 4. Actual configuration of both internal and external leveling networks.

Measurements are always carried out twice a year (in May and November), which is a fundamental requirement to obtain data that can be compared (environmental conditions tend to become similar). However, the external network is measured only in May.

The reference point fixed during data processing is the benchmark of pillar n.39, which was always chosen as a reference since the first stages of data acquisition in 1961. Special small leveling rods were constructed in 1969, providing a magnetic connection with the benchmarks on the pillars. Measurements are adjusted via least squares using in-house software developed in MATLAB, obtaining elevation values with a precision of ± 0.1 mm, i.e., the expected precision of a high precision leveling network for structural monitoring.

It is worth mentioning that initial measurements in 1965–1978 inside the cathedral were acquired with a higher frequency (4 seasons). Continuity with first measurements is available till 1969, notwithstanding a subset of benchmarks has continuous measurements till 1965. The authors carried out the creation of the joint digital archive. We could estimate the discontinuity due to the transition between old and new systems for the internal network in a rigorous way thanks to their 'predecessors' who always performed double measurements with both old and new systems.

Continuous maintenance is carried out to ensure posterity of the system. Since measurements are periodically taken twice a year, the measurement phase is also the opportunity for inspection and conservation activities of the system itself, especially for those parts installed on other buildings around the Duomo.

3   The Archive of Measurements Spanning Half a Century

The acquisition of measurements and their processing is always followed by the preparation of a report including the adjusted measurements in tables showing relative and total differences, which are also represented using graphs, and commented with respect to past values [8]. The report is not limited only to the vertical movements captured by geometric leveling. It includes several other systems that are not deeply described in this manuscript (the reader is referred to the next section for just a general introduction of the active monitoring systems of the Duomo). A new report is delivered every six months to Veneranda Fabrica del Duomo staff, and it is then shared among several specialists and consultants.

More recently (2021), the archive of data was also restructured in digital environments that allow simplified access to information and the opportunity for advanced visualization and analysis. The system is composed of a Web-GIS application, which is already operative and under validation.

The GIS-based part covers the Duomo and its surrounding. The system has a Web-GIS interface (Fig. 5) relying on a single data repository that allows different users to access the monitoring data. The configuration of the leveling network and the availability of multi-temporal data makes a Web-GIS ideal to store and represent such information, which is turned into a geospatial database.

The system relies on different open-source libraries for the different components: (i) geographical server (Geoserver), (ii) data visualization, rendering and interaction (OpenLayers), and (iii) graph generation (Plotly).

A typical use of the system is on-site data retrieval (e.g., using a tablet) and visualization of time series of specific benchmarks in specific periods. For this case, the user can log in to the system, access the main page and select the points on the pillars, or trace a polygon on the map. The user can visualize the data in three ways:

- looking at the tabular data;
- plotting the data with a time series scatterplot;
- computing annual or semestral variations and plotting charts.

All the graphs can be dynamically updated, selecting the period for the analysis.

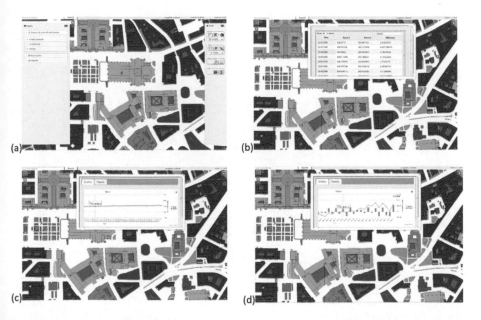

Fig. 5. Web-GIS interface (a) and data visualization: tabular data (b), general displacement data graph (c) and annual/semestral variation graph (d).

The creation of the digital archive also had to face some limits related to missing data and the transition between the different systems and the presence of outliers. For this reason, the different time series were also imported processed using software for time series analysis. Combing the data restructured in a GIS environment allows the creation of additional products that can simplify data interpretation, remotely and on-site.

An example is shown in (Fig. 6), which illustrates the differential movements (in mm) from 1970 to 2020 using a surface-based representation achieved through rational basis functions, which can be obtained with a geospatial interpolation using cartographic $(East, North)$ coordinates and displacement values. The graphs are related to two orthogonal lines of pillars in the West-East (facade-apse) and North-South (transept) directions. As mentioned, the historical reference is pillar n. 39 and variations must always be intended as relative values. The most significant negative displacements are located in the transept area, close to pillar n. 74, i.e. one of the pillars of the tiburium. The curves indicate that movements were significantly more rapid in 1970–1980/85.

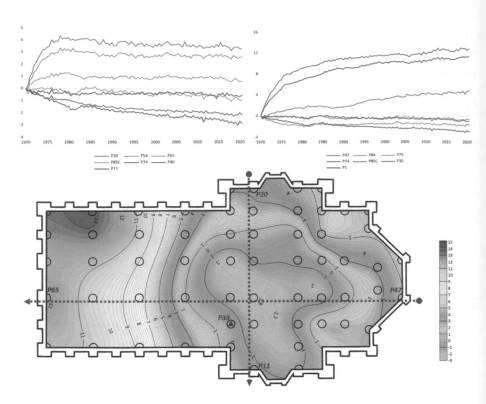

Fig. 6. Interpolated relative displacement map and curves for some pillars on orthogonal directions. Values are reported in mm. The considered period is 1970–2020.

4   Conclusions

The paper briefly presented the monitoring system of vertical movements of the Duomo di Milano, which was installed during the '60 for the settlements caused by sudden variations of the water table level. The system is still operative and was renovated and integrated during more than half a century. Particular attention has always been paid to preserving continuity, guaranteeing the posterity of such a precious archive of data, which more recently was digitized to provide facilitated access with innovative visualization solutions.

Interpretation of the data remains a complicated task in monitoring applications. It requires solutions able to assist the work of the specialists as wells as for the prevention of structural risk.

Conservation activities in the cathedral require daily maintenance and major interventions carried out by the Veneranda Fabrica del Duomo di Milano staff. The collaboration related to monitoring activities started in the '60 between VFD and Politecnico di Milano is still active and will continue using both traditional measurements and the integration of new sensors and methods.

Acknowledgments. The work is carried our within the framework of the collaboration between Veneranda Fabrica del Duomo di Milano and Politecnico di Milano. The authours want to thank Francesco Canali, director of Cantieri del Duomo di Milano for the Veneranda Fabbrica del Duomo.

References

1. Ferrari da Passano, C.: Il Duomo rinato: Storia e tecnica del restauro statico dei piloni del tiburio del Duomo di Milano [in Italian]. Veneranda Fabbrica del Duomo (Diakronia), Milan (1988)
2. Ferrari da Passano, C.: La Gran Guglia, il Duomo e l'acqua. Una piccola storia sulle vicissitudini e i misteri lungo i secoli ed i loro strani rapporti (2019). http://www.risorsa-acqua.it/tag/ferrari-da-passano/. Published online in December 2019
3. Alba, M., Roncoroni, F., Barazzetti, L., Giussani, A., Scaioni, M.: Monitoring of the main spire of the Duomo di Milano. In: Joint International Symposium on Deformation Monitoring, 2–4 November, Hong Kong, China, p. 6 (2011)
4. Gentile, C., Canali, F.: Continuous monitoring the Cathedral of Milan: design, installation and preliminary results. In: The Eighteenth International Conference of Experimental Mechanics, vol. 2, p. 5354 (2018). https://doi.org/10.3390/ICEM18-05354
5. Gentile, C., Ruccolo, A., Canali, F.: Long-term monitoring for the condition-based structural maintenance of the Milan Cathedral. Constr. Build. Mater. **228**, 117101 (2019)
6. Cigada, A., Dell'Acqua, L., Castiglione, B., Scaccabarozzi, M., Vanali, M., Zappa, E.: Structural health monitoring of an historical building: the main spire of the Duomo Di Milano. Int. J. Architect. Heritage 11 (2016)
7. Croce, A.: Questioni geotecniche sulle fondazioni del Duomo di Milano. In: Il Duomo rinato: Storia e tecnica del restauro statico dei piloni del tiburio del Duomo di Milano, vol. 2 (1970)
8. Barazzetti L., Roncoroni, F.: Relazione sulle misure eseguite per il controllo delle deformazioni del Duomo di Milano. Semestral report for the Veneranda Fabrica del Duomo di Milano, p. 39 (2020)

An Integrated Method to Assess Flood Risk and Resilience in the MAB UNESCO Collina Po (Italy)

Carlotta Quagliolo[1]([✉]) [iD], Vanessa Assumma[2] [iD], Elena Comino[3] [iD], Giulio Mondini[2] [iD], and Alessandro Pezzoli[1] [iD]

[1] Interuniversity Department of Regional and Urban Studies and Planning, Politecnico di Torino and Università Degli Studi di Torino, 10125 Turin, Italy
{carlotta.quagliolo,alessandro.pezzoli}@polito.it
[2] Interuniversity Department of Regional and Urban Studies and Planning, Politecnico di Torino, 10125 Turin, Italy
{vanessa.assumma,giulio.mondini}@polito.it
[3] Department of Environment, Land and Infrastructure Engineering, Politecnico di Torino, 10129 Turin, Italy
elena.comino@polito.it

Abstract. Climate change is exacerbating vulnerabilities in cities and territories and compromising the value of the whole heritage. In this context, the UNESCO sites should be prepared to preserve and manage their Outstanding Universal Value (OUV) by the uncertainty posed by flood-related risk. Therefore, this paper deals with the development of an integrated method to assess the biophysical impacts' scenarios through the employment of the Urban Flood Risk Mitigation model (InVEST) and evaluates them through a A'WOT technique. The proposed method considers a cluster of Municipalities of the MAB UNESCO Collina Po (Italy) as case study.

To support the development of the Urban Flood Risk Mitigation model, a number of parameters have been fitted according to defined Nature-Based Solutions scenarios (NBS). The A'WOT employs a set of indicators aiming at assessing the performance of the study area with the aim to find the best NBS solution to facilitate the adaptation of the case study to pluvial flood risk. The output provided by the integration of such models is to support the Decision Makers in risk and resilience assessment of heritage and design long-term strategic guidelines and recommendations to mitigate the flood extremes and preserve and monitor the integrity and authenticity of the MAB bioreserve.

Keywords: MAB UNESCO · Nature-based solutions · A'WOT

1 Introduction

Nowadays, landscape is universally accepted as the outcome of both natural and human actions and of their interrelationships (Consiglio d'Europa 2000), and hence it cannot

© The Author(s), under exclusive license to Springer Nature Switzerland AG 2022
F. Calabrò et al. (Eds.): NMP 2022, LNNS 482, pp. 2545–2555, 2022.
https://doi.org/10.1007/978-3-031-06825-6_243

be crystallized. Everything can be conceived as landscape, from ordinary to outstanding landscapes, including those fragile. Landscape preserves both tangible and intangible cultural and natural values inherited through generations. Many shocks and disturbances affect landscape, primarily as a result of human activity and natural disasters, and their effect is amplified by climate change (Sesana et al. 2021). This can increase the exposure of landscape and of cultural assets and cause a degradation and potential loss of the cultural, environmental, and social values. Landscapes, continuously evolve and adapt across space, time, and social organization (Gunderson and Holling 2002; Bottero and Datola 2020). In this sense, landscape is a component of Socio-Ecological Systems (SES) interested in resilience as both an outcome and a process (Cutter 2021).

The paper focuses on a "Man and Biosphere" (MAB) of UNESCO to be preserved, valorized and managed, thus allowing to next generations to inherit systems of values and innovation. In a context of uncertainty and ambiguity, caused in part by the ongoing climate change and the COVID-19 pandemic, Decision Makers (DMs), planners, CH bodies are ever more interested to employ interdisciplinary and transdisciplinary methods, especially when these can aid the climate adaptation and solutions to respond to extreme events. Today, UNESCO is contributing to increase community awareness, prevention, and preparedness to risk in World Heritage sites. In the context of landscape and urban planning, the expected outcome is to ensure a good governance model for a "good Antropocene" (Bennett et al. 2016). Dynamical modeling and tools can effectively help the assessment and design of adaptation scenarios and thus prioritizing policies and actions.

The paper develops an integrated method that considers on the one hand the assessment of Nature-Based Solutions (NBS) through an Ecosystem Services Approach (ESA) and on the other hand a Multicriteria Decision Analysis (MCDA) called A'WOT to assess the best NBS scenario for UNESCO MAB adaptation. The aim of the integrated method is to help final users to identify suitable interventions to respond to pluvial flood risk and thus increasing the resilience of the MAB bioreserve.

2   Methodology

The integrated method considers the InVEST and A'WOT methods to support decision making process in the design and assessment of scenarios of climate adaptation for a UNESCO MAB strongly characterized by flooding events (UNESCO 2013). This research is focused on pluvial flood risk because water is the most important decay factor for buildings and historic environment (Sabbioni et al. 2008).

2.1   InVEST Modeling

Integrated Valuation of Ecosystem Services and Tradeoffs (InVEST) is an open-source modeling software developed by the Natural Capital Project[1] aimed at including ecosystem services evaluation into decision making processes. Particularly, the Urban Flood Risk Mitigation (UFRM) model is designed to accommodate flood risk reduction during intensive rainfall events (Quagliolo et al. 2021). Such spatial model quantifies the

---

[1] Available at: https://naturalcapitalproject.stanford.edu/software/invest.

potentiality of nature-based measures in reducing runoff production. Even if the bio-physical quantification of runoff in the built environment can be difficult to estimate, due to the sewer systems and the dryness of the soil that may affect the water volume discharge, this model aims at this evaluation by using an empirical simplification (Salata et al. 2021). Indeed, to estimate runoff the model employs the USDA (United States Department of Agriculture) Soil Conservation Service – "SCS curve number" (SCS-CN) method (Kadaverugu et al. 2021). The curve number parameter assumes that highly sealed surfaces have low conductivity with higher runoff production (and vice-versa).

Required data input consider the hydrological aspects (see Table 1) over the study area. The runoff and infiltration aspects are considered by the InVEST model through the employment of LULC (curve numbers) and hydraulic conductivity of the soil (HSG) respectively. Since this work aims to deepen MAB resilience in a Climate Change context, the rainfall data input considered for this analysis is an extreme storm registered for the city of Torino in June 2021[2]. The runoff retention index produced as model output represents the ratio between the quantity of precipitation retained and the total precipitation per pixel. The resulting spatial index elaborated through Geographic Information System (GIS) provides a picture of the greatest flood-prone areas by simulating biophysical impacts of NBS.

Table 1. InVEST data input.

Data type	Description	Source
Administrative boundaries vector	Set of MAB Municipalities	Geoportale Piemonte (2019). https://www.geoportale.pie monte.it/cms/
Depth of rainfall	Amount of water for a single rainfall event (95 mm for 3 h)	ARPA Piemonte and NIMBUS (Società Meteorologica Italiana)
Land cover raster	Integration of Land cover Piedmont; BDTRE (regional topographic geodatabase) and imperviousness map (NHRLC)	Geoportale Piemonte (2010; 2020) and SINANET (2015). https://www.geoportale.pie monte.it/cms/
Soil Hydrological Group (HSG) raster	Map of soil protection capacity	Regione Piemonte (2020)
Sensitivity table	Biophysical values associated to each land use classes and HSG	Computed after USDA (USDA - United States Department of Agriculture 2004)

---

[2] http://www.nimbus.it/eventi/2021/210624NubifragioTorino.htm.

2.2 NBS Scenarios Characterization

The recent "Climate Resilience Plan" from the city of Torino (2020) highlighted the key role of Nature-Based Solutions (NBS) to achieve the hydrological balance in the face of heavy precipitation (Assessorato per le Politiche Ambientali 2020). Indeed, in the Climate Change context, precipitation events are becoming even more short and intense, and producing punctual waterflows. NBS include "Sustainable Urban Drainage System" (SUDS) that enable to reduce the water charge on the sewage drainage system by slowing the runoff flow.

This research proposes the NBS biophysical benefits assessment in mitigating runoff production. Through the definition of two adaptation scenarios, the simulation of changes in city's retention capacity has been developed. The base scenario, which represents the present condition, provides the vulnerability analysis for the prioritization of the suitable areas of intervention. The adaptation scenarios (AS) follow the "Climate Resilience plan" of Torino and the MAB principles by defining measures aimed at reducing the pluvial flood risk while providing social and biodiversity connection. The data input changed in the adaptation scenarios is related to the Land cover while keeping constant all the other input. Figure 1 describes the adaptation scenarios by including the related ecosystem services and MAB effects.

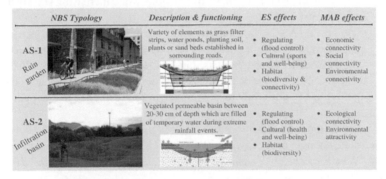

Fig. 1. Adaptation measures' scenarios (Own elaboration)

2.3 Multicriteria Decision Analysis (MCDA)

Multicriteria Decision Analysis (MCDA) can help DMs in solving uncertain problems. Thanks to its quali-quantitative approach, it can be combined within integrated frameworks (Bottero et al. 2021). The A'WOT is retained as suitable to achieve this goal.

2.3.1   Integrating SWOT Analysis and AHP: The A'WOT Method

The A'WOT is a hybrid method that solves some limits of both AHP and SWOT analysis (Kurttila et al. 2000; Kangas et al. 2001). It is developed as follows:

1. Knowledge Analysis: Review of relevant sources on the case study (e.g. Geoportale Piemonte, Climate Resilience Plan of Turin, or MAB dossier);
2. Development of SWOT Analysis: Identification of both internal and external factors;
3. Structuring of the hierarchy tree through AHP (Saaty 2004), where:

    - *Goal*: The best NBS solution for the MAB adaptation to pluvial flood risk;
    - *Criteria*: They are the MAB strengths, weaknesses, opportunities and threats;
    - *Subcriteria*: these are indicators related to the MAB and SWOT components;
    - *Alternatives*: There are three NBS scenarios:

    (1) Scenario 0 (BAU): This is the MAB Business as Usual scenario;
    (2) Scenario 1 (AS-1): It envisions rain gardens installed along public roads to provide multiple benefits to the city before and after a pluvial flood event, such as the increase of soil permeability, psycho-physical benefits from recreative activities, or bio-energy increase amongst urban micro habitats. It can meet both the MAB dimensions while increasing shortly the urban resilience;
    (3) Scenario 2 (AS-2): Infiltration basins are realized within parks to collect a part of water during a pluvial flood event through filtration and regulation. They favour localized habitats, landscape attractiveness and indirect health benefits. These have a constrained accessibility. This scenario meets the natural MAB side while increasing territorial resilience in the medium term.

4. Survey: a questionnaire was proposed to DMs pool with expertise on cultural heritage, climate change, economy, landscape ecology and society. The classical question is "*Which of the two elements A and B is a greater strength? How much greater?*"
5. Weighting of the elements: these are compared pairwise within each SWOT component and then between the SWOT components by using the Saaty's Scale. A performance matrix was provided to the DMs to provide a site-specific asssessment;
6. Determination of local priorities: The comparisons are elaborated within the software Superdecisions[3] (with Consistency Ratio Index – CR ≤ 10%) (Saaty 2004) as a set of weights at sub-criteria and criteria levels through the eigenvalue formula;
7. Priorities synthesis: it means the aggregation of the local priorities into global priorities and thus delivering the final ranking of the NBS scenarios;
8. Sensitivity analysis: the One-at-a-Time approach (OAT) tests the stability of the results through the changing of the criteria's weights (Daniel 1973).

---

[3] Superdecisions by Creative Decisions Foundation http://www.superdecisions.com/.

3 Application

3.1 Case Study: The MAB UNESCO Collina Po, Piedmont (Italy)

Man and Biosphere (MAB) is an intergovernmental program founded in the '70s to preserve and valorize the role of biosphere (UNESCO 2013) by considering three main functions: i) preserving landscapes, ecosystems, species, and genetic diversity; ii) economic and human development to generate benefits on social, cultural, economic, and environmental dimensions; iii) supporting research, monitoring, training, and best practices to strengthen the role of sustainable development within and outside a MAB. The case study is the MAB UNESCO Collina Po and takes the name from the Turin hills crossed by Po River and the initiative of territorial promotion "CollinaPo" (2011) that prepared the strategic basement and partnerships for the UNESCO candidacy (2014, 2016). It extends for about 1,700 Km^2 between the provinces of Asti, Cuneo, Turin, and Vercelli, and where live more than 1.500,000 inhabitants. Therefore, it constituted the first "Urban MAB" of Italy. The MAB involves 86 Municipalities, where interact ecological habitats, cultural assets and landscape, such as the Basilica of Superga and the Royal Residences of the Savoy Dynasty. The MAB perimeter consists of (i) 14 core zones that cover fully the Po River's protected areas (3,853 ha), (ii) a Buffer zone (21,161 ha) adjacent to fluvial and hilly ambits to protect the core zones; (iii) and a Transition area (146,219 ha) as the remaining non-constrained urban and rural areas. The integrated method is employed on four main municipalities included within the MAB to determine the best NBS scenario to reduce the pluvial flood risk: Torino, Pino Torinese, Moncalieri, and Settimo Torinese (Fig. 2).

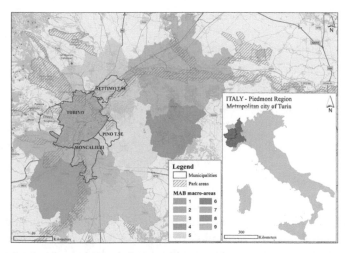

Fig. 2. The study area in the MAB CollinaPo area (Own elaboration)

4   Results and Discussion

4.1   Results of the InVEST Simulation

This section provides a description of the InVEST model's output for the study area by considering a design storm of 95 mm in 3 h comparing the three implemented scenarios. The average values for the four Municipalities considered are between 48% and 70% which means that the catchment area presents a moderate degree of runoff production. Flood mitigation capacity by base- and NBSs-scenarios is quantified in runoff retention volume per hectares (see Table 2). Indeed, to provide the relative runoff retention values, the runoff volume has been calculated in respect to the area covered by each Municipality and it is expressed in m^3/ha. Torino showed the lowest retention capacity (239.4 m^3/ha) in respect to the other municipalities in the base scenario. While Pino T.se is the municipality with the highest retention ability. Through the implementation of Rain Gardens (AS-1) along the most critical streets resulted from the base scenario, all the municipalities increased the runoff retention of 43% in average. The AS-2 which implemented Infiltration Basins in park areas, showed lower retention capacity for Moncalieri followed by Torino. Pino T.se resulted on the highest retention capacity around 476 m^3/ha. In general, all the municipalities increased more the flood mitigation capacity thanks to AS-1 in respect to the AS-2 excepted Pino T.se which has the same improvement level for both adaptation scenarios. Figure 3 shows the runoff retention index per pixels in the four Municipalities within MAB where the pixels' resolution is 5 m. The runoff retention index represents the relative measure of runoff retained related to the precipitation volume and ranges from 0 (low retention) to 1 (high retention).

Table 2. Runoff retention capacity per each municipality

Scenarios	Moncalieri	Pino T.se	Settimo T.se	Torino
Base	295.2 m^3/ha	349.0 m^3/ha	308.1 m^3/ha	239.4 m^3/ha
AS-1	426.6 m^3/ha	475.5 m^3/ha	450.5 m^3/ha	351.0 m^3/ha
AS-2	422.8 m^3/ha	476.1 m^3/ha	445.9 m^3/ha	344.8 m^3/ha

4.2   Results of the A'WOT

Once having collected the DMs responses, these were merged in a unique shared assessment (see Tables 3 and 4). According to the DMs, the most relevant criterion is "Opportunity" (0.384), followed by "Strengths" (0.344), due to the several elements that can enter in synergy and maximize the benefits, thus favoring the MAB adaptation. However, the criteria "Weaknesses" (0.174) and "Threats" (0.097) should be considered as those elements that need to be nullified and limit the potential occurrence and/or concomitance of extreme events. The most important sub-criterion is "PNRR projects" (0.417) because represent a fruitful opportunity to orient properly the MAB adaptation, that means investing in solutions for a better quality of life. The indicator "Population density" (0.354)

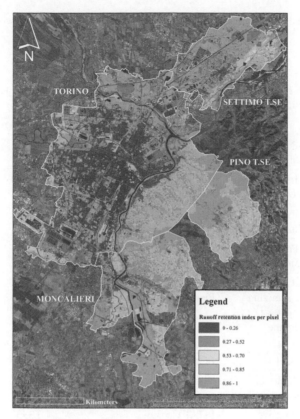

Fig. 3. Runoff retention index for Base scenario (Own elaboration)

is closely related to both the AS scenarios, because these assume value in presence of settlements and thus reducing the exposure of inactive subjects (0.333), protected areas and cultural assets (0.189 and 0.165), or economic income (0.293), among others. The update of Municipal Plans (0.187) whether in synergy with the PNRR can envision resilient actions and with attention to the disaster risk management. The runoff retention and weather drought indicate the soil permeability status and an indirect measure of its response in case of pluvial flood events (0.265 and 0.251, respectively). Finally, the most suitable scenario is AS-2 "Infiltration Basin" (1), followed by AS-1 "Rain gardens" (0.807) and BAU (0.337). A sensitivity analysis was performed through the OAT approach, thus confirming the stability of the results.

Table 3. Set of indicators and their priorities to achieve the best scenario

A'WOT model					
Set of indicators			DMs evaluation		
Criteria (SWOT)	Sub-criteria (Indicators)	Source	Criteria priorities	Sub-criteria	
				Norm priorities	Limiting priorities
S	s.1 Population density (ab/Km^2)	ISTAT (2021)	0.344	0.354	0.097
	s.2 Incidence of protected areas (%)	Geoportale Piemonte		0.189	0.059
	s.3 Incidence of Cultural assets (%)	Geoportale Piemonte		0.165	0.056
	s.4 Pro-capita income (%)	INTWIG on MEF data (2019)		0.293	0.080
W	w.1 Loss of characterizing elements of landscape	Geoportale Piemonte	0.174	0.267	0.052
	w.2 Linear rockslides	Geoportale Piemonte		0.185	0.036
	w.3 Tourism pressure (%)	VisitPiemonte DMO (2020)		0.215	0.045
	w.4 Inactive population (%)	ISTAT Census (2021)		0.333	0.068
O	o.1 PNRR projects (€)	Regione Piemonte (2021)	0.384	0.417	0.166
	o.2 Green initiatives (no.)	Municipalities websites		0.181	0.065
	o.3 Hydrogeological interventions	RendisWEB, ISPRA		0.215	0.076
	o.4 Update of Municipal Plans (Y/N)	Municipalities websites		0.187	0.068
T	t.1 Runoff retention (%)	Output of the InVEST model	0.097	0.265	0.033
	t.2 Extreme rainy events (no.)	ArpaPiemonte		0.166	0.022

(*continued*)

Table 3. (*continued*)

A'WOT model						
Set of indicators			DMs evaluation			
Criteria (SWOT)	Sub-criteria (Indicators)	Source	Criteria priorities	Sub-criteria		
				Norm priorities	Limiting priorities	
	t.3 Weather drought (%)	ArpaPiemonte		0.251	0.029	
	t.4 Lack of funds (€)	–		0.318	0.047	
				Total	1.000	

Table 4. Synthesis of final priorities

Scenarios	Final ranking
BAU (Business as Usual)	0.337
AS-1 (Rain gardens)	0.807
AS-2 (Infiltration basins)	1.000

5 Discussion of Results and Conclusion

The integrated method assessed the best adaptation scenario for the MAB Collina Po to the pluvial flood risk. The ESA approach employed climate and land-use change data and provided a bio-physical valuation, where both the ASs recorded similar future trends ($t_0 - t_1$). The A'WOT model, starting from endogenous and exogenous factors through indicators (t_0), assessed their importance and prioritized the scenarios, thanks to a multidisciplinary panel of DMs. Moreover, the parameters of runoff retention and weather drought represent the link between the ESA approach and the A'WOT model as element of novelty and integration in both the disciplines. This method can be applied for example in procedures such as the Strategic Environmental Assessment (SEA), or the Assessment of Ecological Incidence for Natura 2000 sites, or even to integrate the economic evaluations of properties in proximity of valuable natural and cultural sites (Dell'Anna et al. 2022). This can also support the monitoring of the State of Conservation (SOC) of MAB and other UNESCO World Heritage Sites. It will be replicated by involving local actors and stakeholders actively involved in the MAB. The application will be extended to all the MAB Municipalities to provide a comprehensive overview. Additional scenarios and other relevant risks would be explored in this MAB such as the fire risk, due to the presence of relevant forests and protected areas. As future development, more refined and alternative procedures will be considered (Kangas et al. 2001). A dynamic extension of A'WOT will be developed through a network configuration to refine the assessment of problems with interdependences. The integrated method will be also replicated to other bioreserves of the MAB network, such as Po Grande or the Appennino Tosco-Emiliano, among others.

References

Assessorato per le Politiche Ambientali: Piano di Resilienza Climatica, Torino (2020)

Bennett, E.M., Solan, M., Biggs, R., et al.: Bright spots: seeds of a good anthropocene. Front Ecol. Environ. **14**, 441–448 (2016). https://doi.org/10.1002/fee.1309

Bottero, M., Assumma, V., Caprioli, C., Dell'Ovo, M.: Decision making in urban development: Tthe application of a hybrid evaluation method for a critical area in the city of Turin (Italy). Sustain. Cities Soc. **72**, 103028 (2021). https://doi.org/10.1016/j.scs.2021.103028

Bottero, M., Datola, G.: Addressing social sustainability in urban regeneration processes. an application of the social multi-criteria evaluation. Sustainability **12** (2020). https://doi.org/10. 3390/su12187579

Consiglio d'Europa: Convenzione Europea del Paesaggio, Firenze (2000)

Cutter, S.L.: Urban risks and resilience. In: Shi, W., Goodchild, M.F., Batty, M., Kwan, M.-P., Zhang, A. (eds.) Urban Informatics. TUBS, pp. 197–211. Springer, Singapore (2021). https:// doi.org/10.1007/978-981-15-8983-6_13

Daniel, C.: One-at-a-time plans. J. Am. Stat. Assoc. **68**, 353–360 (1973). https://doi.org/10.1080/ 01621459.1973.10482433

Dell'Anna, F., Bravi, M., Bottero, M.: Urban green infrastructures: how much did they affect property prices in Singapore? Urban Urban Green **68**, 127475 (2022). https://doi.org/10.1016/ j.ufug.2022.127475

Gunderson, L.H., Holling, C.S.: Panarchy: understanding transformations in systems of humans and nature (2002)

Kadaverugu, A., Nageshwar Rao, C., Viswanadh, G.K.: Quantification of flood mitigation services by urban green spaces using InVEST model: a case study of Hyderabad city, India. Model. Earth Syst. Environ. **7**(1), 589–602 (2021). https://doi.org/10.1007/s40808-020-00937-0

Kangas, J., Pesonen, M., Kurttila, M., Kajanus, M.: A'wot: Integrating the AHP with Swot Analysis, pp. 189–198 (2001). https://doi.org/10.13033/isahp.y2001.012

Kurttila, M., Pesonen, M., Kangas, J., Kajanus, M.: Utilizing the analytic hierarchy process (AHP) in SWOT analysis - A hybrid method and its application to a forest-certification case. For. Policy Econ. **1**, 41–52 (2000). https://doi.org/10.1016/s1389-9341(99)00004-0

Quagliolo, C., Comino, E., Pezzoli, A.: Experimental flash floods assessment through urban flood risk mitigation (UFRM) model: the case study of ligurian coastal cities. Front Water **3**, 1–16 (2021). https://doi.org/10.3389/frwa.2021.663378

Saaty, T.L.: Decision making—the Analytic Hierarchy and Network Processes (AHP/ANP). J. Syst. Sci. Syst. Eng. (2004). https://doi.org/10.1007/s11518-006-0151-5

Sabbioni, C., Cassar, M., Brimblecombe, P., Lefevre, R.A.: Vulnerability of cultural heritage to climate change. Pollution Atmospherique, pp. 157–169 (2008)

Salata, S., Ronchi, S., Giaimo, C., et al.: Performance-based planning to reduce flooding vulnerability insights from the case of Turin (North-West Italy). Sustainability **13**, 1–25 (2021)

Sesana, E., Gagnon, A.S., Ciantelli, C., et al.: Climate change impacts on cultural heritage: a literature review. Wiley Interdiscip. Rev. Clim. Chang. **12**, e710 (2021). https://doi.org/10. 1002/WCC.710

UNESCO: Biosphere Reserve Nomination Form, vol. 33, pp. 1–37 (2013)

USDA - United States Department of Agriculture: Hydrologic Soil-Cover Complexes. In: National Engineering Handbook: Part 630 – Hydrology (2004)

Commissione Nazionale Italiana per l'UNESCO. Man and Biosphere program - MAB. https:// www.unesco.it/it/italianellunesco/detail/186 (Accessed May 2022)

Innovative Tools for Green Heritage Management: The Case of the Historic Gardens of Savoy Royal Residences of Piedmont (Italy)

Vanessa Assumma[✉] ⓘ, Daniele Druetto, Gabriele Garnero ⓘ, and Giulio Mondini ⓘ

Interuniversity Department of Regional and Urban Studies and Planning, Politecnico di Torino, Viale Mattioli 39, 10125 Turin, Italy

{vanessa.assumma,gabriele.garnero,giulio.mondini}@polito.it, daniele.druetto@gmail.com

Abstract. New environmental, economic and social challenges are enlightening the need for innovative systems and approaches for the management of green areas. The green sector, in particular green heritage, is directly involved in this issue. On the one hand, green heritage is even more intended as more than an aesthetic component, on the other hand, the economic resources available for management and maintenance aspects have been reduced. To aggravate the general situation is the pressing process of climate change affecting the shrinking of natural resources and impacting negatively on green heritage. Hence the need to research new management approaches to solve this issue effectively and efficiently, require a vision that can provide immediate results, benefits in the medium and long term, in combination with the knowledge, sensitivity and dynamism of the human component which are considered indispensable today. The goal of the paper is the discussion on emerging technologies and tools to better support the management of green heritage, with regard to the historical gardens of Savoy Royal Residences in Piedmont (Italy). The contribution focuses on specific technologies and tools, supported by applicative examples, aimed at improving the current management of green areas with the purpose of creating a unified system, also in language, which can help the collaboration between actors, stakeholders and specialised subjects actively involved in the green heritage field.

Keywords: Green heritage · MEC classification · Multispectral and green indices

1 Introduction

In the international and national contexts, specific directives and agreements such as the Green Deal and the Italian National Plan for Recovery and Resilience - PNRR [1, 2] highlighted an urgent need to operationalise priority axes for a more optimistic future, in terms of interventions inspired by traditional and innovative paradigms as sustainability, resiliency, or circularity, among others. To achieve this purpose, specific technologies and tools can help Decision Makers (DMs) to define proper policies through

© The Author(s), under exclusive license to Springer Nature Switzerland AG 2022
F. Calabrò et al. (Eds.): NMP 2022, LNNS 482, pp. 2556–2564, 2022.
https://doi.org/10.1007/978-3-031-06825-6_244

the analysis, assessment and manage green areas comprehensively, including also the internal and external relationships that occur between all the elements, both natural and anthropogenic. The elements present in a green area, in particular the plant elements, should not be conceived as individual elements and nor as a group. It is necessary to consider them as a system in a large-scale system. In fact, this can interact with endogenous elements and consider at the same time exogeneous factors belonging to neighboring systems and higher order macro-system [3]. The stability of the system, although it is able to rebalance itself naturally, can nevertheless be strongly compromised by anthropic interventions, internal or external [4].

The present paper continues a research in part developed in previous works [5, 6] with the aim to deepen the role of technologies and information systems to support the management and restoration of green heritage for their preservation, valorisation and management and thus of their resilience. Specifically, the attention is focused on the historical gardens of Savoy Royal residences in Piedmont (Italy), which belong to the famous UNESCO serial site. It should be notices that the paper focuses on a site known for its outstanding value, historical, cultural, and unique in the world. However, the authors open the discussion for valuable green areas, and more in general, for all the typologies of green areas..

This contribution focuses on two main categories of tools: the first is represented by the Topographic DataBase (TDB) developed with the support of Geographic Information Systems (GIS) and based on the classification of the Minimum Environmental Criteria (MEC) and its potential implementation for green heritage; the second category is represented by multispectral technology and annexed devices like cameras, drones able to provide and monitor data (e.g. the vitality of the vegetative component and other features not visible by the human eye).

Information produced by the proposed tools is identified by a univocal code and can be processed both singularly and combined with other data within a TDB. The combination of TDB with multi-spectral technology is retained as suitable to collect documented material on green heritage and to monitor the health status of plant species present within historical gardens. In this way, it is possible to optimize their preservation, enhancement, and management and to guarantee the safety of a wide range of subjects that access the green areas, both public and private, spanning from specialized operators to tourists, and so on.

The paper is articulated into the following sections: Sect. 2 describes the current tools and related proposals, Sect. 3 provides applicative examples followed by some reflections of the authors, Sect. 4 concludes with final remarks and opens to future research perspectives.

2   Methodology

In this section two main groups of tools are considered as suitable to support the preservation, enhancement, and management of green heritage: on the one hand, the Geographic Information Systems (GIS) integrated by the classification of the Minimum Environmental Criteria (MEC), and on the other hand, the multi-spectral technologies that use cameras, which are generally employed by specialized operators, or installed on Remote

Piloting Aircraft (APR) vehicles (or "professional drones"). Multispectral sensors are equipment capable of recording the amount of light mirrored by elements, subjected to reading in different wavelengths of the electromagnetic spectrum. The combination of such wavelengths allows the calculation of specific indices of vegetation, that can record for example an outbreak of cryptogamic diseases, or a nutritional deficiency of the foliage, or even the presence of water stress, among others.

Vegetation indices are based on the reading of reflected light compared to the foliar one and by considering the following bands: Red (R), Green (G), Near-Infrared (NIR), and Red-Edge (RE). The last type of band is transitory and located between the R and NIR bands. In the green sector, with particular regard to precision farming, there is increasing attention on specific indices called Normalized Difference Vegetation Index (NDVI), Green Normalized Difference Vegetation Index (GNDVI) and Normalized Difference Red Edge (NDRE) [7].

2.1 GIS and MEC Criteria: A Novel Extension

The structure of the TDB employs the MEC classification by considering the guidelines of the Italian Decree of 10th March 2020 "*Criteri ambientali minimi per il servizio di gestione del verde pubblico e la fornitura di prodotti per la cura del verde*", and related attachments [8].

The MEC classification organizes the several information into four main categories of geometries of primary and secondary level several information. Each geometry is then codified through a univocal code and georeferenced to be visualized by final users (see Fig. 1). Geometries are implemented by a set of attributes, according to the normative in force, where a wide range of options and combinations are available. Concerning green heritage and particularly the historic gardens, potential implementation of the existing attributes is retained more than ever required.

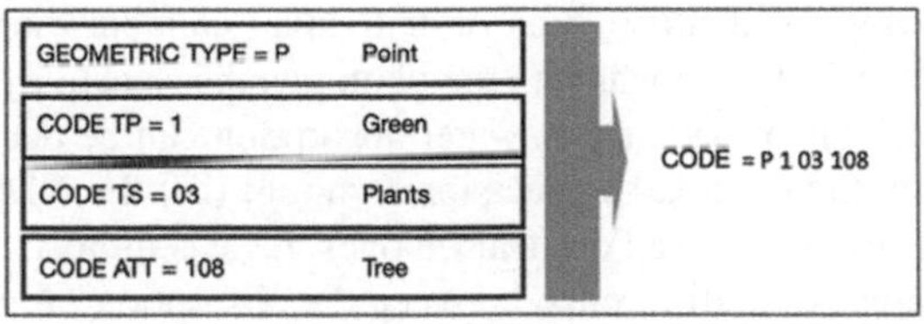

Fig. 1. MEC classification of urban green areas (Elaboration from [8])

Since the MEC classification is considered a tool suitable as necessary. Its possible implementation with attributes related to the historical gardens would be more specific, focusing on an increase of the Level Of Detail (LOD) assigned to each element, and on the aggregation of the individual elements into homogeneous groups when they belong to the same class and composition. More details can be found in Sect. 3 with a focus on the cases study.

2.2 Green Historic Register: VTA and Multi-spectral

Considering the application of the MEC classification developed in previous works for a real case study [5, 6], this paper proposes the integration of TDB with additional information to create a green historic register of the elements that compose historical gardens, from the past to the present ones, with a particular attention for those elements of historic and environmental value that must be preserved (e.g. secular trees). To develop a comprehensive green historic register, it is fundamental the insertion of the following elements:

- Photographic analysis on the current state of the green historic elements to be monitored periodically by considering intervals that do not exceeding one year. According to the authors, it is suggested the semestral monitoring because can compare the health status across the seasons, and in occurrence of interventions of extraordinary maintenance. Both photographic and cartographical materials can enrich the georeferencing and visualization of the elements, for example through the overlay mapping and notes for further analyses and comparisons;
- Data collection related to the classification Visual Tree Assessment (VTA). Data are elaborated in matrices to facilitate the import within TDB by maintaining the values related to the assigned risk class, as well as possible notes and other information provided by the specialized operator during the inspection. In this way, the information of geo-localization of the green element can be enriched and making the information as univocal and easily to be detected;
- Remote sensing through multi-spectral cameras, with the support of specialized operators and/or drones, is finalized to monitor the health status of the vegetative element, spanning from the water stress to the presence of invasive pathogens [9]. The information detected in this stage are efficacy whether a comparison between different vegetation indices is developed, since each of them is propaedeutic to the monitoring of specific components. They are: (i) the NDVI that measures the vigour of a plant component allowing to identify areas or individual elements subjected to a given stress; (ii) the GNDVI index measures the photo-synthesis activity of the plant component from which situations of low water absorption can be detected. In fact, it is often employed in the context of precision farming; (iii) the NDRE index, using a transitory band between the Red e InfraRed ones, reveals more accurate for example in the identification of the difference between the absorption of chlorophyll and the dispersion caused by the leaf structure. Analogously to the photographic analysis, the sensing should be repeated each year, even if for specific situations related to the foliage band, in periods in which deciduous species have begun or not their rest cycle (Fig. 2).

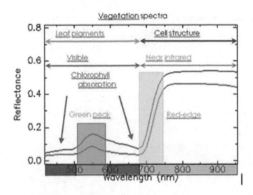

Fig. 2. Electromagnetic spectrum plot of the vegetative component [10]

3   Case Study

The case study of the paper focuses on the UNESCO serial site of Savoy's Royal Residences. Two green heritage components are considered as meaningful for the aim of the paper: the parterre of the Garden of Agliè Castle and the Royal Gardens of the Royal Palace of Turin that deal with the MEC classification and the multi-spectral application.

As an example, a flowerbed can be classified according to ten possible attributes, even if none of the current classification is proper when a parterre is present in a historical garden. The parterre is a group of geometric forms with ornamental function, and each form should be identified as an individual flowerbed. Moreover, the created composition can be realized in meadows, or for topiary shrubs as part of the flowerbed. Similarly, a floral component, which is frequently consisting of herbaceous plants with an annual life cycle, might be present or absent.

The classification of a flowerbed may or may not be devoid of the adjective "flowery" depending on the different periods of the year. Finally, it could be possible that the composition of the parterre can include some architectural elements and decorative elements (e.g. fountains, statues, or vases), in central positions, or following symmetries with respect to the overall geometric design (Fig. 3).

Therefore, the authors propose here an integration of the current MEC classification that leads to greater specificity in the identification of each element present, for example those mentioned above, thus allowing immediate recognition of the macro-group that these elements make up, such as belonging to the design of a parterre.

A last, but not less important consideration, concerns the systems and pipes present in the subsoil of the external spaces and the need to consider them with respect to the vegetation, especially as regards the aspects related to maintenance and the possibility of interventions extraordinary as well as possible new future projects. It is suggested to add these elements in the TDB to be able to identify the overlapping points and create buffer strips to avoid, for example, that a new tree is planted near a pipe that could damage over time or vice versa that the laying of a new pipe can intercept the root system of a tree damaging it.

Secondary type 02 - FLOWERBED		A
NAME	GEOMETRY	CODE
Flowerbed generic	S	S102000
Flowerbed bush scrub	S	S102101
Flowerbed perennial herbaceous	S	S102102
Flowerbed annual herbaceous	S	S102103
Flowerbed roof intensive	S	S102111
Flowerbed roof extensive	S	S102112
Flowerbed vertical green	S	S102113
Flowerbed flowery of value	S	S102121
Flowerbed perennial flowery	S	S102122
Flowerbed with grill	S	S102455

Fig. 3. MEC classification of the element"flowerbed" (**A**); Comparison of the same flowerbed parterre in the Agliè Castele, no classification is suitable for the proposed case (**B, C**).

The Figure below illustrates an orthophoto taken by aerial photography on the Royal Gardens of Turin, the overlay of information regarding the VTA classification provided by the body of the Royal Museums of Turin [5], and the NDVI processing of the same (Fig. 4).

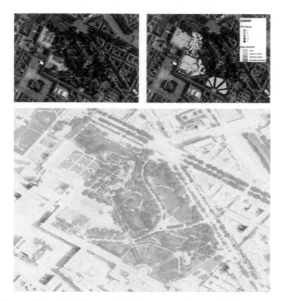

Fig. 4. Royal Gardens of Royal Palace of Turin, comparison RGB image (left) and RGB image with overlay VTA3 information (right) and NDVI elaboration (down) [5]

In the NDVI image, in addition to guessing part of the building bordered in red and the ground in yellow, it is possible to detect different shades of green which are not directly a symptom of low vigour but caused by the different return of light given by an area grass rather than the foliage of a tree. It should be considered that aerial survey, given the distance from the ground at which it is carried out, provides less information than an analysis carried out using a drone, therefore, with reference to the previous topics, the first case can be considered, as in the example proposed, in as a rough indication but it is suggested to deepen to a greater detail and comparing different indices of vegetation. On the other hand, the difference in vitality detected on the lawn areas that has low levels near the fountain of the Nereids is unequivocal, confirmed by the comparison with the orthophoto which shows how the lawn areas in question are almost completely dry [11].

Another important consideration, specifically of the tree element, concerns the comparison with the VTA analysis and the difficulties that may arise from this comparison. As reported in the image below, the VTA analysis indicates some specimens in classes of particular risk, such as class CD and D, which on the other hand in the NDVI image do not have a color tending to yellow or red but have more green tones. or less marked: the motivation lies in the fact that the stability and vitality of a tree can be two disjoint factors. For example, this is the case when foliar apparatus is vigorous while there are important root problems that affect the stability or counter a foliage that presents problems, caused by momentary stress, on a perfectly healthy tree structure.

In conclusion, there is no way believed that one type of analysis can be a substitute for the others. In fact, these analyses should be carried out simultaneously to support each other by producing more accurate results as well as reducing the margins of error: aerial analyses, or by drones, can provide a first screening, in particular over large areas, and bring to the attention sites to which it would be advisable to give priority over others and then proceed with the VTA analysis, following or simultaneously with the which, to carry out another multispectral survey, at a detailed scale considering what previously stated, in order to compare the results obtained before proceeding with the drafting of the final evaluations.

Both methods of analysis are a snapshot of a situation which, although ideally repeated with adequate frequency, cannot reflect the dynamism of a system, such as a plant one, which undergoes continuous variations as a living entity. Add to this the human factor and the relative margins of error and influence that can fall on the plant system, both for the operator who carries out the survey and interprets the data and for who intervenes as maintenance of the green areas, and how these effects, albeit minimal, when added together can cause consistent variations in the results and in the final considerations that then lead to the interventions. However, the combination of the proposed tools, and the mutual sharing of information regarding the neighboring areas, together with a careful and far-sighted management of the green heritage, could certainly solve some current problems and thus favoring its preservation, enhancement, and management.

4 Conclusions

The application of the proposed tools and their combination, whether carefully weighed according to site-specific situations, can provide considerable advantages in the management of prestigious green areas, as regards both the economic aspect, strictly linked to the management itself, and the aspects relating to safety, primarily of the users more in general of all the engaged subjects.

Another fundamental advantage concerns the reduction of the likelihood of information leaks occurring as the inclusion of the same within a unified database limits this possibility as well as allowing greater immediate consultation.

In fact, management situations were found in which, for example as regards the classification of trees, the floor plans are stored on CAD files, the stability assessments on spreadsheets, the history of photographs on digital folders and the documentation relating to the interventions carried out. divided partly on digital files and partly on paper material: consulting different material, located in different locations, inevitably leads to a greater waste of time as well as a greater possibility of errors in the consultation and transposition of information. On the other hand, if this information is collected in an individual system and unified by means of a standard classification, the situation is opposite to the previous one and therefore has considerable advantages. One of the first examples at national level is recalled in this regard: the "Progetto Verde Milano" [12].

Finally, the advantages of a greater knowledge of the environment in which one operates emerge from the proposed methodology, together with new resources that derive from a system whose budget, with the same investment, is more efficient. These resources can be shared to generate additional benefits, but a common language is required.

Sharing resources alone would already be an important milestone in itself, think of the example proposed above regarding the Royal Gardens of Turin, which belong to the complex of the Royal Museums of Turin and are separated from the so-called Lower Royal Gardens which are instead managed by the Municipality of Turin: when one of the two bodies carries out an analysis or survey that a larger area than its own, as in the case of an aerial shot or drone, it could share the information with another neighboring body, which in turn will repeat the "courtesy" on a different occasion and thus increasing mutual advantages.

However, data exchange is not possible if data are not united under the same language, first and foremost because the adjustment may require significant resources, as well as problems arising from conversion errors, but also because even incomplete compatibility during conversion: think in extreme cases of border situations between different regions or different states. Different bodies currently also at a local level, in some cases, differ in the use of methodologies, criteria, software, graphic representations.

From a broad perspective, which therefore looks at national and supranational areas, aimed for example at a European vision, it is necessary to create a unified language which, in addition to what has been said previously, will initially act as a bridge and a first basement for dialogue and comparison between various entities. Individuals and nations that currently do not dialogue, or if they do, do so with difficulty, and this, like others, constitutes a barrier that we are aware of having to overcome in order to create a future management of greenery based on sustainability and on resource efficiency. It is necessary to consider the ecological aspect of green heritage not only as the ability to absorb and adapt over time with respect to internal and external anthropic interventions [3], but rather responding in time in order to properly preserve and enhance the green heritage.

References

1. European Commission: Comunicazione Della Commissione al Parlamento Europeo, Al Consiglio, al Comitato Economico e Sociale Europeo e al Comitato Delle Regioni. Il Green Deal Europeo. Bruxells (2019). https://eur-lex.europa.eu/legal-content/IT/TXT/?uri=CELEX:520 19DC0640. Accessed Dec 2021
2. Italian Government: Commissione Europea, Piano Nazionale di Ripresa e Resilienza, #NextGenerationItalia. Riforme e Missioni, Missione 2 – Rivoluzione verde e transizione ecologica, Roma (2021). https://www.governo.it/sites/governo.it/files/PNRR.pdf. Accessed Dec 2021
3. Holling, C.S.: Engineering resilience versus ecological resilience. In: Schulze, P.E. (ed.) Engineering within Ecological Constraints, pp. 31–43. National Academy Press, Washington DC (1996)
4. Turner, M.G., Robert, H.G.: Landscape Ecology in Theory and Practice - Pattern and Process. Springer, New York (2015). https://doi.org/10.1007/978-1-4939-2794-4
5. Druetto, D.: Modello di analisi e valutazione per la valorizzazione e gestione degli spazi (2021)
6. Assumma, V., Druetto, D., Garnero, G., Mondini, G.: A model of analysis and assessment to support the valorisation and management of green areas: the royal gardens of Turin (Italy). In: Gervasi, O., et al. (eds.) ICCSA 2021. LNCS, vol. 12955, pp. 554–568. Springer, Cham (2021). https://doi.org/10.1007/978-3-030-87007-2_40
7. Gonthier, P., Lione, G., Borgogno Mondino, E.: Tree health monitoring: perspectives from the visible and near infrared remote sensing, Forest - Rivista di Selvicoltura ed Ecologia Forestale 9, 89–102 (2012). https://doi.org/10.3832/efor0691-009
8. Guzzetti, F., Viskanic, P., et al.: Modello dati per il censimento del verde urbano. Version 2.1 (2020). https://www.mite.gov.it/sites/default/files/archivio/allegati/GPP/2020/modello_dati_per_il_censimento_del_verde_urbano_2_1_con_allegati.pdf. Accessed Dec 2021
9. Borgogno-Mondino, E., Lessio, A., Gomarasca, M.A.: A fast operative method for NDVI uncertainty estimation and its role in vegetation analysis. Euro. J. Remote Sens. 49(1), 137–156 (2016). https://doi.org/10.5721/EuJRS20164908
10. Maresi, L., Taccola, M., Kohling, M., Lievens, S.: PhytoMapper – compact hyperspectral wide field of view instrument. In: Sandau, R., Roeser, H.P., Valenzuela, A. (eds.) Small Satellite Missions for Earth Observation. Springer, Heidelberg (2010). DOI:https://doi.org/10.1007/978-3-642-03501-2_30
11. Vaglio Laurin, G., et al.: Potential of ALOS2 and NDVI to estimate forest above-ground biomass, and comparison with lidar-derived estimates. Remote Sens. 9, 18 (2017). https://doi.org/10.3390/rs9010018
12. Di Maria, F., Guzzetti, F., Privitera, A., Viskanic, P.: Progetto verde Milano: il censimento e la gestione del verde con strumenti Web Gis. 9.a Conferenza Nazionale ASITA, Catania (2005). http://atti.asita.it/Asita2005/Pdf/0186.pdf. Accessed Dec 2021

Assessing the Economic Value of the Unmovable Cultural Assets for Improving Their Resilience: The Case Study of the Church of Santa Maria dei Miracoli

Giulia Datola[✉] [iD], Vanessa Assumma[iD], and Marta Bottero[iD]

Interuniversity Department of Regional and Urban Studies and Planning, Politecnico di Torino, Turin, Italy

{giulia.datola,vanessa.assumma,marta.bottero}@polito.it

Abstract. Cultural Heritage is exposed to many stresses and risks due to climate change, natural events, and human actions. During the last decade, the interest in risk reduction and risk assessment is getting significant attention, also for cultural assets. The current challenge is improving their resilience to safeguard and valorize them for future generations. This contribution is related to the European project of ResCult, "Increasing Resilience of Cultural heritage: a supporting decision tool for the safeguarding of cultural assets."(2017; 2018). The present paper focuses on assessing the economic value of the Church of Santa Maria dei Miracoli (Venice, Italy) as a component of the risk assessment. The proposed contribution applies the Travel Cost Method (TCM) to evaluate the economic value of the considered cultural asset. Finally, this contribution discusses the current state of the art of risk assessment in the European context, focusing the attention on its fragmentation. Moreover, the present paper reflects on the possible implementation of the proposed method at the European scale.

Keywords: Cultural heritage · Economic evaluation · Cultural resilience · Risk assessment · Travel cost method

1 Introduction

Cultural heritage (CH) is exposed to several pressures and risks due to natural events and human actions. Therefore, the current challenge is focused on safeguarding and valorizing these assets, both movable and unmovable, with specific attention to improving their resilience [1]. Before going in-depth into actions to increase the resilience of cultural heritage, it is necessary to define CH. In the national context, the first legislative definition of cultural heritage was developed following the "*Convenzione per la protezione dei beni culturali in caso di conflitto armato*", signed in 1954. However, the concept of cultural heritage is very recent. It is defined as "*all the goods of particular historical, cultural and aesthetic importance, which are of public interest and give wealth to a place*" [2]. In this sense, it is possible to outline that this definition has a broad scope that also alludes to the economic value owned by the goods that are part of this ensemble.

© The Author(s), under exclusive license to Springer Nature Switzerland AG 2022

F. Calabrò et al. (Eds.): NMP 2022, LNNS 482, pp. 2565–2574, 2022.
https://doi.org/10.1007/978-3-031-06825-6_245

However, it is necessary to define the typology of economic goods to which CH belongs before introducing its evaluation in monetary terms. CH assets are classified as intermediate between public and private goods [3]. In literature, goods are distinguished into public and private goods according to their consumers' audience and consumption mode. If private goods are characterized by rivalry and excludability variables, public goods are distinguished by non-rivalry and non-excludability variables [4]. Moreover, both public and private goods are distinguished by the reference market's presence or absence. Private goods are characterized by a reference market with a supply and demand for the good for private goods. On the contrary, public goods do not have a reference market [3]. The absence of a market implies different challenges to determine the demand function of a public good. The presence of a reference market permits useful information for economic and monetary valuation. However, when dealing with the public goods without the reference market, the monetary evaluation can be developed through indirect methods or rather the revealed preferences, or by the direct methods represented by stated preferences. The revealed preferences methods evaluate the economic value by observing the market of goods related to the public good to be assessed. The Hedonic Price Method (HPM) and the Travel Cost Method (TCM) belong to this category [4]. The stated preference methods evaluate the economic value of public goods by analyzing consumers' preferences, who are invited to declare their Willingness to Pay (WTP) to utilize the asset under analysis. In detail, this approach is defined as Contingent Valuation Method (CVM) [4]. This contribution focuses on the evaluation of the economic value of the Santa Maria dei Miracoli's Church in Venice (Italy) within the field of the CH risk assessment [1, 5, 6]. In light of the above mentioned motivations, this paper retains the TCM as suitable for this case study [7, 8]. The article is structured as follows: Sect. 2 is dedicated to the description of the risk related to CH and its parameters, Sect. 3 describes the applied methodology, Sect. 4 concerns the illustration of the analyzed case study, Sect. 5 illustrates the TCM application, Sect. 6 discusses the obtained results, and Sect. 7 concerns the conclusion and the reflections on future research perspectives.

2 The Concept of Risk

Over the latest years, Disaster Risk Reduction (DRR) and risk evaluation are getting increasing attention in the political agenda [5]. In this context, three international frameworks have to be considered, the Yokohama Strategy and Plan of Action for a Safer World [9], the Hyogo Framework for action 2005–2015 [10], and the Sendai Framework [11]. In detail, the evaluation of the risk is also getting great importance in the field of CH [1, 12, 13]. This new vision aims at preserving and valorizing cultural assets for future generations as sense of belonging to sites and identity, as well as historical evidence of the past. This paragraph illustrates the concept of risk with its components that have to be considered to develop a comprehensive evaluation [1]. The concept of risk can be divided into its main components, i.e. hazard, vulnerability, and value. The risk is thus described as:

$$Risk = f(p, E, V) \tag{1}$$

where p is the probability of occurrence of a given disaster; E is the exposure;

V stands for the vulnerability.

Vulnerability (V) is defined as the characteristics and circumstances of a community, system, or resources that make it sensitive to hazards [14]. Exposure (E) concerns the totality of elements present in the hazard areas (e.g. people, or buildings), which are subject to potential losses [15, 16].

In detail, the risk for cultural heritage $(Risk_{CH})$ is evaluated considering all these components. The $Risk_{CH}$ is thus characterized:

$$Risk_{CH} = f(P_{CH}, Vu_{CH}, Val_{CH}) \tag{2}$$

where Hazard$_{CH}$ is an expression of the geographical and statistical components of the risk [17], Vu_{CH} describes the intrinsic vulnerability of a building to the considered phenomenon [18, 19], and the Val_{CH} represents the economic, social, and cultural value expressed by the asset [20].

This contribution puts the attention on the value of cultural heritage (Val_{CH}), which is described by Eq. 3:

$$Val_{CH} = f(V_e, V_c, V_s) \tag{3}$$

The economic value (V_e) indicates the estimated value of the building, related to the possible market value and the cost of its hypothetical reconstruction. The cultural value (V_c) is an expression of the importance of the building, also related to the context. The social value (V_s) can be defined as the importance given to the building by the local community and occasional users. In detail, this contribution is focused on the assessment of the economic value of the considered cultural asset.

3 Methodology

As addressed in the previous section, cultural assets refer to a particular category of goods with no reference market [21]. Therefore, their economic evaluation is performed through non-market evaluation techniques that are classified into two categories: revealed preference method and stated preferences method (Sect. 1) [22]. This paper proposes the application of the TCM [8] to assess the economic value of the Church of Santa Maria dei Miracoli in Venice (Italy). This method belongs to the category of the revealed preference, and it uses the consumer surplus as the unit of measurement for valuation. Clawson introduced it in 1959, starting from insights made earlier by economist Harold Hotteling in 1930 to evaluate environmental assets (e.g., parks, forests, etc.). During the last decade, the TCM has been widely applied for the economic evaluation of cultural heritage [21]. The TCM estimates the number of journeys tourists should make to a particular cultural site [21, 23]. It is based on the demand theory, and it assumes that the demand for a site is inversely related to travel costs [24].

The TCM assesses the benefits generated by using the cultural heritage asset. The basing concept of this method is that as the cost of using an asset increases, the number of visitors to the asset decreases. Therefore, the TCM is included in the category of methods based on revealed preferences because it uses tourists' behavior and effective choices to deduce the use-values of tourist sites [25]. In detail, the cost is expressed both in monetary terms and in terms of time spent to reach the good in question. The cost that has the most significant influence on the valuation is the transport costs. Still, other categories of costs are also relevant to the valuation, such as meal costs, accommodation costs, entrance fees, etc. It can be said that the costs incurred by the visitor represent an implicit Willingness to Pay (WTP) for the use of the good itself. The evaluation through the method of travel costs leads to the construction of a demand curve. The curve is constructed concerning the various visitors' costs to enjoy the goods. The considered costs are transport costs (to reach the place under investigation), other costs related to the time dedicated to the trip, or additional costs of various kinds. It can be stated that the objective of the following methodology is to estimate the recreational value connected to the visit of an environmental or cultural good. Five different steps for the recreational experience have been theorized [7]: (i) planning, (ii) outward journey, (iii) recreational experience, (iv) return journey, (v) recollection of the experience.

The most relevant steps for formulating an evaluation through TCM are the outward and the return journey. These steps are of fundamental importance because of the high cost of the economic assessment. The higher the price the visitor is willing to pay to reach the recreational good and enjoy it, the greater the economic value and importance.

4 Case Study

The analyzed case study is the Santa Maria dei Miracoli Church in Venice (Italy). It is located in the historic center of the city, in the Cannaregio district, between the Rialto Bridge and San Marco square (Fig. 1). The Church is one of Venice's first Renaissance-style buildings. The principal artistic value of this Church is to be entirely covered, both outside and inside, by polychrome marble (e.g. veined Tuscan *pavonazzetto*, Istrian stone, or serpentine) [26, 27]. The Church of Santa Maria dei Miracoli has been chosen as one of the pilot cases study for the ResCult project as it represents a representative historic and ecclesiastical building in the city of Venice, which is vulnerable to flooding events, affected by high water events of the Venice lagoon, but also earthquake and fire hazards.

Fig. 1. Territorial framework (Totaro, 2018)

5   Application

The TCM has been applied to evaluate the economic value of the Santa Maria dei Miracoli Church related to the risk assessment, due to its ability to analyze the relationship among its use value and its conservation value, as well to its focus on the market that analyze socio-economic information [21]. The methodology of indirect costs is used here to perform the described analysis. In detail to overcome the issue related to the multiple destination trips, it has been used the cost from the departure location to reach the Santa Maria dei Miracoli Church [23], as in-depth described in the following paragraph. Therefore, the utility obtained from each visitor who uses the asset is thus calculated. Firstly, the data related to tourist flow and their origin are required to carry out this type of evaluation. For this purpose, a questionnaire was carried out to get this information. The questionnaire was performed to be compiled both on-site through a paper-based approach and online through a Google form, specifically created.

The sample of the completed questionnaires amounted to 83 tourists. Moreover, other statistical and non-statistical data have been collected to perform this economic evaluation. These data are the number of Italian and foreign tourists visiting Venice in one year and the number of Italian and foreign tourists staying in one of the Veneto provinces and visiting Venice. This calculation is made by measuring their WTP, a defined amount of money not to renounce the use of the services linked to the good. The WTP for the use of the case study has been calculated by estimating the costs incurred by tourists to reach the asset, from their departure location [23].

The evaluation has been divided into six different steps:

1. *Subdivision of the geographical area of the tourist allocation.* The evaluation assumes that the visitors are in one of the seven provinces of Veneto. Thus, it is possible to estimate the travel costs supported by the tourists to visit the Church. Figure 2 illustrates the subdivision of the geographical area into three zones: 1) the metropolitan city of Venice, 2) the provinces of Padua/Rovigo/Treviso, and 3) the provinces of Belluno/Vicenza/Verona.

Fig. 2. Division of the area into concentric circles (Totaro, 2018)

2. *Calculation of the number of tourists that visit the metropolitan city of Venice and the number of tourists visiting the Church of Santa Maria dei Miracles over a year*

compared to their area of location (Fig. 2). This calculation was based on the ISTAT database, the data provided by the statistical office of the Veneto Region, and the data of the annual Church provided by the Chorus association.

3. Calculation of the visitor frequency essay for each zone assumed in step 1, through Eq. 4

$$S_f = (V_c/V_v) \cdot 1000 \tag{4}$$

where: S_f is the frequency essay; V_c is the number of visits of the Santa Maria dei Miracoli Church in a year; V_v is the total number of the visit to Venice in a year.

4. *Calculation of the travel cost, according to the different zones of departure.* An average distance to the Santa Maria dei Miracoli Church was calculated for each area previously identified. Thus, the travel cost is given by the sum of the economic cost, considering the fuel expenditure for the distance traveled, and the time cost to travel from the place of stay to the Church.

5. *Regression function evaluation.* This function compares the total travel cost with the frequency essay (Fig. 3).

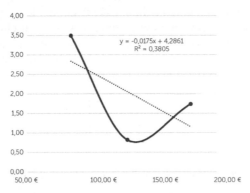

Fig. 3. Regression function (Totaro, 2018)

6. *Estimation of the demand curve.* It is constructed starting with an entry ticket to use the good. Successively, it is assumed that its price increases progressively until the number of visitors to the asset itself. By studying visitors' willingness to pay, we can hypothesize the importance attributed to its users' cultural significance attached to the cultural object. From the study of the price increase, the number of visitors dropped to zero after adding a price increase of about 113€.

7. *Construction of the consumer surplus graph.* From the area below the curve, it is possible to evaluate the economic value of the good through the travel cost methodology (Fig. 4).

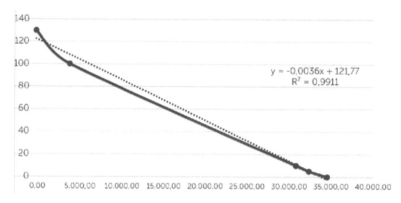

Fig. 4. Costumers' surplus (Totaro, 2018)

6 Discussion of the Result

The Total Economic Value (TEV) of the Santa Maria dei Miracoli Church has been obtained from the consumers' surplus. In this application, the geometric methodology has been applied to get the economic value from consumers' surplus. Table 1 illustrates the Relation between price variation and the number of visits, and the TEV of the Santa Maria dei Miracoli Church [27]. As an example, the calculation of the formula related to the increase of 5 € of the ticket price is here reported by Eq. (5):

$$\left[(0 + 5) \cdot \frac{34.465 - 32.306}{2} \right] = 5.397{,}73€ \tag{5}$$

Table 1. Relation between price variation, number of visits and the TEV of the Church (Totaro, 2018)

Increase of ticket price	Total of visits	Formula
+0 €	34,465	
+5 €	32,306	5,397,7. €
+10 €	30,802	11,282,2. €
+100 €	3,724	1,489,253.19 €
+113 €	0	396,636.15 €
Total Economic Value (TEV)		1,902,569.28 €

From the sum of the obtained results by the geometric method, the TEV of the Santa Maria dei Miracoli Church is equal to 1.902.569,28 €.

7 Conclusion

This paper illustrates the application of the TCM to assess the TEV of the Santa Maria dei Miracoli Church as a component to be included in the risk assessment. In this first and preliminary application, the TCM has been selected considering its ability to assess the economic value through analyzing real behavior (e.g. cost of travel, frequency of visit) [21] of socio-economic reference market. More in detail, the TCM has been chosen with respect to other existing methods, such as the HPM, as the latter requires very specific data, and it bases the evaluation on specific intrinsic characteristics of the asset, e.g. structural and spatial characteristics [28, 29]. Moreover, according to the purpose of evaluating the economic value of the Santa Maria dei Miracoli Church to include it the risk assessment, the TCM has been selected for its ability to relate the use value of the cultural asset and its cost of conservation. In this sense, the TCM has been chosen as a method that is easy to perform and repeatable to insert its result in a more comprehensive risk evaluation framework [1] developed from the ResCult project. Furthermore, it can consider proposing a common risk assessment framework for the European context thanks to its ease of performing. It should be regarded as to ask to the fragmentation that characterizes the protection and the valorization procedures in the EU. This will surely produce benefits on building sustainable and resilient cities and communities [30]. However, according to the current challenge of evaluating the risk of cultural assets to improve their resilience, it is fundamental to analyze the limits of the proposed method, to find a solution in the future implementation. The first vulnerability concerns the multi-purpose trips. In this application, the simplification of calculating the cost from the departure site to the cultural asset has been applied. Moreover, for the future perspective, it should be used the multisite model with random utility maximization, to compare the obtained results [21, 31]. Another weakness detected by the authors to be considered in the future is the substitution, or rather the omission of demand curves that reflect the presence of replaceable sites, which could imply overestimation of the true demand curve [31]. Moreover, a suitable future implementation of the present research is the application of the HPM in order to evaluate the economic value of the cultural asset considering only its recreational use, but according to its different and multiple features [29, 32].

Acknowledgments. The present contribution is based on the thesis work titled "*Un approccio integrato per la valutazione dei beni culturali*" (2018) by Lorenzo Totaro with the supervision of Professor Marta Bottero, Politecnico di Torino.

References

1. Appiotti, F., Assumma, V., Bottero, M., et al.: Definition of a risk assessment model within a European interoperable database platform (EID) for cultural heritage. J. Cult. Herit. **46**, 268–277 (2020). https://doi.org/10.1016/j.culher.2020.08.001
2. Bottari, F., Pizzicannella, F.: L'Italia dei tesori: legislazione dei beni culturali, museologia, catalogazione e tutela del patrimonio artistico. Zanichelli (2002)
3. Roscelli, R.: Manuale di estimo. Valutazioni economiche ed esercizio della professione. UTET Università (2014)

4. Grillenzoni, M., Grittani, G.: Estimo. Teoria, procedure di valutazione e casi applicativi, 2nd edn. Calderini, Milano (1994)
5. Appiotti, F., Assumma, V., Bottero, M., et al.: Un modello di valutazione del rischio per il Patrimonio Culturale. RIV Rass Ital di Valutazione 121–148 (2018). https://doi.org/10.3280/RIV2018-071007
6. Rescult, F., Forum, U.: NEWSLETTER 4/2018 This is the 2nd RESCULT newsletter ! What has been done EID Conceptualization 1–5 (2018)
7. Torres-Ortega, S., Pérez-Álvarez, R., Díaz-Simal, P., et al.: Economic valuation of cultural heritage: application of travel cost method to the national museum and research center of Altamira. Sustainability 10, 2550 (2018). https://doi.org/10.3390/su10072550
8. OECD: Cost-Benefit Analysis and the Environment. OECD (2018)
9. United Nations: Yokohama Strategy and Plan of Action for a Safer World: Guidelines for Natural Disaster Prevention, Preparedness and Mitigation. World Conf. Nat. Disaster Reduct. Yokohama, Japan, 23–27 May 1994 (1994)
10. United Nations International Strategy for Disaster Risk Reduction: Hyogo Framework for Action 2005–2015: Building the Resilience of Nations and Communities to Disasters
11. Sendai framework for disaster risk reduction 2015–2030 (2005)
12. Accardo, G., Giani, E., Giovagnoli, A.: The risk map of Italian cultural heritage. J. Archit. Conserv. 9, 41–57 (2003). https://doi.org/10.1080/13556207.2003.10785342
13. Diaferio, M., Foti, D., Sabbà, M.F., Lerna, M.: A procedure for the seismic risk assessment of the cultural heritage. Bull. Earthq. Eng. 19(2), 1027–1050 (2021). https://doi.org/10.1007/s10518-020-01022-8
14. Alexander, D.: The study of natural disasters, 1977–1997: some reflections on a changing field of knowledge. Disasters (1997). https://doi.org/10.1111/1467-7717.00064
15. UNISDR: Hyogo Framework for Action 2005–2015. In: United Nations International Strategy for Disaster Reduc (2005)
16. UNISDR: Sendai Framework for Disaster Risk Reduction. UN World Conf (2015)
17. Yıldırım Esen, S., Bilgin Altınöz, A.G.: Assessment of risks on a territorial scale for archaeological sites in İzmir. Int. J. Archit. Herit. 12, 951–980 (2018). https://doi.org/10.1080/15583058.2017.1423133
18. Modena, C., da Porto, F., Valluzzi, M.R., Munari, M.: Criteria and technologies for the structural repair and strengthening of architectural heritage. In: International Conference Chennai 2013:16th (2013)
19. Schanze, J.: Flood risk management – a basic framework. In: Flood Risk Management: Hazards, Vulnerability and Mitigation Measures (2007)
20. Drury, P., McPherson, A.: Conservation Principles, Policies and Guidance (2008)
21. Merciu, F.-C., Petrişor, A.-I., Merciu, G.-L.: Economic valuation of cultural heritage using the travel cost method: the historical centre of the municipality of bucharest as a case study. Heritage 4, 2356–2376 (2021). https://doi.org/10.3390/heritage4030133
22. Salvo, F., Dell'Ovo, M., Tavano, D., Sdino, L.: Valuation Approaches to Assess the Cultural Heritage, pp. 1746–1754 (2021)
23. Bedate, A., Herrero, L.C., Sanz, J.Á.: Economic valuation of the cultural heritage: application to four case studies in Spain. J. Cult. Herit. 5, 101–111 (2004). https://doi.org/10.1016/j.culher.2003.04.002
24. Prayaga, P.: Estimating the value of beach recreation for locals in the Great Barrier Reef Marine Park, Australia. Econ Anal Policy 53, 9–18 (2017)
25. Leh, F.C., Mokhtar, F.Z., Rameli, N., Ismail, K.: Measuring recreational value using travel cost method (TCM): a number of issues and limitations. Int. J. Acad. Res. Bus. Soc. Sci. 8, 1381–1396 (2018). https://doi.org/10.6007/IJARBSS/v8-i10/5306
26. Piana, M., Wolters, W.: Santa Maria dei Miracoli a Venezia. La storia, la fabbrica, i restauri, Venezia (2003)

27. Totaro, L.: Un approccio integrato per la valutazione dei beni culturali. Politecnico di Torino (2018)
28. Dell'Anna, F., Bravi, M., Bottero, M.: Urban Green infrastructures: how much did they affect property prices in Singapore? Urban For Urban Green **68**, 127475 (2022). https://doi.org/10.1016/j.ufug.2022.127475
29. Lazrak, F., Nijkamp, P., Rietveld, P., Rouwendal, J.: Cultural heritage: hedonic prices for non-market values. VU Univ. Res. Memo **2009–49**(49), 12 (2009)
30. Berisha, E., Caprioli, C., Cotella, G.: Unpacking SDG target 11.a: what is it about and how to measure its progress? City Environ. Interact. **14**, 100080 (2022). https://doi.org/10.1016/J.CACINT.2022.100080
31. Stellin, G., Rosato, P.: La valutazione economica dei beni ambientali. Metodologia e casi di studio. Città Studi, Torino (1998)
32. Dell'Ovo, M., Dell'Anna, F., Simonelli, R., Sdino, L.: Enhancing the cultural heritage through adaptive reuse. a multicriteria approach to evaluate the castello Visconteo in Cusago (Italy). Sustainability **13**, 4440 (2021). https://doi.org/10.3390/su13084440

Mountain Hamlet Heritage Between Risk and Enhancement

Elisabetta Colucci[✉]

Department of Environmental, Land and Infrastructure Engineering (DIATI), Polytechnic University of Turin, 10125 Turin, Italy
elisabetta.colucci@polito.it

Abstract. This paper focuses on villages and hamlets on the mountain considered historical, cultural heritage. Due to their cultural and historical values, they need documentation, enhancement, and revaluation activities. The research aims to semantically and spatially describe minor historical centres such as hamlets affected by natural hazards and risk phenomena (landslide, avalanches, …). In this domain, many actors and use cases are involved in different activities such as urban and landscape planning, restoration, tourism, etc. For this reason, it is necessary to develop a replicable and sharable workflow able to harmonise and integrate different sources (spatial, structured, unstructured, historical and so on) to document these centres thoroughly. A significant case study has been selected to apply the presented methodology. It is the hamlet of Pomieri, in the municipality of Prali (Piedmont Region, Alps, Germanasca Valley, Italy). Different documents about natural hazards phenomena have been investigated parallelly to semantic and ontologies sources. This method set the basis for developing a knowledge base, including further information on such centres subject to risk and needing documentation, restoration, preservation, maintenance and planning actions.

Keywords: Minor historical centres · Alpine hamlets · GIS · Natural hazards · Geospatial ontologies

1 Introduction and Objectives

It is well known that Cultural Heritage (CH) need to be documented and protected for their intrinsic cultural, historical and artistic values. This work focuses on minor historical centres (MHC) such as hamlets and villages intended as CH, and they need protection, renovation, enhancement, and maintenance activities as built ancient heritage. Moreover, due to their position, morphology, climate conditions and historical characteristics, they could be naturally damaged by time and natural phenomena such as natural hazards. Hamlets and mountain villages are located in marginal and rural inner areas that need revaluation plans and reinhabiting actions. In this scenario, the paper aims to semantic and spatial document these minor historical centres to spread knowledge and cultural values. The methodology focuses on built heritage subject to risks such as avalanches, floods and landslides and their semantic and spatial documentation. This paper is an excerpt of a PhD research aimed to develop a spatial ontology to semantically document

© The Author(s), under exclusive license to Springer Nature Switzerland AG 2022
F. Calabrò et al. (Eds.): NMP 2022, LNNS 482, pp. 2575–2586, 2022.
https://doi.org/10.1007/978-3-031-06825-6_246

MHC. The present work wants to enlarge the doctoral research and populate and enrich the ontology in order to validate the methodology by adding practical example from a real case study in the domain of HC, in this case an hamlet, subject to natural risks. Hence, the core of the methodology aims to semantically, temporal and spatially describe MHC such as hamlets affected by natural hazards and risk phenomena. In this cultural and landscape domain, many stakeholders and use cases are involved in various activities and for this reason, it is necessary to develop a replicable and sharable approach able to harmonise and integrate different sources (spatial, structured, unstructured, historical and so on) to fully describe these centres. For the objectives above mentioned, a significant case study has been selected to apply the presented methodology. It is the hamlet of Pomieri, in the municipality of Prali (Piedmont Region, Alps, Germanasca Valley, Italy). Different documents about natural hazards phenomena have been investigated parallelly to semantic and ontologies sources. This method set the basis for developing a knowledge base, including further information on such centres subject to risk and needing documentation, restoration, preservation, maintenance and planning actions.

1.1 Research Framework and Literature Background

In these last years, a lot of attention was paid to these hinterland areas due to events, such as climate changes conditions and covid isolations, that led people to reinhabit these abandoned places and start thinking about a more "slow living" [1, 2]. As mentioned above, many actors are involved in that domain [3], such as architects, restorers, citizens, civil protection, urban planners, policymakers, etc. All these actors involved in this domain aim to adopt resilient plans for which they need to cooperate. Thus, it is necessary to spatial represent such centres with common standards, a unique shared language and harmonised datasets. Various methods and technologies have been developed to represent CH and their semantic information. In the Geographic Information (GI) domain, Spatial Data Infrastructures (SDIs) and Geographic Information Systems (GIS) helps to store, collect and represent heritage spatial objects. Many examples are reported in the recent academic GI field for restoration purposes, archaeology, 3D city models, and Building Information Modelling (BIM) areas [4–6]. The semantic description of the historical heritage is also a fundamental aspect. For this purpose, spatial ontologies could play an essential role. In computer science, ontology is an information object or computational artefact, "formal, explicit specification of a shared conceptualisation" [7]. Many research and projects adopted an ontological approach to semantic represent cultural and built heritage (see Sect. 1.2). Hence, this methodology focuses on necessary information to document the mountain hamlets at risk by collecting and organising datasets from different sources (structured and unstructured). Then, a step of this workflow is dedicated to describing semantic definitions of objects and characteristics around the domain of minor HC. This work starts from standards and already developed ontologies and conceptualisation in CH and HC (MHC ontology, Colucci, https://w3id.org/lode/owlapi/http://purl.org/net/mh-centre_v1) to widen the knowledge base and specify concepts that connect villages and hamlets with risk phenomena.

The paper is structured as following. The notion of MHC is considered starting from the general concept of Cultural Heritage (CH). Moreover, to spatial analyse and define built heritage and its territory, it is necessary also to give an overview of the existing standards in the geographical and architectural domains. A practical example with a specific case study is investigated, reporting a summary of possible examples (instances) that could be added to an MHC spatial ontology. A significant case study is selected to apply and validate the methodology: the Pomieri hamlet in the Germanasca Valley (Piedmont, Italy) (Sect. 2). 1.2 Cultural Heritage and Minor Historical Centres notions MHC are part of built cultural heritage from which they inherit historical, cultural and social values. Thus, it is necessary to clarify the meanings and the evolution of the *notion of historical centre*. A literature investigation extensively studies historical books and descriptions of architecture and urbanism on the topic to define the ontology domain adequately. In this literature, urban cores, parts of cities and small villages or hamlets are all considered HC. The different semantic definitions of HC have changed over time, and therefore it is necessary to understand the evolution of the concept in the last 50 years. The UNESCO definition of cultural heritage is expressed in the "Convention Concerning the protection of the World Cultural and Natural Heritage: The General Conference of the United Nations Educational, Scientific and Cultural Organization" meeting in Paris from 17th October to 21st November 1972 (UNESCO, 1972). The first article says: "The following shall be considered as cultural heritage: monuments, groups of buildings and sites". Built heritage represents another crucial cultural asset; it is the historical layers of our built environment in places. The term cultural built heritage is, therefore, a broad concept. "According to the UNESCO Universal Declaration on Cultural Diversity (UNESCO, 2002), culture is the set of distinctive spiritual, material, intellectual and emotional features of a society or a social group that encompasses art and literature, lifestyles, ways of living together, value systems, traditions and beliefs. Thus, urban culture covers the notions of culture within an urban setting from both a functional and anthropological perspective". The notion of HC evolved during the centres, including more and more built heritage out of cities. The first definition of Historical centres appears in 1967, in the Commissione Franceschini titolo IV ("Dei beni ambientali"), dichiarazione XL ("Centri storici e loro tutela"). Giovanni Astengo defined HC as "urban settlement structures that constitute cultural unity or the original and authentic part of settlements and testify the characteristics of a lively urban culture. (…) For operational purposes, the protection of historic centres will have to be implemented through preventive measures and definitive through regulatory plans". Alberto Predieri introduces the concept of minor historical centres in his report for the "VI Convegno A. NCSA", 1971 [8]. He classified historical centres into three different categories: Historical centres in cities, Minor historical centres included in developing cities now lapsed, but with great historical-artistic-environmental value and cultural and touristic interest (according to the Venice Charter of 1964 and Disegno di Legge n. 1942 in Italy, minor historical centres are municipalities with fewer than 5,000 inhabitants), and Abandoned Minor historical centres. In 1975 the European Charter of Architectural Heritage defined "abandoned places". Rolli [9] includes in the definition of minor historical centres also small villages and hamlets.

1.2 Material and Methods: GIS, Standards and Geospatial Ontology to Support the Spatial Documentation of MHC

As explained in [10], the interest in ontologies applied to the built heritage and urban domain was initially triggered by the technological challenges linked to the interoperability of urban and territorial databases and the need to interconnect the different databases. GIS spread has characterised many urban and city databases even among urban planning experts. A further update of these databases to make them more readily available and link them to other data sources couldn't be possible without restructuring their content. Given the scale and complexity of the activity, ontological engineering was selected as a necessary step to manage both the conceptual and continuity with previous database versions. Below, definitions of GIS, SDIs and geospatial ontologies are reported. Moreover, some standards to represent built heritage and GI objects are described.

Used Tools: GIS and SDIS. According to the Global Spatial Data Infrastructure Association Cookbook [11], "the term Spatial Data Infrastructure is often used to indicate the collection of technologies, policies and institutional provisions that facilitate the availability and access to spatial data". The best tool to manage data in numeric cartography is GIS, a working method to design a multi-dimensional model to represent the real world with the final aim to create a project by the user. GIS is a computer-based information system that enables capture, modelling, storage, retrieval, sharing, manipulation, analysis, and presentation of geographically referenced data. In the last years, GIS has evolved, including 3D GIS, in which it is possible to visualise, relate, and query 3D geometrics into a geographical environment. Finally, WebGIS is a GIS published on the web.

Methodology Application: Spatial Ontologies. It is possible to refer to an ontology in knowledge communities, especially in Computer Science and Semantic Web, to identify "a kind of information object or computational artefact" [12]. A spatial ontology defines the general concepts of spatial objects and their relations for spatial application domains. "Etymologically, geography means the description of the Earth, while ontology refers to the discourse about existing things. Hence, geographic ontology means describing things existing on Earth, i.e., geographic features." [13]. It is possible to consider the ontologies applied in GIScience as domain ontologies. These "are often called geographic ontologies or geo-ontologies" [14]. The W3C standards also define geospatial ontologies.

Existing Used Knowledge: Standards for Architectural Heritage and Geographic Information. Many existing standards and vocabularies represent built urban and architectural heritage knowledge. The core ontology representing CH is the CIDOC Conceptual Reference Model (CIDOC-CRM). In the framework of the CIDOC-CRM ontology, some extensions have been developed (such as CRMgeo, CRMba, CRMsci and CRMarcheo ontologies). The Getty Vocabularies represent other standards. The Art & Architecture Thesaurus (https://www.getty.edu/research/tools/vocabularies/aat/) contains terms, concepts and vocabularies related to art and architecture. The MIBACT and the ICCD have been developed different projects and ontologies such as ArCo, the

Knowledge Graph (KG) of the Italian CH and Cultural-ON ontology for cultural places. In the geo-information field, there are various spread and used standards. At first, as regards urban content, there is CityGML (http://www.opengeospatial.org/standards/cit ygml), an international standard data model published by the OGC to represent multi-scale 3D information about entities of cities. Secondly, the INSPIRE data model (http://inspire.ec.europa.eu/). It is part of the European Directive to reach interoperable cartography in Europe. GeoSPARQL is an OGC standard that intends to represent and query geospatial data; it defines the geo-classes Spatial Object, Feature and Geometry. Many research [15–17], different methodologies to enrich and populate a geospatial ontology to enable semantic information extraction have been presented. Ontologies for Urban Development are selected to describe ontologies in urban domain areas [18, 19]. Few attempts have been made in architectural built heritage and HC framework [20, 21].

Despite this developed knowledge and standards, none of these examples can completely describe and document hamlets and villages in risk scenarios. Thus, the paper collects various datasets and documentation to describe the hamlet case study. It tries to enlarge the semantic knowledge of such domain by giving the possible example of ontologies integration with classes and properties connected to villages, hamlets, their morphology and natural risks. This investigation represents a possible implementation of the developed MHC ontology mentioned above.

2   The Pomieri Semi-abandoned Hamlet Case Study

The case study here reported is not officially acknowledged and safeguarded for its patrimonial values and it is now in state of abandonment. For this reason the research has an impact in its future preservation. Moreover, it has been selected to validate the methodological approach by checking if the necessary information to describe and document minor historical centres have been considered. This step also represents the possible replicability of the innovative methodology on a MHC subject to risk. It represents semi-abandoned mountain hamlets in Piedmont Alps, damaged by natural phenomena requiring spatial documentation for restoration, valorisation, and sustainable urban resilient planning. This approach also shows the potential reuse of this knowledge, as underlined in one of the ontology tasks of the Ontology Guide of Noy & McGuinness (2001). Moreover, this ontological structure could be applied in a whole or part to other case studies becoming an application ontology. The Italian Mountain hamlet of Pomieri is located in the municipality of Prali in the Piedmont Region. Its old name in dialect is "Li Poumie", which means "the apple trees". It is situated in a mountain area in the Germanasca Valley at 1511 m. A.s.l. Relatively steep slopes characterise Pomieri and the entire municipality of Prali; consequently, they are naturally exposed to snow slides. Some testimonies revealed the existence of Prali even in the 11th century, but the most reliable start from the 15th century. The dimension is around 22.500 sqm with 17 buildings only and 11 inhabitants. In the summer of 2019 (from the 26th to the 28th of June), the geomatics group of Politecnico of Torino and the student Team DiRECT (Disaster Recovery Team) [22] carried out a 3D integrated metric survey in Prali. The team is devoted to acquiring spatial data with rapid systems techniques in damaged areas or places affected by natural hazards. The survey aimed to document, and 3D represent these places to support planning activities and the regeneration of alpine regions (Fig. 1).

Fig. 1. Aerial view of the Pomieri hamlet. Image acquired during the 3D survey.

2.1 Natural Hazards Evidence

The various avalanches that occurred in the Germanasca valley represent a real risk for the population of Prali and their houses. The so-called "springer avalanches" in April month are the ones that damaged more the hamlets. Many villages are now abandoned for this reason, and some bridges are collapsed. In the Maira avalanche of 1832, there were victims of twelve pralines returning from work in the vineyards downstream of Ghigo, Orgiere and Pomieri. Two different dangerous avalanches sites are present in the Pomieri Area, and these are also reported in the SIVA (https://webgis.arpa.piemonte.it/Geovie wer2D/?config=otherconfigs/SIVA_config.json) WebGIS of the Piedmont Region. The sites are:

1. Pomieri site, Orgiere, Prali (Torino) (https://comune.prali.to.it/cgi-bin/prgc/021220 2014373_COMUNE_DI_PRALI.pdf). The code of the site is n° 75_V_TO. The events occurred in: 10th April 1972, 17th March 1978, 16th December 2008 and February 2009.
2. Passo della Scodella site, Pomieri, Prali (Torino) (http://webgis.arpa.piemonte.it/ webval/valdati_new_prod.php?CODECATE=76_V_TO). Code avalanche site n° 76_V_TO. Events occurred in: 11st March 1832, March 1969, 16th April 1972, 17th March 1978, 31st January 1986, 4th April 1987, 9th April 1992, 16th December 2008 and February 2009.

The article in La Beidana [23] reported other risk phenomena, such as the landslide of 1220 that hit the church of San Giovanni in Prali. The paper reported in Fig. 2b described the floods and avalanches that hit the alps in January 1978.

The areas affected by hazards are reported in the "Map of avalanche phenomena" by SIVA. Regarding the avalanche phenomenon that marginally affected the western portion of Pomieri during the winter of 2008–2009, a specific cartographic excerpt has been prepared, shown below, based on the photographic documentation (Figs. 3 and 4).

Fig. 2. (a, b). (a) SIVA WebGIS. In the screenshot, the Pomieri localised danger is queried. (b) Article in the newspaper "La Stampa", 21st January 1978, describing avalanches that hit the area of Prali in those days, http://www.archiviolastampa.it/.

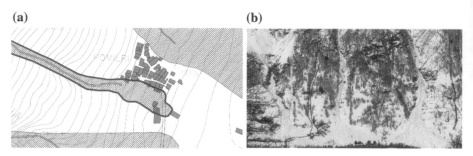

Fig. 3. (a/b) General Inter-municipal Town Plan (PRGI) (Sub Area Val Germanasca Municipality of Prali).

Fig. 4. (a/b) Buildings damaged by the avalanches of 2008 and degradation and abandonment.

3   GIS Projects of Pomieri

3.1   Other Available Sources

Spatial Data Sources. In the Piedmont Geoportal, it is possible to visualise and query spatial data compliant with the INSPIRE Directive. There are various services such as data search, download (WFS service, shapefiles and raster data) and consultation.

The BDTRE (Base Dati Territoriale di Riferimento per gli Enti Piemontesi, s://www.geoportale.piemonte.it/cms/bdtre/specifiche-per-cartografia-di-base) represents all the technical cartography of the Region, structured according to the national "Technical rules for the definition of the content specifications of geotopographic databases". It aims to support planning, governance and protection activities. Other spatial sources, with a high level of detail and accuracy, derived from the 3D metric survey and are the generated DSM (Digital Surface Model) and the orthophoto. All the datasets available for the Pomieri case study GIS are listed in Table 1.

Table 1. Available spatial datasets for the Pomieri case study.

Data description	Scale	Source
Technical Regional Map, the BDTRE 2019 representing toponym, hydrography, borders (of valley and municipalities), buildings, cities, roads, etc.	1:10k	https://www.geoportale.piemonte.it/cms/bdtre/specifiche-per-cartografia-di-base
PPR datasets of mountain areas, paths, peaks, ridges, etc. tables P4 and P5 of PPR	1:50k	https://www.regione.piemonte.it/web/temi/ambiente-territorio/paesaggio/piano-paesaggistico-regionale-ppr
The lithological unit map	1:10k	http://map.chisone-germanasca.torino.it/web/images/ValGermanasca/AdeguamentoPAI/Perrero/4.3_Carta%20caratteri%20litotecnici%20e%20idrogeologici.pdf
Avalanches, floods and landslides derive from Arpa Piemonte maps, Avalanche Information System (SIVA)	Max scale 1:5k	https://webgis.arpa.piemonte.it/geoportale/
DTM of south-western Piedmont – CTRN 1:10k	10 m	http://www.datigeo-piem-download.it/direct/Geoportale/RegionePiemonte/DTM10/DTM10.zip
DTM 2009-2011 ICE, CTR FOGLIO 50-172	5 m	http://geomap.reteunitaria.piemonte.it/ws/taims/rp-01/taimsgriwms/wms_griglie?
Historic cartography of IGM (Istituto Geografico Militare) of 1880, 1930, 1960 years	1:50k, 1:25k	https://www.igmi.org
DSM	1:100	*3D metric survey*
ORTHOPHOTO	1:100	

Documentations and Regulations. In addition to the spatial datasets, the Italian territorial and landscape plans have been analysed. The first important document from the landscape values point of view is the *Regional Landscape Plan* (Piano Paesaggistico Regionale, PPR). The PPR represents the primary tool to guarantee the quality of the landscape and sustainable environmental development of the entire regional territory. The PPR also provides thematic GIS maps for different areas ("Ambiti"). The Germanasca Valley and Prali are located in the "Ambito 41" of PPR. Some other specific existing documents for the municipality of Prali are:

- the General Plan of the Mountain Community of the Chisone and Germanasca Valleys, PRGCM (Piano Regolatore Generale della Comunità Montana Valli Chisone e Germanasca),
- the Intercomuncal general Regulatory Plan the PRGI (Piano Regolatore Generale Intercomunale, Unione Montana dei Comuni Valli Chisone e Germanasca, PRGI),
- the Building Regulation of Prali.

3.2   GIS Project

From the datasets, a GIS project has been designed in QGIS (version 3.16), below the framework map of Prali municipality and avalanches and landslides map from PPR datasets are reported. The available datasets used for the case study are in the cartographic RS WGS84- UTM zone 32N, EPSG:32632 (Figs. 5 and 6).

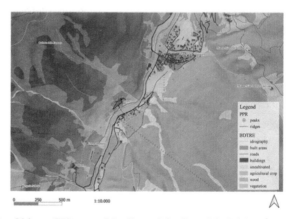

Fig.5. City Objects GIS map of Prali municipality (QGIS OS software, v. 3.16).

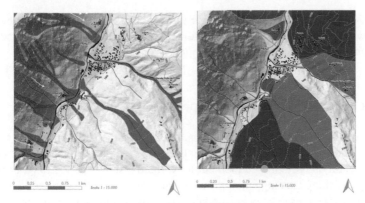

Fig. 6. Avalanches and landslides map (on the left, red show dangerous zone and blue documented avalanches, on the right, in violet scale different types of landslides and red conoids) designed in the framework of the MSc Cours "Riabitare le Alpi" 2019, Biffanti, Crivelli, Caiazzo, Dello Vicario, prof. A.Spanò.

4   Semantic Definition of Classes and Properties

The MHC existing ontology structure has been implemented by adding classes and relations in Protègè software (more detail will be reported in Colucci, Geospatial Ontology to support the Documentation of Minor Historical Centres, forthcoming, 2022). This paragraph presents an excerpt of possible classes, semantic concepts and definitions extracted from texts and documentation related to different activities in which the municipality of Prali and the hamlet of Pomieri are involved. The schema below is an example of the many different thematic areas and domains of the entities in various colours. In grey, existing classes of the ontologies mentioned above or standards already implemented into the MHC ontology are reported (Fig. 7).

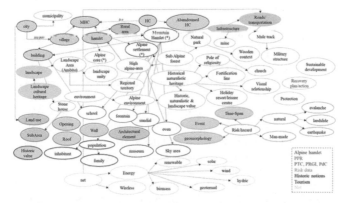

Fig. 7. Classes and relations from plans (black arrows are meronymic relations, blue "is-a" relations).

Moreover, some possible instances for the application ontology could be extracted from the available sources. In the MHC ontology, few examples have been added to prove the possibility of reusing such knowledge in various fields such as restoration actions, historical documentation, regional and city plans, and building permits processes.

5 Conclusions and Future Perspectives

All the information and documentation included and collected in this paper are helpful and essential to semantically and spatial describe the hamlet of Pomieri. This work validates that an ontology-based methodology could also be applied to other mountain villages by enriching existing ontologies and harmonising spatial datasets. The methodology is replicable, reusable, and extendable for further studies and domains. This work is located in the geomatics domain, particularly in the area of GI. As geomatics techniques and methods support the research in different disciplines, this approach wants to help various stakeholders (citizens, researchers, policymakers, etc.) to store and get information on the same domain: the small, historical, and abandoned centres. Thus, the paper gave an overview to develop further work and implement spatial ontologies and GIS or WebGIS with risk data connected with their semantics. A future perspective could be validating the methodology and the ontology by adding instances from other case studies. This process of inserting more information will enhance the documentation of MHC purpose, giving a 360° view of the domain. Other critical reflections and considerations are related to the possibility to improve the idea of "slow living" in inner or mountain area, and to develop digital instruments, twins and tools able to manage risk factors.

References

1. Boeri, S.: Via dalle città, nei vecchi borghi c'è il nostro futuro. La Repubblica (2020)
2. Koolhaas, R.: As dense metropolises become ovewhelmed, the countryside is seen as an escape hatch. Dezeen (2020)
3. Colucci, E., Kokla, M., Mostafavi, M.A., Noardo, F., Spanò, A.: Semantically describing urban historical buildings across different levels of granularity. Int. Arch. Photogramm. Remote Sens. Spatial Inf. Sci. 33–40 (2020)
4. Chiabrando, F., Colucci, E., Lingua, A., Matrone, F., Noardo, F., Spanòa, A.: A European Interoperable Database (EID) to increase resilience of cultural heritage. Int. Arch. Photogramm. Remote Sens. Spatial Inf. Sci. **42**(3/W4) (2018)
5. Brumana, R., Ioannides, M., Previtali, M.: Holistic heritage building information modelling (HHBIM): from nodes to hub networking, vocabularies and repositories. In: 2nd International Conference of Geomatics and Restoration, GEORES 2019, vol. 42, no. 2, pp. 309–316 (2019)
6. Previtali, M., Brumana, R., Stanga, C., Banfi, F.: An ontology-based representation of vaulted system for HBIM. Appl. Sci. **10**, 1377 (2020)
7. Studer, R., Benjamins, V.R., Fensel, D.: Knowledge engineering: principles and methods. Data Knowl. Eng. **25**(1–2), 161–197 (1998). https://doi.org/10.1016/S0169-023X(97)00056-6
8. Predieri, A.: Report for the "VI Convegno A.N.C.S.A." (1971)
9. Rolli, G.L.: Centri storici minori. Ed Alinea (2005)
10. Falquet, G., Mètral, C., Teller, J., Tweed, C.: Ontologies in Urban Devopment Projects. Springer, London (2011). https://doi.org/10.1017/CBO9781107415324.004

11. Nebert, D.D.: The SDI Cookbook–Developing Spatial Data Infrastructures. Global Spatial Data Infrastructure Association, 2 (2004)
12. Guarino, N., Oberle, D., Staab, S.: What Is Ontology? Handbook on Ontologies, International Handbooks on Information Systems, pp. 1–17 (2009). https://doi.org/10.1007/978-3-540-926 73-3
13. Laurini, R., Kazar, O.: Geographic ontologies: survey and challenges. J. Theor. Cartogr. **9**, 1–13 (2016)
14. Fonseca, F., Camara, G., Miguel, A.: A Framework for Measuring the Interoperability of Geo-Ontologies. Perception 5868 (2006). https://doi.org/10.1207/s15427633scc0604
15. Chaves, M.S., Rodrigues, C., Silva, M.: Data model for geographic ontologies generation. In: XML: Aplicações e Tecnologias Associadas (XATA 2007), pp. 47–58 (2007)
16. Kavouras, M.: A unified ontological framework for semantic integration. In: Next Generation Geospatial Information: From Digital Image Analysis to Spatiotemporal Databases, vol. 3, p. 147 (2005)
17. Tomai, E., Kavouras, M.: From 'onto-geonoesis' to 'onto-genesis': the design of geographic ontologies. GeoInformatica **8**(3), 285–302 (2004)
18. Teller, J., Lee, J.R., Roussey, C.: Ontologies for Urban Development. Springer, Heidelberg (2007). https://doi.org/10.1007/978-3-540-71976-2
19. Berdier, C., Roussey, C.: Urban ontologies: The towntology prototype towards case studies. Stud. Comput. Intell. **61**, 143–155 (2007)
20. Kokla, M., Mostafavi, M.A., Noardo, F., Spanò, A.: Towards building a semantic formalisation of (small) historical centres. ISPRS Ann. Photogramm. Remote Sens. Spatial Inf. Sci. **42**(2/W11), 675–683 (2019). https://doi.org/10.5194/isprs-Archives-XLII-2-W11-675-2019
21. Acierno, M., Cursi, S., Simeone, D., Fiorani, D.: Architectural heritage knowledge modelling: an ontology-based framework for conservation process. J. Cult. Herit. **24**, 124–133 (2017)
22. Grasso, N.: DIsaster RECovery Team (DIRECT): formazione e attività del team studentesco del Politecnico di Torino per la gestione delle emergenze. Bollettino della società italiana di fotogrammetria e topografia **2**, 15–21 (2015)
23. Società di Studi Valdesi. Prali e Rodoretto – Una vita in altitudine -. La Beidana. Vultura e Storia Nelle Valli Valdesi n. 74 (2012). https://www.studivaldesi.org/filemanager/pdf/la-bei dana-n-74.pdf

Study of the Cloisters of the Historical Center of Florence: Methodological Approach for the Definition of Restoration Intervention Priorities

Giovanna Acampa[1]([⊠]) [iD], Carlo Francini[2] [iD], and Mariolina Grasso[3] [iD]

[1] Faculty of Architecture and Engineering, University Kore of Enna, 94100 Enna, EN, Italy
giovanna.acampa@unikore.it
[2] Department of Architecture, University of Florence, 50121 Florence, FI, Italy
carlo.francini@comune.fi.it
[3] Department of Architecture, Roma Tre University, 00153 Rome, Italy
mariolina.grasso@uniroma3.it

Abstract. On the VI session of the UNESCO World Heritage Committee, which was held in Paris at the headquarters of the United Nations Educational, Scientific and Cultural Organization between December the 13th and the 17th 1982, the Historic Centre of Florence was included in the World Heritage List. Starting from this recognition, the first objective of the following research is to promote the preservation and management of a specific type of historical architecture, an integral and characteristic part of the religious buildings of the historical center of Florence, namely the Cloister. The research aims to facilitate the development of Heritage Impact Assessment (HIA) procedures through the definition of a computerized and geo-referenced model of the cloisters under examination, through the processing of data contained in the datasheets specifically developed. In particular, a methodological approach has been developed to define the priority of intervention on the Cloisters of Florence. The objective of the paper is thus to highlight the contribution that a multi-method framework can have in supporting public policy decisions, where there is a strong need for transparency, replicability and learning mechanisms.

Keywords: Heritage evaluation · Cloister · Restoration

1 Introduction

Since 1982, the Historic Center of Florence has been inscribed on the UNESCO World Heritage List on the basis of the six distinct criteria for admission expressed by ICOMOS (International Council of Monuments and Sites), an integral part of the Statement of Outstanding Universal Value (OUV) with the statements of integrity and authenticity value for the site.

C. Francini—Site Manager of the UNESCO World Heritage Site "The Historic Centre of Florence".

© The Author(s), under exclusive license to Springer Nature Switzerland AG 2022
F. Calabrò et al. (Eds.): NMP 2022, LNNS 482, pp. 2587–2596, 2022.
https://doi.org/10.1007/978-3-031-06825-6_247

Starting from this recognition, the first objective of the following research is to promote the preservation and management of a specific type of historical architecture, an integral and characteristic part of the religious buildings of the historical center of Florence, namely the Cloister. The achievement of this objective is based on the definition of a methodological procedure to define the priority of intervention among the cloisters under examination according to certain endogenous and exogenous characteristics. In this regard, the research was divided into several phases, including the selection of the scope of investigation through a census of the cloisters of the historic center of Florence, the definition of an ad hoc datasheet for the collection of qualitative and quantitative data for each of them, the study and identification of exogenous and endogenous risk factors and the definition of the methodological procedure for the definition of intervention priorities. The research also aims to facilitate the development of Heritage Impact Assessment (HIA) procedures through the definition of a computerized and geo-referenced model of the cloisters under examination, through the processing of data contained in the datasheets specifically developed.

A methodological approach has been developed to define the priority of intervention of the Cloisters of Florence. The objective of the paper is thus to highlight the contribution that a multi-method framework can have in supporting public policy decisions, where there is a strong need for transparency, replicability and learning mechanisms [1].

2   State of the Art

2.1   Main Heritage Impact Assessment Procedures

Undertaking an assessment of potential risks that could damage cultural heritage should be an action to be included in any planning and management process. In the context of cultural heritage, and particularly cultural centers, risks are generally related to the effects of natural disasters, human intervention and error, and socio-economic conditions.

The UNESCO site of Florence within an urban area that is very problematic and complex is not exempt from critical issues of various kinds: hydro-geomorphological and, more generally, of environmental character that put at risk the very integrity of the architectural and landscape heritage. For these reasons, the site will have to be constantly monitored from the point of view of impact assessment for the effects produced by the physical degradation that threatens its monuments and by the anthropic action of transformation that, if not well controlled, determines disastrous changes and even vandalism.

HIA (also known as CHIA, Cultural Heritage Impact Assessments) is a particular type of risk assessment and is a very important component of heritage planning processes; it is a relatively recent tool (ICOMOS, 2011) [2], whose application methodology is still under development.

The purpose of the HIA is to describe and assess the impacts (or rather, effects) - both positive and negative - that a given process (or action, work, phenomenon, tra-transformation) would have on cultural heritage, from a single asset to an entire built-up area. The assessment must also propose a set of measures to mitigate the negative effects, reducing or eliminating them.

HIA answers three basic questions:

1) What is the asset exposed to risk and what are the attributes of its Outstanding Universal Value?
2) How might a given process affect the heritage values of integrity and authenticity?
3) What mitigation measures, if any, are proposed to ameliorate any negative ef-fect?

The Heritage Impact Assessment is not a compulsory requirement although it rep-resents an objective tool of great usefulness for administrators and planners [3]. HIA evolves from Environment Impact Assessments (EIAs), a mandatory administrative pro-cedure in contrast to the former. The study process and methodology is similar for both assessments; in some states, HIA is an integral part of EIA, in others they are separate [4].

The steps for conducting a HIA are

1) identify and locate the potentially endangered property;
2) describe the significance of the site: OUV (Outstanding Universal Value) and attributes of OUV;
3) define the nature of the potential effects on the integrity and authenticity of the site, both negative and positive, in terms of quality and quantity;
4) recommend mitigation measures for adverse effects, describing their benefits;
5) prepare the scoping report.

HIA is a more widely used and in some cases mandatory tool for archaeological areas, while it is currently only recommended for large-scale heritage, such as historic centers or cultural landscapes, as the presence of numerous elements and factors makes it more complex to implement [5].

An effective approach to drafting the HIA on a large scale is to proceed in stages, introducing the Assessment into studies at different levels (e.g., constraint studies, environmental impact assessments, etc.). The key to the success of the HIA is the production of an adequate inventory of the attributes that confer Exceptional Universal Value to the site; this inventory, organized into a database, must be linked to a GIS (GIS-linked database) and must contain basic location information about the asset and its constituents, information about the regulatory constraint, and information about the value of the asset [6].

The Guidance on Heritage Impact Assessments for Cultural World Heritage Proper-ties (ICOMOS, 2011) provides a methodology to enable heritage impact assessments by considering the attributes that make up the OUV as separate entities and evaluating them in a systematic and consistent manner. Underlying this process, an understanding of the values, meaning, attributes (tangible and intangible) and their relationships becomes critical [7–9]. The link between attributes and spatial components becomes the element on which the assessment of impacts on it is based [10].

2.2 Analysis of the Main Projects for the Preservation of the Historic Center of Florence

On the occasion of the VI session of the UNESCO World Heritage Committee, which was held in Paris at the headquarters of the United Nations Educational, Scientific and Cultural Organization between December the 13th and the 17th 1982, the Historic Centre of Florence was included in the World Heritage List.

Article 11 of the Convention established at UNESCO an Intergovernmental Committee for the Protection of the World Cultural and Natural Heritage, called the World Heritage Committee, which, on the basis of data provided by each participating State, is responsible for the preparation, updating and dissemination of a list of world heritage assets considered to be of exceptional universal value and to draw up a list of world heritage in danger, indicating the assets for the protection of which interventions and maintenance works are necessary and for which international assistance has been requested.

Therefore, the international community, the national community and above all the local community in its most varied expressions is called upon to defend the Outstanding Universal Value of a site in the awareness that its diminution is a very serious loss for the whole of humanity.

The issue of how to respond to a management consistent with the values of inscription has never entered the national and local political agenda until the beginning of the 2000s, thanks to a strong appeal by UNESCO to promote new candidacies with Management Plans and with an explicit call to the States Parties to provide even the old sites with a Management Plan.

In Italy, there was an abrupt acceleration on this issue in 2005, when a process started a few years earlier by the Ministry for Cultural Heritage and Activities and by those responsible for World Heritage sites began to be implemented.

In the peripheral administrations of the State and in the Municipality there was a minimal awareness that Florence, or rather its Historic Center, was included in the World Heritage List. There was no official document, regulation or territorial planning instrument that mentioned this unique characteristic.

In 2005, the City of Florence set up a dedicated office and in 2006, a Management Plan for the Historic Center of Florence was drafted. This plan, inspired by the above-mentioned criteria, involved the City of Florence, the Region of Tuscany and the Ministry of Cultural Heritage and Activities with the Regional Direction for Cultural Heritage and Landscape of Tuscany.

In this decade, the awareness of managing a world heritage site has strengthened by highlighting and trying to go to cover the major emergencies. The most important result is certainly to have provided the Historic Center of Florence with a Buffer Zone, approved in 2015 by UNESCO, able to protect the historic urban landscape.

The second Management Plan for the Historic Center of Florence, mindful of the experience of recent years, is proposed as a strategic and operational instrument capable of combining the many territorial dimensions involved and of identifying objectives and concrete actions to deal with the threats that interfere with the maintenance (of the site's Exceptional Universal Value).

The objective of the new Management Plan is to maintain and raise the Outstanding Universal Value of the site, addressing the critical issues identified by the second cycle of the Periodic Report that involved all World Heritage sites from 2008 to 2015 and that are:

– congestion in the Historic Center due to mass tourism;
– the preservation of the monumental heritage;
– the urban mobility system and air pollution;
– the danger of flooding of the Arno River and risks related to climate change;
– the decrease in the number of residents in the Historic Center.

Having as its goal the reduction of the factors of vulnerability and exposure to risk, the new Management Plan is characterized by the assessment of the main factors of danger and the identification, selection (following meetings with local actors and consultation with the local community) and inclusion within the Action Plan of projects and targeted, realiable and measurable actions that can mitigate the threats and that respond to the 5 lines strategy, the 5Cs: Credibility, Conservation, Capacity Building, Communication, Community.

In order to achieve a comprehensive assessment of the effects that certain development interventions/projects may have on the Outstanding Universal Value of the site, the ICOMOS (International Council of Monuments and Sites) Advisory Body requires that these must be the subject of a Heritage Impact Assessment (HIA) [11].

Through the elaboration of a model HIA for the Historic Center of Florence, it would be possible to provide the administration with information and assessments in a timely manner, which are in line with the provisions of UNESCO/ICOMOS and the Management Plan for the Historic Center of Florence.

Therefore, in order to protect the World Heritage Site, it is important that the HIA becomes an action that can be included within the planning and management processes of the site, and that the administration views the project proposals that take place in and/or around the site, applying the HIA methodology in order to assess the impact of the project itself [12].

3   Cloisters Mapping and Definition of Criteria to Establish Priorities for Intervention

To set up a methodological procedure to define the priority of intervention between the cloisters, the first step was mapping them.

Through the "culture" section of the website of the Municipality of Florence it was possible to select all the ones existing in the city center area. As result of these research 51 cloister were selected. We should point out that many of the cloisters, which are constituent parts of convents, abbeys, churches and monasteries, over time lost this function, changing destination of use (libraries, universities, etc.) assuming therefore more the function of a courtyard. Some of them also lost their typical structure being entirely covered also in the central part by a roofing structure.

Following the mapping phase of the cloisters, the main endogenous and exogenous factors that contribute to the process of deterioration of the assets under examination were identified. The identification of these factors took place following in situ analysis and study of the main bibliographic sources [13, 14].

Exogenous factors	Endogenous factors
Use of the monument (public or private)	Presence of artistic landmarks
Roofing structure (exposure to atmospheric agents)	Percentage of columns decay (LoA < T6 or LoA > T6)
Accessibility (restricted traffic zone or not)	

4   Methodological Approach to Define the Priority of Intervention

To assess the priority of interventions between the cloisters we developed a multi-method process of analysis and evaluation supporting the strategic planning phase for the management of environmental public resources.

The developed process consists of 4 main steps shown in Fig. 1:

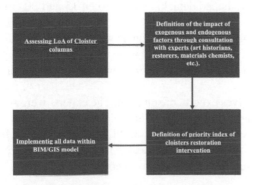

Fig. 1. Methodological procedure workflow

The first step concerns the definition of the Level of Alteration (LoA) of the columns that make up the cloisters. The definition of the level of alteration occurs through a special scale of measurement that allows to quantify the amount of material missing from the shaft of the columns (Fig. 2).

Fig. 2. Level of Alteration (LoA) scale

The scale has an increasing trend from the lowest (T1) to the highest (T8) alteration level (LoA):

- LoA_T1 unchanged conditions;
- LoA_T2 initial state of alteration with circulardetachmentsbetween approximately 0.5 cm and 5 cm in diameter;
- LoA _T3 circular detachments are particularly widespread on the column shaft or in an advanced stage and usually between 5 and 10 cm diameter;
- LoA_T4 irregularly shaped detachments between 10 and 15 cm wide and 1 to 1.5 cm long;
- LoA_T5 irregularly shaped detachments between 15 and 20 cm wide and 1 to 2 cm long
- LoA_T6 detachments of more than 20 cm in width affect the column shaft extensively;
- LoA_T7 detachment more than 20 cm wide with initial erosion of the column;
- LoA_T8 the column shows diffuse *pitting* over most of its surface.

The next step involves expert consultation. The experts were selected based on their skills and considering scientific publications relevant to the subjects under investigation. They were provided with the table in Fig. 3. With reference to the cloister under analysis they were asked to point out whether the exogenous/endogenous factors considered have an impact (in this case a value of 1 is attributed) on it.

	Exogenous factors						Endogenous Factors			
	Use of the monument		Roofing structure		Accessibility		Artistic Landmarks		Percentage of columns' decay	
	Public	Private	Present	Absent	Restricted traffic zone (ZTL)	No Restricted traffic zone (ZTL)	Present	Absent	LoA<T6	LoA>T6
Value	1	0	0	1	0	1	1	0	0	1
Cloister n.										

Fig. 3. Expert form to be filled

The following step is to calculate the Cloister restoration intervention priority index. First of all, for each column of the cloisters we defined its type of degradation. In this sense, the UNI EN 1182/2006 - Macroscopic alterations of stone materials (Ex Recommendations Normal 1/88) distinguish two types of degradation of stone materials:

1) Degradations involving loss of material (such as Disintegration, Detachment, Erosion, Exfoliation, Fracturing, Missing, Pitting, and Spalling)
2) Degradations that do not involve loss of material (such as Chromatic Alteration, Alveolization, Concretion, Crust, Deformation, Surface Deposit, Efflorescence, Encrustation, Stain, Patina, Biological Patina and Presence of Vegetation).

To identify the types of degradation on the cloister columns it is useful to draw up a map as shown in the example (cloister type in Fig. 4) [11].

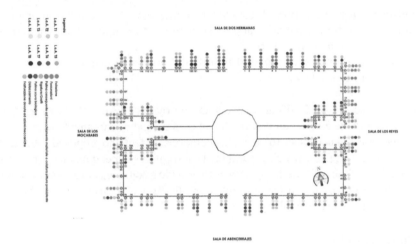

Fig. 4. Map of columns degradation

In this way we lay the groundwork for the next step, which is the calculation of the index of priority for restoration interventions defined as follow:

$$Pi \sum_{i=1}^{n} \left(d_i \sum_{j=1}^{m} \beta_j \right) \qquad (1)$$

where Pi is cloister restoration intervention priority index and:

- d_i is the value indicating the severity of the column degradation that can take any value between 0 and 1. To define it, we distinguish between types of degradation as seen before, consulting expert restorers/chemists of materials. It should also be pointed out that i ranges from 1 to n depending on the number of identified degradations.
- β_j is a dummy variable to highlight or not the presence of the impact of each exogenous and endogenous factor considered. So, there are j dummy variable corresponding to the j factors (exogenous and endogenous) defined. The value of the dummy variable is assigned by the experts through the compilation of the table shown in Fig. 3.

Therefore, we calculate the Index in 2 phases: first the degradation of each column in relation to the exogenous/endogenous factors, then the sum of the degradations of all the columns in the cloister. The index must be calculated for each cloister under examination and allows to define the priority of restoration work considering endogenous and exogenous factors that are responsible for the degradation. Finally, the last step of the methodology is the creation of a H-BIM/GIS model. The H-BIM-GIS model here proposed integrates georeferenced information and combines parametric modeling tools of BIM that can also operate on cartographic information. The BIM-GIS uses objects like buildings, roads, water bodies, vegetation, etc., and offers a 3D environment with parametric modeling tools that are usually used for infrastructure [15].

The proposed BIM-GIS approach is:

- multi-scale: from the territorial scale to the architectural details of the complex;
- multi-sensor: different methods and tools were used to capture the geometric data necessary for the modeling work;
- multi-temporal: data were acquired in different years and can also be used to find changes that occurred over time.

The proposed H-BIM-GIS also covers an intermediate level between cartographic applications and the architectural scale. It can operate and communicate with both systems, which are still necessary for specific operations. In fact, the proposed solution does not replace GIS and BIM, which are still the best solutions depending on the problem to be solved. One of the goals of this work was to provide a better way to access information and coordinate processing instead of replicating those operations which can be carried out with GIS and BIM technology.

5 Conclusion

The methodology described in this paper is set to allows the hierarchization of Florence Cloisters with a similar level of degradation to each other in order to guarantee their safeguard and to facilitate the managing body in the choice procedures concerning restoration interventions.

Specifically, we intend to highlight the aspects of the methodology (Fig. 1) that make it replicable and innovative.

Through the identification of the level of alteration (LA) of the cloisters columns it is possible to evaluate and quantify the evolution of degradation taking into account the variation over time. As for the priority index, it allows to take into account jointly exogenous and endogenous factors and creating a hierarchy among the cloister according to the need for restorations work. Another advantage provided by the application of the methodology concerns the management of the data coming from the analysis phases through the definition of an HBIM (Heritage Building Information Model)-GIS model. The method extends traditional GIS or BIM approaches using a multi-scale approach, in which spatial consistency is provided using georeferencing techniques.

The future objective therefore concerns the definition of an operational tool aimed at the managing team or administration which is user friendly but at the same time capable of assessing the possible balances between the necessary levels of conservation, maintenance and possible uses. Also, it is foreseen to further integrate the proposed methodology with different multi-criteria methods (AHP, ANP) to evaluate different impact of specific degradations on columns.

References

1. Baratta, A., Calcagnini, L., Finucci, F., et al.: Strategy for better performance in spontaneous building. TECHNE J. Technol. Architect. Environ. 158–167 (2017). https://doi.org/10.13128/TECHNE-20797
2. ICOMOS: Guidance on Heritage Impact Assessments for Cultural World Heritage Properties (2011)
3. Forte, F., Del Giudice, V., De Paola, P., Troisi, F.: Valuation of the vocationality of cultural heritage: the vesuvian villas. Sustainability 12, 943 (2020). https://doi.org/10.3390/su12052069
4. Centauro, G., Francini, C.: Progetto HECO (Heritage Colors): metodologie, analisi, sintesi, apparati, valutazione d'impatto sul sito UNESCO Centro Storico di Firenze. DidaPress, Firenze (2017)
5. Acampa, G., Grasso, M.: Heritage evaluation: restoration plan through HBIM and MCDA. IOP Conf. Ser.: Mater. Sci. Eng. 949, 012061 (2020). https://doi.org/10.1088/1757-899X/949/1/012061
6. Baratta, A.F.L., Finucci, F., Magarò, A.: Regenerating Regeneration: augmented reality and new models of minor architectural heritage reuse. VITRUVIO 3, 2 (2018). https://doi.org/10.4995/vitruvio-ijats.2018.10884
7. Ferretti, V., Bottero, M., Mondini, G.: Decision making and cultural heritage: an application of the multi-attribute value theory for the reuse of historical buildings. J. Cult. Herit. 15, 644–655 (2014). https://doi.org/10.1016/j.culher.2013.12.007
8. Appiotti, F., Assumma, V., Bottero, M., et al.: Definition of a risk assessment model within a European interoperable database platform (EID) for cultural heritage. J. Cult. Herit. 46, 268–277 (2020). https://doi.org/10.1016/j.culher.2020.08.001
9. Arrighi, C., Tanganelli, M., Cristofaro, M.T., et al.: Multi-risk assessment in a historical city. Nat Hazards (2022). https://doi.org/10.1007/s11069-021-05125-6
10. Oppio, A., Dell'Ovo, M.: Cultural heritage preservation and territorial attractiveness: a spatial multidimensional evaluation approach. In: Pileri, P., Moscarelli, R. (eds.) Cycling & Walking for Regional Development. RD, pp. 105–125. Springer, Cham (2021). https://doi.org/10.1007/978-3-030-44003-9_9
11. Patiwael, P.R., Groote, P., Vanclay, F.: Improving heritage impact assessment: an analytical critique of the ICOMOS guidelines. Int. J. Herit. Stud. 25, 333–347 (2019)
12. Ashrafi, B., Kloos, M., Neugebauer, C.: Heritage impact assessment, beyond an assessment tool: a comparative analysis of urban development impact on visual integrity in four UNESCO world heritage properties. J. Cult. Herit. 47, 199–207 (2021)
13. Arizzi, A., Rodríguez Navarro, C., Elert, K., Pardo, S.: Bio-consolidation of the marble columns of the lions Couryard in the Alhambra. In: Construction Pathology, Rehabilitation Technology and Heritage Management. Caceres, Spain (2018)
14. Acampa, G., Grasso, M.: Integrated evaluation methods to hbim for the management of cultural heritage: the case study of the colonnade of patio de Los Leones, Alhambra-Granada: Metodi di valutazione integrati all'hbim per la gestione del patrimonio storico: il caso del colonnato del patio de Los Leones, Alhambra-Granada. Valori e Valutazioni 29, 133–153 (2022). https://doi.org/10.48264/VVSIEV-20212910
15. Barazzetti, L., Roncoroni, F.: Generation of a multi-scale historic BIM-GIS with digital recording tools and geospatial information. Heritage 4, 3331–3348 (2021). https://doi.org/10.3390/heritage4040185

Post-fire Assessment of Heritage Timber Structures

Dante Marranzini[1], Giacomo Iovane[1], Veronica Vitiello[2], Roberto Castelluccio[2], and Beatrice Faggiano[1(✉)]

[1] Department of Structures for Engineering and Architecture, University of Naples Federico II, 80, P.le Tecchio, 80125 Naples, Italy
faggiano@unina.it
[2] Department of Civil, Building, Environmental Engineering, University of Naples Federico II, 80, P.le Tecchio, 80125 Naples, Italy

Abstract. Nowadays built heritage is exposed to numerous risks, as earthquakes, floods, landfall, fires, climate changes. The great fire of Notre Dame, as well as other recent fires, testimony that the fire is still one of the most worrying risks for the cultural heritage buildings. Although many causes can lead to the fire trigger, in the context of heritage buildings the timber structure are often involved. After fire, due to the lack of knowledge, funds or time, designers usually decide for the removal of the damaged structures. However, in such a way the heritage value of the building is compromised since the evidence of traditional and ancestral construction techniques is lost.

To address this need, the assessment of the fire damage on timber structures is focused herein. Reference is made to the case study of Palazzo Carafa di Maddaloni, a XVI century monumental building of the historical centre of Naples. On June 2018 the roof structures was involved in a great fire that compromised most of the XIX timber structures. The methodology set up and applied is based on the in-situ application of visual inspection and size measurement, allowing for the identification of the extent of fire damage and residual capacity of structures.

Keywords: Heritage timber structures · Post-fire assessment of timber · Residual capacity of timber structures in fire

1 Introduction

Nowadays irreversible damage to buildings, infrastructures and human facilities, due to natural or artificial disasters, still occur in all the world. Earthquakes, landslides, floods, fires are the most recurrent risks that cause inestimable damage to human heritage. Among the categories of facilities, cultural heritage buildings are among the most exposed ones to the risks, since they have an irreplaceable value for society. The consequences of heritage losses are well known, indeed in addition to functional and economic losses, the social value loss represents an inestimable damage to society [1]. Thus, risk analysis is needed in order to implement the mitigation measures and educational programs in emergency procedures. At the current state of art several methods for the risk

© The Author(s), under exclusive license to Springer Nature Switzerland AG 2022
F. Calabrò et al. (Eds.): NMP 2022, LNNS 482, pp. 2597–2606, 2022.
https://doi.org/10.1007/978-3-031-06825-6_248

analysis of cultural heritage assets have been recognized, they concern multi hazard or specific risk assessment [2, 3].

As evidenced by recent events, fire is one of the most recurrent risks in the field of cultural heritage buildings. Significant examples like the fire of Glasgow School of Arts (2014 and 2018), Brazil's National Museum (2018) and Paris's Notre-Dame Cathedral (2019) are the proof of the power of this action that involves the building and the moveable assets kept inside it (books, paintings, sculptures). In the context of heritage buildings many causes can lead to the fire trigger. It has been seen that about 10% of fires occur when restoration works are in progress [4]. A lot of factors influence the fire vulnerability of heritage buildings. This can be assessed through indicators-based methods and qualitative approaches, available in literature [5–7].

National codes provide rules for prevention and protection of buildings and human facilities against the fire risk. Often protection measures regulated by the fire codes, related to compartmentation and fire protection systems, are difficult to apply in the field of cultural heritage buildings, since works are limited by the conservation needs. Furthermore, fire codes should also include specific rules for the protection of the movable heritage as well as the human life.

Once the fire triggered and spread out into the building, different scenarios can occur as function of the fire characteristics, chemical and physical property of the building materials and furniture inside, as well as time of exposure to fire. Thus, the extent of the damage varies from only smoke to partial or complete carbonization. As part of the building, also structure can be involved in fire. Unlike other building materials, wood is combustible, thus timber structures are susceptible to more severe damages. Regardless the magnitude and the extent of damage, after fire, designers frequently decide for the removal of the damaged structures. However, in the field of cultural heritage buildings this intention should be the last choice into a decisional process, firstly aiming at the restoration and reuse of the structure [8]. Indeed structures have an historic value too and they must be preserved together with the movable heritage. Specifically, the timber heritage structures are the evidence of craftworkers and builders'skill, they are the proof of traditional, cultural and ancestral knowledge [9].

Thus, before making hasty design choices, a careful post-fire assessment is necessary to evaluate the extent of damage. It should be supported by an adequate knowledge of the timber fire behaviour and timber diagnostic techniques, furthermore time and funds are needed to perform the investigation.

In this paper the post-fire assessment of timber historical structures damaged by fire is presented. The methodology set up is focused on damage survey and mapping, aided by *in situ* application of non-destructive techniques. It has been applied to the timber roofs of Palazzo Carafa di Maddaloni, a sixteenth-century monumental building located in the historical center of Naples (Italy), characterized by ancient timber structures that were involved in fire in the latest 2018.

2   Main Issues of Timber Fire Behavior and Diagnostic Techniques

As it is well known, wood is a combustible material. When timber is exposed to high temperature chemical and physical process occurred. Among them pyrolysis consists in the timber main components degradation [10]. However, timber has a predictable and reliable burning behaviour, indeed the rate of charring is approximately constant over time, it being described by the "charring rate" [11, 12]. During the burning process, the temperature distribution inside timber is characterized by high variation. Thus, three main layers can be identified in the cross section: a charred outer layer, a thin pyrolizate layer and a cool interior zone [11]. Given the relationship between mechanical properties and timber temperature it can be stated that the cooler zone retains its structural integrity, thus the cross section preserves a not negligible bearing capacity. The latter property can be estimated exclusively through a careful investigation.

At the state of art, many contributions, in terms of standards, guidelines and procedures, can be found in the field of heritage timber assessment [13]. Visual inspection and non-destructive tests (NDT) represent fundamental methods to carry out the diagnosis of structures, necessary for evaluating safety conditions and appropriate strengthening strategies. Although the purposes of the post fire investigation are quite different, visual inspection and NDT are still applicable. In literature many articles, showing the application of consolidated techniques on materials such as reinforced concrete, masonry and steel, have been recognized [14–16]. The main challenge in post fire assessment of timber structures is to establish the depth of the charred layer in order to estimate the residual cross-section sizes [17–19]. However, different assessment methods are developing: Kasal et al. [20] and recently Cabaleiro et al. [21] proposed the application of the drilling resistance method for the evaluation of the residual cross section, while Deldicque and Rouzaud [22] show the application of "Raman paleothermometry" method on the charcoals collected after the Notre Dame fire, in order to estimate the maximum temperature developed during the fire.

3 The Proposed Method

The assessment of timber structures is an articulated process characterized by several phases, involving study, survey and calculation. A special attention should be given to diagnosis and possible subsequent repairs. The required framework [23] includes the following main steps: desk survey, visual and measured survey, structural analysis and design of interventions. During the desk survey, knowledge about historical aspects and evolution of the structures could be gather from available documents. Useful information to be acquired are related, among other, to the year of construction, possible strengthening interventions, modification of the static scheme and service loads. Visual and measured survey are fundamental steps, which geometrical and mechanical characterization of the structure is carried out through. In order to reach a fully understanding of the structural behaviour, the structural typology have to be properly detected through the identification of the attributes and functions of each member [24]. Thus, the main dimensions and shape of all timber members should be recorded; the boundary condition and static scheme should be correctly identified; wood species should be recognized and strength grade assigned to members through visual inspection analysis, eventually supported by non-destructive tests [25]; the state of conservation and possible degradation should be characterized together with the corresponding causes. After the diagnostic step is completed, structural analysis can be performed and interventions can be finally designed.

In the post fire investigation, unlike the general approach, where visual inspection is used as healthy condition tool for assessment or for strength-grading of wood members [26, 27] in a structure damaged by fire the analysis criteria are quite different. The fire action is generally evidenced by blackened surfaces caused by smoke, while exposure to very high temperatures is evidenced by a wrinkling of the timber surface. About the latter effects, some authors state that shape and size of the cracks are related to the maximum heat flux occurred during the fire [28]. Some other authors claim that it is not possible to obtain quantitative information from the simply visual inspection of the wood surface [11, 22]. However, temperature estimation can be assessed thanks to the observation of the steel connector, indeed high temperatures cause the permanent deformation in steel components [15]. Finally, the most evident sign of prolonged exposure of timber to high temperature is showed by the size reduction of the cross section and change of geometry.

In this context a methodology for the post fire assessment of timber structures has been set up and the general aspects are discussed herein. More details can be found in Marranzini et al. [29]. In particular the visual inspection is performed through a three levels analysis:

1) *Photographic survey:* it is a preliminary study, allowing to distinguish most damaged members from intact ones. It is useful to assess the global damage conditions of the structure.

2) *Member size measurement:* it allows to detect possible size reduction of the structural members. Length of the members and size of the cross section should be measured through simple tools or advanced one, like laser scanning, in order to implement a 3D model [30]. For an immediate evaluation of the damage status the "Size variation index, $\Delta L_{i,j}$" is defined. With reference to a Palladian typology (Fig. 1), commonly used in timber roofs, $\Delta L_{i,j}$ can be evaluated through the Eq. (1):

$$\Delta L_{i,j} = L_{res,i,j}/L_{i,j} \tag{1}$$

where $L_{i,j}$ is the originary length of the i-th member placed on the j-th alignment, $L_{res,i,j}$ is the residual length of the i-th member placed on the j-th alignment ($i = m$, p_1, p_2, where "m" corresponds to the post king, "p_1" and "p_2" corresponds to the left and right truss respectively. $\Delta L_{i,j}$ values close to 1 indicate that members have no reduction in length, conversely $\Delta L_{i,j}$ values close to zero indicate that member has been completely burnt.

3) *Visual mapping:* it is based on a damage scale definition, characterized by increasing levels of damage. The damage scale is defined on the basis of visual criteria, like timber surface color, presence of cracks and geometry variations of the cross section. A damage scale characterized by five level of increasing damage is proposed in Marranzini et al. [29]. The main challenge of the visual mapping consists in defining a dimensionless parameter, namely coefficient of Fire Visual Damage (FVD) useful to quantitatively evaluate the level of damage of timber members. It is a useful tool for planning the subsequent non-destructive tests.

Non-destructive tests (NDT) support visual damage analysis as they provide quantitative information for the identification of timber decay and strength grade. Among the NDT the drilling resistance test is frequently applied for the biological decay evaluation and indirect estimation of the mechanical properties [31, 32]. This technique can be also applied on elements damaged by the fire, as evidenced by applications found in literature [21, 30]. The drilling resistance technique can be used for estimating the residual cross section and thus bearing capacity of members. Therefore a criterion for the char depth prediction can be established by means of the relationship between the levels of damage defined through visual inspection and the depth of the char layers detected through the drilling resistance test [29]. This is indispensable when the structural-safety evaluation of burnt structure is required. Thus, designers acquire the adequate knowledge to prescribe safety or recovery interventions.

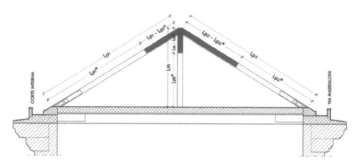

Fig. 1. Member size measurement scheme for the visual inspection of the Palladian truss in Compartment "A" of Palazzo Carafa di Maddaloni in Naples (Italy).

4   Palazzo Carafa di Maddaloni

Palazzo Carafa di Maddaloni (Fig. 2) is a monumental building placed in the historical centre of Naples (Italy) along the "Decumano inferiore", one of the three main roads of the ancient city. Being built in the XVI century, it is one of the main Baroque buildings in Naples. Famous Italian artists, such as Cosimo Fanzago, Fedele Fischetti and Giacomo del Pò, contributed to the construction of the building. Therefore it is protected by the Authority for Archaeology, Fine Arts and Landscape.

Fig. 2. a) Front view of Palazzo Carafa di Maddaloni from Sant'Anna dei Lombardi street; b) the XVII century loggia of the main hall "Sala Maddaloni" (courtesy prof. Castelluccio) [33].

Nowadays the building is the result of subsequent expansion interventions occurred during time [33]. As a consequence, the building is characterized by a complex system of roofs, which structural materials, typologies and different geometries are merged in. Most of the roof structures are made of Palladian timber trusses, whereas the main hall of the building, called "Sala Maddaloni", is covered by "Palladian composite truss", a timber structural type characterized by three king posts and two timber chains. Probably they are made of Chestnut timber (*Castenea sativa Mill.*). From the historiographic analysis, the geometric and photographic surveys, it can be determined that most of the timber roofing structures were built in the XIX century. Moreover, an in-depth analysis of the timber joint between the truss struts and chain (Fig. 3a) has confirmed that the current Palladian trusses are an addition to the previous flat roof supported by chestnut timber beams.

In the last decades the building was partially damaged either by the bombardment of the Second World War, as well as by the Irpinia earthquake in 1980. However the timber roof structures were well preserved. In the 2010 the Palladian roofs were consolidated, furthermore two areas of the roof, which were seriously affected by biological decay, were replaced by a glulam timber structure.

a)   b)   c)

Fig. 3. a) Floor beam jointed to Palladian trusses by means of timber legs and forged iron nails (courtesy of prof. R. Castelluccio); b) The roof of Palazzo Carafa di Maddaloni damaged by the fire of 27th June 2018; c) Localization of the areas of study (compartments "A" and "B").

On 27th June 2018, a fire generated in the building attic, damaging most of the roof timber structures of Palazzo Carafa di Maddaloni. The fire started in the night, triggering in a middle zone of the roof covered by the Palladian trusses. The fire involved about 450 m^2 of the roof surface, it was extinguished after about seven hours, thanks to the work of three teams of firefighters. The result of the disaster was the partial destruction of the Palladian structures placed along the side where fire started, while the glulam structure, placed in the corner of the roof plan, suffered a less significant damage.

Thus, the proposed methodology has been applied to both areas, namely compartment "A" (Palladian type) and compartment "B" (Glulam structure type). In particular, given the best safety conditions of the work area, the combination of visual inspection and instrumental investigations has been carried out only in compartment "B".

5   Application of the Method to the Case Study

5.1   Compartment "A"

The so-called compartment "A" corresponds to the roof area most damaged by the fire, where the XIX century structures were located. Herein the first level of assessment, such as the visual inspection, involved fourteen Palladian trusses. The visual inspection has been performed through the photographic survey and the size measurement. At the first investigation stages, two areas having different magnitude of damage have been found: trusses placed near the ignition area showed a greater damage (trusses alignment "12 to 18"), characterized by the complete carbonization of the timber members and change of members geometry (optical cones "1", "2", "3" in Fig. 4). Besides, low damage, as blackened surface, has been found in the members placed far from the ignition area (trusses alignments "7 to 11").

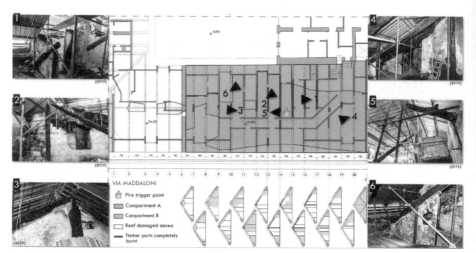

Fig. 4. Compartment "A": Assessment of the fire damage through photographic survey and size measurement.

The size measurement step has been carried out on the fourteen trusses, quantifying the observations of the first step of visual inspection, with regards to the cross section size variation: trusses placed near the ignition area showed a size variation index ranging from 0.4 to 1.0, while trusses placed far from the ignition area showed a size variation index ranging from 0.8 to 1.0 (Fig. 5).

Furthermore, the roof covering area, made of timber boards, which was destroyed by the fire, has been measured as about 240 m^2.

The variability of the damage extent is certainly due to the fire dynamics. The greatest damage reported by the trusses placed in the middle zone of the compartment A was caused by two factors: the proximity to the site of fire ignition and the difficulty of access to the area by the firefighter. Conversely, due to the proximity to the firefighter location, the timber trusses placed along the Maddaloni street showed a smaller damage.

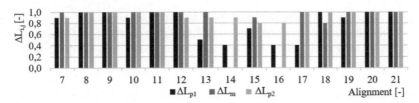

Fig. 5. Compartment "A": Size variation Index $\Delta L_{i,j}$ for Palladian trusses.

5.2 Compartment "B"

With regards to compartment "B", the photographic survey allowed to roughly identify three areas corresponding to different damage levels. Afterwards, for each member, external sizes, height and base of the cross section were measured in three different points placed along the member, corresponding to the both ends and the middle of the member. In this way the cross-section size variation has been determined. The further level of visual analysis (visual mapping) and instrumental investigations through drilling resistance tests were also carried out, allowing to determine the residual cross sections of members. Thanks to this study, the members requiring temporary safety works, as well as the members retaining an adequate residual bearing capacity to be reused, were identified (Fig. 6). Details can be found in Marranzini et al. [29].

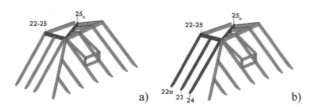

Fig. 6. a) Safety condition assessment (safe and unsafe members in green and red, respectively); b) Reuse in service assessment (reusable and un reusable members in green and red, respectively).

6   Conclusions

This paper briefly illustrates a methodological proposal for the post fire assessment of timber structures and its application to the timber roofs of Palazzo Carafa di Maddaloni in Naples, which was involved in a great fire on June 2018. The procedure based on visual inspection and drilling resistance tests allows to map and quantify the extent of damage, retracing the evolution of fire along the timber structure. To these purposes a visual damage scale has been defined, together with the correspondence between the visual damage levels and the char depths measured through drilling resistance test. Therefore the residual cross section has been quantified and the residual bearing capability of members determined.

The procedure is very useful and efficient for the analysis of timber structures damaged by fire, in order to obtain the adequate level of knowledge required for selecting the more appropriate repair or restoration works, especially for the intervention on heritage structures. Future works are required in order to optimize the methodology. The study sample should be extended in order to make more reliable the visual mapping classification and in order to improve the relationships between visual mapping and drilling resistance measures.

Acknowledgement. Devices and materials were provided by the Department of Structure for Engineering and Architecture of the University of Naples "Federico II". Eng. Emanuele Scaiella support for the in situ experimental campaign is acknowledged.

References

1. Torrieri, F., Oppio, A., Rossitti, M.: Cultural heritage, social value and community mapping Smart Innovation. Syst. Technol. **178**, 1786–1795 (2021)
2. Romão, X., Bertolin, C.: Risk protection for cultural heritage and historic centres: current knowledge and further research needs. Int. J. Disaster. Risk Reduct. **67** (2022)
3. Figueiredo, R., Romão, X., Paupério, E.: Component-based flood vulnerability modelling for cultural heritage buildings, Int. J. Disaster Risk Reduct. **61** (2021)
4. Ferreira, T.M.: Notre Dame cathedral: another case in a growing list of heritage landmarks destroyed by fire. Fire **2**(2), 1–2 (2019)
5. Arborea, A., Mossa, G., Cucurachi, G.: Preventive fire risk assessment of Italian architectural heritage: an index-based approach. Key Eng. Mater. **628**, 27–33 (2014)
6. Granda, S., Ferreira, T.M.: Assessing vulnerability and fire risk in old urban areas: application to the Historical Centre of Guimarães. Fire Technol. **55**(1), 105–127 (2018). https://doi.org/10.1007/s10694-018-0778-z
7. Salazar, L.G.F., Romão, X.: Review of vulnerability indicators for fire risk assessment in cultural heritage, Int. J. Disaster Risk Reduct. **60** (2021)
8. Chorlton, B., Gales, J.: Fire performance of cultural heritage and contemporary timbers. Eng. Struct. **201** (2019)
9. ICOMOS.: Principles for the Wooden Built Heritage, pp. 1–6 (2017)
10. Bartlett, A.I., Hadden, R.M., Bisby, L.A.: A review of factors affecting the burning behaviour of wood for application to tall timber construction. Fire Technol. **55**(1), 1–49 (2018). https://doi.org/10.1007/s10694-018-0787-y
11. Babrauskas, V.: Charring rate of wood as a tool for fire investigations. Fire Saf. J. **40**(6), 528–554 (2005)
12. Friquin, K.L.: Material properties and external factors influencing the charring rate of solid wood and glue-laminated timber. Fire Mater. **35**(5), 303–327 (2011)
13. Riggio, P., D'Ayala, D., Parisi, M.A., Tardini, C.: Assessment of heritage timber structures: review of standards, guidelines and procedures. J. Cult. Herit. **31**, 220–235 (2018)
14. Stochino, F., Mistretta, F., Meloni, P., Carcangiu, G.: Integrated approach for post-fire reinforced concrete structures assessment. Periodica Polytech. Civil Eng. **61**(4), 677–699 (2017)
15. Maraveas, C., Fasoulakis, Z., Tsavdaridis, K.D.: Post-fire assessment and reinstatement of steel structures. J. Struct. Fire Eng. **8**(2), 181–201 (2017)
16. Praticò, Y., Ochsendorf, J., Holzer, S., Flatt, R.J.: Post-fire restoration of historic buildings and implications for Notre-Dame de Paris. Nat. Mater. **19**(8), 817–820 (2020)
17. White, R.H., Woeste, F.E.: Post-fire analysis of solid-sawn heavy timber beams. Struct. Mag. **2013,** 38–40 (2013)

18. Moller, C.: Restoring capacity in compromised glued laminated beams. PhD Thesis, Montana Tech University (Montana, USA) (2016)
19. Kukay, B., et al.: Evaluating Fire-Damaged Components of Historic Covered Bridges. (2016)
20. Kasal, B., Kloiber, M., Drdacky, M.: Field investigation of the 14th century Castle Pernstejn before and after fire damage. In: AEI 2006, Building Integration Solutions - Proceedings of the 2006 Architectural Engineering National Conference, vol. 70 (2006)
21. Cabaleiro, M., Suñer, C., Sousa, H.S., Branco, J.M.: Combination of laser scanner and drilling resistance tests to measure geometry change for structural assessment of timber beams exposed to fire. J. Build. Eng. **40** (2021)
22. Deldicque, D., Rouzaud, J.N.: Temperatures reached by the roof structure of Notre-Dame de Paris in the first of April 15th 2019 determined by Raman paleothermometry. Comptes Rendus Géoscience **352**, 7–18 (2020)
23. Cruz, H., Yeomans, D., Tsakanika, E., Macchioni, N., Jorissen, A., Touza, M., et al.: Guidelines for on-site assessment of historic timber structures. Int. J. Archit. Herit. **9**(3), 277–289 (2015)
24. Faggiano, B., Marzo, A., Grippa, M.R., Iovane, G., Mazzolani, F.M., Calicchio, D.: The inventory of structural typologies of timber floor slabs and roofs in the monumental built heritage: the case of the Royal Palace of Naples. Int. J. Architect. Herit. **12**(4), 683–709 (2018)
25. Faggiano, B., Marzo, A., Mazzolani, F.M.: The Diplomatic Hall of the Royal Palace of Naples: structural characterization of the timber roof by in situ ND investigations. Construct. Build. Mater. **171**, 1005–1016 (2018)
26. UNI 11035-1: Structural Timber – Visual Strength Grading: Terminology and Measurements of Features, Ente Nazionale di Unificazione, Milan, Italy (2010)
27. UNI 11035-2: Visual Strength Grading and Characteristic Values for Italian Structural Timber Population, Ente Nazionale di Unificazione, Milan, Italy (2010)
28. Baroudi, D., Ferrantelli, A., Li, K.Y., Hostikka, S.: A thermomechanical explanation for the topology of crack patterns observed on the surface of charred wood and particle fibreboard. Combust. Flame **182**, 206–215 (2017)
29. Marranzini, D., Iovane, G., Faggiano, B.: Methodology for the assessment of timber structures exposed to fire through NDT. In: 7th International Conference on Applications of structural fire engineering, pp. 242–247, Ljubljana, Slovenia (2021)
30. Cabaleiro, M., Branco, J.M., Sousa, H.S., Conde, B.: First results on the combination of laser scanner and drilling resistance tests for the assessment of the geometrical condition of irregular cross-sections of timber beams. Mater. Struct. **51**(4), 1–15 (2018). https://doi.org/10.1617/s11527-018-1225-9
31. Faggiano, B., Grippa, M.R., Marzo, A., Mazzolani, F.M.: Experimental study for non-destructive mechanical evaluation of ancient chestnut timber. J. Civ. Struct. Heal. Monit. **1**(3–4), 103–112 (2011)
32. Nowak, T.P., Jasieńko, J., Hamrol-Bielecka, K.: In situ assessment of structural timber using the resistance drilling method - evaluation of usefulness. Constr. Build. Mater. **102**, 403–415 (2016)
33. Castelluccio, R., Vitiello, V.: Rilettura critica di interventi di restauro di edifici monumentali ricostruiti nel periodo post-bellico. Il caso del Palazzo Carafa di Maddaloni, Eresia ed ortodossia nel restauro. In: Progetti e realizzazioni 32° Convegno Internazionale Scienza e Beni culturali, pp.431–440, Arcadia Ricerche, Bressanone (2016)

Oil Heritage in Iran and Malaysia: The Future Energy Legacy in the Persian Gulf and the South China Sea

Asma Mehan[1]([✉]) [iD] and Rowena Abdul Razak[2] [iD]

[1] CITTA Research Institute, University of Porto, Porto, Portugal
asmamehan@fe.up.pt
[2] London School of Economics and Political Science (LSE), London, UK

Abstract. The oil industry has played a major role in the economy of modern Iran and Malaysia, especially as a source of transnational exchange and as a major factor in industrial and urban development. During the previous century, the arrival of oil companies in the Persian Gulf, brought many changes to the physical built environment and accelerated the urbanization process in the port cities. Similarly, the development of the national oil industry had a huge impact on post-independence Malaysia, affecting balance sheets, the environment, and society. Oil significantly changed Malaysia's position in the global economy and transformed a predominantly agricultural country into a major producer of petroleum and natural gas. Through implementing the analytical, historical and comparative perspectives, this paper focuses on the legacy of oil cities in the Persian Gulf and the South China Sea as the birthplaces of the oil industry in two regions, whose geopolitical importance along with oil's historical significance has the potential for representing national unity, political memory and collective shared identity. In proposing this grounding, the paper seeks to approach the heritage of oil as a particular form of industrial heritage. This research analyses the future of energy heritage, existing Covid-related challenges, political tensions and examines the various impacts, transitions and capacities associated with the current international relations, post-pandemic urban developments, and the post-oil future to pave the way to these nascent areas of industrial heritage and oil heritage in Iran and Malaysia.

Keywords: Persian Gulf · South China Sea · Oil heritage · Industrial heritage · Post-pandemic urbanism

1 Introduction

The paper situates the Iranian case within a broader Asian and Malaysian perspectives to show that some comparable developments, temporalities, growth processes, spatial practices, international collaborations and exchange may be found. Such processes and practices are often codified in planning documents and enacted by domestic and international actors. Even today, the dynamics and actors of oil in Iran and Malaysia continue to reshape the industry, society, culture, and politics and transform the built environment

© The Author(s), under exclusive license to Springer Nature Switzerland AG 2022
F. Calabrò et al. (Eds.): NMP 2022, LNNS 482, pp. 2607–2616, 2022.
https://doi.org/10.1007/978-3-031-06825-6_249

and urban spaces. In this study, Malaysia represents an important case for comparison, where oil has seeped into different levels of society and politics. Mainly mined off-shore, the presence of the industry altered and affected communities facing the South China Sea. Oil is a source of collective identity at a national level, as represented by the national oil company Petroleum Nasional Berhad (henceforth known as Petronas). From the building of the Twin Towers' rapid development of villages to the sponsorship of students, Petronas is responsible for significant changes in Malaysia's rural landscape and urban cityscape and shifts in traditional society. Its international presence, helped by its sponsorship of the Formula 1 team Mercedes-AMG, has made Malaysia internationally visible beyond the region. In a relatively short period, this multi-faceted nature and leveling of Malaysia's oil industry have ensured that this natural resource is tied into different aspects of the country's image and make-up. Focusing on the contemporary transitions in the Middle East and Iran on one hand and the Asia-Pacific Region and Malaysia on the other, this research explores the legacy of oil in the Persian Gulf and the South China Sea and analyses the future of energy heritage, existing challenges, political tensions and examines the various impacts, transitions and capacities associated with the current international relations, pandemic and the post-oil future in Iran and Malaysia as well as the greater Middle Eastern and southeast Asian regions.

2   Oil (Counter) Narratives: Politics, Transition, Modernization, and Revolution

Over the last 150 years, much of the geopolitical tensions around the world have been directly or indirectly related to energy and, specifically, non-renewable fossil energies. The heritage of Oil Port Cities (OPCs) is the product of natural resources, globally-distributed extractive industries, and vast networks of connectivity and exchange. This is a multi-faceted heritage that is fundamentally spatial and partially expressed in the fabric of the built environment. In these globally interconnected nodes, local geographies of resources coincide with the more complex and fluid labour and knowledge (tech) geographies and the geopolitics of energy. This is an essentially multi-scalar mix. From this perspective and in the world where the shift to renewable energies is of utmost urgency, the heritage of the oil industry in general and OPCs in particular are worthy of consideration both for their direct contribution to the future of energy-producing cities and for the demands of management of these places as heritage localities.

Extraction, refining, transformation, and petroleum consumption have made an extensive impact on OPCs over the past century. In addition, the role and importance of port cities tend to be particularly sensitive to changes arising from larger political, cultural, and societal transitions unfolding around them, as well as the environmental impacts and long-lasting changes to their built environment [1]. The recent devastating Beirut Blast (4th August 2020) shows the importance of the ports in the contemporary globalized world, which calls our attention to the safety, security, governance, connection, and collaboration between port and city regions [2]. In this sense, different layers of the spatial, cultural, and societal memory flows of petroleum, such as physical, represented, and everyday practices, combine into the transnational future petroleumscapes. For analysing the interrelations between oil and global politics in the nineteenth and early twentieth centuries, Timothy Mitchell, in his book-*Carbon democracy*- explores the rise of a certain kind of democratic mass politics and the historical development of energy from fossil fuels [3]. Michael Watt, puts forward the term 'oil complex' (or 'oil assemblage') as the particular territorializing of the oil complex and the technological zone, which is a center of economic, political, and scientific circulation [4]. Rather than the physical representation, the expression, 'oil assemblage', frames the history of oil as the outcome of social and historical 'encounters' between the material world, the actions and the counteractions of different and unequal social workers that underlines the global system of oil provision. By suggesting the term 'oil culture', Barrett and Worden (2012) highlighted the problematic relationships that have taken shape between oil and conceptions of futurity, the profound cultural entanglement of petroleum and apocalypse, and the central role that feeling or effect has played in the interpretation and promotion of oil capitalism [5]. This definition is quite different from the 'oil industry' since the broader social, historical, spatial, and cultural relationships surrounding the rise to oil prominence are excluded from that perspective. Oil production and refining have historically been important sites of labor activism, class relations, social unrest, and political agency. The OPCs were the focal points produced by the actions, interactions and counteractions of the various social actors (such as oil workers, drillers, engineers, and corporate managers, spouses, extended families, urban landlords, bureaucrats, technocrats, political activists, smugglers, beggars, indigenous farmers, merchants, policemen, and migrants who had flooded to the oil complex and involved in the making of this complex history.

In the Middle East, the colonial regimes and rising global corporations greatly influenced the patterns of oil spatializing [6]. Western Europe had no oilfields, so the additional oil would come from the other parts of the world that came under colonial purview, be it the British in Iran or the Dutch in Indonesia [7]. The oil industry in Iran has a close collaboration with national governments and corporate actors to maximize the revenue, develop industrial growth that changed over time, and align with local cultures. After the defeat of the oil nationalization movement (1949–1953), a consortium of multinational oil companies took control with the National Iranian Oil Company (NIOC) (which had replaced the AIOC) divided profits equally between the NIOC and the multinational consortium. This agreement remained the same until the 1979 revolution and the complete nationalization of the Iranian oil industry.

As a result, the growth of OPCs on the southern shores of the Persian Gulf has been phenomenal. As Suleiman Khalaf commented in the mid-2000s, "oil-generated growth has demolished small mud-walled seaports and villages. They have been transformed into glittering commercial capitals and sprawling suburbs integrated within the global economy and culture in just four decades. The speed, pattern, and politics of urban development have been similar across the Gulf [8]." In the Persian Gulf, these transformations have been shaped by the region's strategic importance of its petrochemical resources, leading to the emergence of what may be called "oil urbanization" [9]. In this sense, capitalist expansion has always been closely tied to oil-led urbanization and development.

Similarly, along the Malaysian coasts of the South China Sea, the backwater towns of Miri in Sarawak and Kerteh in Terengganu were transformed by the oil industry. In the case of Miri, when oil was discovered during the colonial period, the town was quickly modernized and westernized to accommodate the British managers. Waterways and roads were improved to transport the oil to the tankers, shop lots were built to provide more luxurious items, and clubhouses were established to entertain the managerial staff. The make-up of the native population changed with the influx of laborers from India and China, with accommodation and facilities installed for their benefit. Miri was industrialized and globalized by the oil industry in less than twenty years, infused by a colonial vision and style. By contrast, the oil changes to the peninsula were different under the purview of an independent country. When oil was struck off the coast of Terengganu, changes came swiftly to accommodate the industry and develop the villages and their people. Schools, hospitals, and modern amenities were built and introduced, seeing the hybrid nature of oil – not only as an economic project but as a social one too. Even with oil rigs far from the coast, its impact is felt on the lives of those living on land. Most recently, the southern city of Pengerang in the state of Johor was touched by the industry. The Pengerang Integrated Petroleum Complex transformed the broader region by Petronas. Petrochemical refineries, storage chambers, and land were replaced by small fishing villages and farmlands for more expansion [10]. As Nelida Fuccaro puts it, "the predominance of studies on oil urbanization precludes an understanding of oil urbanism as a way of life and as a mode of political and socio-economic organization [11]." Therefore, what is needed is a better understanding of "the political and human texture" of the contemporary Persian Gulf and South China Sea cities [12]. This oil-generated growth in the port cities in these two regions has been partly influenced by internal and external dynamics, logistical relationships, regional forces, geostrategic dynamics, and infrastructure development in the region. The policies and priorities of state leaders have been equally important in shaping the overall profile and form of OPCs within and their position in broader regional and global networks.

3   Oil Heritage as National Legacy

The petroleum industry is approximately 160 years old, its origin conventionally dated by historians to the oil wells drilled in Ontario and Pennsylvania in the late 1850s. The International Committee for the Conservation of the Industrial Heritage (TICCIH) (founded in 1978), is an international organisation established to focus on studying industrial heritage in the development period. It aims to explore, protect, conserve and explain the remains of industrialisation. TICCIH has been ICOMOS (The International Council on Monuments and Sites) special advisor for industrial heritage on properties to be added to the World Heritage List, drawing on its advice from TICCIH. The theoretical and practical considerations of these properties as the World Heritage sites are examined in the light of the criteria for Outstanding Universal Value in UNESCO's Operational Guidelines for the Implementation of the World Heritage Convention [13]. In 2020, TICCIH published the first global assessment of the heritage of petroleum production and the oil industry and the places, structures, sites, and landscapes that might be chosen to conserve for their historical, technical, social, or architectural attributes. This TICCIH thematic study on oil heritage also included the proposals for criteria to evaluate this heritage and priorities for conserving the most important sites, ensembles, and landscapes, from regional inventories up to World Heritage sites. In this report by TICCIH, the heritage of the petroleum industry is defined as "the most significant fixed, tangible evidence for the discovery, exploitation, production, and consumption of petroleum products and of their impact on human and natural landscapes" [14]. While the importance of the historical evidence for the oil industry as the tangible cultural heritage is self-evident, it is also challenging to define an integrated and holistic strategy from a conversation point of view [15]. In many cases, the petroleum production sites and historical infrastructures -situated in corrosive and fragile landscapes- are costly to conserve, difficult to re-use and re-function considering their contribution to global warming.

In this sense, for achieving holistic and methodological reuse strategies, it is required to reconsider various factors such as national policies and the economic system. In many cases, the retention and study of documentation and company archives is the best way to conserve the industry's history. Today, there are many museums, memorials, and other historical establishments commemorating the oil industry's contributions to the world's cultural landscape. Bringing together the refinery technology and culture, there are nearly 200 museums in the world that exhibit oil and gas machinery and relics. Due to the lack of an integrated management system, developing new methods to identify and protect industrial heritage with high values and significance [16]. In Iran, starting from January 2014, arrangements began to establish the national Museums and document center of the oil industry set up by the direct order of the Iranian Minister of Petroleum-Bijan Namdar Zangane and under the supervision of Akbar Nematollahi to collect, safeguard and display the old oil industry equipment. Iran's Petroleum Museums and Document Center offer insight into the nation's energy heritage, which began in 1901 when British speculator William D'Arcy received a concession from Iran to explore and develop southern Iran's oil resources which led to the formation of the London-based Anglo-Persian Oil Company (APOC). It tries to collect and display the old oil industry equipment and protect and pass the tangible oil heritages to the next generations.

The launch of Iran's petroleum industry museums started in the OPC of Abadan, which includes old refinery, gas station (as the oldest filling station of Iran has been turned into a museum in Abadan, as well as the 1934 Davazeh Dowlat filling station in Tehran), the oldest oil-related technical training school (as the oldest national training school dedicated for Iranian oil workers in Abadan). Cranes are being preserved in some parts of Abadan's old ports with heavy machinery, such as Ekvan (literally means monster) and Gogerd (literally means Sulfur). There is also an exhibition about the reconstruction process of the refineries after the Iran-Iraq war (1980–1988). The plan also includes the inauguration of the oil museum in other major OPCs such as Masjed Suleiman (located in the southwestern province of Khuzestan as the birthplace of the oil industry in Iran), which includes the oldest oil recovery site in the country. The first thermal power generation plant in Iran- known as Tombi Power Plant (launched in September 1908 and still operational in electricity distribution)- is defined as one of the pilot museum sites in the Masjed Suleiman Petroleum Museum scheme. Based on the editorial report published by the Iran Petroleum Museums and Documents center, "The history of Masjed Suleiman electricity and Tombi Power Plant is directly related to oil eruption from the first well. In 1911, 3 years after, the first barrel of oil was pumped from Well No. 1 of Masjed Suleiman. Oil flow started to the Abadan oil refinery through a pump house in Tombi. Given its oil and gas riches, Masjed Suleiman has always been considered, and it rapidly grew after the discovery of oil. The first station for pumping crude oil from Masjed Suleiman to Abadan was built in 1909, and similar stations started operating in Malasani, Kut Abdullah, and Darkhovin, respectively. Darkhovin station is being operated with a power generator to meet its internal needs [17]."

The plans for two other oil museums in Kermanshah (west part of Iran) and Tehran aim to offer insight to the nation's long oil heritage. In Kermanshah, the previous tin factory is planned to be transformed in to the Petroleum museum. The structure is to be erected at the Tin Factory of Kermanshah Refinery. Given the factory's long history- it has been in place for over a century—and its role in distribution of petroleum products throughout the country in the past several decades, the museum is envisaged to exhibit a rich collection of items belonging to various periods of the factory's operation. Most of the showcased items will be placed in the museum with focus on the industry in Iran's western regions. Tehran-based "Museum of Oil Industry Technology" introduces the nature and importance of oil, gas and petrochemicals in various areas of human life for a long time and the technologies used in it. Unlike other oil museums in other parts of the country, this museum does not have buildings, facilities and content, so a special building will be designed for the Tehran Oil Museum. The museum is expected to take four to five years to set up. Tehran Bureau of Oil Industry Museums and Archives has two major sections including the Treasury of relics and the Archives. The Archives' section of Iran's Oil Industry Museum aims to identifies, gathers, categorizes, organizes, retrieves, repairs, preserves and keeps oil industry documents to provide a comprehensive resource for the oil industry's researchers and the general public [18].

In Malaysia, there is little focus on preserving the oil industry, while discussions remain scattered. One of the main reasons for this is the dominance of off-shore rigs, making them not easily accessible and challenging to convert into an attraction. Another reason for the lack of importance on oil preservation may be the organization of heritage creation in Malaysia, which is tied up with identity and racial politics and is mainly focused on non-industrial aspects (such as food, historical cities and villages, and folklore) [19]. Agricultural heritage is more developed and well thought out, with the creation and planning of Geo-parks, which "support the principles of sustainable development" and, in some cases, are established to preserve some of Malaysia's geological resources [20]. Nonetheless, there have been some efforts to protect Malaysia's oil history and educate and inform. Coinciding with the unveiling of the Petronas Twin Towers, the Petrosains Discovery Centre was opened in 1999 to increase public awareness of petroleum. Beginning with a simulation of arriving on an off-shore oil rig, visitors are brought through interactive exhibitions and information boards about the resource and industry [21].

The only other petroleum museum in Malaysia is situated in Miri, Sarawak, on the first drilling rig in the country. In production since oil was struck in the early 1900s, it was retired in 1972. A year later, the land and the oil well were handed over to the state government by Sarawak Shell. In the early 2000s, the area was developed to include visitors to the oil well, a petroleum museum, and a café. Similar to the Petrosains center, visitors are provided with information on the oil industry. However, in Miri, a concerted effort was made to conserve the old oil well and protect it from fires and decay. Both museums were funded by Petrosains, indicating a governmental interest in preserving oil's historical presence in Malaysia. The focus is scientific and on the environment. Such appears disconnected from other efforts by the Department of Museums, including the conservation of organic and non-organic sites (but not including oil), preservation of culture, and education [22]. In this regard, Malaysia differs sharply from Iran's efforts to preserve the history of the country's oil industry.

In Malaysia, the reach of the oil industry has seeped into many aspects of society, which arguably preserves the oil industry differently. Petronas is a national oil management company and has other ventures in sectors such as education and the automobile industry. It is recognized internationally as a Formula One team with Mercedes AMG, which owns a private university. The company also owns a private university, the Universiti Teknologi Petronas (Technology University Petronas), which, as the name suggests, focuses on technical sciences such as engineering, geoscience, and computer studies. In addition, the company provides scholarships through its Petronas Education Sponsorship Program (PESP), funding talented students for further university education and bonding them to the company after. Such an arrangement creates a sustainable cycle, ensuring that the company constantly gains well-educated employees every year. On the other hand, we see how this results in education's commodification, capitalization, and marketization [23]. By looking at the multi-faceted nature of Petronas' ventures, oil is preserved in several ways in the public space.

The oil industry has not been examined comprehensively from a global standpoint or how it impacts society. The increasing international attention toward the oil heritage as part of the industrial heritage is also a reminder of the coming of a new 'Post-oil future.' In Iran, the management, adaptive reuse, and conservation policies over the oil heritage and its related regulations are still developing. While in Malaysia, efforts are slim, with oil's legacy and heritage being preserved differently. The new holistic framework for achieving sustainable adaptive reuse must be integrated with the social, political, and economic contributing factors. Considering the oil's complex history, more detailed studies are required to assess the cultural impact of the oil industry through the many universities, national museums, and natural parks, which owe their existence to the wealth it has generated.

4   Concluding Notes: Towards Post-pandemic, Post-sanctions, Post-oil Futures?

Fossil fuels are gradually becoming relics of the past. In response to environmental challenges, many countries worldwide are now rallying for climate-friendly green and cleaner energies and new renewable resources as alternatives for oil. As the climate emergency deepens, motor vehicles (which account for half of all oil use globally) are switching to electricity instead. In the Middle East and Asia, the OPEC (Organization of the Petroleum Exporting Countries) states, with an economy firmly rooted in oil production, are moving into a new era and are trying to diversify their economy to secure growth in the long term.

It is essential to identify the moments of decisive change toward new energy values, green transitions, and resilient policies in post-oil future cities. Considering the current de-urbanization, de-growth, and emigration of post-industrial sites, the oil industry's industry's very past and present size needs to shrink dramatically. Moreover, in the aftermath of decolonization and the transformations of neoliberal global order, it is essential to understand the significance and dynamics of the worldwide oil complex [24]. Malaysia and Iran have taken different approaches to their post-oil future when preserving oil as part of their national heritage. In his book, Christopher Dietrich examines how cultural history forms an essential part of the struggle of post-colonial countries turning into international states [25]. As demonstrated here, there is an element of both Malaysian and Iranian policy to decolonize the oil industry through the preservation of historical sites and to narrate them as part of the national story. Through the efforts of Petronas, oil permeates society and dominates education, producing new elites and changing the landscape of coastlines facing the South China Sea. In this narration, oil has resulted in the improvement of society, the upgrading of backwater villages, the advancement of educational opportunities, and drastic changes to landscapes and cityscapes. The relic that is the Miri oil drill is transformed into a tourist site under the purview of Petronas, sanitizing the past and decolonizing British presence from Sarawak's Sarawak's early oil history.

The Covid-19 pandemic outbreak has shown that the industry is not infallible and that countries with significant oil industries are vulnerable to changes in consumer demand. The economic effects of the pandemic saw a drastic drop in consumer demand, which will likely continue to depress Iranian and Malaysian exports for the months to come. In the case of Iran, the Covid-19 crisis and the near-future fluctuations in oil prices coincided with the maximum pressure campaign of the United States against Iran. Tensions between the United States and Iran relaxed somewhat after the nuclear deal of 2015. But, due to the recent US withdrawal from the Joint Comprehensive Plan of Action (JCPOA) in May 2018, Iran's oil exports declined dramatically [26]. Before that, China and India (two giant Asian customers) were purchasing Iranian oil. In response to these and control liquidity and direct it to the right path and production cycle, Iran is considering letting residents invest in oil on the domestic energy exchange [27]. With its rich oil and gas resources, Iran needs new technology investments and development plans to be prepared for the post-Oil future, but that will be hard to achieve without resolving US-Iran tensions and an easing of US sanctions. To balance the future economic growth with social development and environmental protection, Iran also needs to invest more in plans for sustainable development, smart cities, and a smooth transition to less environmentally harmful sources of energy.

Thus far, Malaysia has not been subject to international sanctions and can trade on the world market with few barriers. Its diplomatic relations with Iran have been consistent since the Pahlavi era through to the current presidency of Ebrahim Raisi. Iranians can travel to Malaysia without a visa. Although Malaysian Prime Minister Mahathir Mohamed had condemned the US sanctions against Iran, the country still bowed down to Washington's pressure to enforce them [28]. Petronas has not shied away from interacting with Iran when it comes to the oil industry. Over the years, it has explored oil fields, signed a memorandum of understanding with the NIOC, and expressed interest in Iran's oil and gas projects. But interest has been inconsistent, as seen by its recent withdrawal from a liquified natural gas project in Iran, reflecting Malaysia's adherence to international concerns attached to economic closeness to Iran as well as Petronas' limited scope. Malaysia and Iran have much potential to collaborate over their oil industries, especially in light of good bilateral relations, historical ties, and cultural similarities. However, in the age of sanctions, this remains a barrier for the two countries to pursue economic (and social) projects related to oil.

The Malaysian oil industry played another critical role in light of the global chaos caused by the recent pandemic crisis. Petronas took on a pastoral part by alleviating the difficulties caused by the virus, again blurring the lines between corporate and national responsibility. The company provided medical supplies, monetary support, and daily essentials to frontlines and affected communities. Similarly, Shell in Malaysia launched a similar campaign to show its efforts during the pandemic by channeling funds to local charities and to support medical practitioners. Such measures place oil at the heart of Malaysian society, thus preserving its importance and relevance.

Moreover, Petronas's usage of oil profits challenges the paradox where oil wealth has resulted in inequality [29]. Nonetheless, the Malaysian petroleum industry is not immune to volatilities in the market and international outcries over the ongoing climate crisis. In the last years, Malaysia's palm oil industry has come under much pressure due to its unsustainability and has developed strategies to improve environment management [30]. Malaysian petroleum companies have had to follow suit. They have also developed programmes to safeguard the environment and play an essential role in ensuring sustainability and transparency. The Yayasan Petronas (Petronas Foundation) runs forest and mangrove conservation initiatives and cleaner reloading services for shipping in the South China Sea. As a developing country that still relies on its oil reserves, sustainability here is closely linked with corporate responsibility, which is intertwined with a focus on preserving oil's place within Malaysia's national legacy.

References

1. Hauser, S., Zhu, P., Mehan, A.: 160 years of borders evolution in Dunkirk: petroleum, permeability, and porosity. Urban Plan. **6**(3), 58–68 (2021). https://doi.org/10.17645/up.v6i3.4100
2. Mehan, A., Jansen, M.: Beirut blast: a port city in crisis. The Port City Futures Blog. In: Mehan, A., Mostafavi, S. (eds.) Building Resilient Communities Over Time. The Palgrave Encyclopedia of Urban and Regional Futures. Palgrave McMillan (2022). https://doi.org/10.1007/978-3-030-51812-7_322-1
3. Mitchell, T.: Carbon Democracy: Political Power in the Age of Oil. Verso, London (2011)
4. Watts, M.J.: Crude politics: life and death on the Nigerian oil fields. Niger delta economics of violence, working Papers 25, pp. 1–27 (2009)
5. Barrett, R., Worden, D.: Oil culture: guest editors' introduction. J. Am. Stud. **46**(2), 269 (2012)
6. Fuccaro, N.: Histories of oil and urban modernity in the Middle East. Thematic issue of comparative studies of South Asia. Afr. Middle East **33** (1), (2013)
7. Mitchell, T.: Carbon democracy. Econ. Soc. **38**(3), 406 (2009)
8. Khalaf, S.: The evolution of the Gulf city type, oil, and globalization. In: Fox, J.W., Mourtada-Sabbah, N., Al Mutawa, M. (eds.), Globalization and the Gulf, p. 247. Routledge, London (2006)
9. Fuccaro, N.: Visions of the city: urban studies in the Gulf. Middle East Stud. Assoc. Bull. **35**(2), 177–178 (2001)

10. Rahman, S.: Developing Eastern Johor: The Pengerang Integrated Petroleum Complex, vol. 16, p. 2. Cambridge University Press, Cambridge (2018)
11. Fuccaro, N.: Visions of the city: urban studies in the Gulf. Middle East Stud. Assoc. Bull. **35** (2001)
12. Ibid, 185
13. Bazazzadeh, H., Ghomeshi, M., Mehan, A.: The Trans-Iranian Railway: A UNESCO world heritage site. TICCIH Bull. **95**, 31–33 (2022)
14. Douet, J.: The Heritage of the Oil Industry: TICCIH Thematic Study, p. 8 (2020)
15. Repellino, M.P., Martini, L., Mehan, A.: Growing environment culture through urban design processes 城市设计促进环境文化. Nanfang Jianzhu **2**, 67–73 (2016)
16. Mehan, A., Behzadfar, M.: The forgotten legacy: oil heritage sites in Iran. CONGRESO XVII TICCIH —CHILE (Patrimonio Industrial: Entendiendo el pasado, haciendo el futuro sostenible), pp. 897–900. Universidad Central de Chile, Santiago (2018)
17. http://www.petromuseum.ir/en/content/32/Editorial/4764/Tombi-Power-Plant-in-Masjed-Soleyman-in-1971. Accessed 21 Oct 2021
18. http://www.petromuseum.ir/content/32/Editorial/695/Iran-Petroleum-Museum-Introduct ion-and-Goals. Accessed 21 Oct 2021
19. Gabriel, S.H. (ed): Making Heritage in Malaysia: Sites, Histories, Identities. Palgrave Macmillan, Singapore, pp. ix–xi, 3–4. (2020)
20. Such as the Sarawak Delta Geopark. Badang, D., Aziz Ali, C., Komoo, I., Shafeea Leman, M.: Sustainable geological heritage development approach in Sarawak delta, Sarawak, Malaysia. Geoheritage **9**, 443–444 (2017)
21. Petrosains. https://petrosains.com.my/exhibits-floor-plan/. Accessed 21 Oct 2021
22. Jabatan Muzium Malaysia. http://www.jmm.gov.my/ms/mengenai-jabatan-muzium-mal aysia. Accessed 21 Oct 2021
23. Gunter, H.M., Apple, M.W., Hall, D.: Corporate Elites and the Reform of Public Education, p. 4. Oxford University Press, Bristol (2017)
24. Atabaki, T., Bini, E. (ed): Working for Oil: Comparative Social Histories of Labor in the Global Oil Industry, Palgrave Macmillan, Singapore, p. 27 (2018
25. Dietrich, C.R.W.: Oil Revolution: Anticolonial Elites, Sovereign Rights, and the Economic Culture of Decolonization, p. 7. Cambridge University Press, Cambridge (2017)
26. https://www.iai.it/en/pubblicazioni/covid-19-and-oil-price-crash-twin-crises-impacting-saudi-iran-relations. Accessed 21 Oct 2021
27. https://oilprice.com/Latest-Energy-News/World-News/Iran-Considers-Allowing-People-To-Invest-In-Oil-On-Local-Exchange.html. Accessed 21 Oct 2021
28. Lee, L., Ananthalakshmi, A.: Iranians in Malaysia say banks close their accounts as US sanctions bite. Reuters 30 October 2019
29. Ross, M.L.: The Oil Curse: How Petroleum Wealth Shapes the Development of Nations, pp. 1–2. Princeton University Press, Princeton (2012)
30. Choong, C.G., McKay, A.: Sustainability in the Malaysian palm oil industry. J. Clean. Prod. **85**(15), 258 (2014)

Digital Transformation in the Preservation of Cultural Heritage in Cities: The Example of Love Bank

Katarína Vitálišová [ID], Kamila Borseková[(✉)] [ID], and Anna Vaňová [ID]

Faculty of Economics, Matej Bel University, Tajovského 10, 975 90 Banská Bystrica, Slovakia
{katarina.vitalisova,kamila.borsekova,anna.vanova}@umb.sk

Abstract. The aim of the paper is to demonstrate the implementation of new technologies in the preservation and presentation of unique intangible and tangible culture and historical heritage using the example of Love Bank in Banská Štiavnica in Slovakia. The article presents the selected research results based on the analysis of museum documents, the structured interview with the museum's PR manager and the questionnaire survey among museum visitors. The unique offer of the museum has become one of the key attractions of the city, located in the historic center belonging to the UNESCO World Heritage and has a significant economic and social contribution to the development of the city.

Keyword: Digital technologies · Intangible and tangible heritage · Innovations

1 Introduction

The rapid development of digital technologies significantly changes traditional forms of presentation and preservation of historical and cultural heritage. These changes provide new ways of interaction with the end user and due to the Covid-19 pandemic, not only in the physical environment but also in the virtual space. The great impact of the pandemic during the last two years is evident in the explosion of virtual routes, interactive presentations, videos, etc. in museums, galleries, or even libraries of the world. However, it does not replace the unique experience of physical attendance in these cultural facilities and the authentic experience. The role of culture as a tool of human cultivation has partially lost its importance because of the hard fight with the world pandemic [1].

The respond of regions to the impact of Covid-19 in the cultural sectors depends on the differences in the framework of national cultural policies, data provision, legal framework and social protection [2]. The report from KEA (2020) expects that "the impact of COVID-19 on cultural and creative sectors (CCS) will be more severe in the Balkans, Central and Eastern Europe as often these countries overlook the importance of CCS in the economy and in territorial attractiveness" [3].

In case of museums, the most visible impact on the activity correspond to their educational and social work. In many of them, their knowledge dissemination work was

© The Author(s), under exclusive license to Springer Nature Switzerland AG 2022
F. Calabrò et al. (Eds.): NMP 2022, LNNS 482, pp. 2617–2627, 2022.
https://doi.org/10.1007/978-3-031-06825-6_250

innovated by setting up new devices, by using other media or by the development of digital tools [4]. These impacts of the pandemic influence also Love Bank, the private museum in the Slovak Republic. Because of the postponed national supporting policy for cultural sectors, the museum had to face with lot of challenges. The well-designed strategy to recovery from Covid-19 pandemic, confirms in case of the Love Bank the nomination for the Innovative European Museum of the Year Award and its presentation as Slovak "best" in Expo exhibition in Dubai, in 2022.

The article presents in a form of case study the unique Slovak museum Love Bank situated in Banská Štiavnica. The presentation of the museum combines the intangible heritage – the longest love poem in the world- Marina based on the true love story of Marina and Andrej Sládkovič (the author of the poem) with the historical heritage of the city Banská Štiavnica enlisted in the UNESCO World Heritage. The case study summarizes the selected results of secondary and primary research mapping the IT solutions used in the museum and identifying the role of the museum in city development.

2   Digital Transformation in Cultural Sector

Digital transformation is a new phenomenon evident in all economic sectors. It can be defined as a change in the scope and direction of governance supported by technologies and electronic processes to ensure better value creation for the benefit of customers and companies [5, 6]. Vial adds that important elements to achieve this change are information, computing, communication, and connectivity technologies [7].

The outputs of the digital transformation are usually innovations in the delivery mode of services, forms of direct interactions with customers, as well as the proliferation of smart products that enable real-time monitoring and updating, and services that transform production processes and customer relationship [5]. Innovations, including digital transformation in the preservation of cultural heritage, are crucial to the development of the tourism sector and to ensure competitiveness in tourist destinations [8].

In the cultural sectors, the impacts of digital transformation are reflected in facilitating imaginative engagement with spaces and objects, in affording innovative forms of participation, and in drawing new kinds of value from otherwise inaccessible archives [9]. The new technologies innovated cultural services by "challenging/overcoming shared cultural codes of the product category, and proposing cultural meanings not previously exploited by incumbents that resonate with final customers" [10, p. 432]. In the cultural sector, innovation can be characterized as a soft innovation in goods and services that primarily impacts sensory perception, aesthetic appeal, or intellectual appeal rather than functional performance [11]. NESTA [12] differentiates the innovations in products that are aesthetic or intellectual in nature (music, books, film, fashion, art) and the aesthetic innovations in goods and services that are primarily functional in nature which can be found in other industries where products may also have many non-functional characteristics (sight and touch of a new car, for example, sound of its engine, etc.).

Although technological product and process innovations are widespread within the cultural sector, an important part of innovative activities here is based on novelty instead of functionality and involves a change that is more aesthetic or intellectual in nature [11, 13].

According to Agostino, Constantini the digital transformation in museums has received much attention in the last decade, even there were realized various studies of digital innovations in museums. The objects of research were all activities of the museum value chain, from the digitalization of the heritage assets to the digitalization of the experience offered to visitors [14] (e. g., augmented reality, interactive panels, and mobile technologies to administrative activities, see for example [15–22]).

The impacts of digital transformation in cultural institutions are reflected not only in empowering the customer; enabling staff to think 'beyond my service', encouraging staff to explore new and more efficient ways of working or empowering and supporting staff to continuously improve, encouraging customer-focused thinking and focusing on developing organizational culture [23], but it can also bring new stimuli for city development, for example, higher demand for additional services for tourists. The innovations in cultural and creative industries can rise into new ideas, mobilizes the creative potential of places in the form of new products, services, information, technological innovations, non-technological processes, and outputs that generate creative capital that is increasingly important for the growth of cities and regions [24–26]). Innovations can about bring also the new way of utilization the historical and cultural heritage in other economic activities (e. g., old abandoned historic buildings rebuilt to hotels, restaurants, business offices in a form of co-working space or incubators, etc.). However, all implemented innovations, especially in cultural industries, should be carefully prepared with respect to local identity, acceptable by the local community and its shared values [27, 28]).

In our paper, we focus on digital innovations used in the preservation and presentation of the unique intangible and tangible cultural and historical heritage in a unique world museum, Love Bank.

3   Data and Methodology

The present paper in the form of case study explores the digital innovations realized by the Love Bank, the museum dedicated to the longest love poem in the world, written in 1844 by Slovak writer A. Sládkovič named Marina. It is only one museum of this type in the world. The paper was processed based on the data collected by the analysis of museum documents and a structured interview with the museum's PR manager. These data were supplemented by the questionnaire survey among museum visitors conducted in the March of 2021. In the survey 167 respondents took parts. 160 respondents visited the museum. The research sample includes 88,62% women and 7,19% men. 67,5% of the respondents were in the age range 26–49 years and 21,88% from 18 to 25 years. 92,81% of respondents come from Slovakia, the rest of them from abroad (Switzerland, Czech Republic, Croatia and other countries). The questionnaire survey was conducted in an electronic form, distributed on Facebook, and published in the Love Bank profile. The aim of this research was to identify the unique selling points of the museum and the main reasons people visit the museum. The research was realized as a part of international research oriented on the implementation of smart solutions in various areas of smart cities.

4 The Love Bank in Banská Štiavnica

The Love Bank, an interactive museum, is situated in the manor house and located in Banská Štiavnica. During the last few years, it has become a unique selling point of the city, attracting many domestic and foreign visitors. During Covid-19 pandemic, the museum has had to face with new challenges that needed also innovative solutions.

4.1 History and Present of the Love Bank

The most identified competitive advantage in Slovak regions is cultural and historical heritage, the main tourist attraction in Slovakia [29]. One of the most interesting tourist sites in Slovakia is the city of Banská Štiavnica. Banská Štiavnica, the oldest mining town in Slovakia, was established in the 13th century, although evidence of mining dates to the late Bronze Age. The distinctive form of the property was created by the symbiosis of the industrial landscape and the urban environment resulting from its mineral wealth and consequent prosperity. For centuries, Banská Štiavnica has been the centre of mining and education in Europe. The urban and industrial complex of Banská Štiavnica is an outstanding example of a medieval mining centre of great economic importance. Since 11. December 1993, Banska Štiavnica has been the UNESCO World Cultural Heritage Site. Banská Štiavnica is firmly anchored among established tourist destinations in Slovakia. Tourism has become one of the key economic sectors of the city and region. It continues to progress for several years [28].

The history of the manor house, where the museum is located, is associated with Marina Psichlová who lived in the 16th century and was a love of Andrej Sládkovič, one of the most famous Slovak poets. As a manifestation of his love for Marina, he wrote Marina, the longest love poem on the world, 176 years ago. The story of Marina Pischlová and Andrej Sladkovič could be comparable to Romeo and Juliet, but the difference is that the story is much more powerful and truly happened in the past, in Banská Štiavnica.

In spring 2015, the manor houses on the main square was repurchased by the current owner of the museum. This purchase protected the house before demolition and conversion to a block of apartments. The roof of the house was leaking and in desperate need of repair, several walls were damaged due to the lack of state funds to renovate it. Therefore, the house had to be totally renovated, and the design of the museum was gradually developed and realized.

The main objective of the museum is to preserve the historic house of Marína, to develop the exceptional 'experience house' as a new attraction of lovers (as a tourist attraction) and to promote Sládkovič's Marína around the world as the longest love poem in the world. The aim of this research was to identify the unique selling points of the museum and the main reasons people visit the museum.

The project is realized as a private one. It is under umbrella of the newly established on-profit organization to collect donations from various entities. It is also used as a crowdfunding platform for financing, where museum supporters can contribute financially. The founders covered 54% of total costs until 2018 (504,900 €). The 42% of costs were covered by the Tatrabanka loan, and the rest of the costs were covered by contributions from the Ministry of Culture of the Slovak Republic (4%). Total costs until 2018 were 935,000 € [30].

The uniqueness of the museum lies in the interactive presentation in the longest love poem of the world. The concept of the museum is original and familiar to the real conditions of the house. The development of the project lasted two years and drew inspiration in digital design from abroad.

As was already mentioned, the main idea is to present the poem Marina. The presentation lasts approximately 60 min in a new creative way with support of visualisation, filming, and history. The presentation uses various forms of technologies: 3D visualization; digitalisation of the poem in ultra-high definition. All data and process are backed up to servers.

The presentation in the "experience house" includes different kinds of attractions (Fig. 1). The whole implementation process was in the hands of its creators. Special techniques and technologies were provided by Epson company. The success of the experience house about the poem Marína also lies in the non-traditional approach to the traditional theme.

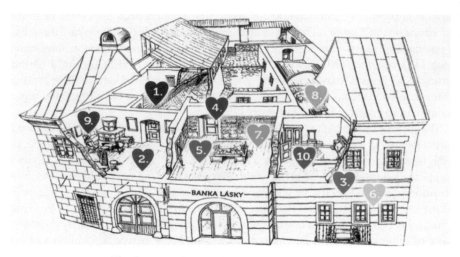

Fig. 1. Attractions of the Love Bank. Source: [32]

The main attraction is the first love vault in the world, where you can store "love" (1). The love vault has been created from the verses of the poem MARÍNA - The Longest Love Poem in the World. Each verse consists of love boxes, in which people can store a symbol of your love or desire for love. Each box contains a piece from Sládkovič's original manuscript of a poem by Marina. The love vault is for everyone because everyone loves someone or something. The Love Box in a love vault is often offered as a gift for occasions such as weddings, engagements, a birth of a child, wedding anniversaries, birthdays, and many others. People can have their love box for a one year for contribution of 50 Euro, or for forever for a contribution of 100 euros. The entire love bank with its love vault is a fundraising project to save and promote the national cultural heritage, The House of Marína.

The old telling paintings (2) thanks to superb high technology and cooperation with the famous Slovak actors and actresses (e.g. Emília Vášáryová, Táňa Pauhoffová, Robert

Roth and Luboš Kostelný (9)) are the next attraction of the museum. Paintings come to live and tell a story of incredible love and learn what really happened between Marína and Sládkovič. For talking paintings, Epson technology and techniques were used. The Epson EB-585W projectors have been used to create an extraordinary interactive experience for visitors. The Epson EB-585W projectors create the illusion of ancient oil paintings. The four portraits in historical wooden frames are portraits of Marina, Sládkovi, Marina's mother, and Marina's husband Gerzsa. During the oil painting show, they suddenly come to life unexpectedly and start communicating with each other. In this unconventional high-tech form, they tell the audience how it was in one of the most famous love stories. The four images are projected by only three projectors. The ultra-short projection distance of the Epson EB-585W projector made it possible to place the projectors just below the images, even eccentrically, thanks to which they are sensitively incorporated into the space so that even technically capable visitors cannot detect them. The three Epson EB-585W projectors, which are installed in the Love Bank, do not disturb the viewer due to their ultra-short distance, and they also have high contrast and brightness, so the projection is clearly visible even in daylight.

Other possibilities are to measure the love of couple by lovemeter (3), expressed in a splendid verse from the poem Marína; to touch the magic handles (4), or to see interactive revival of the poem by new IT technology (5). By the combined creativity with cutting-edge technologies, a unique way of poetry presentation is prepared. Epson Perfection V750 Pro scanners have allowed Marina's original manuscript to be preserved for future generations. Advanced technology recorded the manuscript at 15 times higher quality than before. Scanning technology has combined cold light with high-resolution scanning. The Epson Perfection V750 Pro scanner was able to scan the original handwriting without damaging it, up to 9600 dpi. Epson is an expert in restoring and editing old images with a combination of DIGITAL ICETM technology and Epson Easy Photo Fix. The entire poem was digitalised in cooperation with the Slovak National Library in Martin.

The part of the museum is a Love Post Office as a souvenir shop (6) and Sky of Love (8) that was born in the courtyard of the House of Marina as a thank you to hundreds of people who have a box in the love vault and thus support the fundraising project.

Furthermore, in the Sládkovič library it is possible to see the facsimile of the Marina poem (7), as well as its first edition in 1846 and the latest published in 2017. There is an exhibition of all 50 editions of the poem in 8 languages. Here, visitors experience in an unconventional way how the power of love can miraculously trigger a projection of the most beautiful verses in the poem Marína. This unique experience is made possible by 5 Epson EB-535W projectors. These are cleverly placed under the ceiling and, in addition to spectacular projections, can evoke the authentic atmosphere of an ancient library. The base of the Epson EB-535W projectors is characterized by its ability to cover the entire wall area. Visitors are literally drawn into the story; thanks to 3LCD technology, the device displays clear images in top quality and more realistic colours even in a lighted room without being disturbed by the placement of the device directly in front of the viewers.

Thanks to the financial support from the Ministry of Investment, Regional Development, and Informatization of the Slovak Republic (MIRRI), museum managed in an extremely short time in December 2020 implement a project Professional guide audio device for 10 world languages.

The Covid 19 pandemic has a huge economic impact on the museum, most of 2020 and the first third of 2021, again in November and December 2021 it was closed. However, this situation forced them to look for non-traditional solutions to stay in touch with customers (existing and potential). The museum introduced the online version of love banks, so people can come to the love vault from all over the world. The idea is that for each physical box of love, an electronic version would have the character of the so-called "relationship account" to which owners could upload small symbols and expressions of mutual love online on a daily basis. This online relationship account is closely linked to the physical box of love in the only love vault in the world.

In addition to the possibility of setting up an online box of love and quickly setting up an e-shop, the museum introduces a new product - an online service that allows people to send each other words of support or declaration of love in these difficult times, written in the authentic handwriting of the most romantic Slovak poet Andrej Sládkovič [31].

In the short future, the museum is preparing the new online version of the love box attraction, new presentations in the rest of the Marina House with 3D holograms, etc.).

Due to the uniqueness of the museum, Love Bank was nominated as a candidate for the Innovative European Museum of the Year Award 2021, and also foreign media have been interested in its presentation.

The Love Bank has been open for 4 years and was visited by more than 65,000 people from all continents. The main final users are domestic and foreign tourists who come to Banská Štiavnica. Due to the great interest of the public supported by the promotion in the media of 80 countries, there is a segment of city visitors who come there just for the visit of the museum. Visitors in the form of entrance ticket, as well as fee for the box of love, contribute to the financing and ensure the long-term sustainability and development of the entire project.

4.2   Uniqueness of Love Bank from the Visitors' Point of View

The Love Bank Museum has gathered a lot of good reviews on various platforms: 4.9/5 on Facebook (over 252 reviews), 4.8/5 on Google (over 1 113 reviews) and 5/5 on TripAdvisor (84 reviews) and second recommended thing to do in Banská Štiavnica.

To identify the unique selling points of the museum, we conducted a short questionnaire among museum visitors in March 2021. The survey consists of identification questions (sex, age, country of origin). Subsequently, the research questions were related to the source of information about Love Ban, the motives of the visitor's visit and the evaluation of factors that are the core of Love Bank uniqueness.

The main source of information for visitors is the Internet (38,13% of respondents) and social networks (34,39% of respondents), as well as references from family and friends (23,75% of respondents). As we already mentioned, museum data are available for all potential visitors on widely used communication platforms such as Facebook, Google, and Tripadvisor.

To the main reasons people visit the museum belong the new experience; spending time with family or friends and knowing the history of the poem (Fig. 2).

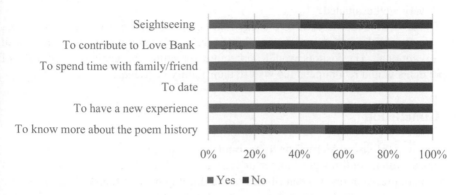

Fig. 2. The reasons why visitors coming to the museum

Subsequently, we asked the respondents to evaluate the various aspects of the museum visit on the scale 1–5 (1- the worst, 5 – the best), and their average evaluation is presented in Fig. 3.

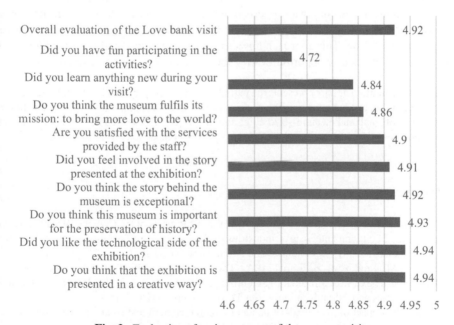

Fig. 3. Evaluation of various aspect of the museum visit

All aspects evaluated were marked with high marks, as the best or almost the best. The highest evaluated aspects of museum exhibition are a creative way of presentation and the technological side of the exhibition. Also, very appreciated is the preservation of the history and the historical story behind the main idea of the museum. As a space to improve the presentation, we introduce more funny activities, even though this aspect is still very well evaluated.

To the most favourite parts of presentation belong to telling paintings, interaction, and interface with technology. They evaluated the whole exhibition as very successful and exceptional. The visitors would appreciate spreading the exhibition. 98,13% of respondents would recommend the visit to their family or friends.

5 Conclusion

The renewed "House of Marina" or "Love Bank" became the best-selling point of Banská Štiavnica, which confirms the number of visitors. The Love Bank as a creative project of museum presentation also contributes to the development of Banská Štiavnica. The effects are possible to identify in increase of revenues of local businesses providing the complementary services (restaurants, hotels, local shops, etc.).

Although the project is private one and there is not a direct link to the municipal strategy. The Love Bank contributes significantly to the preservation of the historical heritage in the centre of the city Banská Štiavnica, which belongs to the UNESCO world heritage. Moreover, the main mission of the museum – to save the poem Marina for future generations – helps to build awareness and save the unique features of the Slovak literature and history in a very creative and modern form. The museum concept is based on the well-designed strategy connected the historical heritage with the innovative forms of presentation. This approach has showed as very relevant also during Covid 19 pandemic and opens new opportunities for the museum (e.g. virtual love boxes).

Acknowledgement. The paper presents the partial outputs of the project VEGA 1/0213/20 Smart Governance in Local Municipalities, the project VEGA 1/0380/20 Innovative approaches to the development of small and medium-sized cities and the project APVV SK-FR-19-0009 Financing cultural policies and creative industries. A Franco-Slovak comparison.

References

1. Vitálišová, K., Borseková, K., Vaňová, A., Helie, T.: Impacts of the COVID-19 pandemic on the policy of cultural and creative industries of Slovakia. Scientific Papers of the University of Pardubice, Faculty of Economics and Administration. Series D, vol. 29, pp. 1–11 (2021)
2. Šebová, M., Révészová, Z., Tótová, B.A.: The response of cultural policies to the Covid-19 Pandemic: The Case of Slovakia. In Scientific papers of the University of Pardubice : Faculty of Economics and Administration. series D, vol. 29, pp. 1–12 (2021)
3. KEA: The impact of the Covid-19 pandemic on the cultural and creative sector. https://kea net.eu/publications/the-impact-of-the-covid-19-pandemic-on-the-cultural-and-creative-sec tor-november-2020/. Accessed 25 Feb 2022
4. UNESCO: Museums Around The World. In the Face of Covid-19. UNESCO, Paris (2021)
5. Mergel, I., Edelmna, N., Haug, N.: Defining digital transformation: results from expert interviews. Gov. Inf. Q. **36**(4), 1–16 (2019)

6. Margiono, A.: Digital transformation: setting the pace. J. Bus. Strateg. **42**(5), 315–322 (2021)
7. Vial, G.: Understanding digital transformation: a review and a research agenda. J. Strateg. Inf. Syst. **28**(2), 118–144 (2019)
8. Gajdošík, T., Gajdošíková, Z., Maráková, V., Borseková, K.: Innovations and networking fostering tourist destination development in Slovakia. Quaest. Geograph. **36**(4) (2017)
9. Arrigoni, G., Schofield, T., Pisanty, D.R.: Framing collaborative processes of digital transformation in cultural organisations: from literary archives to augmented reality. Museum Manag. Curator. **35**(4), 424–445 (2020)
10. Pedeliento, G., Bettinellil, C., Andreini, D., Bergamaschi, M.: Consumer entrepreneurship and cultural innovation: the case of GinO12. J. Bus. Res. **92**, 431–442 (2018)
11. Subottina, N.: Innovation in cultural sector – definition and typology. Zarzadzanie w Kulturze **16**(4), 379–388 (2015)
12. NESTA: Soft Innovation: Towards a More Complete Picture of Innovative Change. NESTA, London (2009)
13. Vitálišová, et al.: Innovations in the cultural and creative local development. Acta aerarii publici **15**(2), 41–55 (2018)
14. Agostino, D., Costantini, C.: A measurement framework for assessing the digital transformation of cultural institutions: the Italian case. Meditari Accountancy Research. 1–28
15. Bertacchini, E., Morando, F.: The future of museums in the digital age: new models of access and use of digital collections. Ebla Working Papers, pp. 1–14 (2011)
16. Borowiecki, K.J., Navarrete, T.: Digitization of heritage collections as indicator of innovation. Econ. Innov. New Technol. **26**, 227–246 (2017)
17. Chiaravalloti, F.: Performance evaluation in the arts and cultural sector: a story of accounting at its margins. J. Arts Manag. Law Soc. **44**, 61–89 (2014)
18. De Bernardi, P., Gilli, M., Colomba, C.: Unlocking museum digital innovation. Are 4.0 Torino museums? In: Smart Tourism, pp. 453–471. McGraw Hill Education, New York (2018)
19. Gombault, A., Allal-Chérif, O., Décamps, A.: CT adoption in heritage organizations: crossing the chasm. J. Bus. Res. **69**, 5135–5140 (2016).
20. Pierroux, P., Krange, I., Sem, I.: Bridging contexts and interpretations: mobile blogging on art museum field trips. Mediekultur: J. Media Commun. Res. **27**(18), 1–18 (2011)
21. Chung, N., Han, H., Joun, Y.: Tourists' intention to visit a destination: the role of augmentedreality (AR) application for a heritage site. Comput. Hum. Behav. **50**, 588–599 (2015)
22. Coman, A., Grigore, A.M., Ardelean, A.: The digital tools: supporting the "inner lives" of customers/visitors in museums. In: Meiselwitz, G. (ed.) Social Computing and Social Media. Design, Human Behavior and Analytics. HCII 2019. LNCS, vol. 11578, pp. 182–201. Springer, Cham (2019). https://doi.org/10.1007/978-3-030-21902-4_14
23. Curtis, S.: Digital transformation – the silver bullet to public service improvement? Public Money Manag. **39**(5), 322–324 (2019)
24. Batabyal, A.A., Nijkamp, P.: Creative capital in production, inefficiency, and inequality: a theoretical analysis. Int. Rev. Econ. Financ. **45**, 553–558 (2016)
25. Borseková, K., et al.: The nexus between creative actors and regional development. Land **10**(3), 276 (2021)
26. Florida, R.: Cities and the creative class. City Commun. **2**, 3–19 (2003)
27. Martinat, S., et al.: Sustainable urban development in a city affected by heavy industry and mining? Case study of brownfields in Karvina, Czech Republic. J. Clean. Prod. **118**, 78–87 (2016)
28. Vitálišová, et al.: Creative potential in the cities and its exploitatiton in the sustainable development. Belianum, Banská Bystrica (2019)

29. Borsekova, K., Vaňová, A., Vitálišová, K.: Smart specialization for smart spatial development: Innovative strategies for building competitive advantages in tourism in Slovakia. Socioecon. Plann. Sci. **58**, 39–50 (2017)
30. Love Bank: About the project. https://bankalasky.sk/o-projekte/. Accessed 21 Dec 2021
31. Love Bank. Declaration of love. https://bankalasky.sk/vyznanie-lasky/. Accessed 21 Dec 2021
32. Love Bank. Attractions of the Love Bank. https://reallovebank.com/visit-the-love-bank/#attractions. Accessed 21 Dec 2021

Circular Economy of the Built Environment in Post-pandemic Era; A Disignerly Proposal for the Future Generation of Workspaces

Hassan Bazazzadeh[1]([X]) (iD), Masoud Ghasemi[2], and Behnam Pourahmadi[1] (iD)

[1] Poznan University of Technology, 60-965 Poznan, Poland
Hassan.bazazzadeh@doctorate.put.poznan.pl
[2] Art University of Isfahan, Isfahan, Iran

Abstract. Considering the impact of COVID-19 outbreak on the foundation of our socio-economic and environmental systems, it is imperative to apply multi-faceted sustainability approaches for the current and post-pandemic era. The built environment plays a key role in the spatial engagement of humans and their workspace in the urban environment. Proposing new concepts for the post-pandemic era that combine the built environment and sustainability techniques may provide an opportunity for better integration into the essence of sustainability. In this regard, this paper recommends applying circular economy idea in adaptive reuse practice of industrial heritage to create circular workplaces for the post-pandemic period. As an example of this given proposal, a scheme for a textile factory in Isfahan, Iran is presented.

Keywords: COVID-19 · Post-pandemic · Circular economy · Built environment · Workspace

1 Introduction

While the global COVID-19 pandemic has made a huge impact on the structure of the world on all walks of life [1], the complexity of the socio-economic environment has been increased to a level that governments and business leaders need to shift toward more resilience approaches to respond to crises [2]. This global emergency has changed the shape of people's lifestyles in a way that an invisible hostile endangers close contacts and reshapes social and economic activities all around the world [3]. Moreover, the pandemic situation has transformed the social and working conditions to a new level of complexity with many challenges, from obligatory quarantine to social distancing, reduction in work productivity due to mental trauma and working adjustment processes. Furthermore, it has negatively affected plans for achieving UN sustainable development goals (SDGs) projects by slowing down the ongoing progress and also worsening some certain situations connected to the quality of human living [4].

In the socio-economic sphere, many areas have faced complex challenges from competition to cooperation, redesigning the current systems and reshaping the facilities to

© The Author(s), under exclusive license to Springer Nature Switzerland AG 2022
F. Calabrò et al. (Eds.): NMP 2022, LNNS 482, pp. 2628–2637, 2022.
https://doi.org/10.1007/978-3-031-06825-6_251

be more adaptable to sudden changes [5]. The urban environment as one of the important sections in the pandemic era got heavily affected by the significance of resilience infrastructure and quick response to changes in emergent times. The importance of developing of new designs that sync up with the resiliency and being environmentally friendly criteria of urban planning in the post-pandemic era [6].

Meanwhile, one key insight from the pandemic crisis is the urgent need to shift from traditional linear economic models to a more dynamic, ecologically friendly, and circular designs [7]. Besides, COVID-19 has destabilized the process of growth and development in "Small and Medium Enterprises" (SMEs), various industries as well as large corporate companies. The financial stability of SMEs got heavily affected by the pandemic due to the market fluctuation, consumer behaviour change, and public healthcare, regulatory intervention [8]. Securing their survival in challenging conditions, enterprises must establish new business models, incorporate adaptive resilience features, and provide flexible workspace.

Concurrently, the working condition has altered differently, due to the necessity of occasional remote working [9] and other required considerations such as respecting social distancing the workspace [10]. Overall, all pandemic scenarios present a slew of issues for the environmental, social, and economic sectors (sustainable development pillars), all of which require a coordinated response based on sustainable ways that address the critical demands of the current situation and the post-pandemic age.

In this regard, circular economy (CE) is one of the concepts that has lately attracted the attention of policymakers, corporate, and government leaders owing to its potential to respond to present and post-pandemic difficulties [11]. The conjunction of CE and the workspace may provide several aspects of consideration from various perspectives. Firstly, remote working in an effective manner may help environment protection and reduce transportation emissions [12]. Secondly, the importance of people's healthcare (mental and medical) state in the working environment should be in consideration in both the current and the post-pandemic era [13], which requires attention in developing new designs for working spaces [14]. Therefore, both concepts of CE and the architecture (the physical aspect) of workspaces in post-pandemic era may find a line of connectivity in the complex system of sustainability.

As far as the architecture of workspaces is concerned, the CE concept may act as a response to current unsustainable business and industrial models that provide environmental challenges [15]. Furthermore, CE approach may improve the quality of circular strategies for creating long-term positive impacts and offering a new spectrum for architects [16]. This transformation could provide an opportunity to synchronize CE and Circular design to create high-quality workspaces to meet both environmental consideration and work adjustment processes [6, 17].

The practicality of remote working shows positive signs in providing a hybrid working style in a variety of industries and businesses. Hence, this experience as well as other post-pandemic work-related challenges highlight the significance of reforming in our current workspaces based on the CE strategies that focus on more resiliency and sustainable goals. For instance, coworking environments with modular design and flexible or movable structure may be a good example of answering to required adaptability of the post-pandemic architecture for working spaces [7]. To engage with CE principles, architectures need to focus on existing CE targets which categorize primarily in recycling, resource efficiency (energy, water, material), recovery (waste, water, energy), and reduction of waste and emissions [18]. To this end, abandoned industrial heritage may be a suitable object for applying modular designs to create a proper workspace and address the sustainability challenges of workspace design in the post-pandemic era. Altogether, connecting CE objectives to the principles of design in workplace architecture in the form of a coworking space may give a cohesive answer to present and post-pandemic concerns.

2 Impacts of Covid-19 on Global Economy

The conditions of last two years were unexpected due to a surprising COVID-19 disruption in all aspects of our life. In order to reduce its transmission rate and isolate the cases, countries around the world implemented a series of actions such as boarder closures or national lockdown [7]. While the 2020s was supposed to be "the Decade of Action" to deliver the SDGs by 2030, COVID-19 has left its mark in the global progress towards these goals so far [19, 20] as achieving two important pillars of SDGs (globalization and sustainable economic growth) has slowed down [21].

The contraction of the global economy as the result of COVID-19 pandemic situations has been caused by withholding of business investments and a sharp decline in international trade because of various reasons [22]. Indeed, pandemic situations indicated how vulnerable and fragile our prosperity is irrespective of the degree of our society's development or status of GDP as it is evident that during Covid-19 pandemic the G7 countries saw 7.8–12.8% decrease in their GDP [23] and 8.8% of global working hours (equal to 255 million full-time jobs) were lost. For example, global lockdown restrictions led to a noticeable reduction in product supply by industries and factories, while self-isolation, quarantine reduced public demand, consumption, utilization of services and products. That is why the drastic fall of economic growth in both global scale and national scale for almost all countries was witnessed (Fig. 1).

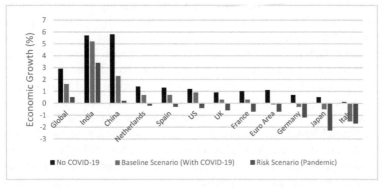

Fig. 1. Global economic impact of COVID-19 on global economy and on most powerful economic countries and areas [24].

During the pandemic situations businesses faced a wide range of problems such as the necessity of remote working which was very difficult for some sectors or the financial problems that happened as the result of decreased consumption of productions as well as disinfection measures. All these challenges highlight the significance of the reform in the workspaces and a paradigm shift towards more resilient workspaces by considering CE principles that can recover the financial loss of businesses and be proper for post-pandemic conditions with all complexity.

3 Circular Economy in the Built Environment in Post-pandemic Era

The concept of Circular Economy (CE) is the most recent effort to integrate economic activities and wellbeing (for both environment and human) in the most sustainable way [25]. The circular economy as an alternative approach to the current linear economy, tend to limit the usage of resources and efficiently make use of already existed structures or using waste (or in AEC industry abandoned structures) as new sources [26]. As a developing field, the CE idea is gaining attraction, with a focus on reducing, reusing, recycling, and recovering from production to consumption via circulation to balance the input and output of economic cycles [27]. At the same time, pandemic situation has changed the working conditions in a global scale [3], from advancement of digitalization process to enhancement of infrastructures for remote working. Circular-economy principles can be categorized into two main groups: (1) fostering reuse practices and extend life cycle of the product through different ways such as repair, upgrades, and retrofits. (2) Turning old goods into as new resources through recycling or reusing the elements and materials [28].

Circular economy strategies can provide a pathway towards a resilient and low-carbon economic recovery for post-pandemic era especially for the building stock. This attitude can be an answer to financial problems the we have been experiencing and it can be a remedy for companies in the building stock as they all have seen massive problems during this time (see Fig. 2). It can help to reshape a pathway towards a more resilient and low-carbon economic that can help achieving mutually reinforcing not only economic, but also societal and environmental objectives [29]. The pandemic has highlighted the shortcomings of our surrounding built environment including low-quality buildings, which lack the adaptability of our current needs. These problems, coupled with the ever-increasing concerns about our well-being have forced us to think about the transformation of the buildings through reusing, upgrading, and renovating buildings along circular principles [30]. This in turn will lead to more comfortable and adaptable buildings, and can have positive environmental impacts. Indeed, reusing infrastructure (such as industrial buildings) could allow a circulation potential for effective use of resources [31]. These circular principles and strategies represent key investments that can promise a more sustainable and resilient future built environment which is comfortable, cost-efficient, safe, and aligned with SDGs.

In this situation, coworking space can reduce the cost of working environment for businesses that have faced financial problems and need to economize on their work place. Moreover, since these places have a high rate of flexibility and adaptability, they

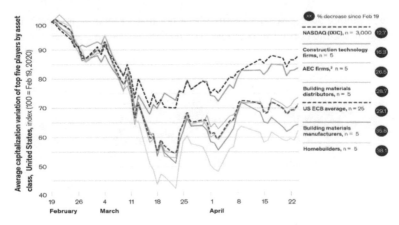

Fig. 2. The impact of COVID-19 on top companies related to the building stock [30].

can serve as a temporary place for businesses during their contract, after which there would be no obligation for them to renew their contract and pay any fee. Thus, this environment could be a smart answer to this challenge and if this attitude would be applied to abandoned industrial heritage the benefits would be even larger and more aligned with the principles of circular economy.

4   Traditional Workspace and Coworking Space

Real estate has usually acted as a serious constraint for almost all entrepreneurs since traditional office spaces are usually located in the downtowns and it means that they are often too expensive, especially for young businesses. Furthermore, traditional office spaces normally need tenants to sign a multi-year lease contract (3 to 5 years at minimum). Due to unpredictable nature of business and the fact that it is impossible to predict the exact degree of growth, entrepreneurs should constantly upgrade their workspace. While using spaces like dorm rooms, kitchen, or even garage could reduce the cost of the workspace, on the one hand they are not ideal places for working and on the other hand working people who work from home may experience loneliness and feelings of isolation [32]. That is why there is a critical need for reforming the architecture of traditional workspaces.

In the past decade, a relatively new phenomenon so called *"coworking spaces"* or *"subscription-based workspaces"* has emerged, *in which teams and individuals from different firms and companies work in a communal, shared space.* This approach can be applied through different ways such as (1) renting buildings from owners under short-term leases, (2) transforming an existing space (and sometimes changing the function) by adding common areas, and (3) renting a part of a space (or desks) and work next to other companies who have rented desks too under more flexible lease terms such as month-to-month contract. A systemically revision of the way the workspace is being planned and designed and even perceived needs a reform in the post-pandemic era to be suitable for our current needs. Even before COVID-19 outbreak these spaces had been very popular among businesses across the globe because of their adaptability to sudden changes which is quite normal in business world (Fig. 3).

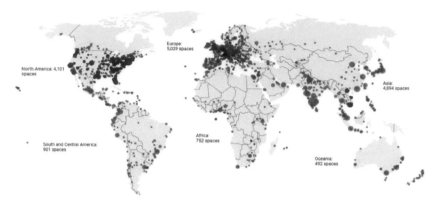

Fig. 3. Coworking spaces around the world in May 2021 [32].

Admittedly, even after the pandemic era, the world will never be the same and the trend of "work from home" and "work smart" will sustain according to announcements of numerous organizations and companies such as Twitter, Facebook, Slack, and Shopify and many other companies will follow them. In this situation coworking spaces would be a smart answer that make the best use of existing built spaces which would be according to CE principles. What make this idea more circular is to use existing buildings and alter the function through adaptive reuse practice in order to create coworking spaces. To that end, abandoned industrial heritage seem an appropriate location due to their physical features for this attitude.

5   Industrial Heritage as the Object of CE Application

The roots of attention to the heritage values of industrial architecture can be traced back to the 1967 seminar on industrial archaeology in Washington. These structures have become the matter of several meetings, studies and debates ever since and the establishment of various national and international organizations (such as "Society for industrial archaeology (SIA)" and "International committee for the conservation of industrial heritage TICCIH") to conserve and use them well confirms it.

According to the *"Nizhny Tagil Charter"*, Industrial heritage consists of the remains of industrial culture which are of historical, technological, social, architectural, or scientific value which can consist of the industrial building such as workshops, mills and factories, warehouses and stores [33–36]. These buildings are currently at risk in most cases mainly because of lack of public awareness about their potential. As a result, using these heritage buildings will be a great opportunity to get them back to life where they can serve perfectly to a wide range of people through adaptive reuse practices be forgotten and abandoned [37]. Industrial heritage buildings usually have some certain features that make them highly suitable for adaptive reuse approaches namely stable structures, existing and modifiable infrastructures and facilities, large and modular plan with high potentials to serve different needs, and in most cases very good access due to its location in urban centres. This feature gives them a great potential for converting to different uses with minimal time, cost, and change. In a way, the changes are temporary and there is the ability to change again. Today, these industrial complexes have good opportunities to establish the required uses of the urban fabric that can be used and re-planned to create vibrant and dynamic urban spaces and to improve the per capita of services, commercial, cultural, recreational, and green space.

6 Case Study; Risbaf Textile Factory, Isfahan, Iran

Risbaf factory in the southeast of *"Si O Se Pol"* Bridge is one of the most important and beautiful examples of industrial architecture in Isfahan which was established in 1932 in a 69,000 square meters land. It is believed that it was built by architect Motamedi under the supervision of Schumann, a German engineer. It has been shut down and abandoned since 2000. Risbaf has an open plan, and a modular structure which gives it a great potential for redesign (Fig. 4). The exterior appearance is more than a factory due to the Slight back and forth of the body, and the Ornament which gives this place a unique look.

Fig. 4. Current situation of Risbaf textile factory.

According to our scheme Risibaf factory will be reused as a coworking space with additional public open space, commercial spaces, cafes, and restaurants, as well as Isfahan's industrial museum. Considering the pioneer role of Risbaf during industrial movement of Iran, it would be a right place to start this another new phase of working environment too. In this design due to the location of four main halls perpendicular to the one of the main, historic, and busiest streets of Isfahan (Chahar Bagh Boulevard) the walls around the complex will be removed and large negative spaces between the halls will be used as green space, public space, and spaces for urban events. Each of the four main halls will be dedicated to coworking spaces and the central space will serve as the Museum of Industry, which will connected by two bridges to increase the interaction of the complex with its urban surroundings (Fig. 5). In addition, there will be spaces for games and relaxation, restaurants and coffee shops, cultural galleries, a library, a training class, and an outdoor and indoor amphitheatre. Through the proposed design, considering remote working opportunities, businesses can reduce their cost of workspace by sharing it with others and also using existing structures instead of building a new one may lead to circular use of these structures.

Fig. 5. Proposed development plan of Risbaf textile factory.

7 Conclusion

Undoubtedly, COVID-19 has left its mark on our perception of our environment and its impact seems never leave us and that is why our economy and consequently other aspects of our life must be adaptable to reach resilient and a more sustainable environment. Workspaces as one of areas that has undergone massive changes in post-pandemic era need a reform in their design approach due to the problems that businesses have faced. Circular economy in this respect is the right approach to be applied for redesigning these spaces to recovery what has been lost both financially and psychologically. Therefore, reusing industrial heritage sites into coworking spaces can be a smart and comprehensive answer to put CE principle into practice and reach sustainability in all senses of the word. The proposed framework may lead to the circularity of the workspaces during the post-pandemic era by reducing the cost for businesses (using shared workspaces) and utilizing the existing industrial heritage as an alternative for constrcuting new buildings.

References

1. Abu-Rayash, A., Dincer, I.: Analysis of the electricity demand trends amidst the COVID-19 coronavirus pandemic. Energy Res. Soc. Sci. **68**, 101682 (2020)
2. Zarghami, S.A.: A reflection on the impact of the COVID-19 pandemic on Australian businesses: toward a taxonomy of vulnerabilities. Int. J. Dis. Risk Reduct. **64**, 102496 (2021)
3. Megahed, N.A., Ghoneim, E.M.: Indoor air quality: rethinking rules of building design strategies in post-pandemic architecture. Environ. Res. **193**, 110471 (2021)
4. Sharma, H.B., et al.: Circular economy approach in solid waste management system to achieve UN-SDGs: solutions for post-COVID recovery. Sci. Total Environ. **800**, 149605 (2021)
5. Bazazzadeh, H., et al.: The importance of flexibility in adaptive reuse of industrial heritage: learning from iranian cases. Int. J. Conserv. Sci. **12**(1), 1–16 (2021)
6. Sharifi, A., Khavarian-Garmsir, A.R.: The COVID-19 pandemic: impacts on cities and major lessons for urban planning, design, and management. Sci. Total Environ. **749**, 142391 (2020)

7. Ibn-Mohammed, T., et al.: A critical analysis of the impacts of COVID-19 on the global economy and ecosystems and opportunities for circular economy strategies. Resour. Conserv. Recycl. **164**, 105169 (2021)
8. Chiu, I.H.Y., A. Kokkinis, and A. Miglionico, Addressing the challenges of post-pandemic debt management in the consumer and SME sectors: a proposal for the roles of UK financial regulators. Journal of Banking Regulation, 2021
9. Uherek-Bradecka, B.: Architectural and aesthetic aspects of workspace at home – a new design challenge in the age of pandemic. IOP Conf. Ser. Mater. Sci. Eng. **1203**(2), 022005 (2021)
10. Bergefurt, A.G.M., et al.: The influence of employees' workspace satisfaction on mental health while working from home during the COVID-19 pandemic. In: Healthy Buildings 2021, Europe, Oslo, Norway (2021)
11. Nandi, S., et al.: Conceptualising circular economy performance with non-traditional valuation methods: lessons for a post-pandemic recovery. Int. J. Logist. Res. App. 1–21 (2021)
12. Fabiani, C., et al.: Sustainable production and consumption in remote working conditions due to COVID-19 lockdown in Italy: an environmental and user acceptance investigation. Sustain. Prod. Consum. **28**, 1757–1771 (2021)
13. Giorgi, G., et al.: COVID-19-related mental health effects in the workplace: a narrative review. Int. J. Environ. Res. Public Health **17**(21), 7857 (2020)
14. Megahed, N.A., Ghoneim, E.M.: Antivirus-built environment: lessons learned from Covid-19 pandemic. Sustain. Cities Soc. **61**, 102350 (2020)
15. Schröder, P., Lemille, A., Desmond, P.: Making the circular economy work for human development. Resour. Conserv. Recycl. **156**, 104686 (2020)
16. Dokter, G., Thuvander, L., Rahe, U.: How circular is current design practice? Investigating perspectives across industrial design and architecture in the transition towards a circular economy. Sustain. Prod. Consum. **26**, 692–708 (2021)
17. Strauser, D., et al.: A tool to measure work adjustment in the post-pandemic economy: the Illinois work adjustment scale. J. Vocat. Rehabil. **54**, 1–8 (2020)
18. Morseletto, P.: Targets for a circular economy. Resour. Conserv. Recycl. **153**, 104553 (2020)
19. Guan, D., et al.: Global supply-chain effects of COVID-19 control measures. Nat. Hum. Behav. **4**(6), 577–587 (2020)
20. Sachs, J., et al.: Sustainable Development Report 2020: The Sustainable Development Goals and Covid-19 Includes the SDG Index and Dashboards. Cambridge University Press, Cambridge, UK (2021)
21. Naidoo, R., Fisher, B.: Reset sustainable development goals for a pandemic world. Nature **583**, 198–201 (2020)
22. Zhang, D., Hu, M., Ji, Q.: Financial markets under the global pandemic of COVID-19. Financ. Res. Lett. **36**, 101528 (2020)
23. International Monetary Fund: World economic outlook, June 2020: The great lockdown. International Monetary Fund (2020)
24. Rushford Business School: COVID-19 effects global economy (2021)
25. Murray, A., Skene, K., Haynes, K.: The circular economy: an interdisciplinary exploration of the concept and application in a global context. J. Bus. Ethics **140**(3), 369–380 (2015). https://doi.org/10.1007/s10551-015-2693-2
26. Teymourian, T., Teymoorian, T., Kowsari, E., Ramakrishna, S.: Challenges, strategies, and recommendations for the huge surge in plastic and medical waste during the global COVID-19 pandemic with circular economy approach. Mater. Circ. Econ. **3**(1), 1–14 (2021). https://doi.org/10.1007/s42824-021-00020-8
27. Kirchherr, J., Reike, D., Hekkert, M.: Conceptualizing the circular economy: an analysis of 114 definitions. Resour. Conserv. Recycl. **127**, 221–232 (2017)

28. Stahel, W.R.: The circular economy. Nature **531**(7595), 435–438 (2016)
29. Ellen MacArthur Foundation: The circular economy: a transformative Covid-19 recovery strategy: How policymakers can pave the way to a low carbon, prosperous future, Ellen MacArthur Foundation (2020)
30. Biörck, J., et al.: How construction can emerge stronger after coronavirus (2020)
31. Clair, A. Homes, health, and COVID-19: how poor housing adds to the hardship of the coronavirus crisis (2020)
32. Howell, T.: Coworking spaces: an overview and research agenda. Res. Policy **51**(2), 104447 (2022)
33. Ticcih, P.: The Nizhny Tagil Charter for the industrial heritage. in TICCIH XII International Congress (2003)
34. Bazazzadeh, H.: Truth of sincerity and authenticity or lie of reconstruction; whom do the visitors of cultural heritage trust? In: International Conference of Defining the Architectural Space Cracow Poland (2020)
35. Bazazzadeh, H., et al.: Requirements for comprehensive management of industrial heritage sites and landscapes. In: The International Conference on Conservation of 20th Century Heritage from Architecture to Landscape, Tehran, Iran (2018)
36. Bazazzadeh, H., et al.: Promoting sustainable development of cultural assets by improving users' perception through space configuration; case study: the industrial heritage site. Sustainability **12**(12), 5109 (2020)
37. Moshaver, A.: Re Architecture: Old and New in Adaptive Reuse of Modern Industrial Heritage. Ryerson University (2011)

A Healthy Approach to Post-COVID Reopening of Sugar Factory of Kahrizak, Iran

Mohsen Ghomeshi[1]($\boxtimes$), Mohamadreza Pourzargar[2],
and Mohammadjavad Mahdavinejad[3]

[1] Lodz University of Technology, 90-924 Lodz, Poland
mohsen.ghomeshi@dokt.p.lodz.pl
[2] Islamic Azad University Central Tehran Branch, Tehran, Iran
moh.pourzargar@iauctb.ac.ir
[3] Tarbiat Modares University, Tehran, Iran
mahdavinejad@modares.ac.ir

Abstract. Post-pandemic architecture and planning is one of the most crucial and controversial issues in the feel of academic discipline and professional practice. This research is to outline the characteristics of a healthy approach toward the industrial heritage buildings, monuments and sites. The research aims to propose a framework for the post-COVID reopening of the industrial heritage monuments and sites. One of the most famous monuments, among the shared-heritage of Iran and Belgium and France - the Sugar Factory of Kahrizak – is adopted as the case study of the research which is an important site for reopening nationally and internationally. The methodology of the research is based on the high-performance architecture theory and the well-building framework. The results of the research show the capabilities of the well-building framework for outlining the pathway for the facilitating the post-COVID reopening of the Sugar Factory of Kahrizak.

Keywords: Post-pandemic era · High-performance architecture theory · Industrial heritage · Healthy to architecture and planning · Shared-heritage of Iran and Belgium

1 Introduction

Architecture, urban planning and health are more interconnected in the post-epidemic world than ever before. The new concept, called "healthy design" in world architecture and urban planning, is a multidisciplinary and interdisciplinary topic [1]. The built environment must adapt to the new reality that Covid-19 will not be the last Corona virus epidemic. Therefore, architecture and urban planning in the post-Crown era is tied to the concept of health of residents and visitors to the building in the pandemic period. The post-pandemic period requires architecture and urban planning in harmony with the introduction of an innovative variable building layer system [2]. This need in addition to new designs, is very important when facing the works of contemporary and industrial architectural heritage [3–7]. Production and development of new building materials is

© The Author(s), under exclusive license to Springer Nature Switzerland AG 2022
F. Calabrò et al. (Eds.): NMP 2022, LNNS 482, pp. 2638–2647, 2022.
https://doi.org/10.1007/978-3-031-06825-6_252

an important part of recent studies in the field of health of residents and visitors to the building in the pandemic period [8]. The Covid epidemic poses a serious challenge to the environment, at all levels, to reduce potential risks or prevent the spread of the virus.

Architecture and urban planning in the post-Corona era are considered the era of architecture with "light", "air" and "green space", what we call the style of post-Corona architecture. The theory of superior architecture more than any other theory can explain a kind of magnificent architecture for the 21st century [9]. The goal of this comprehensive approach to architecture is to prevent the virus from spreading or spreading to humans, along with many other architectural and urban approaches that may increase the protection of our built environment. Health of residents and visitors to the building in the pandemic period is a top priority for trends interested in healthy architecture [10]. In terms of health standards, the promotion of circulation in harmony with open spaces; It is a design initiative to embrace the style of "21st century architecture".

A review of recent studies shows that the harmony of architecture and "nature" is an important issue in the process of designing and executing an architectural work. Harmony with nature as a new approach in healthy architecture; In interaction with the preservation of natural resources is one of the teachings of architecture and urban planning in the post-Crown era [11]. Architects like Steven Hall, reviewing the latest acclaimed projects in the world of architecture, talk about combining buildings with green space and landscape design. Architecture in the post-Corona era is a manifestation of the increasing use of natural light and the flow of fresh air in architectural spaces. Architectural priority in the pandemic period requires a strategy to control the corona. These strategies will continue in the future.

New trends in sustainable economic, cultural, social, and environmental development in the present, in addition to the familiar concepts of the past, have the color and smell of health of residents and visitors to buildings in the pandemic period [12]. Recent studies and researches in this field emphasize that architecture and urban planning in the post-Crown era have been considered as a kind of collective design with ecological innovations that emphasize "health" as a key concept.

2 Conceptual Framework

The coordination of architectural design with "Sun" has been considered by architects and designers since ancient times [13]. In such a way that the harmony of historical monuments in the traditional architecture of Iran and the world cannot be separated from the optimal use of "sunlight" [14]. Iranian Islamic architecture has been considered as an architecture in harmony with daylight. The concept of "Nordic architecture" has once again come to the fore. New research has shown that the corona virus is highly sensitive to rising ambient temperatures and can be easily destroyed by direct or indirect sunlight. As the ambient temperature rises, for example at 65 °C, the environment is easily cleaned. Destruction of the corona virus at a temperature of 50 to 55 °C for 20 min; And at 75 °C in just about 3 min, all the viruses are destroyed. Therefore, in traditional Iranian architecture, a semi-open porch is a desirable space in summer.

Traditional Iranian architecture is an example of design based on a healthy building pattern [15]. Again, design elements in Iranian Islamic architecture such as porches,

porches and front stairs of the famous traditional Iranian architecture have been considered. In addition to input, these elements were designed for friendly conversations and informal interactions with friends and acquaintances. In older homes, there was a toilet next to the yard to improve the health of the residents [16]. In Islamic and Iranian culture, people take off their shoes as soon as they enter the house so that the healthy atmosphere of the house is away from any pollution outside the house. The Persian rug was an allegory of paradise, which no one stepped on with shoes. Shoe maker in traditional Iranian architecture and valuable cultural and historical houses is a useful opportunity to filter out pollutants and remove contaminants.

The health of residents and visitors to the building is rooted in the history of contemporary architecture in Iran and the world [17]. Bauhaus architectural style in the history of contemporary architecture in Iran and the world, it was a leading step in standardizing the health of residents in the process of designing and executing architectural works. This cannot be considered unrelated to the Spanish flu. Wilhelm Chris, for example, paid close attention to health when he designed the Health Museum in Dresden, Germany in 1927.

The international style with the same slogan attracted the attention of everyone in the world. The Marseille and Villa Sava residential complex, designed by Lucor Bouzier, are examples of people who paid serious attention to residents' health and well-being about a hundred years ago. Hence the interaction of architects and designers with Corona is not related to a temporary period, although this pandemic end.

Post-Crown architecture is not just a new style, but a new way of thinking about the architectural design process. The re-emergence of the coronavirus as a worldwide disease has changed the way we live and use the available space. The new design patterns have created a mix of home and workplace, whether personal or public, governmental or nongovernmental. Architectural critics, referring to the history of contemporary architecture, have pointed to the importance of a pandemic in the evolution of architecture. In the meantime, Theodora Filcox [18] in an article entitled "How did the universal diseases transform architecture?" It showed that innovation in engineering and design has been one of the famous strategies of architects against global diseases.

Contemporary architects have once again turned to poetic spaces that are in contact with the open air. These types of spaces can be described as "magical intermediate zones". Large and spacious porches can be a place for people to work and meet guests. The future of world architecture and urban planning awaits new styles and trends, known as architecture and urban planning in the post-Crown era. When the German futurist Matthias Horks published The World After the Corona with his colleagues at the German Institute for Future Research, [19] no one imagined that the disease would affect the global economy and the future to such an extent. Affect. According to Matthias Horks, the relationship between technology and culture will be closer in the future. However, a return to nature and natural resources shows that the era of domination of the quantity and greatness of technique is over.

At both the domestic and industrial scales, especially in the design of communal spaces, health and the removal of pollutants are becoming increasingly important [20]. Architects and designers need to design a place to use the relevant and required elements, for example the use of disinfectant spray. In the process of designing and implementing new offices, a place for the use of ultraviolet radiation is considered at the entrance. Recent studies recommend using shorter wavelengths of 207 nm to 222 nm, which is safer for human tissues. Ultraviolet radiation makes the influenza and corona viruses reasonably ineffective (Figs. 1, 2, 3 and 4).

Fig. 1. The main building with the style of Belgian architecture that became famous in Iran

Fig. 2. Belgian architects managed the size of the openings according to the intensity of the radiation

Fig. 3. The main building of the complex, which became a pioneer for modern architecture

Fig. 4. Kahrizak Sugar Factory The first sugar factory in the Middle East

3 Materials and Method

The present study is a descriptive-analytical case study in terms of applied purpose and research method used. The modernization of valuable and historical buildings is an effective model in preserving them for future generations. Hence, research has a rich paradigm. The analytical method of "healthy building" [21–23] has been selected in the research. This method creates a special analytical model by creating a program on the health of residents, along with key design considerations to meet the unique needs of each building. Based on the selected research methodology, the seven basic research variables are: (1) The variable "air" is measured based on the micro-variables of ambient air quality, humidification and humidification of air, and purity and health of air. (2) The variable of water is measured based on the micro-variables of quality, method of purification and promotion of drinking. (3) The "light" variable is measured based on the natural variables of natural access, color, and intensity control. (4) «The variable "mental condition" is measured based on the variables of cooperation, silence of the work room, the existence of a place for child care on site and the availability of a health library. (5) The "comfort" variable is measured based on ergonomics, sound reduction and olfactory comfort. (6) «The "fitness" variable is measured based on the variables of the existence of improvement or training centers for fitness, stairs, bicycle rooms and incentive programs. (7) «The "nutrition" variable is measured based on the variables of selection/availability, nutritional value of foods and information.

4   Case and Site Analysis

Analytical findings about Kahrizak Sugar Factory as an example of joint heritage of Iran and Belgium, can pave the way for other works of contemporary and industrial architectural heritage. Familiarity with domestic examples such as Isfahan wool weaving factory, Khosravi leather factory in Tabriz, Iqbal Yazd factory, Versak bridge, etc.; Along with familiarity with foreign examples such as the Austrian Samering Railway, the Chatrapi Terminal in India, the city of Oroperto in Brazil, the Carlton Gardens in Australia, etc.; Examples are for a deeper understanding of the subject. The analytical sample selected in this study is Kahrizak Sugar Factory as an example of the joint heritage of Iran and Belgium.

One of the special features of this building is the crystallization of the Belgian architectural style, which in the following years quickly influenced other works of contemporary and industrial architectural heritage. The emphasis on design elements such as Belgian studs in later years in Tehran showed the prosperity and attention to this style of design, which was known in Tehran as the Belgian style.

Stylistically, this building can be considered a successful example of the pattern of gable roofs, whose structure is executed correctly. Prior to this, the gable structure was often executed incorrectly. For this reason, the Belgian gable has been considered by architects and designers as a successful model in the following years.

In historical documents, Jacques Joseph Mornard, a Belgian consultant, is identified as the main designer. Parts of the ruined sugar factory building are so dilapidated that the building is based on a wooden structure, and with two-story and gabled roofs now in the south of Tehran and the old road from Tehran to Qom, attracts the attention of passers-by. The architecture of this factory is one of the first works of Iran's industrial heritage and is over a century and a half old. It was built with the consent of Nasser al-Din Shah Qajar and by Belgian companies with the help of the French and Dutch, and continued to operate as a successful factory for many years.

Construction on a plot of land owned by Mirza Ali Asghar Khan Amin al-Dawla began in 1273 (March 1894) under the supervision of Jozef Van Der Weerd (Belgium). Kahrizak sugar factory went bankrupt in 1291 in competition with Russian imported sugars. The factory remained abandoned until 1308; just when German companies came to Iran between the two world wars and equipped Iranian factories. Kahrizak sugar factory, as an example of the common heritage of Iran and Belgium, France and Germany, is highly important for understanding the developments of modern Iran.

Joseph Naus, better known as Monsieur Noz (1849–1920), was one of the Belgian customs figures who reached the Ministry of Customs of Iran during the reign of Mozaffar al-Din Shah Qajar. From 1277 AD, he began to try to interact as much as possible with Iran and Belgium. According to his plan, Belgian experts conducted an extensive study for five years, which included the location and establishment of new industries in Iran. The report on the "plan to establish a sugar factory in Iran" was an important factor in starting the construction of the factory. A few years later, in 1891, a committee called the "Paris Committee for Industrial Studies" won the sugar concession in Iran. The report on the plan to establish a sugar factory dates back to 1884 to 1889 (1307 to 1312 AH). From this period the study, construction and then operation of the factory continued jointly between Belgium and France. The Paris Committee for Industrial Studies was attended by a group of French-speaking advisers, historically referred to as "Pellet", "Kerchel" and "Rikma Grace". Various historical documents, such as the report of "Monsieur Baron Darp", the then ambassador of Belgium in Tehran in 1891, mention the formation of the Paris Committee for Industrial Studies, and the acquisition of sugar concessions in Iran. The consortium sent these people to Iran for field studies on the development conditions of industrial factories in Iran and the future, as well as the study, construction, implementation, maintenance and construction of a sugar factory. On May 13, 1907, Joseph Naus was expelled from the country with the victory of the Constitutional Revolution. In practice, after this period, Russian and British advisers replaced German, Austrian, Belgian and French labor and engineers. In his famous book, Three Years in Iran, Dr. Fourier, a French physician specializing in Nasser al-Din Shah, states that the reason for the failure of Belgian companies in Iran is their ignorance of Iranian culture and cultural conditions, along with underestimation of their international competitors. Has been from Russia and Britain.

From about 72,000 square meters area of factory, about 1380 square meters was allocated to the location of the factory and other parts belonged to the location of workers and additional sections. The activity of the non-economic factory was stopped again in 1964 due to the growth of imports and also the wear and tear of machinery. Currently, Kahrizak Sugar Factory is registered as the oldest sugar factory in the Middle East with the number 4635 on January 4, 2001 in the National Heritage List of Iran. News agencies reported that the owner of this factory used the incompetence of people in the body of decision-making systems and lack of public awareness. In early 2016, he filed a complaint with the Administrative Court of Justice, requesting to leave the building. But in practice, due to the great importance of this complex in the history of contemporary Iranian architecture and the reaction of the people, the demolition of this building was stopped as part of the initial legacy of modernity entering Iran.

5   Results

Research findings based on methodology and research variables provide important points for forecasting the future of Kahrizak Sugar Factory as an example of the joint heritage of Iran and Belgium (Table 1).

Table 1. Analysis of the capacities of Kahrizak Sugar Factory in the realization of the pattern of architecture and urban planning in the post-pandemic era

Item	Division	Specification	Priority
Air	Air Quality	Access to outdoor space and the ability to maintain ambient air quality by providing healthy air from the environment	✓✓✓
	Humidification and air humidification	Presence of green plants in the range and possibility of natural air humidification and air humidification by indirect methods	✓✓
	Purity and healthy air	Fresh air from the site enters the building complex and its purity is a factor to ensure the health of the complex air.	✓✓✓
	Quality	The new Tehran water treatment plant near the complex somehow guarantees the quantity and quality of water available and suitable for drinking	
	Humidification	An independent treatment plant should be provided for the complex to allow water to be recycled and reused.	
	Purity	The collection somehow provides mobility for the audience to visit different spaces, and this promotes drinking more healthy water.	
Water	Quality	Belgian architectural style has a good relationship with daylight and this has provided natural access to light in buildings	
	Treatment	Light is available in white and natural daylight with the help of built-in skylights and combination skylights	
	Drinking Promotion	The depth of the window, along with the use of intensity control coatings and light level control in practice, leads to glare control	
Light	Natural access	Belgian architectural style has a good relationship with daylight and this has provided natural access to light in buildings	
	Color	Light is available in white and natural daylight with the help of built-in skylights and combination skylights	
	Diming	The depth of the window, along with the use of intensity control coatings and light level control in practice, leads to glare control	
Mind	Collaboration	The vast environment as a shared space in Ali makes the environment encouraging cooperation and interaction	
	Quiet Room	Shared open space in a quiet office environment and unwanted noise is practically possible with large-scale integrated	
	On-site child care	Considering the people who come to the complex, a place for child care in the place of the complex can be anticipated	
	Health and Wellness library	Consider a library in the complex that encourages people to live a healthier life by making available a health library.	
Comfort	Ergonomics	A kind of Belgian engineering of the complex is designed in full harmony with the ergonomics of the work environment and the welfare conditions of the people	
	Sound Reduction	The use of vegetation and green trees is a factor in reducing noise in the environment and noise control	

(*continued*)

Table 1. (*continued*)

Item	*Division*	Specification	Priority
	Olfactory Comfort	The presence of green plants and the reduction of annoying elements in practice leads to easy smell and control and management of odor flow in the yard.	
Fitness	Fitness Centers	Walking from the subway site to the complex is a form of mobility and in practice an effective exercise for fitness	
	Stairs	Two-story buildings with attractive design somehow encourage the use of stairs, is a factor of mobility among people	
	Bike Room	Spaces for physical training in the complex, such as a bicycle room or gym with the aim of mobility and for training are provided.	
	Incentives Programms	Green space and proper access to the environment and infrastructure of the metro is in practice a kind of incentive program for mobility	
Nourishment	Selection / Availibility	Recommend to replace small and cheap home restaurants with fast food in order to select and make healthy food available	
	Serving Size	Introduce the nutritional value of foods by vendors and encourage the use of healthy nutrition in target groups	
	Information	Availability of useful nutrition information available to people with Huff Encouraging healthy, low-calorie, affordable nutrition	

Capacities of Kahrizak Sugar Factory, as an example of the joint heritage of Iran and Belgium, in implementing the idea of modernizing the complex and turning it into a sugar training center at the national level, can create special capacities to realize this concept. The scales are:

1- Very uncompetitive: Newer sets work much better.
2- Unbeatable: Newer sets work better.
3- No competitive priority: The modernized complex acts like other newly built centers.
4- Competitive: Contemporary sets work better.
5- Highly competitive: Contemporary sets work much better.

6 Discussion and Conclusion

Transformation of contemporary and industrial architectural heritage into mental and behavioral sites, regardless of post-coronary requirements, in addition to depletion of physical value and due to specific mental or eventual characteristics; It is also a factor threatening the heritage values of this building.

A comparative study shows that the capacity of Kahrizak Sugar Factory, as an example of the joint heritage of Iran and Belgium, for modernization in the post-Corona era is increasing. In other words, in the pre-Corona era, the chance of success of conventional designs for this site (3.48) was 69.6%, while this number is less than the competitive capabilities of the site on a national scale (3.78) 75.6%. In terms of architecture and urban planning in the post-Crown era and prioritizing the health of residents and visitors to the building in the pandemic period, its competitive values compared to newly built models reach (4.35) 87%, which indicates such a project in the post-Crown era. Finds higher economic and executive values.

References

1. Rasoolzadeh, M., Moshari, M.: Prioritizing for healthy urban planning: interaction of modern chemistry and green material-based computation. Naqshejahan Basic Stud. New Technol. Architect. Plann. 11(1), 94–105 (2021). https://dorl.net/dor/20.1001.1.23224991.1400. 11.1.7.0. (Persian)
2. Mohtashami, N., Mahdavinejad, M., Bemanian, M.: Contribution of city prosperity to decisions on healthy building design: a case study of Tehran. Front. Architect. Res. 5(3), 319–331 (2016). https://doi.org/10.1016/j.foar.2016.06.001
3. Megahed, N.A., Ghoneim, E.M.: Antivirus-built environment: lessons learned from Covid-19 pandemic. Sustain. Cities Soc. 1(61), 102350 (2020). https://doi.org/10.1016/j.scs.2020. 102350
4. Bazazzadeh, H., Mahdavinejad M., Ghomeshi, M., Hashemi safaei, S.: Requirements for comprehensive management of industrial heritage sites and landscapes. In: The International Conference on Conservation of 20th Century Heritage from Architecture to Landscape, Tehran, Iran (2018)
5. Bazazzadeh, H., Nadolny, A., Attarian, K., Safar ali najar, B., Hashemi Safaei, S.: Promoting sustainable development of cultural assets by improving users' perception through space configuration; case study: the industrial heritage site. Sustainability 12(12), 5109 https://doi.org/10.3390/su12125109
6. Bazazzadeh, H., Mehan, A., Nadolny, A., Hashemi, S.: The importance of flexibility in adaptive reuse of industrial heritage: learning from Iranian cases. Int. J. Conserv. Sci. 12(1), 1–16 (2021)
7. Mahdavinejad, M., Didehban, M., Bazazzadeh, H.: Contemporary architectural heritage and industrial identity in historic districts case study: Dezful. J. Stud. Iranian-Islamic City. 6(22), 1–10 (2016)
8. Bazazzadeh, H.: Truth of sincerity and authenticity or lie of reconstruction; whom do the visitors of cultural heritage trust? In: International Conference of Defining the Architectural Space, Cracow, Poland (2020)
9. Pakdehi, S.G., Rasoolzadeh, M., Moghadam, A.S.: Barium oxide as a modifier to stabilize the γ-Al2O3 structure. Pol. J. Chem. Technol. 18(4), 1–4 (2016). https://doi.org/10.1515/pjct-2016-0062
10. Mahdavinejad, M., Hosseini, S.A.: Data mining and content analysis of the jury citations of the Pritzker architecture prize (1977–2017). J. Archit. Urban. 43(1), 71–90 (2019). https://doi.org/10.3846/jau.2019.5209
11. Mahdavinejad, M., Javanroodi, K.: Natural ventilation performance of ancient wind catchers, an experimental and analytical study–case studies: one-sided, two-sided and four-sided wind catchers. Int. J. Energy Technol. Policy 10(1), 36–60 (2014). https://doi.org/10.1504/IJETP. 2014.065036
12. Talaei, M., Mahdavinejad, M., Azari, R.: Thermal and energy performance of algae bioreactive façades: a review. J. Build. Eng. 1(28), 101011 (2020). https://doi.org/10.1016/j.jobe.2019. 101011
13. Bazazzadeh, H., Pilechiha, P., Nadolny, A., Mahdavinejad, M., Hashemi, S.S.: The impact assessment of climate change on building energy consumption in Poland. Energies 14(14), 4084 (2021). https://doi.org/10.3390/en14144084

14. Bazazzadeh, H., Nadolny, A., Hashemi Safaei, S.: Climate change and building energy consumption: a review of the impact of weather parameters influenced by climate change on household heating and cooling demands of buildings. Eur. J. Sustain. Dev. **10**(2), 1 (2021)

15. Eltaweel, A., Yuehong, S.U.: Parametric design and daylighting: a literature review. Renew. Sustain. Energy Rev. **1**(73), 1086–1103 (2017). https://doi.org/10.1016/j.rser.2017.02.011

16. Goharian, A., Mahdavinejad, M.: A novel approach to multi-apertures and multi-aspects ratio light pipe. J. Daylighting. **7**(2), 186–200 (2020). https://doi.org/10.15627/jd.2020.17

17. Pourzargar, M., Abedini, H.: Explaining the components of contemporization and quality improvement of Emamzadeh Saleh's (AS) adjacent texture. Naqshejahan-Bas. Stud. New Technol. Architect. Plann. **10**(1), 63–74 (2020). https://dorl.net/dor/20.1001.1.23224991.1399.10.1.7.3. (Persian)

18. Bahramipanah, A., Kia, A.: Quranic interpretation of holy light idea in Islamic and Iranian ar chitecture of Safavid Era. Naqshejahan Basic Stud. New Technol. Architect. Plann. **10**(4), 287–293 (2020). https://dorl.net/dor/20.1001.1.23224991.1399.10.4.7.9. (Persian)

19. Bates, V.: 'Humanizing' healthcare environments: architecture, art and design in modern hospitals. Des. Health. **2**(1), 5–19 (2018). https://doi.org/10.1080/24735132.2018.1436304

20. César, P.D., Tronca, B., Marchesini, T.Z.: COVID-19 and epidemic diseases transforming lodging facilities: a study of Brazilian cities. In: Virus Outbreaks and Tourism Mobility, 6 Sep 2021, Emerald Publishing Limited (2021).https://doi.org/10.1108/978-1-80071-334-520 211011

21. Horx, M.: Die Welt nach Corona. Zugriff am. 19 Mar 2020. https://www.markt-bechhofen. de/fileadmin/Dateien/Website/Bilder/News/Die_Welt_nach_Corona.pdf

22. Sarada, B.V., Vijay, R., Johnson, R., Rao, T.N., Padmanabham, G.: Fight against COVID-19: ARCI's technologies for disinfection. Trans. Indian Natl. Acad. Eng. **5**(2), 349–354 (2020). https://doi.org/10.1007/s41403-020-00153-3

23. Al-Shammari, B., et al.: Real-time application of integrated well and network models through smart workflows for optimizing and sustaining production targets in an area of Kuwait's greater Burgan oil field. In: SPE/IATMI Asia Pacific Oil and Gas Conference and Exhibition, 17 Oct 2017. https://doi.org/10.2118/186205-MS

Approaches Proposal for Tools Coordinating in Maintenance and Reuse of Architectural Heritage. A Case Study on Urban Complexes of Modern Architectural Heritage

Marco Zerbinatti and Sara Fasana[✉]

Politecnico di Torino, C.so Duca degli Abruzzi, 24, Turin, Italy
{marco.zerbinatti,sara.fasana}@polito.it

Abstract. Resilience assumed at a building scale often refers to efficiency and safety aspects, while extended to the urban and territorial scale it involves (among others) safety, energy efficiency and infrastructure aspects. Actually, it seems rather more complex imagine to bridge these goals together to a reference framework for Cultural Heritage, where technical and regulatory requirements must be usefully balanced with those of enhancing and conservation and, not at least, social involvement. In this perspective, authors deal with an on-going research, referred to an emblematic example of urban environment, recently added to UNESCO's World Heritage List: Ivrea Industrial City of XX century. Here an innovative effort in maintenance program can interpret actual urgent needed in terms of conservation, but, at the same time, it can represent an instrument to govern and coordinate future sustainable transformative and regenerative planning. This paper presents the methodological approach and first results of the research program, which final aim is to develop an integrated BIM-GIS-based tool for coordinated and sustainable redesign and maintenance of complex built heritage environments. Original identity, strictly related to local resources, are considered, with the aim to reach a renewal of perspective, to promote and enhance a circular society, not far from Olivetti's ideals, but also consistent with goals proposed in Agenda 2030.

Keywords: Ivrea Industrial City of XX century · Innovative maintenance program · Digital tools for re-design architectures · Digital-Twin

1 Approaching Maintenance and Reuse for Future City: Critics and Perspective for Current Tool

The age of complexity, in which we live, offers, from the most diverse and often contrasting points of view, numerous points for reflection on the perspective towards which to direct common efforts to reconcile the many needs to be satisfied with the transformation of cities: a transformation that is increasingly necessary, but still difficult to achieve. In this regard, significant challenges are posed, on the knowledge plane, by the

© The Author(s), under exclusive license to Springer Nature Switzerland AG 2022
F. Calabrò et al. (Eds.): NMP 2022, LNNS 482, pp. 2648–2658, 2022.
https://doi.org/10.1007/978-3-031-06825-6_253

current availability of innovative technologies and tools: the so-called digital transition, the theoretical possibility of a refined modelling, able to return digital twins apparently depositaries of sophisticated and comprehensive information storage and sharing systems, often risks to convey a false conviction according to which the "digi tization" and the "interoperability" can together, and on their own, meet the needs of predefined solutions, transversally and universally valid.

These themes are partially related to the so called City Information Modelling [1], aimed to achieve a mapping and consequent collection of technical and consumption information data, for example in order to plan sustainable refurbishment and/or new interventions for energy saving. These tools, when based on/referred to quantitative data, can support project choices with the results obtained by comparative analysis of different intervention scenarios [2].

Such an approach, however, cannot be exhaustive with regard to a more comprehensive transformation program, which must consider qualitative aspects, difficult to be "measured".

Starting from this point of observation, the authors have recently started a specific research on the case study of Ivrea Industrial City of XX Century. This is an emblematic example of complex Architectural Cultural Heritage: that is, a set of buildings that are relevant both if considered as architectural individualities, and as the unitary product of a careful and organic planning program, implemented with coherence and continuity over several decades. In fact, *Ivrea, industrial city of the 20th century*, is listed as a UNESCO World Heritage Site not so much, or not only, for the quality of its individual buildings, but above all for the value of its entire connective environment, road infrastructure and urban green spaces, which make it a unique example. This recognition, however, has not preserved it from a slow process of degradation, mainly caused by the fragmentation of the properties, following the closure of Olivetti's original activities, and the consequent lack of central coordination of administration and management activities (care, maintenance, programmed transformation).

Ivrea, therefore, represents a significant challenge to put in place new medium-term planning strategies for urban regeneration, and to experiment with an innovative application of consolidated tools in parallel project fields for the management, maintenance, reuse of the architectural heritage and urban regeneration.

The consolidated availability of digital tools for modelling and information management in the field of construction, boasts significant and profitable examples of application to the architectural cultural heritage. With particular reference to the planning of restoration interventions and subsequent monitoring activities, consolidated methods and applications are generally referred to as HBIM (standing for Historical BIM, that is BIM applied to historical assets). These also increased thanks to development of accurate and efficient 3D surveying and modeling techniques in the field of geomatics, used to support multidisciplinary and multi-scale accessibility to architectural assets. Most examples refer to the management and sharing of the results of mapping and survey activities for restoration work planning. It generally consists of processing and integrating results of knowledge activities into simplified parametric models highly detailed, obtained from the effective application of survey methods (laser scanning, e.g.) [3]. More recently, the need to use digital technologies to support coordinated actions for the care and maintenance of Cultural Heritage, has oriented many researches and experiments to the development of integrated tools to support decision-making processes towards accessible and feasible conservation strategies [4].

With regard to the field of Cultural Heritage, an interesting development of method is that offered by the case of historical architectural heritage complexes, for which maintenance, safeguarding and preservation activities must consider several factors. These include the integration and management of different types of data, but also their easy manipulation by managers or users [5], who may have different types of interest and skills. In this sense, the HBIM methodology offers an excellent support tool for the management of built heritage [6–10].

Furthermore, there is also an increasing interest in inserting HBIM models in their urban-environmental context, thus allowing more articulated, complex and complete analyses [11–15].

The aim of this research is to propose a methodological approach useful to support the dialogue between the different and numerous actors involved in actions of recovery, reuse and transformation of the built heritage, paying particular attention to the centrality of the Knowledge Level [16]. Knowledge regarding buildings, which is necessary so that every current and future intervention is compatible and coherent with it; but also knowledge and interpretation of the context, understood as *hambitus*, that is of the environment in its meaning of result of transformations governed by man, where the forms "introduced" by man in their manifestation and communication go beyond the simple functional meaning, to become cultural language.

On this perspective, it is clear that experiences and studies only referring to the importance of the geometrical detail of a model, in some cases, this can even be counterproductive in the perspective of data knowledge organization and interpretation. On the contrary, the interpretation of the ideational process and of the instances that have generated an urban complex are a necessary and fundamental patrimony of knowledge to promote and implement urban regeneration programs that involve not only the built component, but also, or perhaps above all, the social and cultural heritage that constitute its expression and testimony (Fig. 1).

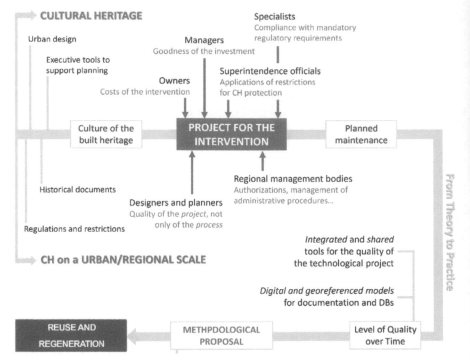

Fig. 1. Overall workflow for the definition of the methodology (Visual Abstract by Fasana S. and Matrone F, 2021).

2   Knowledge as a Basis for Regeneration and Transformation

In order to get to the heart of the structure of the current research, it is necessary to recall some relevant aspects, partly already mentioned in the preceding paragraphs: values, state of preservation and cultural context on the one hand; instruments, actors and management dynamics on the other.

The residential building heritage under study is the result of a building program involving different players, but always with continuity and coherence, guaranteed by the constant supervision and direct dialogue with the Olivetti Presidium; the extraordinary coherence from the compositional and formal point of view was guaranteed by the reference to a simple, but not trivial, vocabulary of a few compositional elements (for example the use of local natural materials, in coherence with the "autarkic" guidelines of the particular historical period), which combined together gave rise to a great variety of solutions, also as a consequence of the interpretation by different designers. In the same way, the "continuity" and coherence in the quality of construction, management, care and maintenance of the heritage has been guaranteed for decades by specific offices, a direct emanation of the Olivetti management and presidency: the permanence of a single and effective "direction" has made it possible to maintain the integrity of the urban environment, even when renewed functional needs have made transformation necessary [17].

Since the presence of Olivetti ceased, and the property has been fragmented, the state of conservation of the artifacts has undergone a rapid compromise.

Faced with the boundless bibliography on Olivetti's work, to say that only with a thorough knowledge we will able to plan adequate transformation and conservation actions, seems paradoxical.

However, to date there has been no organic analysis and interpretation of the technological aspects, and the consequent development of integrated digital tools to be used not only for representation, but also for the management and sharing of dynamic data, scalable according to the object and the actors involved [18].

2.1 Criticalities and Safeguard Actions

Recognition of the exceptional nature of this fabric has led to the implementation, of numerous actions aimed at protection and enhancing. In 2001, the MAAM (Open-Air Museum of Modern Architecture) was created, with the simultaneous drafting of a regulatory document, become part of Municipal Building Regulations, for the "governance" of interventions on buildings belonging to the urban context of what would later become Ivrea, the 20th century industrial city. In the same years, was draft a catalogue of the architecture representing the material legacy of the Olivetti company [7]. Then, in 2013 the Maam Observatory was created: it consists in an information desk to support the owners of the buildings belonging to the Olivetti architecture network, with the role of suggesting methodological guidelines to implement a concrete protection of the buildings and the image of the context. At the same time, the Municipality approved the Regulations for the Safeguarding of Modern Architecture in Ivrea, amended and adapted over the years in view of approved the candidature for the UNESCO list. In the most recent update, which dates back to 2017, it was supplemented with the "Catalogue of Constructive and Decorative Typological Heritage of the City of Ivrea".

In fact, this catalogue is a categorization of artefacts on the basis of indications of permitted interventions of different nature, quality and intensity.

With reference to the current situation, some thoughts have been useful in outlining the research path and, consequently, the development of the tool currently being prepared.

From a general point of view, at least two aspects emerged: 1) the need to harmonize the regulatory aspects, considering that "bound" buildings, protected as recognized monuments, are not contemplated in the enunciation of the categories introduced by the Maam Standard, integrated in the planning tools at municipal level; 2) the need to pursue, over time, what could be defined as homogeneity or "implementation consistency", considering that, in the Standard itself, the definition of permitted interventions and the prescriptions for their implementation contain numerous references to the "discretion" of the municipal technical office.

The study is based on the current situation, with the aim of combining, in a single tool, the different aspects that can contribute to an updated proposal of shared processes for the compatible care and maintenance and regeneration of the heritage in question. The final objective will be the definition of an integrated BIM-GIS-Data Base tool, of which there are two relevant aspects: the first refers to the use of the BIM tool for the interpretation of the connotative values and their restitution in the form of a simplified resolution line for the project [19–21]. The second aspect refers to GIS tool: this is proposed not only as a solution to better interpreting values of spatial relationship (between buildings and the con-text, and between buildings reciprocally); in fact it is settled with the aim of encourage a specific and increasingly in-depth definition of links with the territory, for example with regard to the origin of materials, or the permanence of excellent craftsmanship that over time has helped to shape and characterize the architectural heritage under study (Fig. 2).

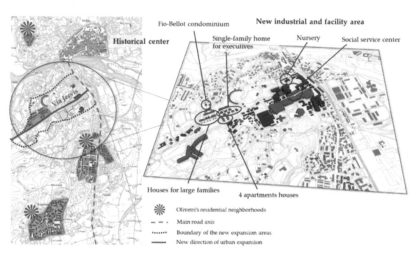

Fig. 2. On the left, it is represented the municipality of Ivrea with the relation between the historical centre and the industrial area and residential neighbourhoods, while on the right, a focus on Olivetti's industrial area with an indication of the HBIM models developed in this study (red circles) (Figure by Fasana S. and Matrone F.).

To this goals, the first step was to define in more detail the set of buildings to which the tool is to be applied (after an experimental phase, it will be applicable to a wider heritage). Although not exhaustive, the set of buildings under study can be interpreted and "classified" according to different aspects: for example, their "size", their vocation for re-use and life-giving transformation, whether they have undergone recent redevelopment, the presence of specific protection constraints, any management strategies and practices that are inherent to the asset and a consequence of the configuration of different scenarios in terms of ownership and, of course, their functional use (original and current). Each of the above aspects takes on a "direct" significance when considered individually, but also, or above all, complementary to the others in conferring on the individual artefact characteristics that may make some buildings more fragile, since they are exposed to a slow but constant over-positioning of "small" uncoordinated interventions. The "classification" deriving from this interweaving of interpretative lines has led to the distinction of certain groups of buildings, and consequently to the identification of which ones to use as reference cases for an initial phase of the study.

3   First Results

3.1   The Structure of the Operational Tool

The interface of the integrated operational tool BIM - GIS - Data Base is structured with the definition of Information Sheets, prepared in coherence with the double reference a) to the phases defined with the methodological approach and b) to the knowledge levels "needed" in each phase. All sheets (Building, Volumetric Unit, Building Part, Building Element and Intervention) are connected and complementary, but autonomous with respect to the others, in order to promote a constant reference to an integral approach to the investigation and the project, without however outlining predetermined paths. These sheets therefore explain the objectives that each level of knowledge contained in the model assumes in relation to a critical approach to the building organism, proposing themselves together as an operational tool, for the methodological address for the program of interventions (maintenance and transformation) and as a synthesis of the corresponding phase (cognitive investigation, ideational conception of the intervention, drafting of the maintenance protocol, execution and control of the interventions) (Fig. 3).

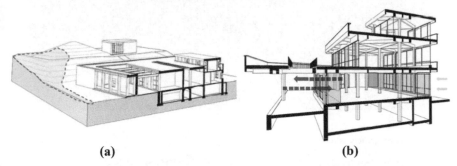

(a)   (b)

Fig. 3. Example of the use of Digital Models to identify buffer strips that interpret spatiality and contribute to compliance with the original design (a - Nursery School) or dialogue with the surrounding environment (b - Social Services Center).

In defining the contents, attention was paid to coordination with documents already available, such as the fact sheets produced and published as part of the Maam program, in order to promote as far as possible a complementary, rather than an alternative, vision, with the aim of propositional synthesis.

Of the individual fact sheets, it is worth noting those defined as Building Part, conceived with particular reference to the perspective of the survey for the recovery and transformation project, of an innovative nature compared to the more descriptive fact sheets, of a more consolidated nature: these fact sheets are aimed at promoting an interpretation of the values of spatiality and the relationship of inter-visibility (both internal and with the context), fundamental for the original project, and consequently essential as decisive lines in the conception of compatible intervention proposals.

Building Information Schedule/Sheet. The building information sheets provide a general view of the building, explaining the "master data" such as the authors, the period of construction, the reference legislation for protection, the ownership, the cadastral data. A section of the sheet contains the DB data, bearing in mind that the following information sheet corresponds to a first level of investigation/knowledge: LOK A. A further section of the sheet is dedicated to the full description of the building.

The purpose of this sheet is to provide a synthesis of technical information, useful for designers, and general information on the consistency and historical value of the building, for which it is intended to be strictly related to Maam Schedule contents. In addition, it is here outlined if the building is composed by Volumetric Units and the general state of conservation (Fig. 4).

DATI

Nome Edificio di riferimento: Fascia Servizi Sociali
Epoca di realizzazione: 1954-1959
Progettisti: Arch. Figini, Arch. Pollini
Valenza architettonica: Riconosciuto
Proprietà: Privata
Concessione: Si
Periodo Concessione: 2020-2030
Funzione originale: Assistenza Sociale
Funzione attuale: Dismesso
Normativa di riferimento: Beni culturali
Dati catastali: Foglio-Part.
Stato di conservazione:

RIFERIMENTI DATABASE

Id numerico: 02
Id alfanumerico: UV_02
Id alfanumerico edificio: A_01
Ente fornitore: UTC Comune Ivrea
Ente produttore: Nome Progettista
Data inserimento dati: 24/11/2020

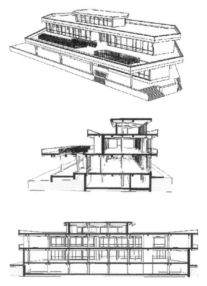

Fig. 4. Example of volumetric unit sheets for the social services building.

Volumetric Unit Schedule/Sheet. This sheet, as its name suggests, has been introduced for buildings that are made up of several volumes, with different and autonomous, but functionally complementary. It contains main data and graphic information about each identified units listed in the Building Schedule. The definition of this sheet is remarkable on methodological perspective: the objective is to drive the apply similar criteria of care and maintenance (propensity to care), but also coherent transformative interventions (ways of implementing, materials, e.g.) on different parts of buildings, often subjected to significant misalignments over time. In fact, despite of they contribute significantly to the image and quality of the environment, at present the application of standards/regulations often shows lacks on this purpose.

4 Discussion

First results obtained are encouraging to deep the applying, as a new perspective of development of results achieved by authors and colleagues in parallel on-going research on Maintenance Programs for Cultural Heritage assets [21, 22]. Current practice confirms the need to offer integrated tools simple to use and the role they can assume in the amelioration of present and future interventions quality. The on-going re-design activities for adaptive re-use of some buildings involved in the research have been developed without a consistent BIM or H-BIM adopting: no use, or use limited to digitalization of buildings, without role in problem oriented solutions and subsequent choices were adopted in these specific cases. As a consequence of this, each intervention is designed independently from the others, and very different outcomes are achieved in terms of material conservation and in terms of quality reflected in the urban context. On the other hand, the mere use of BIM (or HBIM), if considered only as a tool to create a digital twin, would not be sufficient to guarantee coordination on a methodological level, although it would already be useful to integrate the competences of many specialists. Perhaps this is also why it has been so little used so far and in this context.

Starting from these criticalities and assumptions, numerous digital models have been prepared with the work in progress, not with the aim of achieving a refined level of correspondence in terms of geometric precision, but rather of integrating on a common basis, on the one hand, specific knowledge (e.g. about the technological solutions or materials used), and on the other, a common method of interpreting the compositional design choices, which always show an articulated sensitivity and accuracy in creating visual relationships with the surrounding context.

This means, in particular, reproposing, with updated tools, that participatory but effectively coordinated design process that was the basis of Ivrea's experience. This also means laying the foundations for broadening in terms of the transdisciplinary approach innovative methods widely acquired in similar urban examples of adaptive re-use process for the evaluation of alternative solutions, different but united by the imperative of preserving land and protecting heritage [23]. The dialogue with some of the designers who have taken care of interventions in the past, and who are currently involved in the design process for the preservation and reuse of some significant buildings, has found significant and encouraging feedback: the perspective of an organic and reciprocal integration of information and results allows to recognize the value of single "investments" for the innovation of the process and the adoption of such digital tools. In fact, these could become starting points for simplified procedures to be developed in the future, to be applied to similar and complementary interventions on each building. Thus representing an effective advantage for each subject: minor temporal involving for owners (that is minor cost), objective support to technical board of local administration for the approval procedure, useful guidelines for technicians in planning of maintenance and reuse interventions, developed over time, that must be which must be able to be autonomous but coherent pieces in the design of the future city.

Acknowledgements. The authors acknowledge PhD Francesca Matrone, Dr Elisabetta Colucci, Dr Gianvito Ventura and Dr Emmanuele Iacono for a fruitful and constructive exchange of views on parallel research works, as well as Eng. Marco Indolfi for his contribution in archival research and Bim modeling.

References

1. Xu, X., Ding, L., Luo, H., Ma, L.: From building information modeling to city information modeling. J. Inf. Technol. Constr. **19**, 292–307 (2014)
2. Patti, E., Ronzino, A., Osello, A., Verda, V., Acquaviva, A., Macii, E.: District information modeling and energy management. IT Professional. **17**(6), 28–34 (2015)
3. Murphy, M., McGovern, E., Pavia, S.: Historic building information modelling (HBIM). Struct. Surv. **27**, 4 (2009)
4. Castellano–Roman, M., et al.: Dimensions and levels of knowledge in heritage building information modelling, HBIM: the model of the charterhouse of Jerez (Cadiz, Spain). Digital App. Archaeol. Cultural Heritage. **14**, e00110 (2019). https://doi.org/10.1016/j.daach.2019.e00110
5. Tommasi, C., Achille, C., Fanzini, D., Fassi, F.: Advanced digital technologies for the conservation and valorisation of the UNESCO Sacri Monti. In: Daniotti, Bruno, Gianinetto, Marco, Della Torre, Stefano (eds.) Digital Transformation of the Design, Construction and Management Processes of the Built Environment. RD, pp. 379–389. Springer, Cham (2020). https://doi.org/10.1007/978-3-030-33570-0_34
6. Brumana, R., Oreni, D., Raimondi, A., Georgopoulos, A., Bregianni, A.: From survey to HBIM for documentation, dissemination and management of built heritage: the case study of St. Maria in Scaria d'Intelvi. Digital Heritage Int. Congr. IEEE **1**, 497–504 (2013)
7. García-Valldecabres, J., Pellicer, E., Jordan-Palomar, I.: BIM scientific literature review for existing buildings and a theoretical method: proposal for heritage data management using HBIM. Constr. Res. Congr. 2228–2238 (2016)
8. Brumana, R., Georgopoulos, A., Oreni, D., Raimondi, A., Bregianni, A.: HBIM for documentation, dissemination and management of built heritage. The case study of St. Maria in Scaria d'Intelvi. Int. J. Heritage Digital Era. **2**(3), 433–451 (2020)
9. García, E.S., García-Valldecabres, J.O.R.G.E., Blasco, M.J.: The use of HBIM models as a tool for dissemination and public use management of historical architecture: a review. Build. Inf. Syst. Constr. Indust. 101 (2018)
10. Bruno, N., Roncella, R.: HBIM for conservation: a new proposal for information modelling. Remote Sens. **11**(15), 1751 (2019)
11. Vacca, G., Quaquero, E., Pili, D., Brandolini, M.: GIS-HBIM integration for the management of historical buildings. Int. Arch. Photogramm. Remote. Sens. Spatial. Inf. Sci. **42**(2), 1–7 (2018)

12. Matrone, F., Colucci, E., De Ruvo, V., Lingua, A., Spanò, A.: HBIM in a semantic 3D GIS database. Int. Arch. Photogramm. Remote. Sens. Spatial. Inf. Sci. **42**(2), W11 (2019)
13. Bruno, N., Rechichi, F., Achille, C., Zerbi, A., Roncella, R., Fassi, F.: Integration of historical GIS data in a HBIM system. Int. Arch. Photogramm. Remote. Sens. Spatial. Inf. Sci. **43-B4-2020**, 427–434 (2020). https://doi.org/10.5194/isprs-archives-XLIII-B4-2020-427-2020
14. Colucci, E., De Ruvo, V., Lingua, A., Matrone, F.; Rizzo, G.: HBIM-GIS integration: from IFC to cityGML standard for damaged cultural heritage in a multiscale 3D GIS. Appl. Sci. **10**(4), 1356 (2020)
15. Tsilimantou, E., Delegou, E.T., Nikitakos, I.A., Ioannidis, C., Moropoulou, A.: GIS and BIM as integrated digital environments for modelling and monitoring of historic buildings. Appl. Sci. **10**(3), 1078 (2020)
16. Olmo, C., Bonifazio, P., Lazzarini, L.: (a cura di). Le case Olivetti a Ivrea: l'Ufficio Consulenza Case per i Dipendenti ed Emilio A. Tarpino. Ed. Il Mulino, Bologna (2018). ISBN 9788815274625
17. Giacopelli, E., Bonifazio, P.: (a cura di). Il paesaggio futuro. Letture e norme per il patrimonio dell'architettura moderna a Ivrea. Ed. Allemandi, Torino (2007). ISBN 8842215376
18. Fiamma, P.: Il B.I.M. per l'architettura tecnica: ingegno e costruzione nell'epoca della complessita'. In: Ingegno e costruzione nell'epoca della complessità. Forma urbana e individualità architettonica. In: Garda, E., Mele, C., Piantanida, P., (eds.) Proceedings of Colloqui.AT.e, Turin, Italy, 2019, Politecnico di Torino, Torino, pp. 718–727 (2019). ISBN: 978-88-85745-31-5
19. Megahed, N.A.: Towards a theoretical framework for HBIM approach. Historic preservation and management. Int. J. Archit. Res. **9**, 130–147 (2015)
20. Bianchini, C., Attenni, M., Potestà, G.: Regenerative design tools for the existing city: HBIM potentials. In: Andreucci, M.B., Marvuglia, A., Baltov, M., Hansen, P. (eds.) Rethinking Sustainability Towards a Regenerative Economy. FC, vol. 15, pp. 23–43. Springer, Cham (2021). https://doi.org/10.1007/978-3-030-71819-0_2
21. Colucci, E., Iacono, E., Matrone, F., Ventura, G.M.: A BIM-GIS integrated database to support planned maintenance activities of historical built heritage. In: Borgogno-Mondino, E., Zamperlin, P. (eds.) Geomatics and Geospatial Technologies. CCIS, vol. 1507, pp. 182–194. Springer, Cham (2022). https://doi.org/10.1007/978-3-030-94426-1_14
22. De Ruvo, V., et al.: Development of integrated management tools for a maintenance plan of historical heritage. In: 9th ARQUEOLÓGICA 2.0 and 3rd GEORES 2021 Proceedengs, Universitat Politècnica de València, 26–28 April 2021
23. Vardopoulos, J.. Critical sustainable development factors in the adaptive reuse of urban industrial buildings. A fuzzy DEMATEL approach. Sustain. Cities. Soc. **50**, 101684 (2019). ISSN 2210-6707. https://doi.org/10.1016/j.scs.2019.101684

Beyond Culture and Tourism: Conservation and Reuse of Architectural Heritage in a Productive Perspective

Nino Sulfaro[✉]

University Mediterranea of Reggio Calabria, Reggio Calabria, Italy
nino.sulfaro@unirc.it

Abstract. Despite tourism and culture being the most affected by the pandemic, these sectors continue to play a significant role in most Italian economic policies, such as the National Recovery and Resilience Plan. These policy strategies traditionally look at architectural heritage only as a resource in terms of tourism accommodations, museums and facilities, or as an element which is helpful in reinforcing local identity. But is it possible to imagine an alternative role for historical buildings? The present paper proposes a reflection on a change of perspective on architectural heritage that goes beyond pre-established narratives tying historical buildings exclusively to tourist and cultural uses. In addition, the author presents an ongoing didactic experience on conservation and reuse of historical buildings connected to work and production. Mills, oil mills, spinning mills and old factories represent material evidence of an alternative economic system which, if wisely reactivated in a productive way, also through innovative design, can represent a new paradigm of socio-economic development.

Keywords: Culture · Tourism · Conservation

1 Culture and Tourism in the Post-pandemic Economic Relaunch

Tourism and culture represent a strategic combination that traditionally identifies the image of Italy at an international level. Before the COVID pandemic, Italian art cities boasted a trend of continuous growth and were visited by "high spending" foreign tourists passionate about Italian art and culture (60%). Just to mention one striking case, Matera, capital of culture 2019, recorded a 216% increase in foreign visitors in the last seven years[1].

However, the totally deserted historical centers and art cities during the first lockdown demonstrated how much tourism rests essentially on very fragile foundations. At that time, some scholars had highlighted how tourism, «a sector with low added value, low productivity, low innovation, low investments and wages», could hardly be considered strategic for the country's economic growth[2]. Currently, after two years, tourism

---

[1] See ONT (2018).
[2] See Gainforth (2021).

© The Author(s), under exclusive license to Springer Nature Switzerland AG 2022
F. Calabrò et al. (Eds.): NMP 2022, LNNS 482, pp. 2659–2668, 2022.
https://doi.org/10.1007/978-3-031-06825-6_254

and culture continue to be the sectors most affected by the pandemic. Although this combination seems unsustainable, at least in the short- or mid-term period, the idea that Italy can "live on tourism and culture" continues to be an apparently convincing argument. Tourism plays a significant role in most of the country's economic recovery strategies: in Mission 1 of the National Recovery and Resilience Plan (PNRR) tourism and culture are considered as «sectors that most characterize Italy and define its image in the world», and which deserve particular attention «both for their identity role and for the "image" and the "brand" of the country at an international level, as well as for the weight they have in the economic system». The investments planned by PNRR are aimed at "improving the attractiveness" of the country and will be directed both to major attractions and to the more marginal areas[3].

The problem with this strategy is that it directs public spending where it is most profitable not where it is most needed, widening territorial gaps[4]. Just to mention one of the main economic provisions, in August 2020 the Italian government allocated €500 million to support commercial and touristic activities in the art cities affected by the pandemic, assigned only on the basis of the previously high level of local tourism[5]. In addition, we must not forget how, just before the pandemic, the so called "overtourism" phenomenon was one of the most discussed issues in the debate on tourism and cultural heritage[6]. The PNRR solution to overtourism consists of shifting the flows from the more touristic cities to the so-called *borghi* (small, ancient towns) and rural areas of the country, «contrasting depopulation of the territories and promoting the conservation of landscape and traditions» and supporting «production linked to the agricultural world and traditional crafts»[7]. However, this solution seems very generic and conditioned by a representation of Italy based on the contrast between urban dynamism and the immobility of marginal territories, "guardians of traditions", with the *borghi* which, in order to survive, must satisfy the tourist imagination, through a not well specified strategy of regeneration aimed at "enhancing the identity of places". Furthermore, local communities are almost always excluded from the control of resources: rather, the environmental and social costs are passed on to them.

2 A Change of Perspective on Architectural Heritage

In all relaunch economic strategies, architectural heritage is always only considered as a useful resource for tourism – for example in terms of accommodations, museums, itineraries, etc. – and/or as an element which is helpful to reinforce local identity. In this perspective, concepts such as 'identity' and 'culture' generally coincide, and tourism is considered as a 'leverage' for their economic valorization.

---

[3] See PNRR (2021).

[4] See, among the others, - Gainforth (2020).

[5] See DL 14/08/2020.

[6] On the issue of "overtourism" see UNWTO (2018) and Milano, Cheer, Novelli (2019).

[7] See PNRR (2021).

But is it possible to imagine an alternative role for architectural heritage? Is it possible to go beyond the narratives that tie historical buildings exclusively with tourist and cultural uses?

The restoration of architecture, through innovative development processes, for example in the production or artisan sectors or in the context of circular economy, is hardly ever seen as a real resource for economic relaunch. Rather, restoration, born with the aim of defining theories and methods for the protection of monuments, has always been strongly connected with cultural functions. Moreover, the reuse of historical buildings, intended as depositories of finished, immutable values, has constantly been seen as a traumatic operation, as it causes inevitable sacrifices in terms of materials and aesthetics. In this light, converting architectural heritage to cultural functions, such as museums, libraries and galleries, has been traditionally better accepted, also because more 'utilitarian' uses were considered conceptually incompatible[8]. The restoration discipline maintains a controversial relationship with tourism (and with cultural tourism): the conflict between tourism, economic growth, increasing visitor access and conservation, creates an awkward tension between preserving the vitality of places while conserving vulnerable historic fabric and immaterial heritage, which is subject to decay and degradation[9]. In addition, tourism activities activate a commodification of culture, transforming heritage into a product[10].

However, the field of restoration and conservation of architecture has extended its interests over time, not only including urban heritage, industrial archaeology and modern architecture, but imposing a change of perspectives, including issues inherent to relations between cultural heritage and communities that benefit from it for various reasons. Thus, restoration should not be seen as a sector in which to carry out a technical intervention on a historical building for its own sake, but as a set of theories and methods for a process aimed at managing the unavoidable transformation of the territory.

According to the "Territorialist school", born in Italy in the early 90s, the concept of territory as "textiles" derives from the historical bind between nature and culture, which can be respectively assumed as the warp and the weft. Over the period through which territory takes shape, it acquires its own historical depth and peculiar identity made up of a complex system of relationships between local communities (and their cultures) and the environment. However, this interweaving produced by the territory is not static or immutable, it is constantly evolving and can be compared to a "highly complex living organism".

Recently, this approach has encouraged a new way of looking at cultural heritage and, as a consequence, at architectural restoration and the field of conservation[11]. This change means that the final objective of interventions on pre-existing architecture is not transmission to the future of heritage, but its insertion into our daily life system. This means, on the one hand, abandoning the widespread idea that heritage conservation processes are necessarily anti-economic; on the other, as new approaches to the economy of culture demonstrate, if looked at in the medium- or long-term period, the

---

[8] On the relationship between restoration and reuse of architectural heritage see Sulfaro (2018).

[9] See Harney (2019), p. 161.

[10] See Sulfaro (2019), p. 229.

[11] On this issue, see Oteri (2019).

active safeguarding of a historical building can activate development, but only if the preservation programs are inserted in a territorial dimension. Conservation is not only a physical safeguard operation, but an action that includes the building of the past, with its multiple, rich meanings, in the transformation and development processes of a territory.

This implies a new way of involving communities in the reuse and management of architectural heritage. European conventions outline a more active participation of communities in the conservation strategies of cultural heritage. In the Faro Convention (2005) the concept of "Community of inheritance" prefigures an involvement not only in terms of sharing choices in some ways imposed from above, but rather an assumption of responsibility - even if it is the case, also in economic terms - on the part of citizens, businessmen, ONG, companies, in the processes of revitalization of the territory through architectural heritage.

A "history based" approach to the study of the territory becomes the starting point for identifying potentialities and strategies, especially in Italy where there are many productive micro-realities. This approach, however, should be not philological or classificatory, but oriented to a multidisciplinary vision, where the geographical, historical and environmental components assume great importance if aimed at investigating the interconnections between social and cultural phenomena, also regarding changes produced by modernity[12].

3   An Educational Experience

Since the beginning of the pandemic, the author's didactic and research activities have been oriented toward combining the theme of conservation of architectural heritage with that of local development, through the adaptive reuse of buildings connected with work and production[13]. The aim is to develop a vision of cultural heritage that goes beyond pre-established narratives that tie historical buildings exclusively to tourist and cultural uses, thinking also of the training of future architects. Mills, oil mills, spinning mills and old factories represent material evidence of an alternative economic system which, if wisely reactivated in a productive key, including innovative design, may represent a new paradigm of socio-economic development.

These activities led to a first acknowledge phase of this particular heritage in Sicily and in Calabria, to recognize the architectural typologies, chronology, materials and techniques, state of conservation, and last but not least, the susceptibility to be reused in an alternative perspective to tourism.

One category of this rich, varied heritage includes pre-industrial buildings, dating from the late nineteenth century to the first half of the twentieth century - whose activity often continued until a few decades ago - related to agricultural activities, such as oil mills and millstones. It is almost always impossible to reactivate the previous functions in these structures, also due to modern rules connected for example to the production of

---

[12] Oteri (2019), p. 173.

[13] The works presented are the outcomes of the following courses: Corso di laurea magistrale in Architettura-Restauro LM-4 – Dipartimento PAU, Laboratorio di Restauro A.A. 2020–2021 – Docenti: D. Mediati, N. Sulfaro; Corso di laurea a ciclo unico in Architettura – DarTe, Corso integrato di restauro A.A. 2020.2021– Docente: N. Sulfaro.

foodstuffs. For this reason, creative efforts were aimed at identifying alternative uses, paying attention primarily to the conservative aspects, trying to connect new functions with peculiarities of the territory and trying to imagine possible processes of economic and social development for local communities.

In this perspective, for example, an ancient oil mill built in the early twentieth century, in Rosario Valanidi, a small village near Reggio Calabria, was turned into a laboratory for the weaving and spinning of wool, a product which, like oil, belongs to the local productive tradition (see Fig. 1), through a project aimed at preserving the traces of the transformations it has undergone over time.

Fig. 1. Conservation and reuse project of the "Antico Frantoio" in the village of Rosario Valanidi, near Reggio Calabria. Study on the wool cycle process (Students: Polsia Cambareri, Fortunata Marino).

Imagination plays a fundamental role, not only as the architect's free creativity, but also as an ethical exercise in reading and interpreting architectural heritage and responsible restitution in a new key. Likewise, imagination played an important role in the conservation project of two mills called Maltese-Comi and Flesca, dating back to the nineteenth century in the "Vallata del Gallico", province of Reggio Calabria. Through careful study of the remains and of the context, these two buildings were re-imagined as an integrated laboratory for the construction of kites. In particular, structures for the canalization of water were readapted as runways to test the kites produced (see Fig. 2).

Fig. 2. Conservation and reuse project of two mills in the Gallico Valley, Reggio Calabria. The site was transformed into a laboratory for the construction and testing of kites (Students: Nils Burandt, Eliana Catalano, Giorgio Retez).

Another rich category includes buildings and sites for the production of building materials, which very often testify the transition from craft production to industry. An interesting example is the "Fornace Laterizi F.lli Aletti", in Rende, province of Cosenza. Built between 1905 and 1907, it constitutes one of the first examples in Calabria of the use of the so-called "Forno Hoffmann" for the production of bricks. This production system optimized clay firing times, using less fuel and manpower, creating increasingly competitive products. In this case, the proposed reuse project is focused on innovation, transforming the old brick factory into a laboratory for testing and experimenting innovative building materials. In the case of the brick factory called "Sila", in Piano di Bruno, near Mileto, province of Reggio Calabria, it was proposed instead to transform the large spaces for the production of bricks into a contemporary design winery (see Fig. 3).

Fig. 3. Conservation and reuse of the so-called Fornace SILA, in Mileto, province of Vibo Valentia. The brick factory is transformed into a contemporary design winery (Students: Ilenia Salimbeni, Elisabetta Sgroi, Claudia Surace, Elenea Vallone).

In other cases, some buildings bear witness to the phase of early industrial development in Calabria which sought to relaunch some local typical products in a wider market. Among others, the factory of the *"L'Ardoresina"* company, in Ardore Marina, province of Reggio Calabria, was built in 1965 to produce soft drinks - including a drink called Ardoresina. The conservation project of this ruined building, which obviously has modern construction types, materials and structures, adopted the same approach as that used on ancient heritage, trying to preserve traces of time on the building. The proposal reuse consists of returning to the production of soft drinks, with the specific intention

Fig. 4. Conservation and reuse project of "L'Ardoresina", a soft drinks factory in Ardore Marina, province of Reggio Calabria (Students: Rosarina Giardino, Alessia Minniti, Valentina Vitale).

of valorizing typical products of the area (lemons, bergamots, orange, etc.); particular attention was given to innovative production systems and contemporary design (see Figs. 4, 5).

Fig. 5. Conservation and reuse project of "L'Ardoresina", soft drinks factory in Ardore Marina, province of Reggio Calabria (Students: Rosarina Giardino, Alessia Minniti, Valentina Vitale).

4 Conclusions

The valorisation and fruition of architectural heritage are traditionally pursued through cultural and touristic reuse of ancient buildings and sites. However, the COVID pandemic has shown how these two sectors are not sufficient for the economic relaunch of Italy in absence of some sizeable public investments. As a possible consequence, architectural heritage risks being more and more excluded from development policies and, in conclusion, from our daily life. This risk is obviously more serious for the constellation of buildings and sites testifying rural, productive and industrial activities throughout the country and which have no chance of being preserved in a financially sustainable perspective.

Re-thinking the conservation and reuse of rural and industrial heritage – such as mills and factories – in an innovative productive key may represent a new development paradigm activating local economies and, at the same time, preserving important legacies. Exercising and experimenting with this approach in the study of architecture, as proposed in the present paper, may also constitute a stimulus to develop an innovative vision of cultural heritage, which is fundamental for the challenges that architects and other operators in the sector will have to deal with in the coming decades.

References

DL 14/08/2020 - DECRETO-LEGGE 14 agosto 2020, n. 104, Misure urgenti per il sostegno e il rilancio dell'economia. https://www.normattiva.it/uri-res/N2Ls?urn:nir:stato:decreto.legge:2020;104. Accessed 21 Dec 2021

Fiorani, F., Kealy, M.: Calvo Salve 2019. In: Fiorani, D., Franco, G., Kealy, L., Musso, S.F., Calvo Salve, M.A. (eds) Conservation-Consumption. Preserving the tangible and intangible Values, EAAE Transactions on Architectural Education, vol. 66, EAAE, Hasselt, Belgium (2019)

Gainforth 2021 - Gainforth, S.: Oltre la retorica sul turismo. Gli asini, 89 (2021). https://gliasinir ivista.org/oltre-la-retorica-sul-turismo/. Accessed 22 Dec 2021

Gainforth 2020 – Gainforth, S.: Oltre il turismo. Esiste un turismo sostenibile? Eris 2020

Harney, M.: Conservation and cultural tourism: conflicts and solutions. In: Fiorani Franco, Kealy, Musso, Calvo Salve, pp. 161–172 (2019)

Milano, C., Cheer, J.M., Novelli, M. (eds.) Overtourism: Excesses, Discontents and Measures in Travel and Tourism. Wallingford, UK, CABI (2019)

ONT 2018: Osservatorio Nazionale del turismo, Il turismo di ritorno (2018). http://www.ontit. it/opencms/opencms/ont/it/focus/focus/Il_turismo_di_ritorno.html?category=documenti/ric erche_ONT&sezione=focus. Accessed 22 Dec 2021

Oteri, A.M.: Architetture in territori fragili. Criticità e nuove prospettive per la cura del patrimonio costruito. ArcHistoR 11 (2019). http://pkp.unirc.it/ojs/index.php/archistor/article/vie w/432. Accessed 21 Dec 2021

Piano Nazionale di Ripresa e Resilienza. https://italiadomani.gov.it/it/home.html. Accessed 22 Dec 2021

UNWTO 2018: World Tourism Organization; Centre of Expertise Leisure, Tourism and Hospitality; NHTV Breda University of Applied Sciences; and NHL Stenden University of Applied (2018)

Sciences: Overtourism? Understanding and Managing Urban Tourism Growth beyond Perceptions, Executive Summary, UNWTO, Madrid (2018)

Sulfaro, N.: L'architettura come opera aperta. Il tema dell'uso nel progetto di conservazione. ArcHistoR Extra 2 (2018). http://pkp.unirc.it/ojs/index.php/archistorextra/issue/vie w/34. Accessed 21 Dec 2021

Sulfaro, N.: Lost in the (cultural) supermarket: heritage, tourism and conservation practices in the post-globalised world. In: Fiorani Franco, Kealy, Musso, Calvo Salve, pp. 278–288 (2019)

Special Event: Region United Nations 2020–2030

II Edition Rhegion UN 2030: Towards a New Space's ORMA Opportunity and Risks of New Modalities of Anthropization Between Sustainability, Innovation and Fragility for the Territory

Stefano Aragona[✉]

Mediterranea University of Reggio Calabria, Reggio Calabria, Italy
saragona@unirc.it

> *The spirit wanders and from there comes here and from there goes there and slips into any body...*
> *... all over the world there is nothing that lasts. Everything flows, and every phenomenon has erratic forms.*
> Ovid making Pythagoras speak; Book XV, verse 165, in The Metamorphoses.

The reference to Rhegion, the ancient Greek name of Reggio Calabria, intends to highlight the importance of history and memory for the present and to propose scenarios for a sustainable future linked to cultural, material, and intangible heritage. Considering, moreover, that we are in the Region of Bernardino Telesio, author of *De rerum natura iuxta propria principia*, mentor of Tommaso Campanella, coming from Stilo (RC), who wrote *The City of the Sun.*

The Ministry or the Ecological Transition - MiTE, the Association for the development of industry in Southern Italy - SVIMEZ, European Regional Science Association - ERSA, the National Institute of Urban Planning - INU, the Society of Territorialists - SdT, the National Union of Municipalities and Mountain Authorities - UNCEM, the Network of Universities for the Sustainable Development - RUS, the National Council of Engineers - CNI, of the National Agency for New Technologies, Energy and Economic Development - ENEA, the Italian Institute for Environmental Protection and Research - ISPRA, the Italian Alliance for the Sustainable Development - ASviS, the Association Biennial of the Public Space - APS, the Italian Association Youth for the UNESCO - AIGU, the Order of Engineers and the Order of Architects, Planners, Landscape Architects, Conservators - APPC of the Province of Reggio Calabria, and Confindustria Reggio Calabria have already granted their patronage to the Event, waiting for others as the National Council of Architects, Planners, Landscape Architects, Conservators - APPC, the Italian Alliance for the Sustainable Development - ASviS, the Association Biennial of the Public Space - APS, the Order of Engineers and the Order of Architects, Planners, Landscape Architects, Conservators - APPC of the Province of Reggio Calabria, and

© The Author(s), under exclusive license to Springer Nature Switzerland AG 2022
F. Calabrò et al. (Eds.): NMP 2022, LNNS 482, pp. 2671–2679, 2022.
https://doi.org/10.1007/978-3-031-06825-6_255

Confindustria Reggio Calabria have already granted its patronage to the Event, waiting for others as the National Council of Architects, Planners, Landscape Architects, Conservators - APPC, and of the Italian Institute for Environmental Protection and Research - ISPRA.

1   The Changing Earth Pushed by Several Limits and with Many Chances

Even before pandemic cities shown thermodynamic limits, great social and contradictions (Harvey 1993, 2012) and of meaning less (Augè 1993, 1999): results of uncontrolled globalization (Rodrik 2011) and of an unsustainable artificialization (Munafò 2020). To respond to economic, environmental and health crises, with resilience as key element for environmental and social antifragility (Rifkin 2019; Taleb 2012) are required to face new paths of reterritorialization (Raffenstin 1987): an Ecological Transition.

It must be based on reducing the ecological footprint in anthropization processes, that is, of territories and cities. The first aim is to increase vast area and local resilience considering that climate change imposes the reduction of fragility on a large scale and at the local level. This means a reduction in hydrogeological risk and an increase in elements - such as greenery - which raise local conditions regarding the response to phenomena such as heat waves. The other important objective is the use of choices such as Energy Communities or River Contracts as much as possible using elements based on the "0 km" criterion.

This requires planning and action strategies for both the new and, above all, the existing considering that the basic assumption is "0" land consumption.

The Rhegion 2022 Event has, continuing the line of the first Edition held in 2020, proposed other steps of new territorialization or reterritorialization paths proposing design scenarios and emblematic and operational cases. So participating in a robust theoretical and methodological framework inspired by the autopoiesis (Maturana and Varela 1987) and ecopolitics (Morin 1985, 2020), cultured technology (Del Nord 1991; Zeleny 1985), to go beyond the paradigm of the modern city, based on the goals' partnership required by the UN 2030 Agenda (2015). That is the transition from process control to product control (Nilles 1988) together with the possible breaking of the synchrony between spaces and times, the foundations of the Informational city (Castells 1989, 1996, 1997; Beguinot 1989; Aragona 1993), opens to new opportunities for the territory (Faggian 2020; Torre 2020). In this way, contributing to the elimination of land consumption and to the protection of biodiversity (UE 2019, 2020), developing operational paths to implement the Glocal (Robertson 1995), between internal and central areas, based on the circular economy, combining innovation and environment. Following the indications of the 2007 EU Leipzig Charter and the "New" one of 2020 which calls for integrated planning between the various centres and rural areas of the territories in line with the integrated ecology that characterizes the Encyclical Letter *Laudato Bee for the Cure of the Common House.*

2 Keys to Reading as an Opportunity for the Territory

It is interesting to see how the topics of environmental and social sustainability have been addressed in the various papers. And like all of them, however, they have a common feature: the interconnection between the different aspects addressed. This is in line with the *United Nations Agenda 2030* philosophy which, expressly stated in the last Goal, calls for a partnership between all the previous Goals.

The reference topics, for operational reasons, were divided as follows: "Scenarios of territories between cities and villages for the ecological transition", "The ecological transition into the existing: challenges for the ancient and modern heritage", "Energy Communities, Territory, Re-forestation, Water in the ecological transition: participation and monitoring", and "Small local steps essential for the ecological transition".

Ecopolis vs Megacity: a post-crises regional-urban vision towards 2050 by Sandro Fabbro and *Ecological transition and planning strategies*, author Stefano Aragona, offer basic considerations and structural proposals relating to the ecological transition. Scenarios that in *Big Data and Cultural Heritage*, by Vincenzo Barrile and Ernesto Bernardo, have a very interesting contribution on the relationship between digital technological innovation and the necessary change of mentality.

Change of mentality that Rossella Marzullo faces in the essay *Rethinking the South in humanistic, social and pedagogi-cal perspective. Rethinking the relationship between local and global.* Particularly interesting in declining the topic by emphasizing the speficities of the southern anthropological context.

Theme considered, more broadly and attentively to spatial and urban planning issues, by Fernando Verardi in the paper *Rethinking new communities. Moving towards a science of cities.* Argument then deepened in rural villages that are – or can become – smart as Gabriella Pultrone writes in *Combining the Ecological and Digital Transitions: Smart Villages for New Scenarios in the EU Rural Area.* And that Maria Rossana Caniglia treats from the historical point of view and suggests some possible hypotheses of new developments of sustainable anthropization in her contribution *Small Rural Towns and Farmhouses of the Opera for Valorizzazione of Sila in Calabria. Narrated Memory from the Past and the Present, Research for Possible Sustainable Scenarios.* To then come to the description of the scenario proposed by Nunzio Bruno Palermo in his *Hypothesis Of Recovery And Redevelopment in the Territories of the Stilaro*, whose valley was the site of important industrialization in the Bourbon era.

In the Ecological Transition, water is of great importance. Its role in relation to climate change and as an essential resource is fundamental. Cities have had energy from it and it has been, and is, one of the main communication modality. In relation to it, they have been built and can be rebuilt as Carmela Mariano writes in *Climate-proof planning: water as engine of urban regeneration in the ecological transition era* and Irene Poli with Paola Imbesi emphasize in their paper *Green Infrastructures and Water Management. Urban regeneration strategies to face global change.* In these writings the urban and social aspect begin to emerge with ever more force and are present also in *Climate Change and "local Nature Based Solution" towards resilience*, essay fosussed on an holistic approach in town planning by Fabiola Fratini.

The last two writings deal in a more technological way of the possible, and necessary change, for reasons of sustainability, of buildings and urban centers. The first, *Reversible Building Technologies and Unconventional materials for the Circular and Creative reuse of Small Centers* by Francesca Giglio, Sara Sansotta, and Evelyn Grillo, contextualizes the topic in small towns and suggests the use of particular materials, also with the involvement of population. Citizens who are also the actors of the *Application of crowd sensing for sustainable management of smart cities* by Ilaria Pigliautile and Myriam Carraù, essay in which, thanks to their involvement, there can be various uses of smart services useful for their well-being.

In conclusion, it should be emphasized that the topics covered are all present in the *National Recovery and Resilience Plan* (PCM 2021) in which *greenery* and *digitization* constitute the basic philosophy, even if in it lacks an overall strategic vision of the territory and the city.

The Call was promoted on the website of the European Regional Science Association - ERSA, the National Institute of Town planning - INU, the Network of the Universities for the Sustainable Development - RUS, the National Council of Engineers.

https://ersa.org/events/nmp2022-rhegion-united-nation-2030/

https://www.inu.it/news/rhegion-united-nations-2030-call-for-paper-aperto-fino-al-17-dicembre-l-inu-patrocina/

https://reterus.it/bandi-e-opportunita/

https://www.cni.it/media-ing/news/226-2021/3818-rhegion-united-nations-2030-call-for-paper-aperto-fino-al-30-dicembre

https://www.unirc.it/comunicazione/articoli/25290/ii-edition-rhegion-un-2030-call-paper-deadline

http://www.nmp.unirc.it/focus-sessions/

3 II Edition Rhegion UN 2030 Scientific Working Group

Stefano Aragona (Coordinator) - Department Heritage, Architecture, Town Planning - PAU, Mediterranea University of Reggio Calabria, Eng., Ph.D., Researcher in Town Planning, Master of Science in Economy & Policy Planning, Delegate of the Mediterranea University of Reggio Calabria at the Universities for the Sustainable Development Network - RUS.

Gabriella Pultrone - Department Architecture and Territory - dArTe, Mediterranea University of Reggio Calabria, Architect, PhD in "Planning and Design of the Mediterranea City", Assistant Professor in Urban Planning - Mediterranea University of Reggio Calabria, Member of the University Quality Committee (PQA).

Francesca Assennato - Geological Survey of Italy, Institute for Environmental protection and research, Italian Institute for Environmental Protection and Research - ISPRA, PhD in Energy and environment. Head of Land Monitoring Unit. Geological Survey of Italy - ISPRA. Responsible for ISPRA in relevant EU funded projects (H2020 EJP SOIL, Nellife4drylands). Main topics: land monitoring, impacts of urbanization, land degradation and desertification, ecosystem services, land use planning, urban regeneration.

Sandro Fabbro - Polytechnic Department of Engineering and Architecture - DPIA, University of Udine, Graduated in Urban Planning at the University Institute of Architecture in Venice, PhD (1993), currently is associate professor, Scientific director of international and national research projects, he has almost 200 registered research products.

Fabiola Fratini - Faculty of Engineering "Sapienza" Rome University, Department Civil Construction and Environmental Engineering – DICEA, Architect, PhD, professor of Urban Planning at the Faculty of Engineering "Sapienza" Rome University, DICEA. The main research field developed is sustainable urban regeneration. On this topic from 2015 she experimented methodologies and tools to support citizen empowerment and to release small actions for long-term strategies in the neighbourhood of San Lorenzo (Rome). She is author of papers on this subject: "Oasi Verdi a San Lorenzo (Roma).

Francesca Silvia Rota - CNR IRCrES, Researcher at National Research Council, Research Institute on Sustainable Economic Growth - CNR IRCrES. She holds a Ph.D. in Territorial Planning and Local Development. Her research interests include local development, competitiveness, sustainability and resilience.

Paolo Salonia - Director of Research associate of Institute Sciences Cultural Heritage – ISPC CNR, ICOMOS IT Advisor and Executive Board Member, Roma, Architect, former Director of the Institute for Technologies Applied to Cultural Heritage of the National Research Council, with more than five years of research experience in the field of knowledge for the preservation and enhancement Cultural Heritage.

Currently he is Research Director Associate at the ISPC-CNR and ICOMOS IT Advisor and Executive board member.

The Stilaro Valley, History of Industrialization and Intervention Strategies according to the Goals of the Ecological Transition (Source: B. Palermo 2021).

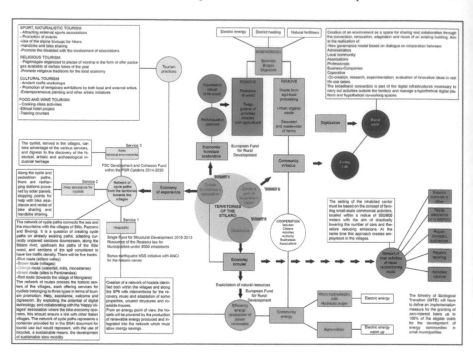

References

Aragona, S.: La città virtuale: Trasformazioni urbane e nuove tecnologie della informazione. Gangemi Editore, Roma - Reggio Calabria (1993)

Augè, M.: Non luoghi. Introduzione a una antropolgia della surmodernità, elèuthera, Milano (or. Ed. Non-lieux: Introduction a une anthropologie de la surmodernite, Editions Seuil, Paris 1992) (1993)

Augè M.: Disneyland e altri non luoghi, Bollati Boringhieri, Torino (L'impossible voyage: le tourisme et ses images, Éditions Payot & Rivages, Paris 1997) (1999)

Beguinot, C. (ed.): La Città Cablata. Un'Enciclopedia, IPiGeT-CNR&DiPiST-Fac. Ingegneria, Giannini, Napoli (1989)

Campanella, T.: La città del sole, Civitas Solis idea republicae philosophica, Friburgo (1623 (Curatori: Ernst G. Salvetti Firpo L.), Laterza, Bari, (2015, IX edizione) (1602)

Castells, M.: The Informational City. Information Technology, Economic Restructuring and the Urban - regional Process. Basil Blackwell, Oxford (1989)

Castells, M.: The Information Age: Economy. Blackwell Cambridge, MA and Oxford, UK (1996)

Castells, M.: The Information Age: Society. Blackwell Cambridge, MA and Oxford, UK (1997)

Del Nord, R.: Presentazione. In: Mucci, E., Rizzoli, P. (a cura di) L'immaginario tecnologico metropolitano, F. Angeli, Milano (1991)

Faggian, A.: Resilience and Inner Areas: is Covid19 an opportunity or a threat? Some preliminary reflections, Keynote Speaker, Closing Ceremony of the ERSA Web Conference 2020 (2020)

Harvey, D.: La crisi della modernità. Alle origini dei mutamenti culturali, Il Saggiatore, Milano, (or.ed., 1990, Blackwell) (1993)

Harvey, D.: Il capitalismo contro il diritto alla città. Neoliberalismo, urbanizzazione, resistenze, Ombre Corte, Verona (2012)

Lettera Enciclica Laudato Sii del Santo Padre Francesco sulla Cura della Casa Comune (2015.05.24), Tipografia Vaticana, Città del Vaticano

Maturana H., Varela L.: L'albero della conoscenza. Garzanti, Milano (1987)

Morin, E.: Le vie della complessità. In: Bocchi, G., Ceruti, M. (a curadi) La Sfida della complessità, Feltrinelli (1985)

Morin, E.: Cambiamo strada. Le 15 lezioni del Coronavirus, Collana Temi, Editore Raffaello Cortina Editore, Milano (2020)

Munafò, M.: Consumo di suolo, dinamiche territoriali e servizi ecosistemici, Edizione 2020. Report SNPA 15/20 (2020)

Nilles J.M.: Managing Teleworking, Centre for Effective Organization, Southern California University, L.A (1988)

PCM - Presidenza del Consiglio dei Ministri: Piano Nazionale di Ripresa e Resilienza - PNRR (2021)

Raffestin, C.: Repers pour une theorie de la territorialite' humaine, in Cahier n. 7, Groupe Reseaux, Parigi (1987)

Rodrik, D.: The Globalization Paradox. Democracy and the Future of the World Economy. W.W. Norton & Company, New York, NY (2011)

Robertson, R.: Globalization: Social Theory and Global Culture. Sage. Newcastle upon Tyne, United Kingdom (1995)

Rifkin, J.: The Green New Deal, 2019. St Martin Griffin, New York (2019)

Taleb, N.N.: Antifragile. Prosperare nel disordine. Il Saggiatore, Milano (2012)

Telesio, B.: (1565, 1570, 1586), De rerum natura iuxta propria principia, libri IX (rist. anast.) (curatore Giglioni G.), Caroccie Editore, Collana Telesiana, Roma (2013)

Torre, A.: Is circular economy a good solution for territorial development?, keynote speaker, International dialogue about good sustainability practices, IV Event "online" of "La Mediterranea e lo Sviluppo Sostenibile: teoria e buone pratche", Festival per lo sviluppo sostenibile, 6th October (2020)

UE: Carta di Lipsia sulle Città Europee Sostenibili (2007)

UE: Il green deal europeo, Bruxelles, 11.12.2019 COM(2019) 640 final (2019)

UE: Strategia dell'UE sulla biodiversità per il 2030, Bruxelles, 20.5.2020 COM(2020) 380 final (2020)

UE: The New Leipzig Charter. The transformative power of cities for the common good. EU20DE, Adopted at the Informal Ministerial Meeting on Urban Matters on 30 November 2020. http://www.bmi.bund.de. Accessed 6 May 2021

Zeleny, M.: La Gestione a Tecnologia Superiore e la Gestione della Tecnologia Superiore. In: Bocchi, G., Ceruti, M. (eds.) La Sfida della complessita', Mondadori, Milano (1985)

Climate Change and "Local Nature Based Solution" Towards Resilience

Fabiola Fratini[✉]

Dipartimento di Ingegneria Civile e Ambientale, Sapienza University Rome, via Eudossiana 18, 00186 Rome, Italy
fabiola.fratini@uniroma1.it

Abstract. "The right to the city" (New Urban Agenda) and the fight against climate change (COP 21, COP 26) have inspired new approaches capable of building bridges between cities and local communities in the search for resilience. In this regard the paper proposes a path, aimed at increasing civic engagement in sustainable regeneration at neighborhood scale.

Keywords: Regeneration · Resilience · Civic engagement

1 Climate Change, Engagement and Actors

The ongoing phenomena will lead to "irreversible mutations that will radically reshape life on Earth in the coming decades, even if humans can tame planet-warming greenhouse gas emissions". This is what the Intergovernmental Panel on Climate Change states. The timing postulated by the COP26 will allow to keep global warming below the +2.4 °C threshold by the end of this century but not to respect the maximum limit of +1.5°, as endorsed by the Paris COP 21.

The Glasgow Climate Pact, approved by over 190 countries, sets the target to cut emissions by 45% compared to 2010, to be achieved by 2030. Experts calculate that if governments decided not to take these measures, emissions will grow by 13.7% in 2030. An outcome at risk. In Italy, between 1990 and today, in thirty years, emissions have been reduced by about 20%. To save the planet, we should double our efforts in a third of the time. Edo Ronchi, president of the Foundation for Sustainable Development and promoter of Italy for Climate, has raised the alarm.

What emerges from this latest global meeting is the seriousness of the emergency, the short time available, the uncertainty about the solutions and their effectiveness. 2030 is upon us and in order to reduce emissions we need urgent plans, actions and investments to accompany the transition, together with the involvement of all global and local actors to implement them.

Not only States but also regions, cities, associations and citizens. They all must be mobilized in favor of the planet in a bottom-up approach. It is necessary to combine great long-term strategies and short-term actions, to invite open and pro-active participation and to review the modalities and timing of the international summits.

© The Author(s), under exclusive license to Springer Nature Switzerland AG 2022
F. Calabrò et al. (Eds.): NMP 2022, LNNS 482, pp. 2680–2691, 2022.
https://doi.org/10.1007/978-3-031-06825-6_256

"The Cop will have to reinvent itself and be integrated with new approaches capable of building bridges with cities, local communities, grassroots movements and ordinary citizens" (Said El Khadraoui 2021).

And if governments struggle to put adequate policies in place, change is already "marching" in local realities. Just think of the over 10,000 cities involved in the Covenant of Mayor for Climate and Energy (CoM) that have adopted the Sustainable Energy and Climate Action Plan (SECAP) and taken action to implement it. Or the network of non-institutional stakeholders who have joined the United Nations "Race to race" campaign and are committed to pursuing the "net zero initiative". They are "733 cities, 31 regions, 3,067 businesses, 173 of the biggest investors, and 622 Higher Education Institutions. These 'real economy' actors join 120 countries in the largest ever alliance committed to achieving net zero carbon emissions by 2050 at the latest. Collectively these actors now cover nearly 25% global CO_2 emissions and over 50% GDP" (https://unfccc.int/climate-action/race-to-zero-campaign).

Or also the networks created by civil society, first of all "Fridays for the Future", which mobilizes young people from all over the world.

"The overall trend is clear: a growing number of actors are signaling an intent to pursue a net-zero trajectory. This momentum represents a first step towards mobilizing much-needed speed and scale, though the ambition and implementation of these efforts varies widely" (Data-Driven EnviroLab & NewClimate Institute 2021).

Society is ready to take its own responsibilities, global strategies must be translated into shared local agendas capable, in the immediate future, of facing the ongoing challenges through social, economic and urban models that are fair, inclusive and sustainable. Local agendas capable to create hope and build a perspective, by mobilizing consensus around the practical decisions that will have to be taken.

2   Rethink the Urban Paradigm

In this global and multifaceted framework, the cities (the increasing urbanization), where 70% of emissions are concentrated and 2/3 of the planet's energy is consumed, must become laboratories for change.

Exposure to pollution, to extreme weather events, the degradation of ecosystems and the erosion of natural capital, high density are already risk factors that undermine the livability (wellbeing) and urban coexistence in the cities, where the 70% of the world population will live in 2050.

The time has come to take care of these territories and to apply restoration measures through "nature". As Mamta Mehra recalls on the "Race to race" website, the key to increase the resilience of habitats are "Natural Climate Solutions".

Ecology comes to the rescue of urban planning, inviting us to rethink the city as a more natural and sustainable, inclusive, reversible and evolutionary system (Clergeau 2020). It is no longer a question of "making the city" but of "making an urban ecosystem" based on biodiversity and on different levels of ecological functioning (Clergeau 2020): a new urban paradigm takes shape.

Under the pressure of a systemic environmental crisis, the question of nature in the city and the re-naturalization of urban spaces is emerging as a priority and a structuring axis (Abbadie 2020) of a new approach that grows in those cities which are linked to global commitments.

Nothing new, if one thinks of the metaphor of *Urbs in Horto*, for the new Chicago plan of 1837, the *Emerald Necklace* project for Boston by Frederick Law Olmsted (built in 1870), the *Garden City* model by Ebenezer Howard (1989), the eco-regional visions made by Lewis Mumford, dotted with urban constellations set within green matrices (Mumford 1953).

Today, the nineteenth-century parks, gardens and parkways become, through European directives and research programs, *Green Infrastructures* (EU 2013) and *Nature Based Solutions* (EU 2015). New visions and urban models blossom in the name of nature, such as the *"forest city"* of Shijiazhuang and Liuzhou in China (2016), the *"national park city"* of London (2017), the *"sponge city"* (a city model transferred from the Netherlands to China), the *"oasis city"* (REBUS 2017).

All of these models are related to the "sustainable city" concept, and propose a "more natural city", capable of offering numerous services, including ecosystem services, and responsible urban planning must integrate the social, economic and functional dimension with the geographical and ecological one. In this framework the value of time, of uncertainty, of the reversibility of choices, of evolution become planning/design tools (Clergeau 2020).

Furthermore, "reintroducing nature in the city is the most effective, complete and affordable 'technology' to face climate change risks, both for the safety of people and activities, and for the healthcare. A "lack of adaptation" leads to very high social and economical costs." (Ravanello 2017).

In addition, "the Climate-Proof cities" are "Cities for people" too (Gehl 2010). The majority of the most effective measures for climate adaptation are based on the reintroduction of nature within the urban area, in the public space. Green-blue infrastructures help to mitigate climate change phenomena and make the cities more healthy, livable, comfortable and attractive.

Therefore, in the outlined model, the size of the neighborhood and the care of the inhabitants play an important role: through the local scale, the inhabitants are able to enhance their territory, both urban and rural, as a common good (Magnaghi 2020). "Awareness of the places" (Magnaghi 2020), active citizenship and individual responsibility are the conditions for carrying out projects in the name of the desired change.

3   Small Actions Can Lead to Change

The New Urban Agenda (UN 2020) identifies the "public space as a place for experimentation and learning, a creative and participatory laboratory to build shared visions and take actions in favor of resilience and sustainability for a convivial and renaturalized city" (SDG 114).

"Public space, the backbone of the local dimension, is the testing ground for enriching the concept of urban quality, integrating it with that of ecological and environmental quality, expanding it to include the theme of mitigating the effects of climate change…

the implementation of these qualities represents the real challenge for the transformation of the city" (REBUS 2017).

In other words, "the systemic integration of social, cultural and nature-based innovation in the design, development and governance of public space has a tremendous potential to transform these spaces into diverse, accessible, safe, inclusive and high quality green areas that increase well-being and health and deliver a fair and equitable distribution of the associated benefits" (Horizon 2020, call SC 14–19).

Many cities have already engaged in this experimental path by declining at a small scale, and in particular in the public space, actions framed in the framework of climate action plans intended to contribute to the mitigation of the effects of climate change to increase urban resilience. These include the cities of Barcelona and Paris.

3.1 Barcelona

Actions for vegetalization and sustainable mobility interventions adopted at the local level can increase comfort, well-being and quality in the "everyday space" and, when replicated, can spread the benefits throughout the city.

This is demonstrated by the case of Barcelona and the *Superilla program* (2012–2015) framed in the *Pla de Mobilitat Urbana 2013–2018*, in the *Compromis de Barcelona pelClima*, in the *Pla del Verdi i la Biodiversitat and the Líniesestratègiques* of the PAM (2016–2019).

The program aims to build a network of *green hubs, green streets* and squares in favor of a resilient, convivial, inclusive city. The strategy takes shape through eleven neighborhoods, starting from the *Eixample*, and five objectives: to improve the livability of the public space; develop sustainable mobility; increase the presence of greenery and biodiversity; promote civic engagement; ensure that inhabitants can access a welcoming public space within a 200-m radius of their homes (Ajuntamento de Barcelona).

The process is developed through an *Action Plan* articulated into local planning schemes, projects and actions. Each project is independent and the planned actions are carried out according to the available budget. This flexibility facilitates implementation, which is done on an action-by-action basis within the overall framework. The process starts from a single part to regenerate the entire "body" of the city.

The planned actions involve the reduction of private vehicle circulation, speed (10 km/h) and parking lots, together with the creation of *green streets* for cycling and walking. All these interventions contribute to composing a welcoming landscape, inhabitant-sized, which improves the local microclimate thanks to greenery.

At least 10% of the available space will be greened up, whereas only 1% of the road section is left to vegetation today. Four thousand trees will occupy the central part of the road, thus becoming new landmarks. Sufficient space is reserved for plants so that they can grow in height and develop an adequate leaf area for the mitigation of temperature and pollution. Replacing the pavement with a permeable surface contributes to the growth and health of plants and promotes proper water management, with the recovery of 30% of rainfall. Furthermore, the space is enriched with urban furniture (seats, fountains and tables) and playgrounds, in order to promote conviviality.

The aim is to move from the concept of the street as a car-friendly space to that of a *Green Infrastructure*: a sustainable, efficient and self-sufficient system designed for people of all ages.

On the basis of the listed criteria, 33 km of green roads and 2,000 m² of new squares were thus created, transforming 3.9 hectares of public spaces into oases of microclimatic comfort, conviviality and play. At the end of the process, the implementation of the Plan in the *Eixample* district will allow to create 33.4 hectares of pedestrian areas and 6.6 hectares of green hubs.

In conclusion, some results: in *Sant Antoni*, the first district where the regeneration was completed, according to the data released by the *Ajuntamento de Barcelona*, vehicular traffic decreased by 82%, NO2 pollution decreased by 1/3 and that of PM10 by 4%.

Research developed by the *Istituto de Salud Global de Barcelona* (ISGlobal) states that if the project were implemented in its entirety (503 Superilles) it would be possible to reduce NO2 pollution by 24% and noise pollution by 5.4%. In addition, there would be a considerable reduction in premature deaths (ISGlobal 2019).

Finally, the actions envisaged also involve impacts on local economy. The survey carried out in the *Sant Antoni* district reveals a growth of 16% per year in the users of retails.

The Barcelona case study demonstrates that strategical environmental goals can be pursued starting from the local dimension, improving the inhabitants' quality of life and wellbeing through a short-medium term and step-by-step planning approach.

3.2 Végétalisons Paris

The *Végétalisons Paris* (2015) program focuses on the concept of "urban micro-naturalization" and the use of a digital platform as a participation tool. The ambition is to actively involve individuals in a collective vegetation program to be implemented at a neighborhood scale. The challenge: to increase green in the dense city, promote a subjective sense of "connection to nature", develop social practices based on citizens' engagement.

The program is part of the *Plan Biodiversité* aimed at greening the city. The objectives of the plan include the realization of +30 hectares of green spaces, +200 hectares of neighborhood gardens, +20,000 trees +100 hectares of roofs (*Objectif 100 hectares* program) and green façades.

A first assessment. As part of this strategy, 18 hectares of gardens were created, 15 hectares of areas for urban agriculture, 38 new shared and pedagogical gardens. Furthermore, to increase the supply of green spaces, the 40 hectares of the former railway belt (*la petite ceinture*) were redeveloped and opened to the public for recreational use. Lastly, 12,000 trees were planted.

In this framework the *Végétalisons Paris* program is set to call each citizen to "spontaneously vegetate the city" through the megaphone of the web. The program stems from the purpose to regenerate the urban landscape through individual micro green actions, favoring the care of the public space.

These micro actions constitute the pieces of a great urban puzzle, and the sum of "many small square meters" turns into a tangible result of multiple hectares. But rather than responding to a question of quantity, the program spreads quality: well-being, social practices and public space quality.

Taking care of the Parisian public space is a commitment that all the citizens could accomplish. Since the 30th of June 2015 it is sufficient to fill in a *Permis de végétaliser*: a "project form" containing the data of the applicant, the location and typology of the project, and a plan. In the case of a special project, such as a collective or pedagogical garden, there must be at least five applicants.

Eligible projects include: trees, flowerbeds around trees, fixed and mobile planters, keyhole gardens, vegetation of parking bollards, fences and signs, vegetated street furniture and "any other form of creativity" (Charte de végétalisation de l'espace public parisien).

The request is submitted on the website, then the project is evaluated and the permit issued within a month. By signing the *Charte de végétalisation*, the applicant undertakes to cultivate with respect for the environment, choose suitable species, to take care of the plants and clean the area. Once the permit has been received, the citizen can benefit from the support of the *Maison du Jardinage*. In 2016, 50,000 packs of seeds and 40,000 of bulbs were distributed by the services of the *Maison du Jardinage* and in 2017 it was the turn of ladybugs and chrysopes to promote the biological control of harmful insects.

2100 *Permis de végétaliser* have been released to date; 20 green streets and 170 green walls created.

According to an initial assessment carried out by the Sorbonne University (2018), the recreational aspect is the main reason why people apply for a permit. Thus 80% of those who joined the program chose to embellish the tree bed in front of their home as if it was a private garden. In second place there is the pleasure of building social relationships, however occasional, and of receiving the appreciation of passers-by and neighbors for the work done.

The satisfaction given by the consent allows urban gardeners to persist despite the difficulties: lack of close access to water, acts of vandalism, cost of replacing plants, garden caring during holidays.

Moreover, the inhabitants strive to build networks through social media in order to overcome adversity. Collaboration between neighbors or with shopkeepers is a determining factor in the survival of green spaces, and it is an interesting clue that reveals the social impact of individual projects.

The lesson learnt from the *Végétalisons Paris* experience can be summarized as follows. Tiny-green spaces are components of the city urban green infrastructures. Increasing their presence allows to release ecosystem services such as developing conviviality, improving the citizens' wellbeing and favoring biodiversity. Furthermore, a flower bed can play the role of "first aid green" less than 15 min far from home.

In conclusion, as illustrated by the examples of Barcelona and Paris, local actions in favor of ecological-environmental quality and wellbeing contribute to pursuing strategic objectives of resilience and, at the same time, increasing the quality of living places.

All those examples lead therefore to experiment, on a local scale, a new "frugal" approach to planning, inspired by *Urban Acupuncture* and *Nature Based Solutions* and capable of involving a multiplicity of actors in favor of change.

4 The "Oases Green Network" in San Lorenzo (Rome)

The "Oases Green Network" project was born with the aim of exploring new *small, green oriented* and *integrated* project strategies inspired by *UrbanAcupunture* (Lerner 2003), *Nature Based Solutions* (APAT 2003; EU 2015; Naturvation 2017) and *Urban Green Infrastructure* - UGI (EU 2013; Green Surge 2017).

The basic component of the model is the Oasis: a green and multifunctional place, of variable size, aimed at regenerating the public open space - streets, squares, parking lots and abandoned places - and the private one- school courtyards, hospital and university grounds, condominium areas.

The Oasis responds to two challenges. The first is ecological: increasing microclimatic comfort, mitigating pollution, improving water management, promoting biodiversity through green and permeable soil. The second is social: creating multifunctional inclusive places, promoting conviviality and developing social practices. The integration of ecological and social aspects contributes to increasing the psycho-physical well-being of citizens and the quality of the urban landscape (Balaÿ et al. 2020).

The Oases realization is framed within a network-system that progressively cover urban spaces from the neighborhood to the city.

The approach that makes it possible is inspired by *Urban Acupuncture* (Lerner 2003), i.e. "urban strategies exclusively applied in public space based on interventions, coordinated with the aim of activating public space on a larger scale by revitalizing urban life, carried out within a short period of time" (Hernandez 2014). These small short-term actions are aimed at achieving long-term changes (Lydon and Garcia 2015).

The neighborhood chosen to explore the Oases model is San Lorenzo in Rome: a "central suburb", born at the end of the 19th century, welcoming railway workers, workers and artisans and which today is affected by dynamics of gentrification. The 50 hectares of San Lorenzo are home to 8,866 census-taken residents (2019), plus a transient population of Sapienza students. The population density is eight times higher than the Roman average (16,969 inhabitants/sqkm compared to 2,213 inhabitants/sq.).

Sapienza university buildings punctuate the fabric of the district along with the headquarters of associations, foundations, theaters, cinemas and artists' studios. This cultural vocation is accompanied by the transformation of San Lorenzo as an entertainment and night-life hub. Among the negative consequences of this process, "new usages" of public space emerge, opposite to those of many locals and to the identity of the place.

A contrast that translates into activities that contribute to the degradation of the neighborhood, highlighting the need of shared solutions.

Criticalities related to usages overlap the negative impact of a typical nineteenth-century grid system where trees and green spaces represent a rare asset. The public green area consists of 20,000 m², about only 2 m/inhabitant against the 9 m² established by the Ministerial Decree 1444/68.

This problematic situation is compounded by the effects of the pandemic and the demand for public space, particularly areas for children's play.

These are the premises on which the "Oases Green Network" project takes shape, a process of co-design and co-construction that began as a green and small-scale sustainable regeneration that spreads to everyday places.

The Oases project stems from the willingness for change advocated by associations (the *Libera Repubblica di San Lorenzo*, the *Grande Cocomero*, the *G.R.U.*), school and university, inhabitants, artisans and artists. The Oases of San Lorenzo are places and projects supporting a sustainable and shared regeneration that is realized through the construction of a network of local actors.

The first experiment to take shape from this collaboration is a temporary oasis created in the *Parco dei Galli* (2016). It is a collective, *short-term vegetable garden* of 10 m^2 built to boost the relationship of children with nature and to encourage the care of shared space. The network that supports and takes care of the project is made up of local and non-local actors: Sapienza University with the students of the Urban Planning II course, Municipio II, the association *Zappata Romana*, *Libera Repubblica di San Lorenzo* and the users of the area (parents, children and grandparents). The *short-term vegetable garden* can be considered as starting point to introduce the topic of green and sustainable engagement within San Lorenzo's community. The results of this first experimentation can be appreciated through the growing willingness by local inhabitants to be part of future green projects.

The second Oasis is the *San Lorenzo Temporary Forest* "built" in 2017 in the former Dogana railways area with 100 potted trees, 18 of which have been donated by the Presidential Reserve of Castelporziano. This unusual forest model, made up of more than 50 species, releases ecosystem services related to environmental and cultural issues. A sensor experimentation project has been set to increase people awareness concerning the important role played by nature, representing chlorophyll photosynthesis, thus making the tree's CO_2 capture activity visible.

The goal is to create a *"smart kind of nature"* capable of catalyzing the interest of a wide public, increasing citizens' knowledge on the multiple functions performed by trees, and spreading a cultural message of reconciliation between nature and the city.

The Oasis has been animated through events, such as *La Giornata dell'Albero* organized with the schoolchildren of the Municipio II, until the end of 2018. Then, some of the trees have been moved to the *Parco dei Caduti,* while others have migrated to other gardens in the Municipio II. The San Lorenzo trees have been adopted by the children of the Saffi school who assist them during the first post-planting period. The *Temporary Forest* has increased the citizens' interest in nature through the creation of an unusual type of *hortus botanicus* which illustrates, with the help of sensors data, the role of trees in urban climate regulation and pollution mitigation (Fig. 1).

The *Oasis of piazzale del Verano* (2019) pertains to the category of the temporary use as well and it is aimed to promote the interest in urban farming and related projects to be created in San Lorenzo. The experimentation arises from the open call issued by the Municipio II *"E-state insieme"* aimed at developing cultural and recreational activities in the public space.

The project, proposed by the HabiCura association (Emanuele Caputo), becomes an opportunity to promote innovative forms of urban farming through the construction of a hydroponic 9 m^2 micro-greenhouse. The Oasis aims to highlight the challenges underlying the supply of cities and to exhibit new techniques for food production at neighborhood scale. The greenhouse thus becomes a concrete action that turns into a reference to be re-proposed shortly as a symbolic building for the proposal of Urban Regeneration of

Fig. 1. Outdoor school class about the ecosystem services delivered by trees during *La Giornata dell'Albero*

the Borghetto dei Lucani (Call for Competition issued by Roma Capitale) developed together with the *Libera Repubblica di San Lorenzo* (2019).

This collaboration has made it possible to convey ecological principles and *Nature Based Solutions* in the design process towards the Urban Regeneration of the Borghetto dei Lucani, such as the vegetable garden, the forest square and the hydroponic greenhouse, revisiting the experiences that took place in the former years.

Then, in 2020, RESPIRA Citizen Science project has taken shape in order to promote evidence based knowledge on climate change and pollution to facilitate a sound understanding of ecological principles and *Nature Based Solutions*. The project intended to increase the awareness of inhabitants and stakeholders on climate change issues, provide open access tools aimed at disseminating data and encourage the creation of Oases. The research, financed with Sapienza university funds, involves the installation of a network of sensors to detect air pollution (PM and CO_2), noise pollution, temperature and humidity. Five low cost sensors have been placed on the roof of the Borsi school, the Faculty of Psychology, the Faculty of Engineering and on two inhabitants' terrace. The project is developed thanks to the collaboration with the *Libera Repubblica di San Lorenzo* and citizens. The sensors were provided by the Barcelona FabLab, which has developed the same Citizen Science tools for the HORIZON2020 iSCAPE project (Improving the Smart Pollution Control in Europe).

The choice of the location of the sensors follows the purpose of comparing the pollution phenomena with the built environment features. Therefore the places chosen to locate the sensors are selected according to a different consistency of vegetation, traffic and density of the building. The data are available on the website https://smartc itizen.me/kits/14604.

The first scientific data collected and discussed highlights that in San Lorenzo the air pollution has exceed the threshold of more than 35 days of the PM10 value over

50 ug/m^3 and CO_2 over 1500 ppm. Furthermore, inside the neighborhood grid (Piazza deiSanniti) the PM10 value is higher than along the border (Largo Passamonti), despite the presence of the main traffic roads. This could be explained with the lack of air circulation between the buildings in Piazza dei Sanniti area and the positive presence of the Verano cemetery's vast green area near largo Passamonti which plays an ecological role, mitigating pollutants and temperatures.

The numbers that demonstrate the high level of pollution can become an evidence to boost the decision making process and request the active concern of local administration (Municipio II) towards green regeneration projects as the one in Piazza dei Sanniti, where citizens demand a considerable reduction in car parking and the replacement with public space and trees.

Moreover, scientific data can be supported by the reference to good practices such as the Barcelona Superilla program which demonstrates that traffic calming zones, car presence reduction and tree planting can mitigate pollution, as in Sant Antoni neighborhood (Fig. 2).

Fig. 2. The nine-square-meters-hydroponic farm in Piazzale del Verano

5   Conclusions

Lesson learnt. The five Oases that have been created are part of a systematic project. The actions that flourish and come to an end are crossed by continuous activity that foster institutions, associations and citizens to discuss the issues of climate, pollution and actions for change, starting from different perspectives. During their existence, the Oases offer citizens the opportunity to participate in the co-construction of a better future.

However, this first cycle of experimentation highlights the need to demonstrate the "immaterial results" due to the Oasis project: the growth in citizen participation and awareness of environmental issues, the recognition of the role played by nature, the increase in the willingness to act to accelerate change.

Therefore the first step of the Oasis experiment ends with a double perspective: multiplying actions and their duration and accompanying new experiments with an impact study aimed at evaluating the increase in civic engagement, well-being and connection to nature in the participants.

This experience is the starting point for *"Oases from school to neighborhood"*, a *Sapienza Terza Missione* project funded in 2022 aimed at implementing learning activities focused on climate change issues on one hand and realizing a Tiny Forest (Afforest 2020, IVN 2021), on the other. The proposal sees the participation of the Municipio II, the Borsi and Saffi schools, the Faculty of Psychology and Engineering, the active engagement of local actors and artists; the involvement of associations "external" to the neighborhood.

On March the 28th 2022 a public presentation was held in the Parco dei Caduti where the schoolchildren drew the border of the Tiny Forest that will be put in place on Novembre the 21st during *La Giornata dell'Albero* (Fig. 3).

Fig. 3. Schoolchildren involved in a Tiny Forest planting action in Paris (March 2018)

References

Abbadie, L.: La nature nous rend et se rend des services. In: Clergeau, P.: Urbanisme et Biodiversité. Vers un paysage vivant structurant le projet urbain, pp. 16–26. Édition Apogées, Paris (2020)

Afforest Homepage. https://www.afforestt.com/. Accessed 15 Oct 2021

Ajuntamento de Barcelona Homepage. https://www.barcelona.cat/pla-superilla-barcelona/es. Accessed 03 Dec 2020

APAT: Gestione delle Aree di collegamento ecologico funzionale. Indirizzi e modalità operative per l'adeguamento degli strumenti di pianificazione del territorio in funzione della costruzione di reti ecologiche a scala locale, Manuali e Linee Guida 26/2003 (2003)

Balaÿ, O., Brossier, J., Lapray, K., Leroy-Thomas, M., Marie, H.: Ménager des Oasis Urbaines: des représentations à la fabrication. In: Marry, S. (ed.) Territoires Durables, p. 56. Èditions Parenthèses, Paris (2020)

Clergeau, P.: Urbanisme et Biodiversité, p. 12. Éditions Apogée, Paris (2020)

Data-Driven EnviroLab&NewClimate Institute Homepage. http://datadrivenlab.org/wp-content/uploads/2020/09. Accessed 05 Sept 2020

Commission, E.: Green Infrastructure (GI) – Enhancing Europe's Natural Capital. EU, Brussels (2013)

European Commission: Towards An EU Research and Innovation Policy Agenda for Nature-Based Solutions & Re-Naturing Cities, Final Report of the Horizon 2020 Expert Group on' Nature-Based Solutions and Re-Naturing Cities', Directorate - General for Research and Innovation, Brussels (2015)

Gehl, J.: Cities for People. Island Press, Washington DC (2010)

Green Surge Homepage. http://www.greensurge.eu. Accessed 1 March 2017

Hernandez, J.: Public space acupuncture. In: Casanova, H., Hernández, J. (eds.) Public space Acupuncture, p. 10. Actar Publishers, New York (2014)

Howard, E.: Tomorrow, a peaceful path to real reform (1989)

ISGlobal Homepage. https://www.isglobal.org/-/el-proyecto-original-de-las-supermanzanaspo dria-evitar-cerca-de-700-muertes-prematuras-anuales-en-barcelona. Accessed 09 June 2019

IVN natuureducatie Homepage. https://www.ivn.nl/tinyforest/tiny-forest-worldwide. Accessed 20 Dec 2021

Lerner, J.: Acupunctura Urbana, pp. 43–56. GrupoEditorial Record, Rio de Janeiro (2003)

Lydon, M., Garcia, A.:Tactical urbanism. Short term Action for Long Term Change. Island Press, Washington DC (2015)

Magnaghi, A.: Sur la bioregion et le territoire comme bien commun, entretien avec Alberto Magnaghi. In: Duhem, L., Pereira de Moura, R.: Design des territoires. L'enseignement de la Biorégion, pp. 43–55. Collection Eterotopia France Parcours, Paris (2020)

Mumford, L.: La cultura della città. Edizioni di Comunità, Milano (1953)

Naturvation Homepage. https://naturvation.eu/atlas. Accessed 1 Dec 2017

New Urban Agenda UN-Habitat Homepage. https://unhabitat.org/about-us/new-urban-agenda. Accessed 7 Mar 2020

Race To Zero Campaign | UNFCCC Homepage. https://unfccc.int/climate-action/race-to-zero-campaign. Accessed 23 Nov 2021

Ravanello, L.: Laboratorio sugli spazi pubblici per la mitigazione e l'adattamento ai cambiamenti climatici. Regione Emilia-Romagna (2017)

REBUS. Dessi, V., Farné, E., Ravanello,L., Salomoni, M.T.: Rigenerare la città con la natura. Maggioli Editore, Santarcangelo di Romagna (2017)

Sorbonne (2018)

The Progressive Post Homepage. https://progressivepost.eu/author/said-el-khadraoui/2021. Accessed 23 Nov 2021

Climate-Proof Planning: Water as Engine of Urban Regeneration in the Ecological Transition Era

Carmela Mariano[✉] and Marsia Marino

PDTA Department, Sapienza University of Rome, Via Flaminia 72, 00196 Rome, Italy
{carmela.mariano,marsia.marino}@uniroma1.it

Abstract. The increasing frequency of climate change-related extreme events observed in recent decades highlights the need to identify new paradigms for the sustainable transformation of the threatened territories, as also recognized in the 17 goals of the Global Agenda for Sustainable Development 2030. From the urban planning perspective, the achievement of the latter can only be pursued through the overcoming of traditionally sectoral approaches, in favour of an integrated one to urban complexity, according to what is called "climate-proof planning". Starting from the analysis of the reference context, the contribution focuses on the need to adopt an "amphibious" approach to urban regeneration actions in flood-prone areas, proposing the case study of *Isola Sacra*, Fiumicino (Italy). This contribution is part of the ongoing research "Urban regeneration strategies for climate-proof territories. Tools and methods for the assessment of vulnerability and for the identification of resilience tactics in coastal urban areas subject to sea level rise" (Scientific coordinator Prof. Arch. Carmen Mariano) carried out by the authors at the PDTA Department, Sapienza University of Rome in collaboration with the Climate and Impact Modelling Lab of ENEA, funded by Sapienza University of Rome within the University funds for scientific research 2020.

Keywords: Climate-proof planning · Urban regeneration strategies · Water landscape

1 Water: From Challenge to Resource

1.1 Ecological Transition of Fragile Territories

Climate change is among the most important issues of the scientific debate on risks connected to the urban transformations of the contemporary cities. Rising global temperatures, largely the product of metropolitan areas, are likely to have severe effects on both the earth's natural systems and human society. Sea level rise and dramatic changes in weather patterns, predicted as a consequence of sustained global warming, could accelerate the weakness of economic systems, moving of coastal communities and port facilities, shortages of food and water supplies, increases in disease, additional health and safety risks from natural hazards, and large-scale population migration [1, 2]. Much

© The Author(s), under exclusive license to Springer Nature Switzerland AG 2022
F. Calabrò et al. (Eds.): NMP 2022, LNNS 482, pp. 2692–2700, 2022.
https://doi.org/10.1007/978-3-031-06825-6_257

of the mitigation and defence debate to date has centred on reducing greenhouse gas (GHG) emissions, but at the same time there is a growing acknowledgement by scientists and policy analysts which claims that a substantial part of the global warming challenge may be met through the urban planning and design of cities, not only through mitigation's strategies but with adaptation's policies [3, 4].

Therefore, the need arose both to improve and extend the skills of urban planners and to update the territorial governance tools currently available, aiming to develop effective and feasible regeneration and resilience strategies [5, 6].

About this, the profound crisis of contemporary cities and territories, and of their environmental, economic and social links and identity, has already been defined for some decades as the *new urban question* [7] and it constitutes the rationale for the renewed awareness in the urban planning community that finally prompted an innovative transdisciplinary approach to the design of an updated notion of urban welfare, focusing on safety, health, economic development, and social innovation [8]. From the urban planning perspective, the achievement of the ecological transition goals for the fragile territories can only be pursued through the overcoming of traditionally sectoral approaches, in favour of an integrated approach to urban complexity [9–11]. In this very direction, the new EU Strategy on Adaptation to Climate Change (2021) promotes policymaking, new investments and urban planning that are climate-informed and future-proofed. The EU Biodiversity Strategy 2030 *Bringing nature back into our lives* (2020) highlights that the promotion of healthy ecosystems, green infrastructure and nature-based solutions should be systematically integrated into urban planning, including in public spaces, infrastructure, and the design of buildings and their surroundings.

The relevance of the topic and the need to identify new paradigms for the sustainable transformation of the threatened territories are also recognized in the Global Agenda for Sustainable Development 2030 [12], which urges to make cities and human settlements inclusive, safe, long-lasting, and sustainable (Objective 11), and recommends that policy makers and institutions take urgent measures to tackle climate change and its consequences (Objective 13). The European Union also recognizes that cities play a central role in the challenge of climate change, with the launch of the Covenant of Mayors in 2009, with the adoption of the Adaptation Strategy in 2013 and with the Mayors' Campaign for Climate Adaptation [13].

This scenario raises the need for an intersectoral and inter-institutional convergence among all the policies that impact on urban regeneration and territorial governance. Moreover, this fits with the most recent Community policies and programming, and also constitutes a cross-cutting objective of the *Piano Nazionale di Ripresa e Resilienza* - PNRR (National Recovery and Resilicnce Plan) (Mission 5 Inclusion and cohesion: Urban regeneration and social housing), and of the PNR (National Research Programme) 2021/2027 (Thematic Area 2 Humanistic culture, creativity, transformations, society of inclusion, in close correlation with Thematic Area 5 Climate and Thematic Area 6 Environment). In this frame of reference, the contribution focuses on territories subject to floods due the effects of climate change, both caused by sea level rise (SLR), a linear and permanent phenomenon [14] and by river floods (transient phenomena).

It is important to highlight the dynamic nature of the water element, which, even historically, has always been considered both a criticality and a resource.

Indeed, from a geographical and strategic point of view, all the greatest civilizations have arisen in the proximities of water basins, rivers, or seacoasts. Water has always been a great ally to urban development, as well as economic and cultural, for the most important cities, becoming an identity element of the place and the landscape [15].

Precisely in these areas, the climate changes highlight the needs to identify preventive and management strategies to prevent the calamitous event. This point of view is inspired to tactical urbanism's actions [16], which integrate the theme of temporality of the interventions and the phased programming, to the thematic of project quality and to the research for innovative solutions [17].

The aim of the authors is to identify theoretical and methodological references and experimental guidelines to adopt specific urban regeneration actions for the ecological transition of areas at risk, which fall into broader urban resilience strategies which the authors, during their joint research path, conceptualized in three main approaches: "defence", "adaptation" and "relocation"[1] [18].

1.2   *Isola Sacra* (Fiumicino): "Amphibious Area" on the Outskirts of Rome

The ongoing research focuses on the municipality of Fiumicino, near the city of Rome in Italy, precisely on the area called *Isola Sacra*[2], characterized by a condition of multi-fragility linked to the water element, which allows a rare opportunity for experimentation.

Fiumicino is one of the thirty-three areas in Italy affected by coastal flooding in a time horizon of one hundred years, due to the progressive sea level rise as effect of climate change [19]. In addition to this aspect, which directly affects the coastal strip, the so-called *Isola Sacra* area is subject to frequent flooding due to heavy rains and overflows of the Tiber River.

The picture (see Fig. 1) shows a first result of the ongoing research, in which a projection (blue polygon) of the areas potentially affected by flooding due to SLR at 2100 is returned according to the RCP 8.5 scenario (the worst), which considers a sea level rise of about 63 cm compared to the current sea level.[3]

Moreover, the internal areas of Fiumicino, especially *Isola Sacra*, in addition to the SLR phenomenon, are subjected to flooding phenomena due to the overflow of the Tiber River, which imply to consider this territory as an "amphibian", between land and sea.

---

[1] For further information see: Mariano, C., Marino, M.: Defense, adaptation and relocation. Three strategies for urban planning of coastal areas at risk of flooding. Planning, Nature and Ecosystem Services INPUT aCAdemy 2019 Conference proceedings (2019).

[2] The name is due to the morphology of the area, bounded to the south-east by the bend of the Tiber River and to the north by Fossa Traianea, a waterway built by the emperor Traiano starting from 42 b.C. as part of the work for the construction of the new port, for further information see: Keay, S., Millett, M., Strutt, K., Germoni, P.: The Isola Sacra Survey Ostia, Portus and the port system of Imperial Rome. Mcdonald Institute Monographs, Cambridge, UK (2021).

[3] Data provided by the ENEA Climate Modelling Lab. For further information on the methodology adopted, see: Mariano, M. Marino, G. Pisacane, G. Sannino: Sea Level Rise and Coastal Impacts: Innovation and Improvement of the Local Urban Plan for a Climate-Proof Adaptation Strategy. Sustainability, 13, 1565 (2021) https://doi.org/10.3390/su13031565.

Fig. 1. Areas at risk of flooding due to sea level rise in *Isola Sacra* – Fiumicino (blue polygon). SLR data compared to today: 63 cm; time horizon: 2100. Drawing by Ph.D. candidate in *Planning. Design, Technology of architecture* Gabriele Pastore (2022).

The following pictures show areas potentially affected by floods (purple and light blue polygons) due to overflow of Tiber River with return periods of 30 (see Fig. 2) and 100 years (see Fig. 3) [20].

Fig. 2. Areas at risk of flooding due to overflow of Tiber River in *Isola Sacra* – Fiumicino. Return period: 30 years (purple polygon). Source: Piano di Gestione del Rischio Alluvione Appennino Centrale – PGRAAC (2016). Drawing by Ph.D. candidate Gabriele Pastore (2022).

Fig. 3. Areas at risk of flooding due to overflow of Tiber River in *Isola Sacra* – Fiumicino. Return period: 100 years (light blue polygons). Source: Piano di Gestione del Rischio Alluvione Appennino Centrale – PGRAAC (2016). Drawing by Ph.D. candidate in *Planning. Design, Technology of architecture* Gabriele Pastore (2022).

Taking into account the vulnerability of the area with respect to the issues briefly exposed, and with reference to the three prevailing strategic approaches to the theme of risk, namely "defence", "adaptation" and "relocation", it is interesting how the criticality is addressed by the public administration of Fiumicino through the sole use of the defensive approach, substantially based on the construction of massive embankments to protect the most vulnerable areas (Fig. 4), denying any communication between urbanized areas and the natural element (the river), to the detriment of place identity. However, the authors believe that integrate the defensive approach with the "adaptation" strategy is the real challenge for urban designers and policy makers to pursue resilient and climate-proof territories and cities, and to educate communities to live together with water.

Adaptation actions and strategies represent a complementary approach to mitigation. This approach implies the population's ability to continue living their habitat while making adjustments that can reduce flood impact to a minimum; and it involves practices of urban regeneration of compromised territories [21], that can effect urban development with a view to sustainability and resilience [22], also by relying on the adoption of nature-based solutions (NbS) as «actions to protect, sustainably manage, and restore natural or modified ecosystems, that address societal challenges effectively and adaptively, simultaneously providing human well-being and biodiversity benefits» or solutions inspired to the Ecosystem-based Approach (EbA) that involve a wide range of ecosystem management activities to increase the resilience and reduce the vulnerability of people and the environment to climate change [23].

To this, the next paragraph is intended to illustrate a project that can encourage reflection on the near future cities.

Fig. 4. One of the embankments protecting the urban fabric of Isala Sacra which completely denies the relationship between land and river. Pictures by Ph.D. candidate in *Planning. Design, Technology of architecture* Maria Racioppi (2022).

2 Resilience Tactics for Urban Adaptation

2.1 *Climate Tiles* Teaches Man How to Walk on Water

For the reasons stated so far, the authors have decided to analyse an emblematic best practice that teaches humans, physically and metaphorically, to walk on water, and which fall within the aforementioned resilient urban regeneration strategy of "adaptation". The project proposed is designed by the Danish firm *Tradje Natur*, based in Copenhagen, it is called *Climate Tiles*, and refers to the need to adapt the public space of cities to flooding. This practice is extremely useful in understanding how, public space can and has the task of becoming an instrument of civil awareness to the effects of climate change [24].

It is about an outdoor tile, designed for future sidewalks capable of handling the extreme and increasingly frequent rainfall that is pouring into our cities.

The pilot project was realized in the summer of 2018 in the Nørrebro area of the city of Copenhagen, along the road adjacent to the headquarters of the *Tradje Natur* studio. It consists of a 50 m long sidewalk to study the effectiveness and functioning of the climate tile and verify the response during the different seasons of the year (see Fig. 5). The purpose of this intervention is to collect rainwater from roofs and sidewalks, to allow its reuse and to manage the overload on the sewage system in periods of extreme rainfall, thus reducing water damage and this is allowed by an underground piping system (see Fig. 6). These tiles are designed for a 50-year cycle, to ensure long-term management of climate challenges. The goal of the project is to create a smart tile with low-tech and high-tech plug-ins. On the merits, sensors could be inserted in the holes that can read and send information on the current water level both to the water network and to the citizens (Fig. 7 and 8). In this, all citizens will be informed and aware of climate adaptation measures in the city.

Copenhagen has more than 700 kms of sidewalks, thus several million square meters. All cities have sidewalks, even the smallest and most dense. The potential of this project, all over the world, is easy to understand [25].

The innovative scope of this project is to guarantee adequate rainwater management, while at the same time adding more value to the city.

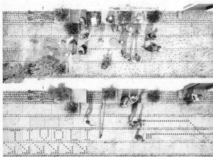

Fig. 5. Detail of Climate Tiles project. Source: Tradje Natur studio (2018).

Fig. 6. Detail of Climate Tiles project. Source: Tradje Natur studio (2018).

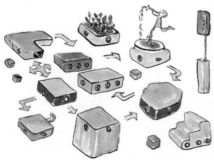

Fig. 7. Detail of Climate Tiles project. Source: Tradje Natur studio (2018).

Fig. 8. Detail of Climate Tiles project. Source: Tradje Natur studio (2018).

3   Conclusion and Future Developments

Thus, the contribution represents the advancement of an ongoing research that aims to respond to the increasing challenges that climate change imposes on our cities.

The case study of Isola Sacra is an emblematic example of "amphibious" territory because of all the different pressures exposed so far, in terms of impacts on the territory, linked to the water element.

The authors wanted to demonstrate how, to date, decision-makers are oriented, when they are, almost only to defensive actions, which deny any relationship between the urban fabric and the natural element. However, the very concept of ecological transition makes this approach obsolete and implies the need to prefigure urban regeneration strategies, aimed at the ecological transition of areas at risk, composed and integrated, which include hard defense actions (mostly infrastructure), associated with soft actions of urban adaptation (considering the contribution of all project disciplines). Indeed, the Climate Tiles project responds to this need, and even if it is mostly an urban design project, its informative purpose makes it a best practice to be considered as a real urban regeneration action for urban metabolism.

For these reasons research developments are aimed at proposing site-specific and climate-proof solutions for Isola Sacra, within an overall vision of urban regeneration in ecological key.

Author Contributions. The contribution is the result of a shared reflection by the authors, however, paragraph 1.1 should be attributed to Carmen Mariano, 1.2 should be attributed to both authors, 2.1 should be attributed to Marsia Marino, conclusion and abstract should be attributed to both authors.

References

1. IPCC: Summary for Policymakers. Climate Change 2021: The Physical Science Basis; Contribution of Working Group II to the Sixth Assessment Report of the Intergovernmental Panel on Climate Change. Cambridge University Press, Cambridge, UK; New York, NY, USA (2021). https://www.ipcc.ch/site/assets/uploads/2018/02/WG1AR5_all_final.pdf. Accessed 3 Feb 2022
2. UNFCCC: United Nations Framework Convention on Climate Change (2021). https://unfccc. int/resource/docs/convkp/conveng.pdf, Accessed 21 Dec 2021
3. Musco, F., Magni, F.: UHI nel contesto ampio del CC: Pianificazione, città e clima. In: Musco, F., Fregolent, L. (eds.) Pianificazione Urbanistica e Clima Urbano. Manuale per la riduzione dei fenomeni di isola di calore urbano. Il Poligrafo casa Editrice, Padova, Italy (2014)
4. Magni, F.: Climate proof planning. L'adattamento in Italia tra Sperimentazioni e Innovazioni. Franco Angeli Editore: Rome, Italy (2019)
5. United Nations Human Settlements Programme (UN-Habitat): Planning for Climate Change: A strategic, values-based approach for urban planners, Cities and climate change initiative tool series, UNON, Publishing Services Section, Nairobi (2014)
6. Maragno, D., dall'Omo, C.F., Pozzer, G., Musco, F.: Multi-risk climate mapping for the adaptation of the venice metropolitan area. Sustainability **13**, 1334 (2021). https://doi.org/10. 3390/su13031334
7. Secchi, B.: A new urban question in Territorio, vol. 53 (2010)
8. Merrifield, A.: The New Urban Question. Pluto Press (2014). https://doi.org/10.2307/j.ctt183 p210
9. Musco, F., Zanchini, E.: Le Città Cambiano il Clima. Corila: Venice, Italy (2013)
10. Mariano, C., Marino, M.: Water Landscapes: From risk management to an urban regeneration strategy. Upl. J. Urban Plan. Landsc. Environ. Des (2018a)
11. de Luca, C., Naumann, S., Davis, M., Tondelli, S.: Nature-based solutions and sustainable urban planning in the european environmental policy framework: analysis of the state of the art and recommendations for future development. Sustainability **13**, 5021 (2021). https://doi. org/10.3390/su13095021
12. United Nation: The 2030 Agenda for Sustainable Development (2015). https://sdgs.un.org/ 2030agenda. Accessed 02 March 2022

13. European Commission: Mayors' Adapt – the Covenant of Mayors Initiative on Climate Change Adaptation (2014)
14. Mariano, C., Marino, M. Pisacane, G., Sannino, G.M.: Sea level rise and coastal impacts: innovation and improvement of the Local Urban Plan for a climate-proof adaptation strategy. Sustainability **13**, 1565 (2021). https://doi.org/10.3390/su13031565
15. Braudel, F.: Il Mediterraneo. Lo spazio e la storia, gli uomini e la tradizione. Milano, IT, Bompiani (1987)
16. European Union's Horizon 2020 project: Generative Commons Living Lab, Tools for generative commons (2022). https://generative-commons.eu/
17. Capon, T., Smith, M., Wise, R.: National climate change adaptation case study: early adaptation to climate change through climate-compatible development and adaptation pathways. In: Baldwin, K., Howden, M., Smith, M., Hussey, K., Dawson, P. (eds.) Transitioning to a Prosperous, Resilient and Carbon-Free Economy: A Guide for Decision-Makers, pp. 365–388. Cambridge University Press, Cambridge (2021). https://doi.org/10.1017/9781316389553.019
18. Mariano, C., Marino, M.: Defense, adaptation, and relocation: three strategies for urban planning of coastal areas at risk of flooding. TeMA Journal – Conference proceedings of INPUT aCAdemy 2019 – Planning, Natur and Ecosystem Services (2019)
19. Antonioli, F., et al.: Sea-level rise and potential drowning of the Italian coastal plains: Flooding risk scenarios for 2100. Quatern. Sci. Rev. **158**, 29–43 (2017). https://doi.org/10.1016/j.quascirev.2016.12.021. Accessed 10 Dec 2021
20. Autorità di Bacino Distrettuale dell'Appenino Centrale: Piano di Gestione del Rischio Alluvioni – PGRAAC (2016)
21. Boateng, I.: Integrating Sea-Level Rise Adaptation into Planning Policies in the Coastal Zone, in Integrating GenerationsFIG Working Week 2008. Stockholm, Sweden (2008)
22. Salata, S., Giaimo, C.: Nuovi paradigmi per la pianificazione urbanistica: i servizi ecosistemici per il buon uso del suolo. In: Un nuovo ciclo della pianificazione urbanistica tra tattica e strategia. Planum publisher, Milan (2016)
23. IUCN: Ecosystem-based Approaches to Climate Change Adaptation (2020). https://www.iucn.org/theme/ecosystem-management/our-work/ecosystem-based-approaches-climate-change-adaptation. Accessed 05 Jan 2022
24. Rockefeller Foundation: 100 Resilient Cities (2019). http://www.100resilientcities.org/about-us/. Accessed 05 Feb 2022
25. Tradje Natur: Climate tiles (2018). https://www.tredjenatur.dk/en/portfolio/climatetile/. Accessed 05 Jan 2022

Rethinking the South in Humanistic, Social and Pedagogical Perspective. Rethinking the Relationship Between Local and Global

Rossella Marzullo[✉]

Mediterranea University, 89124 Reggio Calabria, Italy
rossella.marzullo@unirc.it

Abstract. In the era of globalization of economy and markets, where calculating rationality imposes itself by overwhelming every social dimension, communities seem to no longer serve, they are perceived as anachronistic survivals that are not functional to the new way of interpreting existence and conceiving development. Furthermore, the telematic revolution and the domination of technology impose themselves on everything and ask to reconsider the concepts of spatiality and proximity. But in reality it is precisely the territories, the cities, the inner villages, the places where a process of social recomposition can take place today. Small villages are the primary places for identity formation, the central categories of human existence.

Keywords: Places · Society · Identity · Memory · Community

Rethinking the South. Rethinking the Relationship Between Local and Global.

1 Introduction

A renewed reflection on historical events, cultural behaviors and forms of identity in today's Calabria is destined to go beyond the traditional sphere of reference traced by over a century of southern question. It is now inevitably intertwined with the pervasive mechanism of the global economy and the triumph of technology which, in an increasingly deterritorialized world process, affects individuals, territories and communities (Amoroso and Gomez y Paloma 2007). Rethinking the South today therefore means first of all rethinking the relationship between local and global.

"Economic power is no longer limited to exercising control over the factory and production sites […] it has long since overflowed and invaded society. Control must be total. Society is being remodeled in its structures, in its organization, in its habits of life, in the forms of its consumption and even in its food diets. The result is an artificialization of the world and of life, their distortion in the literal sense of the term" (Alcaro 2006a, b). This distortion concerns the environment and the territories that become amorphous spaces on which the market dominates and also affects human life and those social constructions around which it is formed and gathers.

© The Author(s), under exclusive license to Springer Nature Switzerland AG 2022
F. Calabrò et al. (Eds.): NMP 2022, LNNS 482, pp. 2701–2707, 2022.
https://doi.org/10.1007/978-3-031-06825-6_258

In the global world, "the communities, these centers that have produced an admirable synthesis of nature and culture, that have produced a great work of cultural shaping of the natural data of life, these propulsive centers of the civilization of humanity have now become superfluous" (Alcaro 2006a, b). In the era of globalization of economy and markets, where calculating rationality imposes itself by overwhelming every social dimension, communities seem to no longer serve, they are perceived as anachronistic survivals that are not functional to the new way of interpreting existence and conceiving development. Furthermore, the telematic revolution and the domination of technology impose themselves on everything and ask to reconsider the concepts of spatiality and proximity.

Before, "the neighbor was the neighbor, someone who met and mobilized emotional resources, persuasive resources, who forced us to communicate intensely, did not allow us to disengage. In the current era, paradoxically, we have the feeling of being close to those who are furthest away, and with whom we dialogue, perhaps through the network" (Barcellona 2007). We live in a dimension in which the sense of spatiality and temporality, of before and after, is lost.

The perception of time has always been decisive for the various eras and for the types of culture. "The peasant culture has been built over the millennia as a culture of cyclical time, a time that returns, that follows the rhythm of the seasons [....] Today we live in real time, which is the time of contextuality, for which there is no the possibility of elaboration, of transformation. The present time is apparently a full time, but in reality it is an empty time, made up of moments not linked to each other " (Barcellona 2007).

All this affects the identity of each of us carved over time and linked to the places that have to do with our memory, with our experience, with our daily life, with our network of social and affective relationships.

These characteristics of modernity condition the southern citizen, and the Calabrian one in particular, already involved in a process of disintegration that led him to reject the culture of his fathers to the point of contempt, pushing him to internalize the new universal cultural models that propagate with sirens. of the technological utopia and globalization.

The result is a growing process of anomie in the communities of the South that are increasingly uprooted from the territory and increasingly unable to recognize themselves in the traditions and places around which they were built.

It is therefore necessary to start from the places to rethink the South in modernity.

2   The Old Villages: Places of Identity and Memory, Bridges Between Past and Future

It is precisely the territories, the cities, the inner villages that are "the places where a process of social recomposition can take place today. Villages are the primary places of identity formation, the central categories of human existence. The only opposition to the ongoing processes induced by globalization can therefore only arise from the small size, where people in flesh and blood, and not abstract universal citizens, open conflicts, experience life practices, and consciously oppose each other day after day, and not, with one's own bodies and with one's actions, to the homologative processes that neutralize subjectivities and sterilize passions" (Scandurra 2007).

When the non-local market, the one over long distances, imposes its regime everywhere, the risk of the desacralization of places, reduced to mere space, becomes high, and the living disappears in modern space (Scandurra 2007).

It is through places, in fact, that relationships between the individual and the group are established, it is through the forms of human settlements that relationships between men and the territory are established. "Places and people, intertwining their presences, tell stories, form tangles of relationships until the same places become part of the affective world of men and constitute a vital part of it" (Scandurra 2007).

In this respect, the historical role of the city is decisive, a symbolic place par excellence since it establishes the space of relations between the individual and the group and between the individual and the universe, the space where human specificity is constituted and which always has had, Enzo Scandurra reminds us, "the role of remembering what has been through its physical, material, geometric testimonies, it is the past that becomes present" (Scandurra 2007).

The contemporary city, on the other hand, is the city that thrives on continuous innovation, technology, the tumultuous manifold without shared stories: it does not need places that remember, but an undifferentiated space without memory on which to design a new structure suitable for adapt to the changes of modernity.

Therefore, the function of the city tends to be lost as a specific place that preserves the common memory, that establishes relationships, that helps to build individual and collective identities. The modern city is no longer the repository of the characters of a community, of symbols that speak only to those who are able to decipher them. With devastating consequences because the deconstruction of the city as a specific place represents an epochal break in the tradition of European man and undermines the mental and symbolic space in which the great emotional conflicts, emotions and passions of human beings have been staged (Barcellona 2007).

This is even more evident if we think of the southern towns and cities whose streets, alleys, squares were built and lived in function of social communication, relationships between individuals, families and communities, as places where the community meets, for anniversaries, holidays, mourning. In this sense, the southern city becomes a recognition, an emotion, a belonging, "that special place, topologically singular, where the power of the common intellect is manifested in the production of words, feelings, laws that externalize, so to speak, the specific qualities of the place, the *genius loci*" (Piperno 1997).

3 Losing Focus of Places and Desacralization of Nature

Losing focus of places as constituent factors of identity is accompanied by a process of desacralization of nature that has roots that go back in time, but which is accentuated by the modern mechanisms of the market and the domination of technology.

Mario Alcaro outlined the historical and cultural roots and the "theoretical ingredients that lead anthropocentrism to destructiveness towards the natural context [...] first and foremost the separation between subject and object: on the one hand man as an extra-natural res on the other, nature as an inert and passive object" (Alcaro 2006a, b).

In modernity, nature is no longer thought of, as in pre-modern philosophies, as a living organism rich in potential and symbolic values, but is deprived of all autonomy and independence: "thought in these terms becomes a flexible wax [...] available to appetites some men" (Alcaro 2006a, b).

Nature thus becomes an object from which to obtain everything possible, a resource to be exploited rather than the complex organism in which man is inserted and roots his own identity.

The laws of economics, techniques and the unstoppable logic of development have increasingly accentuated this trend and also in the South, which has not had the advantages of industrialism but has certainly suffered the damage, nature has been experienced as available to any slaughter and abuse, in spite of the centuries-old Mediterranean cultural tradition, for which, as Alcaro recalled recalling Camus, "nature was measure, that is, a fundamental element that dimensioned human existence and gave it a specific role in the cosmos" (Alcaro 2006a, b).

The local dimension, therefore, is the only one that can propose a new relationship between community and environment within which, in a rediscovered harmony with nature, men individualize themselves and make history, they become social individuals when they intertwine relationships with those who experience the specific qualities of the same place.

Reaffirming the importance of the local in the globalized world does not mean a parochial and provincialistic closure or the rejection of mechanisms that are scary because they are uncontrollable. Places are not isolated niches inhabited by communities closed in their own particular, but plural social contexts where the concept of democracy requires to be defined and re-defined from time to time as the relationship between different individuals and the shared common good (Scandurra 2007). In a world of supranational or planetary relations, the local can be the context where one can practice the modern challenge of the conviviality of differences against the abstract universalisms of political rhetoric (Scandurra 2007).

In this way the community can be a factor capable of delineating new compatibilities and new collective needs in a broader scenario, it can provide new tools for evaluating economic-social relations, it can strengthen the "social capital" necessary for development. In a word, the community can be the spring for the reconstruction of a feeling of self-esteem and self-confidence even in a broader geopolitical context.

This is why the local is the other side of the global (Touraine 2008; Sassen 2008; Shiva 2006; Bauman 2005; Stiglitz 2007; Seravalli 2006; Pianta 2001; Sullo 2002). Because the community dimension, the roots in places and the relationships that are constituted in them are indispensable tools to face the logic of a globalization that neglects the social dimension of coexistence.

From this perspective, the venue can become a creative and constructive space where identity, memory and traditions exert forces of liberation from unsustainable lifestyles, imposed by consumerist logic and an unbridled liberal economy. The risk of closure scenarios or barricades against "the different who threatens us", feared by many, vanishes if the identity of local communities is nourished by the exchange and comparison of diversity, in the awareness that only in terms of diversity does it recognizes and expresses itself.

4 Role of Southern Italy in European Economic, Social and Political Dynamics

In light of these considerations, one can try to imagine what role Southern Italy could play in a context that inexorably attracts it to European economic, social and political dynamics.

The economic and social history of Southern Italy in the second half of the twentieth century highlighted the growing dependence on external resources and, at the same time, an ever greater opening of the southern territory to the world market. This has transformed the southern regions into vast areas of consumption in which society and economy are linked, in absolutely dependent forms, to the development model of Northern Italy and Central Europe. In this way, the South was released from the Mediterranean basin (Perna 2008).

Obviously, the South still belongs geographically to the Mediterranean, but the production mechanism and, to a large extent, the lifestyle no longer have as a reference point the models and cultures of the Mediterranean geopolitical area.

All of this, however, took place at the same time as a process of transformation of Europe which took place in that path of westernization of the world which in fact produces a planetary homologation that takes place under the sign of the American way of life (Latouche 2007).

This process of artificialization of the world and distortion of life in the historical forms it has given itself has (Alcaro 2003), therefore, invested the modalities of construction of the new Europe which tends to be realized around the myths of development, of the market and of techno-science, destined to defining its identity characters in a dimension that does not take into account the histories, traditions and cultures of the communities that compose it (Barcellona 2005).

In this way the South of Italy finds itself inserted in a Europe that has broken its historical-cultural roots with the Mediterranean: a Europe separated from the "cradle of Europe". As if a person could form after being deprived of his childhood, his adolescence (Matvejevic 2007).

The identity of the southern communities inextricably linked to the history of the Mediterranean and therefore the unity of a people that has in common the world of nature, but, above all, the world of culture is seriously threatened (Signore 2003).

The perverse effects of globalization and the world market force us to think of the countries of the Mediterranean basin as a Mesoregion, the only institutional and socio-economic scale capable of preventing further marginalization of this area (Perna 2008). The background can be that of polycentric Europe, described by Bruno Amoroso, linked to the four most important meso-regions that surround it - the Baltic, Central Europe, Western Europe and the Mediterranean - which can be represented with "the image of the Olympic circles autonomous in part overlapping and linked together (the "four rings of solidarity")" (Amoroso 2000).

The alternative to catastrophe, wrote Pietro Barcellona, "is the ability to identify a European space in which each of us can reacquire a cultural belonging that is compatible with our geographical belonging" (Barcellona 2006).

In this perspective, Europe can only be Mediterranean. A force, that is, that can resist globalization in the name of a tradition of civilization that is not delivered to individual nations, but is thought of as the symbolic container of a multiplicity of responses (Barcellona 2006).

The risk is that this idea of Mediterranean Europe becomes only a cultured quotation, an ideal reference without consequent choices and political strategies if within it, visibility is not given to the South, its culture, its identity and its ability to become "The bridgehead of another Europe [...] the Europe of a more convivial, more human, more social, more tolerant, more cultural civilization, founded on Mediterranean values that are now derided or repressed: solidarity, the sense of family, an art of living, a conception of time and death" (Latouche 2007).

A Europe that overcomes its Atlantic vocation and in which Mediterranean cultures regain centrality, also in view of the construction of a new economic and social pole in an intercultural dimension (VII Rapporto sul Mediterraneo).

5 Conclusions

In the heart of this Mediterranean Europe, Calabria, with its specific historical-cultural identity, can find itself in a space contiguous with other countries, with which it shares a cultural heritage, and an ethos at the base of which is a conception of the world based on sense of proportion, prudence and balance (Teti 2004).

If the future of Mediterranean Europe depends on the ability to look at the peoples of the other shore, rejecting the temptation to remain locked up in its own banks, Calabria, not only for its geographical location, but, perhaps, even more for its heritage cultural and identity, it can be a bridge between two worlds that a certain conception of the West wants to oppose, in irremediable conflict and divided by a new wall this time constituted by the Mediterranean (Amoroso 2000).

A symbolic role of high ethical content that recalls the passionate words that the Bosnian writer Ivo Andric dedicates to bridges in one of his stories: "Of all that man, driven by his vital instinct, builds and erects, nothing, in my opinion, it is more beautiful and more precious than bridges. Bridges are more important than houses, more sacred, because they are more useful, than temples. They belong to everyone and are the same for everyone, always raised, sensibly, at the point where most human needs meet, more lasting than any other construction, never subservient to dark plots and evil powers. [...] Everywhere in the world, wherever my thought goes and stops, it finds faithful and industrious bridges, a symbol of the eternal and never satisfied desire of man to connect, pacify and unite everything that appears to our spirit, to our eyes, at our feet, so that there are no divisions, contrasts, separations" (Andric 2001).

References

Alcaro, M.: Economia totale e mondo della vita. Il liberismo nell'era della biopolitica, p. 62. Manifestolibri, Roma (2003)

Alcaro, M., Filosofie della natura. Naturalismo mediterraneo e pensiero moderno, p. 106. Manifestolibri, Roma (2006a)

Alcaro, M.: Globale e locale, in M. Cimino (a cura di), Politica e cultura in Calabria. OraLocale (1996–2005), vol. I, p. 229. Cosenza (2006b)

Amoroso B.: Europa e Mediterraneo. Le sfide del futuro, p. 17. Dedalo, Bari (2000)

Amoroso, B., Gomez y Paloma, S.: Persone e comunità. Gli attori del cambiamento, p. 23 ss. Dedalo, Bari (2007)

Amoroso, B.: Politica di vicinato o progetto comune?. In: Cassano, F., Zolo, D. (a cura di): L'alternativa mediterranea, p. 493 ss. Feltrinelli, Milano (2007)

Andric, I.: I ponti, in Id, Romanzi e racconti, pp. 1182–1184. Mondadori, Milano (2001)

Barcellona, P.: Il suicidio dell'Europa. Dalla coscienza infelice all'edonismo cognitive. Dedalo, Bari (2005)

Barcellona, P.: La parola perduta. Tra polis greca e cyberspazio, p. 72. Dedalo, Bari (2007)

Barcellona, P.: Saggio introduttivo a P. Barcellona, F. Ciaramelli, La frontiera mediterranea. Tradizioni culturali e sviluppo locale, p. 8. Dedalo, Bari (2006)

Bauman, Z.: Globalizzazione e glocalizzazione. Armando, Roma (2005)

Husserl, E.: L'idea di Europa, Raffaello Cortina Editore, Milano (1999)

Latouche, S.: La voce e le vie di un mare dilaniato. In: Cassano, F., Zolo, D. (a cura di): L'alternativa mediterranea, p. 122. Feltrinelli, Milano (2007)

Matvejevic, P.: Quale Mediterraneo, quale Europa? In: Cassano, F., Zolo, D. (a cura di): L'alternativa mediterranea, p. 435. Feltrinelli, Milano (2007)

Perna, T.: Mezzogiorno e Mediterraneo: divergenze attuali e convergenze possibili. In: Paesi e popoli del Mediterraneo, n. 0, p. 233 (2008)

Pianta, M.: Globalizzazione dal basso. Economia mondiale e movimenti sociali. Roma (2001)

Piperno, F.: Elogio dello spirito pubblico meridionale. Genius locis e individuo sociale, pp. 89–90. Manifestolibri, Roma (1997)

Sassen, S.: Una sociologia della globalizzazione. Giulio Einaudi Editore, Torino (2008)

Scandurra, E.: Un paese ci vuole. Ripartire dai luoghi, Città aperta, p. 33–34. Troina (2007)

Seravalli, G.: Né facile né impossibile. Economia e politica nello sviluppo locale. Donzelli, Roma (2006)

Shiva, V.: Il bene comune della terra. Feltrinelli, Milano (2006)

Signore, M.: Il Mediterraneo tra memoria e progetto. In: AA. VV., Mediterraneo e cultura europea, Rubbettino, Soveria Mannelli, p. 27 (2003)

Stiglitz, J., La,: Globalizzazione che funziona. Einaudi, Torino (2007)

Sullo, P. (a cura di): La democrazia possibile. Il Cantiere del Nuovo Municipio e le nuove forme di partecipazione da Porto Alegre al Vecchio Continente, Intra Moenia, Roma (2002)

Teti, V.: Il senso dei luoghi, Memoria e storia dei paesi abbandonati. Donzelli, Roma (2004)

Touraine, A.: La globalizzazione e la fine del sociale, Il Saggiatore, Milano (2008)

VII Rapporto sul Mediterraneo curato dal centro Federico Caffè, diretto da Bruno Amoroso, in Paesi e popoli del Mediterraneo

Big Data and Cultural Heritage

Vincenzo Barrile and Ernesto Bernardo(⊠)

Department of Civil, Energetic, Environmental and Material Engineering-DICEAM- Geomatics Laboratory, Mediterranea University of Reggio Calabria, 89123 Reggio Calabria, Italy
ernesto.bernardo@unirc.it

Abstract. The purpose of this research is to help clarify how digital and technological transformation can contribute to improving the protection of the cultural heritage of our nation, through the creation of an innovative monitoring system of cultural heritage, where it is possible to faithfully reproduce the study area in a short period of time, analyzing in detail the problems such as the identification of cracks. Firstly, an innovative methodology will be used for the management of a fleet of automated drones for data acquisition, through the use of charging and data transmission bases. The images will allow to produce a 3D model of the study area and subsequently, this large amount of data acquired by drones will be treated with innovative techniques of segmentation and classification of images to visualize the cracks in the buildings present in the study area. Finally, the information will be transmitted to a GIS platform capable of visualizing buildings showing deterioration. This system can be used in both large cities and small towns (or sparsely populated or rural areas). Its versatility makes it capable of weakening territorial inequalities and the possible socio-economic effects that the loss of cultural heritage would entail, such as the crisis in the tourism sector, layoffs and social exclusion. It is important to recognize the relationship between society and technology because it helps us to understand the willingness of many young people to renew, change and improve the world in order to be able to protect the history of cultural and historical heritage.

Keywords: UAV · Big data · Virtual archaeology · Digital archaeology · Cultural heritage · 3D reconstruction

1 Introduction

For some time the Geomatics Laboratory of the Mediterranean University of Reggio Calabria has been interested in the field of cultural heritage, in particular this note wants to show an innovative system capable of monitoring a large area of cultural heritage (a village) for the identification of various states of deterioration of the houses in the town to be displayed in the GIS. Over the past decade, the advent of electronic computers has revolutionized the work of the surveyor. The way of analyzing and collecting data has changed with the development of new technology which has brought benefits in this sector. Everything that was analyzed until the advent of technology was written in paper, and, now, it is digital; in particular, the measurements were done manually, while the latest technological advances have automated the data acquisition process. Today we can take advantage of new advanced technologies such as, drones, sensors, or even cameras,

© The Author(s), under exclusive license to Springer Nature Switzerland AG 2022
F. Calabrò et al. (Eds.): NMP 2022, LNNS 482, pp. 2708–2716, 2022.
https://doi.org/10.1007/978-3-031-06825-6_259

which allow us to detect images and acquire data at increasingly reduced costs over time. Among the most advanced technologies used in the monitoring of a building are sensors, which detect, through hardware systems, information concerning, for example, the state of a crack or even the geometric elements of a building.

In particular, UAV make it possible to carry out visual inspections of the building continuously and automatically, overcoming the limits that a human operator may have in terms of fallibility and the ability to compare a large quantity. of measures varying over time. The technological revolution has in fact exponentially increased the amount of information about buildings that can be stored in databases [1–6]. Obviously, traditional data analysis and management systems are unable to support a huge amount of data (big data). Despite the effects of this technological revolution dictated by an exponential amount of data, the diversity and structure of information requires a highly cost to be maintained. For this it becomes essential to focus on more effective and automated methods, in order to value only the information useful for the analysis and discard the redundant information to avoid overloading the archives, without the risk of increasing or decreasing the margin of error too much.

Even small systems produce a large amount of data. The concept of "Big Data" is described by 3V: Volume, Variety and Speed. The latter are related by the following relationship: large volumes of data from a variety of data sources are available at high speed. The amount of data flow can be a problem for storing and processing data that can hardly be managed and in most cases being unused, because it do not show anything relevant. On the other hand, large amounts of data increase the chances of having a reliable estimate of the performance parameters, provided that adequate processing of this data is available. A large amount of data can reveal correlations and dependencies that allow for predictions of results and behaviors, thus facilitating informed and rational decision-making for efficient management of the facility. Thanks to improvements in detection capabilities, processing power, storage capacity, software programs and the quality of Internet connections, the ability to acquire, collect, share, store and process huge amounts of data is constantly increasing, offering the opportunity to exploit very large volumes of a wide variety of data collected and analyzed at high speed.

The proposed research activity is based on the study and development of advanced techniques for monitoring, inspection and mapping of building cracks in order to obtain and constantly update the safety status of buildings through a GIS platform. In particular, we have created an innovative automated UAV system for monitoring and acquiring large amounts of data (big data). Data collection, one of the most important phases for monitoring the data process, allows to obtain information about the integrity of buildings, useful for planning future design and intervention choices. We then used Machine Learning techniques to manage the large amount of georeferenced data acquired in order to recognize the presence of cracks in the acquired images and eliminate any photos in which the same crack is present (thus eliminating redundancies in the acquired dataset) As a next step, the system realized recognizes if over time there has been a change in the crack and reports it. Finally, we optimized the representation of the identified elements on a GIS platform in order to obtain an "open and updatable thematic cartography". Our aim is to improve the proposed system, in order to make it an updatable IT tool to store, view, query and manage all the data that the Municipality and the Regions have on their city. It will be possible to represent the elements inherent in the geometric characteristics of the buildings, their relevance, the state of the cracks, the interventions carried out in the most important historic buildings and the systems built, having available databases that allow rapid and selective searches by topic.

1.1 Study Area

We chose to test this experimental system on Cardeto (Fig. 1), an Italian town of 1 473 inhabitants in the metropolitan city of Reggio Calabria in Calabria. The town is located on the right bank of the Sant'Agata stream and has a characteristic stepped structure. The origin of the ancient center probably goes back to the 10th and 11th centuries. In that period, in fact, under the emperor Basil I, the bishopric of Reggio was elevated to the "Metropolis of Byzantine positions of southern Greek church", destination of a continuous influx of Basilian monks, of which we find traces in the ancient female monastery of Sant'Andrea di Mallamaci, then became Santa Maria di Mallamaci. The other hypothesis states that the country hosted the first inhabitants when the Byzantines, around the year 1000, to better face the Arab threat, erected various fortifications in the interland of the cities, building kastre, also called motte, that remember the near Motta Sant'Agata of which Cardeto was a farmhouse until 1783. It was probably the same agatini who, in search of a safer place to escape the continuous raids, founded Cardeto by going inland. The earthquake of 1783 and that of 1908, together with the flood of 1951, caused considerable damage to the community, seriously damaging the two churches present in the place, one of which, that of the patron saint San Sebastiano, will only be definitively reopened in 2000. Cardeto also appears to be the seat of numerous and ancient places of worship.

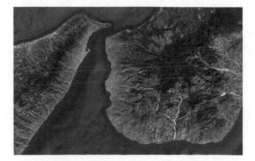

Fig. 1. Study area

2   Materials and Methods

The proposed work includes a hardware part concerning the creation of charging bases and data transmissions, and a software part that allows the correlation of the acquired information and the creation of algorithms to manage a large amount of georeferenced data and the subsequent representation of the results (deteriorations) in the GIS platforms used as an easily updatable thematic cartography. Therefore, this system for collecting information through GIS platforms is a valid tool for updating the Cadastre of unsafe buildings, which can subsequently be proposed to the Authorities responsible for its use.

This system of monitoring and updating the cadastre of unsafe buildings consists of two phases: a phase in the laboratory and a phase in the field.

Laboratory activities:

1) Realization and testing of the 3 neural networks for the recognition of cracks using suitable methodologies for the treatment of Big Data
2) Creation of a charging base and experimental data transmission using suitable methodologies for the treatment of Big Data (Fig. 2)

Fig. 2. Charging base and data transmission

3) Programming of system automations
4) Creation of a flight plan by the operator through the use of the QGround Controll application which is automatically transmitted at set times to the charging and data transmission base
5) Configuring the drone with the charging base
6) Implementation of the system as a whole

Field activities

7) Transmission of the flight plan from the charging base to the drone
8) Execution of the flight plan
9) Sending the images acquired by the drone to the charging base
10) Sending the acquired images from the charging base to the server

At this point we proceed with the processing of the large amount of acquired images (Big Data) in order to jointly process them to identify the cracks in the buildings.

By way of example, an image acquired by the system relating to the identification of cracks is shown (Fig. 3).

In order to populate the database we have suitably processed the images acquired through the described acquisition system, extracting the geometric characteristics of the buildings, the presence of decay in the walls. To better explain the procedure, an example of methodology applied to the classification of a single type of building crack is reported (Fig. 4).

Fig. 3. Image acquired by the drone relating to the identification of cracks.

The acquired images (Fig. 4a) were automatically subjected to a pre-processing and enhancement process (Sobel operator and Prewit), then they were segmented (edge detector, Canny filter, Gaussian filter) and classified (Support Vector Machine) to be able to extrapolate the information we need (Fig. 4b) [7–12].

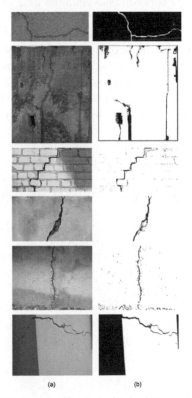

(a)   (b)

Fig. 4. Methodology applied to the classification of a single type of building crack is reported: a) Input Images; b) Crack detection.

In particular, the SVM classification was carried out in two phases: 1) SVM Training. 2) Performance testing. In the first phase, the geometric characteristics of the linked components assigned for SVM training were initially calculated. Then, these features were normalized to a range. Kernel with radial base function (RBF) was chosen as a kernel trick, because the number of instances (connected regions) was not very large, and the size of the space transformed with RBF is infinite. The optimal training parameters for the SVM were found using grid search. During this operation, triple cross-validation was performed to correctly learn the different types of cracks. In this triple cross-validation, the training set was divided into three equal subsets. To ensure proper learning, a subset was tested using the trained classifier on the remaining two subsets. The goal was to identify good parameters so that the classifier can predict test data effectively. After learning that the parameters had been determined, SVM was trained with the "One Against All" approach using the MATLAB LIBSVM library. In the second phase, connected regions were tested that were not used during SVM training [13–20].

We used experimental Machine Learning (ML) methodologies to manage Big Data. ML algorithms and statistical models detect patterns from data based on data mining, pattern recognition, and predictive analytics. They are more effective for addressing uncertainties than traditional algorithms in situations where large and diverse datasets (e.g. Big Data) are available. Due to the large volumes of data, analyzing and detecting correlations and relationships between data could be time exploiting using traditional methods.

In our experimentation, thanks to datasets of about 250,000 images, we then used Machine Learning techniques for the recognition of cracks, we therefore created a model capable of recognizing and identifying in a completely automatic way the presence of a possible crack in the photo, delete any photos in which the same crack is present [21–25]. Over time, our aim for the program is that it will be able to recognize if there has been a change in the crack, if that is the case, it will be reported in the GIS application.

We have therefore created a GIS capable of displaying the map of the buildings where the cracks are present with the trajectories of the drones [26–29]. In particular, by clicking on the trajectory we can see all the images acquired, by clicking on the building we are able to identify the cracks and the most important elements.

We have made a first updated map showing the updated network of buildings created by our experimental system.

The images captured by the drone clearly show the details of the cracks.

The data can be acquired regularly, without the use of any operator, facilitating operations and reducing costs and times. The database reports the coordinates of the buildings which are then identified in the order of study area, type and conditions, geometric information, conservative and functional status.

The study attests that UAV images can capture more information about buildings than field surveys and can be useful in a variety of situations.

Furthermore, we are working on how to make the system compatible with the new EASA regulations (Fig. 5).

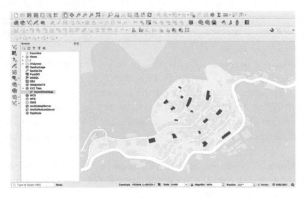

Fig. 5. GIS elaboration

3   Conclusions

Digital and technological transformation can help improve the preservation of our nation's cultural heritage and at the same time be a support for the green transition.

Through the implementation of our innovative monitoring system for cultural heritage, with low environmental impact, we are able to eliminate territorial inequalities and safeguard cities, towns and rural areas from the possible socio-economic effects that the loss of cultural heritage would entail.

From an economic point of view, the system we have created to obtain a Cadastre of unsafe buildings is very expensive; however, if designed for prevention and maintenance functions, it can be an excellent tool for public administrations, which can then plan interventions on the basis of the indications contained within the GIS.

In fact, the Real Estate Cadastre can be used with the sole purpose of taking a census of the heritage of existing buildings in the towns; however, thanks to the acquisition of other data, we are able to extend its function to the planning of maintenance activities such as the identification of buildings with a high level of damage, due to the presence of dangerous cracks and the planning of interventions to increase the security. The databases will then be enriched with a series of possible additional information useful for this purpose. It is therefore decided to study how to implement the cadastre of buildings in order to be able to use it for maintenance planning and restoration interventions. Future developments will aim at improving automatic systems.

References

1. Bernardo, E., Bilotta, G.: Monumental arc 3D model reconstruction through BIM technology. In: Bevilacqua, C., Calabrò, F., Della Spina, L. (eds.) NMP 2020. SIST, vol. 178, pp. 1581–1589. Springer, Cham (2021). https://doi.org/10.1007/978-3-030-48279-4_148
2. Bernardo, E., Musolino, M., Maesano, M.: San Pietro di Deca: from knowledge to restoration. Studies and geomatics investigations for conservation, redevelopment and promotion. In: Bevilacqua, C., Calabrò, F., Della Spina, L. (eds.) NMP 2020. SIST, vol. 178, pp. 1572–1580. Springer, Cham (2021). https://doi.org/10.1007/978-3-030-48279-4_147

3. Barrile, V., Fotia, A., Bernardo, E., Candela, G.: Geomatics techniques for submerged heritage: a mobile app for tourism. WSEAS Trans. Environ. Dev. **16**, 586–597 (2020). https://doi.org/10.37394/232015.2020.16.60

4. Barrile, V., Fotia, A., Bernardo, E., Bilotta, G.: Geomatic techniques: a smart app for a smart city. In: Bevilacqua, C., Calabrò, F., Della Spina, L. (eds.) NMP 2020. SIST, vol. 178, pp. 2123–2130. Springer, Cham (2021). https://doi.org/10.1007/978-3-030-48279-4_200

5. Tonkin, T.N., Midgley, N.G.: Ground-control networks for image based surface reconstruction: an investigation of optimum survey designs using UAV derived imagery and structure-from-motion photogrammetry. Remote Sens. **8**(9), 786 (2016). https://doi.org/10.3390/rs8090786

6. Monteiro, C., Costa, C., Pina, A., Santos, M., Ferrão, P.: An urban building database (UBD) supporting a smartcity information system. Energy Build. **158**, 244–260 (2018). https://doi.org/10.1016/j.enbuild.2017.10.009

7. Gopalakrishnan, K., Gholami, H., Vidyadharan, A., Choudhary, A., Agrawal, A.: Crack damage detection inunmanned aerial vehicle images of civil infrastructure using pre-trained deep learning model. Int. J. Traffic Transp. Eng. **8**(1), 1–14 (2018). https://doi.org/10.7708/ijtte.2018.8(1).01

8. Serna, A., Marcotegui, B.: Detection, segmentation and classification of 3D urban objects using mathematical morphology and supervised learning. ISPRS J. Photogram. Remote Sens. **93**, 243–255 (2014)

9. Chen, C., et al.: Automatic pavement crack detection based on image recognition. In: International Conference on Smart Infrastructure and Construction, (ICSIC), pp. 361–369 (2019). https://doi.org/10.1680/icsic.64669.361

10. Chen, S., Truong-Hong, L., Laefer, D.F., Mangina, E.: Automated Bridge Deck Evaluation through UAV Derived Point Cloud. CERI-ITRN2018, pp. 735–740. Dublin, Ireland (2018)

11. Ameri, A., Dadrass Javan, F., Zarrinpanjeh, N.: Automatic pavement crack detection based on aerial imagery. J. Geomatics Sci. Technol. **9**(1), 145–160 (2019)

12. Dadrasjavan, F., Zarrinpanjeh, N., Ameri, A.: Automatic Crack Detection of Road Pavement Based on Aerial UAV Imagery. Preprints, 2019070009 (2019). https://doi.org/10.20944/preprints201907.0009.v1

13. Barrile, V., Bernardo, E., Fotia, A., Candela, G., Bilotta, G.: Road safety: Road degradation survey through images by UAV. WSEAS Trans. Environ. Dev. **16**, 649–659 (2020). ISSN: 2224-3496. https://doi.org/10.37394/232015.2020.16.67

14. Hoang, N.-D., Nguyen, Q.-L.: A novel method for asphalt pavement crack classification based on image processing and machine learning. Eng. Comput. **35**(2), 487–498 (2018). https://doi.org/10.1007/s00366-018-0611-9

15. Cannistraro, M., Bernardo, E.: Monitoring of the indoor microclimate in hospital environments a case study the Papardo Hospital in Messina. Int. J. Heat Technol. **35**(Special Issue 1), S456–S465 (2017). https://doi.org/10.18280/ijht.35Sp0162

16. Mancini, A., Malinverni, E.S., Frontoni, E., Zingaretti, P.: Road pavement crack automatic detection by MMS images. In: 21st Mediterranean Conference on Control and Automation, pp. 1589–1596. Chania (2013).https://doi.org/10.1109/MED.2013.6608934

17. Ogawa, S., Matsushima, K., Takahashi, O.: Efficient pavement crack area classification using gaussian mixture model based features. In: International Conference on Mechatronics, Robotics and Systems Engineering (MoRSE), pp. 75–80. Bali, Indonesia (2019) https://doi.org/10.1109/MoRSE48060.2019.8998713

18. Sari, Y., Prakoso P.B., Baskara, A.R.: Road Crack Detection using Support Vector Machine (SVM) and OTSU Algorithm. In: 6th International Conference on Electric Vehicular Technology (ICEVT), pp. 349–354. Bali, Indonesia (2019). https://doi.org/10.1109/ICEVT48285.2019.8993969

19. Sekeroglu, B., Tuncal, K.: Image processing in unmanned aerial vehicles. In: Al-Turjman, F. (ed.) Unmanned Aerial Vehicles in Smart Cities. UST, pp. 167–179. Springer, Cham (2020). https://doi.org/10.1007/978-3-030-38712-9_10

20. Barrile, V., Bilotta, G., Fotia, A., Bernardo, E.: Road extraction for emergencies from satellite imagery. In: Gervasi, O., et al. (eds.) ICCSA 2020. LNCS, vol. 12252, pp. 767–781. Springer, Cham (2020). https://doi.org/10.1007/978-3-030-58811-3_55

21. Agrawal, A., Choudhary, A.: Perspective: materials informatics and big data: realization of the "fourth paradigm" of science in materials science. APL Mater. 4(5), 1–9 (2016). https://doi.org/10.1063/1.4946894

22. Bai, S.: Growing random forest on deep convolutional neural networks for scene categorization. Expert Syst. Appl. **71**, 279–287 (2017)

23. Bernardo, E., Barrile, V., Fotia, A.: Innovative UAV methods for intelligent landslide monitoring. In: European Association of Geoscientists and Engineers Conference Proceedings, International Conference of Young Professionals «GeoTerrace-2020», vol. 2020, issue 1, pp. 1–5 (2020). https://doi.org/10.3997/2214-4609.20205713. ISSN: 2214-4609

24. Bernardo, E., Barrile, V., Fotia, A., Bilotta, G.: Landslide susceptibility mapping with fuzzy methodology. In: European Association of Geoscientists and Engineers, Conference Proceedings, International Conference of Young Professionals «GeoTerrace-2020», vol. 2020, issue 1, pp. 1–5 (2020). https://doi.org/10.3997/2214-4609.20205712. ISSN: 2214-4609

25. Deng, J., Dong, W., Socher, R., Li, L., Li, K., Fei-Fei, L.: Imagenet: a large-scale hierarchical image database. In: Proceedings of the 2009 IEEE Conference on Computer Vision and Pattern Recognition (CVPR), pp. 248–255 (2009). https://doi.org/10.1109/CVPR.2009.520 6848

26. Gustavsson, M., Seijmonsbergen, A., Kolstrup, E.: Structure and contents of a new geomorphological GIS database linked to a geomorphological map—with an example from Liden, central Sweden. Geomorphology **95**(3–4), 335–349 (2008). https://doi.org/10.1016/j.geomorph.2007.06.014

27. Barrile, V., Bilotta, G., Fotia, A., Bernardo, E.: Integrated GIS system for post-fire hazard assessments with remote sensing, Int. Arch. Photogramm. Remote Sens. Spatial Inf. Sci. XLIV-3/W1-2020, 13–20 (2021). https://doi.org/10.5194/isprs-archives-XLIV-3-W1-2020-13-2020, 2020

28. Barrile, V., Fotia, A., Bernardo, E., Bilotta, G., Modafferi, A.: Road infrastructure monitoring: an experimental geomatic integrated system. In: Gervasi, O., et al. (eds.) ICCSA 2020. LNCS, vol. 12252, pp. 634–648. Springer, Cham (2020). https://doi.org/10.1007/978-3-030-58811-3_46

29. Barrile, B., Bernardo, E., Fotia, A.: GPS/GIS system for updating capable faults in the Calabrian territory through the use of soft computing techniques. In: European Association of Geoscientists and Engineers. Conference Proceedings, International Conference of Young Professionals «GeoTerrace-2020», vol. 2020, issue 1, pp. 1–5 (2020). https://doi.org/10.3997/2214-4609.20205710

Combining the Ecological and Digital Transitions: *Smart Villages* for New Scenarios in the EU Rural Areas

Gabriella Pultrone[(✉)] [iD]

Mediterranea University of Reggio Calabria, Reggio Calabria, Italy
gabriella.pultrone@unirc.it

Abstract. This paper proposes a reflection on ongoing research activities focused on the concept of *smart village* and its possible operational declinations in helping to implement ecological and digital transitions in a balanced logic of attention to the local dimension, oriented towards the quality and maintenance of territories, the regeneration of minor centres in the rural areas affected by abandonment and depopulation, the creation of new economies based on the enhancement of territorial resources, adequate provision of infrastructures and services for a significant improvement in the quality of life.

From a methodological point of view – in the light of the reference framework outlined by the most recent developments described in various EU reports and international political agendas – the examination of some innovative smart village experiences shows a varied picture in relation to the specificities of the local contexts, and highlights an orientation, already existing before the current policies, which has been accelerated by the pandemic crisis in order to go beyond the emergency aspects, in favour of long-lasting structural interventions, and to "leave no one behind".

The discussion and the final considerations highlight, on the one hand, that the ecological transition is, first of all, a cultural question, based on a new way of thinking, on the recovery of the sense of limits and on a profound rethinking of the scale of values in the search for a durable balance. On the other hand, the integrated approach, methods and tools of urban and territorial planning can and must contribute to pursuing the objectives of territorial rebalancing and cohesion, within new virtuous relational networks between urban and rural areas.

Keywords: Integrated territorial planning strategies · EU smart villages · Ecological and digital transitions

1 Introduction. Towards the Essential Transition: Topic, Aim of the Research and Methodological Approach

The challenge of climate change is a priority issue at a global and local level, also in relation the pandemic crisis from Covid-19. At the EU level, while environmental and climate policies have brought significant benefits in recent decades, many problems persist in the loss of biodiversity, the use of natural resources, the impact of climate change

© The Author(s), under exclusive license to Springer Nature Switzerland AG 2022
F. Calabrò et al. (Eds.): NMP 2022, LNNS 482, pp. 2717–2726, 2022.
https://doi.org/10.1007/978-3-031-06825-6_260

and environmental risks to health and well-being. Furthermore, the health emergency from Covid-19, which has been underway for two years now, has revealed the underlying fragility of large European metropolitan areas and the missing territorial links with rural areas, highlighting the consequent need for a long-term vision, and for closer cooperation to ensure an equitable distribution of the funds from the Recovery Plan between urban and rural areas.

However, for the sake of completeness, it should also be noted that, in recent decades, the European Union and the Member States have issued numerous policies and strategies to take action on city-countryside territorial rebalancing. However, since it is not possible to provide further details here, please refer to the bibliographical references.

There is a need for a holistic approach that includes urban and rural regions, with particular emphasis on the phenomenon of depopulation that threatens them. Rural policies should be linked to other relevant policies sharing the cross-cutting objective of sustainable development, such as those for food, climate and biodiversity, poverty reduction, land use, transport infrastructure, services of general interest or the development of new activities based on the circular economy and the bioeconomy, in order to appropriately localize the SDGs [1, 2]. Indeed, it is equally imperative to address both short-term and long-term problems simultaneously, learning something from crisis management to address the even greater threat of climate change [3].

With reference to the European scenario and to the first results of ongoing research activities, carried out by the author as a researcher of the Mediterranean University of Reggio Calabria, within the Research and Development Network for Southern Europe Sparsely Populated Areas (RDENSESPA) [4], the paper intends to underline the fundamental role rural areas can play in economic and social cohesion policies in increasing the resilience of territories, in the contribution of countless services from various local ecosystems, in the process of equitable and sustainable transition towards innovative models of well-being economy in all sectors [5, 6].

The main aim is to search for new urban, territorial, economic, social, infrastructural, and relational models that could offer alternative possibilities to these places, stopping their tendency to abandonment and decline.

In this context, the concept of *smart village* is gaining ground in the rural development agenda, coinciding with the ongoing reform of the common agricultural policy (CAP), which provides for a new implementation model based on the elaboration of a strategic plan by each Member State, according to the recommendations published by the Commission in 2020 to jointly achieve the objectives of the CAP and those of the European Green Deal, such as access to high-speed broadband internet in rural areas by 2025 [7].

The scientific literature on the concept of "smart" applied to rural areas is also gaining momentum and *smart villages* are an increasingly emerging reality [8]. They can be understood as an extension of the concept of *smartness* to less densely populated territories that can help rural communities unlock potential and opportunities as part of a common strategy for defining territorial development in a renewed balance and complementarity between urban and rural areas. This is also what Bill Slee [9], an active member of the ENRD Thematic Group on Smart Villages, argues about the links between Smart Villages and the European Green Deal and the fact that smart rural communities are already providing many inspiring examples on how to deal with each of the challenges identified in the European Green Deal at the local level, such as climate change and just transition. This is observed in transformations involving various types of smart communities, such as eco-villages and collaborative environmental partnerships addressing water quality and biodiversity, renewable energy communities, local food systems and initiatives for sustainable mobility. The concept of *Smart village* focuses on inclusive development, trying "to leave no one behind" by mobilizing local communities and resorting to collaboration agreements with the world of research and businesses to co-design more sustainable futures. Therefore, considering appropriate to investigate in this direction for the theoretical and operational implications, the paper proposes a reflection on the first results of an ongoing research activity focused on the concept of smart village and on its possible operational declinations in helping to implement ecological and digital transitions in a balanced logic of attention to the local dimension, oriented towards the quality and maintenance of the territories, the regeneration of degraded areas, the prevention of hydrogeological instability and the regeneration of minor centres, especially in the inland areas affected by abandonment and depopulation.

The reference framework is linked to the recent developments described in various reports by the European Commission, the European Committee of the Regions, the European Investment Bank and ESPON, as well as the main recent political agendas, including: the UN 2030 Agenda for Sustainable Development and its 17 SDGs (2015), the Paris Agreement (2015) and the recent outcomes of COP26 (2021), the UN New Urban Agenda (2016), the European Agenda (2016), the document reflection of the Commission "Towards a sustainable Europe by 2030" (2019), the Urban Agenda for the EU (2016), the New Leipzig Charter (2020), the Cork 2.0 declaration on Better Life in Rural Areas (2016), the OECD Principles on Urban Policy and Rural Policy (2019), the European Green Deal (2019), the European Territorial Agenda, the Next Generation EU (2020).

As for the methodology, the examination of some case studies on innovative smart villages at EU level, illustrated in the second paragraph, shows a varied picture in relation to the local specificities of an orientation, already existing before the current policies, to which the pandemic crisis has imprinted the need for acceleration to overcome the emergency aspects in favour of long-lasting structural interventions to "leave no one behind" and live in harmony with nature.

2 European Strategies Towards the Ecological and Digital Transitions for Smarter Rural Regions

According to the New Leipzig Charter – adopted on 30 November 2020 -, which reaffirms the fundamental principles of the 2007 Leipzig Charter and, at the same time, updates them in view of contemporary global challenges, sustainable and resilient urban development takes place within a regional or metropolitan context and is based on a complex network of interdependencies and functional partnerships. This is exemplified by the "functional area", as indicated in the 2030 Territorial Agenda, which partly covers a metropolitan area or a combination of other territorial entities. Cities must cooperate and coordinate their policies and instruments with surrounding suburban and rural areas in relation to important sectors, such as housing, commercial areas, mobility, services, green and blue infrastructure, material flows, local and regional food systems, and energy supply [10].

To ensure that no one is left behind, the digital transformation and the collaboration needed to implement it should be based on common human values, such as inclusiveness, human centrality, and transparency. Equally necessary is the strengthening of the awareness and capacity of local communities to protect, regenerate, and reuse their living environments, landscapes, tangible and intangible cultural assets and other unique values through the instruments of EU cohesion policy, rural development policy, territorial planning or any other instrument that promotes, among others, integrated territorial or local development.

In this context, due to the richness of the cultural heritage, the intensity of ecosystem services, typical products and paths, rural areas have great potential and possible positive and interesting effects of public enhancement policies, and also the smallest towns have the possibility to place themselves on the frontier of innovation, but they need system policies that allow them to project the high quality of life of which they are custodians into the future, enhancing the presence of services and training offers, job and investment opportunities, territory enhancement and maintenance tools.

For instance, in Portugal, the territorial 2014–2020 cohesion policy, with a specific emphasis on the trend towards depopulation, aging and impoverishment in inland regions, particularly those located near the border with Spain, pursued the priority objective to seek suitable solutions. It contains more than one hundred and sixty measures, mostly of government initiative, and an agenda for the inner territories that integrates eight thematic initiatives in a dynamic process based on continuous co-creation, experimentation and revision, which encourage the dynamics of smart specialization at the local and sub-regional level. Covering all the governance areas, whose action is reflected on the inland territories, the envisaged measures are organized around five axes of intervention for a more cohesive, competitive, sustainable, connected, and collaborative inner territory. Among other things, they aim at strengthening the connectivity of inner territories by facilitating their insertion into wider territorial relationships, between coastal-inland and border production bases, thus generating new forms of organization for cohesion, competitiveness, and sustainability. The agenda consists of eight initiatives organized around the following challenges for the sustainable development of inner territories: 1) Aging with quality; 2) innovation of the economic base; 3) Territorial capital; 4) Cross-border cooperation; 5) Rural-urban relationship; 6) Digital accessibility.

In defining and developing a common European vision for 2040, the European Commission has gathered the views of rural communities and businesses through public consultations and stakeholder-led events. Through this collaborative process, it has created a broad vision and comprehensive rural action plan to help rural communities and businesses reach their full potential in the coming decades with four complementary action areas including stronger, connected, resilient and prosperous rural areas by 2040 [5] (Fig. 1).

According to ARC (Agricultural and Rural Convention), smart villages are already significant actors in the response to the Green Deal at the local level, often at the forefront of rural social innovation, considered the key to addressing sustainability challenges.

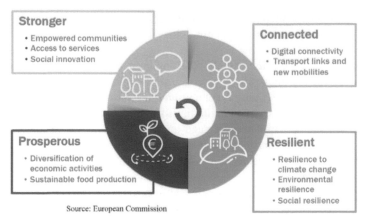

Source: European Commission

Fig. 1. Main drivers shaping the future of rural areas for 2040 and the four complementary areas for action (European Commission 2021:10)

In particular, not only technology, but also institutional factors and, above all, human and social innovation factors must be placed at the heart of the processes of smart cities and territories.

The concept of smart village originates from that of smart city but, in addition to being connected to the opportunities provided by new technologies, it puts more emphasis on the potential of social innovation. To a certain extent, *smart villages* include an application of community-led local development (CLLD) principles on a more local scale than most Leader groups. It might make sense to formally incorporate the smart village principles into the LAG strategy and, in this case, LAGs could become one of the main tools to support Smart Villages. However, in order to play this role, they must be equipped to do so.

The ENRD thematic group meetings on smart villages managed by the European Network for Rural Development have already gathered a wealth of knowledge of good practices on the issues of the Green Deal and the transition to a post-carbon Europe. And even if, in some cases, those who develop the initiatives on a local scale rarely consider their communities as Smart Villages, they can be considered as such because they are able to provide innovative and creative local responses to the long-term challenges and the 10 priorities identified by the Green Deal.

Challenges deriving from the socio-economic context or from access to natural resources very often trigger the radical changes involved in the creation of a smart village. A better understanding of the factors that can favour or hinder the transition paths from traditional villages to smart village status starts with a clear identification of the main challenges affecting rural areas in Europe. These challenges are widely known and have been documented in many policy reports and research papers.

Rural Europe contains many good examples of ecovillages and communities in transition that have implemented more ambitious emission reduction targets than specified in international treaties or European strategies. Their actions go beyond community energy and take a broad look at living with a lighter ecological footprint. For example, Cloughjordan, in Ireland, decided to become an ecovillage in the early 2000s, long before the financial crisis that affected its progress and its inhabitants are still carrying out the project even though the forecast of 130 houses on a plot of 67 acres has been only partially realized, with some self-built and others professionally built, as well as a farm for food production, a district heating plant and a community bakery.

As regards the goal of preserving and restoring ecosystems and biodiversity, even if for about three decades, agri-environmental policies have been oriented towards the protection and improvement of agricultural landscapes and biodiversity, in many areas the problems of ecological damage persist and sometimes they have worsened. Given that individual farm-level initiatives are insufficient to address habitat loss, the Dutch government has actively promoted a new landscape-scale approach using environmental cooperatives run by farmers' collectives, now an option within the Dutch Rural Development Programme (RDP). The same approach has been used in several other Member States, such as Denmark. The coastal village of Karby, Jutland, is an example of the integrated proactive and participatory approaches to landscape governance that are emerging across Europe. The salt flats adjacent to Karby are a Natura 2000 habitat which obliges the competent agency to produce a management plan. Using the agri-environmental measures of the RDP, farmers signed a long-term management agreement on salt marshes, footpaths were created, and arable fields were converted into grasslands. The key to the success of the Karby project was the creation of an effective partnership between the main landscape managers, the municipality and other public bodies and residents in various community groups.

Another challenge for rural areas is the transition to sustainable and smart mobility. Public transport systems in European remote rural areas have been penalized by austerity and the decline in public subsidies, privatizations, and the increase in private mobility. The biggest losers are those who do not have access to a private car, typically young adults, the very old and the poor. The decline has been met with a number of responses that include attempts by the public sector to develop flexible and demand-responsive public transport options. The third sector has emerged as a major player in the provision of community (mini) buses, where the bus is usually owned by a community association and operated by volunteers. Smart service models are being implemented in Smart Villages across Europe, *e.g.*, in France, where the *Rezo Pouce* app is one of the new IT systems to improve the mobility of some of the most disadvantaged groups.

The Green Deal's goal of leaving no one behind may be seen as necessary due also to job losses in coalfields or hydrocarbon communities, but many natural resource dependent communities seek to address job loss problems and a feeling of abandonment, as in the case of the large investment in the forest bioeconomy in the Nordic Countries. The inclusive collaborative approach to revitalize territories with the Smart Villages movement means accepting that the innovations undertaken are always in relation to the context and local needs and that communities with less social capital require more support and construction skills.

Moving to Southern Europe, the population of Artieda, a small mountain village in Aragon (ES), refused to surrender to growing depopulation problems, similar to those of much of Spain's rural inland, and to the added challenge of an expanded reservoir that could take on some of the community's best farmland. They worked hard to strengthen their community, breathed new life into it, and engaged with different sub-groups of the population to ascertain their needs, involving also the elderly and young people. With a combination of citizen action, collective commitment and entrepreneurial talent, Artieda has created a tightly bound partnership of actors fully committed to improving liveability [10].

"Smart Villages" is the sub-theme of the broader ENRD thematic work on "Smart and Competitive Rural Areas" and a specific Thematic Group (TG) has worked on this topic between September 2017 and July 2020 [11]. Following the Cork 2.0 European Conference on Rural Development in Ireland, organized in September 2016, the European Commission published, in April 2017, an "EU Action for Smart Villages" which examines the main EU policy areas that already contribute to the development of smart villages. It also stresses the need to bring together different programs to build a strategic approach that can foster the development of "Smart Villages".

Sixteen concrete actions are also described to promote *smart villages* which are based on a wide range of EU policies, including rural development, regional development, research, transport, energy, and digital policies.

Specifically, the Pilot Project on *Smart Eco-Social Villages*, initiated by the European Parliament and implemented under the responsibility of the European Commission (DG AGRI), aims to explore the characteristics of smart eco-social villages and identify the best practices on which decision-makers and rural communities can build future development strategies [12]. The project was completed by a consortium, including Ecorys, Origin for Sustainability and R.E.D., between January 2018 and April 2019. Its conclusions are important for the future use of the Smart Village concept in the EU. The review of opportunities and challenges provides a solid knowledge base and clarifies the concept of Smart Village and the interactions with villages in the 15 examples of good practice and six case studies gather insights from core experience. In particular, the selection of the final examples of "good practice" of *Smart Eco-Social Villages* projects comes from a pre-selection of 30 examples, of which the 15 most relevant ones were chosen for the analysis: Munderfing (AT); Seeham (AT); Hofheim (DE); Kolga (EE); Infoenergy of Aragon (ES); Eskola (FI); Bras-sur-Meuse (FR); Cozzano (FR), Ceglédbercel (HU); O'Gonnelloe (IE); Pinela (PT), Cluj-Nap (RO); Bohinj (SI); Fintry and Superfast Cornwall (UK).

Among the differences, worth mentioning is the way in which people's involvement is organized: it can concern a specific platform (Agenda 21 in Seeham, AT), existing structures (Local Action Group in Aragón Infoenergía, ES), steering committees (Hofheim), pilot centres (Kolga, EE), informal meetings (Ceglédbercel, HU) or digital communication (Bras-sur-Meuse, FR). In some cases, like in the Romanian example (Cluj-Nap), different means of interaction are combined: monthly events, online communication, and daily business transactions.

As regards funding, most of the sources of funding are European or private; Member States are rarely involved financially, and the role of private actors depends heavily on the local economy and opportunities. If we look more closely at EU funds, we can clearly see that LEADER financed by the EAFRD and ERDF funds are the most frequent in support of Smart Village strategies.

Besides, while carrying out the case studies, the study team observed that most of the selected villages were very advanced in the use of planning and participation tools and have already developed a clear strategy outlining the main needs and objectives to be achieved.

In a nutshell, the main aspects covered by the good practices identified in the 15 target villages were: the identification of innovative solutions to face the challenges of rural areas; the diversity of rural areas; the accessibility to and the use of place-based assets in combination with ICT and other technologies; the ability to access funding and identify opportunities/initiatives in progress; the involvement of the local population in the development of the local development strategy; the identification of the most relevant development opportunities; the very different and complex socio-economic systems of rural regions, but also the existence of many resources in these sparsely populated regions that can and must be exploited endogenously, adopting a bottom-up perspective.

Finally, three main lessons learned can be taken away from the project: 1) a first lesson is that, although the concept of Smart Village is relatively recent, a wide range of initiatives are already underway in EU rural areas; the Pilot Project has identified many examples of villages currently engaged in initiatives to address the challenges or improve the quality of life of the inhabitants with innovative and intelligent solutions covering a wide range of relevant thematic areas, including agriculture, environment, energy, mobility, health, education, culture or tourism; 2) a second lesson is that, despite the diversity of situations across the EU, many Smart Villages share common characteristics which often include the importance of citizen participation, the presence of adequate governance and the use of a "project anchor" to guide the strategy towards a specific objective; 3) the third lesson is that adequate support must be provided for the development of Smart Villages at the three essential levels of government, *i.e.*, community, national and regional [12].

3   Discussion and Final Considerations

Considering the above, the monitoring and evaluation of ongoing experiences are important for the future use of the concept of *smart village* in terms of inspiration, both for stakeholders and for decision-makers in the field of public support.

Smart Villages require rapid and appropriate access to knowledge in a very wide range of fields – not only those that have been discussed above, such as renewable energy and sustainable mobility, but also on sustainable models of service delivery in rural areas, digital transformation and many more. To address these broader rural challenges, several Member States are exploring the possibility of creating teams of "innovation mediators" in these fields and incorporating LEADER groups into a broader rural innovation ecosystem, or "ARKIS" (Agricultural and Rural Knowledge Innovation Systems).

Villages are developing a wide range of creative solutions to overcome challenges and/or improve citizens' quality of life. They innovate in various spheres and in very different ways, depending on the opportunities and challenges arising from their local contexts. The results of the pilot project, which closely match those of the case studies reviewed by the European Network for Rural Development (ENRD), show the wide diversity of scope, scale and type of innovative services developed by villages. Some villages have experienced difficult situations and, in reaction, they have responded by developing activities and services. In other examples, villages have developed new and innovative services to improve the quality of life of their communities. The pilot project found that, in these circumstances, services and activities are often focused on a specific issue, such as energy, tourism or education.

In conclusion, the analysis of opportunities and challenges provides a solid and up-to-date knowledge base that highlights the current situation of wide variation between regions in terms of levels of development. A just transition is needed in the European Green Deal, as is evident in the transformative changes taking place among *smart villages*, for example in eco-villages and collaborative environmental partnerships addressing water quality and biodiversity, in the development of the local food sector, and in renewable energy communities. *Smart villages* also focus on inclusive development that leaves no one behind and rely on collaborative agreements with researchers to co-design more sustainable futures. Collaborative action at the village or small-town level can be seen as a highly suitable entry point for the injection of the principles of the European Green Deal [10, 13].

Despite the diversity of specific local contexts, many smart villages share common characteristics and are providing inspiring examples on how to address each of the identified challenges for a just transition and inclusive development. The wider purpose is not limited to bridge the distance that separates the major urban centres from the villages and rural areas, but it rather aims to integrate their potential to obtain a mutual advantage in a wider territorial dimension. Many experiences make use of new digital technologies which, however, are only one of the many tools available: there are also many examples of social innovation in rural services, new relationships with urban areas that prove to be beneficial for both parties, and activities that strengthen the role of rural areas in the transition towards a greener, healthier and more sustainable society, also in line with the message of the Encyclical "Laudato si'" which highlights the urgency of a true ecological transition and an inclusive economy.

References

1. Cavalli, L.: Agenda 2030 da globale a locale. Report Marzo 2018. Fondazione ENI Enrico Mattei-FEEM, SDSN Italia (2018). https://www.feem.it/m/publications_pages/2018-cavalli-agenda2030.pdf
2. Pultrone, G.: Passato e/è futuro nell'implementazione dell'Agenda 2030. Strategie di valorizzazione del patrimonio culturale per i territori fragili. ArcHistoR EXTRA 06/2019, pp. 488–501 (2019)
3. Pultrone, G.: What planning for facing global challenges? approaches, policies, strategies, tools, ongoing experiences in urban areas. In: Leone, A., Gargiulo, C. (eds.) Environmental and Territorial Modelling for Planning and Design, pp. 577–587. FedOA Press, Napoli (2018)
4. RDENSESPA Homepage. https://www.rdensespa.eu/. Accessed 30 Dec 2021
5. European Commission: A long-term Vision for the EU's Rural Areas - Towards stronger, connected, resilient and prosperous rural areas by 2040, Communication from the Commission to the European Parliament, The Council, The European Economic and Social Committee and The Committee of The Regions, Brussels, 30.6.2021 COM (2021) 345 final (2021). https://ec.europa.eu/info/sites/default/files/strategy/strategy_documents/documents/ltvra-c2021-345_en.pdf

6. ESPON: POLICY PAPER Territorial evidence and policy advice for the prosperous future of rural areas. Contribution to the Long-Term Vision for Rural Areas, 2021PORTUGAL. EU (2021). https://www.espon.eu/rural

7. European Commission: EU Action for Smart Villages. https://ec.europa.eu/info/sites/info/files/food-farming-fisheries/key_policies/documents/rur-dev-small-villages_en.pdf

8. Visvizi, A., Lytras, M.D., Mudri, G.: Smart Villages in the EU and Deyond. Emerald Publishing, Bingley (2019)

9. Slee, B.: Smart Villages and the European Green Deal: making the connections. ENRD (2020). https://enrd.ec.europa.eu/sites/default/files/enrd_publications/tg6_smart-villages_sv-green-deal-bill-slee.pdf

10. EU2020.DE: The New Leipzig Charter. The transformative power of cities for the common good Adopted at the Informal Ministerial Meeting on Urban Matters on 30 November 2020 (2021)

11. ENRD: Borghi intelligenti. Nuova linfa per i servizi vitali, Rivista rurale dell'UE n. 26 (2018)

12. European Commission: Pilot Project. Smart eco-social villages. Final Report 2019, Luxembourg: Publications Office of the European Union (2020)

13. Martinez Juan, A., McEldowney, J.: European Parliament: Smart villages. Concept, issues and prospects for EU rural areas. EPRS-European Parliamentary Research Service (2021)

Rethinking New Communities. Moving Towards a Science of Cities

Ferdinando Verardi[✉]

Pegaso Telematic University, Centro Direzionale, Isola F2, 80143 Napoli, Italy
ferdinando.verardi@unipegaso.it

Abstract. Development, according to a network logic, starts from the peculiar excellences present in the area, but pushes them to invest in new skills, in increasingly qualified human resources. Therefore, immediately thinking of the training of new urban planners and territorial planners, as well as a strategic reorganization of the urban and territorial systems of the country, starting from a conscious integration of the concepts of smart cities and territories, in which cities include the suburbs and other parts of communities with their own specific names that from a functional point of view are part of the wider urban network in which the different components of that community coexist, does not appear as a choice, but on the contrary, as a necessity. This research aims to analyse the multiple correlations and convergences between two different visions. The first is represented by a school of thought on a new science of cities, described by *Geoffrey West*, according to which cities are the product of the integration of the structure and dynamics of social networks with the physical infrastructural networks that constitute the stage on which urban life takes place in all its articulations, and the second conceived by *Stefano Boeri*, who imagines a planet crossed by corridors of biodiversity, in which forests and cities find a new balance, with a model of urban development, that considers the urban and environmental regeneration of new communities, represented by historic villages, as well as an innovative vision of metropolises, understood as archipelagos of self-sufficient eco-neighborhoods. Finally, a further aspect to consider, derives from the new component that will be characterized, given by digital networks, which will have a preponderant role in a post-pandemic phase, also with respect to the realization of applications capable of carrying out different and complex monitoring systems.

Keywords: New urban communities · City science · Digital innovation

1   The City as a Social Incubator

The theories of cities are largely elaborated, starting from studies dedicated to specific cities or groups of them. They rarely integrate infrastructural issues with socio-economic dynamics. *Geoffrey West* proposes theoretical models on cities, inspired by phyics, i.e. quantitative analysis. Cities and the urbanization process are perhaps too complex to be subjected in a useful way to laws and rules that go beyond their individuality (Geoffrey 2018). There are universally recognized problems, such as consciousness as the

© The Author(s), under exclusive license to Springer Nature Switzerland AG 2022
F. Calabrò et al. (Eds.): NMP 2022, LNNS 482, pp. 2727–2736, 2022.
https://doi.org/10.1007/978-3-031-06825-6_261

origin of life and of the universe, and in fact of the cities themselves, which cannot be addressed, within a computational structure, in a mathematical and predictive sense. But both the complexity and the diversity of these themes must not discourage us with respect to an expansion of the scientific paradigm, useful for a greater understanding and knowledge. Several years ago, a research program on the city was started at the *Santa Fe Institute*, which has the long-term goal of identifying and addressing global sustainability issues. Based on these research trajectories, some results will be indicated, which have contributed to the formulation of a possible science of cities. The goal is to develop ideas and models, which will be related to more traditional schemes, on the many aspects relating to cities and their urbanization. We started by asking ourselves some questions: what are cities? How did they arise? How do they work? What's their future? In the university context, there are a multiplicity of scientific disciplinary sectors that represent the different alternative ways of understanding cities (urban geography, urban economics, urban planning, urban studies, urbanomics, architectural studies, etc.). With the advent of big data, the situation is rapidly changing with respect to a vision of smart cities, even if *Geoffrey West* is convinced that the elements just mentioned are naively sold as panaceas for the solution of all our urban problems. It is significant that there are currently no defined departments of urban science. They represent a new frontier when there is an urgent need to understand cities from a more scientific point of view. Geoffrey West, introduces <<*the use of scale considerations as a tool to open a window on the development of a quantitative, integrated and systemic conceptual structure for understanding cities*>> (Geoffrey 2018. It has been asked whether cities, similarly to other natural systems, are roughly scaled versions of each other. Based on their measurable characteristics, cities such as New York, Los Angeles, Chicago, and Santa Fe, are scale versions of each other, and if so, whether their scale relationship is analogous to that which exists between Tokyo, Osaka, Nagoya and Kyoto, despite their very different aspects and characters. To a certain extent, urban studies have generally not been quantitative, also because few computational mechanistic models of cities have been developed. The question arises: is there a scale behavior of cities? Various researches, carried out by scientific institutes, demonstrate a surprising simplicity and regularity in the comparison between different cities and different countries (Kuhnert et al. 2006). As regards the overall infrastructures, cities behave like organisms: they vary in scale in a sublinear way, following a simple trend determined by a power law (logarithmic) thus presenting a systematic economy of scale. Other researches were based on the collection and analysis of data covering a wide range of parameters, relating to urban systems around the world; the result was always the same, showing a sublinear scale trend of the infrastructural quantities. An important question that has conditioned all discussions on cities is to define what a city actually is. We all have an intuitive idea, but to develop a quantitative conception we need something more precise. In general terms, a city in the sense in which it is understood here is not to be identified with its political or administrative definition. Generally the city includes suburbs and other communities with their own specific names which from a functional point of view, however, are part of the larger urban network. This is universally recognized by most planners, administrators and governments by introducing broader categories into which this more realistic representation of the city fits. For example, these functional agglomerations are called metropolitan statistical areas (MSAs) in the United States, metropolitan areas in Japan and large urban areas or functional urban areas in the European Union. Unfortunately, there is no common definition, and therefore caution must be exercised when comparing different countries. *Geoffrey West* argues that cities are a self-organized emerging phenomenon that is the product of interaction and communication between human beings exchanging energy, resources and information. *<<as urban creatures we all, wherever we live, participate in the multiple networks of intense human interaction that manifests itself in the metropolitan buzz of productivity, speed and ingenuity>>* (Geoffrey 2018. It can be said that the larger the city, the greater the social activity, the more opportunities there are, the higher the salaries, the more diversity there is, the greater the availability of good restaurants, with, museums and educational facilities, and the more intense is the sense of activity, enthusiasm and commitment. These aspects of big cities have proved to be a sources of enormous attraction and seduction for people all over the planet, who at the same time neglect, ignore or fail to take into account the inevitable negative aspects and the dark side of the increased incidence of crime, pollution and disease. In addition to the perception of the individual advantages deriving from the larger size of cities, there are enormous benefits deriving from systematic economies of scale. In fact, *Geoffrey West* states that *<<This extraordinary combination of increasing benefits for the individual as the size of the city increases, added to the systematic increasing benefits for the community, constitutes the fundamental driving force of the continuous expansion of urbanization in the whole planet>>* (Geoffrey 2018). But how come urban systems around the world in countries as diverse as Japan, Chile, the United States, and the Netherlands vary in scale in substantially the same way despite these countries having profoundly different geographies, histories and cultures, and despite have evolved independently of each other. But then what is the unifying common factor that transcends these differences and underlies this surprising structural and dynamic similarity? The great unifying element is the universality of social network structures around the world. Cities are made up of people, and to a large extent people are quite similar in the way they interact with each other and in the way they come together to form groups and communities. *Geoffrey West* also states that *<<the fundamental unifying element that is expressed in the surprising universality of urban scale laws is the fact that the structure and dynamics of human social networks are more or less the same everywhere>>* (Geoffrey 2018). Cities are the product of the integration of the structure and dynamics of social networks with the physical infrastructural networks that constitute the stage on which urban life takes place in all its articulations.

2 Thinking of the Plural City

It is also natural to extend this concept to social networks: in the temporal average, each person interacts with a certain number of other people as well as with groups of people in such a way that collectively their network of interactions fills the socio-economic space available. In fact, this urban network of socio-economic interaction constitutes a complex of social activity and interconnectivity, which really defines what a city is and what its borders are. To be part of a city one must be part of this network continuously. And, of course, the invariant terminal units of these networks, the analogue of the capillaries, of the cells, are people and their homes. Compared to biological life, cities have not existed for a very long time, only for a few centuries, while many organisms have been present for millions, if not hundreds of millions of years. Thus, any optimization drive resulting from progressive adaptations and feedback mechanisms, in the course of the growth and evolution of cities, did not have much time to fully achieve its effect and settle down. This is further complicated by the much faster pace with which innovation and change have taken place in cities, compared to the typical rhythms of biological evolution. With the invention of the city and its formidable combination of economies of scale combined with innovation and wealth creation, the great divisions of society were born. Current social network structures practically did not exist in their present form until urban communities developed. Just like biological ones, these networks are hierarchical and fractal in nature. The fractal nature of these networks and the flows that pass through them ensures an efficient distribution of energy and resources and is the basis of the sublinear scale trend as well as of the economies of scale. In reality, things are a little less simple than that, because the cities are not uniform, but generally have a certain number of local activity poles that behave in a semi-autonomous way, even if they are hierarchically interconnected. These local poles are often called central locations following a popular model of urban systems known as the central location theory, which, introduced in the 1930s by the German geographer *Walter Christaller*, gained considerable credit among urban planners and geographers. In essence, it is a static and highly symmetrical geometric model of the physical configuration of cities and urban systems. It was postulated by *Walter Christaller* based on his observations in cities in southern Germany, somewhat like Jane Jacobs when formulating her conception of the city based on her personal experiences of New York. Little or no attention was paid to quantitative calculation and verification, to systematic analysis and comparison with data, or to mathematical formulation with subsequent predictions; so it is not really science. In spirit, it has much more in common with *Ebenezer Howard's* rigid and inorganic garden city designs, which were basically inspired by idealized configurations of Euclidean geometry, with almost no consideration of the role of people other than as unity economic. Nonetheless, it has many interesting features, and was extremely influential in the design and conception of cities during the 20th century. In a curious pun on his own name, *Christaller* postulated that urban systems, and implicitly individual cities, can be ideally represented as two-dimensional crystalline geometric structures based on a highly symmetrical hexagonal reticular pattern that repeats itself forming graininess on an ever smaller scale. The mathematician Mandelbrot in the seventies introduced with his studies the theory of fractal geometry, extending the studies of as many famous mathematicians, such as Peano himself, Cantor and Hausdorff. To understand complex geometric forms, and even more aspects of nature, Mandelbrot developed a theory that, until recently, escaped an adequate mathematical representation. With the famous passage of his book, (The Fractal Geometry of Nature) Mandelbrot, in paraphrasing the new paradigm of fractal geometry, stated the following: <<Why is geome-try often described as cold and dry? One reason lies in its inability to describe the shape of a cloud, a mountain, a coastline or a tree. Clouds are not spheres, mountains are not cones, coastlines are not circles, and bark is not smooth, nor does lightning travel in a straight line.. >> (Mandelbrot 1982). Therefore, the geometric shapes that previously appeared almost like mathematical monstrosities, which for several years have escaped an adequate mathematical representation, can now be measured and studied with fractal theory, like cities. The most famous fractal is certainly the Mandelbrot set, shown in Fig. 1.

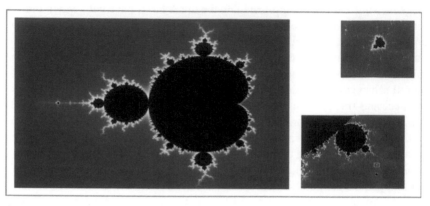

Fig. 1. The Mandelbrot set is one of the most important and famous fractals. It has been reproduced in a wide range of colors, each of which is striking for its particular appearance; an enlargement of it reveals a detail identical to the whole itself.

Although this regularity is not generally found in most urban systems or within cities, Christaller's model for the geometry of urban systems, despite its artificial and unnatural structure, incorporates two very important characteristics that it shares with evolved reticular structures in an organic way. It fills the space and is self-similar (and therefore hierarchical), even if neither of these two concepts had yet been invented. In addition, the model also included other important general characteristics, such as the idea of minimal travel time and distance for the distribution of services. An accurate mathematical analysis of these patterns shows that cities are, in fact, to a good approximation self-similar fractals, just like biological organisms or the coastlines of geography. For example, if the length of the visible boundary of a city is measured at different resolution levels, similar to what Lewis Fry Richardson did for coastlines, and the different measurements are plotted in a logarithmic diagram, we obtain approximately straight lines whose slope is the conventional fractal dimension of the city contrast. Fractal dimensions are the measure of the degree of fragmentation of the outline of an object, which some interpret as a measure of its complexity. In the 1980s, eminent urban geographer *Michael Batty* conducted an extensive statistical analysis of cities to measure their fractal dimensions (Batty et al. 1994, 2005). *Batty*, found values of the order of 1.2 but with a considerable variance that reached values close to 1.8. In addition to providing a parameter for comparing the complexity of different cities, the fractal dimension finds perhaps one of its most interesting uses as a diagnostic barometer of the state of health of a city. In general, the fractal dimension of a robust and healthy city steadily increases as it grows and develops, reflecting ever-increasing complexity to accommodate a growing population engaged in ever more diversified and complicated activities. Conversely, its fractal dimension diminishes when it goes through difficult economic times or when it temporarily contracts. The main proponent of the elaboration of the concept of fractal city and of the integration of ideas drawn from the theory of complexity into traditional urban analysis and planning was Mike Batty, who directs the CASA (*Center for Advanced Spatial Analysis, Center for Advanced Space Analysis*) at University College London. His work has mainly focused on computer models of the physical dimension of cities and urban systems. *Batty* is enthusiastic about the concept of cities as complex adaptive systems and has consequently become a major proponent of developing a science of cities. Over the last twenty years a sub-discipline has developed, the science of networks, which has led to a deeper understanding both of the phenomenology of networks in general and of the underlying mechanisms and dynamics that generate them (Barabasi 2004). Network science studies an enormous variety of objects, including classical community organizations, criminal and terrorist networks, innovation networks, ecological and food networks, health and disease networks, linguistic and literary networks. These studies have provided meaningful insights into a wide range of major societal challenges, including devising the most effective strategies for fighting pandemics and terrorist organizations, addressing environmental issues, empowering and facilitating innovative processes and optimizing social organizations. It appears that the two main components that make up a city, its physical infrastructure and its socio-economic activity, can be conceived as approximately self-similar fractal-type network structures. In fact, *Geoffrey West*, states that <<*fractals are often the result of an evolutionary process that tends to optimize specific characteristics, for example to ensure that all the cells of an organism or all the people of a city are supplied with energy and information, or to maximize efficiency by minimizing transport times or the time required to perform tasks with minimal energy. However, it is obvious what is being optimized in social networks*>> (Geoffrey 2018). Certainly, much remains to be done to build a quantitative theory of social networks, and future investigations will face many exciting challenges. These considerations show that there is a natural explanation for the fact that social connectivity and therefore socio-economic quantities vary in a superlinearly scale with the size of the population. Socio-economic quantities are the sum of interactions or connections between people and therefore depend on how closely they are related. In this sense of variation in scale of the networks, it is no coincidence that the physical and the social are mirror images of each other, so much so that we can think of the physical city, with its networks of buildings, streets, of electricity, gas and water distribution lines, as an inverse non-linear representation of the socio-economic city, with its networks of social interactions. This combination of increased socio-economic activity and infrastructural economies of scale, as a function of the inverse relationship between the two variables, has its mechanistic origins in a similar inverse relationship between their underlying network structures. Although social and physical networks share common general characteristics such as fractal-like nature, space-filling property and the presence of terminal invariant units, there are some essential differences between them. One of these, which is relevant and has enormous consequences, is the way in which the dimensions and flows within networks vary in scale as one progresses downward through their fractal-like hierarchies (Bettencourt 2013).

3   Models of Territorial Governance

It is understood that the involvement and participation of citizens in the government of the city, and in urban planning, is a necessary, but not sufficient, condition for the promotion of Community political strategies. Cities around the world have become the object of particular attention as potential laboratories for new participatory policies. In Italy, the phenomenon was perceived on the occasion of the establishment of metropolitan cities, the operational purposes of which are still being defined, as the Mayor of Milan (Sala 2018) highlighted. Currently, the researches that turn to the definition of innovative models are oriented towards the city-states, such as, for example, Hong Kong, Singapore, etc., and many other emerging urban realities in many areas of the planet. At the height of the Renaissance, in Italy, history has handed down to us an inexhaustible wealth of experience gained in the transition from the medieval urban model, represented by the Municipality, to the lords and principalities, which in the fifteenth-sixteenth century already proposed city-state models, able to have leadership in Europe, in trade, in the economy and in strategic sectors for the development of a community, therefore with the convinced participation of citizens. The evolution of this new model, as *Bertuglia and Vaio* affirm, is subordinated to the establishment of *Urban Laboratories*, in which man is once again at the center of political and administrative attention. <<*The educational processes, able to trigger these new innovative forms of city government, are aimed at the formation of new skills, towards the resolution of urban problems*>> (Bertuglia and Vaio 2019). Here, a strategic mission is certainly represented by the change of skills. Training, research and the environment are the three keywords through which the success of the country and the recovery of the economy pass. Thus, the need to establish real *City Schools* of new conception arises, which, like what Business Schools did for post-Taylorist's enterprises in a development crisis, can train the necessary figures of urban managers. Schools capable of training innovative profiles, scholars, operators, professors who work in the field of project actions concerning the city. The idea is to train operators, tools and models of action to re-read them as an opportunity to address the different themes and various aspects, on which the feasibility of urban projects seems to depend. An organization that brings together multidisciplinary skills necessary to face the development of the territory in an integrated way, with the involvement of public and private actors. An activity that is in line with the most recent reinterpretations of the urban planning profession prompted by the planning theory, where it is increasingly considered necessary to combine skills with respect to the elaboration of visions, the ability to attribute to these concreteness and operability (Pizzorno et al. 2013) working to their feasibility in the strategic combination of resources, problems and contingent opportunities that arise in the city. The territory is therefore probed, in an intentional and strategic way, rather than a field of problems, rather as a field of possible integrations. An orientation that recalls the vision of planning as a trading zone, proposed by *Balducci and Mantysolo* (Balducci and Mantysolo 2013) and on the basis of which it seems to be area-based, ultimately, plus the process of finding a solution to certain problems highlighted by the actors, than the process of defining the problem itself. Also in consideration of the evolution of the health emergency linked to Covid 19, there are various hypotheses on which public institutions and scientific bodies are working. The process will be gradual and timed based on the level of risk assigned to each professional, entrepreneurial and social activity. Every crisis, explains *Richard Florida* (Florida 2010), corresponds to a profound reorganization of the territories and cities because the lifestyles and the adaptability of the population change. What needs to be rebuilt is a sense of belonging to shared values, this requires a long time and good quality public spaces. Digital in the next phases of the post-coronavirus will have a preponderant role, from the creation of applications capable of monitoring a significant sample of the population at risk of contagion to the new functioning of the city. In a similar context, the question that needs to be asked is: in the stages following the chorus-navirus do you need urban planning? Urban planning has always been involved in planning and programming the city and the territories according to medium and long-term logic. Urban planning is needed today, if it will be able to give immediate answers. The new urban planning will have to identify solutions capable of anticipating and governing, today and not tomorrow, future scenarios attributable to the effects of the pandemic. In a subsequent phase, it will be necessary to quickly produce a map of the national territory showing the presence or absence of the relevant local dimension (Dematteis 2015) for each single city or municipality. Is this the dimension that guarantees the self-sufficiency of local and urban communities? In a context in which movements are and will be allowed, by law, only in reference to basic necessities, the question that urban planning must ask is: the Italian territory and the cities are organized according to a logic of the right to access to essential goods? The task of urban planning must be to answer this question. In areas where this right is not guaranteed, urban planning will have to produce a proposal for the rapid reorganization of essential functions on a territorial and urban scale. In this perspective, some authors, including *Stefano Boeri* (Boeri 2021), imagine a planet crossed by corridors of biodiversity, in which forests and cities find a new balance, with an urban development model, which considers the urban and environmental regeneration of new communities, represented by historic villages, as well as by an innovative vision of metropolises, understood as archipelagos of self-sufficient eco-districts. An Ecological Network understood as territorial infrastructure, connects the different areas, endowed with a greater presence of naturalness and with a high degree of integration between local communities and natural processes, mending the landscape quality. The theme of ecological networks has established itself in Europe in the last decade as a central theme of environmental policies. This has led to a radical change of perspective, passing from the idea of conserving specific protected areas to that of conserving the entire structure of the ecosystems present in the area. This change of perspective arises from the consideration that the policies for protected areas aimed at conserving territorial units (Parks and Reserves) are not sufficient to counter the growing environmental pressures and to guarantee processes of conservation of nature and the environment. The elaboration of the various regional strategies for biodiversity is part of the commitments undertaken by the Regions. Since the establishment of the Law on Parks, various and multiple best practices have been created, of Protected Natural Areas, marine reserves, as well as of the ZPS (special protection areas) and SIC (sites of community importance). Parks, terrestrial and marine reserves, historical centers and more generally the elements of the environmental structure, are the subjects of this policy of attention, aimed at combining the objectives of protection and conservation with those of compatible and lasting development, integrating the economic and social issues of the territories affected by the protected areas, with the overall policy of conservation and enhancement of environmental resources. With reference to strategies, we can refer to protection, in the sense of defending, restablishing and connecting resources in a balanced network; development, in terms of restructuring and strengthening of weak areas, as well as balance, aligning living and working conditions between areas of different levels. Strategies linked to the creation of skills, the dissemination of knowledge, the strengthening of project skills, linked to the specificities of individual situations and operating in an integrated and systemic vision, acquire an important role.

4 Digital Innovations. Spaces for Participation and Sharing

The city of tomorrow, increasingly smart, transformed by advances in technology and the spread of networks, and the related digital innovations represent a great opportunity for urban planning, both from the point of view of infrastructure monitoring and the environmental status of the urban environment and both from the point of view of modeling and knowledge of the urban object. A further opportunity is represented by the possibility of visualizing the changes resulting from certain projects and choices. Due to all these characteristics and the ease of collecting contributions and opinions, digital technologies also appear useful as tools for involving and participating in citizens. However some cautions must be used; in the collection and use of data, in the construction of algorithms, and in the awareness that some issues will continue to slip through the mesh of digitalization. This is why it is important to continue to build public participatory moments in real space. Italian urban planning has, for some time now, been questioning its effectiveness and has been trying to equip itself with methods for overcoming the operational limits of its tools. One of the problems that has recently emerged is that of the disconnection between the data used in drawing up the plans and the perception of the daily life of the inhabitants. Some of these data are obsolete, others distorted or in flux, while the questions that the plan should govern change at a rate at which the planning tool cannot respond quickly enough. Others are data, which are not taken into account, because they have origins outside or within the limits of legality, in self-management, in self-organization as well as in abuse. Some practices of self-organization can fall into what Giancarlo Paba called public policies from below. Practices of self-organization for the collective response to social needs, which can represent an important source of widespread urban well-being, through the activation of cultural and recreational spaces capable of responding creatively to the multiple demands of everyday life, through modalities that in some cases can be considered more public than the public. Beyond the issues related to gender planning, it is nevertheless useful to welcome the critique of traditional functionalist planning, especially in a moment like the present one, in which work loses its predefined boundaries of time and space and agile and smart working work practices are spreading changing mobility and consumption habits. A knowledge of the urban and territorial object that is closer to the experience of the inhabitants from the point of view of temporality and complexity and multidimensionality therefore appears strategic for planning. For this reason, a great opportunity is represented by the use of digital technologies both in the context of the collection and arrangement of a large number of data, both in the context of decision support, and as a useful tool for participation.

References

Geoffrey, W.: Le leggi universali della crescita, dell'innovazione, della sostenibilità e il ritmo di vita degli organismi, delle città, dell'economia e delle aziende, pp. 297, 305, 316, 321, 349. Mondadori, Milano (2018)

Kuhnert, C., Helbing e West, G.: Scaling Laws in Urban Supply Networks. In: Physica A, CCCLXIII, pp. 96–103 (2006)

Mandelbrot, B.: The Fractal Geometry of Nature. EPBM Brattleboro LLC, Vermont (1982)

Batty, M., Longley, P.: Fractal Cities, A Geometry of Form and Function. Cambridge (MA), Academic Press (1994)

Batty M., Cities and Complexity. Cambridge (MA), The MIT Press (2005)

Barabasi, A.: Link. La nuova scienza delle reti. trad. it. Torino, Einaudi (2004)

Bettencourt, M.A.: The Origins of Scaling in Cities. In: Science, pp. 1438–1441 (2013)

Sala, G.: Milano e il secolo delle città. La nave di Teseo, Milano (2018)

Bertuglia, S., Vaio, F.: Il fenomeno urbano e la complessità. Bollati Boringhieri (2019)

Pizzorno, A., Crosta, P.L., Secchi, B.: Competenze e rappresentanza, (a cura di) Balducci A. e Bianchetti C., Donzelli, Roma (2013)

Balducci, A., Mantysolo, R.: Urban Planning as a Trading Zone. Springer, Dordrecth (2013)

Florida, R.: The Great Reset. Harper Collins, New York (2010)

Dematteis, G.: Territorialità, sviluppo locale, sostenibilità: il modello sLot. Franco Angeli (2015)

Boeri, S.: Urbania, Editori Laterza (2021)

Hypothesis of Recovery and Redevelopment in the Territories of the Stilaro

Nunzio Bruno Palermo[(⊠)]

University of Mediterranean Studies of Reggio Calabria, Reggio Calabria, Italy
nunziopa@alice.it

Abstract. The proposal concerns the territorial regeneration of a territory in Calabria (Italy) whose history begins from 580 BC. C. up to the early 1900s, a story linked to the mining and processing activities in the foundries in the area. The territory falls between the province of Vibo Valentia and that of Reggio Calabria embracing the Regional Park of the Serre, the industrial archeology sites are now part of nature itself. The territory has many potentials and many resources to be taken into consideration for a hypothetical 'green' territorial development to relaunch both the territory and the villages, on the basis of sustainable tourism with activities and attractions that could create induced activities and jobs. On the basis of what the territory can offer, it is possible to tackle some design themes aimed at sustainable development. The issues touch on some of the objectives of the United Nations Agency 2020–2030 and could be implemented thanks also to the resources of the National Plan for the Recovery and Resilience of Italy. The goal is to adapt the villages and the territory to a Smart Model, considering them points of attraction to create different economies, including renewable energy (micro-hydroelectric, agrovoltaic) to power those that would become communities in full autonomy. Power.

Keywords: Development · Sustainability · Work

1 Introduction

The territory, due to its industrial past, can today be defined as "the cradle of the first national industrialization". The whole area that runs along the river retains very significant remains of its industrial history. The mining exploitation of the resources of the subsoil of the Stilaro dates back to the Greek period, and continued until the fifties of the last century. Favored by mining (limonite, galena, chalcopyrite, molybdenite, etc.), the largest public iron and steel center in the whole of southern Italy was born, the most substantial remains of which can be traced back to the Bourbon period. In just two hundred years (18th–19th centuries), three arms factories were built (the most important ones in Stilo and Mongiana), three foundries and twenty ironworks; Thirty mines were activated between Bivongi and Pazzano and three steel villages were founded. Technological innovation meant that in the late 1800s and early 1900s, in addition to the iron and steel activities, three hydroelectric plants were built, two of which, although no longer

© The Author(s), under exclusive license to Springer Nature Switzerland AG 2022
F. Calabrò et al. (Eds.): NMP 2022, LNNS 482, pp. 2737–2746, 2022.
https://doi.org/10.1007/978-3-031-06825-6_262

active, still exist along the upper Stilaro river; The dams hidden in the vegetation, the hydraulic mills, a flotation plant, together with the foundries and ironworks, complete the cultural heritage linked to the industrial archeology of this area. Today this heritage can be considered an attraction, some sites are being converted into museums, parks, cultural environments, and together with the beauties and natural resources, together with the villages, together with all the potential of the area, it is possible create economies, work, a virtuous community, contribute to reducing CO_2 emissions, creating a series of sustainable scenarios. A first consideration about the regeneration of the area is the evaluation of the contexts both upstream and downstream by evaluating how the elements present (example: Waterfall, Norman Castle, Natural Paths, Historic Villages, Mines, Power Plants, Dams, Foundries) (Fig. 1) can be exploited in a different way and clearly represent strengths. Creating "sustainable attractions" to make the territory known is a possible way to go to revitalize the territory from an economic, social and cultural point of view.

Fig. 1. .

2 The Territory

2.1 Territorial Framework

The territory considered is between the Serre Calabresi and the province of Reggio Calabria, it includes the Stilaro valley with the municipalities of Bivongi, Pazzano, Stilo and Monastrace up to the territories of Ferdinandea and Mongiana (Fig. 2). It preserves its great treasure linked to mining, metallurgical and iron and steel activities which led it to be defined as the "cradle of industrialization in Southern Italy". The valley forms a semicircle around Monte Consolino while the valley floor gives rise to a wide gravel plain. Towards the mountain extends the area of Ferdinandea which is part of the Serre Regional Park, entirely covered with tall woods (fir and beech predominate, but the presence of pine, chestnut, oak, poplar and of other essences), there are testimonies of the industrial past such as the foundry of Ferdinandea, built at the time to support the iron and steel plants of nearby Mongiana. All the territories along the Stilaro river are characterized by sites and natural settings of a certain particularity, by scattered monasteries, and industrial archeology sites immersed in nature that represent a heritage to be exploited for the regeneration of the area.

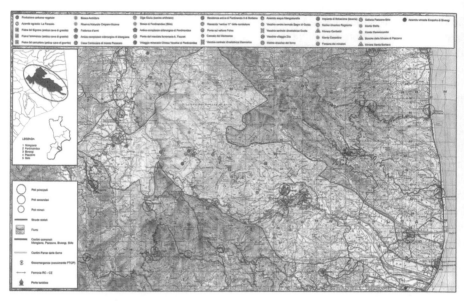

Fig. 2. .

The elements considered are the testimonies of the production processes, the water resource, the vegetation resource, the iron and granite quarries (Fig. 3).

A Production facilities
B The quarries
C Water as an essential source of energy, a natural component
D Flora, a source of energy and an element of naturalistic engineering, soil consolidation and reduction of hydrogeological risk
E Architectural emergencies F Identity and anthropization processes
F Anthropization processes

Fig. 3. .

2.2   Criticalities and Strengths of the Territory

The criticalities in the Stilaro area are those found in many internal areas of Calabria such as the employment situation with modest activity rate (40.1%) and general unemployment (17.1%) and youth (46.1%)) rather high. The area is characterized by a very limited percentage of the population. There is no governance in the management of public transport services. The digital divide is marked for the municipalities furthest from the coast. In the wooded sector, we note: the scarce valorisation of the wood cut product; the inconsistency of the forest-wood supply chain; the failure to approve forest management plans which does not allow for sustainable forest management. There is no governance to manage and make known all the historical cultural heritage and industrial archeology present between Mongiana, Bivongi, Pazzano and Stilo. In addition to the natural beauties that represent a potential, the upstream area is characterized by the enormous amount of water that represents a great resource, and is currently not exploited. The potential of the territory is characterized by numerous factors: The uniform distribution of the protected area of the Regional Park moreover the territory is characterized by a significant biological diversity, also linked to the arid and dry landscapes of the coastal dune formations, which alternate with the Apennine ones of the pine, beech and white fir forests of the Serre Massif. The morphological variety is characterized by the passage from zero altimetric sea level to 1,400 m above sea level. of Monte Pecoraro, in a distance of only 25 km; From the point of view of agri-food production, the quality is high with the production of wine, oil, honey, cheeses, PDO (Protected Designation of Origin) and PGI (Protected Designation of Origin) products. In addition to these factors that characterize the territory, the resources present such as the natural ones, represent strengths for the entire area, together with the historical cultural and artistic heritage which includes the Norman castle, the Byzantine Catholic, religious sites such as monasteries, industrial archeology sites scattered between the mountain and the vicinity of the villages. A heritage that must become an attraction to capture the attention of the tourist. The natural resources that represent a real potential for developing sustainable ideas both upstream and downstream are water, wood from the woods and the sun for the entire valley, to finally arrive at the presence of numerous abandoned buildings scattered both in the historic centers than in other natural areas and which in any case can be recovered and reconverted in a productive way to implement innovative ideas that bring jobs and economies.

3   Strategy

3.1   Strategic Framework

The schematized design strategy (Fig. 4) has three main objectives: to create economies and work, to create a pole of experimentation of innovative ideas, to reduce CO_2 emissions. Based on the environmental contexts and the artistic, cultural, historical, religious, archaeological and industrial heritage, project themes are brought into play to be developed on different scenarios. A sustainable idea that is part of the experience economy concerns the creation of a network of cycle paths, which connect the sea and the mountains by crossing the historic centers of the villages on existing paths, making them passable both with electric bicycles and on foot (Fig. 6). By exploiting the potential of digital technology, and collaborating with the 'Happy Villages' association where the bike-economy operates, this should guarantee a connection with other Italian villages.

The settlement of the inhabited center must be based on the concept of favoring small commercial activities, located within a radius of 500/600 (Fig. 11) meters, with the aim of drastically reducing the number of cars and therefore reducing emissions. At the same time, this approach creates employment in the villages. In the context of the circular economy, the theme of the "energy community" is synonymous with a virtuous local community. Using a natural resource such as water for micro-hydroelectric energy means producing clean energy, reducing CO_2 emissions and using it for the benefit of the territory. Sustainable forestry also aims to use bioenergy to produce clean energy. This takes place within a cooperation that aims to find innovative ideas within a living lab, exploiting digital technology by making the territory intelligent.

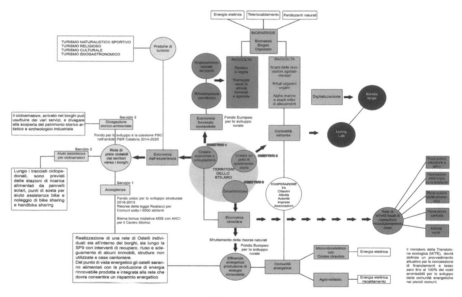

Fig. 4. .

3.2   Financial Instruments

The implementation of the project themes is realized with the investments envisaged by the PNRR (Fig. 5), considering certain financial instruments and making them synergistic at the same time, such as the fund of the Realacci law, the Fund for Development and Cohesion, the European Fund for Rural Development, and other tools that can help make the planned issues concrete. The local strategy takes into account both the data present in the SNAI document (National Strategy for Internal Areas) (Fig. 6) and considers any interventions envisaged by the document and already financed, and the issues present in the SEA document (Strategic Environmental Assessment - Regional Operational Program FESR FSE 2021–2027).

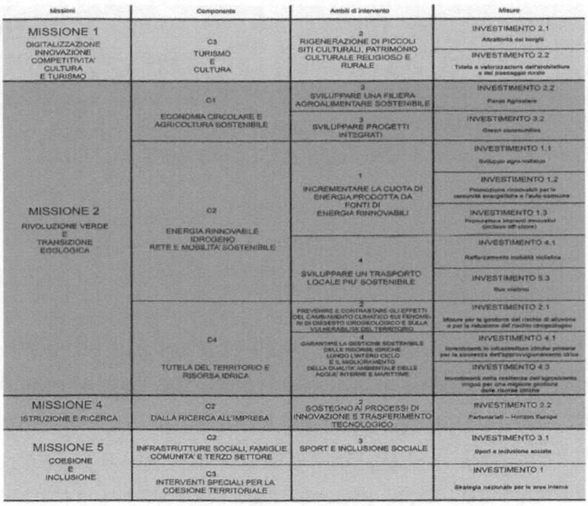

Missioni	Componente	Ambiti di intervento	Misure
MISSIONE 1 DIGITALIZZAZIONE INNOVAZIONE COMPETITIVITA' CULTURA E TURISMO	C3 TURISMO E CULTURA	2 RIGENERAZIONE DI PICCOLI SITI CULTURALI, PATRIMONIO CULTURALE RELIGIOSO E RURALE	INVESTIMENTO 2.1 Attrattività dei borghi
			INVESTIMENTO 2.2 Tutela e valorizzazione dell'architettura e del paesaggio rurale
MISSIONE 2 RIVOLUZIONE VERDE E TRANSIZIONE ECOLOGICA	C1 ECONOMIA CIRCOLARE E AGRICOLTURA SOSTENIBILE	2 SVILUPPARE UNA FILIERA AGROALIMENTARE SOSTENIBILE	INVESTIMENTO 2.2 Parco Agrisolare
		3 SVILUPPARE PROGETTI INTEGRATI	INVESTIMENTO 3.2 Green communities
	C2 ENERGIA RINNOVABILE IDROGENO RETE E MOBILITA' SOSTENIBILE	1 INCREMENTARE LA QUOTA DI ENERGIA PRODOTTA DA FONTI DI ENERGIA RINNOVABILI	INVESTIMENTO 1.1 Sviluppo agro-voltaico
			INVESTIMENTO 1.2 Promozione rinnovabili per le comunità energetiche e l'auto-consumo
			INVESTIMENTO 1.3 Promozione impianti innovativi (incluso off-shore)
		4 SVILUPPARE UN TRASPORTO LOCALE PIU' SOSTENIBILE	INVESTIMENTO 4.1 Rafforzamento mobilità ciclistica
			INVESTIMENTO 5.3 Bus elettrici
	C4 TUTELA DEL TERRITORIO E RISORSA IDRICA	2 PREVENIRE E CONTRASTARE GLI EFFETTI DEL CAMBIAMENTO CLIMATICO SUI FENOMENI DI DISSESTO IDROGEOLOGICO E SULLA VULNERABILITA' DEL TERRITORIO	INVESTIMENTO 2.1 Misure per la gestione del rischio di alluvione e per la riduzione del rischio idrogeologico
		4 GARANTIRE LA GESTIONE SOSTENIBILE DELLE RISORSE IDRICHE LUNGO L'INTERO CICLO E IL MIGLIORAMENTO DELLA QUALITA' AMBIENTALE DELLE ACQUE INTERNE E MARITTIME	INVESTIMENTO 4.1 Investimenti in infrastrutture idriche primarie per la sicurezza dell'approvvigionamento idrico
			INVESTIMENTO 4.3 Investimenti nella resilienza dell'agrosistema irriguo per una migliore gestione delle risorse idriche
MISSIONE 4 ISTRUZIONE E RICERCA	C2 DALLA RICERCA ALL'IMPRESA	2 SOSTEGNO AI PROCESSI DI INNOVAZIONE E TRASFERIMENTO TECNOLOGICO	INVESTIMENTO 2.2 Partenariati – Horizon Europe
MISSIONE 5 COESIONE E INCLUSIONE	C2 INFRASTRUTTURE SOCIALI, FAMIGLIE COMUNITA' E TERZO SETTORE	3 SPORT E INCLUSIONE SOCIALE	INVESTIMENTO 3.1 Sport e inclusione sociale
	C3 INTERVENTI SPECIALI PER LA COESIONE TERRITORIALE		INVESTIMENTO 1 Strategia nazionale per le aree interne

Fig. 5. .

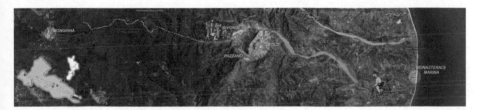

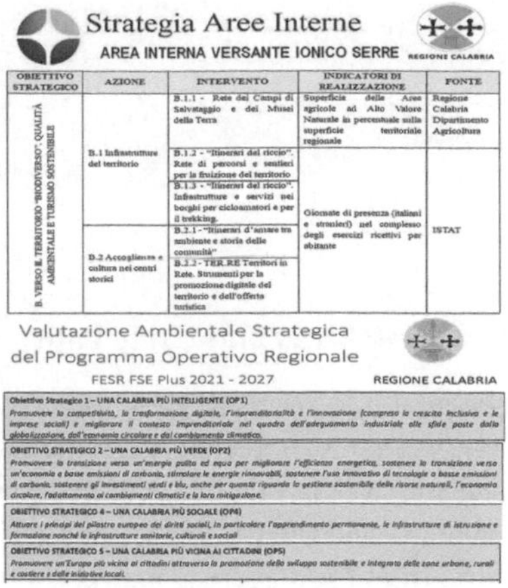

Strategia Aree Interne

AREA INTERNA VERSANTE IONICO SERRE

REGIONE CALABRIA

OBIETTIVO STRATEGICO	AZIONE	INTERVENTO	INDICATORI DI REALIZZAZIONE	FONTE
B. VERSO IL TERRITORIO "BIODIVERSO". QUALITÀ AMBIENTALE E TURISMO SOSTENIBILE	B.1 Infrastrutture del territorio	B.1.1 - Rete dei Campi di Salvataggio e dei Musei della Terra	Superficie delle Aree agricole ad Alto Valore Naturale in percentuale sulla superficie territoriale regionale	Regione Calabria Dipartimento Agricoltura
		B.1.2 - "Itinerari del riccio". Rete di percorsi e sentieri per la fruizione del territorio	Giornate di presenza (italiani e stranieri) nel complesso degli esercizi ricettivi per abitante	ISTAT
		B.1.3 - "Itinerari del riccio". Infrastrutture e servizi nei borghi per cicloamatori e per il trekking.		
	B.2 Accoglienza e cultura nei centri storici	B.2.1 - "Itinerari d'amare tra ambiente e storia delle comunità"		
		B.2.2 - TER.RE Territori in Rete. Strumenti per la promozione digitale del territorio e dell'offerta turistica		

Valutazione Ambientale Strategica

del Programma Operativo Regionale

FESR FSE Plus 2021 - 2027

REGIONE CALABRIA

Obiettivo Strategico 1 – UNA CALABRIA PIÙ INTELLIGENTE (OP1)

Promuovere la competitività, la trasformazione digitale, l'imprenditorialità e l'innovazione (compreso la crescita inclusiva e le imprese sociali) e migliorare il contesto imprenditoriale nel quadro dell'adeguamento industriale alle sfide poste dalla globalizzazione, dall'economia circolare e dal cambiamento climatico.

OBIETTIVO STRATEGICO 2 – UNA CALABRIA PIÙ VERDE (OP2)

Promuovere la transizione verso un'energia pulita ed equa per migliorare l'efficienza energetica, sostenere la transizione verso un'economia a basse emissioni di carbonio, stimolare le energie rinnovabili, sostenere l'uso innovativo di tecnologie a basse emissioni di carbonio, sostenere gli investimenti verdi e blu, anche per quanto riguarda la gestione sostenibile delle risorse naturali, l'economia circolare, l'adattamento ai cambiamenti climatici e la loro mitigazione.

OBIETTIVO STRATEGICO 4 – UNA CALABRIA PIÙ SOCIALE (OP4)

Attuare i principi del pilastro europeo dei diritti sociali, in particolare l'apprendimento permanente, le infrastrutture di istruzione e formazione nonché le infrastrutture sanitarie, culturali e sociali

OBIETTIVO STRATEGICO 5 – UNA CALABRIA PIÙ VICINA AI CITTADINI (OP5)

Promuovere un'Europa più vicina ai cittadini attraverso la promozione dello sviluppo sostenibile e integrato delle zone urbane, rurali e costiere e delle iniziative locali.

Fig. 6. .

4 Project Themes

The territories both downstream and upstream are connected by a cycle-pedestrian (Fig. 7) path that starts from the coast to reach the hinterland and crosses areas where it is possible to hypothesize a design theme, to be developed on the basis of what are the needs and objectives foreseen.

Fig. 7. .

4.1 Bike Sharing and Handbike Sharing

The shared electric bicycle is a service that allows sustainable mobility for all people on a network of cycle and pedestrian paths. A different service is given by the introduction of Handbike sharing. It is a special service which meets the needs of the handicapped. The main objective of the project is to provide disabled people with a service equal to that of bike sharing (Fig. 8). The vehicles used will be like classic bicycles that are equipped with assisted pedaling, obviously have different characteristics, as they are equipped with three wheels and a particular mechanism that allows you to pedal with your hands.

Fig. 8. .

4.2   Rest Areas and Cycle Paths

The goal is to combine sport, nature, culture and socialization between able and disabled people, in one direction: to be the same in particular situations. Hence the hypothesis of inserting a network of cycle-pedestrian (Fig. 9) paths with adjacent signs, for cyclists, amateurs and the disabled, with vehicles such as handbikes on tracks within a path that starts from Monasterace marina. Along the routes there are services such as refreshment points with bike charging stations, powered by solar energy, welcome service in the villages, and bike maintenance service.

Fig. 9. .

4.3   Mines Park

The mines of Pazzano and Bivongi (Figs. 3, 10) are realities of a past history that must emerge, through a reconversion of every single area into what could be defined as the Mines Park (Fig. 10) of an urban center with the addition of a series of services for the visitor, enhancing the importance of the place, promoting mobility and social rebirth.

Fig. 10. .

4.4   Reduce CO_2 Emissions

New Approach, Based on the Concept of Favoring Small Commercial and Repair Businesses (Fig. 11).

Fig. 11. .

Renewable Energies: Micro Hydroelectric, Agrovoltaic and Bioenergy

The territory would allow the production of renewable energy that can be used as an integration to the network in order to guarantee certain services for sustainable mobility or a green reception service. Solar energy and the kinetic force of water guarantee the production of renewables organized within an energy community. The basic idea of agrovoltaics is to ensure that agricultural land can be used to produce clean electricity, leaving room for agricultural crops. Micro hydroelectricity, given the abundance of water, could be an alternative through a small plant along the Stilaro river (Fig. 12).

Fig. 12. .

An energy biomass (Fig. 13) power station fed with urban organic waste, waste from agri-food processing, wood residues from the forest and branches from forestry processes produces clean electricity, guarantees district heating and produces natural fertilizers for agricultural land.

4.5   Marmarico Museum

He old hydroelectric power station of Marmarico (Figs. 3, 14) is currently abandoned to the right of the route that tourists travel on to go and see the waterfalls, a unique attraction in the area. Converting the structure into a museum (Fig. 14) or exhibition environment where the culture of the area will be exalted at the base could be an alternative to restore a degraded structure in complete abandonment.

Fig. 13. .

Fig. 14. .

4.6 Youth Hostel

The presence on the territory of abandoned structures (Fig. 3) gives rise to the idea of reusing them in a tourism key to promote eco-sustainable tourism, creating a network of hospitality hostels (Fig. 17) that will be managed by associations or cooperatives. Each hostel will be characterized by the use of the amount of renewable energy provided, with the presence of timed sensors for the light.

Fig. 15. .

5 Conclusions

The study carried out revealed the importance that these territories have had in past centuries, and the potential that exists for a sustainable regeneration of the entire area. All the topics covered are part of a strategic panorama for the achievement of certain objectives. The strategy is based on a series of targeted interventions and actions to reverse a deadlock. In the case study of the Stilaro territories, some design hypotheses could be a driving force for creating development, work, knowledge, attraction and investments. The theme of renewable energy leads to exploiting one of the natural potential of the area which is the water resource, producing clean energy in the area. The aim is also to guarantee the basis for attracting every person to know and experience the villages, the historical places of the industrial past, the traditions that are highlighted and valued in a different way. In the urban context, changing the layout of the village means creating an economy based on maintenance and on the concept of encouraging small repair activities, while in natural contexts the goal is to create an economy based on attraction and services for tourism. of experience and local sustainability. This strategic panorama in the context of cooperation between citizens, organizations, associations, companies, brings different territorial scenarios and dynamics compared to a situation characterized by depopulation and lack of jobs and economies. It is necessary to consider each financial instrument and make them synergistic, in order to implement the issues addressed and carry out the planned interventions. The creation of a virtuous community is essential to continue along the path of cooperation for the development of the territory, giving life to innovative ideas within a possible pole of experimentation.

References

1. De Stefano Manno, B., Matacena, G.: Le Reali Ferriere ed Officine di Mongiana. Storia, condizioni di lavoro, tecnologie, prodotti, trasformazione del territorio e architettura delle antiche e più importanti fonderie del Regno delle Due Sicilie. Rubbettino, Soveria Mannelli (2008)
2. Rubino, G.E.: Archeologia Industriale e Mezzogiorno. Problemi di architettura degli impianti, tecnologia, società. Mario Giuditta, Isola di Capo Rizzuto (1978)
3. Matacena, G.: Architettura del lavoro in Calabria tra i secoli XV e XIX. s.l., Edizioni Scientifiche Italiane, Napoli (1983)
4. Franco, D.: Il ferro in Calabria. Vicende storiche-economiche del trascorso industriale calabrese. Kaleidon, Reggio Calabria (2003)
5. Aragona, S.: Costruire un senso del territorio. Spunti, riflessioni, indicazioni di pianificazione e progettazione. Gangemi, Roma (2012)
6. Bevilacqua, F.: Il Parco delle Serre. Guida naturalistica ed escursionistica. Rubbettino, Soveria Mannelli (2002)
7. Bevilacqua, F.: Montagne della Calabria. Guida naturalistica ed escursionistica, Rubbettino, Soveria Mannelli (2003)
8. Franco, D.: Le reali fabbriche del ferro in Calabria. Rubbettino, Soveria Mannelli (2020)

Small Rural Towns and Farmhouses of the Opera for Valorizzazione of Sila in Calabria. Narrated Memory from the Past and the Present, Research for Possible Sustainable Scenarios

Maria Rossana Caniglia(⌨)

University of Messina, Messina, Italy
mariarossana.caniglia@unime.it

Abstract. With the enactment of the Sila law in 1950 as a first step towards that land innovation, so much invoked in the years after the Second World War, intended to respond to the precarious economical condition and the backwardness of the agricultural system, the agrarian revolution in Calabria began. While performing infrastructural works and allocation of the farm lands, the institution of the Opera for the Valorizzazione of Sila (OVS) was also constructing the settlement system: the small rural towns and farmhouses.

As for the farmhouses, arranged according to the settlement of the "scattered house", 16 housing types have been defined with respect to the climatic, morphological, landscaped and architectural parameters. Almost simultaneously, the projects of the 29 small rural towns, built between 1951 and 1958, with the exception of those of Turrutio and San Leonardo di Cutro, both in the province of Crotone, were entrusted to the same officials of the OVS technical office, the first one to Saul Greco and the second one to Giovanni Astengo.

It is necessary to reflect about those agrirarian and architectural actions carried out within the Reformation in the fifties, which have not always changed the outlook and the skyline of the Calabrian landscape, but the signs of which are still strongly present and perhaps even the consequences of a late intervention of the social and economical redemption. Today, it is necessary to think about the possible future scenarios for the cultural heritage of the small rural towns system, with an objective to rediscover their identity and peculiar memory, by defining new production models oriented towards sustainability, protection, valorization and safeguard and by investing, in particular, in more efficient and circular production processes.

Keywords: Small rural towns · Sustainable scenarios · Calabria

1 Introduction

«La riforma agraria è cominciata in Calabria […] non perché è avvenuto questo o quello, ma perché non poteva non cominciare, perché – come dice appunto la misteriosa e

© The Author(s), under exclusive license to Springer Nature Switzerland AG 2022
F. Calabrò et al. (Eds.): NMP 2022, LNNS 482, pp. 2747–2756, 2022.
https://doi.org/10.1007/978-3-031-06825-6_263

affascinante espressione biblica – i tempi erano maturi» [1, p. 143] this is what Manlio Rossi Doria (1905–1988), technical advisor at the Opera for Valorizzazione of Sila, wrote, after the enactment of the Sila law has determined the «colpo di rottura» [1, p. 147] of the agricultural reality of the Calabrian latifundium [2].

The institution of the Opera for Valorizzazione of Sila (OVS) was established with the law of 31 of December, 1947 no. 1629, with the goal of promoting and implementing the land and the agricultural transformation plan of the Silan Plateau, and of constructing the infrastructures necessary for economic development and for tourist and industrial enhancement. Although the birth of the OVS represented an act of progress, it was limited to an administrative body whose powers were insufficient to carry out effective planning for implementation of the predetermined transformations.

Since the Second World War, the country in Southern Italy were already facing strong social and economic tensions due to the precarious condition of the agricultural system. In Calabria, these tensions lead to peasant struggles and spontaneous movements of workers ready to fight for the occupation of the uncultivated lands, in particular with the events of Melissa which took place in the summer of 1949 in the Fragalà area, in the province of Crotone.

The implementation of the Agrarian Reformation in Calabria was ratified with the Sila law of 12 of May 1950 no. 230 and ensured the expropriation of properties that exceeded 300 hectares, while the only requirement was to be "susceptible to transformation". It also entrusted to the OVS both «il compito di provvedere alla ridistribuzione della proprietà terriera e alla sua conseguente trasformazione, con lo scopo di ricavarne i terreni da concedersi in proprietà a contadini» [3, p. 157] and the organisation of «servizi di assistenza tecnica ed economica-finanziaria per gli assegnatari. [...] promuovere, incoraggiare, ed organizzare corsi speciali gratuiti di istruzione professionale» [3, p. 159].

The Silano-Jonico district, a territory defined by Article 1 of the Sila law to which the transformations program was applied, was divided into what Rossi Doria defined as the four realities of the Agrarian Reformation, socially, economically and geomorphologically homogeneous areas: «1°) quella dell'Altopiano Silano; 2°) quella delle terre latifondistiche asciutte del Marchesato di Crotone; 3°) quelle delle terre di futura irrigazione della Bassa Valle del Neto ed in parte della Piana di Sibari ed infine 4°) quella vastissima e varia, tra collinare e montana, che gira tutto intorno all'Altipiano silano, da Squillace alla Sila Greca e alle pendici orientali della Piana di Sibari» [1, p. 156].

At first, the "transformation" was focused on the construction of infrastructures and on the organization of the services which were necessary for the development of the agricultural production. But «il coronamento [...] è stato rappresentato, senza dubbio, dallo sforzo di ridimensionamento degli insediamenti rurali, [...]. E, infine, si è proceduto alla costruzione di borghi e centri, che integrassero i servizi degli abitanti persistenti, oppure li fornissero ex novo ai nuovi insediamenti» [4, p. 208].

In the first six years of the implementation of the law, the Works for the Valorisation of Sila intervened on approximately 84,000 hectares distributed among 19,148 families. The institution undertook various hydraulic and forestry works, construction of country roads, land reclamation and forest paths, the construction of about 4,107 farmhouses and fifteen small rural towns and service centers, in addition to rural aqueducts, sheds, cereal storage warehouses, oil mills and vocational school buildings.

On 11 of November, 1955, on the session of the Senate Special Commission for the examination of laws concerning the extraordinary measures for Calabria, Senator Umberto Zanotti Bianco (1889–1963) declared that: «l'opera compiuta è stata talmente grandiosa che ha cambiato il volto di vaste zone della Calabria, in maniera tale che in alcuni punti mi pareva di essere nell'Italia centrale o del Nord. Dove prima c'era un latifondo con pochino di grano stento (lo ricordo perché sono stato molte volte in quelle zone) ora vi sono terreni floridi. Il marchesato di Crotone non si riconosce più…Si tratta di opere grandiose, non possiamo negarlo!» [4, p. 209].

2 In the Housing Project: Small Rural Towns and Farmhouses

In 1952, the Opera for the Valorizzazione of Sila began to implement the first land transformation plans, but during the process, the question of what characteristics the housing system should have was raised. The choice oscillated between a scattered settlement, positive for an agricultural management determined by the farmhouses factor and a centralized one, an instrument for developing a greater social life and reducing the costs for construction of services. In the end, the preferred settlement was the one of the "scattered house", defined as an ideal, self-sufficient model for a resident peasant family, adopted in most of the area, except for the lands in the mountain areas, where the preferred settlement was the "semi-centralized" one. Indeed, scattered houses can be found in the flat areas near the Crati, Neto and Tacina rivers, where the land was easily irrigated and allowed a rapid growth of crops, as well as in the Isola Capo Rizzuto area, with the exception of some centralized settlements such as the one in Cutro. Even in the hilly areas, where the splitting of the terriory was more extensive, scattered settlements were preferred and the criterion of groups of 3–4 houses placed on the crossroads of farm and country roads was adopted only in exceptional cases. In the mountain areas of the Silan Plateau and Apollinara, in the province of Catanzaro, centralized settlements or large residential areas can be spotted, to contrast the excessive isolation and accustom the population to the new housing conditions [5].

Each farm land was defined by a more or less regular grid and the dimension varied with respect to the characteristics of the concerned area. In the flat area, the farm could vary from 2 to 4 hectares, in the hilly area from 4 to 6 hectares, in medium-productive land, especially those of the Marquisate of Crotone, 6 hectares and, on the Silano Plateau, up to 8 hectares. Afterwards, additional integrative quotas were assigned, to enable the assignment of small additional parts be added to the pre-existing properties [6].

The institution in charge of building the housing system was expected to take into consideration various functional requirements that the farmhouse was supposed to have with respect to the agricultural production, as well as to the climatic and morphological conditions of the four homogenous zones. Furthermore, a specificity of an equal importance, the "architectural" monotony was supposed to be avoided, while respecting the characteristics of the landscape. Actually, five typologies of two floor houses were defined for the mountain area, while eleven typologies of one or two floor houses were defined for the hilly and flat areas.

On the Silan Plateau, the houses were built on the flat zones and on the slopes of the Neto and Savuto Valleys, according to the typologies of *Mucone, Cecita, Rovale* and Lorica. All of them are on two floors and differ one from the other by the dimension, the form, the organization of the space and by the types of services attached (stable, barn, warehouse, porch etc.). The construction modes were taking into consideration the climate, while paying attention to the termical protection of the vertical structures (an inner tube with perforated partitions, distanced from the walls 8–10 cm) and of the covers (bricks and reinforced concrete with one layer of Marseillais tiles), as well as to the technical standards for the seismic zones. For the Marquisite of Crotone, a hilly and flat zone, the *Cassano* typology has been selected, with a square plan on two levels, and housing and services connected with an external staircase and the *Fedula* typology, bigger than the first one, with the stable placed outside of the house. As for the territory between the plain of Sibari and the ionic hill, caracterized by repetitive geomorphological changes, the combination of more typologies was necessary. For certain houses, the *Cassano* and *Fedula* typologies were suggested, while for the hilly area a new typology was selected *Bisignano*, a farmhouse with brick walls and with pitched roof (all the materials used are traditional and found on site), constructed on two levels connected with an internal staircase and which could accommodate up to seven people.

Once the construction of the farmhouses began and the working activities increased, the need to build small rural towns arose, the necessary reference points for the new farmers scattered between the farms, where the civil and the social services were to be concentrated.

The selected location for the construction of each small rural town was not accidental, but had to respond to the fundamental parametars such as: location, accessibility and morphology of the territory as well as the exposition. The location usually was close to the main road, on a distance not greater than 500 m and often connected to the inter-farm road. The zone of influence of each small rural towns was calculated based on a radius of 2 to 2,5 km, the maximum distance to reach school or church. These distances were varying, depending on the morphological and climate conditions of the territory.

The general planimetric composition of each small rural town included a church with the parsonage, elementary school and kindergarden with accommodations for the teachers, an ambulance with an accommodation for the doctor, the offices of the municipal delegation and post offices, as well as a grocery store. In certain cases, there were also police station, local artisan shops and taverns. If the distance between two small rural towns was too big and if the radius of a certain small rural town was not reaching some residences, which, therefore could not use the necessary services, the Minor small rural towns or Service Centers were built. Those were different from the former by the number and the dimensions of the buildings, but they were always including a church, an elementary school with an accommodation for the teachers and a small food corner. One variant of the Service Centers are the Detached nuclei, which included a small church and a school.

The projects of the 29 small rural towns, built between 1951 and 1958, with the exception of those of Turrutio and San Leonardo of Cutro, both in the province of Crotone, were entrusted to the same officials of the OVS technical office, the first to Saul Greco and the second to Giovanni Astengo.

The location of the small rural towns in the four homogenous zones of the Silano Jonic region was planned as follows: in the Silano Plateau were planned Caporose, Quaresima and Torre Spineto in Aprigliano (Cs), Cagno-Ceraso and Germano-Serrisi of San Giovanni in Fiore (Cs), Camigliatello Silano and Neto di Ferrara in Spezzano Albanese (Cs), Cecita-Lagarò in Celico (Cs), Lorica in Pedace (Cs), Pantano-Racisi in Taverna (Cz). In Marquisate and in the Valle of the Neto, the small rural towns were Apriglianello, Bucchi, Domine Maria, Salica on the territory of the city of Crotone, Turrutio in Corazzo of Scandale (Kr), Rosito and San Leonardo di Cutro in Cutro (Kr), Campolongo, Sant'Anna, Soverato, Stumio in Isola Capo Rizzuto (Kr), Torre Melissa in Melissa (Kr), Armirò in Santa Severina (Kr). On the other hand, in the plain of Sibari, six small rural towns were planned, all in the provincial territory of Cosenza: Apollinara and Fabrizio in Corigliano, Lattughelle in Cassano Jonio, Mirto scalo in Mirto Crosia, Pietrapaola Scalo in Pietrapaola, San Lorenzo del Vallo in San Lorenzo del Vallo.

2.1 San Leonardo: The Unexpected Small Rural Town of the Agrarian Reformation

In 1928, Hélène Tuzet (1902–1987), a young graduate in literature, was sent to Calabria and Sicily by the Laura Spelman Rockefeller Foundation to verify the activities carried out by the schools of the National Association for the interests of Southern Italy (ANIMI), in relation to the social environment in which they were located. From her narrations that also enhance the beauty of the Calabrian cities and landscape, one wishes to dive deep into the discovery of that part of mysterious and still wild territory. This impression will lead Tuzet to visit San Leonardo too.

> «I dintorni del paese sono suggestivi: vi si accede per un monumentale viale di pini marittimi. [...]. Questo villaggio è uno dei più miserabili d'Italia ed appartiene a un grande latifondista, il barone Barracco, [...]. La parte più antica del borgo si compone di tuguri bassi disposti per quattro in file dritte a formare un quadrato completamente chiuso, dove si accede per una porta ad arco, al centro del quale c'è la famosa torre del tesoro. Chi può dire il segreto di questa disposizione? Alle spalle del fortino, altre due file di case, sempre in linea retta, molto distanziate tra di loro. Una di esse, include il castello del signore locale, la posta, una o due casette nuove dipinte di rosa di proprietà di contadini agiati. [...]. Le famiglie [...] vivono stipate in locali dove la pulizia è pressoché impossibile; e il barone è favorevole alla costruzione di nuove case [...]. La piccola scuola è tutto il superfluo, la sola cosa fatta unicamente per soddisfare i bisogni dello spirito [...], che secondo i contadini sembra una chiesa. Di chiese, d'altronde, gli abitanti non ne hanno altre: è tanto tempo che quella vera è chiusa e nessun prete si premura di venire a dire messa in un posto così sperduto» [7].

In the fifties, the Marquisate was the epicenter of the agrarian reformation, because, with respect to the rest of the region, it was the "reality" in which the first and decisive actions of "rupture" marked the real beginning of the land transformation. In fact, the OVS proceeded with the implementation of massive and numerous expropriations in the Valle of the Neto and on the plateau of Isola di Capo Rizzuto which determined an intense land distribution and the abandoned territory of San Leonardo di Cutro is the most interesting example of its application. Its clayey lands «fra le vallate del Dragone e del Purgatorio» [5, p. 181] were the quite desolate, but the presence of a spring and a large olive grove around the houses of the workers of the baronial families favored its expropriation. Thus, each beneficiary was assigned «un ettaro di oliveto e il resto (3 o 6 ettari) in terreni coltivabili sulle superfici ondulate che scendono verso il mare» [8].

When the construction of the thirteen previously planned small rural towns began in the Punta delle Castella-Capo Colonna reclamation consortium, San Leonardo did not exist. Manlio Rossi Doria did not agree with this decision, on the contrary, he was particularly interested in arranging of the current planimetric layout and in improving the poor hygienic and social conditions of the inhabited center. In fact, in July 1951 after a continuous exchange of letters with Rossi Doria, Giovanni Astengo (1915–1990) himself arrived in San Leonardo, accompanied by Giuseppe Samonà (1898–1983).

On the first inspection, Astengo verified whether the «case a schiera non addensate (apparivano) facilmente migliorabili» [9] and, at the same time, chose the location for the necessary structures (school, office, ambulance-pharmacy, shops and covered market) to transform the small center into a service small rural town. The next year, the architect prepared the transformation plan which included: arrangement and extension of the existing barracks near the olive grove, in new housings with a large vegitable garden; the re-arrangement of Vaglio (a settlement founded by the Jesuit Fathers in 1957, and from 1834 in property of the baron family Barracco), with an intention to redefine the shape of the yard with the two long sides intended for accommodation and the third, where a small pre-existing church was located, was intended for the parochial offices and other collective spaces.

For the arrangement of the core of services, Astengo chose the free place between the square on the left of Vaglio and the one along the entrance street. Planimetrically, the core was formed from four linear elements connected by a porch, but in the next phase, from 1954 and 1955, it went through significant changes and additions.

The block of the buildings, at first homogenous and compact, was comprised of five elements, out of which, the elementary school and the ambulance-pharmacy were distinguishable by their artichetural plan and hierarchical typology. The school included a large aula used as a theater, a space for three didactic aulas and another one, smaller, for the offices; the ambulance-pharmacy, instead, was located in a small room.

San Leonardo was a laboratory where Astengo was working on and was experimenting with an organic architectural language. The design of the new small rural town included a special articulation and an expressive heterogeneity of all buildings which were supposed to respond to the basic characteristics defined by him: functional, constructional and economic essentiality [10].

2.2 Turrutio Small Rural Town in the Countryside of Scandale

Between 1952 and 1954, the architect Saul Greco built the small rural town Turrutio, located in the hilly zone of the countryside of Scandale, along the main road Corazzo-Scandale, in the province of Crotone. The small rural town was intended to host around 1500 residents distributed in various farmhouses on a territory of about three thousand hectares.

The first drawings of the planimetry included a very ambitious project, designed for the inside of a rectangular area of about 56.000 square meters, well articulated and complex, that was not completely implemented. This failure of implementation, is, in particular, due to the ideology of the Reformation which affected the entire project. The small rural town that we see today is profoundly different, the general plan has been modified, but one can still recognize the special quality and the linguistics of certain buildings, built according to the original drawings of the architect.

The church, the most representative building and above all, the most original from architectural and typological point of view, is comprised of an aula, a central octagonal floor plain (11 m on 11 m), designed to host around 200 people, and an elongated body for the altar and the parochial offices. The choice of a central floor plain in that context was a real "experimental" exercise, especially because of the various solutions adopted, despite their extreme simplicity. In the octagonal space, also highlighted by the strengthened concrete structure that supports the roof, we see two entrances, both connected by a small base, one on the main façade and the other one on the right profile. Externally, the white mural mass is interrupted by small openings on the upper side, that perhaps were supposed to simulate the shutter drum of the dome. The bell tower, located in the back side, was built with a metal lattice; the parsonage, which also includes the priest's parish was located next to the church, connected with it by a sacristy room. The school building is represented by a single body divided into three branches where the elementary school, the kindergarden and other services are located. The main entrance, from where different areas can be reached, has a role of a connection element and a convergence point for the whole complex. The two anterior bodies are placed on a single floor, above ground, the posterior one being articulated on two levels and distinguishable for the porch and the accommodation for the teacher. The topic of the porch was reproposed by the architect as well on the opposite side of the square, to connect the market building with the one of the tavern [11].

«L'effetto d'insieme è, quindi affidato al movimento dei volumi, al gioco di chiaroscuro dei vuoti e dei pieni e alle colorazioni delle masse opportunamente ricercate per ottenere particolari accordi di toni col verde della vegetazione e l'azzurro del cielo» [12].

3   Narrated Memory and/or Research for Possible Sustainable Scenarios?

«Il territorio, sovraccarico com'è di tracce e di letture passate, assomiglia piuttosto a un palinsesto. […]. Ciascun territorio è unico, per cui è necessario 'riciclare', grattare una volta di più (ma possibilmente con la massima cura) il vecchio testo che gli uomini hanno inscritto sull'insostituibile materiale del suolo, per deporvene uno nuovo, che risponda alle esigenze d'oggi, prima di essere a sua volta abrogato» [13, p. 27].

The narrated memory has the cultural heritage as a protagonist, a living testimony, precious and undeniable, both an architectural or an environmental good. The landscape, of all the cultural heritage, is the one that contains in itself the bond between man and nature. Actually, it is essential to underline that the landscape is a result of multiple spontaneous processes and changes, where the humans were almost constantly alternating the actions of cancelling with those of rewriting of new layers, a continuous o Silation between the harmony and the conflict, between the sence of belonging and opposing. All this leads us to define the landscape both as a "product", an object of construction, and as a "project", because the «dinamismo dei fenomeni di formazione e di produzione prosegue nell'idea di un perfezionamento continuo dei risultati, in cui tutto è correlato» [13, p. 24].but where the temporal duration, very often, escapes the control of the human. To this dynamism corresponds the semantic narration of the landscape, a collection of images that retells the evolutionary paths and the possible transformations, both those from the past, that contributed to the creation of the identity of the place and those future ones, that have a task to corroborate the realization of new projects.

The different forms of settlement and the development of the transformation activities on the territory actually represent an expression of different social organizations, which include various aspects (from innovation of the used technologies to ensure the survival to the evolution of the collective dynamics). The final result is an overlap of *layers* which determine another "palimpsest" (or more palimpsests), based on the integration of the plurality of potential resources, which, with a proper selection of elements and possible dynamics, can be transformed into real future scenarios.

Reflections that have a special value if applied to the agrarian and architectural actions from the Agrarian Reformation, which did not always change the outlook and the skyline of the landscape, from which some signs remain and perhaps also the consequences of a late intervention of social and economic redemption «ci si è adagiati, [...], in una concezione dello sviluppo economico precedente per compartimenti-stagni, per fasi successive nettamente separate e distinte; quando (ancora peggio) non si è pensato che la valorizzazione della locale agricoltura rappresentasse l'unica prospettiva di sviluppo che si potesse o dovesse perseguire» [7, p. 224].

Today, walking along what once were the borders of the Silano-Jonic District, and particularly of Marquisite of Crotone, it is possible to observe the ruins of the numerous farmhouses, some of which were never inhabited and others abandoned already few years after the construction (Fig. 1). This was due to the effects of the new industrial production models, to the lack of services, to the socio-cultural conditions and to the gradual migration of the population towards the newly built centers. Even if those houses still define the landowner identity landscape, they can be defined as a contemporary archeology, frozen in a space-time dimension. An analysis which leads to two considerations: the first is the duration aspect, as a key element of whatever thought connected to sustainability; the second is the consciousness that a new "ruin" exists.

Regarding the small rural towns, the reflections are different. Among the 29 which were built, some were incorporated in the urban landscape of the neighboring centers and are quite difficult to recognize because they lost their original outlook; others were abandoned; and, the houses of the small rural towns of San Leonardo (today a fraction of Cutro) (Fig. 2) and Turrutio (fraction of Scandale) (Fig. 2) continue to play a role of social connection between the years of their foundation and today, first as a reference place for the farmers scattered on the farms and today as small urban centers.

Fig. 1. The abandoned farmhouses in the territory of San Leonardo di Cutro (photo F. Scarpino, 2019).

Fig. 2. Small rural towns today, left San Leonardo di Cutro (photo F. Scarpino, 2019), right Turrutio (photo M.R. Caniglia, 2017).

The antropic impact of the environment and the related changes, elements of sometimes unstable balance, lead us to rethink new models of production oriented towards sustainability, protection, valorization and safeguard and in particular of investments into more efficient and circular production processes. It is, therefore, necessary to consider possible future scenarios for the cultural heritage of the small rural towns system, and in particular, those of San Leonardo and Turrutio.

In 2017 in San Leonardo was inaugurated the "Museo della Cultura Contadina", sponsored by the Cultural Association Radici in collaboration with the town hall of Cutro, an exhibition space dedicated to preserve and enhance the identity of the territory of the small rural town and the Marquisate of Crotone. The building chosen is the school designed by Astengo, perhaps the most representative of that transformation that in the fifties had defined the new village and, today once again could express that added value in implementing the "design" of all those economic, social and environmental actions necessary. This action has started a path of participatory planning, involving the inhabitants (main resource of knowledge and competence) in a shared survey and, activating a propositional relationship-interaction of comparison elaborating proposals and solutions. However appreciable this example remains isolated and perhaps not completely up to "expectations", both for financial reasons and for the lack of a structured planning of interventions.

The objective is to rediscover the memory of the identity of these small rural towns, by recognizing their "permanent" and "peculiar" characters, by defining a strategy for conscious development. It is necessary either to reconstruct spaces where the correlation between the material and the immaterial elements will be an added value, or to define a society able to recognize an equilibrium between the production, the consumption and the environment, necessary for a circular model made by synergies and good practices.

References

1. Rossi Doria, M.: La riforma agraria in Calabria e l'Opera per la Valorizzazione della Sila, estratto dagli Atti dell'Accademia dei Georgofili. Tipografia Giuntina S.A., Firenze (1950)
2. Caniglia, M.R., Passalacqua, F.: La Riforma agraria degli anni Cinquanta in Calabria. Conoscenza, conservazione e trasformazione del paesaggio del Marchesato di Crotone. In: Mistretta, M., Mussari, B., Santini, A. (eds.) La Mediterranea verso il 2030. Studi e ricerche sul patrimonio storico e sui paesaggi antropici, tra conservazione e rigenerazione. ArcHistoR Extra, 6, supplement of «ArcHistoR», vol. 12, pp. 94–109, VI (2019)
3. Seronde, A.M.: La riforma agraria. In: Meyriat, J. (ed.) La Calabria. Lerici editori, Milano (1961)
4. Galasso, G.: Latifondo, lotte per la terra e riforma agraria nel Marchesato di Crotone. In: Arlacchi, P. (ed.) Territorio e società. Calabria, pp. 1750–1950. Lerici, Cosenza (1978)
5. Rogliano, G.: La casa rurale nel comprensorio dell'O.V.S. Tipografia eredi Serafino, Cosenza (1962)
6. Celani, S.: L'intervento pubblico in agricoltura nel Mezzogiorno dal 1950 ad oggi, con particolare riguardo alla Calabria. In: De Martino, U. (ed.) Pianificazione urbanistica delle aree agricole, pp. 59–60. Gangemi editore, Roma (1986)
7. Napolitano, S. (ed.): Hélène Tuzet-Jules Destrée. In Calabria durante il fascismo. Due viaggi inchiesta, pp. 41–42. Rubbettino, Soveria Mannelli (2008)
8. Carratelli, O.: La povera gente del Marchesato. L'illustrazione italiana 46, 17 (1949)
9. Dolcetta, B., Maguolo, M., Marin A.: Giovanni Astengo urbanista: Piani progetti opere, p. 190. Il Poligrafo, Padova (2015)
10. Caniglia, M.R.: San Leonardo di Cutro nel Marchesato di Crotone: conoscenza narrata tra passato e presente, ricerca per possibili scenari futuri. In: Capano, F., Visone M. (eds.), La Città Palinsesto. Tracce, sguardi e narrazioni sulla complessità dei contesti urbani storici, tomo primo. Memorie, storie, immagini, pp. 499–506. FedOA-Federico II University Press, Napoli (2020)
11. Caniglia, M.R.: Saul Greco. Un architetto contemporaneo in Calabria. In: Martorano, F. (ed.) L'architettura in Calabria dal 1945 ad oggi. Selezione delle opere di rilevante interesse storico-artistico, pp. 297–305. Iiriti, Reggio Calabria (2020)
12. Medici, R.F.: Architettura rurale: esperienze della bonifica, p. 230. Edizioni agricole, Bologna (1956)
13. Corboz, A.: Il territorio come palinsesto. Casabella vol. 156 (1985)

Ecopolis vs Megacity: A Post-crises Regional-Urban Vision Towards 2050

Sandro Fabbro[✉]

University of Udine, 33100 Udine, Italy
sandro.fabbro@uniud.it

Abstract. The paper is concerned with a reflection on spatial visions capable to:

- overcome the territorial inequalities and criticalities that have exploded as a consequence of the recent interwoven global crises (environmental, economic, pandemic and of social justice);
- pursue the so called "ecological transition" towards 2050.

The paper supports the idea that the too big urban agglomerations (and in particular the so called "megacities"), exploded in number during the last decades in the Far and in the Middle East, are at the origin of the current interwoven crises.

Megacities must be decidedly abandoned in order to return, as soon as possible, to an ecological and decentralized city as had already been claimed, in the twenties of the last century, in the thought and action of the planners of the Regional Planning Association of America (RPAA). With reference to this alternative model, the European vision of "Ecopolis", elaborated by Espon in 2012, is recovered together with the "ecopolitical" vision of Edgar Morin (2020) and adapted to the conditions arisen, in particular in the European context, due to the pandemic.

With reference to the Italian situation, the ideal-type of Ecopolis could be introduced, with the name of "Area Ecopolitana" - instead of the old and weakened Provincia -, as an institutional body with special powers aimed at the rebalancing of the territorial structure through the regeneration of the non-metropolitan territories.

Keywords: Ecopolis · Megacity · Regional city

"Starting from the primitive settlements of the Stone Age, due to the ever closer approach determined by the need to satisfy ever increasing needs, we will reach the great contemporary city. In the latter, growth accelerates precisely when the drawbacks of urban concentration appear. This phenomenon is a consequence of the incessant scientific and technical advances which induce men to live more united in order to enjoy the relative advantages, while waiting for the same progress, thanks to the ever greater reduction of distances, to allow them to live more distant from each other, without losing its benefits" (Marcel Poëte, Introduction à l'urbanisme: l'évolution des villes, la leçon de l'Antiquité 1929).

© The Author(s), under exclusive license to Springer Nature Switzerland AG 2022
F. Calabrò et al. (Eds.): NMP 2022, LNNS 482, pp. 2757–2777, 2022.
https://doi.org/10.1007/978-3-031-06825-6_264

1 Introduction

The objectives of the paper are: a. to support an urban vision totally different and alterna-tive to the dominant ones; b. to analyse the technical conditions of its implementation; c. to see how to implement it in the particular geographical and institutional conditions of the Italian territory, also thanks to European funds, and consistently with the deadlines of the European Green Deal.

The paper supports the idea that the too big urban agglomerations (and in particular the so called "megacities"), which exploded in number over recent decades in the Far and Middle East, are among the causes of the current interwoven crises. In less than one hundred years (which is only a very short period in the history of the city), megacities have transformed themselves from the promise to become polar stars of innovation, to places where the violence of urbanization -at the expense of nature, ecosystems and humans- has become one of the causes of the current environmental, social and economic disasters. We wonder if the time has come to free ourselves from the illusion that megacities still represent great opportunities for human prosperity and if it is not the moment to seriously begin to impeach these creations as one of the main causes of the devastating current crises (as well as of their combination): "The reconstruction of the planetary biophysical balances requires a slowdown, better an inversion, of the urbanization process (...) also in consideration of the substantial ungovernability, on an environmental as well as social level, of the large urban conglomerates" (Butera 2021, p. 54).

A totally different urban vision should be strongly proposed and promoted by the scientific and technical culture that deals with cities and regions (Andreotta 2017). In the work "Territorial Scenarios and Visions for Europe" (Espon 2012) the third of the three proposed scenarios/visions is named "Ecopolis"[1] while, in the list of its main inspirers, after the name of Patrick Geddes, the "father" of all ecological cities, also the name of L. Mumford -the head of the RPAA- is indicated. The concept of Ecopolis will be taken up in this paper and considered the ideal-type of the spatial vision proposed. In 1954, Clarence Stein, a member of Mumford's RPAA, proposed a model of an ecological and decentralized urban system with the name "Regional City" that can be considered not only a structural alternative to megalopolis, but also a structural model of reference for the proposed vision of Ecopolis. Recently, Edgar Morin (2020), the theorist of social complexity, in a book that aims to indicate the new ways to be followed after the pandemic crisis, explicitly speaks of an "Ecopolitical" way of which he also outlines the main different aspects. Ecopolis, which comes from the regionalist thought of the RPAA, can therefore represent a possible spatial version of Morin's ecopolitics.

The ideal type of Ecopolis has, therefore, been taken (Fabbro 2021) to define a spatial vision coherent with, at the same time: the humanistic regionalism of the RPAA (who was the first, in the twenties of the last century, to criticize the emerging metropolitanism); the European vision named "Ecopolis" (Espon 2012); and the «ecopolitical» perspective indicated by Morin (2020).

---

[1] "This scenario responds to the challanges of energy scarcity and climate change (...) by promot-ing small and medium-sized cities as centers of self-contained ecological regions and sustainable mobility patterns..." (Espon 2012, p. 151).

The citation that introduces this paper, by the French urban planning historian, Marcel Poëte, frames and addresses the evolution of the contemporary city in a more general evolutionary theory: the urban form, as we know it today, for Marcel Poëte is the result of a process of coevolution between human needs and scientific technical progress. But, as such, it will not have its present form forever. Indeed, after the modern trend of accelerated concentration will have shown all the drawbacks of the same concentration, the human aspirations could change towards a minor density without losing the benefits of proximity (or, in other words, enabling a greater integration between natural ecosystems and cities via the reduction of the dimensions of cities and the increase of the distance between settlements and without losing the benefits of proximity). Different forms of civilization as well as wars, natural disasters, famines and epidemics have framed and selected cities and settlements: some of them have completely disappeared or have been reborn after centuries of abandonment (such as some "megacities" of the past such as Uruk, Babylon, Thebes, Nineveh, Rome, etc.), but others, more "resilient", have emerged and developed over the centuries and even millennia.

This evolutionary but not linear theory of the city in history, developed by historians as Poëte and Mumford (1934, 1961), support our approach to the interpretation of the current relationships between megacities and the rest of the world as a structural conflict between (mega)centres and the wider peripheries.

Ecopolis makes reference to the structure of the "Regional city" of the RPAA, but it is not to be considered as the blueprint of a new town, or of a cluster of new towns, as has often been proposed, in the urban thinking of the last century. Ecopolis is more a territorial ecosystem resulting from a long co-evolution process between human communities, settlements and natural environment which, as such, has generated that degree of variety necessary not only to preserve its structures and identity, but also to pursue that degree of complexity (of its control system), as well as to cope with the complexity of the surrounding environment[2]. The role of urban and regional planning is, therefore, not so much that of inventing new structures -as modernism in architecture and urbanism has supported- as much that of "rediscovering", conserving and empowering this historical territorial ecosystem, in order to deal, at the right level of complexity, with the emerging, both social and environmental, issues. In this perspective, Ecopolis can be considered the most suitable territorial ecosystem for the coming "Era of resilience" (Rifkin 2019).

In order to pursue now, a future not subordinated to the forms and dimensions imposed by megacities, the first step could be, at least in Italy, that of embedding Ecopolis into the reform of the territorial institution, named Provincia, which currently has the duty to manage the Italian "non-metropolitan" territories. In fact, for various reasons, the profile and the duties of this institution arc no longer fit to address the ecological transition in the territories (Fabbro 2021). This is the reason why a radical transformation

---

[2] "Each traditional or vernacular settlement contains, integrated in its morphology, the organized complexity of its inhabitants and the additional levels of complexity of the society that they have built. They reflect a balance between two systems: (i) urban fabric and (ii) human society. Throughout history, a close mutual relationship is implemented: people and society govern their built environment, while the built environment in turn governs people and society" (Salingaros in https://www.ladirezione.it/la-legge-della-varieta-necessaria-e-lambiente-costruito-di-nikos-a-salingaros/).

of the Provincia should be undertaken in the form of a new territorial institution as it happened, in 2001, with the "Metropolitan City", introduced by the Italian Constitution to replace the Provincia in the municipalities belonging to the metropolitan systems.

The five main arguments of the paper will be developed as follows:

1. In Sect. 2, the reasons why megacities developed in recent decades can be considered at the origin of the current crises, are reconstructed.
2. In Sect. 3, a fundamental criticism of the large urban concentration, which had already begun in the twenties of the last century when the metropolitan phenomenon was at its beginning, is discussed. The regionalist movement and Clarence Stein's Regional city can be considered the pioneer proposals towards a city as an alternative to the too large urban concentrations.
3. In Sect. 4, the structural characteristics of Ecopolis and the wider "ecopolitical" con-text which has to support its development-, are illustrated, analysed and outlined.
4. In Sect. 5, it is shown how the European Green Deal (EGD 2019), the Next Generation EU (NGEU 2020) and the Italian National Plan for Recovery and Resilience (PNRR 2021) would already be directed towards the development of territorial systems coherent with Ecopolis, if only a territorial vision had been included in these fundamental European and Italian initiatives.
5. Finally, in Sect. 6, the "Ecopolitan Area" is proposed as a new territorial institution to substitute the existing Italian Province.

2 The Role of Megacities in Causing the Crises

If the current interwoven crises (environmental, economic, pandemic and of social justice) have an anthropogenic origin, then, like other types of «human creations», even Megacities (that are human creations at the highest degree) must be questioned as possible causes of the current interwoven crisis.

How to define, with a unique term, those huge urbanized areas that have spread all over the world in the last century, and that are growing more and more?[3] Metropolis has become ever widening concept that can include a medium (with less than 1 million inhabitants), as well as very big urban agglomerations. Currently, a city of 1 million people isn't remarkable at all. In China alone, there are now over 100 cities with a million people – and as such, our mental benchmark for what we consider to be a "big city" has changed considerably from past times. Megacities, instead, are cities that, according to the current definitions[4], concentrate more than ten million inhabitants. They often also

---

[3] In its study of 2014, the Global Cities Institute (GCI 2014) estimates that, by 2025, the 101 largest cities will concentrate 900 million inhabitants ranging from Tokyo, at the top, with 36.4 million inhabitants, to Dakar with 4.2 million. But, in 2100, the 101 big cities will concentrate 2.3 billion people, will mainly be in Africa and Asia and will range from 88 million in Lagos to 8 million in Rawalpindi.

[4] This quantitative definition is more or less the same both in the Oxford English Dictionary ("A very large city, typically one with a population of over ten million people") and in the Cambridge English Dictionary (a very large city, especially one with more than 10 million people living in it").

correspond to those huge metropolitan systems, called Megalopolis, that, according to the first studies (Gottmann 1961) represent the interconnection of different metropolitan areas. Another definition for a large city with a primary role in the global economic network, (and therefore, with a qualitative more than quantitative definition), is "global city". This term, rather than "megacity", was popularized by Sassen in her 1991 work. Megacities are not all global cities, but they certainly include the largest and most famous of these (London, New York, Tokyo, Shangai, Beijing etc.). Less than ten at the end of the XX Century, megacities have more than tripled in the following twenty years, reaching a number of almost thirty-five.

From now on, we will use the term "megacity" to refer, more widely, to those hundred cities going from the largest megacity of Tokyo (36 million), to that of Addis Ababa with 3.5 million, which are currently home to, more or less, 10% of the world's population. Considering that, by the end of the century, the world population is likely to grow with estimates ranging from 7 billion to 13 billion, and the percentage of people residing in the hundred larger megacities is estimated to be from 15% to 23% (GCI 2014).

Megacities have been considered, since their onset, as drivers of social and economic growth (Gottmann 1961). Moreover, the recent cycle of globalization, with its powerful drivers of innovation and creativity, has also been considered as a big push toward the growth of contemporary cities (Florida 2005, Glaeser 2011). But after the beginning of the Great Depression in 2008 and the current pandemic, maybe this overly optimistic judgment should be revised, because these "catastrophic" and not inevitable events clearly show some structural weaknesses of our current economic and social life systems. Therefore, the megacities generated by the recent hyper-globalization (Rodrik 2011) inevitably share the responsibilities of the causes and effects of these disastrous events, in terms of social inequalities and major environmental imbalances at a planetary scale.

Gottmann himself, four years after the issuing of the *Limits to growth* (Meadows et al. 1972), warned (Gottmann 1976) about the possible risks facing the megacities that he had been studying with great optimism only fifteen years earlier. Those risks, in the following decades, have grown to such an extent that they have become a reality. Megacities, in fact, have fuelled the most advanced but also voracious forms of multiplication of economic value, as well as of exploitation of social, human and natural resources:

1. Due to huge real estate overproduction, they have contributed to generate the structural conditions to globalize the 2008 financial crisis (Harvey 2008; Harvey and Nak-chung 2017).
2. With more or less 10% of the total world population, they produce more than half of the total CO_2 emissions (Wei et al. 2021).
3. In one of them (Whuan), at the end of 2019, the Coronavirus started (with the so called «spill over» from animals to humans or, technically speaking, with the "zoonosis") its deadly spread around the world.
4. More generally, megacities are pushing the "deterritorialisation" of the world in the sense of the uprooting of populations, social groups, and individuals from their cultural and environmental contexts (Vasenev et al. 2018). The avant-garde of this process are the so called "global cities" or the global hubs for trade, finance, banking, technology and economic outlets. Rightly, Sassen herself (2004) asks whether these

realities can still be called cities. In fact, they are globally connected cities but totally disconnected locally, physically and socially, to the point that it no longer makes sense to talk about cities that are, instead, locally, physically and socially embedded (Mumford 1961).

There is nothing really unexpected and unpredictable in the enormous process of change which has occurred in recent decades, in the urban structure of the world and in the way of considering cities. Marxist scholars (Harvey and Nak-chung 2017) consider megacities both the outcome of the spatial process of circulation and expanded accumulation of capital, and the means to preserve the conditions of this process and reproduce it as an increasingly extended growth cycle. At the same time, in fact, megacities: a. promise to large masses of population dis-embedded from their land, a future as "citizens", but in reality make them dependent on wage labour and disproportionate rents; b. concentrate in few hands and in relatively limited spaces new economic and financial value, investing the surplus of capital in ever new infrastructures and ever larger real estate; c. impoverish the value of the assets existing in other peripheral territories, making it possible to easily transfer these assets, at a lower price, into the hands of big investors capable of revaluing them for new business purposes and for new capital accumulation. In other words, the process of concentrated urbanization of the world has produced, at the same time:

1. The creation of tremendous environmental bombs (all the skyscrapers that populate these cities are totally air-conditioned, producing huge amounts of carbon emissions).
2. Social inequalities between centres and peripheries and the impoverishment of entire territories.
3. Enormous real estate bubbles which, when they explode, put a large part of the world in crisis and generate new social and economic imbalances.
4. More in general, totally ungovernable processes and consequences at a planetary scale.

In conclusion, in less than one hundred years[5] (which is a very short period of time in the history of the city), megacities have transformed themselves from polar stars of innovation, to places where violence against nature, global ecosystems and humans has become one of the causes of the current environmental, social and economic disasters. We wonder if the time has come to free ourselves from the illusion that megacities still represent great opportunities for human prosperity, and if it is not the moment to seriously begin to impeach these human creations as one of the main causes of the radical crises (as well as of their combination) that are everywhere disrupting the contemporary social systems. Notwithstanding the dominant narratives, not everything is going well: there is a hidden side of the urbanization of the world that has been irresponsibly forgotten for decades, and even exalted as the best of all possible worlds.

---

[5] The harsh conflict between the environmentally oriented proposals of the Regional Planning Association of America of L. Mumford (as it will be discussed later) and the metropolisation of New York (the first global city outside Europe), driven by the financial power of Wall Street, dates back to the early 1920s (Wesley 2008).

3   The Historical Alternatives to Megacities

In light of what has been said, it is difficult to argue that megacities should be considered radiating centres of prosperity for all. It is easier to argue that, due to their violent concentration of wealth, power, people and activities, they have become a source of social inequality and environmental degradation. As such they had already been stigmatized by L. Mumford when, in his book *The City in History* (1961), he equated megalopolis with necropolis[6]. It is, therefore, at the dawn of megacities, that a thought clearly opposed to that form of urbanization, was born. It is under the pioneering work of the Regional Planning Association of America (the RPAA), and under the guidance of Lewis Mumford, during the twenties of the last century, that these ideas assert themselves in social theory and practice. With a strong criticism to the megalopolis, the planners of the RPAA developed radical alternative ideas and projects (Lubove 1963; Lucarelli 1995). Their communitarian and regionalist approach gave birth to the idea of a multipolar city structured as a polycentric network of medium and small centres well integrated within the rural land, and their surrounding environmental qualities. These ideas, in turn, originated from those pioneering scholars that, at the turn of the nineteenth and twentieth centuries, shaped the principles of modern urban planning to contrast the negative effects of the industrial city of the nineteenth century. Among them it is worth mentioning the Russian geographer Pyotr Kropotkin (who criticized the excess of specialization and division of labour introduced and fostered, in urban and rural areas, by capitalism) (Kropotkin 1974), and the British planner Ebenezer Howard (who criticized the loss, in the industrial city, of rural values and therefore advocated a new town-country alliance through the so called "Garden city") (Howard 1922).

Among the planners of the RPAA, Clarence Stein, urban planner, architect and writer, imported, into the American context, not only the British idea of the garden city (with the plan for Radburn in New Jersey, USA) and of the "neighbourhood unit" (with the "Sunnyside Gardens" in the New York City borough of Queens), but also the idea of a "Regional city" for one million people (see Fig. 1).

Stein's "Regional City" (see diagram in Fig. 1) represents an evolution, on a wider scale and with fewer structural and geometric constraints, of the garden city of E. Howard. It consists of a city of 1 ml inhabitants distributed in 1,000 square miles (about 2500 kmq, or an area of 50 × 50 km). The population of the city is articulated in thirty-six clusters with four neighbourhoods each (with a surface area of one square mile and a population of 6 thousand inhabitants for each neighbourhood). In total, each cluster is about 25 thousand people each. The structural hierarchy of the centres is the following: one town of about 100 thousand inhabitants in an urbanized area of 16 square miles (4000 ha) and a population density of 6250 ab/sqm or 25 ab/ha. Four towns of about 75 thousand inhabitants. 9 towns of 50 thousand inhabitants and a number of small towns of about 25 thousand inhabitants. The land use of the region is about 84% rural or natural, while the remaining 16% is urban. The building density is quite low (about 5000 mc/ha). Having a low building density, the urban land use density is also rather low. But it must be also considered that the urban land use is so low because it comprises wide urban parks and

---

[6] It is worth to mention that, the critique of the large urban agglomerations, had already been formulated in his previous *Culture of Cities* (1938).

REGIONAL CITY – DIAGRAM

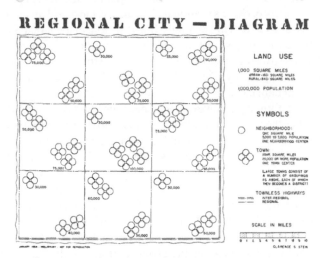

Fig. 1. The diagram of the Regional City by Clarence Stein.

green areas. The Regional City is connected to other regional cities through a network of interregional highways, while the internal network of regional roads is a regular grid with a cell of 10-mile sides. The Regional City is articulated in nine sub-systems of a dimension that ranges from a minimum of 75 thousand to maximum of 125 thousand inhabitants.

The diagram, of course, as any spatial diagram, represents an ideal-type and therefore is more a coherent system of guiding principles than a model to be replicated to the letter and that, as such, leaves a certain freedom of interpretation, adjustment and application to reality. The guiding principles are: a. the basic relationship between the order of magnitude of the resident population (around 1 mL) and the vastness of the entire territory of the region (1 thousand square miles); b. the polycentric structure of the settlements and the fact that, the largest of them, has the size of a medium-small city; c. the relationship between urban land uses and non-urban land uses, where the latter should be at least five times wider than the former; the articulation of the whole system in about ten local sub-systems well connected to each other and to the whole system, in terms of transport.

It may be a coincidence, but this is also the real structure of many "regional cities" still recognizable, despite all the transformations they have undergone in the last hundred years, in the Italian urban geography and in particular in those areas, often mountainous, foothills and hills where urbanization has never reached metropolitan dimensions and has preserved, among the networks of small and medium-sized cities inherited from the Middle Ages, large rural and natural spaces. The population of Stein's Regional City corresponds, with reference to the Italian territorial administrative context, either to a large Provincia within a large Region (such as those of Bergamo or Brescia within Lombardy which has about 10 million inhabitants) or to a small Region (such as Abruzzo, Calabria, Friuli Venezia Giulia, Marche, Liguria, Trentino Alto Adige, Umbria). In terms of land area, on the other hand, Stein's Regional City corresponds to the average area of the Italian Provincia. In any case, several non-metropolitan Italian Provinces correspond to the general parameters of Stein's Regional city, such as those of Treviso and Vicenza in the North, or Lecce and Caserta in the South. It is therefore evident that Stein's Regional City is not an abstract ideal-type, but instead, it seems to find a strong correspondence both to the real urban structure and to the administrative structure of a strongly urbanized and historically shaped territory like the Italian one. These observations will be useful for the final proposal delineated in Sect. 6.

It should not be forgotten, however, that between the sixties and the nineties of the last century, the theme of the regional, dispersed or diffused city was taken up, in the disciplinary debate on urban planning, with great attention. In addition to the international, the Italian debate is also full of suggestions. They range from the pioneering work of Giancarlo De Carlo (1962) on the "city region", to the elaborations of Pier Carlo Palermo on the "urban region" (1996); from the interpretations of recent forms of the Italian urbanization (Astengo and Nucci 1990), to the emerging networked urban structures (Dematteis 1985, 1990). The interpretative and analytical use of the concepts of city region, diffused city or urban network seems, however, rather controversial in its final conclusions: in acknowledging the distance, from the planners' schemes, of the real urban systems realized in Italy along with the post-war economic boom and then by the manufacturing spread of the "third Italy", much emphasis is placed particularly on their negative effects in terms of congestion, waste of land, spread of mobility and, in any case, of the many diseconomies of these semi-spontaneous models of city growth (Indovina 1990, 2009). In this research the regional and reticular approach is used, in any case, more as an interpretative scheme to analyse the transformations of post-war Italian settlements than as design concepts to approach and restructure those urban systems. Nowadays, some of the diffidence brought about by those interpretations against the regional city concept, are probably no longer justified: the population, at least in western countries and certainly in Italy, is no longer growing, and there is, therefore, no demand for new expansions of the cities, especially if dispersed in the countryside. On the other hand, sustainable alternatives in transport, no longer based on fossil fuels, are developing with respect to mobility and private motorization, going from electric cars to regional rail and tram transport systems. Other criticisms to small polycentric urban systems, are related to their supposed insufficient critical mass for an efficient service provision and management, and for an efficient collective transport (Parr 2004), compared to bigger urban concentrations. But the tendency to reduce the cost of distance, thanks to the great development, in many fields, of digital technologies -including in the transport sector- and the potential of new renewable energies to reduce energy costs -also avoiding dependence on overly centralized sources-, is also leading to a reduction of the magnitude of the critical mass necessary to guarantee the efficiency of public services and transport (Rifkin 2019).

Even though it should be recontextualized to our present age (i.e. interregional highways should be complemented by electric collective transport such as interregional railways and interurban tramways), Stein's Regional City can be, in any case, considered one of the first structural alternatives to megalopolis and a reference diagram for the ideal-type of Ecopolis. Moreover, if Stein's Regional City is something that exists in reality and not only in the planner's mind, and if Stein's Regional City is the ideal-type of Ecopolis, then it can also be said that Ecopolis is not a utopian city but a possible meta-model that can serve to relaunch existing non-metropolitan areas as well as to reshape too big urban concentrations.

The cultural path that leads from Clarence Stein's Regional City to Ecopolis is, moreover, made recently explicit in the work "Territorial Scenarios and Visions for Europe" (Espon 2012), where the third of the three proposed scenarios (the "Europe of Regions") is also named "Ecopolis"[7] and in the list of its main inspirers, after Patrick Geddes - the "father" of all the ecological cities -, L. Mumford, - the head of the RPAA -, is indicated.

But before entering a more precise definition and description of Ecopolis, it is necessary to make a reflection on the more general contextual conditions that allow for the realization of Ecopolis.

4 Ecopolitics and Ecopolis

4.1 Ecopolitics

Strangely, despite the large amount of literature produced on the theme of ecology and the city, not many books and papers, at least in the international literature of the last thirty years, have been dedicated to the topic of Ecopolis, a word that integrates the two very dense concepts of *oikos* and *polis* -coming from the ancient Greek culture-, that are are at the basis of Western civilization. With a dominant utopian perspective some articles are included in the book *Urban eco-communities in Australia* (Cooper and Baer 2018). Previously, worth mentioning are the works of Tjallingii (1995, 2004), who considers ecopolis as a "conceptual tool" for integrated urban planning, while Ignatieva (2000), Pearce (2006), Babalis (2006) and Girardet (2014), with the term Ecopolis underline, with different levels of detail, the close relationship between urban planning, the city and the environment in a perspective of sustainability.

In Espon (2012), when the effects of the 2008 crisis had not yet spread widely and the pandemic was still to come, the Ecopolis scenario/vision was considered on a par with two other spatial scenarios/visions: the "Europe of flows" (Metapolis) and the "Europe of cities" (Metropolis). Today, after the great depression started in 2008, after the explosion of global inequalities of economic and climatic origin, after two years of the Coronavirus pandemic, the scenarios based on global flows and on megacities should be seriously questioned. At the structural level certainly, they still are prosecutable. But, from the point of view of citizens and territories that seem now quite disillusioned with the great promises of globalization and "metropolisation" of the world (Fabbro 2020), are they, at the same time, also desirable?

If not, then, among the three spatial scenarios/visions proposed in 2012 by Espon, the one that seems the most desirable is, at the moment, that of Ecopolis. In any case, a choice among three spatial scenarios does not make sense without questioning the more general political conditions of each of them. The "Europe of flows" (Metapolis) and the

---

[7] "This scenario responds to the challanges of energy scarcity and climate change (…) by promoting small and medium-sized cities as centers of self-contained ecological regions and sustainable mobility patterns…" (Espon 2012, p. 151).

"Europe of cities" (Metropolis), represent, in fact, two scenarios (after all, compatible each other) functional to the current development model dominated by the financial capitalism. Ecopolis, instead, is a vision (more than a scenario because it should be, in any case, strongly promoted), alternative to the other two scenarios and whose feasibility conditions seem quite different from the existing dominant models.

The three spatial scenarios/vision of Espon (2012) are, therefore, quite different from each other because they belong to different orders of reality and abstraction: the first two represent two existing realities, but perhaps less desirable than just some years ago. The third is, at least for now, almost only a desirable vision for a quite distant future. But what particular political conditions does it need?

Edgar Morin, the centennial scholar of «social complexity», in his recent book (2020) written to look beyond the pandemic and to overcome the negative social and environmental consequences of the dominant financial capitalism, introduces and develops an eco-political paradigm explicitly called "ecopolitics".

Morin, after having criticized the devastating global effects of neoliberal politics, proposes a new way, that he names ecopolitics, to combine: "globalization and de-globalization, growth and de-growth, development and envelope". In particular, he proposes the "resilience" of territories as the new way to deal with all the new risks and with both sides of those dichotomies.

Morin, as other authors such as Rodrik (2011), Crouch (2015) or Rifkin (2019), certainly does not ignore the role and importance, in terms of human rights and human solidarity, of globalization but does not hesitate to declare, looking inside the recent crisis of globalization, its failure at least at the techno-economic level. A "humanized globalization", according to Morin, involves partial de-globalization as well, to ensure, in particular, food, energy and health autonomy for nations and local territories. When Morin evokes a regeneration of local and regional communities, horticulture in the urban periphery, food at zero kilometres, the return of small commerce and craftsmanship in the cities etc., does not just make a recall, which can appear nostalgic, to "Fields, factories and workshops", the book where P. Kropotkin, in 1898, with a detailed analysis of the economic life in the European territories during the industrial era, lashed out against the excesses of specialization of territories and the international division of labor (at the time at its beginning), but he seeks, with realism, a remedy to the structural weaknesses of the recent cycle of globalization. On the other hand, Rifkin, one year before the pandemic and the European decisions to contrast its long-term effects, inaugurates the concept of a Green New Deal to promote "an infrastructure for an Age of Resilience", where, for resilience, he means the capacity to deal with a nature that, from pacified and tamed, is becoming wild again. When he publishes its book, the spill-over, from animals to humans, of a new deadly virus -that, in the space of two years will produce more than 5 million deaths-, is spreading from Wuhan in China to the rest of the world in just a few weeks. The consequences of the pandemic brutally highlight, therefore, the fragility of the last cycle of globalization from at least the following points of view: a. the fragility demonstrated by the evident incapacity of the global system to control the perverse - and even lethal - outcomes of an economy of flows not only of goods and people but even of viral agents; b. the fragility of an economy of flows connecting countries too distant and different from each other in terms of the relationship between humans and nature; and c. the fragility of value chains even thousands and thousands of kilometres long and that, when they are interrupted by some global shock, produce the collapse of entire economic sectors, as well as of local economies dependant on those chains.

Ecopolitics, for Morin, can be summarized as a future process strongly implied with a partial deglobalization, with less international division of labor and more locally integrated economic activities, all in the context of a different relationship between humans and nature. What better interpretation also of the integrated and global concept of resilience!

4.2   Ecopolis

Ecopolis is probably the most resilient urban eco-system.Moreover, in terms of resilience, between Ecopolis and Megacity, there seems to be no comparison at all as, above all, the history of the ancient megacities seems to demonstrate.

The ideal-type of Ecopolis has been taken (Fabbro 2021) to define a spatial vision coherent with the "Ecopolis" proposed by Espon (2012) and with the «ecopolitical» perspective indicated by Morin (2020). Furthermore, Ecopolis makes reference to the structure and model of the "Regional city" of the RPAA. But Ecopolis is not to be considered as the construction of a new town (or a cluster of new towns), as often has been proposed in the XXth century to deal with the population growth. Ecopolis is, to some extent, the historical outcome of the co-evolution between human communities and the physical space. As such it is a "naturally resilient" territorial ecosystem. The role of urban and regional planning is, therefore, that of conserving and enhancing its resilience through continuous efforts to adapt the system to the environmental, social and technological issues of the present. In this perspective, Ecopolis can be considered the proper urban ecosystem for the "Era of resilience" because its principles of order (it is multifunctional, redundant, modular, based on different organizational principles, polycentric, multiscalar) are opposed to those generated by the simplifying techno-economic rationality that has guided the development of cities in the last century and that has led to megacities via high levels of concentration, specialization, uniformity, maximization of utilities, etc.

An eco-system, as any complex biological system (Salingaros 2014, gives the example of the tropical forest), is able to cope with unpredictable upheavals, while preserving for a long time structure and identity, because it has a structure that embodies inter-connections, diversity, redundancy, wide distribution of structures at different scales, adaptability and self-organization. An urban or, more in general, a territorial eco-system (as a region, for instance) that arose from historically long co-evolutionary processes of adaptation between human, social and environmental entities (Holling 2001) is, like-wise, a complex system capable of integrating physical, ecological, social and decisional dimensions at the same time. Ecopolis is, therefore, a "territorial ecosystem" because it is a territorial system that can exhibit qualities (such as a structure interconnected as a network, diversity and redundancy, wide distribution of structures at different scales, adaptability and self-organization) that are typical of biological resilient systems (Salingaros 2014) and, as such, are more resilient than others types of too artificial urban system.

Resilience, therefore, is neither a performance to be artificially implemented and imposed from above, nor is it the result of the implementation of technologies, however advanced. Resilience cannot be planned as a totally new artefact or service. Instead, it has to be nourished, fostered, conserved starting from the recognition of the historical ecosystemic qualities that already characterize the territorial systems in terms of resilience (Holling 2001, Olazabal et al. 2012). A territorial system that has evolved over the centuries in spite of wars, natural disasters, famines and epidemics, and that, notwithstanding, has maintained its basic order and identity, is resilient (Alexander 2004). Centuries of adaptive constructions, of balancing human activities and forces of nature, of energy savings and economic sobriety, of respect for the sense of places and communities, without however, giving up the technologies available today to improve our lives, they are certainly sources of integrated resilience (Salingeros 2014).

According to the aforementioned characteristics, Ecopolis, in structural terms, can be sketched as a polycentric, self-governed, regional network of small-medium towns and rural areas and villages:

- complementary with each other in terms of service provision;
- with strong provision of ecosystem services at different scales;
- integrated with natural and seminatural ecosystems;
- with limited dependence on fossil energy sources;
- with a short supply chain among manufacturers;
- with a sustainable mobility among the service centers based on an electric intercity railway network;
- with a high level of digital interconnections all over the regional areas;
- and with a certain level of self-sufficiency in energy and food supply.

Ecopolis, therefore, should rely on the virtuous integration between some powerful drivers as:

- the *resilience driver* that reinterprets the territorial justice in terms of anti-fragility and the ability to react to climate change thanks to a greater integration between society and ecosystems;
- the *political driver* that integrates, in a multilevel governance, the micro dimension of the local government with the macro dimension of an international cooperation;
- the *economic-energy driver* that integrates clean energies - and the new ways of producing and distributing them on a local basis - with digital technologies;
- the *driver of the reorganization of the times and spaces of the territory* to make the associated life of the ecopolitan contexts, more liveable and attractive than in the metropolitan cities;
- the *social driver* based on the reorganization of key public services -such as hospitals, schools, cultural centers and public transport- for a more cohesive and humane society.

4.3   The Critical Mass Weak Point

The main criticism that has probably been made against the "Regional City" -as well as to other polycentric and networked cities- consists in the fact that it would not be able to generate the critical mass that allows the emerging of those services, particularly in the field of amenities, leisure and culture, comparable with those of a city, of the same dimension, but structurally very concentrated. In other words, due to the so-called "agglomeration economies", two cities with the same dimension (for example of one million inhabitants) but radically different in their structural form (one as a polycentric network of small and medium centers and the other as a monocentric agglomeration), would appear very different in terms of presence and performance in the higher level of services offered (Meijers 2008). If the planning objective is, therefore, that of pursuing a high level of services, in a region with dispersed small towns, the principle of the critical mass is, according to the "agglomeration economies" principle, to be kept in serious consideration. A complementarity services network among the urban centers is considered a spatial strategy capable to integrate, in a unique structure, all the different centers and, consequently, to approximate the necessary critical mass capable to generate higher level services (Camagni 1993). But it must be recognized that this strategy has at least three main criticalities: a. it is not so easy to realize, particularly in a short time, a complete complementary structure; b. consequently it cannot reach the desired integration; b. in any case it generates a large number of physical interactions between the different service centers and, consequently, the private mobility, within the network, risk to become higher than in the more concentrated urban centers.

Ecopolis, anyway, may be insufficient in guaranteeing very high-level services in sectors as high education, health, and culture (top level universities, very specialized hospitals, famous theatres). Therefore, for these services, it risks being dependant on some metropolitan system. But it is also true that, normally, the most of a person's life does not take place in the classrooms of an important university or by attending the performances of a famous theatre or even, at least hopefully, in a well-known highly specialized hospital. Instead, he spends a lot of time living in a neighbourhood with the family, working in an office, shop or laboratory, walking, cycling, using a car or public transport, doing sports outdoors or in a gym, meeting friends at the bar and so on. Ecopolis, on the other hand, excels in providing ecosystem services and in being very attractive for new concepts of housing, mobility and home working (Tjallingii 2004). One solution, therefore, may be that of developing, as much as possible, neighbourhood services in the smaller centers and to limit to some particular categories, the services that generate high mobility in the network (as high schools and health services). The latter, in turn, should be interconnected by rail and tram networks in order to limit, as much as possible, the environmental and social externalities of the mobility between centers.

Another positive consideration for Ecopolis is that it has to be contextualized in this epoch of crises when resilience has become, as said before, the central issue. Resilience is not a question of critical mass but of a much more complex relationship between the human, settlements and nature. The objective is, consequently, a greater resilience to crises and climate change. As such, Ecopolis can guarantee shorter economy cycles; shorter interconnections between built-up areas; natural amenities and resources; smaller neighbourhoods and with a more human scale; greater ease of concentration of essential services and travel on the short distance; a greater proximity to environmental amenities and ecosystem services along with a lower incidence of environmental impacts and also a greater identity and self-organization of the various local communities.

Last but not least, Ecopolis, being a medium-low density settlement, certainly needs, in relative terms, much larger surfaces than megacities. This inevitably implies the recovery of the wider anthropized territory (the "Ecumene"), as well as of that which has been progressively abandoned (because considered not efficient according to the parameters of modern capitalism) but which today exhibits great qualities of resilience in particular towards climate change.

5 Ecopolis, the European Green Deal (2019), the NGEU (2020) and the Italian PNRR (2021)

The «European Green Deal» (EGD, issued by the European Commission in December 2019)[8] is the frame of reference of the Next Generation EU initiative (approved 2020) that, through the "recovery and resilience facility"[9], represents the great European response to the pandemic. This package of initiatives (starting from a questionable idea of growth), cannot be considered a proper form of "ecopolitics". It is, instead, decidedly a new European "political economy" and, as such, it questions neither the current forms of globalization, nor the ways the international division of labor is organized (through too long and vulnerable supply chains). A fundamental ingredient of an effective ecopolitics should be even a spatial perspective aimed at overcoming the structural inequalities between centers and peripheries and therefore close, to a certain extent, to that of Ecopolis (also outlined by Espon in 2012), but, this perspective, for the moment, is lacking. Important purposes of the EGD, anyway, starting from the 2050 objective of net zero emissions, are aimed at a different relationship between human activities and nature. Therefore, even if without scratching the ecopolitical premises (including an adequate European spatial vision[10]), some important features of the so-called "ecological transition" seem to have been addressed.

Italy is currently facing the start of the «National Recovery and Resilience Plan» (in Italian PNRR) issued in the context of the NGEU and in coherence with the EGD'

---

[8] "The European Green Deal (…) is a new growth strategy that aims to transform the EU into a fair and prosperous society, with a modern, resource-efficient and competitive economy where there are no net emissions of greenhouse gases in 2050 and where economic growth is decoupled from resource use. It also aims to protect, conserve and enhance the EU's natural capital, and protect the health and well-being of citizens from environment-related risks and impacts" (from the Communication of the European Commission to the European Parliment, Brussels, 11/12/2019).

[9] "The Recovery and Resilience Facility is the centerpiece of Next Generation EU with €723.8 billion in loans and grants available to support reforms and investments undertaken by EU countries. The aim is to mitigate the economic and social impact of the coronavirus pandemic and make European economies and societies more sustainable, resilient and better prepared for the challenges and opportunities of the green and digital transitions" (https://ec.europa.eu/info/strategy/recovery-plan-europe_en).

[10] Before the 2008 crisis, European spatial visions and policies were explicitly addressed, starting with the ESDP of 1999, towards the strengthening of polycentrism as an alternative to the major European urban agglomerations (often represented with the images of the Blue Banana or the «Pentagon»), seen as possible sources of too big imbalances with non-metropolitan European territories.

objectives. It is certainly a huge and concrete opportunity not only to relaunch the national economy after the pandemic, but also to restructure, in the territories, the relationship between settlements and ecosystems in order to get a more advanced resilience (which is also provided in the name of the PNRR). In this perspective, the PNRR contains many appropriate measures that should also be implemented by Regions and Comuni (municipalities).

Let's take a closer look at the PNRR. It foresees 134 investments (235 if sub-investments are also taken into account) and 63 political-administrative reforms, for a total of 191.5 billion euros of funds. Of these, 68.9 billion are non-repayable grants and 122.6 billion are loans. To these allocations, the resources of the European React-EU funds and a National Plan for Complementary investments (PNC), for a total of approximately 235 billion euros (which correspond to approximately 14% of the current Italian gross domestic product) are added. The PNRR consists of six Missions and sixteen Components, which are articulated around three strategic axes shared at the European level: digitalization and innovation, ecological transition, social inclusion. 59.5 billion of the total amount of the PNRR, are destined to the "Green Revolution and Ecological Transition". The implementation of the PNRR is in charge of all levels of the Italian public administration. Local authorities play a central role in this process of implementation: about 36% of the PNRR resources will be entrusted to Regioni, Provinces, Comuni (municipalities), Metropolitan Cities or other local administrations. 28.3 billion will be managed only by Municipalities and Metropolitan Cities, 10.8 billion by Regioni alone and another 10.8 billion by Regioni, Provinces and Municipalities together. The only territorial restriction on the allocation of resources is that at least 40% of the resources should be reserved for the southern regions (34% of the Italian population). The 59.7 billion allocated to the "Green Revolution and Ecological Transition" are divided into: 5.27 for circular economy and sustainable agriculture; 23.78 for renewable energy; 15.36 for the energy efficiency of buildings; 15.06 for the protection of soil and the territory. 33% of the 59.7 billion (around 20 billion) will be managed directly by the Municipalities.

Even assuming that only the resources of the PNRR destined to the "Green Revolution and Ecological Transition" and directly under the responsibility of Regions and Municipalities (20 billion), were allocated for the realization of local planned models of Ecopoli, we would be in the presence of a truly unique amount of public investments in the territory, which had not occurred in Italy since the post-war reconstruction and which will not occur again for a long time. It goes without saying, therefore, that this mass of money should be aimed, in perfect coherence with the objectives of the PNRR and, upstream, of the NGEU, at the regeneration of the local territories towards ecologically new and resilient territorial systems. But this will probably not be the case due to the fact that the implementation timetable of the PNRR is not compatible with Ecopoli's medium and long-term planning and programming.

The PNRR appears, therefore, caged by a serious paradox: it is aimed at the implementation of the 2050 EGD objectives, but its operational deadlines are too limited (by the end of 2026 all the interventions must be concluded) to pursue those objectives and, moreover, do not have any spatial perspective to address the huge amount of resources, if not in the very general terms of a rebalancing between North and South of the Country. A strict deadline and the absence of a spatial vision can favour only those big national agencies that, in terms of operational capacities on big investments, are particularly well equipped[11]. Local and small municipalities will be obliged to quickly pull old projects out of dusty drawers without being able to frame them in a new and resilient spatial vision. The European constraints of the PNRR, therefore, make it very difficult to address investment strategies according to new territorial projects and to new spatial visions. Without having the possibility to adopt and implement new spatial visions, plans and projects, it seems, therefore, very hard to pursue an effective and real «ecological transition» towards 2050. It seems even more difficult, therefore, to be able to pursue an Ecopolitical spatial vision.

6   The "Area Ecopolitana"

Even if the PNRR deadline of 2026 is very restrictive and compliance with it requires recourse to already existing projects, everything possible will have to be done to ensure that Municipalities and Regions propose projects close to an "ecopolitical" perspective. Ecopolis could work as a general framework operating towards an "ecological transition" strongly aimed at resilience and the contrast to climate change, and to guide, at least from a political-cultural point of view, the selection of projects. Many of them should be addressed, by Regions and Comuni, to the territorial scale that seems the most effective towards a well rooted ecological transition. This is the intermediate scale that stays between small and medium Municipalities and the Regioni, and that is also the right scale for Ecopolis (as supported in Sect. 3). Currently, this territorial scale, in the Italian non-metropolitan territories is, in many cases and especially along the Apennines and the Alps, affected by the growing phenomena of demographic decrease and socio-economic shrinkage. To this, must be added the fact that these territories are presided over by the Provinces which, in the last years, have become politically weaker and neglected. So, any planning strategy interesting the intermediate territorial scale today means, in Italy, to deal with the weak role and powers of the Provincia[12] and, therefore, to renounce *a priori* any effective planning. If, instead, planning efforts at the intermediate scale are considered necessary, the only solution is that of also rethinking the role and functions of the Provincia from its roots. Instead of the old and weakened Provincia, the role and functions of an "Area Ecopolitana" could, then, be promoted -in complementarity with the existing "Città Metropolitane"[13]-, as the institutional body equipped with special powers for the regeneration and resilience of non-metropolitan territories.

---

[11] This is the case of national large public transport companies (such as the "Ferrovie dello Stato") which, with the resources of the PNRR, will be able to realize the already planned and projected high-speed railway sections. But the high-speed rail is the most distant strategy that can be thought of, to bring about a rebalancing of the local territories and to help the development and resilience of the systems of small towns distributed in the wide Apennine and Alpine areas (Fabbro 2017).

[12] The Provincia is a political-administrative body of prefectural origin, recognized by the Italian Constitution, and centered on the local capital of the territory. It has been deprived, by the law n. 56 of 2014, also of its direct election powers.

[13] In Italy, the Constitution already recognises the metropolitan areas as territorial institutions named "Città metropolitane".

The Area Ecopolitana can be interpreted as a polycentric federation of Municipalities, at the intermediate territorial scale, which has four major advantages that the old Italian Provincia do not has: it can ensure a wider political participation to all the territories of the region, even in the minor and marginal ones (in the Provincia, centered on the local capital city, this does not happen); 2. it can also ensure that geographic and ecosystemic diversities are worthily recognized and represented (this is not the main mission of the current Provincia); 3. it can ensure such strength and size to be able to face and support economic and ecological policies of significant impact (the Provincia, given its current powers, is unable to support them); 4. last but not least, it can introduce an innovation, in the Italian and perhaps European institutional landscape: an 'ecopolitan' territorial federation that can go beyond the out-dated administrative model of the 'Westphalian state' (Faludi 2018). The Area Ecopolitana, furthermore, postulates not only development policy processes but also the creation of an institutional body reinforced by powers of territorial justice and then potentially capable of dealing with the inversion of those negative demographic trends and of the socio-economic shrinkage that currently distinguish non-metropolitan territories in Italy. As the Area ecopolitana is not currently provided by the Italian Constitution nor by the ordinary law, the desired institutionalization of the Area ecopolitana would necessarily require a specific law and perhaps also an integration of the Italian Constitution. Changing the Constitution to replace the Provincia with the Area ecopolitana is certainly not an easy thing.

However, if the current devastating and cumulative crises push towards a profound re-discussion of the principles upon which our associated life is based, why should we not also be able to think about the revision -or introduction- of new principles in the Italian Constitution, or to believe that the Provincia must always remain the same as the past? In any case, if it were too difficult to change, with a Constitutional law, the name and the role of the Provincia towards the "Area ecopolitana", at least the mission and functions of the existing Provincia should be substantially changed -towards a new mission coherent with "Ecopolis" -, through the enforcement of an ordinary law.

7 Conclusions

After two years of a pandemic that has already produced 5.5 million deaths worldwide (as of January, 2022) and the succession of increasingly frequent and intertwined crises of different types (economic, financial, climatic, of social justice, etc.), due not to chance but to the persistent and systematic anthropogenic action of altering global balances, there are very few people who do not believe we are facing a breaking point of the old system, as well as the need to rethink the paradigms of our associated life at the root. But with respect to this situation, what role should we attribute to cities and regions?

To try to address the question, it is necessary to approach, in some way, the great evolutions that have profoundly marked the history of cities and regions, and also their more recent transformations. This backward mapping can allow us to assume the right critical distance towards the problems we are facing and to avoid two mirror errors in designing the possible changes: on the one hand, the error to neglect the great transforming forces operating in history and, on the other, the error of imagining changes that contrast with the latent but permanent structure of the territory and, consequently, exposing us not only to the loss of historical identity but also to the generation of unprecedented new fragilities and vulnerabilities.

According to these more general assumptions, three arguments have been discussed in the paper:

1. megacities must be questioned because they are at the origin of the most recent crises, as well as of the current territorial inequalities and criticalities;
2. strong criticism and real alternatives to megacities have been firmly promoted since the beginning of the last century.
3. Ecopolis is an ideal-type deeply rooted in that criticism as well as a spatial vision alternative -and more valid than ever-, to megacities.

The EU's most recent new political economy -as well as its Italian consequence, the National Recovery and Resilience Plan (PNRR in Italy)- are pursuing an «ecological and digital transition», but without the adoption of a spatial anti-crises vision. The absence of an explicit spatial alternative opens the way and makes possible a new concentration of resources and opportunities prevalently in the European metropolitan core areas and this seems, in light of the previous arguments, not acceptable.

At the European level, there is a possible way to avoid this negative perspective:

1. On the one hand by adopting, as soon as possible, a European spatial vision address-ing spatial forces beyond the metropolitan concentration perspective. The 2012 Espon vision named Ecopolis, is, in general, still valid.
2. On the other, by seeking and promoting, in the context of the EU Recovery and Resilience policies, some «pilot projects» strongly addressed towards the ecological and digital transition at the spatial scale proper of the Ecopolis scenario.

In coherence with these European perspectives, the paper also proposes Ecopolis as a vision to regenerate the Italian territories as well as a possible institutional reform to empower the territorial government. In this perspective it could be;

1. A feasible spatial vision to address, at the regional and urban scale, the new territorial projects of the PNRR aimed at the «ecological and digital transition»;
2. In the context of the desired reform of the Italian territorial administration, a new institutional body (with the name of "Area Ecopolitana") to replace, with the mission of planning Ecopolis and with particular reference to the non-metropolitan Italian territories, the obsolete existing «Province».

References

Alexander, C.: The Nature of Order. Center for Environmental Structure, Berkeley, California (2004)

Andreotta, C.: Visioneering Low Carbon Futures, PHD Dissertation, Vienna University of Technology (2017)

Astengo, G., Nucci, C.: It. Urb. 80. Rapporto sullo stato dell'urbanizzazione in Italia, Urbanistica Informazioni, Quaderni, n. 8. Inu Edizioni, Roma (1990)

Babalis, D.: Sustainable Planning and Design Principles. Alinea, Firenze (2006)

Butera, F.: Piccolo è meglio per l'uomo e per l'ambiente. Urbanistica Informazioni, pp. 298–299. Inu Edizioni, Roma (2021)

Camagni, R.: From city hierarchy to city network: reflections about an emerging paradigm. In: Lakshmanan, T.R., Nijkamp, P. (eds.), Structure and Change in the Space Economy, Springer, Heidelberg (1993). https://doi.org/10.1007/978-3-642-78094-3_6

Cooper, L., Baer, H.: Urban Eco-Communities in Australia. Springer, Singapore (2018)

Crouch, C.: The Globalization Backlash. Polity Press, Cambridge (2015)

De Carlo, G.: Relazione di sintesi. In: ILSES – Istituto Lombardo per gli Studi Economici e Sociali, Relazioni del Seminario «La nuova dimensione della città – La città regione, ILSES, Milano (1962)

Dematteis, G.: Contro-urbanizzazione e strutture urbane reticolari. In: Bianchi, G., Magnani, I. (a cura di) Sviluppo multiregionale: teorie, metodi, problemi. FrancoAngeli, Milano (1985)

Dematteis, G.: Modelli urbani a rete. Considerazioni preliminari. In: Curti, F., Diappi, L. (a cura di) Gerarchie e reti di città: tendenze e politiche, FrancoAngeli, Milano (1990)

EGD: Communication from the Commission to the European Parliament, the European Council, the Council, the European Economic and Social Committee and the Committee of the Regions, Com (2019) 640 final, Brussels (2019)

Espon: Territorial Scenario and Visions for Europe, Project 2013/1/19, Intermin Report I 31/05/2012 (2012)

Fabbro, S.: An alternative approach to the construction of the European space: transport corridors as complex chains of territorial projects. Crios **13–1**, 61–71 (2017)

Fabbro, S.: Ecopoli vs Cosmopolis: un'altra strada contro e oltre la deterritorializzazione. Una riflessione a partire dal caso italiano. In: Archivio di Studi Urbani e Regionali, vol. 127, pp. 122–147. FrancoAngeli, Milano (2020)

Fabbro, S.: ECOPOLI, Visione Regione 2050. INU Edizioni, Roma (2021)

Faludi, A.: The Poverty of Territorialism: A Neo-Medieval View of Europe and European Planning. Edward Elgar Publishing (2018)

Florida, R.: Cities and the Creative Class. Routledge (2005)

GCI Global cities Institute: Socioeconomic Pathways and Regional Distribution of the World's 101 Largest Cities, Working Paper n. 04. University of Ontario, Institute of Technology (2014)

Geddes, P.: Cities in Evolution. Ernest Benn, London (1968)

Girardet, H.: Ecopolis: the regenerative city. In: Lehmann, S. (ed.) Low Carbon Cities. Routledge, London (2014)

Glaeser, E.: Triumph of the City. How our Greatest Invention Makes us Richer, Smarter, Greener, Healthier, and Happier. New York, Penguin (2011)

Gottmann, J.: Megalopolis. The urbanized northeastern seaboard of the United States, The twentieth century fund, New York (1961)

Gottmann, J.: Magalopolitan systems around the world. Geografski Glasnik **38**, 103–111 (1976)

Harvey, D.: The right to the city. In: New Left Review, vol. 53. London (2008)

Harvey, D., Nak-chung, P.: How capital operates and where the world and China are going: a conversation between David Harvey and Paik Nak-chung. Inter-Asia Cultural Studi. **18**(2), 251–268 (2017)

Holling, C.S.: Understanding the complexity of economic, ecological and social systems. Ecosystem **2001**(4), 390–404 (2001)

Howard, E.: Garden Cities of To-morrow. Swan Sonnenschein & Co., Ltd., London (1922)

Ignatieva, M.: Ecopolis-towards the holistic city: lessons in integration from throughout the world (2000)

Indovina, F.: (a cura di), La città diffusa, Daest-IUAV, Venezia (1990)

Indovina, F.: Dalla città diffusa all'arcipelago metropolitano. Franco Angeli, Milano (2009)

Kropotkin, P.: Campi, Fabbriche, Officine. Edizioni Eléuthera, Milano (1974)

Lubove, R.: Community Planning in the 1920's: The Contribution of Regional Planning Association of America. Un. of Pittsburgh Press (1963)

Lucarelli, M.: Lewis Mumford and the Ecological Region: The Politics of Planning. Guilford Press, New York (1995)

Meadows, D.H., Meadows, D.L., Randers, J., Behrens, W.W.: The Limits to Growth: A Report for the Club of Rome's Project. Universe Books, New York (1972)

Meijers, E.: Stein's 'Regional City' concept revisited. Town Plan. Rev. **79**(5), 485–506 (2008)

Morin, E.: Cambiamo strada. Le 15 lezioni del coronavirus. Raffaello Cortina Editore, Milano (2020)

Mumford, L.: Technics and Civilization. Harcourt, Brace & Company Inc, New York (1934)

Mumford, L.: The City in History: Its Origins, Its Transformations, and Its Prospects. Harcourt, Brace & World, New York (1961)

NGEU: Special meeting of the European Council (17, 18, 19, 20 and 21 July 2020), European Council (2020)

Olazabal, M., Chelleri, L., Waters, J., Kunath, A.: Urban resilience: towards an integrated approach. Paper presented at 1st International Conference on Urban Sustainability & Resilience, London, UK (2012). ISSN 2051-1361

Palermo, P.C.: Interpretazioni di forme. In: Clementi, A., Dematteis, G., Palermo, P.C. (a cura di) Le forme del territorio italiano, Laterza, Bari (1996)

Parr, J.B.: The polycentric urban region: a closer inspection. Reg. Stud. **38**, 231–240 (2004)

Pearce, F.: Eco-cities special: ecopolis now. New Sci. **191**, 8–9 (2006)

PNRR: Piano Nazionale di Ripresa e Resilienza (2021). https://www.governo.it/sites/governo.it/files/PNRR.pdf

Poëte, M.: Introduction à l'urbanisme: l'évolution des villes, la leçon de l'Antiquité. Boivin, Paris (1929)

Rifkin, J.: Un Green new deal globale. Mondadori, Milano (2019)

Rodrik: The Globalization Paradox. Democracy and the Future of the World Economy. Norton & Company, New York (2011)

Salingaros, N.: (2014). https://www.ilcovile.it/scritti/COVILE_807_Architettura_resiliente_4.pdf

Sassen, S.: The Global City. Princeton University Press, Princeton (1991)

Sassen, S.: The Global City: introducing a concept. Brown J. World Aff. **XI**(2), 27–43 (2004)

Tjallingii, S.P.: Ecopolis. Bakhuys Publishers, Leiden (1995)

Tjallingii, S.P.: Sustainable and Green: ECOPOLIS and urban planning. IUFRO World Series **14**, 43–63 (2004)

Vasenev, V.I., Dovletyarova, E., Chen, Z., Valentini, R. (ed.): Megacities 2050: Environmental Consequences of Urbanization. Springer (2018)

Wei, T., Wu, J., Chen, S.: Keeping track of greenhouse gas emission reduction progress and targets in 167 cities worldwide. Front. Sustain. Cities (2021). https://doi.org/10.3389/frsc.2021.696381

Wesley, J.: Regional divide: the Regional Planning Association of America and the Regional Plan of New York and its Environs, July 2008. AESOP-ACSP Joint congress, Chicago (2008)

Reversible Building Technologies and Unconventional Materials for the Circular and Creative Reuse of Small Centers

Francesca Giglio[✉], Sara Sansotta, and Evelyn Grillo

Department of Architecture and Territory, Mediterranea University of Reggio Calabria, Reggio Calabria, Italy

{francesca.giglio,sara.sansotta,evelyn.grillo}@unirc.it

Abstract. The paper describes the research experience transferred to the teaching activity of the authors through the description of design experiments carried out in small Italian centres, in which emergency conditions become drivers of innovation in the context of the Circular Economy and application of circular and creative reuse strategies of unused spaces. The post-pandemic conditions have accelerated the rethinking and shaping of neighbourhoods, cities and small centres, which have become places of social innovation and cultural growth, to all intents and purposes. At the same time, the key role of Design, widely reiterated by the European strategies related to the European Green Deal, draws attention to the activation of Open Innovation, Circular Design, Design for Disassembly dynamics, highlighting the social, economic and environmental impact of innovative and sustainable measures that focus on people and the environment, involving citizens in urban regeneration activities. The intervention strategies concern the rethinking of open spaces and meeting places in our cities through temporary, modular, off-site, mountable, demountable, aggregable structures to provide adequate services to citizens. Two case studies will be described as best cultural practices of design experiments oriented to the principles of Circular Building Technologies, Reversible Building Design, Design Thinking, providing students with scientific and technical competences on the topic, and transferring them to the construction sector.

Keywords: Creative reuse strategies · Small centres · Circular building technologies

1 Urban Reuse of Unusued Spaces: Circular Building Technologies as Enabling of Culture Creativity Processes

Cohesion and inclusion, green transition, innovation and digital technology are the themes that Europe is placing at the centre of its policies for tackling the pandemic and the climate crisis. The architecture design, at the different scales of intervention, is more and more stressed by a generalized condition of permanent environmental emergency that concerns different aspects, to which is added a post-pandemic condition that

© The Author(s), under exclusive license to Springer Nature Switzerland AG 2022
F. Calabrò et al. (Eds.): NMP 2022, LNNS 482, pp. 2778–2789, 2022.
https://doi.org/10.1007/978-3-031-06825-6_265

represents the current moment in which we live and to which the construction sector is called to respond. The emergency conditions represent a stimulating ground for experimentation, outlining new trajectories of innovation and culture creativity, as an enabling factor for the circular economy and culture processes [1] relating competitiveness, environment, social cohesion, innovation, new technologies, human capital and connections with the territory.

This theoretical aspect is particularly significant and relevant if transferred to the realities of small urban centres at national level, which represent a capital of humanitarian relations to be regenerated in material and immaterial terms. Of the approximately 8,000 Italian municipalities, about 70% do not exceed the threshold of 5,000 inhabitants and more than 10 million people live there, about 15,58% of the national population. This is not, therefore, a minor Italy, but the backbone, in demographic, economic and cultural terms, of our national community: small and medium-sized urban and rural centres characterised by environmental and artistic assets, productive and social activities [2, 3]. The challenges of the next decades will identify the innovative dynamics that are triggered in the design, to respond to the process of transition to the Circular Economy, in the construction sector, involving all stakeholders of the production / construction processes and promoting reuse and regeneration as a strategic model of sustainable urban development. The last year's emergency has accelerated the rethinking and shaping of neighbourhoods, cities and small centers, which have become places of social innovation to all intents and purposes.

Disused and underutilized spaces and buildings in all European cities and in small italian centers, can become opportunities for new jobs, the promotion of a collaborative economy and social innovations, through urban regeneration and re-use of buildings, in line with SDG 11 and SDG 15[1] [4]. The attention at European level on the cultural and creative sector is demonstrated within Horizon Europe 2021–2027, by a specific action (Cluster 2: Culture, Creativity and Inclusive society) dedicated to culture, European arts, European values and traditions as well as cultural and creative industries and technologies for cultural heritage, also supported by a dedicated research and innovation programme. The opportunities offered by European programmes can also be found in other programmes such as the regionally managed Structural Funds and InvestEU 2021–2027 to support the economy and investment in Europe. At the same time, the key role of Design, widely reiterated by the European strategies related to the European Green Deal [5], draws attention to the activation of Open Innovation, Circular Design Technologies, Design for Disassembly dynamics, highlighting the social, economic and environmental impact of innovative and sustainable measures that focus on people and the environment.

The intervention strategies concern the rethinking of open spaces and meeting places in our cities through temporary, modular, off-site, mountable, demountable, aggregable structures to provide adequate services to citizens. The redevelopment of spaces also concerns social innovation activities, involving citizens in urban regeneration activities [6]. What emerges is a vision in which it is necessary to redesign the life cycle of buildings and urban spaces, concerning their functional temporariness, considering the choice of

---

[1] SDG 11 Agenda 2030: Make cities and human settlements inclusive, safe, resilient and sustainable and SDG 15 Agenda 2030: (Life on land) concerning land saving.

materials compatible with the available resources and in relation to their possible reuse, thus pushing the need to opt for components to be assembled and disassembled, with a view to the circularity of resources, with a "Cradle to Cradle" approach [7].

With regard to this social and cultural context, the aim of the paper is to describe the research experience transferred to the teaching activity of the authors, through the description of design experiments carried out in small centres, with the intention of leading students to a reflection in which emergency conditions (including post-pandemic conditions) become drivers of innovation in the context of the Circular Economy and strategies of circular and creative reuse of unused spaces. The resulting need is to design a concept that can generate an identity and not just be a mere space, becoming a real place with positive externalities on the community through social innovation, regeneration and urban requalification projects. Sustainability, Technology and Flexibility are three common keys to the approach that the design experiments present, making a contribution not only to SDGs 11 and 15, but also to SDG 12[2]. The design experiments are therefore oriented towards the principles of Circular Building, Reversible Building Design, Design Thinking, providing students with scientific and technical skills on the subject, investigating the areas in evolution and applying them to the construction sector.

The themes of the projects have been chosen on the basis of the analysed planning documents and the needs highlighted, also addressing different uses. The projects are controlled throughout their executive process, through the use of innovative dry technologies, unconventional materials (recovered, recycled, reused), modular components, with the aim of proposing eco-efficient design and construction solutions with respect to the environmental characteristics of the places of intervention [8]. The working method is based on two key documents: the White Paper "Design tech for future" [6], a programmatic document - for the restart after Covid 19 - drawn up by a Task Force of designers, companies, enterprises, which highlights how design, combined with technological innovation, can play a central role in overcoming the health and economic crisis; - "Sustainable and Circular reuse of spaces & buildings", an Handbook that identifies new strategies for the integrated management of urban reuse, following the principles of circular economy [4].

2   Reversible Building Technologies for Design Experiments as Cultural Best Practices

In the face of the health and economic emergency, the world of design and architecture must develop innovative design models with a view to sustainability and the circular economy. It is tracing a path of economic growth that takes account of experience to date and redesigning new production and distribution guidelines close to the reference markets. In this scenario, transforming the current crisis into a new great beginning for humanity, enhancing the design culture, and giving it an important innovation function, is the ambitious challenge set by the working group that generated this document. Whatever the type of intervention, in this case in small towns and cities of a national character, the great challenge of urban regeneration will be to combine the need for density and

---

[2] SDG 12 Agenda 2030: responsible consumption and production.

efficiency with the need to create spaces and environments that are also safe from a health point of view [9]. In general, to anticipate and foresee users' needs, using current emergencies as opportunities for innovation, dialogue between clients, architects, designers, and planners will be indispensable. The latter will be called upon to find innovative solutions to solve practical problems, redefine aesthetics, respond to regulations, and suggest quality proposals and management systems for new architecture.

Architecture and design must once again have a social function that has been lost, becoming a self-referential tool. Attributing a symbolic and identity value to the built environment, in addition to its functional role, makes it possible to open up a new way of thinking, designing, creating, and communicating spaces. Specifically, the intention is to give more space and time to the design and study of materials and constituent components to optimize the quantities, processes, forms used, and their impact on management, maintenance, and dismantling. Even when buildings become obsolete or non-functional, the design of their life cycle must provide that they can be 'dismantled' instead of being demolished, using technologies that facilitate this process [10].

In addition, the Covid-19 health emergency raises the delicate and necessary question of rethinking urban spaces from the point of view of promoting a greater sense of tranquillity, security and liveability. In particular, the year 2020 has inaugurated a new sociality, and therefore the idea of public space has undergone a radical change. Therefore, the didactic and design challenge focuses on the innovation of the types of transformation that the public places and unused spaces of our small and medium-sized urban centres will undergo.

The result of this approach is the definition of a new framework of social innovation, in which space, environment and community become interconnected ecosystemic realities. They feed off each other in terms of positive values, creating a common language where social relations, technological trends, ethical and sustainable principles become the pillars of a new circular development model. The definition of an 'age-friendly city' [11] is taken as a starting point to manage and drive these design experiments. It rethinks public spaces concerning a plurality of uses, avoiding segregation by age group; it encourages the creation of conditions of accessibility, safety, and walkability of its urban structure; encourages the provision of daily opportunities for socialization and contact with nature (such as green and pre-treated footpaths, urban gardens, neighborhood micro-parks); enhances the knowledge of elderly people, who preserve the historical memory of traditional places and activities, also to the protection of cultural heritage (material and non-material) and the promotion of sustainable lifestyles in our cities. Cities are understood here as complex systems where transformations take place. Changes in society and different production needs redesign territories and urban spaces, often favouring some areas and marginalizing others. Until a short time ago, the same useful buildings could become empty and abandoned. For example, the solution of consuming new land for new buildings, together with the production of waste materials from the demolition of buildings, contributes to the negative balance of the city system in terms of consumption of resources and creation of waste.

Aware of the fact that statements such as 'zero waste' and 'zero emissions' are more often than not just slogans. It is necessary to start taking action at all levels, in an integrated and conscious manner, so that at micro and macro-level, behaviour that increasingly tends towards the principles of the circular economy [12]. Therefore, accelerating the transition to the circular economy is the current challenge for institutions, cities, production districts, organizations, and individuals. Alongside the many virtuous initiatives at the micro-level, systemic and integrated circular models must increasingly develop and assert themselves. The local and urban dimension is where the transition process can be favoured. The following paragraph describes some paradigmatic examples of implemented design experiments as cultural best practices for small centres' circular and creative reuse.

3 Design Experiments

To test its applicability and effectiveness, the proposed reading method was applied to two case studies, intended as good practices of circular experimentation to revitalize historic city centres, assuming that they could constitute examples of social, managerial, and technological innovation induced by Emergential Innovation, about the quality of the design approach, the implementation methodologies and the application experiments. In order to do this, innovation and regeneration policies had to be taken into account at the urban and local level; innovating processes means rethinking them in a circular way, considering the entire life cycle of products. Regenerating the city, its unused or underused spaces and buildings means rethinking their use and new productive, recreational and social functions. The approach to urban regeneration can be bottom-up for cities or, vice versa, top-down, adopting different models for dealing with specific situations [13]. In addition to these practices, it is advisable to combine adequate communication and storytelling of the regeneration process, also to make citizens an active part of the change.

3.1 Akro(C)s – Polis Historic Centre of Cosenza (Italy)

The first experiment in temporary design, called "Akro(C)s - Polis"[3], deals with the theme of public space to guarantee a more generalised, but at the same time precise, project that could univocally embrace most of the interstices of which the historic centre of Cosenza is composed (see Fig. 1). Imagining a post-pandemic project situation helped choose the location for the project, but especially in identifying the objectives to be achieved through it. It was decided to focus on one of the sectors that suffered most from the crisis during the pandemic period, namely the entertainment sector. Starting from this assumption, a network of interstitial spaces has been defined, which make up the historic centre and which, connected, generate an itinerant, mobile and temporary theatre, capable of hosting various activities linked to the world of art and entertainment. The objective of the project management concerns the design of spaces that are flexible in their use and space to be able to host all the activities that are inherent to the theatrical arts.

The idea for the project stems from the choice of some spaces in the historic centre which are very different from each other in terms of shape and size and which have

---

[3] Project team: F. Armocida. Project coordinators: F. Giglio with S. Sansotta, E. Grillo.

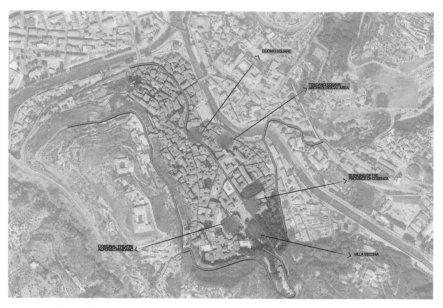

Fig. 1. Plan of the Historic Centre of Cosenza and points of attraction.

been regularised through a square grid capable of assigning a geometric measure and therefore a structural step for the project itself (see Fig. 2). The mesh is the generator of a grid of sliding guides for rotating panels which, through the simple operations of moving and rotating, circumscribe spaces, scenes, interchangeable moments. Many of these panels are updated with certain functions, which are also flexible, at the discretion of the user and directly linked to the planned activities.

Finally, two micro-architectures are inserted, also generated by the same mesh and the same panels, which act as the heads of the entire route and contain services and relaxation areas. The aspects of circularity are used not only for the temporary use of these micro-architectures, thinking about a limited time of the equipment but also through the materials that constitute the structural and buffering elements of them. "Akro(C)s - Polis" can be understood as good practice for circular design as it respects the above-mentioned criteria both for the use of unconventional materials and because all the materials used can be easily found in other production chains and can be reused in other fields, even different from the original one, once disassembled.

The principle of single-materiality is also promoted through the use of simple, reversible connections and single-material packages that do not generate high levels of production waste (see Fig. 3). The reference design for the composition of the load-bearing structure is inspired by the 'Brasserie 2050 Restaurant, Frits Ham by Overtreders W Architecture Office', a project in which the structures of industrial shelving become a load-bearing element of the architecture itself. The metal supporting structure also has the task of housing the set of guides for the upper sliding panels and the projecting roof of the pavilion. The generating principle of the system on which the panels slide and rotate comes from the technological operation of the roto-translating doors, which,

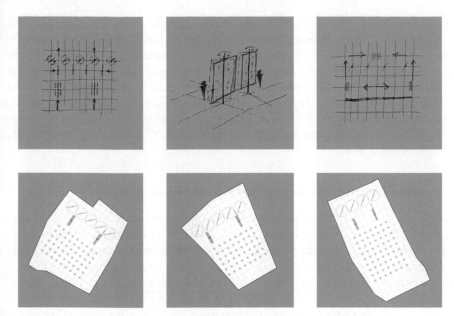

Fig. 2. Square Mesh grid and guide rails for rotating panels.

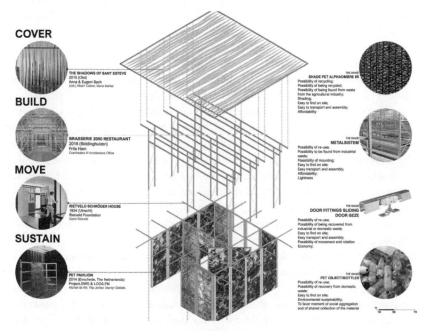

Fig. 3. Technological detail of the microarchitecture structure.

through a central pin on nylon wheels, allow both sliding and rotation around the same. The sliding guides are connected starting from the crosses positioned at the intersections of the mesh on which the other rectilinear sliding guides are connected to form a grid-connected in turn to the supporting structure (above) and the platform through special bearings (below). The roto-translating panels, made of waste plastic materials, are assembled directly on-site starting from the central load-bearing structure consisting of a circular metal profile with a hollow section previously welded to two fins connected to pre-drilled metal plates and connected, in turn, to the 'C' profiles 100 that form the frame of the panel itself.

3.2 Urban Gardens Share for Villaggio Gesso in Messina (Italy)

If in "Akro(C)s - Polis", the issue of the health emergency is understood as good practice for the revitalization of the historic centre at the service of the entertainment sector, in the second design experiment, the declination of the transformation of urban space, as a response to the practices of social life that we are experiencing, calls into question the consolidated models of habitat, bringing out the criticality and limitations of the environments we inhabit. 'Villaggio Gesso: evolution of public spaces in the post-pandemic era'[4] deals with the theme of public space concerning the needs of the village in the province of Messina, Italy, to trigger mechanisms of sharing and social integration, making people responsible for looking after their places and learning about nature through the design of a small urban gardens share. Meeting spaces need to be rethought to create a new way of living. Small villages are particularly important in this climate of rebirth: most affected by the emergency, as they are often disadvantaged by the lack of services, they become potential hubs of social innovation as cradles of space, safety, and nature, far from metropolitan concentrations.

The project idea stems from the identification of a degraded area in the village that connects the Mother Church of Gesso (ME) to an ancient medieval church in a state of abandonment (see Fig. 4). The design of the urban gardens share placed along the path linking the two focal points of the project aims to rehabilitate the area on the one hand and on the other to trigger a mechanism of attraction and repopulation of the public space. (see Fig. 5). The circular aspects of the design process concern the temporary use of the structure that defines and regularises the garden and the use of materials that come from waste materials, otherwise destined to be considered waste. In this sense, the last step of the case study reiterates the concept of circular economy in which waste becomes a resource for a 'new' project. In addition, the use of materials dry-assembled through-bolted or interlocking connections from Japanese culture is the translation of dictates from Circular Design.

Considering the redevelopment phase of the old town centre, the elements that will have to be dismantled for the laying of the pavement define two scenarios also related to the Circular Design if at least 70% of the pavement remains intact, the recovered material will be reused for the renovation of the medieval church. Otherwise, the material will be processed in a new form, laying the basis for the construction of the wooden plank pavement. About the basic grid, the path is composed of a structural frame in laminated wood, which takes inspiration from the 'Boxing Club project by FT Architects' against

---

[4] Project team: M. Carlier. Project coordinators: F. Giglio with S. Sansotta, E. Grillo.

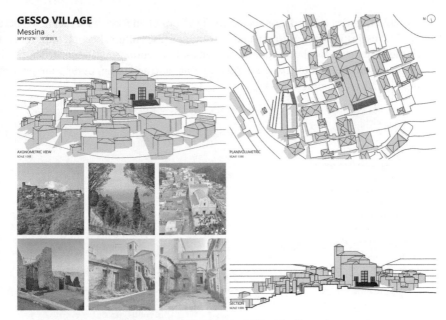

Fig. 4. Identification at plan level of the Villaggio Gesso intervention area.

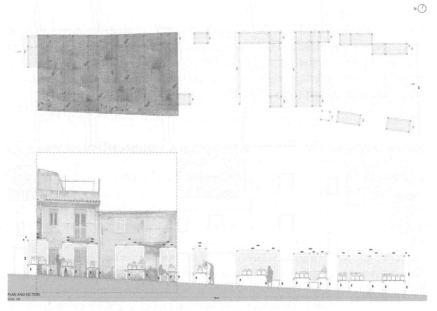

Fig. 5. Plan and Section of the urban garden.

which a double frame is created, connected by a system of anchors with metal washers and bolts. The bearing structure system, taken from Japanese culture, becomes an element on which the containers for the vegetable garden and the corten steel seats are grafted. The bamboo canopy is used to shade the path on one side and to encourage the growth of climbing plants on the other (see Fig. 6).

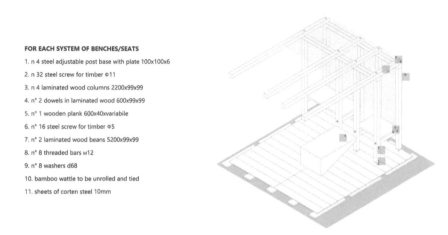

FOR EACH SYSTEM OF BENCHES/SEATS

1. n 4 steel adjustable post base with plate 100x100x6
2. n 32 steel screw for timber Φ11
3. n 4 laminated wood columns 2200x99x99
4. n° 2 dowels in laminated wood 600x99x99
5. n° 1 wooden plank 600x40xvariabile
6. n° 16 steel screw for timber Φ5
7. n° 2 laminated wood beans 5200x99x99
8. n° 8 threaded bars M12
9. n° 8 washers d68
10. bamboo wattle to be unrolled and tied
11. sheets of corten steel 10mm

Fig. 6. Technological detail of the bench/seating system.

Finally, it is emphasized that the methodology of the two experiments focuses on disassembly, transformation, and reuse as a means to bring the construction closer to the continual use loops of resources in its systems, products, and materials [15]. A world where buildings and materials are circulating in continuous reuse loops represent multiple value propositions and form a base for Circular Economy (CE) in the construction sector.

4 Social Impact: The Benefits to Social Communities

The common thread linking these experimentations and others carried out is based on the concept of "Reversibility", which is defined as a process of transforming buildings or dismantling their systems, products and elements without causing damage [14]. The principles of reversibility applied to construction technologies can generate a virtuous circle in order to promote RBD the rebirth of small towns of our territory.

The active and participatory involvement of the population becomes an integral part of a virtuous process of cultural and social reactivation of the villages through temporary and reversible activities which allow the entire population to contribute to the active transformation of the places. Therefore, intervention models combine cultural and material identities with simple building systems, interpreting reversibility as a paradigm in the relationship with the context, in the urban connections, in the revitalization of dead areas. The need to intervene in emergency conditions and for emergency conditions is a principle driving the objectives of constructive and productive innovation in a circular economy perspective, intending to use material and immaterial resources through socially and materially sustainable models for intervention contexts. Replicable experimental models allow for possible future developments in other contexts where similar problems require technological and operational responses on the territory.

5 Conclusions

The common parameters to the two experiments, which led to the transformation of urban centres, can be summarised and identified as elements that can be repeated in other contexts:

1. Volume dimensions that are compatible with desired scenarios;
2. Position of the core elements that is not restricting number of use options;
3. Capacity to carry loads and provide space for services for desired upgradability and use scenarios;
4. Disassembly potential that takes care of separation of main building functions.

The need to respond to the needs posed by an emergency, in a short time, finds answers through reversible, low-cost technologies and construction processes with reused and/or recycled materials, to ensure the circularity of processes, with reference to the sustainable consumption of non-renewable resources and waste reduction. The urban reuse of unused spaces, especially in smaller centres, through circular building technologies, becomes an enabler cultural creativity processes and regeneration strategies. Minimising waste, designing flexible spaces, keeping products and materials in use for longer, regenerating natural systems, converting the linearity of transformation processes towards their circularity, are some of the possible actions of the Circular Design approach through the application of circular building technologies.

Transferring research activities concerning strategies for creative urban reuse through circular building technologies to teaching activities, contributes to a necessary and fundamental dissemination of knowledge for the actual transition towards a circular economy also in construction sector. Repeatable results are also hoped for in different contexts but respecting the diverse needs of the users and the place, where temporary architecture will be needed in small and medium-sized villages. Once again, the intervention of municipalities and all other stakeholders is fundamental.

References

1. Antonini, E., Boeri, A., Giglio, F.: Emergency Driven Innovation. Low Tech Buildings and Circular Design. Springer, Cham (2020). https://doi.org/10.1007/978-3-030-55969-4
2. Beccatini, G.: La coscienza dei luoghi. Il territorio come soggetto corale. Donzelli, Roma (2015)
3. Bulsei, G.L.: Essere comunità in condizioni averse. Sociologia urbana e rurale, n 10 (2016)
4. Partnership on Circular Economy and Sustainable Land Use: Sustainable and circular re-use of spaces and buildings – Handbook (2019)
5. COM: 640 final "The European Green Deal" (2019)
6. White Paper "Design tech for Future". Design e tecnologia per progettare il mondo dopo il COVID -19, Design Tech

7. McDonought, W., Braungart, M.: Cradle to Cradle: Remaking the Way We Make Things, North Point Pr (2002)
8. Giglio, F.: Low Tech and unconventional materials. Measure, Time, Place. Techne. FUP Press, Firenze (2018)
9. Verso il Xv Rapporto: Report I Snpa 09/2019- Qualità Dell'ambiente Urbano. 10 AZIONI E STRUMENTI PER LA SOSTENIBILITÀ LOCALE ISBN 978-88-448-0973-7 (2019)
10. Durmisevic, E.: Circular Economy in Construction Design Strategies for Reversible Buildings. University of Twente, The Netherlands (2019)
11. OMS: Network of cities tackles age-old problems. Bull. World Health Organ. **88**, 406–407 (2010)
12. Altamura, P.: Costruire a zero rifiuti: strategie e strumenti per la prevenzione e l'upcycling dei materiali di scarto in edilizia. Costruire a zero rifiuti (2015)
13. Tricarico, L., De Vidovich, L.: Economie di prossimità post Covid-19. Riflessioni con alcuni riferimenti al contesto urbano italiano (2021)
14. Lespagnard, M.J.P., Cambier, C., Vandervaeren, C., Galle, W., De Temmerman, N.: Understanding the design process and the impact of reversible design tools and strategies through timeline development. In: Proceedings of the IBA Crossing Boundaries conference. Zuyd University of Applied Sciences (2021)
15. Askar, R., Bragança, L., Gervásio, H.: Adaptability of buildings: a critical review on the concept evolution. Appl. Sci. **11**(10), 4483 (2021)

Green Infrastructures and Water Management. Urban Regeneration Strategies to Face Global Change

Irene Poli[✉] and Paola N. Imbesi

PDTA Department, Sapienza Università di Roma, Rome, Italy
{irene.poli,paola.imbesi}@uniroma1.it

Abstract. The global crisis that exacerbates pathologies of the contemporary city and territories directs research and urban planning experimentation towards new approaches aimed at defining integrated and multi-scalar strategies of regeneration, based on the concept of resilience. At a global level, Green and Blue Infrastructures constituite a shared and consolidated field of experimentation to pursue the rebalancing of urban and territorial systems through place-based policies and approaches and a strong involvement of socio-economic actors and local communities in decision-making and management processes, designing new sustainable and resilient assets and encouraging the emergence of virtuous and ecologically oriented behaviors. In this context, *water* resource is a structural and strategic component for renaturalization, adaptive and resilient management of risks, and revitalization of territories, but also a leitmotiv of social-innovation and social-inclusion processes. Water management allow to restore ecological functioning of natural cycles, to integrate virtuous forms of recycling and reuse, to adopt sustainable mobility, and to activate eco-friendly uses of public spaces with recreational and inclusive character. The cases illustrated represent, on the international scene, emblematic and consolidated examples: the case of Copenhagen with regards to the pivotal role of urban areas in addressing climate change; the case of Philadelphia, both in terms of private and community involvement in pursuing a tangible improvement of the *cadre de vie*.

Keywords: Green infrastructures · Water management · Urban regeneration

1 Urban Regeneration and Global Challenges

The current growing attention of national and urban European Agendas and great international organizations towards issues of *Sustainable Urban Development* (WCED 1987) is closely related to the achieved awareness of the pathologies induced on the territory by unlimited urban growth. Pathologies such as fragmentation of ecosystems, loss of biodiversity, progressive artificialisation of hydrographic networks, waterproofing of soils together with degradation of landscapes that hold the identity of places. These global phenomena are increasingly exacerbated by the emerging environmental challenges related to climate change, whose complex dynamics generate considerable and

© The Author(s), under exclusive license to Springer Nature Switzerland AG 2022
F. Calabrò et al. (Eds.): NMP 2022, LNNS 482, pp. 2790–2799, 2022.
https://doi.org/10.1007/978-3-031-06825-6_266

now unavoidable disciplinary implications for the planning of urban systems. Intensification of extreme climate events is, in fact, generating further significant impacts on ecosystems and human health, exacerbating the already precarious geo-environmental balances, especially in those territories characterized by high levels of socio-economic fragility and affected by high levels of urbanization (Poli and Uras 2018).

The pandemic has increasingly exacerbated these structural issues, bringing to global attention the impacts on human well-being provided by ecosystem services (MEA 2005) and the presence of "nature in the city". The current phase of global crisis that aggravates the imbalances of the city and territories has directed Urban Planning research and experimentation towards new approaches aimed at defining integrated and multiscalar strategies of urban and territorial regeneration (Gasparrini 2016; Oliva and Ricci 2017, Arcidiacono 2017). In the wake of the disciplinary tradition addressed to analytical and design approaches in an ecological-environmental perspective, these must address contemporary urban pathologies and risks related to natural cycles and innovate socio-economic structures in the real areas of their phenomenology and their forms of interaction (Ravagnan et al. 2019).

Urban and territorial regeneration strategies move simultaneously on several levels, physical-morphological, functional, ecological-environmental and landscape as well as socio-economic, in a *resilience* perspective which prefigures specific integrated lines of intervention to get out of the ongoing environmental, economic, health and social crisis, in a logic of mitigation of environmental risks related to land use, adaptation to climate change in urban transformation interventions and promotion of new ecologically oriented urban collective values (Ravagnan 2018). In this sense, on a global level, Green and Blue Infrastructures (GI) constitute a shared and by now consolidated field of experimentation for deployment of strategies to rebalance urban and territorial systems through place-based policies and approaches and a strong involvement of socio-economic actors and local communities in decision-making and management processes, designing new sustainable and resilient assets and encouraging the emergence of virtuous and ecologically oriented behaviors (EC 2015; EPA 2011, 2013).

GI used as an alternative or in synergy with traditional *gray infrastructures*, overcome traditional monofunctional conception of ecological networks (Gasparrini 2018a, b) and contribute to the improvement of environmental risk management and urban and community resilience. Such infrastructures, in addition to constituting reference for ecosystems with different levels of naturalness and resilient structure to reduce vulnerability to environmental risks and climate change (EC 2013), configure a multifunctional and continuous frame of open spaces that contributes to the construction of a sustainable, adaptive and resilient public city (Gasparrini 2016), reorganizing the economic base of cities, triggering new urban metabolisms and new lifestyles (Poli and Ravagnan 2016), as it constitutes the "backbone of the urban and environmental design of the contemporary public city" (Arcidiacono 2017).

2   Green Infrastructures and Water Management

The integrated and multiscalar construction of GI, structural components with a strong strategic value, directs planning innovation towards new references aimed at formulating of proactive solutions to overcome the crisis phase.

In line with objectives of the *European Urban Agenda* (EU 2016), these solutions constitute inputs for more sustainable and "reliable" physical and socio-economic assets (Ravagnan and Poli 2017), which combine, through implementation of Nature-Based Solutions, protection of fundamental resources and natural capital with social inclusion and economic development (Andreucci 2017). At the same time, in line with guidelines of the *New European Bauhaus* (2020) for the achievement of *Green New Deal*, GI allow improvement of quality of living, focusing on sustainability, aesthetics and inclusion, through enhancement of open spaces, promotion of soft mobility, participation and sharing in the reduction of environmental impact caused by human activities in the Anthropocene era, based on ethical value of common goods. This objectives converge towards lines of action that require a cultural and technical sensitivity of urban planning towards management of environmental networks and resources in design terms (Gasparrini 2018a, b), but also new building, urban planning and technological regulatory frameworks, to be conceived at the vast scale and implemented at all levels of intervention. To these ends, inter-institutional governance systems of variable geometry are also needed, capable of elaborating and implementing construction of GI as part of partnerships between territorial authorities, planning agencies, institutions and other stakeholders (park authorities, productive clusters, associations, etc.). GI can be constituted as places of convergence of projects and funding at international, national, regional and local level (Arcidiacono 2017), but also experiences of participatory and shared planning and management of common goods, based on re-appropriation by communities of degraded or abandoned areas, also through innovative types of reuse of public assets, such as urban agriculture or other eco-friendly temporary uses.

In this context, *water* resource represents par excellence "a vector of sustainability" (Masboungi 2012), a component at the same time structural and strategic for renaturalization, adaptive and resilient management of risks and resources, and the revitalization of territories, but also a leitmotif of social-innovation and social-inclusion processes (Ravagnan et al. 2019). Projects related to it allow to restore ecological functioning of natural cycles, to integrate forms of recycling and virtuous reuse, to adopt sustainable forms of mobility, and to activate eco-friendly uses of public spaces from recreational and inclusive character. The strategic role of water management is reflected in most recent international references such as the *Sustainable Development Goals*, which crosses all, but in particular goals *15 Plant a tree and help protect the environment, 6 Avoid wasting water, 7 Educate young people on climate change.* Operationally, these goals converge in the design, multiscalar and multidimensional, of GI oriented to the coordination and linking of tangible measures and interventions to contain climate-changing emissions, mitigate risks and adapt to changing climate conditions, through specifically nature-based and local-based approaches (EEA 2015; MEA 2005). Increasingly, therefore, water management becomes an integral part of Territorial Government, going to be a privileged field of innovation for the reconfiguration of the strategic, regulatory and programmatic aspects of urban planning. This is reinterpreted, at the global level, primarily in a new integrated approach between planning and risk management, for the construction of new sustainable and resilient urban assets, differently declined with reference to the existing city, characterized by density and compactness, or to the metropolized territories, characterized by extensive urbanization and discontinuity (Gralepois and Guevara

2015). The reports of the Intergovernmental Panel on Climate Change (IPCC 2021), in fact, highlight the direct relationships between climate change and risks (Galderisi 2014) of urban and territorial systems, already connoted by precarious geo-environmental and socio-economic balances and high levels of urbanization. This requires the need to govern these relationships with *Climate Plans*, brought back within the framework of the tools of Territorial Government (Reckiena et al. 2018), in order to implement integrated mitigation and adaptation strategies, considered synergistic, respectively, to contain the causes of climate change and to address its consequences and socio-economic and environmental costs (Castellari et al. 2014; Poli and Uras 2020). At the same time, however, instances related to water management, related to saving and reuse, and the related Sustainable Urban Drainage Systems solutions (standards and technologies for drainage, treatment and recycling of rainwater and wastewater), are combined with instances of regeneration of marginal and degraded areas both within dense urban fabrics and in contemporary territories (Sgobbo 2018), to raise the quality of living and, more generally, the quality of life. This has led, in many cases, to identify virtuous forms of partnerships with various stakeholders involved in implementation of interventions, always in the face of a public direction, as well as the involvement of comunity in their management and care starting precisely from the concept of water as a common good. A process of awareness and inclusion that brings new lifestyles, social innovation, site-specific creativity, but also virtuous forms of circular economy, marked by the concept of environment as a unifying and universal value (Poli and Ravagnan 2016).

The two cases illustrated below represent, in the international panorama, emblematic and consolidated examples of the aspects mentioned above: the case of Copenhagen regarding the pivotal role of urban areas in addressing climate change; the case of Philadelphia both in terms of the involvement of private individuals and the community in pursuing a concrete raising of the *cadre de vie*.

3 Risk Management for Mitigation and Adaptation: The Copenhagen Case

The Danish context appears to be at the forefront in terms of awareness of the relationship between cities and environmental risks. The adaptation plans, mandatory since 2013, have been integrated into the urban planning system by national law in 2018 (Reckiena et al. 2018). Copenhagen is one of the most livable cities in the world. With one of the lowest CO_2 emissions per capita among European cities (at 2.5 t/inhabitant per year), it has embarked, with tangible measures and actions, on a path of complete transition to a sustainable economy (CO_2 neutrality by 2025). The attention to environmental issues and to the balance between natural and anthropic space traditionally characterizes the approach to the territory of the City, starting from the famous Fingerplaner of the metropolitan area in 1949.

City Administration has launched a process of cultural, technical and managerial updating aimed at the implementation of tools, measures and actions for adaptation to climate change, with the aim of making the City safer and more livable, which also led to its recognition as European Green Capital in 2014. Particularly after the rainfall of extraordinary intensity that hit the city in the years 2010/2011, causing serious inconvenience and costly damage especially within the tissues of the consolidated city, morphologically and structurally unprepared to handle persistent emergency.

The *Copenhagen Climate Adaptation Plan*, approved in 2011, identifies the challenges posed by climate change that the City will have to face in the medium term and the most appropriate solutions as an opportunity to start processes of urban regeneration and to define more sustainable urban assets. The flexible and modular character of the strategy is expressed in the identification of different risk gradients (articulated by probability and costs) that determine three levels of adaptation, declined in several scales of intervention (region, municipalities, districts, roads, buildings). A first level with prevention initiatives (construction of dams, adaptation of the sewerage system and stormwater management system); a second level with interventions to limit the extent of damages (implementation of warning systems and monitoring of rainfall levels, construction of flood basins); a third level with measures to reduce vulnerability (artifacts to cope with flooding, pumps and drainage systems). The Plan identifies among the main adaptation measures the construction of a *Green structure* as a structural invariant consisting of public and private green areas within the most compact tissues, together with public spaces and roads, as well as natural and agricultural areas. It is aimed at reducing and preventing flooding due to rainfall; balancing and moderating temperature; creating shade and air circulation, to reduce energy consumption and increase comfort; reducing air and noise pollution; preventing stress and creating leisure opportunities for the population; protecting and developing biodiversity. The *Green structure* is conceived and realized at different scales of intervention: at the large scale, providing green wedges and environmental corridors that connect the free areas more internal to the urban system with the natural and agricultural areas; at the urban scale, with the aim of optimizing the existing high ecological-environmental quality, the Plan provides for the creation of an integrated system of interconnected components through continuous links. In the areas of development, however, the *Green structure* should be the reference structure for the new assets (Poli and Uras 2020). The adaptation measures are modulated, then, in correspondence of the different urban tissues, through the inclusion of new functions compatible with the protection and historicity of green areas of the parts of the oldest plant (i.e. fountains, water features); the reconfiguration and enhancement of free areas within the more compact and consolidated tissues, also using innovative technological solutions (i.e. green roofs, green walls), as well as reconnecting and inserting in the system even the most interstitial areas (i.e. courtyards, parking lots, tree rows along the road system).

The *Cloudburst Management Plan* of 2012 puts in coherence more than 300 projects for adaptation and mitigation to extreme weather phenomena, for the 8 most vulnerable urban basins, providing the integration between "traditional" interventions and blue-green solutions. The implementation plans of the pilot projects (*Copenhagen Concretization Plans*) are multidisciplinary tools aimed at bridging the gap between urban planning and site-specific solutions through the identification of a *Typology-based Cloudburst Toolkit*. Designers, planners, sociologists, economists, biologists, geographers, information specialists and communication experts are involved in the definition of the projects, which are mainly aimed at the redesign and configuration of public spaces, interacting with service companies, public and private entities, investors and citizens. Both of these

Plans have been established as references for the drafting of the new Territorial Government system for Copenhagen and its metropolitan region. In particular, the new *Finger Plan* of 2013, at the metropolitan scale, and the new *Københavns Kommuneplan* of 2019 have operationally integrated measures and interventions aimed at adapting to climate change within the urban strategies and regulatory frameworks of reference (Fig. 1).

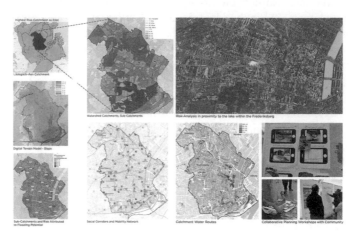

Fig. 1. Cloudburst Management Plan. Source: https://www.asla.org/2016awards/171784.html

4   Circular Economy for Quality of Life: The Case of Philadelphia

In the tradition of the well-established experimentation of GI in the United States urban settings (Clean Water Act 1972; NRC 2009; EPA 2011) in Philadelphia, one of the oldest municipalities and the sixth-most-populous North American cities, since 2011 the Water Department (PWD) has reached an agreement with the Pennsylvania Department of Environmental Protection (PADEP) and the Environmental Protection Agency (EPA) to develop a green, sustainable stormwater management infrastructure plan that reconstructs the natural water cycle in order to reduce wastewater overflows. Instead of choosing for expand traditional gray infrastructure, they invest in GI to restore nature's ability to capture water and treat it as a resource. The same year, PWD launched the *Green City, Clean Waters program* (GCCW) to come up with federal Clean Water Act regulations (PWD 2011).

The City of Philadelphia is crossed by two major rivers that historically collected water from an extensive network of surface and underground streams. In urban expansion over the past 100 years, many of these streams have been tombed and converted to concrete sewer systems, and most of the city's naturally porous surfaces have been paved over and built upon, making it impossible for stormwater to be absorbed, thus placing barriers to natural processes. Such often aging infrastructures are not always able to accommodate water that overflows directly into rivers and streams collecting most surface pollutants (Shade and Kremer 2019).

The GCCW program aims to primarily use GI to convey and reuse stormwater, decreasing runoff into traditional gray infrastructure. The environmental success of the program is measured using the concept of *"Green Acres"*, defined as the amount of GI sufficient to handle one inch of stormwater from one acre of drainage area (PWD 2016). The creation of these "Acres" represents an opportunity to return a diverse range of green areas (rain gardens, green roofs, draining sidewalks, etc.) to the city on both public and private lands. The private sector is involved in investing in the "greening" of the public city and in contributing to the creation of paths of continuity, through investments in Nature-Based Solutions in new development and redevelopment projects. This also includes providing for re-permeabilization of private land through a combination of incentive devices of tax relief and cost-sharing of implementation. The GCCW was the first program in the United States to use GI in parallel with gray infrastructure to manage stormwater runoff in critical weather conditions. Construction of GCCW projects began in 2011 and will continue through 2036. In an effort to test a wide range of GI types, GCCW is proposing eight different management models to reduce the amount of impervious surfaces within city: green streets, green schools, green public facilities, green parking, green outdoor spaces, green industry, business, commerce and institutions, green alleys, driveways and walkways, and green homes. A variety of GI creation and redevelopment practices have been implemented through the development of GCCW projects throughout the City, such as rain gardens, artificial wetlands, tree trenches, planters, swales, and green roofs, as well as non-vegetative practices such as cisterns, trenches, bump outs, basins, and permeable pavement. Managing water resources in an environmentally friendly and sustainable way has involved a multi-pronged approach, from distributing free rain barrels to residents for home watering, to planting rain gardens in strategic locations in parks, sidewalks, and rooftops, to the more challenging and expensive task of replacing 30% of the city's concrete streets and sidewalks with porous ones (PWD 2018). PWD is also working to leverage private capital to support the shift from gray to green by developing compelling financial products to encourage investment by demonstrating long-term energy and resource savings (Econsult Solutions 2016).

In June 2021, the GCCW program successfully marked a decade of activity: to incentivize this green transition in "America's poorest big city", interventions in part were funded through tax leverage, such as through water bill tax relief. The most important challenge the GCCW is pursuing has been a shift in the cultural paradigm of thepopulation, introducing themes of resource conservation and recycling into local communities and building the broad support necessary to sustain a water resource management plan for the area with a 25-year horizon, creating strong partnerships with an ever-widening network of stakeholders.

As the PWD Commissioner states "we continue to be dedicated to our vision: combining widespread, neighborhood-based green stormwater infrastructure investments and targeted improvements to our robust sewer system to strategically reduce combined sewer overflows" (Statement to the Greater Philadelphia Sustainable Business Network, 2020). The goal of creating green acres in the first 10 years was successfully pursued: despite major obstacles to the COVID-19 pandemic, the department worked with its partners on the PADEP and to implement the first 2,148 *Acres*, a figure that to date has equaled nearly 800 GI sites with nearly 2,800 individual rain gardens, stormwater trees, ponds, and other green systems. By zooming out, these systems represent an estimated

2.7 billion gallons reduction in annual overflows, as well as a significant boost in community green spaces by improving the livability and sustainability of neighborhoods throughout the city (Fig. 2).

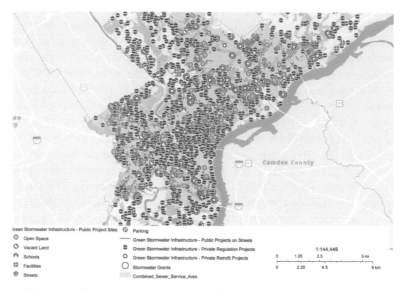

Fig. 2. GCCW interactive map. Source: https://water.phila.gov/green-city/

The gradual implementation of the program is expected to achieve for urban community not only environmental and economic benefits but also benefits of social equity. As far as environmental and economic benefits are concerned, the PWD is carrying out a project to monitor environmental effects of interventions carried out. The GCCW has in fact constituted an enabling platform for investments and funding sources, which has so far catalyzed more than 9 million dollars of investment to support biodiversity. In terms of social equity, the majority of investments under the program have focused on implementing public and private services and facilities to improve the environmental and physical health of urban habitats with a particular focus on low-income communities. This has also involved the development of educational projects, for awareness and involvement, but also cultural activities, related to the theme of water and environmental resources (street art, art contests, cultural events, etc.), of dissemination of best practices and re-appropriation of places involved in the projects by the local community (public assemblies, educational projects, joint public-private initiatives, etc.). The program has also allowed local architecture, engineering and landscape companies, to test and codify innovative stormwater management technologies and services and propose them in other urban realities such as Washington, DC and New York City, going on to identify an exportable and replicable water management "model" (Meerow and Newell 2017). As a testament to the open, processual and emblematic nature of the program, the PWD has created an inetrative map, progressively augmentable, including by the community, with all the projects implemented and underway in the municipal territory.

5 A Brief Conclusion

If in recent years events related to climate change have grown in number and intensity,, the areas that have suffered and will suffer the greatest impacts are large urban metropolises.In this perspective, urban planning must and can (re)play an important role, taking into account the challenges posed by climate change, starting from the assumption of strategic environmental and ecological dimension to recompose the design of urban and metropolitan territories. The multiplicity of risks, as well as their dynamic and cumulative interaction, requires planning strategies guided by adaptive logic in such a way as to structurally rethink space of daily life without limiting itself to the simple "safety" but redefining goals and themes of the urban discipline focusing on concepts such as blue and green infrastructure, regeneration and raising the quality of life. Dynamics of climate change require a profound revision, not only of approaches, but at the same time of tools at the service of planning activities: cases presented in this paper represent the need/ ability to take a closer look and anchored to the physical and social reality of the places, with an attention to the spatial project but also social and economic to recognize the peculiarities, opportunities and to ensure not only quality urban landscapes, but also externalities and interdependencies that only green, efficient and safe territories and cities can offer.

References

Andreucci, M.B.: Progettare Green Infrastructure: tecnologie, valori e strumenti per la resilienza Urbana. Wolters Kluwer Italia, Assago (2017)

Arcidiacono, A.: Nuove priorità per il progetto urbanistico, pp. 273–272. Le infrastrutture ambientali nel progetto di piano, Urbanistica informazioni n (2017)

Castellari, S., et al.: Analisi della normativa comunitaria e nazionale rilevante per gli impatti, la vulnerabilità e l'adattamento ai cambiamenti climatici. MATTM (2014)

Clean Water Act (1972). https://www.epa.gov/laws-regulations/summary-clean-water-act

EC: An EU Strategy on adaptation to climate change (2013)

EC: Towards an EU Research and Innovation policy agenda far Nature-Based Solutions & Re-Naturing Cities, Final Report Horizon 2020, Luxembourg (2015)

Solutions, E.: The Economic Impact of Green City, Clean Waters: The First Five Years. Sustainable Business Network of Greater Philadelphia, Philadelphia (2016)

EEA: Exploring nature-based solutions. The role of green infrastructure in mitigating the impacts of weather and climate change-related natural hazards, Report n. 12 (2015)

EPA: A Strategic Agenda to Protect Waters and Build More Livable Communities Through Green Infrastructure (2011) and EPA: Green Infrastructure Strategic Agenda (2013)

EPA Environmental Protection Agency Report (2011). http://www.epa.gov/compliance/resources/cases/civil/caa/ethanol/index.html

EU: Urban Agenda for the EU. Pact of Amsterdam. Agreed at the Informal Meeting of EU Ministers Responsible for Urban Matters. Amsterdam, The Netherlands (2016)

Galderisi, A.: Cambiamento climatico, rischi e governo delle trasformazioni urbane: quali prospettive per l'integrazione?. Urbanistica Informazioni n. 257 (2014)

Gasparrini, C.: Un'urbanistica del paesaggio per città resilienti. In: Storchi, S. (ed.) La qualità nell'urbanistica. MUP Editore, Parma (2016)

Gasparrini, C.: Infrastrutture verdi e blu. Una priorità nazionale per la pianificazione urbanistica e la coesione territoriale nei prossimi anni. Urbanistica Informazioni n. 282 (2018a)

Gralepois, M., Guevara, S.: L'adaptation aux risques d'inondation façonnée par les métiers de la ville. Tensions à l'échelle du projet d'aménagement, Developpement durable et territoire vol. 6, n. 3 (2015)

Gasparrini, C.: Una buona urbanistica per convivere con i rischi. Urbanistica n.159 (2018b)

IPCC: Climate Change 2021: The Physical Science Basis. VI Assessment Report of the Intergovernmental Panel on Climate Change. Cambridge University Press (2021)

Masboungi, A.: Projets Urbains Durables: Strategies. Moniteur, Parigi (2012)

MEA Millennium Ecosystem Assessment (2005). http://www.millenniumassessment.org/en/Global.html

Meerow, S., Newell, J.P: Spatial planning for multifunctional green infrastructure: growing resilience in Detroit. Landscape Urban Plann. **159** (2017)

NRC, National Research Council. Science and Decisions: Advancing Risk Assessment. The National Academies Press, Washington, DC (2009). https://doi.org/10.17226/12209

Oliva, F., Ricci, L.: Promoting urban regeneration and the requalification of built housing stock. In: Antonini E., Tucci F. (eds.) Architettura, Città, Territorio verso la Green Economy. Edizioni Ambiente (2017)

Poli, I., Ravagnan, C.: The urban plan within sustinability and resilience. New operational concepts and new collective values. Urbanistica n. 157 (2016)

Poli, I., Uras, S.: Pianificare le infrastrutture verdi in ambito urbano. Il caso spagnolo di Vitoria-Gasteiz. In: Ricci L., Battisti A., Cristallo V., Ravagnan C. (eds.) Costruire la città pubblica. Tra storia, cultura e natura. Urbanistica Dossier n. 15, Inu Edizioni, Roma (2018)

Poli, I., Uras, S.: Il ruolo delle green infrastructure nella costruzione di strategie adattive resilienti. In: Talia, M. (ed.) La città contemporanea: un gigante dai piedi d'argilla. Planum (2020)

PWD: Green City, Clean Waters: The City of Philadelphia's Program for Combined Sewer Overflow Control, Philadelphia Water Department, Philadelphia, PA, USA (2011)

PWD: Green City, Clean Waters-Evaluation and Adaptation Plan. Philadelphia Water Department, Philadelphia, PA, USA (2016)

PWD: Green Stormwater Infrastructure Projects. Philadelphia Water Department, Philadelphia, PA, USA (2018)

Ravagnan, C.: Prospettive di resilienza per la città e i territori contemporanei. Il ruolo delle reti verdi e blu nelle strategie di rigenerazione. Urbanistica Informazioni n. 278 (2018)

Ravagnan, C., Poli, I.: Green and blue networks: towards a safe future within risk management and strategic vision. Urbanistica n. 160 (2017)

Ravagnan, C., Poli, I., Uras, S.: The role of water management in European regeneration strategies. From problem to opportunity. UPLanD n. 4 (2019)

Reckiena, D., et al.: How are cities planning to respond to climate change? Assessment of local climate plans from 885 cities in the EU. J. Clean. Prod. **191** (2018)

Sgobbo, A.: Water sensitive urban planning. Approach and Opportunities in Mediterranean Metropolitan Areas. INU edizioni, Roma (2018)

Shade, C., Kremer, P.: Predicting Land Use Changes in Philadelphia Following Green Infrastructure Policies. in Land 8, no. 2 (2019)

WCED: Our Common Future. Brundtland Report (1987)

Application of Crowd Sensing for Sustainable Management of Smart Cities

Ilaria Pigliautile[1,2](✉) ⓘ and Myriam Caratù[3] ⓘ

[1] CIRIAF – Interuniversity Research Centre, University of Perugia, Via G. Duranti 67,
16125 Perugia, Italy
`ilaria.pigliautile@unipg.it`
[2] Engineering Department, University of Perugia, Via G. Duranti 93, 06125 Perugia, Italy
[3] Faculty of Economics, UNINT – Università degli Studi Internazionali di Roma, Via delle Sette
Chiese 139, 00147 Rome, Italy
`myriam.caratu@unint.eu`

Abstract. Cities are key contributors to the climate system alteration and the urban environment is, in turn, significantly affected by the climate change effects, such as the increasing occurrences of extreme weather events, which compromise citizens' life quality and health status. Nevertheless, hubs of economic and innovation and have the potential of leading the needed green transition by implementing effective mitigation and adaptation strategies. The high heterogeneity of the urban environment requires the implementation of site-specific and targeted actions that need to be supported through fine-grain data collection of the most significant environmental parameters such as air temperature and pollutants concentration. Recent advances in sensor technology and data communication capability allow to conceive the active participation of citizens in the urban environmental monitoring process, moving the analysis to a human-centered approach: citizens become, at the same time, vectors of data and target of data-processing that could be used also for boosting the required behavioral change of individuals and increasing resiliency of territories and communities. Wearables are already implemented for the smart management of the healthcare sector and their usage can be easily extended in the coming future for a better management of other sectors involving citizens' everyday life.

Keywords: Urban resilience · Crowdsensing · Data-driven innovation · Smart city · E-management · Policy-making

1 Introduction

Urban areas are responsible for 70% of the Green House Gas emissions and of 2/3 of the energy consumption worldwide, despite covering only 3% of the world surface [1]. These peculiarities of urban metabolism highlight the great impact of cities on the climate system. Furthermore, cities host more than half of the world population and their structure, along with their land cover and land usage changes, increase their vulnerability to climate change effects [2]. Extreme weather events, like massive precipitation patterns

© The Author(s), under exclusive license to Springer Nature Switzerland AG 2022
F. Calabrò et al. (Eds.): NMP 2022, LNNS 482, pp. 2800–2808, 2022.
https://doi.org/10.1007/978-3-031-06825-6_267

and heat waves, are going to be more frequent and more intense in the near future due to the on-going climate change and its negative impacts on human life quality: in particular, these events are going to be further exacerbated in urban areas, where they synergically interact with some phenomena typical of the urban environment, such as the Urban Heat Island, and the concentration of air, noise, and light pollution [3]. Therefore, for reaching a sustainable development of urban areas, it is urgent to intervene through both mitigation and adaptation actions for enhancing urban resilience (in terms of both its physical heritage and communities' organization) while reducing human contribution to the climate system alteration [4].

Tailored strategies must be planned and implemented accounting for peculiarities of each area of intervention through a multiple-scales approach, ranging from the international down to the neighborhood and the single-building scale. A resiliency plan that simultaneously accounts for the required green transition and economic growth (as required by the European Green Deal) must consider geographical and climatic aspects of the region but also morphological and metabolic characteristics of a specific settlement which cause intra-urban microclimate variation, different levels of urban outdoors environmental quality, and thus different environmental risks for citizens [5].

Along with the heterogeneity of urban forms, materials, and activities, an efficient resiliency plan must account also for the urban social dimension, considering economic, and cultural discrepancies of its community: the finer the scale of the analysis, the better tuned and the more effective will be the planned and implemented strategy.

In this view, it is important to properly decline the concept of Smart City towards the definition of an efficient infrastructure for a fine grain collection of a multitude of data belonging to different fields of knowledge and considering several dimensions for the individual wellbeing: environment, health and possibility to adopt wearable technologies that enable a smart assessment and monitoring of possible life patterns of citizens into smart cities. The infrastructure, along with the data analytics, is going to become the skeleton for the development of sustainable communities and territories increasing the efficiency in service provision for both public and private companies. In this sense, here a proposal for an innovative data collection infrastructure is presented: this focuses on individuals – single citizens – as the vector for data gathering and, at the same time, as the main target of the data analysis. This work moves from the description of the climate change effects on the urban environment and citizens' health to progressively define what are the proper tools for critically addressing the main challenges for a sustainable development of urban settlements under the Smart City definition. Therefore, potentialities in optimizing resources consumption through citizens involvement and engagement are stated.

2   Climate Change Impact on Citizens Well-Being

The majority of the world population currently lives in urban areas, where most of the economic activities, employment, and wealth are concentrated. Despite the great opportunities of living in cities, most of the social inequalities arise in urban areas, which are very heterogeneous places in a socio-economic perspective [6]. Furthermore, the urban environment could threaten citizens' health in different ways: the urbanization process, combined with the concentration of anthropogenic activities in urban areas, alters the urban microclimate (characterized by the well-known phaenomenon of the Urban Heat Island (UHI), exposing citizens to temperatures higher than those of the rural areas) and leads to high level of air, noise, and light pollution in cities. Peculiarities of urban morphology and metabolism (its physical form and socio-economic activities) alter environmental stressors exposure of citizens at local scale and further exacerbate weather extreme events due to climate change, like Heat Waves (HWs) that act synergically with UHI, or extreme precipitation patterns leading to high water flood risks due to lack of greenery and, more generally, of permeable surfaces in cities. Citizens' exposure to heat and cold extremes (HWs and cold spells), high concentration of pollutants (ground-level O3, particulate, volatile organic compounds, NOx, SOx), and flooding are key impacts of climate change causing negative health outcomes [7]. These are both related to the exacerbation of pre-existing chronic diseases and to the onset of acute effects. Prolonged exposure to extremely high temperatures has an impact on human health, increasing morbidity and mortality due to heat-related illnesses with syndromes that vary from less severe (heat syncope), to severe and lethal (heat stroke) [8].

Extreme heat also worsens the effects of chronic diseases (including respiratory diseases, cardiovascular diseases, and kidney problems), while it has been observed that high temperatures and air pollution lead to synergistic negative health effects [9]. High temperatures negatively impact also emotional and psychological health [10] and correlations have been found with air pollutants concentration and psychiatric emergency visits [11]. Extreme weather events, in general, affect mental health in several ways. Following disasters, acute mental health impacts (including anxiety, depression, post-traumatic stress disorder (PTSD) and substance abuse) increase, both among people with no history of mental illnesses, and those at risk [12].

Other climate-extreme impacts on mental health are recognized in literature like climate and ecological anxiety and grief [13] and exacerbated psychosis [14]. Negative health outcomes of urban climate extremes vary according to individuals' vulnerabilities. Characteristics that affect the individual vulnerabilities include age, pre-existing medical conditions, economic conditions (including energy poverty and poor housing), level of instruction, accessibility to urban services and facilities (such as the Public Leisure Spaces, PLSs), health education and awareness, all affecting the adaptive capacity of people to environmental stressors. Although the nexus between human health and environmental quality of the built environment (both indoors and outdoors) has been only partially disclosed, and despite the influence of socioeconomic variables on people adaptation capacity has been recognized, an evidence-based and transparent procedure capable of estimating health-related costs of exposure to environmental stressors is still missing. Furthermore, the impact of intervention primary aimed at limiting cities contribution to climate change (mitigation actions for decarbonizing cities) has to be highlighted also in terms of health risk reduction and socio-economic co-benefits for comprehensively depicting the effectiveness of those actions and properly guide policymakers and public authorities in the implementation of the more effective strategies [15].

Therefore, there is a strong link between human health, well-being, and the surrounding environmental quality which is in turn strictly related to the progression of climate change [16]. Nevertheless, urban ecosystems and citizens' lifestyles are leading to the emergence and sprawl of new diseases. Indeed, much of the global mortality in the coming future will be due to diseases attributable to risks related to urbanization and the growing sedentary lifestyle according to the World Health Organization (WHO). All these diseases must be - if not prevented - at least managed in a smart way and e-Health represents the application of the Smart City concept to the healthcare sector.

3 Sustainable Technological Development in Smart Cities

Despite its captivating and futuristic name, the concept of Smart City is not new: it is intended as a place where, "through the availability and quality of dedicated infrastructures to communication and social participation, the city becomes smart to the extent that it intelligently manages economic activities, mobility, environmental resources, relationships between people, housing policies and its own administration model. All this, favoring the sustainable economic development of the city, guarantees its inhabitants a high quality of life" [17].

Sustainable development means a type of development that does not negatively impact either the environment, society or economic profit, according to the model of the Three Ps- People, Profit, Planet [18]: a model that is currently not respected, if we think that a large part of global emissions is due to electricity and heat production systems both in urban and rural areas.

In this scenario, (ICTs) become crucial in order to achieve the sustainable objectives outlined by the Intelligent Community Forum of New York in 2014: less pollution and traffic thanks to the use of electric cars, savings on domestic consumption thanks to the use of renewable energy and/or from energy recovery from waste ("waste to energy" is a process developed by systems that the most advanced smart cities are equipped with), presence of urban green that contributes to lower the level of carbon dioxide in the air and thus counteract global warming (an example, in this sense, is the well-known "vertical forest" of Milan). ICTs further facilitate and optimize the provision of public services to citizens: from irrigation of parks or street lighting to more complex challenges including the healthcare management, the energy grid management, or the urban resiliency planning which need a well-structured use of different kind data to be collected and processed real-time also thanks to the collaborative participation of citizens.

In recent decades, the integration of the most advanced technological and communication tools in every area of daily life, as well as the introduction of the Smart City paradigm, brought to great changes in urban lifestyles and in the management of city public services. Being invested by a huge flow of information, new mobility options, environmental criticalities and energy efficiency issues, municipalities have become the place of social change par excellence, in their organizational (social order, work) and structural dimensions (family, school, politics, culture). Changes in the urban dimension are due to a series of complex phenomena: the urban population growth, the increase in social inequalities, the unfair access to public services between the various areas of the city, etc.

In this frame, the main question is: *what is the real innovative potential of the "technological city" in a socio-urban paradigm? How could public authorities take advantage of ICTs for a better management of cities and what are the expected benefits for citizens?*

The transformations of our collective life, which are taking place in modern cities, pose complex questions of political nature on the "governance of social space". In this perspective, an intelligent city needs to be associated with a new form of urban space management, based on a multidisciplinary perspective: a new place facing the challenges that globalization and climate change pose in terms of competitiveness and sustainable development (economic, environmental, and social sustainability). Additionally, the implementation of the best performing ICTs modifies the physical boundaries of the urban identity that have become more blurred, and re-shape the housing, commercial and production densities, and the associated support services. The urban space management and public service provision thus requires to programming choices that focus on the economic component (in the sense of "efficient") as essential for the achievement of the "smartness": from the commercial services that are being revolutionized by the affirmation of e-commerce, to the logistics of individuals and goods accounting for personal services (primarily health and education, but also social, cultural, and recreational activities). Public authorities need to identify proper tools maximizing the data availability and the relevance of data-processing associated information while minimizing the infrastructure costs in terms of ICT systems installation, maintenance, and physical impact on the urban ecosystem.

Advances in electronic and communication fields led to the introduction of many technical apparata (e.g., wearables for personal health status tracking, Internet of Things, 5G urban Wi-Fi, self-mapping and forcing real-time positioning) in everyday life. The spread of these systems has contributed to changing how social individuals perceive and interact with the organizational structures. As a matter of fact, it is common practice among citizens to evaluate services within the city or to share personal progress in physical activities on dedicated platforms. In this view, it is worth presenting the Laney introduction to the big data concept, as follows: "every activity carried out online and in particular on social media or through communication technologies connected to GPS or corporate sensors and devices, such as the Intranet, generates a series of records that are classified as "metadata" or "big data" based on their volume, variety and speed with which they proliferate" [19]. Not by chance previous research studies, have investigated the influence of big data on business intelligence [20], strategic management [21], organization change [22], organization engagement in social media [23] and human resources management [24]. However, the analysis of big data is fundamental also to streamline the interaction between systems, within which the themes of social organization, social order, and urban resilience and renaissance represent interesting targets of the analysis. More specifically, data retrieved by citizens, via crowdsourcing, are essential and reliable data for a deep analysis of the urban ecosystem due to their spatial and temporal collection scale: fine-grain spatial resolution and a continuous data stream.

3.1 e-Health Management Systems

The post-Covid-19 era is a difficult time for the management of patients with psychiatric disorders, as social distancing and blockages in many countries have threatened the mental health of individuals and families: not surprisingly, the WHO has stated that mental health systems in all nations must be massively strengthened to cope with the impact of the restrictive measures adopted by governments due to the health emergency [25].

In this scenario, this paragraph focuses on the implementation of the ICTs for a smart management of urban services, which is framed into the specific sector of the smart healthcare management (e-health). In fact, the digitalization of the healthcare sector by means of the 5G, along with Artificial Intelligence (AI) and Big Data Analytics, is the promising future for the field of smart healthcare (or e-healthcare) since this will support and improve the service provision. Indeed, we are moving towards the matching of sensing and actuating: a process of broadening our skills in knowing how to use information to make decisions that are useful to the individual and the community. The sensing component takes advantage from the above-mentioned achievements in the electronic field that have brought to the opportunity of introducing wearable, not invasive, apparata in our daily life. The sprawl of these technologies allows a rising number of people to constantly monitor their status in terms of fitness advances, in most of the applications, but also in terms of health status. These data are available to the final user but could be used for a broader scope supporting a decision-making process (the actuating): a conscious reaction to the retrieved information. The benefits are twofold: the final user can enhance its quality of life and service providers can optimize their work putting in place new practices: an example of effective implementation of ICTs in the e-Health sector is the H2020 project BRAINE ("Big data pRocessing and Artificial Intelligence at the Network Edge", [26]) devoted in developing Edge Computing solutions implementing AI for Big Data processing at local scale thanks to potentials of the 5G. The developed technological solutions cover diversified fields of action: from healthcare assisted living, to the hyper-connected smart city, robotics and the supply chain for the industry 4.0 in the energy efficiency field. Specifically, concerning healthcare, the BRAIN project evaluates the opportunity of adopting machine learning algorithms and AI for the smart hospital and caregiving with the aim of optimizing disease treatment by predicting the subjects' health status according to their daily monitoring.

The healthcare assisted living became a need due to the rising demand of healthcare automatized services, remotely enabled and capable of providing real-time outcomes for their smart management. In this view, the Internet of Things (IoT) provides a further incentive towards data-driven and real-time patients management. The role of Data Science is to preliminary skim, process, and store a huge amount of information. The proper development, application, and smart management of the healthcare would result in a more sustainable system putting in place a digital economy, community-based, that will actively contribute in promoting healthy lifestyles. Indeed, an effective implementation of ICTs in the healthcare sector may result in the promotion of best practices and virtuous behaviors among final users (citizens). Furthermore, AI is going to be more than needed due to the huge amount of data availability in order to explore their complex interactions.

Nowadays, the main categories for the e-Health application concern some macro-areas such as: diagnosis, therapeutic consultations, involvement in experiments and patient consent management (in line with General Data Protection Regulation [30]), administrative processes relating to therapies and patient medical history. As a matter of fact, AI algorithms already overcame outperformed radiologists in detecting malignant tumours and this procedure could be extended to many other health diagnoses in the coming future, along with the ability to find the best solution for the treatment of various diseases.

Finally, it is worth mentioning the possibility of monitoring the individuals' exposure to environmental variables by means of the same wearable devices already adopted in the e-Health sector [27]. The simultaneous collection of personal and environmental data would provide further insights on the existing correlation between the environment and the human health. These data could be integrated into dedicated platforms that would represent a common informatics and conceptual ground, where inputting a variety of historic and real-time information sprawling from: (i) environmental contextual factors (UHI intensity, air pollution, weather forecast, etc.), (ii) medical treatments, (iii) information collected by patients via wearables, i.e., personal exposure and physiology. Systematizing all these information will contribute toward the creation of the smart urban resilience platform against climate change stress on citizens.

In particular, citizens with chronic complex disease, i.e. mental disorders, would mostly benefit from these systems of smart healthcare, since these subjects may not be truly investigated and treated without a continuous long-term smart and low-disturbance monitoring and care, because of the dynamic evolution and fluctuating conditions of their disease, e.g. recurrent depressions, seasonal affective disorders, etc.) [28, 29].

4   Conclusions

Urban settlements host most of the word population and are characterized by peculiar environmental conditions that are further exacerbated by the on-going climate change process. Environmental stressors impact citizens' life quality and well-being. In order to reduce the urban contribution to climate system alteration and to enhance its resilience to climate change (including urban communities' resiliency) it is fundamental to define tailored strategies relying upon data-driven evidence. A fine-grain data collection is suggested for planning the most effective solutions. Achievements in electronic and ICTs are leading towards the adoption of human-centered solutions that would lead to the twofold aim of (i) providing a fine-spatial data resolution and (ii) being tailored on the final target of the data analysis whenever it concerns individuals' adaptation to external stressors or enhancement of current lifestyles and behaviors.

Examples in e-Health are thus presented as real applications of the Smart City paradigm to a specific sector but such case studies can be extended to other fields of citizens life and urban services management. The usage of real-time and human-centered data can be also pursued for a better tuning of the public mobility services, including electric car sharing in a sustainable framework. Other applications can involve the implementation, management, and development of energy communities where renewable resources, distributed on a neighborhood scale, are wisely managed among prosumers for limiting grid overloads and distribute daily peaks of production thanks to energy consumption data shared among members of the same community.

This work has the aim of highlighting the role of crowd sensing data on a variety of smart city services, specifically suggesting their usage for citizens' lifestyle improvement by coupling personal data (physiological, behavioral, etc.) to environmental data.

Urban planners, policymakers, and medical doctors can take advantage of big data to define urban resiliency plan tailored on site-specific criticalities and further supported by individuals' engagement provided as personalized alters (aimed at limiting personal exposure to harmful environmental stressors) and triggers (promoting virtuous behaviors in a sustainable and healthy framework).

References

1. Chen, S., Chen, B.: Urban energy consumption: different insights from energy flow analysis, input-output analysis and ecological network analysis. Appl. Energy **138** (2015). https://doi. org/10.1016/j.apenergy.2014.10.055

2. Garschagen, M., Romero-Lankao, P.: Exploring the relationships between urbanization trends and climate change vulnerability. Clim. Change **133**(1), 37–52 (2013). https://doi.org/10. 1007/s10584-013-0812-6

3. Zhao, L., et al.: Interactions between urban heat islands and heat waves. Environ. Res. Lett. **13**(3) (2018). https://doi.org/10.1088/1748-9326/aa9f73

4. Rani, W.N.M.W.M., Kamarudin, K.H., Razak, K.A., Asmawi, Z.M.: Climate change adaptation and disaster risk reduction in urban development plans for resilient cities. In: IOP Conference Series: Earth and Environmental Science, vol. 409, no. 1, p. 12024 (2020). https:// doi.org/10.1088/1755-1315/409/1/012024

5. Pioppi, B., Pigliautile, I., Pisello, A.L.: Human-centric microclimate analysis of Urban Heat Island: wearable sensing and data-driven techniques for identifying mitigation strategies in New York City. Urban Clim. **34** (2020). https://doi.org/10.1016/j.uclim.2020.100716

6. Vaziri, M., Acheampong, M., Downs, J., Majid, M.R.: Poverty as a function of space: understanding the spatial configuration of poverty in Malaysia for sustainable development goal number one. GeoJournal **84**(5), 1317–1336 (2019). https://doi.org/10.1007/S10708-018-9926-8/TABLES/7

7. Bai, X., Nath, I., Capon, A., Hasan, N., Jaron, D.: Health and wellbeing in the changing urban environment: complex challenges, scientific responses, and the way forward. Curr. Opin. Environ. Sustain. **4**(4), 465–472 (2012). https://doi.org/10.1016/J.COSUST.2012.09.009

8. Heaviside, C., Vardoulakis, S., Cai, X.: Attribution of mortality to the Urban Heat Island during heatwaves in the West Midlands, UK, vol. 15 (2016). https://doi.org/10.1186/s12940-016-0100-9

9. De Sario, M., Katsouyanni, K., Michelozzi, P.: Climate change, extreme weather events, air pollution and respiratory health in Europe. Eur. Respir. J. **42**(3), 826–843 (2013). https://doi. org/10.1183/09031936.00074712

10. Mullins, J.T., White, C.: Temperature and mental health: evidence from the spectrum of mental health outcomes. J. Health Econ. **68**, 102240 (2019). https://doi.org/10.1016/J.JHEALECO. 2019.102240

11. Bernardini, F., et al.: Air pollutants and daily number of admissions to psychiatric emergency services: evidence for detrimental mental health effects of ozone. Epidemiol. Psychiatr. Sci. **29**, e66 (2020). https://doi.org/10.1017/S2045796019000623

12. Makwana, N.: Disaster and its impact on mental health: a narrative review. J. Fam. Med. Prim. Care **8**(10), 3090 (2019). https://doi.org/10.4103/JFMPC.JFMPC_893_19

13. Cunsolo, A., Ellis, N.R.: Ecological grief as a mental health response to climate change-related loss. Nat. Clim. Chang. 2018 84 **8**(4), 275–281 (2018). https://doi.org/10.1038/s41 558-018-0092-2

14. Clayton, S., Manning, C.: Psychology and Climate Change: Human Perceptions, Impacts, and Responses. Elsevier Academic Press (2018). https://doi.org/10.1016/C2016-0-04326-7

15. Aldinger, F., Schulze, T.G.: Environmental factors, life events, and trauma in the course of bipolar disorder. Psychiatry Clin. Neurosci. **71**(1), 6–17 (2017). https://doi.org/10.1111/PCN.12433
16. EURISPES: Non solo lavoro: città healthy, la salute come bene comune. In: 31mo Rapporto Italia. p. Scheda 25 (2019)
17. EURISPES: Le Smart City in Italia tra successi e ritardi. In: 30mo Rapporto Italia, Minerva Publisher, p. Scheda 54 (2018)
18. Elkington, J.: Towards the Sustainable Corporation: Win-Win-Win Business Strategies for Sustainable Development, vol. 36, no. 2, pp. 90–100 (1994). https://doi.org/10.2307/41165746
19. Doug, L.: 3D data management: Controlling data volume, velocity and variety, META Gr. Res. Note (2001). https://www.bibsonomy.org/bibtex/742811cb00b303261f79a98e9b80bf49%0Ahttps://www.zotero.org/rudedude/items/IBTGZ7TK%0Ahttp://blogs.gartner.com/doug-laney/files/2012/01/ad949-3D-Data-Management-Controlling-Data-Volume-Velocity-and-Variety.pdf
20. Dutta, D., Bose, I.: Managing a big data project: the case of ramco cements limited. Int. J. Prod. Econ. **165**, 293–306 (2015). https://doi.org/10.1016/J.IJPE.2014.12.032
21. Bhimani, A.: Exploring big data's strategic consequences Article (Accepted version) (Refereed). J. Inf. Technol. **30**(1), 66–69 (2015). https://doi.org/10.1057/jit.2014.29
22. Rifkin, J.: The Zero Marginal Cost Society: The Internet of Things, The Collaborative Commons, and the Eclipse of Capitalism. St. Martin's Press (2014)
23. Dijkmans, C., Kerkhof, P., Beukeboom, C.J.: A stage to engage: social media use and corporate reputation. Tour. Manag. **47**, 58–67 (2015). https://doi.org/10.1016/J.TOURMAN.2014.09.005
24. Scullion, H., Linehan, M.: International Human Resource Management. Macmillan, London (2006)
25. Ghebreyesus, A.T.: Addressing mental health needs: an integral part of COVID-19 response. World Psychiatry **19**(2), 129 (2020). https://doi.org/10.1002/WPS.20768
26. BRAIN project. https://www.braine-project.eu/project/
27. Pigliautile, I., Pisello, A.L.: A new wearable monitoring system for investigating pedestrians' environmental conditions: development of the experimental tool and start-up findings. Sci. Total Environ. **630** (2018). https://doi.org/10.1016/j.scitotenv.2018.02.208
28. Rosenthal, N.E., et al.: Seasonal affective disorder: a description of the syndrome and preliminary findings with light therapy. Arch. Gen. Psychiatry **41**(1) (1984). https://doi.org/10.1001/archpsyc.1984.01790120076010
29. Shapira, A., Shiloh, R., Potchter, O., Hermesh, H., Popper, M., Weizman, A.: Admission rates of bipolar depressed patients increase during spring/summer and correlate with maximal environmental temperature. Bipolar Disord. **6**(1) (2004). https://doi.org/10.1046/j.1399-5618.2003.00081.x
30. EU, Regulation (EU) 2016/679 of the European Parliament and of the Council of 27 April 2016 on the protection of natural persons with regard to the processing of personal data and on the free movement of such data, and repealing Directive 95/46/EC (General Da, 2016)

Ecological Transition and Planning Strategies

Stefano Aragona[✉]

Dipartimento Patrimonio, Architettura, Urbanistica, University Mediterranea of Reggio Calabria, via dell'Università 25, 89124 Reggio Calabria, Italy
saragona@unirc.it

> *We destroy the beauty of the landscape because the splendors of nature,*
> *freely available, they have no economic value.*
> *We would be able to put out the sun and the stars because they don't pay a dividend.*
>
> J.M. Keynes

Abstract. Starting from environmental and social unsustainability, the paper proposes the rethinking of territorialization or reterritorialization paths. Highlight some key elements to build the ecological transition. Knowing that implementing the ecological approach takes time, remembering that the industrial city has taken over 350 years to establish itself. It highlights that planning and urban planning are increasingly land and city management and that the object of interest is the existing one, its maintenance or transformation having the common good as its guiding star.

Keywords: Re-territorialization processes · Resilience · Ecological approach · Integrated planning · Common goods

1 Ecological Transition is Much More …

The Ecological Transition (ET) must be based on reducing the ecological footprint in anthropogenic processes i.e. of territories and cities and of the development model. This is the only way to deal with the limitation of non-renewable resources, both those used as production inputs and as land that is increasingly "consumed" or urbanized. The processes of anthropization have always used natural resources to build human spaces. With the modern city this use has grown exponentially and unsustainably in the near future. One of the main threats, global warming, is highly dependent on overbuilding and desertification. With regard to both these phenomena, territorial and urban planning can play a non-secondary role.

Digitization, together with greenery, is the other key element of ET. The reference is the *New Green Deal* of Europe, inspired by what Rifkin wrote in 2017. Capra and Mattei in the same year ask for a necessary great *Copernican revolution* in the relationship

© The Author(s), under exclusive license to Springer Nature Switzerland AG 2022
F. Calabrò et al. (Eds.): NMP 2022, LNNS 482, pp. 2809–2821, 2022.
https://doi.org/10.1007/978-3-031-06825-6_268

between man and nature and which also changes the legal, social and economic conditions of modern society and its spatial expressions. But even the components of digital transformation, the raw materials for its electronic components, are limited. As planners we cannot intervene on this availability, but we can and must indicate territorial and city structures that are less dependent and fragile regarding this danger (Butera 2021).

And the theme is much broader and concerns the use of the territory. This guides the spatial arrangements. Assets that can also be in competition with each other. One of the most important, which has always existed, is the competition between greenery and agricultural activities. Agricultural activities that are in competition with those of animal breeding for food purposes. This competition is one of the reasons for the enormous deforestation of the largest green lung on Earth, the Amazon. We are talking about activities - agricultural and livestock - which are among the main causes of both emissions that increase the greenhouse effect and water consumption. In many areas, very relevant in Amazonia, there is still another competition: that between the harvesting of wood for various purposes and the protection of green areas.

These are essential lungs of the Earth that help counteract the greenhouse effect. The current deforestation capacity is not even comparable to that which existed in the past. Thus, emblematic case, the Romans destroyed forests, transformed landscapes and made woods disappear to make building material, ships, energy, etc. but certainly they could not eliminate, in a year, as many hectares as the surface of Belgium as happened in the aforementioned Amazonia. Furthermore, those ancient transformations occurring over the centuries, however, allowed a sort of adaptation both to natural components and also to human ones: processes that were at the basis of the construction of the landscape, as recalled by the EU *Landscape Charter* (2000).

Regarding these issues, territorial and urban planning has some spaces for action. But, as required by the UN Agenda 2030, a truly integrated approach is needed. In this case e.g. hypothesize that the activities to produce agricultural products take place in vertical greenhouses and not in the field, thus minimizing the transformation of the soil, the consumption of water, and transport. On the other hand, agricultural productions no longer even offer many jobs as they are increasingly mechanized. It is no coincidence that in these activities there are many multinational companies that apply economies of scale and automation. For the most part, they use chemicals and do not respect the rotation of crops thus impoverishing the land.

Forest management is another field in which spatial planning, in a coordinated way with rural disciplines, can play a significant role. In this sense, *River Contracts* are among the instruments that can be used. The more these work well, that is, they offer economically and socially valid solutions, the more local communities can be convinced and involved not to transform the territories into pasture and not deforest them.

The basic purpose of ET must be to increase resilience on both large and local scale, considering that climate change first of all imposes a decrease in fragility at both levels. This, first of all, translates into an increase in their resilience. The reason is because natural events have no administrative boundaries. That means a reduction in hydrogeological risk and an increase in elements – such as greenery – which raise local conditions regarding the response to phenomena such as heat waves. Choices that require actions to improve the "passive" response of spaces and lifestyles.

The other important objective is the use of tools such as *Energy Communities* or *River Contracts* mentioned above and using as much as possible elements based on the "0 km" criterion. Hence opportunities for local development, construction of a circular economy, recovery and proposal of the meaning and identity of places. This requires planning and action strategies for both the new and, above all, the existing considering that the basic assumption is "0" land consumption.

2   Difficult Changes for the Territory in Ecological Transition

In the windows of time that open between one peak and another of the various variants of Covid, many are hoping and asking for "going back" to before the pandemic.

But they do not consider that already, when it begins in 2019, the paradigm – using the term used by Khun in 1962 for scientific revolutions – of development manifested great problems both environmental and social. The reason is at the root of it.

In recent decades, international and national organizations have focused on large urban centres so that they become the main engines of a globalization which, according to *neo-liberal* theories, had to redistribute wealth on a planetary level. This is done using *economic efficiency* as a key element. And the growth of GDP and per capita income were the essential indicators, in fact the only ones, even if since the Rio Conference in 1992 – if not since 1972 with *The Limits to Growth* – the question of environmental and social sustainability has always been more and more emerging.

Europe chooses to face globalization with strategies resulting from this overall vision. Thus proposes the *European Space Development Scheme* of 1999 and launches the transport policy based on the *Transeuropean Network for Transport, TEN-T* with High Speed, between large European urban and metropolitan areas. Since this choice, in many cases, was not accompanied by the reinforcement of local public transport, the result was to "bring closer who are far away", that is the already strong poles and "make farer those who are closer", that is the smaller centres and/or internal areas. The situation is even worse in Italy where in 1992 there was the so-called "dry branches cutting" in rail transport which meant cutting off networks and services especially for what will later be defined as "internal areas".

EU Urban Agenda is the attempt to focus on some European cities to create poles capable of competing on a global level. The economic function of urban centres is also one of the essential components of them. They are born for security reasons, to have trading markets and places of production. They are one of the three *territorial invariants* – together with the areas and networks – well identified by Raffestin (1985). Their weight changes in relation to the different societies/eras. With the modern city, formed with the industrial revolution and engine of growth, then with the metropolis (Gottman 1961, 1982, Sassen 1991; McKinsey 2011; Glaser 2013), but from the beginning of the '70 of the twentieth century there is the progressive withdrawal of politics.

Initially this happens in Great Britain (the era of M. Thatcher), then in the United States (the years of Regan and then Clinton), and gradually throughout Europe with liberalization and privatization. They are increasingly vast and have led to the transformation of many services, previously considered rights, into goods to be purchased. The city is left in the hands of the economy, global digital companies and large international finance (Rodrik 2011; Crouch 2018). Attention to balanced territorial and social development disappears. The integrated planning strategies between rural, urban, small, medium, large, metropolitan areas, required the Leipzig Charter in 2007, are almost everywhere forgotten.

These neoliberal policies have been particularly negative from a social, territorial, and environmental point of view. On the one hand, in the large cities have increased their economic competitiveness - for the advantage of few actors -, on the other, the rest of the territory has lost its attractiveness, wealth and population (Roses and Wolf 2019; Rodriguez-Pose 2018). The large megalopolises have shown a strong increase in the various types of emissions, pollution and related diseases (WHO 2016), a great growth in forms of social inequity (Harvey 2007, 2012), all with an ever greater loss of identity and sense of places (Augè 1993, 1999). The set of these consequences - social, environmental, and spatial - is absolutely consistent with the development philosophy of neoliberal thought. The bases of which are to consider: a) work as a production input to be purchased and which can be "moved" or where there is a need, or demand, for labor; b) the territory as another production input, "quarry" for materials; c) a development that means aiming at the expansion of the demand for goods including those consisting of real estate and material infrastructures.

In Italy they have determined an increase in territorial vulnerability to disasters (PCM 2017), a loss of infrastructure safety (remember the Morandi bridge disaster in Genoa), the deterioration of a landscape unique in the world (Munafò 2020).

This development model, consolidated over more than thirty years, "exploded" with the Covid 19 pandemic of 2020. It first hit health systems severely weakened in their resilience (Olazabal et al. 2012): in previous years in Italy many hospitals have been closed. There have been and are devastating consequences on people's lives as well as on cities, with their *forms*, *networks* and *behaviours*.

Lockdowns and social distancing have "demolished" the foundations of the modern "mass city" built on economies of scale and agglomeration (Weber 1909/22). The large business centres, the shopping centres, the Quaternary areas and all the service activities connected to them, the tourist centres, etc., have lost much of their dominant position and their even symbolic attractiveness. The overall arrangement of the space, the growing urban concentration seems to be useful to avoid. Thus, the forecasts of urbanization of about 70% of the world population forecast for 2050 from multiple WTO, World Bank, UN sources, and supported by these with strategies and policies, need to be revised and modified.

However, in addition to the fundamental thrust due to the aforementioned competitiveness, there are other reasons related to sustainability and the use of resources in favour of the "compact city". It is the least expensive compared to travellers. So, the denser it is, the less resources are needed to accomplish them. Land consumption is also lower since the connecting infrastructures serve more users and are therefore more efficient in terms of use. But the higher the concentration of the population, the more there is a situation of fragility as shown by the ongoing Covid2 Sars pandemic. It is no coincidence that research on the quality of life (Mercer 2017) shows that at the top of the charts there is a city like Vienna, ca. 2 million inhabitants and not megacities, and it was made before the pandemic.

Very often gentrification mechanisms are associated with urban concentration. The increase in the density of demand and the increase in spending capacity are the basis of these urban and social transformations. They are made through changes of use or transfer of ownership of residences. This logic also guides the various service sectors: water, electricity, sewage networks, both physical – as mentioned above – and intangible communications. The latter are a particularly sensitive element since they are the basis of what is now called *digital divide* and which, in 1985, Goddard and Gillespie had defined as a *competitive disadvantage*. A condition highlighted in relation to the increasingly important role that the so-called *value-added services* were playing. They are born thanks to the new opportunities of remote telecommunication means, i.e. *telematics*: word composed of the suffix "tele" that is "distance" and "matics" from "informatics" (Aragona 1993). Thus advanced networks and services have been offered only where the demand is densest and with the highest spending capacity, i.e. the main centres. The creation of these is absolutely convenient for network and service providers: the higher and richer the demand per square meter, the lower the unit investment costs and the higher the profits (Aragona and Pietrobelli 1989).

These considerations also concern networks for water, sewage, gas etc. Regarding these services, it is important to highlight that, despite the failure of neoliberal policies, in Italy the recent (2021) Competition Bill under art. 6 of the Government Draghi calls into question the public and social function of the Municipalities. It reduces them to the role of entities solely responsible for putting on the market the public services of their own ownership, with serious prejudice to their duties as guarantors of the rights of the reference community. And this despite being the fundamental resolution of the UN General Assembly of 2010 on the universal right to water and, in our country, the 27 million Italian citizens who in 2011 expressed themselves in the referendum saying that the water had to go out of the market and that no profit can be made on this good.

These considerations are analogous to those regarding natural resources in their aspect as an opportunity to be energy sources: the current National Recovery and Resilience Plan (PNRR) allocates a significant part of the funds to many of them, i.e. 59.47 billion euros out of the total of 191 (PCM 2021), without considering the additional ones of REACT-EU and national programming. However, the beneficiaries and the primary actors of these opportunities must be the local communities and the inhabitants. In wind power, which has also been present in Italy for years, the income was mostly in the hands of a few, in various cases even in illegal situations.

However, the problem of the consumption of materials to build equipment for the use of renewable energy remains. Materials that in large part, also in this case, are not renewable. This is the main reason for having the diffusion of the circular economy and "closing cycles" as well as focusing on the use of natural elements as basic components: see the example of research on the use of vegetables for photovoltaic panels. And this is because the paper focuses on the *resilience increasing* on a large and local scale, that is, on reducing the *ecological footprint* (Fig. 1).

There are consolidated economic interests that would like to hinder this philosophy. They argue that technology will solve those limitations. They do not understand, or do not agree, to admit that what is needed is "cultured technology" (Del Nord 1991). Thus,

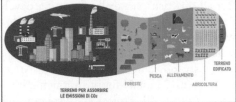

Fig. 1. The ecological footprint (source: Redazionale Ecosviluppo 2021)

in the Call of the PNRR relating to the Energy Communities it is desirable that these aspects are taken into due consideration.

3 Ecological Planning Strategies

With the spread of the pandemic, scenarios and new opportunities open up to propose an ecological, socially and environmentally more sustainable approach (Morin 2020) for the territory and the city. that supported for years in Italy by various scholars and in particular by the school of territorilists (Magnaghi and Paloscia 1992). Based on the integration of the different areas that contribute to the different qualitative and quantitative aspects of urban quality as proposed by ISTAT-CNEL, 2013, with the *BES - Fair and Solidarity Well-being* and as it arises from the monitoring of land consumption by of the National Environmental Protection System - ISPRA (Munafò 2020). Recovering also the proposals for sustainable urban development that emerged a few years ago (Camagni 1996) in a very different framework.

All this was emerging from the first decade of 2000 with the *National Strategy for Inner Areas - SNAI* (2013) inspired by the "place based" theory (Barca 2009) and the *Glocal* proposals, conjugation of the global with the local (Robertson 1995, Rifkin 2019). *Thinking global and acting locally* was the model on which the IV Report on National Planning of the Netherlands of 1985 was inspired by the first emergence of "advanced telecommunications", what we now call *telematics,* before mentioned. In it there was a planning of the necessary infrastructures which, starting from the 4 main centers of the "heart of the country", then extended to a provincial and then municipal scale. In this way, over time, the digital divide was overcome and the territory developed in a more balanced way.

Integrated development also between being and becoming, between matter and energy, in a changing relationship between space and time that is the basis of in-formation and memory, as Saragosa (2016) initially reports in *The path of Biopoli*. Development based on rethinking the relationship with nature, which must aim at the re-naturalization of cities, i.e. *Planning Cities with Nature*, text by Lemes De Oliveira and Mell I., as if it were an extension of McHarg's thought of *Design with Nature*. Texts that tend to focus attention in any case on urban development, while *Ecopoli* by Sandro Fabbro (2021) aims at a much more territorialist reading that raises very radical criticisms of the social, spatial and environmental settlement methods and suggests new tools and forms of anthropization.

But all this risks not really changing development trajectories and, once the pandemic is over, there is the danger of returning to the consolidated previous model. Some innovations were already present but they only occurred sporadically. This is the case of the tele-assistance already provided for in the *Social Master Plans*, created in 1999 to connect socio-health needs of the area (Aragona 2003).

As from 2017 in Italy for the Municipalities of over 50,000 inhabitants, the law ask for *Green Plans*, with Project and Program managed by them (MATTM). And documents from the Ministry of the Environment, Land and Sea Protection, now Ministry for Ecological Transition, speak of the need to combine greenery with water and ecosystem services. Green and blue infrastructures have been possible for years. And only with the Conte2 government were there the first loans for reforestation experimentation in metropolitan areas (l.n.141/2019). In the PNRR there is a significant endowment in this sense – indeed together with digitization – green is the basis of the *New Green Deal*, an idea launched in 2019 by Jeremy Rifkin, and adopted by the EU just before the start of the pandemic. But then how to keep and cur it? Already the one existing in many Italian cities in the parts built from the 30s of the twentieth century onwards, under the influence first of Howard's Garden City and then Mumford and then of zoning, has grown and makes its way, indeed occupies the streets and sidewalks. New Bahuaus and New Urban Agenda, the many experiences of "nature-based solutions" indicate certainly valid paths of sustainability but how is it possible to create the city of 15 min if the neighbourhood shops are increasingly disappearing? They are crushed by commercial centres and by changes in use. Both of these linked to economies of scale, therefore lower prices, with the "presumption" that the quality of life of the inhabitants results, first of all, in the offer of cheapest products and not, primarily, in a wider life quality.

To reverse the situation, political choices are needed to be made through territorial and urban planning options. Certainly, it is difficult to have territorial and urban policies that are challenging the aforementioned liberalizations and privatizations. But the Region Authority and the Municipalities must take responsibility for combating the social and spatial damage, the loss of identity, which their territories are suffering also due to neoliberal choices. Otherwise, the ET cannot be realized except in exceptional, even if significant cases. Thus, the example of the 3 million trees planned by the Municipality of Milan or the creation of a public company for the management of water by the Municipality of Naples is emblematic.

The ET in historical centres is a theme that needs particular attention. Already fourteen years ago the Order of Architects, Planners, Landscape Architects, Conservators of the Province of Pistoia was trying to address this topic (Aragona 2008) in the *Conference 3 Days of Architecture, Thematic Seminar: Population centres and energy*. The question is of great importance for Italy given the great importance and widespread presence of them. In this regard, the countries of central and northern Europe naturally have an advantage first of all because they have always required the greatest possible efficiency due to the climatic conditions: thus the use of wood has been widespread and this has implied an equally effective process in the production chain and its programming in the conservation of the resource. But also because of the conception of natural resources as common goods. It is no coincidence that in England there is the House of Commons. It is probably a legacy of the civic uses. These were created by the Romans and then had further declination in Central European lands.

There is also a difference in the relative weights of the different components of the ET between large cities and medium or small centres. Among the main differences are those related to mobility, both in relation to flows and modalities of it. As a relevant example is car-sharing or car rental which is almost absent in medium or small cities while it is widely present in larger ones. Although it should be noted that even in the latter the service often does not cover the entire municipal territory but only the densest, central areas – that is, those where presumably the spending capacity is greater – and not the more peripheral ones, even if it is lived in these areas the majority of the inhabitants.

The pandemic is also being faced thanks to the possibilities given by telematics. During the pandemics of past centuries there were no such opportunities, so Boccaccio's Decameròn, mid-14th century, has the narrative plot that takes place outside the city of Florence since the plague was here in 1348. Telematics can allow the breaking of the synchronicities between spaces and times, thus undermining one of the presuppositions of the industrial city, that is, of the modern city (Ernesti 1995). As pointed out in 1983 (Clementi), a large part of its structure has been designed over the centuries on the basis of what were then called public services, hospitals, schools. But it, as mentioned above, can support an integrated widespread territorial development as required by the *Leipzig Charter*. The inland areas and villages can be "reserve" – idea launched by Bussone, President of the National Union of Community Municipalities and Mountain Bodies (UNCEM), in 2019 at the INBAR assembly and it was before the pandemic and recently taken up by subjects like ISPRA – for territories and urban areas at risk including Paris, Manhattan in New York, Jakarta and many others built at river or sea level and which are already fighting with the current rise in water caused by the greenhouse effect. Consider that some islands of the Polynesian archipelagos have already been abandoned due to rising sea levels.

Thus, consistent with the aforementioned New Green Deal in the PNRR, another great financial endowment is for digitization, as mentioned above, i.e. 40,32 billion euros which is something less than a fifth of the total of 191 billion euros mentioned above, also here without taking in account the additional ones from REACT-EU and the national programming. But it must be used in a way that is really useful to people. Teleactivity, teleworking, requires the *overcoming of the logic of process control and to pass to product control* (Nilles 1988). However, it seems that the current Minister of Public Administration does not share all this and is linked to the consolidated logic of control. Certainly, this passage implies the loss of control of the managers and gives more freedom to the employees. And equally certainly this requires the definition of rights and duties as the late Stefano Rodotà had already anticipated for some decades, even asking for a "Constitution of the Internet". And there are great differences between the activities of the Public Administration and many others such as tele-teaching or tele-health.

Telework was invented by Jack Nilles, a NASA engineer in the early 1980s. The Mayor of Los Angeles said to him "Could you, who send people to the moon, help us solve the traffic problem in LA?". The "remote work" was experienced thanks to the spatial activities between astronauts and the Earth Base. So also the "group ware", form of remote cooperative work, was born. The many experiences of distance learning or tele-services in this year and a half of pandemic are its widespread diffusion. Around the same time, the Southern California Air Quality Commission imposed a tax on companies about the number of employees for car in downtown locations: the minor this latter was the minor was the taxation. The aim was to reduce traffic and pollution through tele-working or carpooling. All this, of course, in a vast mobility plan for the city. The same reasons for the reduction of traffic and pollution were the motivation for the teleworking experience made in Rome in the mid-90s by the Mayor Rutelli: TraDe or Traffic Decongestion. Experimentation carried out as part of the EU LIFE Projects (Aragona 2000) that is about Life Quality. Already in 2018 there were contracts that anticipated digitization in such as the SPC Consip framework contract in the Public Administration. Thus emblematically the Digital Agenda reported the image of Fig. 2.

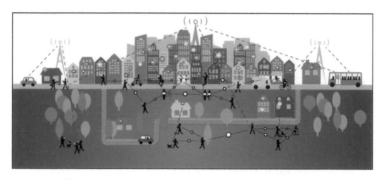

Fig. 2. Innovation and territory (Source: Amati 2018)

In this context there are the digital nomads. Example Vicari (CT) one of the municipalities that has sold empty houses for 1 euro and where thanks to both things the canter is reborn. So UNCEM, emphasizing the possibilities that really exist, talking about PNRR asks some questions (2022) <<*If ItaliaDomani, the Italian Next Generation EU, is truly capable of transforming the country. To make it more cohesive and to generate opportunities and solutions for the new generations. Without putting new debt on them. Certainly, in these first months of implementation of the PNRR we have highlighted as Uncem that there are lights and shadows. We have never made a criticism without making proposals. On some calls released so far, there are several things that we would have liked different. Starting with that on schools, in which municipalities are prevented from using the projects that had already been assembled and set up thanks to the announcement of the state planning fund, for example. Or the one on the villages in which the Regions must identify, it is not clear how, 20 villages with 20 million euros each and then another 229 villages that will have "only" 1.6 million Euros (thousands of potential candidates). It happens, for example, for the announcement on integrated development projects of metropolitan cities, that in these days a list of works is requested from the capitals to the municipalities, when instead the ratio is quite different. And the urban-centric logic actually excludes small municipalities. On urban regeneration, with the applications reopened, many make confusion with the "Integrated urban plans". On too many components, we have so far a resource switch between budget laws and PNRR. As on urban regeneration itself, on works for municipalities, on high-speed railways. An overall change of pace is needed. But for now, we remain confident. In addition to ten, one hundred private (and non-private) subjects who are inventing an "accompaniment" for the Municipalities, UNCEMU works on the political front to avoid urban-centric drifts and a destination of resources entirely towards the big cities, towards the hundred capitals. So far they are hardly recognizing the "metro-mountain" logic as the only solution to the crises (pandemic, climatic, economic). A change of pace is necessary and the municipalities together can, want and must do more.*

Regarding *"the notices published so far… never act running run after all the calls, but according to a plan of growth, inclusion, well-being that together the Municipalities define for their territories and their communities. With a long look and the logic that "everything is connected", and "walking together", which we have learned in decades of collaboration between Bodies in the same valley or in the same homogeneous area.>>.*

Exemplary is what, before the pandemic and also the PNRR, the Municipality of Pegognaga (MN) proposed together with the Italian Biogas Consortium in collaboration with Energia Media. In other words, a model for the enhancement of the Italian territory ready to relaunch itself through a new spatial, social and economic vision: i.e. a Smart Land (Fig. 3).

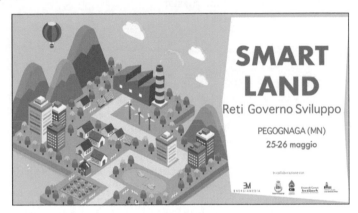

Fig. 3. Smart Land proposed in Pegognaga (MN)

4   Concluding Suggestions in Progress

The ET was born as a push to change the model which proved to be environmentally unsustainable, and which has become more and more socially unequal. For several decades now there has been an awareness of the limits regarding the relationship between man and nature and of the social effectiveness that is increasingly disappearing, crushed by technical and economic efficiency. The pandemic is making these limitations more evident and jarring. Even virtual accessibility, since its manifestation, has followed this scenario of economic efficiency and not of social effectiveness.

The ET, when it succeeds, triggers gentrification processes if it is not managed by politics. Thus the beautiful example of environmental sustainability of the High Line in Manhattan meant an increase in value in the sale or rental prices. While in San Francisco, the "sustainable" bus transport of companies in Silicon Valley to the city, solicited by the City with tax-reduction of these busses, has led to the expulsion of residents in historic Frisco neighbourhoods and those who remain have organized themselves into struggle committees.

However, there are examples of local "resistance" and autopoiesis processes (De Rossi 2018). As the Diversity Inequality Forum (2018) has been saying for some time, as well as the SNAI Strategy (Lucatelli 2015) suggests, as required by UNCEM and finally also by the EU, such experiences need to be supported. By doing so accepting that the dynamics of anthropization must really go towards a passage, that is, make a transition towards an ecological approach – more than a model, a paradigm – based on an integrated vision of territories and cities, as requested by the quoted *UN 2030 Agenda*, the *EU New Green Deal*, and Pope Francis' Encyclical Letter *Laudato Be for the Care of the Common Home*.

Acknowledgement. Thanks to the notes and suggestions of Sandro Fabbro - University of Udine, Fabiola Fratini - University of Rome La Saoienza, Francesca Silvia Rota - CNR Turin, and Paola Pittaluga - University of Sassari, Heads of the Research Units and their members of the PRIN - Project of Relevant National Interest "ORMA - Opportunities and Risks for the New Paths of Anthropization", led by me as Principal Investigator.

References

Amati, A.: Smart city, così l'SPC Consip sta cambiando città e territori col digitale (2018). https://www.agendadigitale.eu/cittadinanza-digitale/smart-city-cosi-lspc-consip-sta-cambiando-citta-e-territori-col-digitale/. Accessed 15 Jan 2022

Aragona, S., Pietrobelli, M.: Innovazione tecnologica e trasformazioni territoriali. Il caso italiano: politiche, strategie, sviluppi, pubblicazione del Dipartimento di Tecnica Edilizia e Controllo Ambientale, Facoltà di Ingegneria, Università La Sapienza, Roma (1989)

Aragona, S.: La città virtuale: trasformazioni urbane e nuove tecnologie della informazione, Gangemi Editore, Roma - Reggio Calabria (1993)

Aragona, S.: Ambiente urbano e innovazione. La città globale tra identità locale e sostenibilità, Gangemi Editore, Roma - Reggio Calabria (2000)

Aragona, S.: Piano Urbanistico e Piano Regolatore Sociale, in (a cura di) Bonsinetto F., Il Pianificatore Territoriale. Dalla formazione alla professione, Quaderni del DSAT, Gangemi Editore, Roma (2003)

Aragona, S.: L'evoluzione nei rapporti tra centri storici e riqualificazione energetica: possibili proposte. In: 3 Days of Architecture, Thematic Seminar: Population centres and energy, Order of Architects, Planners, Landscape Architects, Conservators of the Province of Pistoia, Pistoia 11–13 September (2008)

Augè, M.: Non luoghi. Introduzione a una antropolgia della surmodernità, elèuthera, Milano (ed or. Non-lieux: Introduction a une anthropologie de la surmodernite, Editions Seuil, Paris) (1993)

Augè, M.: Disneyland c altri non luoghi, Bollati Boringhieri, Torino (ed. or. L'impossible voyage: le tourisme et ses images, Éditions Payot & Rivages, Paris 1997) (1999)

Barca, F.: Un'agenda per la riforma della politica di coesione. Una politica di sviluppo rivolta ai luoghi per rispondere alle sfide e alle aspettative dell'Unione Europea. Introduction, capp I and V Rapporto "An Agenda for a Reformed Cohesion Policy", 2010 (original full edition 2009). http://europa.formez.it/content/unagenda-riforma-politica-coesione-0. Accessed 26 May 2021

Boccaccio, G.: Decameròn (1350)

Bussone, M.: Speech at the meeting "Il Manifesto per la Pianificazione territoriale integrata", National Insitute of BioArchitecture – INBAr, CNAPPC, Rome, January 29th (2019)

Butera, F.: Affrontare la complessità. Per governare la transizione ecologica, Edizioni Ambiente, Milano (2021)

Camagni, R. (ed.) Lo sviluppo urbano sostenibile: le ragioni e i fondamenti di un programma di ricerca, Il Mulino, Bologna (1996)

Capra, F., Mattei, U.: Ecologia del diritto. Scienza, politica, beni comuni, Aboca Edizioni, Sansepolcro (AR) (original Ecology of Law Elsevier, Amsterdam (NL) (2017)

Clementi, A.: Pianificare i servizi, Casa del libro, Reggio Calabria - Roma (1983)

Crouch, C.: The Globalization Backlash. Polity Press, Cambridge, UK (2018)

De Rossi, A., et al.: Riabitare l'Italia. Le aree interne tra abbandoni e riconquiste, Progetti Donzelli, Roma (2018)

Del Nord, R.: Presentazione. In: Mucci, E., Rizzoli, P. (eds.) L'immaginario tecnologico metropolitano. Franco Angeli, Milano (1991)

Encyclical Letter Laudato Be for the Care of the Common Home of the Saint Father Francisco, Tipografia Vaticana, Città del Vaticano (VA)

Ernesti, G.: Tempo pubblico e tempo della soggettività: discilplina e società oggi, Urbanistica n.104 (1995)

Fabbro, S.: ECOPOLI. Visione Regione 2050, INU Edizioni, Roma (2021)

Forum Diseguaglianze e Diversità: Aree interne e il problema delle distanze: le proposte della SNAI (2018). https://www.forumdisuguaglianzediversita.org/aree-interne-distanze-proposte-snai/. Accessed 15 May 2018

Glaser, E.L.: Il trionfo della città. Come la nostra più grande invenzione rende più ricchi e felici, Bompiani Milano (2013)

Goddard, J.B., Gillespie, A.E.: Advanced telecommunications and regional economic development. Geograph. J. **152** (1986)

Gottmann, J.: Megalopolis. The urbanized northeastern seaboard of the United States, Twentieth Century Fund (1961)

Gottman, J.: (introduction of C. Muscarà). La città invincibile. Una confutazione dell'urbanistica negativa. Franco Angeli, Milano (1982)

Harvey, D.: Breve storia del neoliberismo, Il Saggiatore, Milano (2007)

Harvey, D.: Il capitalismo contro il diritto alla città. Neoliberalismo, urbanizzazione, resistenze, Ombre Corte, Verona (2012)

Lucatelli, S.: La strategia nazionale, il riconoscimento delle aree interne, Franco Angeli, Milano (2015)

Keynes, J.M.: Royal Economic Society, Collected Writings 1971–1989, vol. XXI, p. 242. Macmillan (London), St. Martin's Press (New York) (1989)

Legge 12.12.2019, n.141, Conversione in legge, con modificazioni, del DL 14.10.2019, n. 111, Decreto Clima misure urgenti per il rispetto degli obblighi della direttiva 2008/50/CE sulla qualità dell'aria e proroga del termine di cui all'articolo 48, commi 11 e 13, del DL legge 17.10.2016, n. 189, convertito, con modificazioni, dalla legge 15.12.2016, n. 229

Lemes De Oliveira, F., Mell, I. (eds.): Planning Cities with Nature: Theories, Strategies and Methods, Cities and Nature. Springer (2019)

Magnaghi, A., Paloscia, R.: Per una trasformazione ecologica degli insediamenti, Franco Angeli, Milano (1992)

MATTM - Ministero dell'ambiente e della tutela del territorio e del mare, Comitato per lo sviluppo del verde pubblico "Strategia nazionale del verde urbano. Foreste urbane resilienti ed eterogenee per la salute e il benessere dei cittadini" (2017). https://www.minambiente.it/sites/default/files/archivio/allegati/comitato%20verde%20pubblico/strategia_verde_urbano.pdf. Accessed 05 Dec 2021. "Linee guida per la gestione del verde urbano e prime indicazioni per una pianificazione sostenibile". http://www.minambiente.it/sites/default/files/archivio/allegati/comitato%20verde%20pubblico/lineeguida_finale_25_maggio_17.pdf. Accessed 6 Dec 2021

Meadows, H.D., et al.: I limiti dello sviluppo, Club di Roma, Mondadori, Milano (1972)

Meadows, D.L., et al.: The Limits to Growth. Universe Books, New York (1972)

Mercer: Quality of Living City Rankings (2017). https://mobilityexchange.mercer.com/Insights/quality-of-living-rankings. Accessed 27 Dec 2021

McKinsey Global Institute: Urban world: Mapping the economic power of cities, March 1, 2011 Report (2011). https://www.mckinsey.com/featured-insights/urbanization/urban-world-mapping-the-economic-power-of-cities. Accessed 09 Dec 2021

Morin, E.: Cambiamo strada. Le 15 lezioni del Coronavirus, Collana Temi, Editore Raffaello Cortina Editore, Milano (2020)

Munafò, M. (ed.) Consumo di suolo, dinamiche territoriali e servizi ecosistemici. Edizione2020. Report SNPA 15/20 (2020)

Nilles, J.M.: Traffic Reduction By Telecommuting: a Status Review, Transportation Research, vol. 22a, 4 (1988)

Olazabal, M., Chelleri, L., Waters, J., Kunath, A.: Urban resilience: towards an integrated approach. I International Conference on Urban Sustainability & Resilience, London, UK (2012). https://www.researchgate.net/publication/236236994_Urban_resilience_towards_an_integrated_approach. Accessed 15 Dec 2021

PCM - Presidenza del Consiglio dei Ministri, Struttura di Missione Casa Italia. Rapporto sulla Promozione della sicurezza dai Rischi naturali del Patrimonio abitativo (2017)

PCM - Presidenza del Consiglio dei Ministri: Piano Nazionale di Ripresa e Resilienza, #NEXTGENERATIONITALIA, Italia domani (2021)

Raffestin, C.: Repers pour une theorie de la territorialite' humaine, Cahier n. 7, Groupe Reseaux: Parigi (1987)

Redazionale Ecosviluppo: Verso un mondo a impatto zero: la transizione all'economia circolare (2021). https://www.ecosviluppo.it/verso-un-mondo-a-impatto-zero-la-transizione-alleconomia-circolare/. Accessed 6 Dec 2021

Rifkin, J.: Un Green New Deal Globale. Il crollo della civiltà dei combustibili fossili entro il 2028 e l'audace piano economico per salvare la Terra, Mondadori (original edition The Green New Deal: Why the Fossil Fuel Civilization Will Collapse by 2028, and the Bold Economic Plan to Save Life on Earth, 2019, St. Martin's Press, New York) (2019)

Robertson, R.: Globalization: Social Theory and Global Culture. Sage, Newcastle upon Tyne, UK (1995)

Rodotà, S.: Una Costituzione per Internet. Giangiacomo Feltrinelli Editore, Milano (2005)

Rodriguez-Pose, A.: The revenge of the places that don't matter (and what to do about it). Cambridge J. Reg. Econ. Soc. 11(1), 189–209 (2018). (Oxford University press, Oxford)

Rodrik, D.: La globalizzazione intelligente, Laterza, Bari (I ed. or. The Globalization Paradox. Democracy and the Future of the World Economy. W.W. Norton & Company, New York, NY (2011), Oxford University Press Oxford, GB (2012)

Rosés, J.R., Wolf, N.: The Economic Development of Europe's Regions: A Quantitative History since 1900. Routledge, London-New York (2019)

Saragosa, C.: Il sentiero di biopoli. L'empatia nella generazione della città, Saggi. Natura e artefatto, Donzelli Eitore, Mentana, Roma (2016)

Sassen, S.: The Global City. Princeton University Press, Princeton (1991)

Smart Land. Reti Governo Sviluppo, Centro Culturale Livia Bottardi Milani, Pegognaga, 25 – 26 maggio 2017. http://www.centroculturalepegognaga.it/smart-land/. Accessed 27 May 2021

UE Landscape Chart: Florence (2000)

UNCEM, News Letter: Piano Nazionale Ripresa e Resilienza. Riepilogo dei bandi del PNRR usciti e in uscita, January 6th (2022)

Weber, A.: Über des Standort der Industrien. Part. I. Reine Theorie des Standorts, Tübingen. Trad ingl. Alfreds Weber's Theory of Location of industries (C.J. Friedrich), The Universty of Chicago press (1909/1922)

WHO: Ambient Air Pollution, a global assessment and exposure and burden of disease, Geneve (2016)

Urban Happiness Planning Through Interactive *Chorems*

Marco Romano(✉) (iD)

Università di Salerno, Fisciano, Salerno, Italy
marromano@unisa.it

Abstract. The Sustainable Development Goals of 2030 Agenda and in particular Goal 11, "*Make cities and human settlements inclusive, safe, resilient and sustainable*", encourage researchers, businesses, and public administrations to make efforts to deal with the growing difficulties in urban planning and to use it as a useful tool for improving citizens' wellbeing. This work shows how the concept of chorematic maps can be useful to make easier the understanding of urban planning and its connections with mental wellbeing of citizens. To do so, a new set of chorems is introduced, designed to describe the qualities and elements of the urban environment that can influence the happiness, and consequently, the mental wellbeing. The work shows an example of the application of the proposed chorems in an urban environment, highlighting how they can be used both by experts and ordinary citizens, making it possible to exploit them as a means to encourage both administrators, technicians, and citizens to live and deal with their city with greater awareness and participation.

Keywords: Geovisual analytics · Smart communities · Chorems · Urban planning

1   Introduction

Nowadays, urban planning is a more and more relevant field of interest, indeed, it is not only capable of improving citizens' lives, but it is also a constantly expanding market.[1] Moreover, companies, researchers, and municipal administrators work together encouraged by the 17 Sustainable Development Goals of the United Nations 2030 Agenda.[2] In particular, Goal 11, "*Make cities and human settlements inclusive, safe, resilient and sustainable*" points it out the growing difficulties of modern urban planning and encourages to work for more healthier cities using urban planning as a means to reach the goal.

Modern cities try to deal with such issues by capturing the vast and variegate number of data coming from the urban fabric and making them available to all the stakeholders in the form of services and open data [15]. Indeed, in [10, 11] the authors highlight that

---

Partially supported by MIUR PRIN 2017 grant number 2017JMHK4F 004.

[1] https://tinyurl.com/2p8ezp2b.

[2] https://sdgs.un.org/goals.

© The Author(s), under exclusive license to Springer Nature Switzerland AG 2022
F. Calabrò et al. (Eds.): NMP 2022, LNNS 482, pp. 2822–2832, 2022.
https://doi.org/10.1007/978-3-031-06825-6_269

an active and continuous collaboration between citizens, organizations and institutions through information technology is essential component for the wellbeing of the community and the city itself. However, understanding or interpreting such data and the various phenomena related to them can be difficult both to ordinary citizens, administrators, and companies, and for such reason, data should be processed and presented in an understandable way [1]. In [1], the authors showed how chorematic interactive maps can be used to process urban data and allow administrators and citizens to better understand the complex connections between them and the development of environmental phenomena such as urban heat islands.

Chorems are a schematic territory representation, which eliminates details not useful to the map comprehension [2]. They can be defined as visual syntheses of geographical elements and phenomena developing on a territory, presenting these in an understandable way despite the limitations of a map and the multitude of data [4, 5]. Moreover, because of their intuitive nature chorems are generally appropriate to both technical and non-technical people [1].

Chorems have been used mainly to represent geographical characteristics and the related environmental phenomena, notwithstanding other phenomena, such as social, emotional and wellbeing ones, which normally develop in an urban area, may benefit from them. Indeed, in [3, 12, 16] the authors argue that the specific characteristics of the urban environment, such as pedestrian or cycling orientedness, urban greenery, and furniture, may influence the mental wellbeing of citizens. Moreover, the elements participating to the mental wellbeing of inhabitants such as happiness and perception of the surrounding beauty can be effectively measured starting from the analysis of data coming from urban environment [3, 12].

This work shows how chorematic maps can be used to represent the happiness phenomenon deriving from the urban environment and how they can be useful to show understandably its links with urban planning, at the same time, encouraging both administrators, technicians, and citizens to live and deal with their city with greater awareness and participation.

The rest of the paper is so structured: Sect. 2 describes the chorem characteristics and introduces an initial proposal of urban happiness index deriving from the environmental characteristics. Section 3 describes the specific chorems developed to deal with the urban environmental characteristics and the urban happiness index. In Sect. 4, conclusions and future work are given.

2 The Chorem Composition and the Happiness Index

This section describes the structure of the chorems and introduces an initial proposal of urban happiness index based on the environmental characteristics.

Chorems. The chorems are synthesized visual representations of geographical elements and environmental phenomena [4, 5]. They eliminate any element not necessary for the understanding of a geographical aspect or a phenomenon to make the representation more understandable and intuitive. There are three types of chorems: *Geographical*, *Phenomenal*, and *Annotation*. The first represents geographical elements, the second spatio-temporal phenomena involving one or more geographic chorems. The Phenomenal type can be further divided into *Flow, Tropism,* and *Spatial Diffusion.* The *Flow* chorem summarizes the flows of movements between two or more geographic chorems, such as migratory flows. The *Tropism* chorem represents the attractiveness or repulsiveness of a space associated with a geographical chorem, such as the capacity of an area to attract people. *Spatial Diffusion* chorem represents a spatial progression or regression, from a geographic chorem along a given direction. Finally, the *Annotation* chorem shows information useful for understanding a chorem, such as the degrees of the temperature of an area. Finally, chorems are structured by two properties which are: the chorem type and its source, and by its graphic representation composed by a graphic component and its meaning.

Chorematic maps have some specific operations that can be applied, both of a geographic and semantic nature such as the zoom. A *geographic zoom* corresponds to a traditional map zoom and is applied to the chorem's aspect concerning the map, keeping unchanged its meaning or structure. A *semantic zoom* acts only on the meaning of the Chorem allowing access to a different level of information.

Urban Happiness Index (UHI). In [9], the authors state that happiness comes not only from the person but also from the elements and characteristics of the surrounding environment. Also in [8] the authors state that the shape of the city and the buildings, lead to a change in human behaviors and relationships in public spaces and can increase the level of happiness. In [12] the authors declare that there is a strong correlation between happiness and mental health and that urban planning should contribute to mental health. They collected evidence from academic studies and practical cases to build a model of citizen happiness deriving from the physical stimuli present in the urban environment. The model was then validated through a questionnaire assigned to the inhabitants of the Tehran downtown. This allowed the authors to understand also the weight of the correlation of the physical stimuli with happiness with the downtown environment.

In the model, happiness is influenced, in order of relevance, by the *Pedestrian-orientedness* (PO), *Environmental elements* (EE), *Spatial cohesion* (SC), *Bicycle-orientedness* (BO), *Quality of the space* (QS), and *Good vegetation* (GV). Moreover, *Environmental elements* are composed by the following elements in order of relevance: *Cafés and restaurants* (CR), *Benches to sit and talk* (BST), *Works of art* (WA), *Fountains* (F), *Illumination* (I), *Access to lavatories* (AL). Finally, *Quality of the space* is composed by the following elements in order of relevance: *Variety* (V), *Physical penetrability* (PP), *Place identity* (PI), *Flexibility* (F) and *Legibility* (L).

Because the model was synthetized for the city of Tehran, the authors warn that each city may require small variations of it, depending, for example, on the climate, geography, and culture of the area.

A widespread happiness index in literature is the one in [13]. It is based on a collection of answers to a questionnaire ranging from 1 to 10. Similarly, the urban happiness index (UHI), which we propose, varies from 1 to 10 but is measured starting from the characteristics of the urban environment. UHI is designed to be applied for neighborhoods of a city, defining the neighborhoods as the smallest and cohesive urban area, which could correspond to the administrative districts, depending on the urban topology. This is to give flexibility to a municipal administration because of the variety of urban characteristics.

UHI is the result of the weighted average of the elements that compose it. The weight (W_x) ranging from 0.0 to 1, depending to the specific necessities of a city. Each element is measured by a value ranging from 1 to 10:

The following formula expresses the composition of the UHI:

$$UHI = (PO * W_1 + EE * W_2 + SC * W_3 + BO * W_4 + QS * W_5 + GV * W_6)$$
$$\sum_x W_x = 1 \tag{1}$$

Similarly, the *EE* and *QS* indexes can be calculated using the following formulas

$$EE = (CR * W_7 + BST * W_8 + WA * W_9 + F * W_{10} + I * W_{11} + AL * W_{12})$$
$$\sum_x W_x = 1 \tag{2}$$

$$QS = (V * W_{13} + PP * W_{14} + PI * W_{15} + F * W_{16} + L * W_{17})$$
$$\sum_x W_x = 1 \tag{3}$$

3 The New Set of Urban Chorems

This section describes the chorems proposed to represent and synthesize the urban happiness in districts and the elements that compose it.

Figure 1 shows the iconography and the corresponding meaning for the chorems representing the UHI in the district and the elements that contribute to its value.

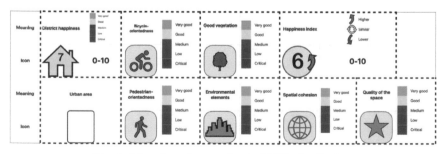

Fig. 1. The urban chorems for the UHI in the districts.

When indicated, the color background of the icons follows chromatic scale of 5 elements going from brown to green. Where brown means an index between 1 and 2 and green between 9 and 10. Table 1 describes in detail each chorem in Fig. 1.

Table 1. The descriptions of the chorems for the UHI in the districts.

Chorem	Type	Description	Representation
District happiness	Spatial diffusion	It is associated with the *Urban Area* chorem. It represents the UHI in the specific area. The UHI computation is computed using formula (1). Its source is the centroid of the associated *urban area*	Its iconic representation is a house whose color scale of five levels indicates the level of happiness, and a numeric annotation which corresponds to the computed UHI
Urban area	Geographical	An *urban area* represents the visual synthesis of a territory in which several urban phenomena develop. Its source is a municipal database containing the coordinates of the smallest and cohesive urban areas	Its iconic representation is a rectangle with rounded corners
Pedestrian-orientedness	Spatial diffusion	It is associated with a specific urban area. It indicates the ease level of walking through an area, taking into consideration factors such as the shape and continuity of the sidewalks or pedestrian areas, the presence of crossings of congested streets, unauthorized parking, and difficult slopes	It is represented by a little man walking set in a colored rectangle. The rectangle has a chromatic scale of five levels that indicates the ease level of walking in the area
Environmental elements	Spatial diffusion	It refers to environmental elements such as urban furniture and services that allow people to enjoy the district. Such elements work as urban stimuli to invite people to spend time outdoors in a place. It is computed using formula (2)	Its icon is a city inscribed in a rectangle whose color scale of five levels indicates the presence of such elements
Spatial cohesion	Spatial diffusion	It is associated with a specific urban area. It refers to a combination of spatial structure, relationships, processes, and integration in the urban area. It can be evaluated using both quantitative or qualitative techniques performed by experts [7]	Its icon is a globe composed of a network of connected nodes and inscribed in a rectangle with a color scale of five levels corresponding to the level of cohesion

(*continued*)

Table 1. (*continued*)

Chorem	Type	Description	Representation
Bicycle-orientedness	Spatial diffusion	It is associated with a specific urban area. It indicates the ease level of cycling through the area Such level can be calculated by experts based on the presence of bike paths, mixed areas with pedestrians or motor vehicles, the presence of unauthorized parking, the presence of vehicles, and their average speed	It is represented by a little man biking set in a colored rectangle. The rectangle has a color scale of five levels that indicates the degree of cycling
Green spaces	Spatial diffusion	It indicates the presence of good vegetation in the Urban area. It depends on the seasons and the care of the plants. Its source is a database frequently updated with a quality index ranging from 1 to 10 and estimated with data from satellite images or field inspections	Its icon is a tree set in a colored rectangle with a color scale of five levels corresponding to the good vegetation level
Quality of the space	Spatial diffusion	It is associated to the Urban area and is calculated using formula (3) with a value ranging from 1 to 10. It represents a visual synthesis of different quality characteristics of the area, that are the variety, the physical penetrability the place identity, the flexibility, and the legibility	The iconic representation is a five-pointed star in a colored rectangle with a color scale of five levels corresponding to the computed quality level of the area

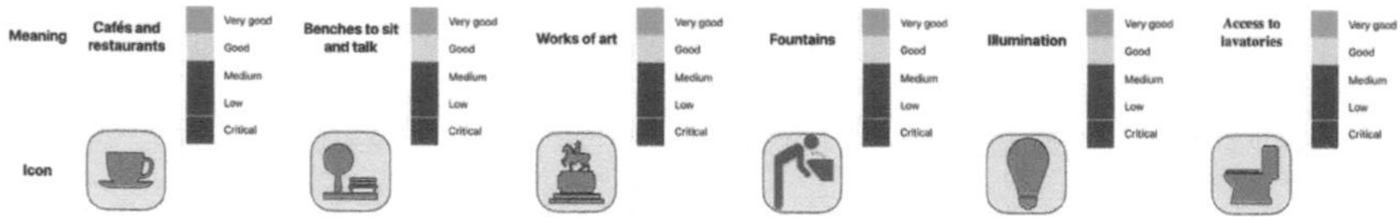

Fig. 2. The urban chorems for the *Environments elements* of an *Urban Area*.

Figure 2 shows the chorems proposed to represent the elements that form the *Environmental elements*. As in the previous cases, when indicated, their background is based on a chromatic scale that indicates their level of presence in the given urban area.

Table 2 describes in detail the chorems in Fig. 2.

Table 2. The descriptions of the chorems for the *Environments elements* of an *Urban Area.*

Chorem	Type	Description	Representation
Cafés and restaurants	Geographical	It represents the visual synthesis of the number of cafés and restaurants in the associated area. Its value ranges from 1 to 10 and is calculated as the ratio between the area's inhabitants, regular visitors and these facilities	The icon is a café set in a colored rectangle with a color scale of five levels corresponding to the presence level of these facilities
Benches to sit and talk	Geographical	It represents the visual synthesis of the number of public places where people can stay spending time to interact with other people in the associated Urban area. Its value ranges from 1 to 10 and is calculated as the ratio between the area's inhabitants, regular visitors and these facilities	The icon is a tree close to a bench set in a colored rectangle with a color scale of five levels corresponding to the presence level of these facilities
Works of art	Geographical	It represents the ability of the associated urban area to stimulate inhabitants and regular visitors through artistic works, such as temporary exhibitions, murals, or sculptures. Its value ranges from 1 to 10 and is defined by art experts and sociologists and updated as time goes by	The icon is a statue set in a colored rectangle with a color scale of five levels corresponding to the level of the stimuli of works of art
Fountains	Geographical	It represents the availability and accessibility of public fountains in the associated urban area. They allow people to stay outdoors for longer. Its value ranges from 1 to 10 and is calculated as the ratio between the area's inhabitants, regular visitors and these facilities	The icon is a public fountain set in a colored rectangle with a color scale of five levels corresponding to accessibility level of the public fountains in the area
Illumination	Spatial Diffusion	It represents the ability of the urban area to be adequately illuminated. The data comes from a frequently updated municipal database fed from environmental sensors, or smart streetlamps, or, alternatively, by public operators	The icon is a bulb set in a colored rectangle with a color scale of five levels corresponding to quality level of the public illumination

(continued)

Table 2. (*continued*)

Chorem	Type	Description	Representation
Access to lavatories	Spatial diffusion	It represents the availability and accessibility of lavatories in the associated urban area. They allow people to stay outdoors for longer. Its value ranges from 1 to 10 and is calculated as the ratio between the area's inhabitants, regular visitors and these facilities	The icon is a toilet set in a colored rectangle with a color scale of five levels corresponding to the availability and accessibility level of the toilets in the area

Figure 3 shows the chorems proposed to represent the elements that compose the Quality of the space. As in the previous cases, their background is based on a chromatic scale that indicates their level in the given urban area.

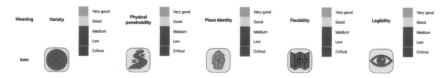

Fig. 3. The urban chorems for the *Quality of the space* in an *Urban Area.*

Table 3 describes in detail each chorem in Fig. 3.

Table 3. The descriptions of the chorems for the *Quality of the space* in an *Urban Area.*

Chorem	Chorem type	Description	Representation
Variety	Spatial diffusion	It represents the level of variety of urban stimuli in the area under examination	The icon is a pie chart set in a colored rectangle with a color scale of five levels corresponding to the presence level of these facilities
Physical penetrability	Spatial diffusion	Penetrability is the characteristic of an urban area that allows ease of movement within it, avoiding cutting off the neighborhood	The icon is a path set in a colored rectangle with a color scale of five levels corresponding to the level of penetrability

(*continued*)

Table 3. (*continued*)

Chorem	Chorem type	Description	Representation
Place identity	Spatial diffusion	Place identity is the capability of an urban space to spread its meaning to the inhabitants and visitors by contributing to the sense of community [14]	The icon is a fingerprint set in a colored rectangle with a color scale of five levels corresponding to the level of capacity of spreading place identity
Flexibility	Spatial diffusion	It is the capability of an urban space to be flexible by offering the possibility of holding a variety of events such as ceremonies, competitions, or art exhibitions	The icon is a flexed map set in a colored rectangle with a color scale of five levels corresponding to flexibility level the area
Legibility	Spatial diffusion	Legibility is the capability of the space to provide an understanding of itself by helping people to create cognitive maps and way-finding [6]	The icon is an eye set in a colored rectangle with a color scale of five levels corresponding to the level of legibility of the area

Image 4 shows an example of the use of the urban chorems to analyze happiness in urban areas. The map system used is the Open Street Map[3] and shows a portion of the city of Rome. The data used to perform the computations are fictitious and used only by way of example. Figure 4a shows 4 urban areas of interest represented by the house icons. Each icon corresponds to a color of the chromatic scale and an annotation of the UHI in the area. Figure 4b shows a semantic zoom performed on the *District happiness* icon with UHI 6 to reveal the factors that contribute to the computation of the UHI. Figures 4c and 4d respectively show a semantic zoom on the *Environmental elements* and the *Quality of the space*.

As it is shown in the examples, the chorematic maps are intuitive despite they represent various phenomena and geographical factors that make part of the complexity of an urban environment. They are suitable for both technical experts and administrators or ordinary citizens. For such reason, they can be used to share knowledge about the urban environment with the various actors that form the community of a city. They can be a tool for making more informed urban planning decisions, and on the other hand, a tool for improving communication and collaboration with citizens.

---

[3] https://www.openstreetmap.org.

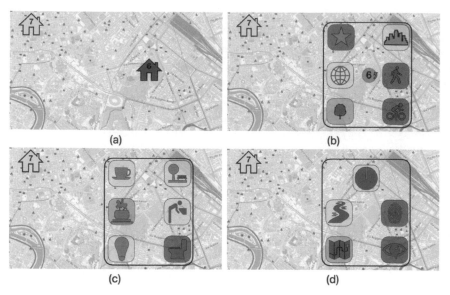

(a)   (b)

(c)   (d)

Fig. 4. An example of application of the new set of urban chorems overlapping a map of Rome

4   Conclusions and Future Work

This work introduces a new set of urban chorems used to represent the factors that contribute to the development of happiness in the different areas of a city. The work also describes an initial proposal for an urban happiness index elaborated from existing models in the literature and associated with the chorems. These chorems are also a tool useful to plan or rethink a city considering the elements and stimuli that support the psychophysical state and social interactions of citizens. At the same time, they are an effective way to allow the different actors of a community to understand the urban data and related phenomena and to encourage them to live and deal with their city with greater awareness. In the next future, working closely with municipal administrators, this research will experiment the proposed chorems with real data to measure the effectiveness of visual synthesis and compare the effects of urban planning on mental wellbeing calculated through UHI with those extracted from field studies.

References

1. Battistoni, P., Grimaldi, M., Romano, M., Sebillo, M., Vitiello, G.: Interactive maps of chorems explaining urban contexts to align smart community's actors. In: Gervasi, O., et al. (eds.) ICCSA 2021. LNCS, vol. 12953, pp. 549–564. Springer, Cham (2021). https://doi.org/10. 1007/978-3-030-86976-2_37
2. Brunet, R.: La carte-modèle et les chorèmes. Mappemonde **86**(4), 2–6 (1986)
3. Calafiore, A.: Measuring beauty in urban settings (2020)
4. De Chiara, D., Del Fatto, V., Laurini, R., Sebillo, M., Vitiello, G.: A chorem-based approach for visually analyzing spatial data. J. Vis. Lang. Comput. **22**(3), 173–193 (2011)

5. Del Fatto, V., Laurini, R., Lopez, K., Sebillo, M., Vitiello, G.: A chorem-based approach for visually synthesizing complex phenomena. Inf. Vis. **7**(3–4), 253–264 (2008)
6. Herzog, T.R., Leverich, O.L.: Searching for legibility. Environ. Behav. **35**, 459 (2003)
7. Kołata, J., Zierke, P.: Assessment of spatial cohesion in suburban areas based on physical characteristics of buildings. Annals of Warsaw University of Life Sciences–SGGW. Hort. Lands. Archit. **41**, 37–49 (2020)
8. Montgomery, C.: Happy City: Transforming Our Lives Through Urban Design. Macmillan, New York (2013)
9. National Research Council: Neem: A Tree for Solving Global Problems. The Minerva Group Inc., Lakewood (2002)
10. Romano, M., Díaz, P., Aedo, I.: Emergency management and smart cities: civic engagement through gamification. In: Díaz, P., Bellamine Ben Saoud, N., Dugdale, J., Hanachi, C. (eds.) ISCRAM-med 2016. LNBIP, vol. 265, pp. 3–14. Springer, Cham (2016). https://doi.org/10.1007/978-3-319-47093-1_1
11. Romano, M., Díaz, P., Aedo, I.: Gamification-less: may gamification really foster civic participation? A controlled field experiment. J. Ambient Intell. Humaniz. Comput., 1–15 (2021). https://doi.org/10.1007/s12652-021-03322-6
12. Samavati, S., Ranjbar, E.: The effect of physical stimuli on citizens' happiness in urban environments: the case of the pedestrian area of the historical part of Tehran. J. Urban Des. Mental Health **2**(2), 1–37 (2017)
13. Veenhoven, R.: World database of happiness tool for dealing with the 'data-deluge.' Psihologijske teme **18**(2), 221–246 (2009)
14. Smith, M.P., Bender, T.: City and Nation: Rethinking Place and Identity. Routledge, Milton Park (2017)
15. Vives, A.: Smart City Barcelona: the Catalan quest to improve future urban living. Int. J. Iberian Stud. **33**(1), 103–104 (2018)
16. Zeile, P., et al.: Urban Emotions–tools of integrating people's perception into urban planning. In: REAL CORP 2015. PLAN TOGETHER–RIGHT NOW–OVERALL. From vision to reality for vibrant cities and regions. Proceedings of 20th International Conference on Urban Planning, Regional Development and Information Society, pp. 905–912. CORP–Competence Center of Urban and Regional Planning, May 2015

A Preliminary Model for Promoting Energy Communities in Urban Planning

Roberto Gerundo$^{(\boxtimes)}$, Alessandra Marra⬥, and Michele Grimaldi⬥

Department of Civil Engineering, University of Salerno, 84084 Fisciano, SA, Italy
{r.gerundo,almarra,migrimaldi}@unisa.it

Abstract. Renewable energy communities (REC) represent an important opportunity to promote the use of energy from renewable sources and the buildings efficiency, to reduce greenhouse gas emissions and, more generally, to combat energy poverty, exacerbated by the recent pandemic and rising energy prices. Although the topic of energy communities is still little explored in municipal planning, the latter can make a significant contribution in their promotion. The aim of this work is the definition of a preliminary model for the construction of a priority areas map, in which to plan the development of REC on an infra-urban scale, in the Italian geographical context. This map makes it possible to identify critical urban areas both in terms of energy performance and income vulnerability, in order to maximize the benefits brought by the REC establishment, with particular reference to the reduction of energy poverty. To this end, in the absence of open data for the estimation of the buildings energy performance, a bottom-up approach is proposed that exploits the technological advances offered by GIS and BIM software, requesting input data normally available in urban planning processes.

Keywords: Energy communities · Energy poverty · Urban planning

1 Introduction

Renewable Energy Communities (REC) represent coalitions of citizens, small and medium-sized enterprises and local authorities, including municipal administrations, who are able to produce, consume and exchange energy produced from renewable sources, with the main aim of providing environmental, economic or social benefits to the community itself or to the local areas in which it operates [1].

The topic is of growing interest, due to the results of some experiments conducted in Europe and the United States in terms of reducing energy poverty and, more generally, contributing numerous environmental benefits, including: the energy efficiency of existing buildings; the promotion of energy use from renewable sources (RES); the reduction of GHG emissions and the consequent contrast to climate change in urban areas [2–4].

Multiple agreements and international agendas establish the need to introduce actions to contain the global average temperature [5, 6]. Furthermore, the worsening of poverty,

Author Contributions—All authors contributed to the design of the research and the definition of the methodology. A.M. prepared a draft manuscript. R.G., A.M. and M.G. revised the manuscript.

© The Author(s), under exclusive license to Springer Nature Switzerland AG 2022
F. Calabrò et al. (Eds.): NMP 2022, LNNS 482, pp. 2833–2840, 2022.
https://doi.org/10.1007/978-3-031-06825-6_270

due to the recent pandemic, and the rise in energy prices are significantly increasing the risk of experiencing conditions of energy poverty [7].

In Italy, the National Recovery and Resilience Plan aims at the energy transition, understood as the construction of a new model of social organization based on the production and consumption of energy from renewable sources, reserving a specific investment channel for renewable energy communities [8].

The issue of REC is still little explored in urban planning, as emerged both from the review of international literature and from the analysis of the municipal urban planning tools of the main Italian cities [9–14]. The few experiences of Italian energy communities are disconnected from planning processes, in which the issue has not been addressed, with the exception of the recent SECAP (Sustainable Energy and Climate Action Plan) of Bologna and Genoa, as well as the 'Air and Climate Plan' of Milan. Even in such cases, the relationship of these instruments with the general urban plan is not explicit with regard to REC, while the general plan cognitive framework could be integrated with documents representing the priority urban areas in which to favor the establishment of energy communities. In fact, a significant contribution in their promotion can be played by planning, in particular municipal planning, also considering the key role that local authorities can have in the construction of renewable energy communities [15, 16].

This work is part of a broader research project, aimed at promoting the development of energy communities in the municipal urban planning process. Specifically, the paper aims at defining a preliminary methodological proposal for the construction of a map of priority areas for the development of REC on an infra-urban scale, in the Italian geographical context. This map makes it possible to identify those neighbourhoods or groups of critical neighbourhoods, where a maximization of the benefits brought by the establishment of renewable energy communities is expected, with particular reference to the reduction of energy poverty. For this purpose, the following Sect. 2 explains the methodological approach underlying the proposed model, presented in Sect. 3; Sect. 4 illustrates the main conclusion of the work, its potential implications for municipal urban planning and future research development.

2 Methodological Approach

In the absence of an agreement in the international scientific community on the methods and techniques for measuring energy poverty at a sub-municipal scale, this paper refers to the definition adopted by the European Commission, according to which energy poverty represents "a situation in which a family or an individual is unable to pay for the primary energy services needed to ensure a decent living standard, due to a combination of low income, high energy expenditure and low energy efficiency in their homes" [17]. In fact, these causal factors represent the most frequently investigated dimensions for the construction of composite indices of energy poverty [18–20].

However, in the Italian geographical context, it is not always possible to measure the aforementioned variables through simple indicators at an infra-urban level, so for the purposes of municipal planning, due to the partial availability of public data.

The cost of energy, if the latter derives from traditional centralized non-renewable sources, depends on the economy and on the global and national market, on taxation, so it can be extremely variable. Nevertheless, local operators are able to provide information on official prices or on consumption based on the bills issued.

The data relating to per capita income are provided by the Ministry of Economy and Finance, mostly on a municipal basis, except for the major cities, for which data is available at the sub-municipal level. However, the localization of low-income areas, can be supported by the knowledge, usually in support of municipal planning tools, of the territorial distribution of existing Public/Social Residential Building districts, in which the housing is assigned to the beneficiaries on the basis of certain income thresholds. It is particularly complex to find open data relating to building energy performance certificates, even where this documentation has actually been drawn up: the national cadaster of energy performance certificates (SIAPE), established on the basis of Ministerial Decree 26/06/2015, provides information, on a municipal basis, only about some Italian regions.

In the absence of data that make it possible to directly measure the consumption and the consequent energy performance of buildings, the technical-scientific literature on the subject offers many examples of evaluation, with particular reference to the residential fabric, relating to two most widely used approaches: top down, based on the disaggregation of data referring to the least detailed level, typically on a municipal basis; bottom-up, starting from data available at the same spatial level or at a detailed building scale [21–23]. The top down approaches are indicated for a first reading, on an infra-urban scale, of the territorial data distribution provided at a higher level of detail, however, they are affected by a high uncertainty degree, deriving from the disaggregation process, to keep in count in the evaluation of the obtained results. On the contrary, the bottom-up approaches are more suitable for the detail scale investigated in this paper. The most recent studies on bottom-up approaches exploit the technological advances offered by GIS software [24], but few works integrate the potential offered by 3D models in a BIM environment, which allow to obtain more realistic results.

In this paper, the energy performance is estimated with a bottom-up methodological approach, which can be implemented by exploiting the technological advances offered by recent GIS and BIM software.

3   Description of the Model

The proposed methodology is divided into three main macrophases, which consist in the construction of the following maps, respectively (Fig. 1):

1) *Energy Poverty Map*, representative of the spatialization of urban areas characterized by energy poverty;

2) *Constraints Map*, summarizing all the existing constraints to the establishment of renewable energy communities;

3) *Priority Areas Map*, built on the previously obtained maps, indicative of the urban areas in which to promote the establishment of energy communities through municipal urban planning.

The aforementioned macrophases are better explained below.

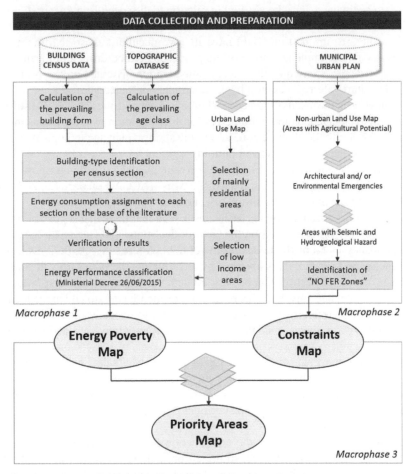

Fig. 1. Workflow of the proposed methodology.

The first macrophase consists in estimating the energy performance of the residential-building stock in terms of heating and domestic hot water production, on a census basis. This is possible by tracing the stock of residential buildings in each section to typical buildings, representative of the entire section, of which the energy consumption, expressed in kWh/m^2 per year, is known on the basis of the relevant literature. More precisely, the reference buildings-types are those identified within the TABULA project, indicative of the typical construction and plant engineering characteristics of the Italian building stock, considering an average climatic zone [25].

The association to a building-type can be carried out on the basis of the building typology and the average age class per section, calculated in a GIS environment, according to a method already proposed and tested by the authors in previous contributions, in support of large area planning [26, 27].

In this work, in order to make the proposed methodology more suitable for municipal urban planning purposes, the data relating to the energy consumption of residential buildings, thus obtained, must be verified in relation to the specific characteristics of the municipal area assumed as case study.

In fact, the area in question could have its own peculiarities, so that the parameters necessary for the implementation of the proposed method differ significantly from the national average. Consequently, energy consumption, assessed on the basis of the technical literature referred to, could be overestimated or underestimated.

A verification of the results obtained can be carried out with the aid of a BIM software, which allows to change the input parameters for the estimation of energy consumption, both with reference to the climatic zone and to the stratigraphic, technological and plant characteristics of the identified typical buildings.

More precisely, among the commercial software available for this purpose, in this work we propose to use the BIM software 'TerMus' (ACCA Software S.p.A.), for its moderate cost and, mainly, for its simplicity and speed of use.

Once the energy consumption for each census section is known, it is possible to associate the energy performance class on the basis of the Ministerial Decree of 26/06/2015, which establishes a total of 10 classes, from A4 to G (A4, A2, A1, A, B, C, D, E, F, G).

The energy performance map, obtained by recalibrating the values from the literature on the basis of this verification, can be crossed with the urban land use map, where it is necessary to extrapolate only the areas for mainly residential use, to which the identified performance classes refer. In this map it is also possible to select the districts made up of existing Public and/or Social Residential Buildings, representative of low-income urban areas.

Subsequently, the Energy Poverty Map is obtained by selecting the urban areas characterized by unfavorable energy performance, identified in classes E, F and G. Similarly, among all low-income urban areas, those falling in the above-mentioned classes are selected, in order to identify the most critical urban areas from an energy and social point of view.

In the second macrophase, for the construction of the Constraints map, it is necessary to identify the factors that constitute an obstacle to the technical feasibility of renewable energy communities, according to current legislation. In order to develop a general method, regardless of specific regional regulations, reference is made to the guidelines drawn up by the national energy service manager [28]. The latter specifies that constraining factors characterize the so-called "No FER zones", where it is not allowed the installation of RES systems: protected areas; forest areas; water bodies; areas and buildings burdened by constraints, such as buildings of documented historical, artistic and architectural value; areas with high agricultural potential; areas with high seismic and hydrogeological hazard.

The data source for the spatial localization of the aforementioned factors is represented by the maps produced in the ordinary municipal urban planning activity, with reference to the system of protections, architectural and/ or environmental emergencies and the use of agricultural land.

In the last macrophase, the map of the most critical urban areas is obtained through an *overlay mapping* between the Energy Poverty map and the Constraints map. The criterion for identifying the areas in which to test the ERC implementation interventions is the following: the most critical areas correspond to urban areas characterized by the worst energy performance, identified in performance classes E, F, G, and, at the same time, not affected by the areas in which the installation of RES plants is not allowed.

4   Conclusion and Research Developments

The map of priority areas, obtained by implementing the proposed model on a given municipal area, allows to support the urban planning process regarding the location of critical areas in which to primarily incentivize the development of energy communities, with a view to guaranteeing the reduction of energy poverty.

The implications of the presented work, in terms of policies and interventions for the promotion of the transition through renewable energy communities, are readable on several levels.

On the one hand, the described methodology makes it possible to support municipal administrations in the application to benefit from dedicated funding, most recently the investment "C2- Inv. 1.2 - Promotion of renewables for energy communities and self-consumption" as part of the mission "M2 - Energy transition and sustainable mobility" [8], envisaged by the National Recovery and Resilience Plan: municipalities can demonstrate that they have preceded an adequate analytical phase to the planning phase, thus showing a more effective potential allocation of public resources where it is expected a maximization of the benefits provided.

At the local level, the priority areas map represents a tool available to the municipal administrations to promote, also by allocating specific resources obtained from the aforementioned funding, the establishment of REC in certain urban areas characterized by energy poverty, under the strictly energy profile linked to buildings performance, or even from a social point of view, with reference to the income situation of residents [29, 30]. To this end, the cognitive framework of municipal urban planning tools, in particular of the general municipal urban plan, possibly consistent with the objectives of the Sustainable Energy and Climate Action Plan, could be integrated with documents representative of the identified critical urban areas, in which to favor the establishment of energy communities.

In order to test its validity, potential limits and possible transferability to other geographical contexts, the proposed methodology will be applied to the case study of the Municipality of Pagani, in Campania Region (Italy), within the framework of an institutional agreement between the Municipality and the Department of Civil Engineering of the University of Salerno for the technical-scientific support to the drafting of the Municipal Urban Plan. Furthermore, the method here illustrated may be further refined in the future, by investigating the technical-energy and economic feasibility aspects of renewable energy community projects in the identified priority areas.

References

1. EU-European Commission: Directive (EU) 2018/2001 of the European Parliament and of the Council of 11 December 2018 on the promotion of the use of energy from renewable sources

(2018). https://eur-lex.europa.eu/legal-content/EN/TXT/?uri=CELEX%3A02018L2001-201 81221. Accessed 28 Feb 2022

2. Brummer, V.: Community energy – benefits and barriers: a comparative literature review of community energy in the UK, Germany and the USA, the benefits it provides for society and the barriers it faces. Renew. Sustain. Energy Rev. **94**, 187–196 (2018)

3. McCabe, A., Pojani, D., van Groenou, A.B.: Social housing and renewable energy: Community energy in a supporting role. Energy Res. Soc. Sci. **38**, 110–113 (2018)

4. Koltunov, M., Bisello, A.: Multiple impacts of energy communities: conceptualization taxonomy and assessment examples. In: Bevilacqua, C., Calabrò, F., Della Spina, L. (eds.) NMP 2020. SIST, vol. 178, pp. 1081–1096. Springer, Cham (2021). https://doi.org/10.1007/978-3-030-48279-4_101

5. UNFCCC: The Paris agreement. In: Paris Climate Change Conference – November 2015, COP 21 Paris (2015). FCCC/CP/2015/L.9/Rev.1

6. UN-United Nations General Assembly: Transforming our world: The 2030 Agenda for Sustainable Development. A/RES/70/1 (2015)

7. EU-European Commission: State of the Energy Union 2021 – Contributing to the European Green Deal and the Union's recovery (2021). https://eur-lex.europa.eu/legal-content/EN/TXT/?uri=CELEX:52021DC0950&qid=1635753095014. Accessed 28 Feb 2022

8. Governo Italiano: Piano Nazionale di Ripresa e Resilienza. Next Generation Italia. (2021). https://italiadomani.gov.it/it/home.html. Accessed 28 Feb 2022

9. Colombo, G., Ferrero, F., Pirani, G., Vesco, A.: Planning local energy communities to develop low carbon urban and suburban areas. In: IEEE International Energy Conference (ENERGY-CON), Dubrovnik, Croatia, 13–16 May 2014, pp. 1012–1018 (2014). https://doi.org/10.1109/ENERGYCON.2014.6850549

10. Huang, Z., Yu, H., Peng, Z., Zhao, M.: Methods and tools for community energy planning: a review. Renew. Sustain. Energy Rev. **42**, 1335–1348 (2015). https://doi.org/10.1016/j.rser.2014.11.042

11. De Pascali, P., Bagaini, A.: Energy transition and urban planning for local development. A critical review of the evolution of integrated spatial and energy planning. Energies **12**(1), 35 (2018). https://doi.org/10.3390/en12010035

12. Walnum, H.T., Hauge, A.L., Lindberg, K.B., Mysen, M., Nielsen, B.F., Sørnes, K.: Developing a scenario calculator for smart energy communities in Norway: Identifying gaps between vision and practice. Sustain. Cities Soc. **46**, 101418 (2019). https://doi.org/10.1016/j.scs.2019.01.003

13. Brunetta, G., Mutani, G., Santantonio, S.: Pianificare per la resilienza dei territori. L'esperienza delle comunità energetiche. Archivio di Studi Urbani e Regionali LII **131**(suppl.) 44–70 (2021)

14. Curreli, S., Zoppi, C.: Carbone e pianificazione del territorio: retorica del declino e criticità della transizione energetica in Sardegna. Archivio di Studi Urbani e Regionali, LII **131**(suppl.), 166–185 (2021)

15. De Mare, G., Di Piazza, F.: The role of public-private partnerships in school building projects. In: Gervasi, O., et al. (eds.) ICCSA 2015. LNCS, vol. 9156, pp. 624–634. Springer, Cham (2015). https://doi.org/10.1007/978-3-319-21407-8_44

16. Friends of the Earth Europe, REScoop.eu, Energy Cities: Municipalities & Local Authorities: an ideal partner, in Community Energy. A practical guide to reclaiming power (2020). https://energy-cities.eu/publication/community-energy/. Accessed 28 Feb 2022

17. Rademaekers, K., et al.: Selecting Indicators to Measure Energy Poverty. European Commission, DG Energy, Brussels, Belgium (2016)

18. Thomson, H., Bouzarovski, S.: Addressing Energy Poverty in the European Union: State of Play and Action. European Commission Energy Poverty Observatory, Luxembourg (2018)

19. Fabbri, K.: Building and fuel poverty, an index to measure fuel poverty: an Italian case study. Energy **89**, 244–258 (2015). https://doi.org/10.1016/j.energy.2015.07.073
20. Halkos, G.E., Gkampoura, E.C.: Coping with energy poverty: measurements, drivers, impacts, and solutions. Energies **14**(10), 2807 (2021). https://doi.org/10.3390/en14102807
21. Gerundo, R., Fasolino, I., Grimaldi, M.: ISUT model. A composite index to measure the sustainability of the urban transformation. In: Papa, R., Fistola, R. (eds.) Smart Energy in the Smart City. GET, pp. 117–130. Springer, Cham (2016). https://doi.org/10.1007/978-3-319-31157-9_7
22. Kavgic, M., Mavrogianni, A., Mumovic, D., Summerfield, A., Stevanovic, Z., Djurovic-Petrovic, M.: A review of bottom-up building stock models for energy consumption in the residential sector. Build. Environ. **45**, 1683–1697 (2010). https://doi.org/10.1016/j.buildenv.2010.01.021
23. Torabi Moghadam, S., Delmastro, C., Corgnati, S.P., Lombardi, P.: Urban energy planning procedure for sustainable development in the built environment: a review of available spatial approaches. J. Clean. Prod. **165**, 811–827 (2017). https://doi.org/10.1016/j.jclepro.2017.07.142
24. Torabi Moghadam, S., Coccolo, S., Mutani, G., Lombardi, P.: A new clustering and visualization method to evaluate urban energy planning scenarios. Cities **88**, 19–36 (2018). https://doi.org/10.31224/osf.io/b9znk
25. Ballarini, I., Corgnati, S.P., Corrado, V.: Use of reference buildings to assess the energy saving potentials of the residential building stock: the experience of TABULA project. Energy Policy **68**, 273–284 (2014). https://doi.org/10.1016/j.enpol.2014.01.027
26. Gerundo, R., Marra, A., Giacomaniello, O.: A methodology for analyzing the role of environmental vulnerability in urban and metropolitan-scale peripheralization processes. In: La Rosa, D., Privitera, R. (eds.) INPUT 2021. LNCE, vol. 146, pp. 459–468. Springer, Cham (2021). https://doi.org/10.1007/978-3-030-68824-0_49
27. Gerundo, R., Marra, A., Giacomaniello, O.: Environmental vulnerability to peripheralization risk in large area planning. Sustain. Mediterr. Constr. **14**, 75–83 (2021). http://www.sustainablemediterraneanconstruction.eu/SMC/The%20Magazine%20n.14.html. Accessed 28 Feb 2022
28. RSE-Ricerca sul Sistema Energetico: Le comunità energetiche in Italia. Note per il coinvolgimento dei cittadini nella transizione energetica. Editrice Alkes, Milan, Italy (2021)
29. Fasolino, I., Grimaldi, M., Zarra, T., Naddeo, V.: Implementation of integrated nuisances action plan. Chem. Eng. Trans. **54**, 19–24 (2016). https://doi.org/10.3303/CET1654004
30. Grimaldi, M., Sebillo, M., Vitiello, G., Pellecchia, V.: Planning and managing the integrated water system: a spatial decision support system to analyze the infrastructure performances. Sustainability **12**(16), 6432 (2020). https://doi.org/10.3390/su12166432

Author Index

© The Editor(s) (if applicable) and The Author(s), under exclusive license
to Springer Nature Switzerland AG 2022
F. Calabrò et al. (Eds.): NMP 2022, LNNS 482, pp. 2841–2847, 2022.
https://doi.org/10.1007/978-3-031-06825-6

Printed by Printforce, the Netherlands